MOSFET MODELS
FOR SPICE SIMULATION,
INCLUDING
BSIM3v3 AND BSIM4

MOSFET MODELS FOR SPICE SIMULATION, INCLUDING BSIM3v3 AND BSIM4

WILLIAM LIU
Texas Instruments

A Wiley-Interscience Publication
John Wiley & Sons, Inc.
New York · Chichester · Weinheim · Brisbane · Singapore · Toronto

This book is printed on acid-free paper. ∞

Copyright © 2001 by John Wiley & Sons, Inc. All rights reserved.

Published simultaneously in Canada.

No part of this publication may be reproduced, stored in a retrieval system or transmitted in any form or by any means, electronic, mechanical, photocopying, recording, scanning or otherwise, except as permitted under Sections 107 or 108 of the 1976 United States Copyright Act, witbout either the prior written permission of the Publisher, or authorization through payment of the appropriate per-copy fee to the Copyright Clearance Center, 222 Rosewood Drive, Danvers, MA 01923, (508) 750-8400, fax (508) 750-4744. Requests to the Publisher for permission should be addressed to the Permissions Department, John Wiley & Sons, Inc., 605 Third Avenue, New York, NY 10158-0012, (212) 850-6011, fax (212) 850-6008, E-Mail: PERMREQ@WILEY.COM. For ordering and customer service, call 1-800-CALL-WILEY.

Library of Congress Cataloging-in-Publication Data is available.

ISBN: 0-471-39697-4

Printed in Great Britain.
10 9 8 7 6 5 4 3 2 1

To: *my wife, Lee-Ping Chong*
my daughters, Yiling Ashley Liu, and Yilan Audrey Liu
my parents, Chien-Shun and Li-Yue Liu
the Liu family, abroad in US or back in Taiwan
the coach, Prof. James Harris

and . . .
my idosyncratic chess pals, who master both

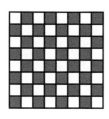 *and*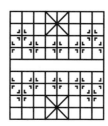

CONTENTS

Preface   xi

1 Modeling Jargons   1

    1.1   SPICE Simulator and SPICE Model, 1

    1.2   Numerical Iteration and Convergence, 8

    1.3   Digital vs. Analog Models, 11

    1.4   Smoothing Function and Single Equation, 27

    1.5   Chain Rule, 36

    1.6   Quasi-Static Approximation, 38

    1.7   Terminal Charges and Charge Partition, 44

    1.8   Charge Conservation, 50

    1.9   Non-Quasi-Static and Quasi-Static y-Parameters, 62

    1.10   Source-Referencing and Inverse Modeling, 70

    1.11   Physical Model and Table-Lookup Model, 77

    1.12   Scalable Model and Device Binning, 86

           References and Notes, 98

2 Basic Facts About BSIM3 — 101

2.1 What Is and What's Not Implemented in BSIM3, 101

2.2 DC Equivalent Circuit Model, 104

2.3 BSIM3's y-Parameters, 116

2.4 Large-Signal Equivalent Circuit, 122

2.5 Small-Signal Model, 133

2.6 Noise Equivalent Circuit, 140

2.7 Special Operating Conditions: $V_{DS} < 0$, $V_{BS} > 0$, $V_{GS} < 0$, or $V_{BD} > 0$, 149

References and Notes, 156

3 BSIM3 Parameters — 158

3.1 List of Parameters According to Function, 158

3.2 Alphabetical Glossary of BSIM3 Parameters, 162

3.3 Flow Diagram of SPICE Simulation, 275

References and Notes, 283

4 Improvable Areas of BSIM3 — 284

4.1 Lack of Robust Non-Quasi-Static Models: Transient Analysis, 285

4.2 Problem with the 40/60 Partition: The "Killer NOR Gate", 299

4.3 Lack of Channel Resistance (NQS Effect; Small-Signal Analysis), 303

4.4 Incorrect Transconductance Dependency on Frequency, 313

4.5 Lack of Gate Resistance (and Associated Noise), 316

4.6 Lack of Substrate Distributed Resistance (and Associated Noise), 323

4.7 Incorrect Source/Drain Asymmetry at $V_{DS} = 0$, 329

4.8 Incorrect C_{gb} Behaviors, 333

4.9 Capacitances with Wrong Signs, 339

4.10 C_{gg} Fit and Other Capacitance Issues, 341

4.11 Insufficient Noise Modeling (No Excess Short-Channel Thermal Noise), 356

4.12 Insufficient Noise Modeling (No Channel-Induced Gate Noise), 362

4.13 Incorrect Noise Figure Behavior, 366

4.14 Inconsistent Input-Referred Noise Behavior, 375
4.15 Possible Negative Transconductances, 376
4.16 Lack of GIDL (Gate-Induced Drain Leakage) Current, 382
4.17 Incorrect Subthreshold Behaviors, 387
4.18 Threshold Voltage Rollup, 390
4.19 Problems Associated with a Nonzero RDSW, 391
4.20 Other Nuisances, 392
References and Notes, 396

5. Improvements in BSIM4   399

5.1 Introduction, 399
5.2 Physical and Electrical Oxide Thicknesses, 400
5.3 Strong Inversion Potential for Vertical Nonuniform Doping Profile, 402
5.4 Threshold Voltage Modifications, 403
5.5 $V_{GST,\text{eff}}$ in Moderate Inversion, 408
5.6 Drain Conductance Model, 409
5.7 Mobility Model, 412
5.8 Diode Capacitance, 414
5.9 Diode Breakdown, 417
5.10 GIDL (Gate-Induced Drain Leakage) Current, 425
5.11 Bias-Dependent Drain–Source Resistance, 427
5.12 Gate Resistance, 431
5.13 Substrate Resistance, 434
5.14 Overlap Capacitance, 435
5.15 Thermal Noise Models, 436
5.16 Flicker Noise Model, 450
5.17 Non-Quasi-Static AC Model, 451
5.18 Gate Tunneling Currents, 459
5.19 Layout-Dependent Parasitics, 465
References and Notes, 472

Appendixes 473

- A BSIM3 Equations, 473
- B Capacitances and Charges for All Bias Conditions, 507
- C Non-Quasi-Static y-Parameters, 510
- D Fringing Capacitance, 515
- E BSIM3 Non-Quasi-Static Modeling, 519
- F Noise Figure, 522
- G BSIM4 Equations, 528

Index 583

PREFACE

Here is a little secret about BSIM3. You cannot find what BSIM stands for in the official BSIM3 manual, nor the official BSIM3 web site (at least as of March, 2000). The reason is simple—BSIM is a household name in the SPICE simulation community. It is so much used that it has become a standard vocabulary, in a similar way that we tend to forget laser stands for *l*ight *a*mplification by *s*timulated *e*mission of *r*adiation. To save you from scratching your head, I state that BSIM is an acronym for *B*erkeley *S*hort-Channel *I*GFET *M*odel. As the full name suggests, it is a MOSFET model capable of handling modern CMOS devices with pronounced short-channel effects. But this description also begs the question exactly what short-channel effects are handled by BSIM3. Or, more generally, does the model account for various leakage currents and terminal resistances?

These questions as well as others are answered in Chapter 2, which summarizes the key components of BSIM3. Because circuit designers tend to view devices with an equivalent circuit, I also represent the BSIM3 model with various equivalent circuits, for various operating conditions. Although Chapter 2 serves as an entry for those who are eager to learn the basics of BSIM3, some modeling background is presented in Chapter 1. This introductory chapter is where the readers can find the meanings and connotations of various modeling jargons, such as charge conservation, quasi-static approximation, and so on. For those who really have no time to understand the details (and thus the beauty) of modeling, I have prepared Chapter 3. It contains an alphabetical listing of the BSIM3 parameters, as well as their meanings and relevant equations.

No model is perfect. BSIM3 certainly shares many of the problems that haunt other models. In Chapter 4 several areas of improvement for the BSIM3 model are pointed out, whether it is used in a dc, ac, transient, or noise analysis. Most of these improvements have been made in BSIM4, the to-date newest member of the BSIM family introduced in 2000. The improvements of the BSIM4 are described in Chapter 5.

Many people cry foul when I praise BSIM3. They contend that it is just a biased opinion. Well, the time has come for me to clarify one point. Although I have used BSIM3 quite often and personally liked BSIM3, I am not one of the BSIM3 authors. The misconception can arise because the chief researcher in charge of the BSIM3 (and BSIM4) development happens to have the same first name initial (W.) and the same last name (Liu) as mine. This coincidence is evident particularly when the BSIM3 manual is referenced. See, for example, Ref. [14] of Chapter 1. I surely did not write any portion of the BSIM3 manual, although the reference seemingly suggests so. I merely write this book to complement the manual. Our last name, incidentally, is the *royal* last name of the most powerful Chinese dynasty. The Han dynasty, which existed some 2000 years before the Republic of China, is credited for the naming of the Chinese people as the Han race. And for those who aspire to be a millionaire (such as in the popular television show), here are some trivia for you— paper was invented in Han dynasty by a chamberlain/eunuch serving Emperor Liu, and the newly elected president of Republic of China is a National Taiwan University graduate named Shui-bian Chen.

The development of this book has been benefited by my colleagues at or formerly at Texas Instruments, as well as the BSIM team from the University of California at Berkeley. I would especially like to thank Keith Green, Karthik Vasanth, Tom Vrotsos, Paul Koch, James Hellums and David Zweidinger for their written comments and suggestions, and Professor Cheming Hu, Weidong Liu, Xiaodong Jin, Mark Cao, and Yuhua Cheng for their outstanding efforts in providing the BSIM models and for answering my questions about BSIM3 and BSIM4. Professor Hu and Yuhua's book, *MOSFET Modeling & BSIM3 User's Guide,* has been very helpful in the preparation of this work.

The discussions with Rajni Aggarwal, Ajith Amerasekera, Jonathan Brodsky, Britt Brooks, Mi-Chang Chang, Ming Chiang, Paul Cox, Charvaka Duvvury, Paul Ehnis, Ulvi Erdogan, Ranjit Gharpurey, Vinod Gupta, John Krick, Brian Mounce, Ismail Oguzman, Shridar Ramaswamy, Jerry Seitchik, Robert Steinhoff, Sreenath Unnikrishnan, Robert Virkus, Bruce Thornton, Doug Weiser, and Ping Yang, who have helped shape the final form of this book, are gratefully acknowledged. My special gratitude goes to "Coach", Professor James Harris of Stanford University, who, since my graduate studies there, has inspired me to do great research and seek physical understanding.

My fellow Stanford colleagues, Darrell Hill, now at Motorola, Chong-Hong Dai, a manager at Intel, as well as Ken Shepard, a Columbia University professor, also helped me with some simulator questions. I would also like to thank Christian Enz of Conexant, Dr. Klaassen, and Luuk Tiemeijer of Philips for clarifying rf issues in modeling. The book editor at Wiley, George Telecki, as well as the anonymous reviewers, have all been extremely helpful in shaping this project to publication. Finally, I thank my wife, Lee-Ping Chong, and my parents, Chien-Shun and Li-Yue Liu, for their encouragement and support.

WILLIAM LIU

1

MODELING JARGONS

1.1 SPICE SIMULATOR AND SPICE MODEL

Just about all electrical engineers have some form of encounter with SPICE, $\underline{S}$imulation $\underline{P}$rogram with $\underline{I}$ntegrated $\underline{C}$ircuit $\underline{E}$mphasis. It could be in a homework assignment during the undergraduate studies, or as an essential part of circuit design at work. As its full name suggests, SPICE is a computer program that accepts a circuit schematic as input and outputs the simulated circuit behaviors. The simulation can be performed under the nonlinear dc, nonlinear transient and linearized ac operating conditions. The circuit may contain resistors, capacitors, inductors, mutual inductors, independent voltage and current sources, dependent sources, lossless and lossy transmission lines, switches, uniform distributed RC lines, and various semiconductor devices including MOSFETs (metal oxide semiconductor field-effect transistors). The original SPICE program, SPICE1, was developed at University of California, Berkeley, and released for public use in May, 1972. By 1975, after the next major release, called SPICE2, SPICE was in widespread use and adopted by most integrated circuit manufacturers. SPICE2 was written in Fortran. With the advent of UNIX computers in the 1980s it became increasingly obvious that SPICE would benefit from the C-shell utilities that come with UNIX computers. This prompted the rewriting of the SPICE program in the C-language, although the basic algorithms remain largely intact. The revised SPICE program, called SPICE3, was released in public domain in March, 1985.

The free distribution by Berkeley is a key factor contributing to the universal acceptance of SPICE. Accompanying with this free lunch, however, is the inevitable lack of "product" support. Several large companies have written their own version of the circuit simulation program to better serve their companies' particular interests. One notable proprietary SPICE, which pioneered the use of charge-based device

2   MODELING JARGONS

model as the solution to the charge non-conservation problem, is the TI-SPICE, still in use in Texas Instruments today [1]. In a charge-based approach, the charge (instead of voltage) becomes the state variable in calculating the transient behavior of MOSFET. This ensured charge conservation, as discussed later in Section 1.8. There are also vendors who develop commercial versions of SPICE to tailor the needs of the small companies without a CAD (computer aided design) group of their own. One commercial version is HSPICE. It is a robust SPICE program combined with an excellent graphic interactive interface. HSPICE is used extensively on UNIX-based workstations. Another example is PSPICE, introduced in 1984, which runs on PC and Macintosh platforms (but at a significantly slower speed than UNIX-based SPICE programs). Because of the proliferation of personal computers, this PC-based SPICE program has attracted many new users and helped SPICE make another leap in popularity since the introduction of SPICE2. These various versions of the SPICE program provide improvements in user interfaces, user support, numerical convergence, and device modeling. Although these (non-Berkeley) versions contain several differences in the implementation of the numerical routines or device models, they nonetheless retain the original SPICE's programming structure. The following lists some notable SPICE programs that are (were) readily available for the public's use (free or for a fee).

Berkeley SPICE   The original SPICE. It is in the public domain and can be run on UNIX platforms. See http://infopad.eecs.berkeley.edu/~icdesign/SPICE and http://hera.eecs.berkeley.edu/~software/spice3f5.html. The latest version, SPICE3f5, supports BSIM3v3.1, but not higher versions of models. BSIM3v3.2 and higher are supported by SPICE3e2.

I-SPICE   Interactive SPICE, developed in the late 1970s. This is the first commercial version of SPICE. As stated in [2], "it probably would have been a success were it not for the impending failure of the time-sharing service concept."

HSPICE   Created by Meta-Software and now owned by Avant!. It is popular within UNIX-based users, and known for its interactive user interfaces. BSIM2, BSIM3v2, and BSIM3v3 are implemented as levels 39, 47, and 49, respectively. Its proprietary MOSFET model (level 28) remains popular today.

PSPICE   PC-based version SPICE created by Micro-Sim, which was recently acquired by Orcad. It has evolved to support BSIM models as well as complete IC models.

SPECTRE   Created by Cadence. It can be thought of as an improved Berkeley SPICE that addresses several numerical problems and the inadequacies in simulation for r.f. circuits. It supports all Berkeley MOS models.

Today there are more than 100,000 copies of SPICE in active use at universities and industry [3], and some trial versions can be found online [4]. There are currently about 20 companies providing SPICE products and support commercially [3].

1.1 SPICE SIMULATOR AND SPICE MODEL 3

In the previous discussion of various derivatives of the Berkeley SPICE, we did not make a clear distinction between the SPICE simulator and the SPICE device model. Together, these two parts form the overall SPICE program. The SPICE simulator is the mathematical engine of SPICE, consisting of several basic subroutines to perform numerical analyses. One exemplar numerical subroutine is matrix inversion, which is the backbone algorithm to solve n linearly independent equations with n unknowns. This routine, when combined with the Newton-Raphson iteration technique, allows the solution of nonlinear equations which govern the static nodal voltages and the branch currents of a given circuit.

Another basic SPICE subroutine solves ordinary differential equations. This type of routine is invoked in a transient analysis involving time as the independent variable. The transient analysis is a one-dimensional problem in which the voltages and currents are functions of time only. Other simulation programs, notably PISCES [5] (or its commercial version—MEDICI), which simultaneously solve the Poisson equation and current continuity equations, address four-dimensional problems involving space (x, y, z) and time (t) as the independent variables. As far as numerical analysis is concerned, it is inherently more difficult to solve a multi-dimensional partial differential equation than a one-dimensional ordinary differential equation. Therefore, SPICE simulation of a circuit is generally faster and enjoys less convergence problems than PISCES simulation of a device. On the other hand, PISCES (or MEDICI) simulation solves the fundamental physical equations governing the device. Once the doping profiles and physical device structure are supplied to the simulator, the solution represents the actual characteristics expected from the device. Often the phrase "MEDICI results" is mentioned in the book. It is implied that the MEDICI simulation results are the exact solution against which the SPICE simulation results can be compared. If the results of a SPICE simulation deviate significantly from a MEDICI simulation, we conclude that either the SPICE model used in the simulation is imperfect, or its model parameters are not properly extracted.

Besides its number-crunching ability, the SPICE simulator also handles the input and output details of the overall SPICE program. It takes the input SPICE deck and constructs the circuit for simulation. It prepares results in a data format specified by the user.

The second part of a SPICE program is the device model. There can be many semiconductor devices in a circuit, such as a diode, a bipolar transistor, or a capacitor. Since this book concerns MOSFET, we shall concentrate on the MOSFET device model in particular. A model mathematically represents the device characteristics under various bias conditions. In dc and ac analyses, the inputs of the device model are the drain-to-source, gate-to-source, bulk-to-source voltages, and the device temperature. The outputs are the various terminal currents. The model parameters, along with the equations in the SPICE model, directly affect the final outcome of the terminal currents. In a transient analysis we can think that the SPICE model accepts the time derivative of the bias voltages, in addition to the absolute values of the biases themselves at an instant of time. The output of the SPICE model is then the terminal currents at that particular instant of time. If a noise analysis is specified in the input deck, the SPICE model also computes the noise voltages at a particular set of bias condition and frequency.

We illustrate the interaction between the SPICE simulator and the SPICE model using the dc circuit shown in Fig. 1-1. Although the input bias current is given, the nodal voltages as well as the branch currents are not known. The SPICE simulator first guesses a solution set for all the unknown nodal voltages (at node 1, 2, and 3). A subset of these guessed voltage values which bias up the MOSFET (V_2 and V_3) are passed to the SPICE model as inputs. Subsequently, the SPICE model evaluates the currents at the four terminals of the MOSFET and feeds the information back to the SPICE simulator. The SPICE simulator then checks whether the Kirchoff current law is fulfilled at each node, that the sum of currents at a node is zero. (Strictly speaking, the SPICE simulator does not really check the Kirchoff current law. It uses a slightly different form of convergence test. We will elaborate on this fine point in Section 1.2.) More than likely, the first guess does not prevail and the Kirchoff current law is violated. Subsequently, the derivative of the currents with respect to the voltages (such as $g_m = \partial I_D/\partial V_{GS}$, $g_d = \partial I_D/\partial V_{DS}$ and $g_{mb} = \partial I_D/\partial V_{BS}$) are evaluated. These derivatives are used in the Newton-Raphson subroutine to make an educated guess of the nodal voltages for the next iteration. These newly guessed voltages are again fed to the SPICE device model, and the terminal currents are re-evaluated. With each successive iteration, the guessed set of voltage values approaches the actual solution. This process iterates until the Kirchoff current law becomes satisfied at every node. (Or, as more correctly put in Section 1.2, the iteration stops when the nodal voltages and the branch currents are within a prescribed tolerance in two consecutive iterations.)

There are two points relating to the convergence process of SPICE simulator. First, even for a dc analysis, the small-signal conductances (such as g_m, g_d, and g_{mb}) are required from the model. These small-signal quantities calculated from the SPICE model provide valuable information about the circuit itself, aiding the SPICE simulator to make educated guesses in subsequent iterations. Therefore, suppose that a circuit is designed to operate only at dc. A MOSFET model still needs to supply the small-signal quantities. Second, these small-signal quantities aid the speed at which the dc convergence is met, but does not preclude convergence when their values are incorrect (such as due to coding errors). The SPICE simulator may still

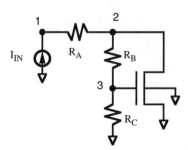

Fig. 1-1 A circuit used to illustrate the interaction between a SPICE simulator and a SPICE model.

converge, albeit at a much slower pace. There are occasions where a model produces negative conductances. In this case, the convergence for the dc analysis may be significantly hampered, and in extreme cases, prevents a solution from ever being reached (Section 1.2). In an ac (or noise) analysis, however, the accuracy of the small-signal parameters are critical to the accuracy of the solution itself.

We have mentioned several SPICE simulators before, including Berkeley SPICE, HSPICE, and PSPICE. They are not complete without the SPICE models. Some notable SPICE models for MOSFET are listed in the following, in somewhat a chronological order of the time the model was introduced.

Level 1 — Also known as the Shichman-Hodges model [6], this is the original model since the dawn of Berkeley SPICE. The model equations are simple, resembling those used in undergraduate textbooks, and are applicable mainly to long-channel devices. The C-V portion of the model is the Meyer model [7], which is not a charge-conserved model (Section 1.8).

Level 2 — This model addresses several short-channel effects such as the velocity saturation [8]. However, the mathematical implementation of the model was complicated, leading to many convergence problems. For C-V calculation, either the Meyer model of Level 1 or the Ward-Dutton model [9] can be used. The Ward-Dutton model is a charge-conserved model, which forms the backbone of all present models.

Level 3 — This semi-empirical model is regarded as a simplified version of Level 2 [8]. This model has proven to be robust and is popular for digital circuit design. However, the model is not very scalable and binning (Section 1.12) is almost always required. Discontinuities in the first derivative of the drain current exist.

BSIM — It is the Berkeley Short-Channel IGFET Model [10], sometimes referred to as Level 4. The model places less emphasis on the exact physical formulation of the device, but instead relies on empirical parameters and polynomial equations to handle various physical effects. This generally leads to improved circuit simulation behavior compared to previous models, although its accuracy degrades in submicron FETs. Furthermore, the polynomial equations can behave poorly, causing negative output conductance and convergence problems.

HSPICE Level 28 — This proprietary model developed by Meta-Software is similar to BSIM [11]. However, with proper modification in the binning strategy and mathematical description in the transition region (Section 1.12), Level 28 has been made suitable for analog design and remains popular to date.

6 MODELING JARGONS

BSIM2   This is an extension to BSIM, with comprehensive modifications which make it suitable for analog circuit design [12]. Although BSIM2 improves upon BSIM in terms of model accuracy as well as convergence behavior in circuit simulation, it still breaks the transistor operation into several regions. This leads to discontinuity in the first derivative in I-V and C-V characteristics, a result that can cause numerical problems in simulation.

BSIM3   With the help of smoothing functions (Section 1.4), BSIM3 adopts a single-equation to describe device characteristics in various operating regions [13]. This eliminates the discontinuity in the I-V and C-V characteristics. BSIM3 has evolved through three versions. BSIM3v1 and BSIM3v2 contain many mathematical problems so that they are largely replaced by the third version; BSIM3v3. There are several variations within BSIM3v3 itself, including BSIM3v3.1 and BSIM3v3.2. The latter eventually grew into two further variations: BSIM3v3.2.1, and BSIM3v3.2.2 [14]. A procedure to identify the exact version of a BSIM3v3 model is found in Section 3.2, under the entry of VERSION. These variations of BSIM3v3 have minor differences, and have been demonstrated for accurate use in 0.18 μm technologies. The manuals, codes, and news about BSIM3 can be found online at http://www-device. eecs.berkeley.edu/~bsim3. BSIM3 is a level 8 model. At the time of this writing there has been discussion to release BSIM3v3.3 [15].

Model 9   MOS Model 9 is the primary non-Berkeley model available for public use [16]. The model also employs smoothing functions to achieve continuity in device characteristics. The model is accurate for sub-quarter micron technologies and exhibits good behaviors in circuit simulation. Model 9 is probably as good as BSIM3. However, companies generally opted for BSIM3 because it was not obvious whether there were intellectual property issues associated with MOS 9, a model developed at Philips Laboratories. The manuals, codes, and news about Model 9 can be found online at http://www-us.semiconductors.com/Philips_Models/mosmodel9.stm.html

EKV Model   This model is unique in its use of bulk-referencing [17], while all other mentioned model employs source-referencing (Section 1.10). This fundamental philosophical change allows the EKV model a greater hope of fundamentally eliminating the asymmetry problems unavoidable in the source-referencing models. Despite its adoption of a more physical modeling approach, the EKV model is not yet popular, partly because of its relatively late arrival compared to other models. The manuals, codes, and news about the EKV model can be found online at http://legwww.epfl.ch/ekv/.

BSIM4   The newest addition to the BSIM family, made public in the year 2000. BSIM4 offers several improvements over BSIM3, not just in the traditional I-V modeling of the intrinsic transistor, but also in the transistor's noise modeling, and in the incorporation of extrinsic parasitics. BSIM4, a level 14 model, is discussed in Chapter 5.

Level 1, 2, and 3 models are generally referred to as the first-generation models, in which the models emphasize the device physics. The attention given to physically accurate representation without equal consideration to mathematical representation often creates numerical problems during circuit simulation. The second-generation models, epitomized by BSIM, BSIM2, and HSPICE Level 28, corrects this problem with greater focus on mathematical implementation. Several empirical parameters without clear physical meanings are incorporated in the model. This approach gains the advantage of improved convergence properties, but at the cost of complicating the parameter extraction process as well as weakening the link between model parameters and fabrication process. The third-generation models, represented by BSIM3 and Model 9, seek to reintroduce a physical basis to the model, while maintaining the mathematical fitness of the model equations. These models rely on smoothing function to result in a single-equation that describes the I-V and C-V characteristics. In this book we will concentrate on the BSIM3 model, the Compact-Model-Council-supported standard model and the *de facto* industry standard model being used today. Details of other mentioned models can be found in Ref. [18]

Generally the term SPICE is used without a clear distinction between the SPICE simulator and the SPICE models. For example, a statement "the simulation is done with the Berkeley SPICE" is ambiguous. Depending on the context, the sentence can mean that the simulation is performed with the Berkeley SPICE simulator (such as SPICE3f), although the MOSFET model is non-Berkeley, for example, the level 28 model created by MetaSoftware. Conversely, the sentence could mean that the simulation was done with the HSPICE simulator, but with a Berkeley SPICE model (such as BSIM). Finally, it can also mean that the simulation was done with both the Berkeley SPICE simulator as well as the Berkeley SPICE model. Not all MOSFET models are supported by all SPICE simulators. Obviously, while TI-SPICE simulator supports both TI's proprietary SPICE MOSFET model as well as BSIM3, the Berkeley SPICE simulator does not support the TI-SPICE model. Moreover, a SPICE simulator often supports several SPICE models for the MOSFET, using the parameter LEVEL to distinguish between various models. For example, HSPICE supports BSIM2, BSIM3v2, and BSIM3v3 models, and are implemented as levels 39, 47, and 49, respectively.

8 MODELING JARGONS

Some SPICE programs are basically known for their SPICE simulators. For example, PSPICE is well known because its SPICE simulators run on the PC platform. PSPICE's SPICE models for MOSFETs are those public domain models created by Berkeley over the years. However, occasionally the SPICE model is as well known as its SPICE simulator. Take HSPICE, for example. Its Level 28 MOSFET model was introduced at a time when there was no good analog model. This proprietary model is used as a vehicle to gain widespread acceptance of the HSPICE simulator.

1.2 NUMERICAL ITERATION AND CONVERGENCE

SPICE is a numerical program constructed to solve the voltages and currents of a given circuit. Consider the simple 3-node circuit shown in Fig. 1-1. According to the Kirchoff's current law (KCL) for node 2, we have

$$\frac{v_1 - v_2}{R_A} - \frac{v_2 - v_3}{R_B} - \frac{W}{L}\frac{\mu_n C'_{ox}}{2}(v_3 - V_T)^2 = 0. \qquad (1\text{-}1)$$

For simplicity, we have modeled the drain current of the MOSFET with the well-known saturation current equal to $W/L \cdot \mu_n \cdot C'_{ox} \cdot (V_{GS} - V_T)^2/2$ [19], where W, the channel width, L, the channel length, μ_n, the channel mobility, C'_{ox}, the device's oxide capacitance per unit area, and V_T, the threshold voltage, are known quantities (as specified in or calculated from the SPICE model). Equation (1-1) can be rewritten in a more abstract fashion,

$$f_2(v_1, v_2, v_3) = 0, \qquad (1\text{-}2a)$$

where f_2 is the function relating to the various circuit elements which tie at node 2. Similar KCL statements imposed at nodes 1 and 3 result in the following equations:

$$f_1(v_1, v_2, v_3) = 0; \qquad (1\text{-}2b)$$
$$f_3(v_1, v_2, v_3) = 0. \qquad (1\text{-}2c)$$

The SPICE simulator solves these three equations of three unknowns by the Newton-Raphson algorithm [20], which seeks the solution of a set of nonlinear equations through the iterative solutions of a sequence of linear equations. The algorithm starts out by taking an initial guess of the solution of the circuit. Let us denote the zero-th iteration solution as, $v_1^{(0)}$, $v_2^{(0)}$, and $v_3^{(0)}$. The guessed solution is checked against a set of convergence criteria. If the guessed solution is not close to the true solution, then the algorithm calls for an evaluation of the derivatives of $f_n(v_1, v_2, v_3)$ for all nodes $n = 1, 2$, and 3. These derivatives point to the general direction in which the next solution should be guessed to maximize the chance of convergence. An intuitive understanding of how the derivatives help convergence is shown in Fig. 1-2, which illustrates the 1-dimensional analog of the present problem.

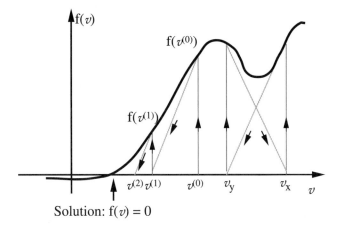

Fig. 1-2 Iterative process used in the Newton-Raphson technique to solve for the root v that satisfies $f(v) = 0$.

We intend to solve for v such that $f(v)$ crosses the x-axis. $v^{(0)}$ is the initial guess, and obviously not the solution. At this particular guess, both $f(v^{(0)})$ and its derivative, $\partial f(v^{(0)})/\partial t$, are evaluated. We draw a dotted line from $f(v^{(0)})$ toward the x-axis, with a slope equal to the derivative. The intersection of the dotted line with the x-axis is the next best guess, denoted as $v^{(1)}$. We evaluate the function f with the new solution, finding it equal to $f(v^{(1)})$. Because $f(v^{(1)})$ differs from 0, it is not the solution, and the next solution is guessed, again based on the information of the derivative, which is $\partial f(v^{(1)})/\partial t$ this time. This iterative process continues until some convergence criteria are satisfied.

While most often the Newton-Raphson method converges to a solution, occasionally the iteration fails to converge. The right portion of Fig. 1-2 illustrates this possibility. Suppose instead that the initial guess is v_x. From $f(v_x)$ and the derivative at that point, we follow the algorithm and trace the next guess of solution to be v_y. Because $f(v_y) \neq 0$, we have not converged to a solution and another iteration resumes. From $f(v_y)$ and the derivative at that point, we follow the algorithm and trace the next guess of solution to be v_x. This is obviously not a solution, either. From here onward, the subsequent guesses of solution oscillate back and forth between v_x and v_y, without any hope of ever converging to the root. In this particular example, two causes are responsible for the nonconvergence. The first is a bad initial guess, and the second, the existence of negative conductance (equal to the slope at $f(v)$), which leads to the oscillations in the guessed solutions. The first problem can be overcome by the user, with either the .NODESET or the .FORCE functions of the SPICE simulator. The .NODESET statement allows for the user-determined initial voltage values for all or part of the nodes, whereas the .FORCE statement fixes the voltage to some particular values. The second problem, however, is intrinsic to the device model, and usually cannot be modified by the user. The user, however, may modify the bias voltage slightly so the transistor is biased at a different operating point, with the hope that somehow a new set of conductance values facilitates the exit of the oscillation mode.

Figure 1-2 is a graphical illustration of the Newton-Raphson algorithm in one dimension, for the solution of one variable and one unknown. Let us return to the discussion of the SPICE solution for the circuit of Fig. 1-1, which is a multi-dimensional problem. As mentioned, at each iteration the SPICE simulator guesses a set of solutions, $v_1, v_2, \ldots, v_n$, where n is the number of nodes. If the solution set does not meet a certain convergence criteria, then a new set of solution is obtained from the derivatives at $f(v_1), f(v_2), \ldots, f(v_n)$, and the iteration continues. To simplify the discussion of the convergence criteria used in the SPICE simulator, let us denote the solution set at the j-th iteration as $v_1^{(j)}, v_2^{(j)}, \ldots, v_n^{(j)}$. (Again, n refers to the node number.) In SPICE, this set of solution is considered the desired solution when the following two conditions are met simultaneously for all nodes n:

$$|v_n^{(j)} - v_n^{(j-1)}| < \text{vntol} + \text{reltol} \cdot \max(|v_n^{(j)}|, |v_n^{(j-1)}|); \quad (1\text{-}3\text{a})$$

$$|f_n(v_1^{(j)}, v_2^{(j)}, \ldots, v_n^{(j)}) - f_n(v_1^{(j-1)}, v_2^{(j-1)}, \ldots, v_n^{(j-1)})| < \text{abstol} + \text{reltol}$$
$$\cdot \max(|f_n(v_1^{(j)}, v_2^{(j)}, \ldots, v_n^{(j)})|, |f_n(v_1^{(j-1)}, v_2^{(j-1)}, \ldots, v_n^{(j-1)})|). \quad (1\text{-}3\text{b})$$

The vntol, abstol, and reltol are the tolerance levels, having default values of 1 µV, 1 pA, and 0.001, respectively. Their values can be modified in the .OPTION statement. (*Note*: Some SPICE simulators may have different names for these tolerances.) The first criterion requires the voltage of a node at a given iteration to be close to its predecessor, within a specified tolerance level. This guarantees that the nodal voltages of a circuit settle at some stable values. The tolerance is composed of two components. The relative tolerance (reltol) allows the simulation of the high-voltage and low-voltage circuits without adjusting the tolerance levels. The absolute nodal voltage tolerance (vntol), in contrast, is necessary in cases where the voltage and current of the circuit are fairly close to zero. Without the latter, the updated solution can be within the computer rounding error or simply too small to be ever considered converged. While the first criterion (Eq. 1-3a) focuses on the voltage, the second criterion (Eq. 1-3b) ensures the functions obtained from the KCLs at all the nodes are relatively the same at each successive iteration. We emphasize that, although Eq. (1-3b) seems to impose the KCL, it really just checks whether the function values settle down. In other words, even if a solution meets both the requirements of Eq. (1-3), the KCL is not necessarily satisfied. It turns out that checking the Kirchoff's current law as a part of convergence criteria is not a straightforward task. To ensure that KCLs are satisfied, we should have checked whether $f_n(v_1^{(j)}, v_2^{(j)}, \ldots, v_n^{(j)})$ is close to zero, rather than close to $f_n(v_1^{(j-1)}, v_2^{(j-1)}, \ldots, v_n^{(j-1)})$, which needs not be zero. In the event that the convergence criteria are met at the j-th iteration but $f_n(v_1^{(j)}, v_2^{(j)}, \ldots, v_n^{(j)}) \neq 0$, then at each node the currents do not quite sum to zero. Equivalently, a small current source, with a value smaller than that allowed by the convergence criteria, is connected to each of the nodes. The additional injected current can become problematic at a high impedance node, since it would contribute to an appreciable amount of voltage error. The requirement of Eq. (1-3a) curtails the degree of this problem somewhat, by forcing the convergence to take place only after the node voltages settle down. Nonetheless, there is always a possibility that false convergence occurs, and the SPICE simulator reaches the wrong solution.

The accuracy and speed of the overall circuit simulation depend critically on both the SPICE simulator and the device model. If the numerical algorithm implemented in the SPICE simulator is not robust, then we have no hope of expediency in the numerical convergence process, even though the SPICE model provides a realistic representation of the device. Conversely, if the SPICE simulator is well implemented but the SPICE model produces negative conductances or kinks in its device characteristics, then the convergence will be slow (or never).

The above description of the algorithm to solve for the voltages and currents of a 3-noded circuit is a gross simplification of what is actually implemented in the SPICE simulator. The actual implementation is more robust in the construction of the nodal equations to be solved, and incorporates several subroutines to facilitate convergence. However, the presented concepts of the algorithm are correct and allow us better to appreciate the inner working of the SPICE simulator. The convergence criteria of Eq. (1-3) are encountered in typical SPICE simulators, especially those derived from the Berkeley SPICE [21]. As such, Kirchoff's current law is not checked during the simulation (although the to-be-solved equations are formulated based on KCL).

1.3 DIGITAL VS. ANALOG MODELS

Having differentiated the SPICE simulator and the SPICE model portions of the overall SPICE program, we shall concentrate exclusively on the SPICE model from here onward. We begin with a classic oxymoron expected of the device modeling engineers:

A model shall be accurate *and* simple.

Certainly, a model should be accurate enough so that the results produced in a simulation are trustworthy. The model should also be simple so that the simulation time is minimal and the process for parameter extraction can be easily implemented. However, creating a model that is both accurate and simple is by no means a simple task.

A balance between the model simplicity and accuracy needs to be attained. The exact equilibrium point is primarily determined by the end application for which the model is used. Over 95% of the world's semiconductor sales is in silicon, the majority of which are digital CMOS circuits. With this level of sales, it is no wonder that the traditional MOSFET modeling efforts focus on digital circuits. These models, loosely called *digital models*, are used in circuitries with many transistors

12 MODELING JARGONS

(as opposed to only a few, as in analog circuits). It is crucial that the time evaluating the model equations be short, to minimize the overall SPICE simulation time. Therefore, as far as the digital models are concerned, the delicate balance had favored model simplicity in the past, sacrificing accuracy in the device model. Although the accuracy at the device level is compromised, the simulation at the overall circuit level can still be accurate.

The I-V characteristics produced by one extreme example of a digital model are schematically shown in Fig. 1-3. Although the fitted I_D-vs.-V_{DS} curve is mostly inaccurate and the subthreshold characteristics are way off (even in a logarithmic plot), the drive current is fitted fairly accurately. (Drive current is the device current when the gate and drain biases are at the maximum supply voltage of a given technology.) This kind of fitting is ghastly to analog circuit designers, but it may just be good enough for some types of digital circuits. In fact, experiences have shown that if the drive currents and the parasitic capacitances are modeled well, an inverter's switching speed is well predicted by the model. The relaxed requirement on accuracy in fitting the device I-V characteristics allows the digital model to be fairly crude, describable with only the least amount of model parameters. Consequently, in the past, when modeling activity was predominantly for the prediction of the inverter delay time in digital circuits, a model capable of putting forth the I-V characteristics shown in Fig. 1-3 was deemed good enough. An elaborate model which can fit I-V characteristics in all operating regions would necessarily involve more model parameters, making the parameter extraction process more time-consuming. During a SPICE simulation, this more elaborate model would also require more mathematical evaluations, thus slowing down the speed of the simulation. The elaborate model, in the interest of digital circuits, is both overly constructed and inefficient.

However, when the same digital model is applied to analog circuit design, gross simulation error results. This is because there are a lot more concerns in analog circuits than just the drive current and the parasitic capacitances. With the recent trend toward mixed analog-digital chips, low-voltage operation, and higher speed, the concerns of analog circuit designers progressively receive more attention. Thus the delicate balance for the model acceptance has gradually shifted toward model

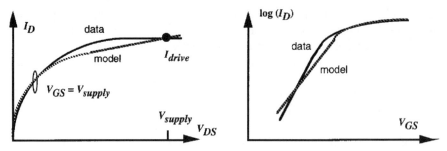

Fig. 1-3 I-V characteristics produced by a particular digital model. The drain current is plotted in the normal and logarithmic scales.

accuracy. The resulting model, being analog-circuit-friendly, is often referred to as an *analog model*. One particular interest of analog circuit design is the accurate fit of the small-signal parameters such as g_m, g_d, and g_{mb}. In the MOSFET, the drain current is a function of V_{GS}, V_{DS}, and V_{BS}. The *mutual transconductance* measures the amount of drain current increase caused by the increment in the gate bias:

$$g_m = \left.\frac{\partial I_D}{\partial V_{GS}}\right|_{V_{DS}, V_{BS} = const.} \tag{1-4a}$$

The *drain transconductance* measures the amount of drain current increase caused by the increment in the drain bias. It is defined as

$$g_d = \left.\frac{\partial I_D}{\partial V_{DS}}\right|_{V_{GS}, V_{BS} = const.} \tag{1-4b}$$

Finally, the *bulk transconductance* reveals the effect of back-gate bias on the drain current conduction:

$$g_{mb} = \left.\frac{\partial I_D}{\partial V_{BS}}\right|_{V_{GS}, V_{DS} = const.} \tag{1-4c}$$

We use Fig. 1-4 to illustrate the stringent requirement on the analog models. From the I-V characteristics on the left, we are tempted to think that the fit is good. The corresponding g_d, equal to the slope of I_D with respect to V_{DS}, is shown on the right. It is clear that although the model produces fairly good fit in the I-V characteristics, the model is nowhere close to the fitting of g_d. A good analog model is required to produce accurate fitting in both the I-V characteristics as well as the g_d characteristics. In order to accommodate the increased requirement, an analog model usually consists of a lot of parameters. Besides, the model equations need to be carefully derived so that the simultaneous fit of I-V and small-signal quantities is possible. From a pragmatic viewpoint, we cannot really say that the analog model is *better* than the digital model, since the digital model can produce accurate circuit

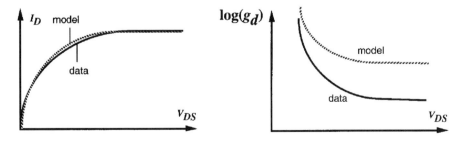

Fig. 1-4 A model that produces a good fit to the I-V characteristics may still yield a bad fit to the small-signal quantities.

simulation for its intended applications. We can only conclude that the digital model is simpler yet less accurate. This is the recurring theme—that the delicate balance between accuracy and simplicity depends on the application.

We have stated that the requirements for digital and analog circuit models are quite distinct. In the rest of this section, we list some of the distinguishing features of these models. We should recognize that the exact requirements for the digital model differ for designers working in various applications. For example, the fitting of subthreshold characteristics is crucial to some digital circuit designers (e.g., for the determination of the off current), while being totally irrelevant for other digital circuitries. Therefore, some generalizing remarks about the characteristics of the digital and analog models, to be made in the following, are inevitable. The digital model can be thought as a model sufficient for the design of low-end consumer product, and the analog model, for the relatively higher-performance higher-speed analog parts. By the natures of the applications, these two models result in sharply different device characteristics, although they both satisfy their respective circuit simulation needs. Incidentally, the analog model employed for the following discussion is based on BSIM3v3, while the digital model is not. The two model parameters were initially created from the same device measurement results. However, we intentionally change some parameter values so that the I-V characteristics are displaced from each other, allowing for a clear comparison between the simulation results of the two models.

1. *Discontinuity in g_d* Figure 1-5 shows the calculated g_d as a function of V_{DS} for the digital and analog models, for $V_{GS} = 2$ and 3 V. $V_{BS} = 0$. The analog model produces smooth transition as V_{DS} varies from 0 to 3 V, within which the transistor moves from linear to saturation region of operation. The drain conductance decreases monotonically, not saturating at a constant value. In contrast, the digital model has some undesirable features. Upon transiting from the linear to saturation region, the drain conductance exhibits a kink, a reminiscent of the regional equations employed in the digital model. That is, in a typical digital model the model prepares several sets of equations, with each set devoted to a particular operating region such as the saturation, linear, subthreshold, depletion, and accumulation regions. While each set of equation works well within its intended region of operation, the transition from one set to the other as the bias sweeps across different operating regions is not always smooth.

Besides the kink feature, there is another inaccuracy of the digital model revealed in Fig. 1-5. After the transistor enters the saturation region, the digital model produces a constant g_d, independent of V_{DS}. This implies that, in this digital model, the equations are such that the drain current can be expected to increase with V_{DS} only at a constant slope. In reality, g_d continues to be a function of V_{DS}, even during saturation.

2. *Discontinuity in g_m* Figure 1-6 exhibits the mutual transconductances simulated by the two models, at $V_{DS} = 0.1$ V and $V_{BS} = 0$ and -1.5 V. We shall focus on the curve obtained at $V_{BS} = 0$. Just as in the discussion concerning g_d, the analog model gives rise to smooth g_m across the subthreshold region ($V_{GS} < V_T \sim 0.7$ V), the saturation region ($V_T < V_{GS} < 0.8$ V), and the linear

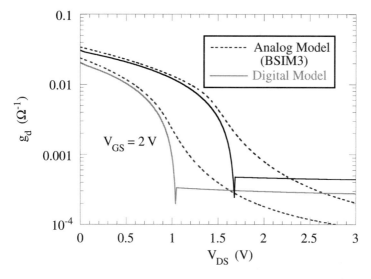

Fig. 1-5 Calculated g_d as a function of V_{DS}, at $V_{GS} = 2$ and 3 V. The discontinuity is a characteristic of the digital model, while BSIM3 produces smooth characteristics.

region ($V_{GS} > 0.8$ V). The digital model, on the other hand, produces kink characteristics between the subthreshold and the saturation region. In addition, the calculated g_m in the digital model changes abruptly as the transistor moves from the saturation to the linear region, with a discontinuity in the slope of g_m. These inaccuracies are again the direct result of the regional modeling approaches adopted in the digital model. The use of the to-be-discussed smoothing equations to smooth out the transitions between regions are key to nice characteristics.

3. *Kink in I_D* A discontinuity in the small-signal quantities is translated into kink problem in the drain current. Figure 1-7 plots the drain current as a function of V_{GS}, with $V_{DS} = 0.01$ and $V_{BS} = 0$. The left side reports I_D in linear scale, while the right side, in logrithmic scale. The analog model shows that the drain current increases exponentially with V_{GS}, without any kink behavior. The characteristics produced by the digital model, however, display a wiggle, right when the transistor enters into the saturation region from the subthreshold region. This is yet another manifestation of the digital model's adoption of the regional equations. The analog model's use of smoothing functions adds a fair amount of complexity into the model equations, but successfully removes the kink problems.

4. *Discontinuity in g_m/I_D* The mutual transconductance can be used to estimate the magnitude of the voltage or power gain of a transistor. It is desired that it be as large as possible. A transistor with a large area has a higher g_m, but at the cost of incurring more dissipation current. A more meaningful figure of merit than g_m itself is g_m/I_D, which normalizes the mutual transconductance to the current dissipation. We bear in mind that the g_m/I_D ratio is not the only parameter that a circuit designer cares about. At the maximum g_m/I_D point, the current mirror matching can be so

16 MODELING JARGONS

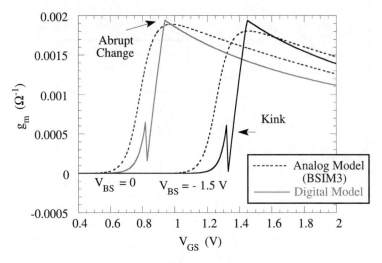

Fig. 1-6 Calculated g_m as a function of V_{GS}, at $V_{DS} = 0.1$ V and $V_{BS} = 0$ and -1.5 V. The digital model produces a kink at the transition between the subthreshold and the saturation regions.

poor or the cutoff frequency can be so low such that the circuits are not operated there. Nonetheless, it is a good figure of merit upon which several technologies can be compared.

The quantity g_m/I_D as calculated from the analog and digital models are shown in Fig. 1-8, plotted as a function of V_{GS}, and in Fig. 1-9, as a function of I_D. The transistor is biased with a V_{DS} of 3.3 V and $V_{BS} = 0$ and 3 V. When the transistor is in saturation, g_m/I_D is proportional to $1/(V_{GS} - V_T)$. As V_{BS} becomes more negative,

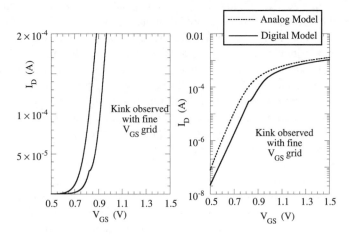

Fig. 1-7 Calculated I_D as a function of V_{GS}, with $V_{DS} = 0.01$ V and $V_{BS} = 0$. The drain current is plotted in the linear and logarithmic scales.

1.3 DIGITAL VS. ANALOG MODELS   17

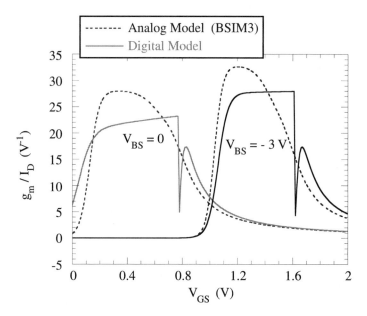

Fig. 1-8 Calculated g_m/I_D as a function of V_{GS}, at $V_{DS} = 3.3$ V and $V_{BS} = 0$ and -3 V. The reasons why the ratio decreases as V_{GS} approaches zero are elaborated in Section 4.17.

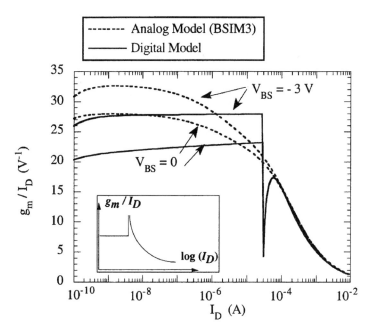

Fig. 1-9 Same plot as Fig. 1-8, except that the g_m/I_D ratio is now plotted as a function of I_D. The inset shows the calculated g_m/I_D ratio simulated with Level 2 and Level 3 models.

the threshold voltage increases because of body effect. Therefore, Figs. 1-8 and 1-9 reveal that the peak of g_m/I_D is higher for the negative V_{BS} case than when $V_{BS} = 0$. The peak value occurs at a V_{GS} near the turning point between the subthreshold and strong inversion regions. As V_{GS} increases so that the transistor moves away the saturation region and into the linear region, the ratio decreases monotonically.

As previously demonstrated, the analog model does not produce a kink or discontinuity in either g_m or I_D. Hence, the g_m/I_D ratio calculated from the analog model is well behaved, without any abrupt change of value. The digital model, again due to the regional modeling approach, suffers from a discontinuity in g_m/I_D near the boundary of the subthreshold and the strong inversion regions. Another inaccuracy in the digital model relates to that fact that g_m/I_D stays relatively constant in the entire subthreshold region, as more evident in the plot of Fig. 1-9. In reality, the ratio should show some degree of bias dependence. In some even more simplified digital models (not the one whose characteristics are shown), the g_m/I_D value is, in fact, fixed as a constant value, due to the use of simplistic modeling in the subthreshold region.

The g_m/I_D ratio starts out with low values when I_D is about 10^{-12} A. This result comes from the fact that, for both the analog and digital models under consideration, there is an extra resistor placed in parallel to the source-bulk and the drain-bulk junctions (Section 2.2). The conductance of these resistors is equal to GMIN, which is defaulted to 10^{-12} mho but modifiable in the .OPTION statement. These resistors are placed mainly to improve the convergence of circuit simulation. In this simulation these resistors allow a leakage path for the drain current to pass to the source through the bulk terminal. So, these resistors intended for numerical stability do, in fact, model the finite leakage current flowing in the transistor. If GMIN was set to be 10^{-16}, the g_m/I_D ratio would remain high when I_D is at the 10^{-12} A levels, rather than having a low value as indicated in Figs. 1-8 and 1-9. In a real-life circuit, whether the g_m/I_D ratio should start to increase at 10^{-12} or 10^{-16} A (or other current levels) is determined primarily by the amount of leakage current between the drain and source terminals.

The inset of Fig. 1-9 shows schematically the g_m/I_D simulated from Level 2 or Level 3 models [22], which are even more primitive than our digital model, whose results are shown in the figure itself. The g_m/I_D calculated by Level 2 or 3 is notorious among analog circuit designers. Because of the unphysical spike feature, some circuit designers, believing everything they got from the simulation, thought that they could gradually increase the bias current from 0 and reach the "sweet spot" where g_m/I_D peaks at its maximum. The simulated maximum value, unfortunately, is not real and the designed circuit often does not work properly with this unrealistic value of g_m/I_D.

5. *Discontinuity in subthreshold slopes* The inaccurate modeling of the subthreshold region as well as the transition between the subthreshold and the inversion regions can be identified from the aforementioned g_m/I_D plot. Sometimes an alternative quantity is used to examine the modeling accuracy in the subthreshold region—the *subthreshold-slope ratio*. Suppose the drain current at a small V_{DS} value, V_{DS1} (such as 0.01 V), is I_{D1}, and the drain current at the next V_{DS} increment,

V_{DS2} (such as 0.02 V), is I_{D2}, then the subthreshold-slope ratio is defined as $(I_{D2} + I_{D1})/(I_{D2} - I_{D1}) \times (V_{DS2} - V_{DS1})/(V_{DS2} + V_{DS1})$. This ratio is the average of (a) the slope of the line joining the origin and the midpoint between (V_{DS1}, I_{D1}) and (V_{DS2}, I_{D2}); and (b) the slope of the line joining (V_{DS1}, I_{D1}) and (V_{DS2}, I_{D2}). In the strong inversion when V_{DS} is small, the transistor operates in the linear region and the channel region can be treated as a linear resistor (linear with respect to the terminal voltage of the channel, i.e., V_{DS}). Therefore, the defined slope ratio approaches unity. In the subthreshold region, the drain current increases exponentially with V_{GS} and exhibits a V_{DS} dependence as $(1 - \exp(qV_{DS}/kT))$. Therefore, the ratio is some number greater than unity, with the value determined by the temperature as well as V_{DS1} and V_{DS2}.

Figure 1-10 shows the ratios calculated from the analog and digital models. It is obtained with $V_{DS1} = 0.01$ V, $V_{DS2} = 0.02$ V, and $V_{BS} = 0$. The analog model exhibits smooth characteristics, while the digital model displays a discontinuity as the transistor enters the strong inversion from the subthreshold region. This discontinuity feature is characteristic of the digital model, which uses separate models to describe the transistor characteristics at different operating regions.

6. *Incorrect Noise Modeling* Noise is the random signal variation about its average value, in the presence or the absence of externally applied biases. The fluctuation in voltage or current can interfere with the weakest signals of an analog circuit. Hence, the modeling of noise is of great importance to an analog model. A digital model, concerned mostly with switching between two extreme voltage levels, does not place great emphasis on the accuracy of noise modeling. Consequently, the noise model adopted in a digital model is primitive and inaccurate in many instances.

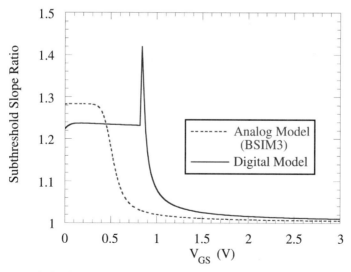

Fig. 1-10 Calculated subthreshold-slope ratio with $V_{DS1} = 0.01$ V, $V_{DS2} = 0.02$ V, and $V_{BS} = 0$. The subthreshold-slope ratio is defined as $(I_{D2} + I_{D1})/(I_{D2} - I_{D1}) \times (V_{DS2} - V_{DS1})/(V_{DS2} + V_{DS1})$, where I_{D1} and I_{D2} are the drain currents corresponding to the application of V_{DS1} and V_{DS2}, respectively.

An exemplar circuit highlighting the deficiency in the digital model's noise modeling is shown in the inset of Fig. 1-11. The zero current source connected at the drain terminal ensures that $V_{DS} = 0$ for the transistor. With a gate voltage of 1.5 V, the transistor operates in the linear operating region. Furthermore, the current source behaves as a high impedance point in a small-signal (noise) analysis. Therefore, the thermal noise associated with the drain current all flows through the output conductance g_d, and at high frequencies, through the output drain-bulk junction capacitance as well. As shown in Fig. 1-11, the analog model yields a nonzero noise voltage at the drain node, and its frequency dependence is qualitatively correct. The digital model, in contrast, defies the intuitively obvious with a result of null noise. (If the calculated mean square noise voltage spectral density is below 10^{-20} V^2/Hz in this SPICE simulator, the value of 10^{-20} is outputted, as shown in Fig. 1-11.) Although the dc current of the transistor is zero, the noise is always present. After all, the noise is considered as a small-signal quantity, and the device's drain conductance is nonzero. The digital model formulates the drain noise to be proportional to the current, thereby producing a zero noise. (It is not because we set the noise parameters to zero.) This result is especially unacceptable when the MOS transistor is intentionally used as a resistor. In the linear region, the MOS transistor's channel is characterized by a resistance equal to $1/g_d$ at low frequencies. We expect the thermal noise voltage to be related to $4kT\Delta f/g_d$, as approximately calculated by the analog model.

7. *Piecewise Continuity in C-V characteristics* In the digital model, the same methodology of regional equations to model the I-V characteristics is adopted in the

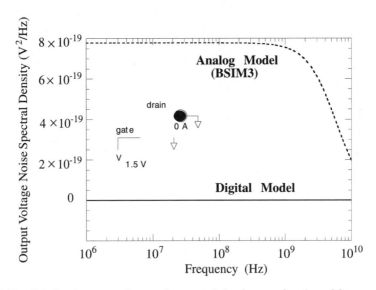

Fig. 1-11 Calculated output voltage noise spectral density as a function of frequency. The circuit is shown in the inset. The absence of proper noise modeling in the digital model leads to the zero calculated result.

1.3 DIGITAL VS. ANALOG MODELS   21

capacitance-voltage (C-V) modeling. Different capacitance expressions are developed for the accumulation, depletion/subthreshold, linear, and saturation regions. Despite the fact that the capacitance values are continuous across different regions of operation, they are only piecewise continuous, with discontinuity in the derivative at the boundaries of different regions. Figure 1-12 illustrates the calculated C_{gg} as a function of V_{GS} from the analog and the digital models at $V_{DS} = 0$. C_{gg} is the total gate capacitance, equal to the sum of the gate-to-drain (C_{gd}), gate-to-source (C_{gs}), and the gate-to-bulk (C_{gb}) capacitances. (The exact definitions of the capacitances will be discussed in Section 1.9.) Figure 1-13 illustrates the calculated C_{gg} and C_{gd} as a function of V_{DS} when $V_{GS} = 2$ V. Both figures reveal the piecewise continuous nature of the C-V characteristics produced by the digital model. In contrast, the analog model, with the same smoothing techniques adopted in the dc I-V characteristics, produces smooth behavior without abrupt transition.

8. *Lack of Subthreshold Capacitance* Subthreshold characteristics do not affect a digital circuit's performance much. While the dc value of the subthreshold drain current is somewhat important because it determines the off current of the device, the transient subthreshold current arising from the device capacitances can well be neglected. To reduce the overhead calculation during a transient simulation, the device capacitances associated with the channel (notably C_{gs}) are simply made zero in a digital model. Figure 1-14 compares the C_{gs} of an analog and a digital model. Whereas the C_{gs} difference in the strong inversion region is caused by the differences

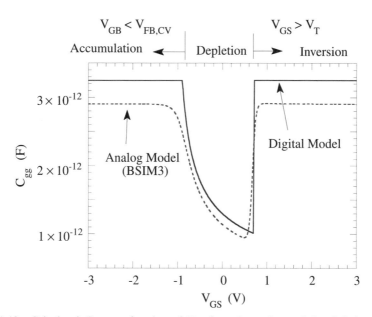

Fig. 1-12 Calculated C_{gg} as a function of V_{GS} from the analog and the digital models at $V_{DS} = 0$. C_{gg} is defined as $\partial Q_G/\partial V_G$, as discussed in Section 1.7. The digital model produces discontinuity in the slope, whereas the analog model is continuous in both the value and the derivative.

22 MODELING JARGONS

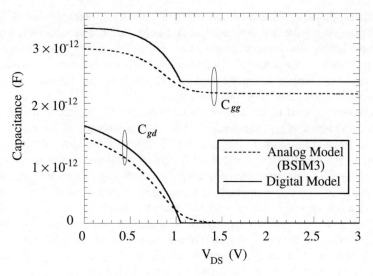

Fig. 1-13 Calculated C_{gg} and C_{gd} as a function of V_{DS} when $V_{GS} = 2\,V$.

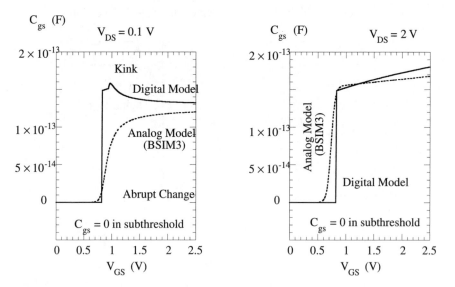

Fig. 1-14 Calculated C_{gs} as a function of V_{GS} when $V_{DS} = 0.1$ (linear region) and $2\,V$ (saturation region). In the digital model, the capacitance immediately goes to zero when V_{GS} is below the threshold voltage.

between the models, the difference in the subthreshold region is attributed to modeling practice. In the digital model, C_{gs} immediately goes to zero when V_{GS} falls below V_T. This characteristic is especially prominent in Fig. 1-14b, for $V_{DS} = 2$ V. The analog model capacitance, in contrast, gradually tapers off in the subthreshold region. In Fig. 1-14a, for $V_{DS} = 0.1$ V, we also highlight two additional problems of the digital model. One, the capacitance turns abruptly from zero to a large value right at $V_{GS} = V_T$. Second, there is a kink behavior, again at the boundary of different operating regions (between the linear and the saturation region, in this case).

The gross approximation of C_{gs} by zero in the subthreshold region can cause problems in analog circuits, although it is generally acceptable for digital circuits. Figure 1-15 shows the output voltage of a passgate circuit. Initially, when the gate voltage is still high, the device is in strong inversion, and both the digital and analog models yield the correct results. However, as the device enters the subthreshold region, the current flow in the drain of the device is found to be zero in the digital model, because the subthreshold capacitance is modeled to be zero. The sudden halt of current fall as V_{GS} falls below V_T prevents the output voltage from decreasing further. The analog model, in contrast, continues to calculate a finite drain current associated with the subthreshold capacitances.

9. *Reciprocal Capacitances* In a parallel-plate capacitor, there is an equal amount of charges on the top and the bottom plates. Its capacitance, $\Delta Q/\Delta V$, has the same value independent of whether ΔQ is measured from the top or the bottom plate. In a four-terminal MOS transistor, the device capacitance no longer behaves as a parallel-plate capacitor. For example, C_{dg}, defined as $-\partial Q_D/\partial V_G$, is not identical to C_{gd}, defined as $-\partial Q_G/\partial V_D$, where Q_D and Q_G are the drain and gate charges,

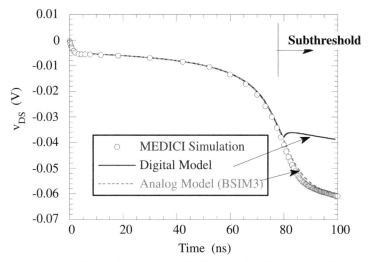

Fig. 1-15 Output voltage of a passgate circuit operating from strong inversion to the subthreshold region. The MOSFET has a channel length of 5 μm and the loading capacitance is 3 times the oxide capacitance. The details of a passgate circuit are shown in Fig. 4-8.

respectively, and V_D and V_G are the drain and gate voltages, respectively. (A minus sign is inserted in the above definitions to make the capacitance positive.) We shall elaborate on the definition of capacitances and charges in Section 1.9. For now, we just need to realize that a reciprocal relationship is not upheld in MOSFETs. That is, $C_{xy} \neq C_{yx}$.

The reciprocal relationship is unphysical. Nonetheless, the first-generation models such as Level 1, 2, and 3, based on the Meyer capacitance model [7], all adopt the reciprocal relationship. This is likely because it simplifies the derivation, but certainly at the cost of producing erroneous device characteristics. It turns out that for digital circuits, the unphysical nature associated with the reciprocal relationship does not lead to significant error in circuit simulation. Besides, the simplistic equating of C_{xy} to C_{yx} has the advantage of simplified model equations. Therefore, the models based on reciprocal relationship are quite numerically efficient, and still find some use today (e.g., for simulation of a large digital circuit).

BSIM3's C-V model is derived from device physics. As such, it does not assume $C_{xy} = C_{yx}$. Figure 1-16 compares the calculated C_{dg} and C_{gd} from the Level 1 model and BSIM3. While C_{dg} is identical to C_{gd} in the Level 1 model, they differ in BSIM3. We shall revisit the reciprocal relationship in the discussion of charge conservation (Section 1.8).

10. *Different V_T Expressions for I-V and C-V Characteristics* An exact formulation of the threshold voltage (V_T) is critical in the calculation of drain current. The charge-sharing, drain-induced barrier-lowering, and narrow-width effects, for example, all need to be carefully accounted for in order for the modeled drain current to

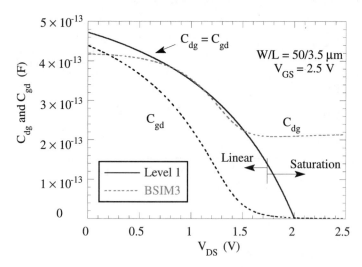

Fig. 1-16 Calculated C_{dg} and C_{gd} from the Level 1 model and BSIM3. Because the Level 1 model assumes the reciprocity relationship, its simulated C_{dg} is identical to C_{gd}. These two capacitances differ in BSIM3, as they should.

display proper voltage and geometrical dependencies. If V_T is incorrectly calculated, then the drive current will be off, and even a simplistic simulation of the inverter delay time will yield wrong results. In contrast, the device capacitance in saturation is not a strong function of V_T. For example, $C_{gs} \approx 2/3 \cdot WLC'_{ox}$, independent of V_T, as long as the transistor is in saturation. V_T is an important parameter for the calculation of capacitances primarily in the linear operation region. Digital models take advantage of this property. Because the switching of digital circuit often takes place when the transistor is in or nearly in saturation, the exact formulation for V_T is not critical for the C-V modeling, although it is for I-V modeling.

Figure 1-17 illustrates the simulated I-V characteristics and capacitances from an analog and a digital model, for a short-channel device biased at $V_{DS} = 0.1$ and 4.5 V. The drain currents produced by both models at the two voltages are displaced from one another at the two V_{DS} values. This separation in I_D correctly indicates that the threshold voltage at $V_{DS} = 4.5$ V is smaller due to the short-channel effects. In Fig. 1-17a, simulated by the analog model, C_{gs} at $V_{DS} = 4.5$ V is again seen to turn on at a smaller V_{GS} than at $V_{DS} = 0.1$ V. This means that the analog model accounted for the short-channel effects in the C-V models. In Fig. 1-17b, however, the digital model predicts the same C_{gs} turn-on behavior for both V_{DS} values. It is clear that although V_T's at the two V_{DS} values are different in the I-V model, they are identical in the C-V model.

As far as the digital model is concerned, arbitrarily equating the V_T of a short-channel device to the V_T of a long-channel device for C-V calculation has the following merits: First, the V_T in the C-V model is simplified. The mathematical development of charges and capacitances becomes less algebraically intensive compared to a C-V model using the full-blown V_T expression. This simplification allows the model to be efficient. As we have argued before, for the simulation of digital inverter delay, the exact value of threshold voltage for the calculation of device capacitance is nonconsequential. Second, without the consideration of short-channel effects, the V_T expression used in C-V modeling is independent of V_{DS}. This property turns out to be crucial in preserving capacitance symmetry with respect to drain and source at $V_{DS} = 0$ (Section 1.10). Indirectly, this property can also be shown to avoid a C_{gb} problem encountered in analog models. The problem is that the simulated C_{gb} in the analog C-V model is nonzero during linear operating region (see Section 4.8), while it is experimentally (and also verified with MEDICI simulation) found to be zero.

As we have stated at the beginning of the section, it is not correct to say that a digital model is inferior to an analog model. As far as its application is concerned, the digital model is able to deliver accurate simulation with the least amount of mathematical details. The analog model is more accurate, but more complicated, in terms of implementation of the model as well as setting up a strategy for model parameter extraction. However, with the present trend of increasingly more analog content in the circuit, the desire to have an accurate, more mathematically involved analog model is mounting. BSIM3 is an analog model capable of meeting the expectation of analog designers as well as the digital designers. It offers continuous

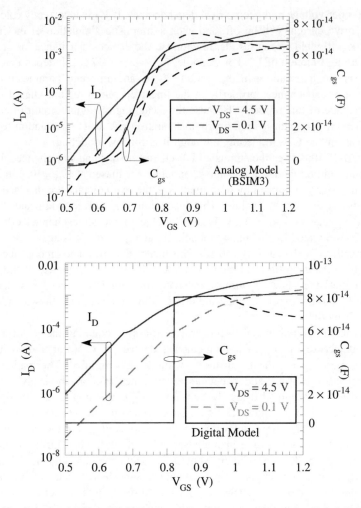

Fig. 1-17 Calculated I-V and C-V characteristics from an analog and a digital model, for a short-channel device biased at $V_{DS} = 0.1$ and 4.5 V. Part (a) is the analog model result. In the digital model, shown in part (b), the threshold voltages used for the I-V and C-V calculation differ, resulting in the different turn-on characteristics.

I-V, C-V, and conductance characteristics, not merely in their first derivatives. The device characteristics are smooth in the transition between the accumulation, subthreshold, linear, and saturation regions. There are no kink behaviors which are prevalent in a digital model with a regional approach. These characteristics are key to BSIM3's success as the industry's *de facto* standard model. The promotion and monitoring of BSIM3 as the standard compact MOSFET model is also under way [23].

1.4 SMOOTHING FUNCTION AND SINGLE EQUATION

BSIM3 is an analog-friendly model that retains all of the good model characteristics described in Section 1.3. The features of continuous and smooth I-V and C-V characteristics are essential to its popularity. The continuity property is a natural consequence of BSIM3's adoption of the smoothing function, which allows the use of a single equation to describe BSIM3's device characteristics across all operating regions.

We illustrate the concept of the smoothing function (or single equation) by citing a simple equation, which governs the drain current in the linear region:

$$I_D = \frac{W}{L}\mu_n C'_{ox}\left[(V_{GS} - V_T)V_{DS} - \frac{1}{2}(1+\delta)V_{DS}^2\right]. \tag{1-5}$$

V_T is the threshold voltage and the parameter δ is the bulk-charge factor accounting for the back-gate (bulk) effects on the transistor behavior. V_T is taken to be equal to the value at $V_{BS} = 0$ plus a term due to the body effect:

$$V_T = V_{T_o} + \gamma\left(\sqrt{2\phi_f - V_{BS}} - \sqrt{2\phi_f}\right), \tag{1-6a}$$

where γ is the body-effect coefficient, and ϕ_f is the Fermi potential with respect to the midgap in the substrate [22]. At strong inversion, the surface potential is assumed in BSIM3 to be equal to $2\phi_f$, given by

$$2\phi_f = 2\frac{kT}{q}\ln\left(\frac{\text{NCH}}{n_i}\right) \tag{1-6b}$$

NCH is a BSIM3 parameter, which is the concentration of the substrate doping near the channel region. Throughout this book, parameters that can be specified by the user, such as a BSIM3 parameter, are printed written in the `Courier` font.

$2\phi_f$ for a substrate doping of 1×10^{17} cm^{-3} is ~ 0.81 V. In a long-channel device, γ is a function of only the oxide thickness and the substrate doping. The functional dependence is shown in Fig. 1-18a. In a short-channel device, γ is smaller because of the charge-sharing between the bulk and the drain/source.

The bulk-charge factor relates to γ as

$$\delta = \frac{\gamma}{2\sqrt{2\phi_f - V_{BS}}} = -\frac{dV_T}{dV_{BS}}. \tag{1-7}$$

Once γ is determined from Fig. 1-18a, δ can be identified from Fig. 1-18b.

Equation (1-5) states that I_D is a parabolic function of V_{DS}. The functional behavior is plotted in Fig. 1-19. The drain current first increases with V_{DS}, eventually reaching a maximum value when V_{DS} is equal to the saturation voltage, $V_{DS,\text{sat}}$. The

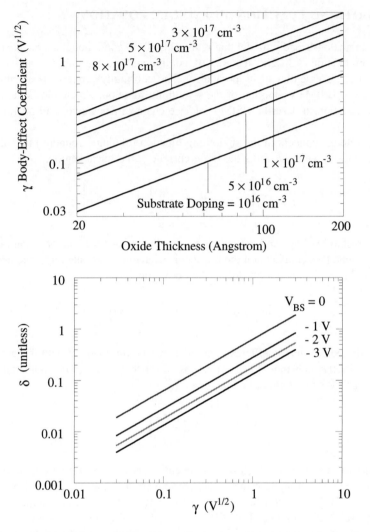

Fig. 1-18 (a) Calculated body-effect coefficient (γ); (b) bulk-charge factor (δ) as a function of the oxide thickness and substrate doping. In the calculation of the bulk-charge factor (δ), the strong-inversion potential appearing in Eq. (1-7) is assumed to be 0.81 V (the value when the substrate doping is 1×10^{17} cm^{-3}).

saturation voltage is obtained by taking the derivative of I_D with respect to V_{DS} and setting the derivative to zero. It is given by

$$V_{DS,\text{sat}} = \frac{V_{GS} - V_T}{1 + \delta}. \tag{1-8}$$

1.4 SMOOTHING FUNCTION AND SINGLE EQUATION

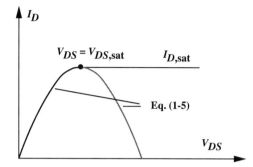

Fig. 1-19 Comparison of measured I-V characteristics and the calculated characteristics with Eq. (1-5). The saturation voltage is typically defined as the voltage at which $\partial I_D / \partial V_{DS} = 0$.

Once V_{DS} surpasses $V_{DS,\text{sat}}$, Eq. (1-5) predicts the drain current to decrease with further increase of V_{DS}. The behavior that I_D decreases at high values of V_{DS} clearly contradicts the experimental observation that I_D becomes relatively constant when V_{DS} reaches a certain value. The reason for the discrepancy is that Eq. (1-5) is derived for the *linear region*, when $V_{DS} < V_{DS,\text{sat}}$. It ceases to be valid when the transistor enters *saturation*; that is, when $V_{DS} > V_{DS,\text{sat}}$. The appropriate equation to simultaneously describe the transistor in the linear and saturation regions is obtained from an observation of the measured characteristics:

$$I_D = \begin{cases} \dfrac{W}{L}\mu_n C'_{ox}\left[(V_{GS}-V_T)V_{DS} - \dfrac{1}{2}(1+\delta)V_{DS}^2\right] & V_{DS} \leq V_{DS,\text{sat}}; \\ \dfrac{W}{L}\mu_n C'_{ox}\dfrac{(V_{GS}-V_T)^2}{2(1+\delta)} & V_{DS} > V_{DS,\text{sat}}. \end{cases} \quad (1\text{-}9)$$

The lower equation in Eq. (1-9) is derived by substituting Eq. (1-8) into Eq. (1-5); it is the maximum value for a particular V_{GS}.

The first generation of SPICE models used equations similar to Eq. (1-9) to describe the transistor's I-V characteristics across the linear and the saturation regions. It soon becomes obvious that this equation, a regional-model equation consisting of two distinct equations for different operating regions, can lead to kinks and discontinuities in the device characteristics, and thereby numerical difficulty during a circuit simulation. A fundamental problem daunting the regional model is that, although both I_D and dI_D/dV_{DS} are continuous at $V_{DS,\text{sat}}$, d^2I_D/dV_{DS}^2 is not. In order to ensure the numerical robustness, the derivatives of arbitrary order must be continuous at all voltage values of interest. This property is sometimes referred to as ∞-*differentiability*.

One solution to guarantee ∞-differentiability, adopted in BSIM3, is to use a single equation to describe the drain current, rather than with two separate equations given by Eq. (1-9). This is achievable by defining an effective drain-source bias, $V_{DS,\text{eff}}$, given by

$$V_{DS,\text{eff}} = V_{DS,\text{sat}} - \tfrac{1}{2}\left(V_{DS,\text{sat}} - V_{DS} - \Delta + \sqrt{(V_{DS,\text{sat}} - V_{DS} - \Delta)^2 + 4\Delta V_{DS,\text{sat}}}\right). \quad (1\text{-}10)$$

$V_{DS,\text{eff}}$ is a *smoothing function*, a function that gradually changes a variable between two extreme values. The exact value of the parameter Δ determines the degree of smoothness in the transition. (Note: Δ is essentially the BSIM3 parameter `DELTA`, modifiable by the user.) In this case, $V_{DS,\text{eff}}$ approaches V_{DS} (the applied bias) when V_{DS} is small, yet approaches $V_{DS,\text{sat}}$ when V_{DS} exceeds $V_{DS,\text{sat}}$. A graphical representation of Eq. (1-10) for $V_{DS,\text{sat}} = 1.5\,\text{V}$ is shown in Fig. 1-20 for various values of Δ's.

With Eq. (1-10), we now express I_D as

$$I_D = \frac{W}{L}\mu_n C'_{ox}\left[(V_{GS} - V_T)V_{DS,\text{eff}} - \tfrac{1}{2}(1+\delta)V_{DS,\text{eff}}^2\right] \qquad V_{GS} > V_T. \quad (1\text{-}11)$$

When V_{DS} is small, $V_{DS,\text{eff}}$ given by Eq. (1-10) is approximately equal to V_{DS}. Hence, Eq. (1-11) degenerates to the top equation of Eq. (1-9). In the other extreme, when V_{DS} greatly exceeds $V_{DS,\text{sat}}$, $V_{DS,\text{eff}}$ is approximately equal to $V_{DS,\text{sat}}$. The same Eq. (1-11) then degenerates to the bottom equation in Eq. (1-9). The advantage of

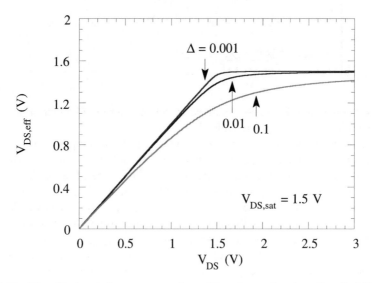

Fig. 1-20 The effective drain-to-source voltage ($V_{DS,\text{eff}}$) as a function of applied V_{DS}. Δ, a parameter in Eq. (1-10), determines the degree of smoothness in the transition between two extreme values.

Eq. (1-11) is that it describes the drain current behaviors in both the linear and saturation regions. It is a continuous function without an abrupt change of value. The derivatives of I_D with respect to V_{DS}, taken to an arbitrary order, are continuous throughout all bias ranges. The numerical problem caused by discontinuity in derivatives disappears in models based on the single-equation approach.

The aforementioned smoothing function is a quadratic function. It is conveniently used to smooth out the transition between the linear and the saturation region occurring in the V_{DS} domain. By now, the readers already have a good idea about what is meant by the smoothing function. As another example of the use of single equation, we analyze the transition from the subthreshold to strong inversion, which occurs when V_{GS} increases from values below V_T to values above. However, the analysis becomes more algebraically involved. The readers can skip the rest of this section without losing generality.

When $V_{GS} \ll V_T$, the transistor operates in the *subthreshold region* and the drain current can be expressed as [13]

$$I_D = I_0 \left[1 - \exp\left(-\frac{qV_{DS}}{kT}\right) \right] \exp\left(\frac{q(V_{GS} - V_T - V_{\mathit{off}})}{nkT}\right) \qquad V_{GS} \ll V_T, \quad (1\text{-}12)$$

where V_{off} is related to the off current (the drain current at zero V_{GS}) and I_0 is a constant specific to a technology. In the *strong inversion region* where $V_{GS} \gg V_T$, the drain current was that given by Eq. (1-11). Each of Eqs. (1-11) and (1-12) describes the drain current in one particular region. Neither of them remains applicable in the entire range of V_{GS}. For example, while Eq. (1-12) is accurate at $V_{GS} \ll V_T$, the same expression leads to excessively high current when $V_{GS} \gg V_T$ (due to the exponential term.) Likewise, whereas Eq. (1-11) applies to $V_{GS} \gg V_T$, it may lead to a negative current when it is unwisely applied at $V_{GS} < V_T$. From the modeling perspective, it is desirable to have a single equation working for all regions of operation. One popular but problematic method to integrate the two equations is to write the drain current as

$$I_D = I_{D,\text{sub}} + I_{D,\text{inv}}. \qquad (1\text{-}13)$$

$I_{D,\text{sub}}$ is identical to Eq. (1-12) when $V_{GS} < V_T$. For $V_{GS} > V_T$, $I_{D,\text{sub}}$ is defined to be the value at which $V_{GS} = V_T$, as shown in Fig. 1-21. $I_{D,\text{inv}}$, in contrast, is given by Eq. (1-11) when $V_{GS} > V_T$, but is made to be zero when $V_{GS} < V_T$. As demonstrated in Fig. 1-21, I_D as defined in Eq. (1-13) is continuous for all V_{GS} values, even when $V_{GS} = V_T$. The continuity feature is attractive for several digital models. The drawback of Eq. (1-13), however, is that its derivatives are discontinuous, a fact that analog-model users may find problematic.

Just as we utilized the $V_{DS,\text{eff}}$ function to smooth out the transition between the linear and saturation regions in the V_{DS} domain, we can define some effective V_{GS} function to smooth out the transition from the subthreshold to the strong inversion. Experiences have shown that it is easier to define an effective $(V_{GS} - V_T)$ function,

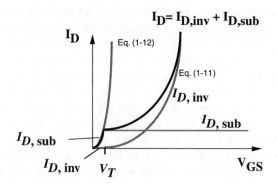

Fig. 1-21 A schematic illustration of a popular but problematic method to smooth the transition between the subthreshold and the strong inversion regions.

rather than an effective V_{GS} function directly. An effective $V_{GS} - V_T$ function, or $V_{GST,\text{eff}}$, is given by

$$V_{GST,\text{eff}} = \frac{2nkT/q \ln\left[1 + \exp\left(\dfrac{V_{GS} - V_T}{2nkT/q}\right)\right]}{1 + 2n \exp\left(-\dfrac{V_{GS} - V_T - 2V_{\text{off}}}{2nkT/q}\right)}, \quad (1\text{-}14)$$

where n is the *ideality factor* of the subthreshold current increase as V_{GS} increases, having a value larger than unity, but smaller than 2. We will revisit the ideality factor shortly, after we derive a simplified drain current expression in the subthreshold region. (*Note*: Just as many equations presented in this chapter, the above $V_{GST,\text{eff}}$ equation is not identical to that used in BSIM3, in order to reduce complexity. For the actual BSIM3 equations, see Appendix A.)

Besides the introduction of $V_{GST,\text{eff}}$, BSIM3 also modifies the $V_{DS,\text{sat}}$ expression of Eq. (1-8) to

$$V_{DS,\text{sat}} = \frac{V_{GST,\text{eff}} + 2kT/q}{1 + \delta}. \quad (1\text{-}15)$$

We have replaced $V_{GS} - V_T$ in Eq. (1-8) by $V_{GST,\text{eff}}$. The reason why we add the small voltage $2kT/q$ in the numerator will become obvious shortly. Independent of the exact value of $V_{DS,\text{sat}}$, the $V_{DS,\text{eff}}$ function remains that of Eq. (1-10).

To incorporate the effective V_{GST} function, the drain current (in the single equation form adopted by BSIM3) is evolved from Eq. (1-11) to

$$I_D = \frac{W}{L} \mu_n C'_{ox} V_{GST,\text{eff}} \left[1 - \frac{(1+\delta)V_{DS,\text{eff}}}{2V_{GST,\text{eff}} + 4kT/q}\right] V_{DS,\text{eff}}. \quad (1\text{-}16)$$

We show that this equation degenerates into either Eq. (1-11) or (1-12) under appropriate conditions. When $V_{GS} \gg V_T$, $V_{GST,\text{eff}}$ given by Eq. (1-14) approaches $V_{GS} - V_T$. Since kT/q is a small number, Eq. (1-16) then degenerates to Eq. (1-11).

At the other extreme, when $V_{GS} \ll V_T$, $\exp[q(V_{GS} - V_T)/(2nkT)]$ is a small number. Knowing that $\ln(1 + x) \approx x$ when x is small, we find that

$$\ln\left[1 + \exp\left(\frac{V_{GS} - V_T}{2nkT/q}\right)\right] \approx \exp\left(\frac{V_{GS} - V_T}{2nkT/q}\right). \quad (1\text{-}17)$$

Hence, when $V_{GS} \ll V_T$, $V_{GST,\text{eff}}$ of Eq. (1-14) approaches $kT/q \cdot \exp(q(V_{GS} - V_T - V_{\text{off}})/nkT)$. Because $V_{GS} - V_T$ is smaller than zero and because V_{off} is roughly zero, $V_{GST,\text{eff}}$ is much smaller than $2 \cdot kT/q$. $V_{DS,\text{sat}}$ given by Eq. (1-15) is roughly equal to $2kT/q/(1 + \delta)$. This $V_{DS,\text{sat}}$ value is also on the order of kT/q, a small value. To proceed with our demonstration that Eqs. (1-16) and (1-12) are equivalent, we need to consider two cases. If $V_{DS} \gg 2kT/q/(1 + \delta)$, then $V_{DS,\text{eff}}$, computed with the help of Eq. (1-10), approaches $V_{DS,\text{sat}}$ when we are considering large values of V_{DS}. On the other hand, if $V_{DS} \ll 2kT/q/(1 + \delta)$, then $V_{DS,\text{eff}}$ is approximately equal to V_{DS}. We briefly summarize the final results of these effective functions. When $V_{GS} \ll V_T$:

$$V_{GST,\text{eff}} \rightarrow \frac{kT}{q} \times \exp\left(\frac{V_{GS} - V_T - V_{\text{off}}}{nkT/q}\right) \quad \text{all } V_{DS} \quad (1\text{-}18)$$

$$V_{DS,\text{eff}} \rightarrow \begin{cases} \dfrac{2}{1+\delta}\dfrac{kT}{q} & V_{DS} \gg \dfrac{2}{1+\delta}\dfrac{kT}{q} \\ \\ V_{DS} & V_{DS} \ll \dfrac{2}{1+\delta}\dfrac{kT}{q} \end{cases} \quad (1\text{-}19)$$

Substituting Eqs. (1-18) and (1-19) into Eq. (1-16), we find that the drain current becomes

$$I_D = \begin{cases} \dfrac{W}{L}\mu_n C'_{ox}\left(\dfrac{kT}{q}\right)^2 \dfrac{1}{1+\delta} \times \exp\left(\dfrac{V_{GS} - V_T - V_{\text{off}}}{nkT/q}\right) & V_{DS} \gg \dfrac{2}{1+\delta}\dfrac{kT}{q}; \\ \\ \dfrac{W}{L}\mu_n C'_{ox}\left(\dfrac{kT}{q}\right)^2 \dfrac{qV_{DS}}{kT} \times \exp\left(\dfrac{V_{GS} - V_T - V_{\text{off}}}{nkT/q}\right) & V_{DS} \ll \dfrac{2}{1+\delta}\dfrac{kT}{q}. \end{cases}$$
(1-20)

We find that indeed under the appropriate extreme conditions (i.e., $V_{DS} \rightarrow 0$ or V_{DS} large), Eqs. (1-20) and (1-12) are equivalent. In fact, we can deduce that the premultiplication factor, I_0 of Eq. (1-12), is given by

$$I_0 = \frac{W}{L}\mu_n C'_{ox}\left(\frac{kT}{q}\right)^2 \quad \text{(neglecting } 1 + \delta \text{ on top equation of Eq. 1-20).} \quad (1\text{-}21)$$

We briefly revisit the parameter n, which was previously introduced as the ideality factor. We rewrite Eq. (1-20) as

$$I_D \propto \exp\left(\frac{V_{GS} - V_T - V_{\textit{off}}}{nkT/q}\right); \quad \text{or equivalently,} \quad I_D \propto \exp\left(\frac{V_{GS}}{nkT/q}\right). \quad (1\text{-}22)$$

We see that in the subthreshold region the drain current increases exponentially with V_{GS} divided by nkT/q. From undergraduate device physics, we know that the diode current under an external bias V is proportional to $\exp(qV/nkT)$. The n in the exponent is called the *ideality factor*; its ideal value is unity, but can take on a value slightly larger than 1, such as 1.1. The ideality factor gives an indication to the degree the measured device characteristics deviate from the ideal. Because Eq. (1-22) has the same functional form as the diode current, the factor n inside the exponent of Eq. (1-22) is also referred to as the ideality factor. Generally, the MOS's ideality factor will not be identical to unity (although the diode would), but larger than 1. To make the matter even more complicated, n is really not a constant, but continuously varying with V_{GS}. It approaches the ideal value of unity only after $|V_{BS}|$ increases. These experimental observations are used to test the model accuracy in the subthreshold region (Section 4.17).

Figure 1-22 illustrates the smoothing functions used in BSIM3. Along the V_{DS} axis, we can classify the operating regions to be either saturation or linear, depending on the relative magnitude of $V_{DS,\text{sat}}$ compared to V_{DS}. $V_{DS,\text{eff}}$, the effective V_{DS} function discussed previously, allows us to use a single equation to present the I-V characteristics for all V_{DS}. The second axis on Fig. 1-22 is the V_{GS} axis. The transistor operates either in the subthreshold or the strong inversion, depending on the value of V_{GS} compared to V_T. The smoothing function $V_{GST,\text{eff}}$ is used to represent the I-V characteristics for all V_{GS}. These two smoothing functions, $V_{DS,\text{eff}}$ and $V_{GST,\text{eff}}$, together cover the entire I-V characteristics. For C-V calculation, however, another effective function needs to be defined. In I-V calculation, whether the device is in accumulation or depletion region (both belong to the subthreshold region), the current equation is the same. However, for C-V calculation (in CAPMOD = 2 or 3; the model which supports continuous capacitance), the capacitances in accumulation differ drastically from those in the depletion. The third smoothing equation employed by BSIM3, aimed to have a unified C-V equation, is the flatband smoothing function, $V_{FB,\text{effCV}}$. $V_{FB,\text{effCV}}$ is made to approach $V_{FB,\text{CV}}$ when $V_{GB} \ll V_{FB,\text{CV}}$, and V_{GB} when $V_{GB} \gg V_{FB,\text{CV}}$. (The exact flatband voltage for C-V calculation, $V_{FB,\text{CV}}$, depends on the CAPMOD and VERSION values specified in the parameter set. See the entry of VFBCV in Section 3.2 for details.) This smoothing function has the property that the channel charge becomes zero during accumulation, yet it gives rise to the right depletion charge in the depletion region. The exact equation for $V_{FB,\text{CV}} > 0$ is given by

$$V_{FB,\text{effCV}} = V_{FB,\text{CV}} - \tfrac{1}{2}\Big(V_{FB,\text{CV}} - V_{GB} - 0.02 \\ + \sqrt{(V_{FB,\text{CV}} - V_{GB} - 0.02)^2 + 4 \cdot 0.02 \cdot V_{FB,\text{CV}}}\,\Big) \quad (1\text{-}23a)$$

1.4 SMOOTHING FUNCTION AND SINGLE EQUATION

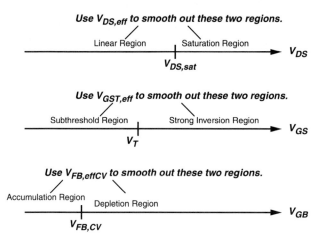

Fig. 1-22 The three smoothing functions used in BSIM3 at various bias axes.

or, for $V_{FB,CV} < 0$,

$$V_{FB,\text{effCV}} = V_{FB,CV} - \tfrac{1}{2}\Big(V_{FB,CV} - V_{GB} - 0.02$$
$$+ \sqrt{(V_{FB,CV} - V_{GB} - 0.02)^2 - 4 \cdot 0.02 \cdot V_{FB,CV}}\Big) \quad (1\text{-}23b)$$

These equations have the similar structure as Eq. (1-10) which calculates the effective drain-to-source voltage. A noticeable change is that Δ in Eq. (1-10) is

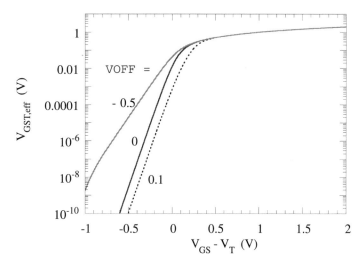

Fig. 1-23 The $V_{GST,\text{eff}}$ smoothing function used in BSIM3, plotted as a function of $V_{GS} - V_T$. The BSIM3 parameter VOFF controls the amount of leakage current at $V_{GS} = 0$.

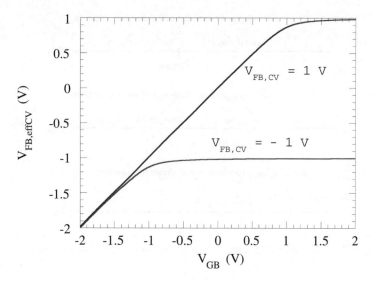

Fig. 1-24 The $V_{FB,\text{effCV}}$ smoothing function used in BSIM3, plotted as a function of V_{GB}.

replaced by 0.02 in Eq. (1-23). The Δ serves to adjust the curvarture of the transition of the effective function between two values. The Δ of Eq. (1-10) is basically DELTA, a SPICE parameter modifiable by the user. However, in Eq. (1-23), the same coefficient is hard-coded to 0.02.

We have shown $V_{DS,\text{eff}}$ as a function of V_{DS} in Fig. 1-10, at several values of Δ (i.e., DELTA). $V_{GST,\text{eff}}$ is shown in Fig. 1-23, as a function of $V_{GS} - V_T$, at several values of VOFF. The simulation results are obtained with $n = 1.2$ and $kT/q = 0.0258$ V. The figure indicates that a close-to-zero value of VOFF is desirable. A fairly negative value tends to distort the curves, and the subthreshold slope would exceed unity. Figure 1-24 reveals how $V_{FB,\text{effCV}}$ depends on V_{GB}, with $V_{FB,\text{CV}} = -1$ and 1 V. Although the exact equation used to calculate $V_{FB,\text{effCV}}$ depends on the sign of $V_{FB,\text{CV}}$, the calculated $V_{FB,\text{effCV}}$ curves look as though they came from the same equation, without an obvious dissimilarity.

1.5 CHAIN RULE

Chain rule is a fundamental concept in calculus. Let us take the drain current function of Eq. (1-16) as an example. The drain current can be expressed more fundamentally as $I_D(V_{GST,\text{eff}}(V_{GS}, V_{DS}, V_{BS}), V_{DS,\text{eff}}(V_{GS}, V_{DS}, V_{BS}))$. Both $V_{GST,\text{eff}}$ and $V_{DS,\text{eff}}$ are function of all three biases because V_T and $V_{DS,\text{sat}}$ are. If we want to take the partial derivative of I_D with respect to V_{GS}, we apply the chain rule and write

$$\frac{\partial I_D}{\partial V_{GS}} = \frac{\partial I_D}{\partial V_{GST,\text{eff}}} \frac{\partial V_{GST,\text{eff}}}{\partial V_{GS}} + \frac{\partial I_D}{\partial V_{DS,\text{eff}}} \frac{\partial V_{DS,\text{eff}}}{\partial V_{GS}}. \qquad (1\text{-}24)$$

$\partial I_D/\partial V_{GS}$ is the definition of g_m. Using Eq. (1-24) to evaluate g_m is actually not as easy as it seems. For example, $\partial I_D/\partial V_{GST,\text{eff}}$ is not a straightforward function given I_D's dependence on $V_{GST,\text{eff}}$ in Eq. (1-16). A lot of patience is required to ensure the derivative is derived correctly. Once $\partial I_D/\partial V_{GST,\text{eff}}$ is found, we need to pursue $\partial V_{GST,\text{eff}}/\partial V_{GST}$. This derivative again cannot be straightforwardly evaluated because $V_{GST,\text{eff}}$ is a convoluted function of V_{GS} in Eq. (1-14). Afterwards, we need to determine $\partial I_D/\partial V_{DS,\text{eff}}$, which is again evaluated from Eq. (1-16). Finally, $\partial V_{DS,\text{eff}}/\partial V_{GS}$ needs to be determined from Eq. (1-10), knowing that $V_{DS,\text{sat}}$ is some function of V_{GS}, such as that given in Eq. (1-8). We caution that the starting equations, Eqs. (1-16), (1-14), (1-10), and (1-8), are all relatively simplified equations used to simplify our discussion. BSIM3 employs more complicated equations, thus making the evaluation of the derivatives even more challenging.

BSIM3 indeed goes through the trouble of deriving expressions of these chain-rule derivatives. The small-signal quantities such as g_m, g_d, and g_{mb} are hard-coded in terms of SPICE parameters after the derivatives are derived. BSIM3 does not determine, for example, g_m, in the following manner. First, $I_{D,1}$ at a particular V_{GS}, V_{DS}, and V_{BS} was determined. Then, $I_{D,2}$ at a related bias condition with $V_{GS} + \Delta V_{GS}$, V_{DS}, and V_{BS} was determined. Finally, g_m was taken to be $(I_{D,2} - I_{D,1})/\Delta V_{GS}$. This way of finding g_m is conceptually satisfying because it is basically the numerical equivalent of the definition of the derivative. However, this method leads to serious numerical consequences because $(I_{D,2} - I_{D,1})/\Delta V_{GS}$ is a ratio of two small numbers. Computer rounding error may cause the so-defined g_m to be discontinuous at a particular bias.

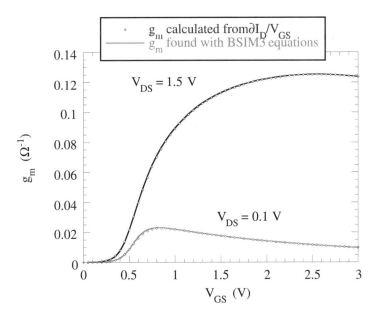

Fig. 1-25 Comparison of the g_m directly outputted from BSIM3 and the g_m evaluated from the derivative of the BSIM3-simulated I_D-vs-V_{GS} curves.

Instead, BSIM3 faithfully evaluates all of the derivatives and expresses the small-signal quantities in terms of the derivatives. Because the derivation of the derivatives is a painstaking process, it is rather easy to make mistakes. In fact, several bug fixes of BSIM3 concern the correction of the expressions of the derivatives. Knowing that there is a lot or room for algebraic error during the derivative-evaluation, we wonder if BSIM3's expressions for the small-signal quantities are correct. We first use a set of BSIM3 parameters to calculate both the drain current and the mutual transconductance, as a function of V_{GS}. Both I_D and g_m so-obtained are calculated from BSIM3 equations. Let us denote this g_m as g_{m_BSIM3}, to emphasize that the g_m here is calculated in BSIM3. We could have gotten g_m through other means. We could have taken the I_D-vs.-V_{GS} curves and found the derivative of I_D with respect to V_{GS}, numerically. This then constitutes $g_{m\_\partial ID/\partial VGS}$, which is not the same g_m outputted by BSIM3. We plot out both g_{m_BSIM3} and $g_{m\_\partial ID/\partial VGS}$ in Fig. 1-25. The agreement indicates that the g_m equations used in BSIM3 are correctly derived, such that g_{m_BSIM3} is nearly identical to $g_{m\_\partial ID/\partial VGS}$.

1.6 QUASI-STATIC APPROXIMATION

Solving transient problems is an important capability of a SPICE simulator. To address the time-varying nature of the currents and voltages of a MOS transistor, a device model needs to find the capacitances and charges that characterize the transient properties of the device. To date, practically all of MOSFET models used in SPICE simulation have been quasi-static models. These are models based on the *quasi-static approximation*, which will be discussed shortly. We should emphasize that BSIM3 offers both a quasi-static (QS) and a non-quasi-static (NQS) model. However, in practice, its non-quasi-static option is hardly ever used because (1) the quasi-static approximation is, in fact, a good approximation for most of the simulations; (2) BSIM3's non-quasi-static model corrects some deficiencies in the QS model, but may produce other errors not seen in its QS model; and (3) there are remedies with the BSIM3 QS model when the NQS effects are pronounced (Section 4.1).

Describing the quasi-static approximation in words alone is a difficult task. A typical explanation states that quasi-static approximation is one wherein the channel charge is assumed to respond instantaneously to any change in the bias voltage. The channel charge profile at a moment in time depends only on the bias voltage values at that instant, not the history leading to that value at that instant.

Is this wording explanation ambiguously unclear? Perhaps it is best to comprehend the quasi-static approximation through an example. Consider the transient problem posed in Fig. 1-26. Both the gate-source bias, $v_{GS}(t)$, and the drain-to-source bias, $v_{DS}(t)$, vary with time. We digress to elucidate our notation convention. A dc quantity, such as V_{DS}, is expressed with all letters capitalized. A small-signal quantity is expressed in all small letters, such as v_{ds}. Their sum represents the

1.6 QUASI-STATIC APPROXIMATION

instantaneous total value, and is given by a small letter with capitalized subscripts. Therefore, v_{DS} appearing in Fig. 1-26 is a instantaneous value, hinting that we are concerned with a transient problem. If we were only concerned with dc, we would have used V_{DS} instead.

With these bias voltages as given, we are interested in finding the time evolution of the terminal currents: $i_D(t)$, $i_S(t)$, $i_G(t)$, and $i_B(t)$. (*Note:* Although the gate and bulk currents are zero at dc when leakage currents are neglected, they can still be significant at high frequencies, due to the presence of device capacitances.) It turns out that all four of these terminal currents can be determined once the channel charge density, $c'(x, t)$, in the unit of coulombs per area, is known, where x denotes the channel position from the source and t is time. The prime in $c'(x, t)$ is used to emphasize that this charge is a per-unit-area quantity. The equation governing $c'(x, t)$ in a long-channel MOSFET with a constant mobility μ_n is a partial differential equation given by [22]

$$\frac{\mu_n}{C'_{ox}} \frac{\partial}{\partial x}\left(c' \frac{\partial c'}{\partial x}\right) = -\frac{\partial c'(x, t)}{\partial t}. \tag{1-25}$$

To avoid unnecessary complexity, we have assumed $\delta = 0$ in the above equation, thereby omitting the bulk-charge effects. The channel charge density $c'(x, t)$ is a negative quantity in NMOS whose conduction carriers are the electrons. If we knew how to solve this nonlinear partial differential equation, $c'(x, t)$ would be known and all four terminal currents could be readily established. In the interest of keeping up the physical picture but not the mathematical details, we concentrate on the source and the drain currents:

$$i_D(t) \propto \left.\frac{\partial}{\partial x} c'(x, t)\right|_{x=\text{drain}} \quad ; \quad i_S(t) \propto -\left.\frac{\partial}{\partial x} c'(x, t)\right|_{x=\text{source}}. \tag{1-26}$$

Despite that both currents are proportional to the spatial derivative of the charge, they are not diffusion currents. They are drift currents, with $\partial c/\partial x$ happening to be equal to the electric field in a MOSFET channel.

We hinted that Eq. (1-25) is more complicated than it appears; there is no general solution of $c'(x, t)$ for arbitrary $v_{GS}(t)$ and $v_{DS}(t)$. (There can be some particular solutions for a specific kind of boundary condition.) In order to make progress, we must make some approximation.

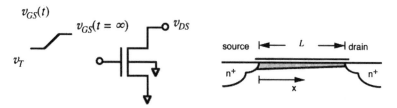

Fig. 1-26 A transient bias condition used to illustrate the concept of quasi-static analysis.

As a first trial, we make the *static assumption*. The static assumption neglects the fact that the bias voltages are functions of time. Instead, it focuses at $t = \infty$, when $v_{GS}(t)$ and $v_{DS}(t)$ have ceased varying and settled to $V_{GS,\infty}$ and $V_{DS,\infty}$, respectively. Because there is no longer a time dependence, $\partial c'/\partial t$ appearing in Eq. (1-25) is zero. Suddenly, Eq. (1-25) degenerates into an ordinary differential equation and an analytical solution is available. The solution of $c'(x, t)$, equal to $c'(x)$ as the time dependence drops out in this static approximation, is given by

$$c'(x, t) = c'(x) = c'(0)\sqrt{1 - \frac{x}{L}(1 - \alpha^2)} \quad \text{(static solution)}, \quad (1\text{-}27)$$

where α, the *saturation index*, under the biases of $V_{GS,\infty}$ and $V_{DS,\infty}$ are given by

$$\alpha = \begin{cases} 1 - \dfrac{V_{DS,\infty}}{V_{DS,\text{sat}}} & \text{for } V_{DS,\infty} < V_{DS,\text{sat}} \\ 0 & \text{for } V_{DS,\infty} \geq V_{DS,\text{sat}} \end{cases}; \quad V_{DS,\text{sat}} = \frac{V_{GS,\infty} - V_T}{1 + \delta}. \quad (1\text{-}28)$$

When the device is in saturation, $\alpha = 0$. When α is between 0 and 1, the device operates in the linear region. $c'(0)$, the charge density at the source, is known once the gate-source voltage is known. In this static analysis, the gate-source voltage is just $V_{GS}(t = \infty)$, and $c'(0)$ appearing in the above equation is given by

$$c'(0) = -C'_{ox}(V_{GS,\infty} - V_T). \quad (1\text{-}29)$$

If we stopped our work here, we would have made little progress. After all, this static solution has absolutely no use in a practical situation, which, by nature, involves some time transient. Our next logical step is the *quasi*-static assumption, in which we assume the overall transient to consist of a series of static events. The approximate solution of $c'(x, t)$ obtained this way is referred to as the *quasi-static charge density*, denoted as $c'_{QS}(x, t)$. In contrast, the actual solution of $c'(x,t)$, obtained by purely numerical analysis for example, is referred to as the *non-quasi-static solution*.

As shown in Fig. 1-26, the gate bias is initially at a value equal to V_T, such that there is no charge inversion in the channel. Because the entire channel layer is depleted, the transistor is initially at the cutoff mode, without a current flow. (The minor component of the drain current, the diffusion current, is neglected.) Let us consider the transistor operation at $t = t_1$, at which instant $v_{GS}(t) = v_{GS,1}$. Although only at the instant $t = t_1$ does $v_{GS}(t)$ attain the value $v_{GS,1}$, we pretend that $v_{GS}(t)$ had been equal to $v_{GS,1}$ for some time prior to reaching to the instant $t = t_1$. In this thought experiment, the device had attained the dc operation state by the time $t = t_1$

1.6 QUASI-STATIC APPROXIMATION

was reached, and that $v_{GS}(t)$ had been $v_{GS,1}$. The channel charge profile under such a scenario would be that of the static solution given in Eq. (1-27):

$$c'_{QS}(x, t_1) = c'(0, t_1)\sqrt{1 - \frac{x}{L}(1 - \alpha^2)}. \quad (1\text{-}30)$$

Now consider $t = t_2$, at which time $v_{GS}(t) = v_{GS,2}$. Again, although only at the instant $t = t_2$ does $v_{GS}(t)$ reach the value $v_{GS,2}$, we imagine that $v_{GS}(t)$ had been equal to $v_{GS,2}$ for a long time prior to reaching to the instant $t = t_2$. Basically, we assume the device had been in the dc operation long before the time $t = t_2$, and that $v_{GS}(t)$ had been $v_{GS,2}$. The channel charge profile at t_2 would then be governed by the following equation:

$$c'_{QS}(x, t_2) = c'(0, t_2)\sqrt{1 - \frac{x}{L}(1 - \alpha^2)}. \quad (1\text{-}31)$$

This procedure is repeated at other times. In each instant, $c'_{QS}(x, t)$ is obtained by assuming that its steady-state distribution had been obtained right before the instant t of concern. That is, $c'(x, t)$ at any moment obeys this equation:

$$c'_{QS}(x, t) = c'(0, t)\sqrt{1 - \frac{x}{L}(1 - \alpha(t)^2)}. \quad (1\text{-}32)$$

In the quasi-static approximation, both $c'(0, t)$ and α appearing in the above equation are functions of time because $v_{GS}(t)$ and $v_{DS}(t)$ are functions of time. They are given by equations similar to those of Eqs. (1-28) and (1-29), with appropriate modifications for the time dependence:

$$\alpha(t) = \begin{cases} 1 - \dfrac{v_{DS}(t)}{v_{DS,\text{sat}}(t)} & \text{for } v_{DS} < v_{DS,\text{sat}} \\ 0 & \text{for } v_{DS} \geq v_{DS,\text{sat}} \end{cases} ; \quad v_{DS,\text{sat}}(t) = \frac{v_{GS} - V_T}{1 + \delta}; \quad (1\text{-}33)$$

$$c'(0, t) = -C'_{ox}(v_{GS}(t) - V_T). \quad (1\text{-}34)$$

We added the subscript QS to $c'(x, t)$ to emphasize that the so-obtained solution represents the quasi-static solution. Despite the similarity in forms, $c'_{QS}(x,t)$ is not identical to $c'(x)$ of Eq. (1-27), which represents the channel charge density at dc. There is a time dependence in $c'_{QS}(x, t)$, mainly through the time variations of $v_{GS}(t)$ and $v_{DS}(t)$. If we drop the time dependencies, then $c'(x)$ and c'_{QS} appear in the same form.

In the quasi-static approximation, the instantaneous channel charge density given by Eq. (1-32) is a sole function of the instantaneous values of the applied v_{GS} and v_{DS}. The charge responds instantaneously to obey the above equation the moment when v_{GS} or v_{DS} changes to a different value. The charge density profile as a function of time for the circuit of Fig. 1-26, is as shown in Fig. 1-27. For simplicity, we assume that $v_{DS}(t)$ is at values above the saturation voltage ($V_{DS,sat}$) during the entire transient; hence, α is always zero. At $t = t_1$, $c'_{QS}(0, t_1)$, the channel charge density at the source edge, is equal to $v_{GS}(t_1) - V_T$. At the drain side, $c'_{QS}(L, t_1)$ is zero because the drain bias is assumed to be high, exceeding $V_{DS,sat}(\alpha = 0)$. The quasi-static channel potential at $t = t_1$ basically follows the static solution given by Eq. (1-27). As t increases to t_2 and $v_{GS}(t)$ increases to $v_{GS,2}$, the channel potential at the source increases, and the overall charge density at $t = t_2$ continues to have the square-root shape given by Eq. (1-27). As input voltage continues to increase over time, the channel charge per unit area at $x = 0$ increases. The curves shift upward, as shown in the figure. In the quasi-static approximation the channel charge's time evolution consists of a series of square-root-like curves of Eq. (1-27), which would be truly obtained only in the steady state. Once the approximation is made, the drain current given by Eq. (1-26) attains finite values, even at a time t very close to 0. In reality (when the quasi-static assumption is dropped), as discussed in the following, the drain current remains zero for a while during the initial part of the transient. The time duration at which i_D stays zero is called the *intrinsic channel transit time* (τ_{tr}).

We now describe the actual device behavior when the QS assumption is dropped. At $t = 0$ when the input bias is equal to V_T, the transistor is just about to turn on. Since $v_{GS}(t) = V_T$, $c'(x, t)$ is equal to 0 at the source. There is no charge inversion anywhere in the channel. At $t > 0$ such that v_{GS} becomes greater than V_T, the electron sheet concentration at $x = 0$, $c'(0, t)$, increases to a certain value. Propelled

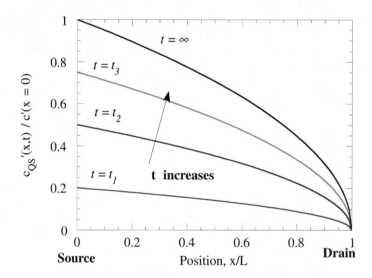

Fig. 1-27 Quasi-static charge density profile, $c'_{QS}(x, t)$, as a function of location in the channel (x), plotted at several time frames (t). The source corresponds to $x = 0$ and the drain, $x = L$, the channel length. The result, which applies to the bias condition shown in Fig. 1-26, is normalized by the constant charge density at the source. (From Ref. [24], p. 375, © Wiley 1999).

by the electric field, these electrons at the source start to move toward the drain. The moment these charges move, the source current at the source contact is nonzero. However, the charges have not reached the drain. Hence, the drain current remains zero initially. Since it takes a certain time for the charge to traverse through the channel, we cannot expect the steady-state profiles of Fig. 1-27 to be set up instantaneously. We numerically solved the partial differential equation of Eq. (1-25) for a particular case wherein α is always 0 and wherein $v_{GS}(t)$ increases linearly with time. The charge profile $c'(x, t)$, plotted in Fig. 1-28, reveals the existence of a *charge front*, which progresses from the source to the drain as time elapses. There is no drain current before the charge front reaches the drain (before $t = t_3$). Although Fig. 1-28 applies to the special case wherein $v_{GS}(t)$ increases linearly with time, the central theme that the drain current is initially zero remains applicable for other functions of $v_{GS}(t)$.

We briefly examine Fig. 1-28 to show that the figure makes sense. With $\alpha = 0$, the boundary condition specified in Eq. (1-27) is indeed satisfied, that throughout the entire transient, $c'(L, t) = 0$. At the source side, $c'(0, t)$ increases linearly with time, since the gate voltage increases linearly. The source current inferred from Eq. (1-26) is finite because $\partial c'(x, t)/\partial x$ is finite at $x = 0$. When t is small, the drain current is zero because $\partial c'(x, t)/\partial x = 0$ at $x = L$. The drain current rises above zero only after the charge front arrives at the drain, an event that gives rise to a nonzero $\partial c'(x, t)/\partial x$ at $x = L$.

Comparing the quasi-static curves of Fig. 1-27 and the non-quasi-static curves of Fig. 1-28, we see that only one curve in Fig. 1-27 is exactly accurate. It is the $t = \infty$ curve. This is because only at $t = \infty$ can we truly obtain the steady-state solution, consistent with the very premise of a quasi-static analysis. However, we do not need

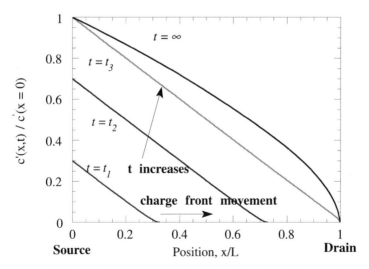

Fig. 1-28 Exact (non-quasi-static) charge density profile without the quasi-static approximation, $c'(x, t)$. The result applies to the bias setup of Fig. 1-26. (From Ref. [24], p. 377, © Wiley 1999).

to wait untill $t = \infty$. The quasi-static approximation is a pretty good approximation as soon as the time scale of concern exceeds the intrinsic channel transit time. The intrinsic channel transit time is given by [24]

$$\tau_{tr} = \frac{1}{\omega_0} \times \left[\frac{4}{3} \frac{1 + \alpha + \alpha^2}{(1 + \alpha)(1 - \alpha^2)} \right], \qquad (1\text{-}35)$$

where ω_0, a popular normalization frequency for MOSFETs, is

$$\omega_0 = \frac{\mu_n(v_{GS} - V_T)}{L^2}. \qquad (1\text{-}36)$$

The factor within the square brackets of Eq. (1-35) is equal to 4/3 at saturation ($\alpha = 0$). If an ac analysis is performed instead of a transient analysis, then the quasi-static analysis is considered valid if the frequency of operation is smaller than $1/(2\pi \cdot \tau_{tr})$.

We shall mostly be using the quasi-static model. Although it is just an approximation, it is remarkably accurate for many practical situations that take place at a time scale longer than the intrinsic channel transit time. Besides, if we want to pursue the non-quasi-static solution in a rigorous fashion, then we must solve the partial differential equation of Eq. (1-25), which does not have a general analytical solution. A non-quasi-static solution almost always requires numerical analysis. Despite that there is a non-quasi-static model in BSIM3, it is a compact model based on some physical intuitive understanding of the device transient. By not solving the partial differential equation, the BSIM3's NQS model suffers from several deficiencies that naturally come about when the governing equation is bypassed. For example, the bulk current is identically zero at any time and under any bias condition. The drain current is sometimes artificially made zero to avoid nonphysical results. While it is agreed that the NQS BSIM3 model represents a good modeling effort, and is perhaps the most accurate NQS model that winds up in a SPICE simulator, the model has not been widely used. When a BSIM3 model is used in circuit simulation, it is almost certain that the quasi-static model is being used.

Thus far our quasi-static analysis has led us to the solution of $c'_{QS}(x, t)$, given in Eq. (1-32). Although we have described what a quasi-static approximation entails and how $c'_{QS}(x, t)$ is obtained, we have not demonstrated how the $c'_{QS}(x, t)$ solution can give us the device's terminal currents, which are the ultimate quantities of interest. As will be discussed in Section 1.8, we can relate the transistor terminal currents to the bias voltages after we derive the terminal charges, such as the gate charge (Q_G), drain charge (Q_D), source charge (Q_S), and bulk charge (Q_B). Hence, we must somehow relate $c'_{QS}(x, t)$ to the terminal charges, a task performed in the next section.

1.7 TERMINAL CHARGES AND CHARGE PARTITION

MOSFET is a four-terminal device. In order to accurately calculate the dynamic behavior of the device, a model must provide the charge expressions for each of the

1.7 TERMINAL CHARGES AND CHARGE PARTITION

terminal. This means, from the $c'_{QS}(x, t)$ obtained from the last section's quasi-static analysis, we need to establish the gate charge (Q_G), drain charge (Q_D), source charge (Q_S), and bulk charge (Q_B). For the clarity of presentation, we shall now assume Q_B to be zero. This is tantamount to assigning δ to zero. Once the relationships among Q_G, Q_D, Q_S, and $c'_{QS}(x, t)$ are determined, we will present all four terminal charges' expressions given a nonzero δ. We believe that with this way of presentation the mathematics is minimized and the physical picture is highlighted.

Before delving into Q_G, Q_D, or Q_S immediately, let us find the total inversion charge in the channel (sometimes referred to as the *channel charge*), Q_I. It is the integral of the charges per unit area in the entire channel:

$$Q_I(t) = -W \int_0^L c'_{QS}(x, t)dx = -WC'_{ox} \int_0^L (v_{GS}(t) - V_T)\sqrt{1 - \frac{x}{L}(1 - \alpha(t)^2)}dx, \quad (1\text{-}37)$$

where $c'_{QS}(x, t)$ was given by Eq. (1-32). The channel charge is negative because the inverted charges, the electrons, have negative charges. It is clear that the time dependence in $Q_I(t)$ comes from the time dependencies of $v_{GS}(t)$ and $v_{DS}(t)$. We shall from now on drop (t), remembering that Q_I, v_{GS}, v_{DS}, and α are all time-dependent quantities. Carrying out the above integration, we find

$$Q_I = -WLC'_{ox}(v_{GS} - V_T)\frac{2}{3}\frac{1 + \alpha + \alpha^2}{1 + \alpha}. \quad (1\text{-}38)$$

Charge conservation, stating that the sum of the net charge in the entire device is zero, must be upheld at any instant of time. With $Q_I + Q_G = 0$ at all times, Q_G is equal to

$$Q_G = -Q_I = WLC'_{ox}(v_{GS} - V_T)\frac{2}{3}\frac{1 + \alpha + \alpha^2}{1 + \alpha}. \quad (1\text{-}39)$$

The gate current, defined positive when the positive charges are flowing *into* the gate, is just the derivative of the gate charge with respect to time:

$$i_G(t) = \frac{d}{dt}Q_G(v_{GS}(t), v_{DS}(t)). \quad (1\text{-}40)$$

By now we have identified the gate charge and the gate current. Naturally, we would like to determine the drain charge and the drain current next. However, thus far we have determined the total channel inversion charge only, given in Eq. (1-38), but not the individual drain and source charges. We know that the sum of the two charges is equal to the overall channel charge:

$$Q_I = Q_D + Q_S. \quad (1\text{-}41)$$

However, we do not exactly know which portion of the channel inversion charge belongs to the drain and which portion belongs to the source, at least not yet. The assignment of the channel charge to source and drain charges is called *charge partition*. A very simplistic partitioning scheme is the so-called *50/50 partition*, which arbitrarily assigns each of Q_D and Q_S to be 50% of Q_I. This partition scheme has no physical basis. It is correct only when v_{DS} is small such that the device is operated in the linear region. With $v_{DS} \sim 0$, the MOSFET device is symmetrical, and it is reasonable to assume $Q_D \sim Q_S \sim Q_I/2$. As soon as v_{DS} deviates from 0, the 50/50 partition scheme becomes progressively more erroneous, with the largest error occurring when the device enters the saturation region.

The 50/50 charge partition is the partition scheme of convenience when only Q_I is known, but not the source and drain charges individually. After all, it is not straightforward to determine Q_D (or Q_S) from physical principles. The first derivation of Q_D (or Q_S) involves a double integral [9], with a slight mathematical complexity which sometimes confuses the physical nature of the quasi-static charges. (However, in a later publication [25], it was found that the drain and source charge expressions can be developed from a simpler method called the *first-moment* technique, to be described shortly.)

To establish Q_D, we first ponder the meaning of Q_D. (This seems circuitous, but really, what is Q_D anyway?) In this regard, we make a distinction between the gate current just discussed and the drain current. The drain current is composed of charges which flow out of the drain contact, through the drain-to-source voltage source, through the source contact, and then into the channel. There is a continuous loop of current path for the electrons to travel. The charges constituting the gate current, in contrast, flow into the gate metal and stay there. There is not a loop path (discounting the displacement current); the electrons cannot proceed further after entering the gate metal, because the metal-semiconductor barrier height (denoted as $q\phi_B$ in typical textbooks) is sufficient to block the electrons from entering the semiconductor. As the gate voltage increases with time, more charges flow into and accumulate at the gate metal. Conversely, as the gate voltage decreases with time, the number of the charges in the gate decreases as the charges flow out of the gate. If we neglect any possible leakage current between the gate and the semiconductor, we can then write the gate current as the time derivative of the gate charge, as done in Eq. (1-40).

If we were to omit the difference between the gate and drain current flow, and blindly apply Eq. (1-40) to formulate I_D, then the drain current would be

$$i_D(t) = \frac{dQ_D(t)}{dt} \quad \text{(not useful)}. \tag{1-42}$$

This drain current definition is quite awkward in a dc analysis. In dc, $i_D(t) = I_D$, so $Q_D(t)$ as defined in Eq. (1-42) would be required to increase linearly with time, even though the external bias voltages v_{GS} and v_{DS} are constant with time. This is in violation of the concept embodied in the quasi-static gate charge, whose time dependence originates from those of the bias voltages as discussed in Eq. (1-37). An

1.7 TERMINAL CHARGES AND CHARGE PARTITION

alternative definition of the drain charge to circumvent the problem is to define $Q_D(t)$ as

$$i_D(t) = I_D + \frac{dQ_D(t)}{dt}. \tag{1-43}$$

In a non-quasi-static analysis, the definition of the drain charge in Eq. (1-43) will not be useful either. $Q_D(t)$ will turn out to be an intractable expression, depending on the time evolution of the boundary conditions. However, the above formulation becomes quite meaningful in a quasi-static approximation. By applying the first-moment technique [24,25], it has been demonstrated that the instantaneous drain current is related to the dc drain current as:

$$i_D(v_{GS}(t), v_{DS}(t)) = I_D(v_{GS}(t), v_{DS}(t)) + \frac{\partial}{\partial t} \frac{W}{L} \int_0^L x c'_{QS}(x,t) dx, \tag{1-44}$$

where I_D is the dc current that will flow when the gate and drain biases are $v_{GS}(t)$ and $v_{DS}(t)$, respectively. An expression for I_D can be as simple as Eq. (1-9) or, as in BSIM3, I_D can incorporate the smoothing functions and appear like Eq. (1-16). By comparing Eqs. (1-44) to (1-43), we find that a useful definition of the drain charge is

$$Q_D = \frac{W}{L} \int_0^L x c'_{QS}(x,t) dx. \tag{1-45}$$

Again, x is the location from the source (Fig. 1-26) and c'_{QS} is the quasi-static charge density in the channel. By substituting $c'_{QS}(x,t)$ given in Eq. (1-32) into Eq. (1-45), we find that

$$Q_D = -WLC'_{ox}\left[(V_{GS} - V_T)\frac{6\alpha^3 + 12\alpha^2 + 8\alpha + 4}{15(1+\alpha)^2}\right]. \tag{1-46}$$

The source charge (Q_S) is defined in a similar fashion as the drain charge. At dc, the source current I_S is the negative of I_D (we are defining a positive current as the current which flows *into* a given node in the device). Therefore,

$$i_S(t) = -I_D + \frac{dQ_S(t)}{dt}. \tag{1-47}$$

The same first moment technique used to establish Q_D can be used to find Q_S. It is given by

$$Q_S = \frac{W}{L}\int_0^L (L-x)c'_{QS}(x,t)dx$$
$$= -WLC'_{ox}\left[(V_{GS}-V_T)\frac{4\alpha^3+8\alpha^2+12\alpha+6}{15(1+\alpha)^2}\right]. \quad (1\text{-}48)$$

When the MOSFET operates in the saturation region, $\alpha = 0$, and the Q_D/Q_S ratio calculated from Eqs. (1-46) and (1-48) is 2/3. Consequently, this partition scheme, in which Q_S and Q_D are given by Eqs. (1-46) and (1-48), is referred to as the *40/60 partition scheme*. Although so named, the 40/60 partition does not mean that Q_D is always 40% of Q_I (which is equal to $Q_S + Q_D$). Q_D is 40% of Q_I only in saturation. During linear operation region, the integrals of the drain and source charges produce a ratio different from 40/60.

Besides the 50/50 and the 40/60 partition schemes, another charge-partition scheme that remains popular today is the *0/100 partition*. The 0/100 partition *arbitrarily* assigns all the channel charge in saturation to the source, leaving the drain charge equal to zero. The logic is that, since the channel is pinched off at the drain, the drain charge is zero. In reality, the fact the channel is pinched off merely suggests that $c'_{QS}(x=L) = 0$. It does not necessarily imply $Q_D = 0$. After all, Q_D is defined such that Eq. (1-43) relating the instantaneous and dc currents is valid. It is impossible to establish the value of Q_D based solely on the information of $c'_{QS}(x=L)$. Despite that the 0/100 partition is physically unsound and leads to wrong delay time in a transient analysis, it is quite often used. This is because at times the physically sound 40/60 partition scheme can produce numerical problems in the initial part of a transient simulation during which the quasi-static analysis is not correct. One particular simulation problem with the 40/60 partition, found in the so-called Killer NOR-gate circuit, will be examined (Secţion 4.2). However, in a small-signal analysis, the 40/60 partition should always be used.

There is a gross misconception that any partition scheme, be it 40/60, 50/50, or 0/100, is an arbitrary assignment of the channel charge to the source and the drain. This misconception arises from the old days, when the methodology to seek the charge equations based purely on device physics was not straightforward, that the modeling engineers may have had a tendency to arbitrarily assign charges. However, as first demonstrated with a double integral technique [9] and later with the first-moment technique [24,25], the 40/60 partition scheme is based entirely on device physics rather than the hunches of device engineers. The moment we make the quasi-static assumption, the 40/60 partition becomes *exactly* correct. [If we were to abolish the quasi-static assumption, then Q_D and Q_S defined in Eqs. (1-46) and (1-48) become meaningless. It does not make sense to partition charge under the non-quasi-static condition anyway.] The 40/60 is not an arbitrary partition scheme. Other partition schemes, such as 50/50 or 0/100, are arbitrary partitioning schemes, at least in the sense that they cannot be rigorously shown to obey physical laws.

In the development of the terminal charges, we have assumed that the bulk effects are negligible and made δ zero (particularly in Eq. 1-25). Deriving Q_G and Q_B with finite δ becomes more complicated, although Q_D and Q_S are still given by Eqs. (1-46) and (1-48). For convenience, we list without proof the equations of Q_G and Q_B, in the event that $\delta \neq 0$:

$$Q_G = WLC'_{ox}\left[\frac{V_{GS} - V_T}{1+\delta}\left(\delta + \frac{2}{3}\frac{\alpha^2 + \alpha + 1}{1+\alpha}\right) + 2\delta(2\phi_f - V_{BS})\right]. \quad (1\text{-}49)$$

$$Q_B = -WLC'_{ox}\left[\frac{\delta}{1+\delta} \cdot (V_{GS} - V_T)\left(1 - \frac{2}{3}\frac{\alpha^2 + \alpha + 1}{1+\alpha}\right) + 2\delta(2\phi_f - V_{BS})\right]. \quad (1\text{-}50)$$

How do the equations developed in this section compare to those used in BSIM3? BSIM3's equations, once stripped of those smoothing functions, are written in a manner found in Ref. [1]. The BSIM3 equations in the 40/60 partition scheme, are given by [26]

$$Q_G = WLC'_{ox}\left[V_{GS} - V_T - \frac{V_{DS}}{2} + \frac{A_{bulk}V_{DS}^2}{12(V_{GS} - V_T - A_{bulk} \cdot V_{DS}/2)} + \gamma\sqrt{2\phi_f - V_{BS}}\right]. \quad (1\text{-}51)$$

$$Q_B = -WLC'_{ox}\left[\frac{(A_{bulk} - 1)V_{DS}}{2} + \frac{(1 - A_{bulk})A_{bulk}V_{DS}^2}{12(V_{GS} - V_T - A_{bulk} \cdot V_{DS}/2)} + \gamma\sqrt{2\phi_f - V_{BS}}\right]. \quad (1\text{-}52)$$

$$Q_D = -WLC'_{ox}\left[\frac{V_{GS} - V_T}{2} - \frac{A_{bulk}V_{DS}}{2}\right.$$
$$\left. + \frac{(V_{GS} - V_T)^2/6 - A_{bulk}V_{DS}(V_{GS} - V_T)/8 + A_{bulk}^2 V_{DS}^2/40}{(V_{GS} - V_T - A_{bulk} \cdot V_{DS}/2)^2} A_{bulk}V_{DS}\right]. \quad (1\text{-}53)$$

These equations are slightly modified from the actual BSIM3 equations to conform with the symbols used in this book. Further, these BSIM3 equations (Eqs. 1-51 to 1-53) apply only to the linear region. At saturation, V_{DS} is replaced by $V_{DS,\text{sat}}$. Although these equations appear to be very different from those of Eqs. (1-49), (1-50), and (1-46), respectively, they are actually the same. The equality can be demonstrated by substituting δ by $A_{bulk} - 1$ and α by $1 - V_{DS}/(V_{GS} - V_T)$, and γ by Eq. (1-7). Although the forms in BSIM3 can strain eyes (especially the drain charge term), those forms are more consistent with the rest of BSIM3 development, which determines the modification to the charges in short-channel devices. Nonetheless, for a first-order calculation, sufficient for many circuit engineers' applications, Eqs. (1-49), (1-50), and (1-46), which apply to both linear and saturation regions, are more convenient.

1.8 CHARGE CONSERVATION

It is often said that BSIM3 is a charge-conserved model. We ask, what does the charge conservation mean, in the context of modeling? Shouldn't charge be always conserved anyway, and so why are we making a big deal out of charge conservation? A simple answer is that a model branded as "charge-conserved" does not necessarily conserve charge.

Early MOSFET models were charge-conserved in the derivation of their equations, but are not charge-conserved during the implementation of the model into SPICE. We will revisit this fine point in the latter part of this section. For now, we demonstrate how the charges can be non-conserved. The input capacitance in this example, C_{gs}, is hooked directly to the input voltage V_{GS}, as shown in Fig. 1-29. We consider a rather ideal case where the output node of the transistor (the drain) is isolated from the input capacitance C_{gs}. In spite of this idealization, C_{gs} is still made to be a function of the output voltage, V_{DS}, besides the terminal voltage difference, V_{GS}. (We disregard the possible dependence on V_{BS} in this simple example.) In the interest of having less algebra, we shall assume $C_{gs}(V_{GS}, V_{DS})$ to be given by

$$C_{gs}(v_{GS}, v_{DS}, v_{BS}) = 3 - v_{DS} \quad \text{(capacitance in F and voltage in V)}. \quad (1\text{-}54)$$

In a real device, C_{gs} is a complicated functions of v_{GS}, v_{DS}, and v_{BS}. The voltage waveforms of $v_{GS}(t)$ and $v_{DS}(t)$ for this example are shown in Fig. 1-30. The initial state and the final state of the circuit are identical, both being at $v_{GS} = v_{DS} = 1$ V. Without the presence of any lossy elements in the circuit, we expect that, between $t = 0$ and t_5, the net transfer of charge through the input voltage source (v_{GS}) is zero; that is, the charge is conserved. However, as demonstrated in the following, charge non-conservation can occur if improper procedure is used to compute the charge transfer during transitions. Let us focus at time $t = t_1$ when $v_{GS}(t)$ steps up from 1 V to 2 V while $v_{DS}(t)$ maintains at 1 V. According to Eq. (1-54), $C_{gs} = 2$ F. An incremental charge $\Delta Q = \int i \cdot dt = \int C_{gs} \cdot dv_{GS} = 2$ coulombs is delivered to the

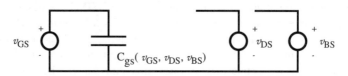

Fig. 1-29 A voltage-dependent capacitor used to illustrate the concept of charge non-conservation. Its capacitance value depends on v_{GS}, v_{DS}, and v_{BS} in accordance with Eq. (1-54). (From Ref. [30], © IEEE 1995, reprinted with permission. The original figure is modified to conform with our discussion).

top plate of C_{gs} from v_{GS}. We now consider $t = t_2$. As v_{DS} increases from 1 V to 2 V, v_{GS} is held at 2 V and $\Delta v_{GS} = 0$. During this transition, $\Delta Q = \int i \cdot dt = \int C_{gs} \cdot dv_{GS} = 0$ and there is no charge transfer. At $t = t_3$, v_{GS} shifts back down from 2 V to 1 V and v_{DS} remains steady. At this moment, C_{gs} is still 1 F. Hence, $\Delta Q = \int i \cdot dt = \int C_{gs} \cdot dv_{GS} = -1$ coulomb. At $t = t_4$ when only v_{DS} changes value while V_{GS} remains unmodified, ΔQ is again zero as at $t = t_2$. At the end of the transitions, the overall change of the charge is $\Delta Q = 2 + 0 - 1 + 0 = 1$ coulomb. The important point is that $\Delta Q \neq 0$, even after both v_{GS} and v_{DS} have long settled at their initial values. The unphysical outcome of the nonzero net charge transfer is termed the *charge non-conservation*.

This anomalous result is traced back to the very use of a branch relationship used in SPICE simulators: $i = C \cdot dv/dt$. In reality, when the capacitor is *nonlinear* (i.e., capacitance value is not constant), the proper relationship should have been $i = C \cdot dv/dt + v \cdot dC/dt$. In particular, in the case of $C_{gs}(v_{GS}, v_{DS})$, the correct expression is

$$i = C_{gs}\frac{dv_{GS}}{dt} + v_{GS}\frac{dC_{gs}}{dt} = C_{gs}\frac{dv_{GS}}{dt} + v_{GS}\frac{\partial C_{gs}}{\partial v_{GS}}\frac{dv_{GS}}{dt} + v_{GS}\frac{\partial C_{gs}}{\partial v_{DS}}\frac{dv_{DS}}{dt}. \quad (1\text{-}55)$$

This equation properly accounts for the capacitance's variation on v_{GS} and v_{DS}. Because the way the program is constructed, most SPICE simulators do not correctly evaluate the current in nonlinear capacitors. A SPICE program considers only the

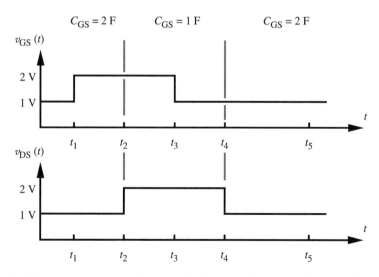

Fig. 1-30 The v_{GS} and v_{DS} waveforms applied to the voltage-dependent capacitor of Fig. 1-29.

$C_{gs} \cdot dv_{GS}/dt$ term in Eq. (1-55), neglecting all other contributions. Based on Eq. (1-55), the proper amount of charge transfer at a given time is

$$\Delta Q = \int i\, dt = C_{gs}\Delta v_{GS} + v_{GS}\frac{\partial C_{gs}}{\partial v_{GS}}\Delta v_{GS} + v_{GS}\frac{\partial C_{gs}}{\partial v_{DS}}\Delta v_{DS}. \qquad (1\text{-}56)$$

In this particular example $\partial C_{gs}/\partial v_{GS} = 0$ and $\partial C_{gs}/\partial v_{DS} = -1$, $\Delta Q = C_{gs} \cdot \Delta v_{GS} - v_{GS} \cdot \Delta v_{DS}$ (where capacitances are in farads and voltages are in volts). We repeat the above exercise to track the charge increments across each transition of v_{GS} and v_{DS}, except this time we use the modified ΔQ expression. At $t = t_1$, $C_{gs} = 2$ F, $\Delta v_{GS} = 1$ V, and $\Delta v_{DS} = 0$. ΔQ at this moment is equal to 2 coulomb, just as calculated the last time. However, at $t = t_2$ when $C_{gs} = 2$ F, $v_{GS} = 2$ V, $\Delta v_{GS} = 0$ and $\Delta v_{DS} = 1$ V, ΔQ based on this modified expression gives $\Delta Q = 2 \cdot 0 - 2 \cdot 1 = -2$ coulomb, rather than zero as calculated previously. At $t = t_3$, $\Delta Q = C_{gs} \cdot \Delta v_{GS} - v_{GS} \cdot \Delta v_{DS}$ is again -1 coulomb, whereas at $t = t_4$, ΔQ is 1 coulomb. Summing up these incremental charges across each transition, we find after $t = t_5$, $\Delta Q = 2 - 2 - 1 + 1 = 0$. This time, the charge is conserved.

We see that the first cause of charge non-conservation is fundamental to the algorithms adopted in a SPICE simulator that, in this particular example, the simulator calculates ΔQ from only the first term in Eq. (1-56). One way to fix this problem is to modify the SPICE simulator formulation to include the remaining terms in Eq. (1-56). With the use of a linear capacitor, a voltage-dependent resistor, some current sources, Ref. [27] demonstrates the charge conservation can indeed be restored. Despite the demonstration, this fix has not made its way to the mainstream SPICE simulators. We describe three likely reasons in the following.

The first reason is a fundamental numerical issue. We want to find the stored charge after a period of time. According to Eq. (1-56), we write

$$Q = \int_{t_1}^{t_2} C_{gs}(v_{GS}(t), v_{DS}(t), v_{BS}(t))\, dv_{GS}(t) + \int_{t_1}^{t_2} v_{GS}(t)\frac{\partial C_{gs}(v_{GS}, v_{DS}, v_{BS})}{\partial v_{GS}}\, dv_{GS}(t)$$

$$+ \int_{t_1}^{t_2} v_{GS}\frac{\partial C_{gs}}{\partial v_{DS}}\, dv_{DS} + \int_{t_1}^{t_2} v_{GS}\frac{\partial C_{gs}}{\partial v_{BS}}\, dv_{BS}. \qquad (1\text{-}57)$$

Let us focus on just the first term (Q_1), which is sufficient to reveal the origin of the problem. The function $C_{gs}(v_{GS}, v_{DS}, v_{BS})$ is not exactly known in the entire time interval between t_1 and t_2. During the numerical evaluation of the integral, the SPICE simulator is able to calculate only $C_{gs}(v_{GS}, v_{DS}, v_{BS})$ at $t = t_1$ and t_2, but not other time points in between. The approach to perform the numerical integration in typical SPICE simulators is to rewrite Q_1 (the first term of Q appearing in Eq. 1-57) as,

$$Q_1 = \int_{t_1}^{t_2} C_{gs}(v_{GS}, v_{DS}, v_{BS})\, dv_{GS} \approx \int_{t_1}^{t_2} \overline{C_{gs}}\, dv_{GS} = \overline{C_{gs}}[v_{GS}(t_2) - v_{GS}(t_1)], \qquad (1\text{-}58)$$

1.8 CHARGE CONSERVATION

where $\overline{C_{gs}}$ is the average of $C_{gs}(v_{GS}(t_1), v_{DS}(t_1), v_{BS}(t_1))$ and $C_{gs}(v_{GS}(t_2), v_{DS}(t_2), v_{BS}(t_2))$. As demonstrated in the above integral, the charge calculation, as implemented in the SPICE simulators, calculates the terminal charges from the corresponding average capacitance whose value depends only on the values of the terminal voltages at the beginning and the end of a given time interval. This non-exact approximation (Eq. 1-58) to the actual charge integral (Eq.1-57), when carried over a period of time, can give rise to a nonzero net transfer of charge as the bias voltages restore to their initial values after some ramp-ups and ramp-downs. There is always some amount of error associated with this non-exact numerical integration, although the problem can be mitigated with the use of a stringent tolerance (by setting .TOL to a smaller value than the default of 10^{-5} in the SPICE input deck.) Tightening the .TOL value forces the SPICE simulator to takes finer time steps. Consequently, the change in voltage across the capacitor over a time step is smaller, and the approximation of the averaged capacitance becomes more accurate. When carried to an extreme, we could perceive that if .TOL were made to be infinitesimally small, then the error would approach zero. This statement can, in fact, be proven with a rigorous analysis of vector calculus. However, due to computer limitations, setting .TOL to an arbitrarily small number prevents the SPICE from ever converging to a solution. (In fact, if the tolerance is too tight, SPICE can converge to a wrong answer far from the true solution.) In this regard, we see that the charge non-conservation problem is not solely an issue of the device model. The charge non-conservation can also come from incorrect numerical integration within the SPICE simulator.

We recapitulate. If a SPICE device model does not take special precaution to the treatment of the nonlinear capacitance, such as accounting the extra terms in Eq. (1-55) in addition to $\int C_{gs} \, dv_{GS}$, then the model by itself is a non-conservative model. The aforementioned integration problem associated with the SPICE simulator will only make matters worse. When the model is made to uphold charge conservation, the overall simulation can still appear to be non-conservative because of the numerical integration problem. However, we should stress that there are some device models which are naturally less susceptible to the above numerical problem than others. This is the second reason why the model of Ref. [27] is not as popular as we might think. In 1983, Yang proposed a model that uses charge as the state variable rather than voltage [1]. [For our purposes here, the state variable can be thought of as the integration variable which appears in the charge integral, e.g., in Eq. (1-58)]. It was recognized that the ultimate source of the integration inaccuracy in early simulators lay in the approximation of the capacitance at various time points. The capacitances were needed in the evaluation of the charge because the state variable (or equivalently, the integration variable) was voltage. If the state variable is charge instead, then the overall charge can be found from $Q = \int dQ$ and no capacitance would be required to complete the calculation. The capacitance information, which represents the derivatives of the charge, will be used only in the numerical subroutines to speed up the convergence of the solution. Even if the capacitances are wrongly calculated in charge-based models, a correct solution is still possible after convergence, although wrong values of capacitance will delay the convergence. This is similar to what we discussed about the dc convergence in Section 1.2. A charge-conservative model refers to one that specifically calculates the charges as the state variable. Let us demonstrate the charge conservation through the use of charge as the state variable, for the aforementioned example. We establish an equation for $Q(v_{GS}, v_{DS}, v_{BS})$ explicitly. To better relate the development here with that of Section 1.7, we shall denote the charge as Q_G, the gate charge. It is given by $Q_G = C_{gs}(v_{GS}, v_{DS}, v_{BS}) \times v_{GS}$:

$$Q_G(t) = (3 - v_{DS}(t)) \times v_{GS}(t) \qquad (Q \text{ in coulombs and } v \text{ in volts}). \qquad (1\text{-}59)$$

We emphasize that in the charge conservative model, we set out to determine the charge expression first, as is done above. From the equation, we can then find $C_{gs} = \partial Q_G / \partial v_{GS} = 3 - v_{DS}$, which is exactly that given in the example. The proper sequence is to find charge first, from which the capacitances are derived. This approach is also that used in BSIM3.

We go through the timing diagram of Fig. 1-30. At $t = 0$ when $v_{GS} = v_{DS} = 1$ V, Q, according to Eq. (1-59), is equal to 2 coulombs. At $t = t_1^+$, v_{GS} has changed from 1 V to 2 V while v_{DS} maintains at 1 V. Q is thus 4 coulombs. At $t = t_2^+$, $v_{GS} = v_{DS} = 2$ V, and Q calculated from Eq. (1-59) is 2 coulombs. Similar calculations are performed and we find Q to be 1 and 2 coulombs at t_3^+ and t_4^+, respectively. As expected, the charge at the beginning and the end of the transient, after which v_{GS} and v_{DS} settle to their original values, is the same. ΔQ at the transition times t_1, t_2, t_3, and t_4 are 2, -2, -1 and 1 coulombs, respectively. These results agree exactly with the calculation following Eq. (1-56). They both give an overall ΔQ of zero. The main point about a charge-based model is that the charge expression is explicitly expressed as a function of the terminal voltages. Even though the bias voltages may undergo some large excursion, the charge calculated from the charge expression regresses to exactly its original value whenever the bias voltages settle to their initial values.

With charge as the state variable, we only need to know the voltages at a given time, not the amount of the voltage increment, as in the previous case. In this unrealistic example where the capacitances and voltages assume integer values, charge calculations based on Eq. (1-56) with voltage as the state variable, and on Eq. (1-59) with charge as the state variable, both lead to the correct charge-conservative result. In real devices whose capacitances and bias voltages vary with non-integer values, we expect the charge-based approach to incur fewer numerical errors because the charge-based approach does not require the evaluation of $\int C_{gs} \cdot dv_{GS}$. However, numerical truncation error is unavoidable in any computer simulator. At every numerical iteration or time point, a current tolerance δI, a voltage tolerance δV, and a charge tolerance, δQ, are always present. When the convergence tolerance is improperly set, these small errors made on each time step of the voltage trajectory accumulate and the final charge is not equal to the initial charge despite the bias voltages' return to their original values. Basically, a charge-conservative model does not automatically guarantee the charge conservation at the circuit simulator level. The proper convergence criteria need to be carefully chosen. In fact, with a bad set of

1.8 CHARGE CONSERVATION

convergence criteria, Ref. [1] showed that charge non-conservation can appear in a linear capacitor whose capacitance values is a constant, independent of any terminal voltages.

For a MOSFET with four terminals, a charge-based model needs to express the terminal charges in terms of terminal voltages. In Section 1.7 we have identified these charges Q_G, Q_D, Q_S, and Q_B. As intuitively obvious, these equations must be such that the sum of these charges be zero:

$$Q_G + Q_D + Q_S + Q_B = 0. \quad (1\text{-}60)$$

All charge-based models satisfy Eq. (1-60), although the equation is not sufficient to guarantee that the model is charge-conservative. Sometimes a model may not establish explicit expressions for Q_D and Q_S, but instead lump them together as one quantity, the channel charge Q_I. In this kind of models, the charge is conserved (i.e., net charge is zero after bias voltages return to initial state) because Eq. (1-60) is upheld in these models. However, because there are no explicit relationships for the drain and source charges, the charge can mysteriously appear or disappear on the drain and source. According to Ref. [21], one example is a pioneering GaAs MESFET (metal-semiconductor field-effect transistor) model which develops only the gate charge but not the source and the drain charges explicitly [28]. In MESFETs, the channel charge density $c'(x, t)$ does not appear as simplistic as Eq. (1-30) for the MOSFETs. In fact, $c'_{QS}(x, t)$ in MESFETs does not have an analytical form, a result that impedes the drain and source charges from being easily derived, though it is still possible. The equations of Q_D and Q_S for MESFET can be found in Ref. [24].

The third reason that Ref. [27]'s non-charge-based approach is not widely adopted to solve the charge non-conservation problem is the following: According to Eq. (1-56), the non-charge-based model would require the knowledge of $\partial C_{gs}/\partial v_{GS}, \partial C_{gs}/v_{DS}$, and $\partial C_{gs}/\partial v_{BS}$. At the time when Ref. [27] was introduced, the smoothing function approach discussed in Section 1.4 was not yet popular. Most of the MOSFET models then can be characterized as the digital models discussed in Section 1.3. The models, based on regional analyses carried out separately in the inversion, subthreshold, and accumulation regions, unavoidably introduce discontinuity in the slopes of the capacitances. As evident in Figs. 1-12 and 1-13, these models produce a sudden jump or decrease of capacitances as the bias voltage varies. These abrupt features cause $\partial C_{gs}/v_{GS}$ and $\partial C_{gs}/\partial v_{DS}$ to be undefined at the boundaries of operating regions. A remedy to this dilemma is possible by defining some intermediate region during which the capacitance values at the two regions are interpolated. However, the remedy comes with a hefty price of complexity.

We have described two ways to modify the device model to ensure charge conservation. The first one is exemplified in Eq. (1-56), which accounts for the current flow associated with the derivative of capacitance with respect to time. The second approach is the charge-based approach, which sought after the charge as the state variable. There is a third approach, the so-called transcapacitance approach [29]. This method eliminates the charge non-conservation problem of equating ΔQ

to just $\int C_{gs} \cdot dv_{GS}$, by the addition of the two other transcapacitance in parallel. The transcapacitive scheme is illustrated in Fig. 1-31. The current through the original capacitor, in this approach, will be taken to consist solely of $C_{gs} \cdot \partial v_{GS}/t$, as accounted for in typical SPICE simulators. The other current components (the other terms in Eq. 1-55), are accounted for by the added transcapacitances. The current through the transcapacitance C_{gd} is $C_{gd} \cdot \partial v_{DS}/\partial t$. Likewise, the current through C_{gb} consists of only $C_{gb} \cdot \partial v_{BS}/\partial t$. This approach can be understood as follows: Instead of writing the overall current as $Cdv/dt + vdC/dt$ as in Eq. (1-55), we write

$$i = \frac{dQ_G}{dt} = \frac{\partial Q_G}{\partial v_{GS}} \cdot \frac{\partial v_{GS}}{\partial t} + \frac{\partial Q_G}{\partial v_{DS}} \cdot \frac{\partial v_{DS}}{\partial t} + \frac{\partial Q_G}{\partial v_{BS}} \cdot \frac{\partial v_{BS}}{\partial t}. \quad (1\text{-}61)$$

Therefore, in terms of the capacitances, the current is given by

$$i = C_{gs} \frac{\partial v_{GS}}{\partial t} + C_{gd} \frac{\partial v_{DS}}{\partial t} + C_{gb} \frac{\partial v_{BS}}{\partial t}. \quad (1\text{-}62)$$

We intentionally write down Eq. (1-61) prior to Eq. (1-62). This way, we clearly show that C_{gd} is $\partial Q_G/\partial V_{DS}$, and that C_{gd} can be obtained once we know the functional form of Q_G. In the transcapacitance approach reported in the literature, C_{gd} in particular is not obtained in this manner [30]. (C_{gb} is irrelevant in our example since $\partial V_{BS}/\partial t$ is zero anyway.) Instead, C_{gd} is determined from the relationship that, since $\partial^2 Q_G/(\partial v_{GS}\partial v_{DS}) = \partial^2 Q_G/(\partial v_{DS}\partial v_{GS})$,

$$\frac{\partial C_{gs}}{\partial v_{DS}} = \frac{\partial C_{gd}}{\partial v_{GS}}. \quad (1\text{-}63)$$

For the $C_{gs}(v_{GS}, v_{DS}, v_{BS})$ given in Eq. (1-54), the relationship governing C_{gd} is therefore, $\partial C_{gd}/\partial v_{GS} = -1$. There are many possibilities of C_{gd} satisfying this relationship. For example, the capacitance can be C_{gd} (in F) $= -v_{GS}$ (in V). We go through the timing diagram of Fig. 1-30 once more, using the transcapacitance

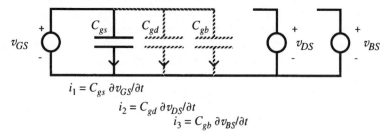

Fig. 1-31 With an addition of two capacitors in parallel to the original capacitor in Fig. 1-29, this transcapacitive approach can eliminate the charge non-conservation problem. (From Ref. [30], © IEEE 1995, reprinted with permission. The original figure is modified to conform with our discussion.)

method represented by Fig. 1-31. Based on Eq. (1-62), the incremental charge due to an incremental time change is: $\Delta Q = C_{gs} \cdot \Delta v_{GS} + C_{gd} \cdot \Delta v_{DS} + C_{gb} \cdot \Delta v_{BS}$. In this example, Δv_{BS} is always zero, hence ΔQ is determined from the first two terms only. $C_{gs} = 3 - v_{DS}$, as given in Eq. (1-55), and $C_{gd} = -v_{GS}$ as one expression which satisfies Eq. (1-63). At $t = t_1$ when v_{GS} increases from 1 V to 2 V and v_{DS} remains at 1 V, $C_{gs} = 2$ F; $C_{gd} = -1$ F; $\Delta v_{GS} = 1$ V and $\Delta v_{DS} = 0$. Therefore, ΔQ at this transition is $2 \cdot 1 - 1 \cdot 0 = 2$ coulombs. At $t = t_2$, $C_{gs} = 2$ F; $C_{gd} = -2$ F; $\Delta v_{GS} = 0$; and $\Delta v_{DS} = 1$ V. This leads to a ΔQ of $2 \cdot 0 - 2 \cdot 1 = -2$ coulombs. At $t = t_3$, $C_{gs} = 1$ F; $C_{gd} = -2$ F; $\Delta v_{GS} = -1$ V; and $\partial v_{DS} = 0$. This leads to a ΔQ of $1 \cdot (-1) - 2 \cdot 0 = -1$ coulombs. Finally, at $t = t_4$, $C_{gs} = 1$ F; $C_{gd} = -1$ F; $\Delta v_{GS} = 0$; and $\Delta v_{DS} = -1$ V. This leads to a ΔQ of $1 \cdot 0 + 1 \cdot 1 = 1$ coulomb. Overall, the sum of the calculated ΔQ's across the various transitions is equal to zero, as expected from charge conservation.

In this transcapacitance method, the transcapacitance C_{gd} needs to be determined from Eq. (1-63). We had used $C_{gd} = -v_{GS}$ to result in charge conservation. However, this C_{gd} expression is not unique. Another expression, $C_{gd} = v_{DS} - v_{GS}$, still satisfies Eq. (1-63) and can be shown to result in charge conservation as well (although the amount of energy transfer will differ in the two cases [30]). The nonuniqueness in the transcapacitance can be resolved if we use the charge-based model discussed previously. In a charge-based model, we need to find Q_G, among other charges. Once Q_G is identified, C_{gd} can then be uniquely determined from the partial derivative of Q_G with respect to v_{DS}. Therefore, even the transcapacitance method points to the charge-based model as the preferred approach toward device modeling.

We end this section by showing some circuits which expose the problems associated with charge non-conservation. For this purpose, we compare results simulated with BSIM3, a charge-conservative model, and MOS Level 1, a non-charge-conservative model. The first circuit, a charge-pump circuit, is shown in Fig. 1-32. The transistor size is such that the total gate capacitance is 1 pF. Hence, when the transistor turns on, and at $v_{DS} \approx 0$, as is the case for this circuit, $C_{gd} = C_{gs} \approx 0.5$ pF. These intrinsic transistor capacitances is roughly equal to the loading capacitors attached to the source and drain terminals. (The parasitic capacitances, such as the overlap and the junction capacitances, are made zero by adjusting the model parameters.) When the input voltage ramps to a high value, the device turns on. The drain node is connected at the one side with the 0.5 pF to ground, and another, with $C_{gd} \approx 0.5$ pF to the input voltage. Therefore, the drain voltage increases to a value roughly equal to $(v_{IN} - V_T)/2$, or about 1.1 V. The ramp time of 0.02 μs is long enough such that the source voltage settles to its stationary value when v_{IN} reaches 3 V. Conversely, when v_{IN} decreases to 0 V, the transistor turns off and its C_{gd} becomes zero. Since the downward ramp takes a time of 0.02 μs, we expect that the drain voltage to reach 0 V when v_{IN} reaches 0 V. This is physically expected behavior is indeed simulated with a charge-conserved BSIM3 model with a .TOL of 1×10^{-12}.

However, when the same charge-conserved BSIM3 model is used with a default .TOL of 1×10^{-5}, a sign of charge non-conservation becomes obvious. As shown

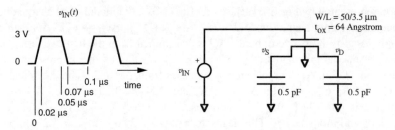

Fig. 1-32 A charge-pump circuit capable of exposing the problems of a charge non-conservative model. (From Ref. [1], © IEEE 1983, reprinted with permission. The original figure is modified to conform with our discussion.)

by the long dash line in Fig. 1-33, the drain voltage after one cycle of v_{IN} variation does not reach its initial state of zero volts, although the input voltage has returned to its initial state at $t = 0$. The drain voltage is negative, inconsistent with the circuit configuration whose most negative voltage is the ground (0 V).

Figure 1-33 also shows that, with a non-charge-conservative model such as Level 1, the simulated result is unphysical even under stringent .TOL of 1×10^{-12}. At the time when the input voltage has returned to its initial state, the drain voltage produced by the Level 1 model takes on positive value in some cycles, and negative values in other cycles.

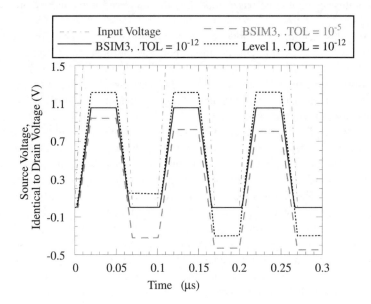

Fig. 1-33 Simulation results of the circuit in Fig. 1-32, with a charge conservative model BSIM3 and a non-conservative model MOS Level 2. Even with a charge conservative model, a wrong simulation result can still occur if the tolerance value is not carefully chosen.

Figure 1-34 is a switch capacitor circuit, another problematic circuit for the non-charge-conservative models. The transistor size is small enough such that the device capacitance is much smaller than the load capacitors of 0.5 and 5 pF. Therefore, when ϕ_A turns high, v_1 is immediately charged to 3 V, the constant voltage maintained at the source of the input transistor. The total charge transferred into the drain of the input transistor is 1.5 pico-coulombs. Because the junction leakage current as well as the parasitic capacitances are made zero in the device model, the transferred charges remain undiminished at the brief period when both ϕ_A and ϕ_B are zero. When ϕ_B turns on, a certain fraction of the 1.5 pico-coulombs flows into C_2, charging up C_2 to a voltage value equal to that at C_1 after the charge transfer. Let us express the statement in the mathematical form. Assume at a particular time t that v_2 is equal to $v_2(t)$. We are interested in finding $v_2(t+1)$, the v_2 value at the next cycle after the charge transfer has taken place. We denote the amount of charge transfer in pico-coulombs from C_1 to C_2 as x, then equating the voltages at the two capacitors leads to

$$v_2(t) + x/C_2 = (1.5 - x)/C_1. \qquad (1\text{-}64)$$

Substituting $C_2 = 5\,\text{pF}$ and $C_1 = 0.5\,\text{pF}$, we find that $x = 1.364 - 0.455 \cdot v_2(t)$. Because $v_2(t+1) = v_2(t) + x/C_2$, we have

$$v_2(t+1) = 0.273 + 0.909 \cdot v_2(t) \qquad \text{(voltages in V)}. \qquad (1\text{-}65)$$

During the zero-th cycle (at $t = 0$), the charge transfer has not taken place. Thus, $v_2(0)$ is initially 0. After one cycle of ϕ_A and ϕ_B, Eq. (1-65) states that $v_2(1) = 0.273$ V. After one additional cycle, $v_2(2)$ according to Eq. (1-65) is $0.273 + 0.909 \cdot 0.273 = 0.521$ V. The subsequent v_2 values after each additional

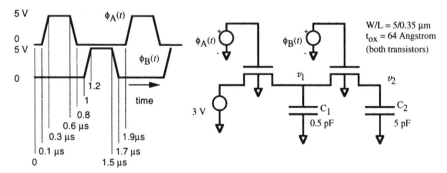

Fig. 1-34 A switch-capacitor circuit capable of exposing the problems of a charge non-conservative model. (From Ref. [1], © IEEE 1983, reprinted with permission. The original figure is modified to conform with our discussion.)

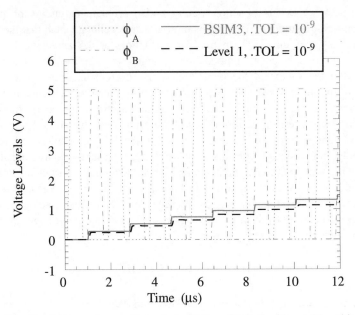

Fig. 1-35 Simulation output voltage (v_2) of the circuit in Fig. 1-34, with a charge conservative model BSIM3 and a non-conservative model MOS Level 2. The transistors have a W/L of $5/0.35\,\mu m$.

cycle are 0.746 V, 0.952 V, 1.102 V, 1.275 V, ... These values are correctly simulated with the BSIM3 model, as shown in Fig. 1-35. Despite that the non-charge-conservative Level 1 model predicts a similar voltage waveform, the quantitative value of v_2 after each cycle is incorrect. The magnitude of the error increases as the number of cycles increases, a feature shared by the non-charge-conservative models.

Figure 1-35 illustrates the simulation results when the transistors in Fig. 1-34 have a W/L of $5/0.35\,\mu m$. When both transistors' size is enlarged to $50/3.5\,\mu m$, the simulation results are shown in Fig. 1-36. Because the transistor capacitances are of similar values as the loading capacitors, there are charging and discharging of current when ϕ_B is ramped up and down. However, the v_2 values specified in Eq. (1-65) remain accurate at the portion of a cycle when ϕ_A is high. Again, we see that the BSIM3 model predicts v_2 to be 0, 0.273 V, 0.521 V, and so on, etc during those times. These values agree the theoretical prediction of Eq. (1-65). In contrast, the Level 1 model gives wrong voltage values as before.

The Level 1's C-V model is the Meyer model, which equates C_{xy} with C_{yx}. This is an erroneous simplification because, for example, C_{dg} is $-\Delta Q_D/\Delta V_G$ while C_{gd} is $-\Delta Q_G/\Delta V_D$. C_{dg} and C_{gd} would have been equal if the gate and drain are the two terminals of a parallel-plate capacitor. A plot of the two capacitances was shown in Fig. 1-16. To explain why physically these two capacitances in the MOS transistor

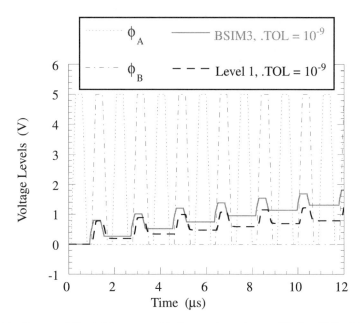

Fig. 1-36 Repeat of Fig. 1-35, except with the transistor size modified to W/L of $5/3.5\,\mu$m.

should differ, we examine the saturation operation region in which the channel charge is pinched off at the drain end. The source is tied to the ground so that $V_G = V_{GS}$ and $V_D = V_{DS}$. By definition, C_{dg} is proportional to the change in the drain charge when the gate bias is changed. Stated in another way, C_{dg} relates to the amount of drain current variation as the gate bias is modified. When V_G increases while the source is grounded, we expect a substantial increase in the channel charge. The additional charges are supplied, in part, through the drain. We therefore expect C_{dg} to be a large value. The situation for C_{gd} in saturation, however, is drastically different. C_{gd} is equal to the variation of the gate charge as V_D is modified. During saturation, the channel is pinched off at the drain, regardless of the exact value of V_{DS}. As long as V_{DS} exceeds the saturation voltage value, $V_{DS,\text{sat}}$, the channel charge is not modified. The channel potential right at the pinchoff point is related to $V_{DS,\text{sat}}$, but unaffected by V_{DS}. Therefore, during saturation, $c'(x, t)$ remains to be the same at all locations and for all the time, independent of the V_{DS} variation. The unchanging $c'(x, t)$, in turn, brings about a constant Q_G. This insensitivity of the gate charge with respect to V_D leads to a zero C_{gd} in saturation.

We have seen that, physically, $C_{gd} \neq C_{dg}$. The Meyer's capacitance model, however, stipulates that $C_{xy} = C_{yx}$. It is thus not surprising that this reciprocity relationship is the root cause of the non-charge-conservation problem associated with the Meyer model [31].

1.9 NON-QUASI-STATIC AND QUASI-STATIC y-PARAMETERS

The instantaneous gate, drain, and source currents of a MOSFET were given by Eqs. (1-40), (1-43), and (1-47), respectively. The bulk current can also be written in a fashion similar to the gate current. We write all four terminal currents as

$$i_G(t) = \frac{dQ_G}{dt} = \frac{\partial Q_G}{\partial v_G} \cdot \frac{dv_G}{dt} + \frac{\partial Q_G}{\partial v_D} \cdot \frac{dv_D}{dt} + \frac{\partial Q_G}{\partial v_S} \cdot \frac{dv_S}{dt} + \frac{\partial Q_G}{\partial v_B} \cdot \frac{dv_B}{dt}; \quad (1\text{-}66a)$$

$$i_D(t) = I_D + \frac{dQ_D}{dt} = I_D + \frac{\partial Q_D}{\partial v_G} \cdot \frac{dv_G}{dt} + \frac{\partial Q_D}{\partial v_D} \cdot \frac{dv_D}{dt} + \frac{\partial Q_D}{\partial v_S} \cdot \frac{dv_S}{dt} + \frac{\partial Q_D}{\partial v_B} \cdot \frac{dv_B}{dt}; \quad (1\text{-}66b)$$

$$i_S(t) = -I_D + \frac{dQ_S}{dt} = -I_D + \frac{\partial Q_S}{\partial v_G} \cdot \frac{dv_G}{dt} + \frac{\partial Q_S}{\partial v_D} \cdot \frac{dv_D}{dt} + \frac{\partial Q_S}{\partial v_S} \cdot \frac{dv_S}{dt} + \frac{\partial Q_S}{\partial v_B} \cdot \frac{dv_B}{dt}; \quad (1\text{-}66c)$$

$$i_B(t) = \frac{dQ_B}{dt} = \frac{\partial Q_B}{\partial v_G} \cdot \frac{dv_G}{dt} + \frac{\partial Q_B}{\partial v_D} \cdot \frac{dv_D}{dt} + \frac{\partial Q_B}{\partial v_S} \cdot \frac{dv_S}{dt} + \frac{\partial Q_B}{\partial v_B} \cdot \frac{dv_B}{dt}. \quad (1\text{-}66d)$$

Most often the FETs are biased in the common-source configuration, in which $dv_S/dt = 0$, $dv_G/dt = dv_{GS}/dt$, $dv_D/dt = dv_{DS}/dt$, and $dv_B/dt = dv_{BS}/dt$. Therefore,

$$i_G(t) = C_{gg} \cdot \frac{dv_{GS}}{dt} - C_{gd} \cdot \frac{dv_{DS}}{dt} - C_{gb} \cdot \frac{dv_{BS}}{dt}; \quad (1\text{-}67a)$$

$$i_D(t) = I_D - C_{dg} \cdot \frac{dv_{GS}}{dt} + C_{dd} \cdot \frac{dv_{DS}}{dt} - C_{db} \cdot \frac{dv_{BS}}{dt}; \quad (1\text{-}67b)$$

$$i_B(t) = -C_{bg} \cdot \frac{dv_{GS}}{dt} - C_{bd} \cdot \frac{dv_{DS}}{dt} + C_{bb} \cdot \frac{dv_{BS}}{dt}. \quad (1\text{-}67c)$$

The source current can be inferred from the Kirchoff's current law:

$$i_S(t) = -i_G(t) - i_D(t) - i_B(t). \quad (1\text{-}67d)$$

In digital circuits where the transistor terminal voltages toggle between two extreme values, the transistor characteristics change drastically during the transient. For this situation, Eq. (1-67) can be used. However, in some analog circuits, the external bias changes represent only a small perturbation to the prior operating condition. In these small-signal situations, some approximations can be made to simplify Eq. (1-67). A small-signal operation of particular interest is a sinusoidal perturbation at a frequency f. If the frequency is small, then the d/dt terms can be neglected. Equation (1-67b), in particular, can be written as

$$i_D = I_D(V_{GS} + \Delta v_{gs}, V_{DS} + \Delta v_{ds}, V_{BS} + \Delta v_{bs})$$
$$\approx I_D(V_{GS}, V_{DS}, V_{BS}) + \frac{\partial I_D}{\partial V_{GS}} \Delta v_{gs} + \frac{\partial I_D}{\partial V_{DS}} \Delta v_{ds} + \frac{\partial I_D}{\partial V_{BS}} \Delta v_{bs}. \quad (1\text{-}68)$$

1.9 NON-QUASI-STATIC AND QUASI-STATIC

In the interest of understanding the high-frequency device physics in MOSFET, we limit our consideration to long-channel MOSFET with constant mobility. At the end, the results will be extended to all kinds of MOSFETs, a popular approach used in compact models. The drain current appropriate for the long-channel transistor was given by Eq. (1-9). With the definition of the saturation index (α) of Eq. (1-33), the drain equation can be written as

$$I_D = \frac{WC'_{ox}\mu_n}{L} \frac{(V_{GS} - V_T)^2}{2(1+\delta)} (1 - \alpha^2). \tag{1-69}$$

Knowing that $\partial \alpha / \partial V_{GS} = (1 - \alpha)/(V_{GS} - V_T)$, and that $\partial \alpha / \partial V_{DS} = -(1 + \delta)/(V_{GS} - V_T)$, we write the small-signal conductances (from Eq. 1-4) as

$$g_m = \frac{WC'_{ox}\mu_n}{L} (V_{GS} - V_T) \times \frac{1 - \alpha}{1 + \delta}; \tag{1-70}$$

$$g_d = \frac{WC'_{ox}\mu_n}{L} (V_{GS} - V_T) \times \alpha. \tag{1-71}$$

We also state without proof that [19],

$$g_{mb} = \delta g_m. \tag{1-72}$$

In the case of small-signal perturbation at low frequency, we can then write $\Delta I_D = i_D(v_{GS}, v_{DS}, v_{BS}) - I_D(V_{GS}, V_{DS}, V_{BS})$ as

$$\Delta I_D = g_m \Delta V_{GS} + g_d \Delta V_{DS} + g_{mb} \Delta V_{BS}. \tag{1-73}$$

Further, since there is no dc current component in the gate and the bulk (and we are neglecting transistor capacitances due to low frequency), we have

$$\Delta I_G = \Delta I_B = 0. \tag{1-74}$$

When the small-signal voltages applied to an MOS transistor vary rapidly with time, there are (capacitive) current associated with the dQ_G/dt, dQ_D/dt, and dQ_B/dt terms, in addition to the current given by Eq. (1-73). In this section, we consider the capacitive currents originating in the intrinsic portion of a MOSFET, shown in Fig. 1-37. As shown in part (a) of the figure, the d.c. bias voltages V_G, V_D, V_S, and V_B are applied across the gate, drain, source, and bulk terminals, respectively. When we place an incremental voltage across the gate as shown in Fig. 1-37b, there will obviously be a corresponding change in the gate charge. Because the total charge of the overall device sums up to zero, the increased gate charge induces an equal amount of negative charge in the channel.

Some of the negative charge appears as drain charge, some as the source charge, and the rest, as the bulk charge. If ΔV_G is positive, then ΔQ_G is positive while ΔQ_D, ΔQ_S, and ΔQ_B are negative. There are two sign conventions in defining the

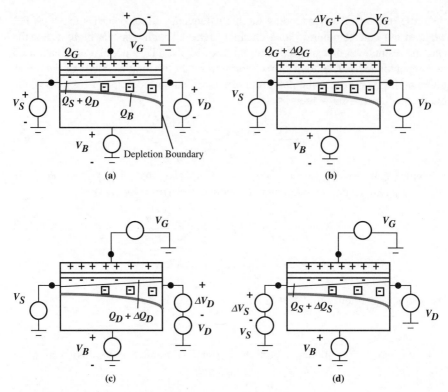

Fig. 1-37 Schematic charge distributions in the gate, channel and bulk when (a) only dc biases are applied; (b) an incremental voltage is applied at the gate; (c) the incremental voltage is applied at the drain; and (d) the incremental voltage is applied at the source. At all times, $Q_G + Q_B + Q_S + Q_D = 0$.

capacitances. We shall adopt the convention in which the capacitance C_{xy} is defined as $\delta_{xy} \times \partial Q_x / \partial V_y$, where δ_{xy} is equal to 1 if $x = y$, and -1 if otherwise. That is,

$$C_{gg} = \frac{\Delta Q_G}{\Delta V_G}; \quad C_{dg} = -\frac{\Delta Q_D}{\Delta V_G}; \quad C_{sg} = -\frac{\Delta Q_S}{\Delta V_G}; \quad C_{bg} = -\frac{\Delta Q_B}{\Delta V_G}. \quad (1\text{-}75)$$

In this convention, C_{gg}, C_{dg}, C_{sg}, and C_{bg} are all positive quantities, in agreement with our intuitive understanding of a capacitor. For example, with Q_G defined in Eq. (1-49) and α defined in Eq. (1-33), $\partial \alpha / \partial V_G = (V_D - V_S)/(V_G - V_S - V_T)^2 = (1 - \alpha)/(V_{GS} - V_T)$, $\partial \delta / \partial V_G = 0$, we find

$$C_{gg} = \frac{\partial Q_G}{\partial V_G} = WLC'_{ox} \times \left\{ \frac{2}{3} \frac{1 + 4\alpha + \alpha^2}{(1 + \alpha)^2} + \frac{\delta}{3(1 + \delta)} \left(\frac{1 - \alpha}{1 + \alpha} \right)^2 \right\}. \quad (1\text{-}76)$$

1.9 NON-QUASI-STATIC AND QUASI-STATIC

When we place a positive incremental voltage ΔV_D at the drain (Fig. 1-37c), the device operates closer to saturation and there will be more depletion of the channel charges at the drain end. The magnitude of the channel charge decreases. However, since the channel charge, made of electrons, is a negative quantity, the change of channel charge in response to a positive ΔV_D is positive. The source charge and the gate charge also adjust themselves in a way such that the total incremental charge inside the device is zero. Following the aforementioned convention, we define the drain-related capacitances as

$$C_{gd} = -\frac{\Delta Q_G}{\Delta V_D}; \quad C_{dd} = \frac{\Delta Q_D}{\Delta V_D}; \quad C_{sd} = -\frac{\Delta Q_S}{\Delta V_D}; \quad C_{bd} = -\frac{\Delta Q_B}{\Delta V_D}. \quad (1\text{-}77)$$

Likewise, when we place an incremental voltage ΔV_S at the source (Fig. 1-37d), we can determine the capacitances from the amount of charge variations:

$$C_{gs} = -\frac{\Delta Q_G}{\Delta V_S}; \quad C_{ds} = -\frac{\Delta Q_D}{\Delta V_S}; \quad C_{ss} = \frac{\Delta Q_S}{\Delta V_S}; \quad C_{bs} = -\frac{\Delta Q_B}{\Delta V_S}. \quad (1\text{-}78)$$

Lastly, for a ΔV_B at the bulk (not shown in Fig. 1-37),

$$C_{gb} = -\frac{\Delta Q_G}{\Delta V_B}; \quad C_{db} = -\frac{\Delta Q_D}{\Delta V_B}; \quad C_{sb} = -\frac{\Delta Q_S}{\Delta V_B}; \quad C_{bb} = \frac{\Delta Q_B}{\Delta V_B}. \quad (1\text{-}79)$$

We revisit Fig. 1-37b. When a ΔV_G is applied, all four terminal charges change in magnitude. However, the sum of the induced charges remains zero. Because $\Delta Q_G + \Delta Q_D + \Delta Q_S + \Delta Q_B = 0$, a relationship among the gate-related capacitances exists:

$$C_{gg} - C_{dg} - C_{sg} - C_{bg} = 0. \quad (1\text{-}80\text{a})$$

The same rationale is applied to other terminals, resulting in the following relationships:

$$-C_{gd} + C_{dd} - C_{sd} - C_{bd} = 0; \quad (1\text{-}80\text{b})$$

$$-C_{gs} - C_{ds} + C_{ss} - C_{bs} = 0; \quad (1\text{-}80\text{c})$$

$$-C_{gb} - C_{db} - C_{sb} + C_{bb} = 0. \quad (1\text{-}80\text{d})$$

Let us consider the situation when four voltage sources of equal magnitude are connected to the four terminals. Namely, $\Delta V_G = \Delta V_D = \Delta V_S = \Delta V_B = \Delta V$, as shown in Fig. 1-38. Because effectively the bias condition is the same before and after the application, $\Delta Q_G = \Delta Q_D = \Delta Q_S = \Delta Q_B = 0$. Further, because all four

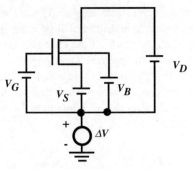

Fig. 1-38 Schematic showing the bias condition when $\Delta V_G = \Delta V_D = \Delta V_S = \Delta V_B = \Delta V$.

terminals are subjected to the same incremental voltage, the total incremental charge ΔQ is equal to

$$\Delta Q_G = C_{gg}\Delta V_G - C_{gd}\Delta V_D - C_{gs}\Delta V_S - C_{gb}\Delta V_B; \quad (1\text{-}81a)$$

$$\Delta Q_D = -C_{dg}\Delta V_G + C_{dd}\Delta V_D - C_{ds}\Delta V_S - C_{db}\Delta V_B; \quad (1\text{-}81b)$$

$$\Delta Q_S = -C_{sg}\Delta V_G - C_{sd}\Delta V_D + C_{ss}\Delta V_S - C_{sb}\Delta V_B; \quad (1\text{-}81c)$$

$$\Delta Q_B = -C_{bg}\Delta V_G - C_{bd}\Delta V_D - C_{bs}\Delta V_S + C_{bb}\Delta V_B. \quad (1\text{-}81d)$$

From the facts that $\Delta Q_G = \Delta Q_D = \Delta Q_S = \Delta Q_B = 0$ and that $\Delta V_G = \Delta V_D = \Delta V_S = \Delta V_B = \Delta V$, we conclude

$$C_{gg} - C_{gd} - C_{gs} - C_{gb} = 0; \quad (1\text{-}82a)$$

$$-C_{dg} + C_{dd} - C_{ds} - C_{db} = 0; \quad (1\text{-}82b)$$

$$-C_{sg} - C_{sd} + C_{ss} - C_{sb} = 0; \quad (1\text{-}82c)$$

$$-C_{bg} - C_{bd} - C_{bs} + C_{bb} = 0. \quad (1\text{-}82d)$$

These four relationships, together with those given by Eq. (1-80), allow all of the sixteen capacitances to be determined once a linearly independent set of nine capacitances are known.

Equations (1-75) to (1-79) can be used to find all the sixteen capacitances. They are listed in Appendix B, for all operation regions.

We proceed to determine the y-parameters of a MOSFET whose operating point is V_D, V_G, V_S, and V_B. The small-signal perturbations on top of these biasing voltages are sinusoidal at a radian frequency of ω (equal to $2\pi f$):

$$\begin{aligned} v_D(t) &= V_D + \tilde{v}_d e^{j\omega t}; & v_G(t) &= V_G + \tilde{v}_g e^{j\omega t}; \\ v_S(t) &= V_S + \tilde{v}_s e^{j\omega t}; & v_B(t) &= V_B + \tilde{v}_b e^{j\omega t}, \end{aligned} \quad (1\text{-}83)$$

Each of the resultant terminal currents consists of a quiescent dc value and a small-signal value varying at the same frequency ω:

$$i_D(t) = I_D + \tilde{i}_d e^{j\omega t}; \qquad i_G(t) = \tilde{i}_g e^{j\omega t};$$
$$i_S(t) = I_S + \tilde{i}_s e^{j\omega t}; \qquad i_B(t) = \tilde{i}_b e^{j\omega t}. \qquad (1\text{-}84)$$

For the intrinsic transistor without leakage currents, $I_G = I_B = 0$. I_S, of course, is equal to $-I_D$.

We digress briefly to clarify a point. In Section 1.6 we stated that the partial differential equation of $c'_{QS}(x,t)$ has no analytical solution. However, our focus in this section is small-signal operation. When the bias voltages in the time domain can be expressed in a sinusoidal form as that in Eq. (1-83), then the time derivative in Eq. (1-25) drops out, and the partial differential equation becomes a second-order ordinary differential equation. Suddenly, an exact solution is possible. Once the exact solution is found, we can then find the small-signal y-parameters characterizing the transistor. This is the fundamental reason why there is often an apparent disconnect between the MOSFET's small-signal and large-signal models.

A formal definition of the y-parameter is

$$y_{xy} = \left. \frac{\partial \tilde{i}_x}{\partial \tilde{v}_y} \right|_{\text{node voltages other than } \tilde{v}_y \text{ are set to zero}}, \qquad (1\text{-}85)$$

where x and y refer to any of the for transistor terminals: gate (g), drain (d), source (s), and bulk (b). For example, y_{dg} denotes the ratio of the small-signal current at the drain node with respect to a small-signal voltage at the gate, while the drain, and source small-signal voltages are maintained at zero. With this y-parameter definition, we express the terminal small-signal currents as a function of the small-signal bias voltages in a compact form:

$$\begin{bmatrix} \tilde{i}_g \\ \tilde{i}_d \\ \tilde{i}_s \\ \tilde{i}_b \end{bmatrix} = \begin{bmatrix} y_{gg} & y_{gd} & y_{gs} & y_{gb} \\ y_{dg} & y_{dd} & y_{ds} & y_{db} \\ y_{sg} & y_{sd} & y_{ss} & y_{sb} \\ y_{bg} & y_{bd} & y_{bs} & y_{bb} \end{bmatrix} \begin{bmatrix} \tilde{v}_g \\ \tilde{v}_d \\ \tilde{v}_s \\ \tilde{v}_b \end{bmatrix}. \qquad (1\text{-}86)$$

There are a total of sixteen y-parameters. Just like the capacitances, only nine of them are linearly independent. The remaining seven y-parameters can be generated once nine linearly independent parameters are known. Similar to the equations for the capacitances, we write

$$\begin{aligned}
y_{gg} + y_{gd} + y_{gs} + y_{gb} &= 0; & y_{dg} + y_{dd} + y_{ds} + y_{db} &= 0; \\
y_{sg} + y_{sd} + y_{ss} + y_{sb} &= 0; & y_{bg} + y_{bd} + y_{bs} + y_{bb} &= 0; \\
y_{gg} + y_{dg} + y_{sg} + y_{bg} &= 0; & y_{gd} + y_{dd} + y_{sd} + y_{bd} &= 0; \\
y_{gs} + y_{ds} + y_{ss} + y_{bs} &= 0; & y_{gb} + y_{db} + y_{sb} + y_{bb} &= 0.
\end{aligned} \qquad (1\text{-}87)$$

In the *common-source* configuration, all voltages are referenced to the source, which is grounded. The biasing voltages consist of quiescent values as well as the small-signal perturbations. These voltages result in terminal currents which also consist of quiescent values and the small-signal variations. The voltages and currents for a common-source transistor are expressed as

$$v_{DS}(t) = V_{DS} + \tilde{v}_{ds}e^{j\omega t}; \qquad v_{GS}(t) = V_{GS} + \tilde{v}_{gs}e^{j\omega t}; \qquad v_{BS}(t) = V_{BS} + \tilde{v}_{bs}e^{j\omega t}; \tag{1-88}$$

$$i_D(t) = I_D + \tilde{i}_d e^{j\omega t}; \qquad i_G(t) = \tilde{i}_g e^{j\omega t}; \qquad i_B(t) = \tilde{i}_b e^{j\omega t}. \tag{1-89}$$

With the gate terminal being port 1, drain terminal, port 2, and bulk terminal, port 3, the small-signal relationships between the port voltages and currents are

$$\begin{bmatrix} \tilde{i}_g \\ \tilde{i}_d \\ \tilde{i}_b \end{bmatrix} = \begin{bmatrix} y_{11} & y_{12} & y_{13} \\ y_{21} & y_{22} & y_{23} \\ y_{31} & y_{32} & y_{33} \end{bmatrix}_s \begin{bmatrix} \tilde{v}_{gs} \\ \tilde{v}_{ds} \\ \tilde{v}_{bs} \end{bmatrix}. \tag{1-90}$$

The subscript s in the y-matrix emphasizes that it applies to the common-source configuration. Comparing Eq. (1-90) with Eq. (1-86), we find the common-source y-parameters to be

$$\begin{bmatrix} y_{11} & y_{12} & y_{13} \\ y_{21} & y_{22} & y_{23} \\ y_{31} & y_{32} & y_{33} \end{bmatrix}_s = \begin{bmatrix} y_{gg} & y_{gd} & y_{gb} \\ y_{dg} & y_{dd} & y_{db} \\ y_{bg} & y_{bd} & y_{bb} \end{bmatrix}. \tag{1-91}$$

Often, especially in the rf circuits, the bulk is tied to the source, making the transistor a two-port network, with port 1 referring to the gate and port 2, the drain. The y-parameters characterizing this configuration with reference to bulk and source are

$$\begin{bmatrix} y_{11} & y_{12} \\ y_{21} & y_{22} \end{bmatrix}_{s-b} = \begin{bmatrix} y_{gg} & y_{gd} \\ y_{dg} & y_{dd} \end{bmatrix}. \tag{1-92}$$

We briefly go through the procedure to determine the common-source y-parameters of MOSFETs. Once the nine common-source y-parameters are established, all the sixteen y-parameters can be identified, through the application of Eq. (1-87). We start with the partial differential equation which dictates the time evolution of channel charges (Eq. 1-25). Just as the currents and voltages which are composed of quiescent values and small-signal perturbations, we can write the channel charge as

$$c'(x,t) = c'_{QS}(x) + \tilde{c}'(x)e^{j\omega t}. \tag{1-93}$$

The small-signal properties of the MOSFET are governed by an ordinary differential equation, which is obtained by substituting Eq. (1-93) into Eq. (1-25):

$$WC'_{ox}\mu_n \frac{d^2}{dx^2}[c'_{QS}(x) \cdot \tilde{c}'(x)e^{j\omega t}] = -j\omega \tilde{c}' e^{j\omega t}. \tag{1-94}$$

1.9 NON-QUASI-STATIC AND QUASI-STATIC

The time dependence can be dropped out from the above equation. After $\tilde{c}'(x)$ is determined from solving the differential equation, the small-signal terminal currents i_g, i_d and i_b can then be established as a function of the v_{gs}, v_{ds}, and v_{bs} specified in Eq. (1-88). The common-source y-parameters can then be evaluated in accordance with the definition given in Eq. (1-85).

This procedure of finding the y-parameters is conceptually straightforward, but quite algebraically intensive. Without proper software, the derivation easily takes days and is error-prone. A good technique to solve the nonlinear differential equation (Eq. 1-94) involves the Bessel functions [24]. The results are listed in Appendix C, from which, y_{gg} can be put in the following form:

$$y_{gg,\text{NQS}} = \frac{N_{gg,0} + (j\omega)N_{gg,1} + (j\omega)^2 N_{gg,2} + \cdots + (j\omega)^k N_{gg,k}}{D_0 + (j\omega)D_1 + (j\omega)^2 D_2 + \cdots + (j\omega)^k D_k}, \quad (1\text{-}95)$$

where $N_{gg,k}$ and D_k are coefficients identified in Appendix C. We place the subscript NQS for y_{gg} to emphasize that the y-parameter is determined by solving Eq. (1-94), without any simplification such as the quasi-static assumption. The solution in the form of Eq. (1-95) is exact, and is called the non-quasi-static y-parameter.

Appendix C points out that $N_{gg,0} = 0$, $D_0 = 1$, and $N_{gg,1} = C_{gg}$, where C_{gg} is precisely equal to that given by Eq. (1-76). Note that y_{gg} in Eq. (1-95) represents the exact solution, yet the C_{gg} was derived from Q_G, the quasi-static gate charge. We rewrite $y_{gg,\text{NQS}}$ in the following form:

$$y_{gg,\text{NQS}} = j\omega C_{gg} \left[\frac{1 + (j\omega)N_{gg,2}/C_{gg} + (j\omega)^2 N_{gg,3}/C_{gg} + \cdots}{1 + (j\omega)D_1 + (j\omega)^2 D_2 + \cdots} \right]. \quad (1\text{-}96)$$

The term enclosed in the square brackets represents the higher-order modification on $j\omega C_{gg}$. If we drop these higher-order terms, the simplified y-parameters result. These simplified y-parameters shall be referred to as the *quasi-static y-parameters* because the remaining term, C_{gg}, are identical to that derived from quasi-static charges. For example, the quasi-static $y_{gg,\text{QS}}$ corresponding to Eq. (1-96) is

$$y_{gg,\text{QS}} = j\omega C_{gg}. \quad (1\text{-}97)$$

At small frequencies, ω approaches zero. The NQS y-parameters then reduce to the QS y-parameters.

We listed out only y_{gg} above. The other eight common-source y-parameters can be determined in a similar way. We summarize the quasi-static y-parameters for the common-source configuration in the following. From now on we will drop the subscript QS. The y-parameters without a subscript implicitly means they are quasi-static y-parameters without the higher-order terms. Since BSIM3 employs a quasi-static analysis, the y-parameters calculated in BSIM3 are quasi-static y-parameters:

$$y_{gg} = j\omega C_{gg} \tag{1-98a}$$
$$y_{gd} = -j\omega C_{gd} \tag{1-98b}$$
$$y_{gb} = -j\omega C_{gb} \tag{1-98c}$$
$$y_{dg} = g_m - j\omega C_{dg} \tag{1-98d}$$
$$y_{dd} = g_d + j\omega C_{dd} \tag{1-98e}$$
$$y_{db} = g_{mb} - j\omega C_{db} \tag{1-98f}$$
$$y_{bg} = -j\omega C_{bg} \tag{1-98g}$$
$$y_{bd} = -j\omega C_{bd} \tag{1-98h}$$
$$y_{bb} = j\omega C_{bb}. \tag{1-98i}$$

The remaining seven y-parameters are determined from Eqs. (1-87).

$$y_{gs} = -j\omega C_{gs} \tag{1-99a}$$
$$y_{ds} = -g_d - g_m - g_{mb} - j\omega C_{ds} \tag{1-99b}$$
$$y_{sg} = -g_m - j\omega C_{sg} \tag{1-99c}$$
$$y_{sd} = -g_d - j\omega C_{sd} \tag{1-99d}$$
$$y_{ss} = g_d + g_m + g_{mb} + j\omega C_{ss} \tag{1-99e}$$
$$y_{sb} = -g_{mb} - j\omega C_{sb} \tag{1-99f}$$
$$y_{bs} = -j\omega C_{bs}. \tag{1-99g}$$

Because the capacitances denote the intrinsic device capacitances, these y-parameters do not include the parasitics. They are not exactly the y-parameters used in BSIM3, which include both the intrinsic and parasitic components. However, the general equation form is correct, and remains valid in all operating regions of MOSFET. We will incorporate the parasitic capacitances in Section 2.3.

1.10 SOURCE-REFERENCING AND INVERSE MODELING

MOSFET is often built as a symmetrical device with respect to the middle of the gate (or bulk). The source and drain are interchangeable. Whichever terminal is labeled as drain is somewhat arbitrary, although the drain generally refers to the terminal at the higher potential. When the device's V_{DS} is at 0, we expect from symmetry that the device's $C_{dx} = C_{sx}$, and $C_{xd} = C_{xs}$, $C_{ds} = C_{sd}$, and $C_{dd} = C_{ss}$, where x refers to either the gate or the bulk terminal. As an example, we expect $C_{gd} = C_{gs}$, for all V_{GS} values, as long as $V_{DS} = 0$. Figure 1-39 illustrates BSIM3-calculated C_{gd} and C_{gs} for a long-channel and a short-channel device. It is obvious that, while C_{gd} is roughly equal to C_{gs}, they are not identical. This asymmetry found in BSIM3, besides being a concern in the accuracy of modeling, can cause numerical difficulties during circuit simulation (Section 4.7). To facilitate understanding of the cause of the asymmetrical

1.10 SOURCE-REFERENCING AND INVERSE MODELING

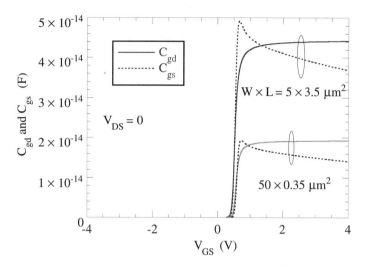

Fig. 1-39 Calculated C_{gd} and C_{gs} for a long-channel and a short-channel device, at $V_{DS} = 0$. The asymmetrical results are due to the source-referencing used in BSIM3.

capacitances, we shall refer the right node in Fig. 1-40 as the drain of the MOS, whether V_{DS} is positive or negative. Conversely, the left node is always the source. (This practice is followed in SPICE. When the MOSFET specified in the SPICE input deck is, "MN 1 2 3 4 ···", then nodes 1 and 3 are always the drain and source nodes, respectively, regardless of whether v_1 is larger than v_3 or not.) We know from symmetry, the transistor's drain current (current flowing into the drain is counted as positive) should obey:

$$I_D(V_{DS}) = -I_D(-V_{DS}) \quad \text{(The drain current symmetry test)} \quad (1\text{-}100)$$

Let us determine if the drain current expression used in BSIM3 has the symmetry property specified in Eq. (1-100). Equation (1-5), on which BSIM3's drain current equation is based, is one that we have taken for granted and appears in lots of textbooks. Unfortunately, it fails to obey drain-current symmetry relationship. The drain current equation specified in Eq. (1-5), as well the actual (and much more complicated) equation used in BSIM3, is applicable only when $V_{DS} \geq 0$.

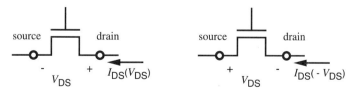

Fig. 1-40 Schematic illustrating the swapping of the source and drain nodes and the resulting device currents.

72 MODELING JARGONS

The fundamental reason why Eq. (1-5) fails the I_D-symmetry test is that the equation is derived with voltages referenced to the source. During strong inversion, most of the MOSFET current is carried by majority carriers through the drift motion. The drain current is $-\mu_n \cdot (qn) \cdot \varepsilon = \mu_n \cdot c'_{QS}(x) \cdot dV_{CB}/dx$, where V_{CB} is the channel potential with respect to the bulk. According to Ref. [22, p. 120],

$$I_D = \mu_n \frac{W}{L} \int_0^L c'_{QS}(x) \frac{dV_{CB}}{dx} dx = \mu_n \frac{W}{L} \int_{V_{SB}}^{V_{DB}} c'_{QS}(V_{CB}) dV_{CB}. \tag{1-101}$$

We first encountered $c'_{QS}(x,t)$ in Section 1.6. It is the channel-charge density distribution as a function of both time and position. In this dc situation, we drop the time dependence in Eq. (1-32), making $c'_{QS}(x)$ to be that given in Eq. (1-27).

When *source referencing* is used, all the voltages will be made with respect to the source. The channel-source potential (V_{CS}) is the variable of choice. We recognize $V_{CS}(x) = V_{CB}(x) - V_{SB}$ where V_{SB} is independent of x. Hence, the drain current, based on the more general equation above, is now expressed as

$$I_D = \mu_n \frac{W}{L} \int_0^{V_{DS}} c'_{QS}(V_{CS}) dV_{CS}. \tag{1-102}$$

Carrying out the integration eventually leads to Eq. (1-5). As mentioned, the very act of source-referencing in Eq. (1-102) causes V_{DS} to appear in the drain current expression, a fact which in turn, results in the failure of the symmetry test. Once the drain current equation fails, all the other equations such as charge expressions for C-V modeling, are doomed.

We will describe a patch employed in BSIM3 to pass the symmetry test shortly. But we caution that, just like most patches, the BSIM3 patch also creates lots of other problems. The fundamental way to ensure symmetrical drain current is to use *bulk-referencing*, in which all voltages are made to be referred to the bulk terminal. Thus far, there is only one (known to author) public-domain model supported by major SPICE simulators which employs bulk referencing: the EKV model [17]. The basic framework in the EKV model also begins with Eq. (1-101). However, care is taken to make voltages referenced to the bulk, preserving the symmetry of the source and the drain terminals. Starting from Eq. (1-101), we write

$$I_D = \mu_n \frac{W}{L} \int_{V_{SB}}^{\infty} -c'_{QS}(V_{CB}) dV_{CB} - \mu_n \frac{W}{L} \int_{\infty}^{V_{DB}} -c'_{QS}(V_{CB}) dV_{CB} \equiv I_F - I_R. \tag{1-103}$$

In the linear region, the forward and reverse current components simplify to [32]

$$I_F = \frac{W}{L} \mu_n C'_{ox} \cdot \frac{1}{2n} \cdot (V_{GB} - V_T - nV_{SB})^2; \tag{1-104a}$$

$$I_R = \frac{W}{L} \mu_n C'_{ox} \cdot \frac{1}{2n} \cdot (V_{GB} - V_T - nV_{DB})^2. \tag{1-104b}$$

1.10 SOURCE-REFERENCING AND INVERSE MODELING

The factor n is used in the EKV model to account for the bulk charge effects. To compare with Eq. (1-5) employed in BSIM3, we may equate n to $1 + \delta$. EKV model separates the drain current into I_F and I_R. Each component depends solely on either V_{SB} or V_{DB}. Therefore, the I_F and I_R equations given in Eq. (1-104) are valid, independent of whether V_{DS} is greater or smaller than zero. A little bit of algebraic exercise shows that EKV's I_D given by Eqs. (1-103) and (1-104) can be expressed as Eq. (1-5) as used in BSIM3. However, the I_D of EKV continues to work at $V_{DS} < 0$, while Eq. (1-5) becomes inaccurate. We demonstrate this observation with a current calculation under bias conditions schematically represented in Fig. 1-41. In a symmetrical device, we expect the drain current to behave according to Eq. (1-100). Figure 1-42 compares the calculated drain currents from Eq. (1-104), on which the EKV model is based, and Eq. (1-5), on which the BSIM3 is based. It is clear that while $I_D(-V_{DS})$ is indeed $-I_D(V_{DS})$ for the EKV model, such a symmetry relationship is not upheld in the BSIM3 model.

Without any patchup, the BSIM3 model fails the I_D-symmetry property specified in Eq. (1-100). The model equations, such as those based on Eq. (1-5), are applicable only at $V_{DS} \geq 0$. The additional modeling activities required specifically to take care the $V_{DS} < 0$ situation is called *inverse modeling*. The general strategy of BSIM3's inverse modeling is the following. When V_{DS} is less than zero, BSIM3 internally swaps the drain and the source nodes. We shall call the actual device's drain and source as D and S, respectively. Inside BSIM3, the flipped drain and source are denoted as D' and S', respectively. Hence, D = S' and S = D' (if $V_{DS} < 0$). The gate and bulk nodes are not swapped. In our notation, this means G = G' and B = B'. With the interchanged nodes, $V_{D'S'}$, equal to $-V_{DS}$, is now a positive number. $V_{D'S'}$ is then substituted into Eq. (1-5), for example, to calculated I'_D. For clarity, we rewrite Eq. (1-5) with the symbols under the inverse modeling:

$$I_{D'} = \frac{W}{L}\mu_n C'_{ox}\left[(V_{GS'} - V_T)V_{D'S'} - \frac{1}{2}(1+\delta)V_{D'S'}^2\right]. \quad (1\text{-}105)$$

$I_{D'}$, the current flowing at the D' node, is positive. Now that redefined $V_{D'S'}$ is a positive value, Eq. (1-5) is valid. At the end of the BSIM3 calculation, every quantities involving D' is swapped back to S, and S', back to D. Hence, $I_{D'}$ calculated from Eq. (1-105) becomes I_S of the actual device. Similarly, I_D of the actual device is

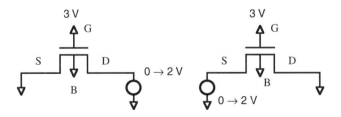

Bias configuration when $V_{DS} > 0$ Bias configuration when $V_{DS} < 0$

Fig. 1-41 Bias conditions and transistor definition used to highlight the difference between the EKV and the BSIM3 models.

74 MODELING JARGONS

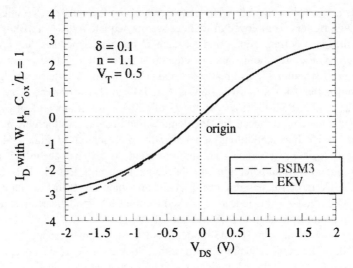

Fig. 1-42 Calculated channel currents for the circuit of Fig. 1-41, from the EKV and the BSIM3 models.

just $I_{S'}$, which is $-I_{D'}$. In this manner, the symmetry property of the drain current required by Eq. (1-100) is recovered. (However, the symmetry of C_{gd} and C_{gs} are still not preserved in the model.)

There are more details to BSIM3's inverse modeling. We used only the I_D calculation as an example. Other quantities, such as g_m, all need to be properly modified when $V_{DS} < 0$. The details are given in an algorithm:

0. If ($V_{DS} < 0$), then do the following.
1. Let $V_{D'S'} = -V_{DS}$, $V_{G'S'} = V_{GD}$, $V_{B'S'} = V_{BD}$.
2. Use the normal mode equations (those derived with the mind-set that $V_{DS} > 0$, such as Eq. 1-5), except substitute V_{DS}, V_{GS}, V_{BS} of the normal mode equations by $V_{D'S'}, V_{G'S'},$ and $V_{B'S'}$. We apply the normal mode equations to find

$$I_{G'}, I_{D'}, I_{S'}, I_{B'}, Q_{G'}, Q_{D'}, Q_{S'}, Q_{B'}, g_{m'}, g_{mb'}, g_{d'}$$

$$y_{g'g'}, y_{g'd'}, y_{g's'}, y_{g'b'}, y_{d'g'}, y_{d'd'}, y_{d's'}, y_{d'b'},$$

$$y_{s'g'}, y_{s'd'}, y_{s's'}, y_{s'b'}, y_{b'g'}, y_{b'd'}, y_{b's'}, y_{b'b'},$$

(*Note*: In the process of calculating y-parameters, all the capacitances C_{xy} are also calculated. The above list of dc and ac quantities are all the variables which are supposed to be calculated in a model. In reality, BSIM3 calculates only 3 of the four currents (excluding I_S), since the last one is obtainable from the Kirchoff's current law. BSIM3 calculates only 3 of the charges (excluding Q_S), since the last one is obtainable from the charge conservation principle.

1.10 SOURCE-REFERENCING AND INVERSE MODELING

Finally, BSIM3 calculates only nine of the sixteen y-parameters, because only nine of them are linearly independent.)

3. Finally, we swap nodes, replacing S' by D, D' by S, G' by G, and B' by B.

Let us take the calculation of C_{gd} for example. When V_{DS} is positive, C_{gd} is calculated with BSIM3's C_{gd} expression, which is valid for $V_{DS} \geq 0$. As V_{DS} becomes negative, BSIM3 internally uses a positive intermediate variable, $V_{D'S'}$, which is equal to $-V_{DS}$. C_{gd} is then evaluated with BSIM3's C_{gs} expression, which is valid for $V_{D'S'} > 0$. The inverse modeling procedure therefore requires that C_{gd} be equal to C_{gs} right at $V_{DS} = 0$. However, as shown in Fig. 1-39, $C_{gd} \neq C_{gs}$ at $V_{DS} = 0$ in BSIM3, due to the source-referencing embodied in its fundamental equations. This inequality causes a discontinuity in the C_{gd}, as displayed in Fig. 1-43, which is simulation result of a transistor biased at $V_{GS} = 2$ V, $V_{BS} = 0$, as V_{DS} is scanned from -0.1 V to 0.1 V. During the SPICE simulation, the parameters relating to parasitic capacitances are made zero so that only the intrinsic capacitances are shown. For convenience, C_{ds} and C_{bd} are also plotted. Although C_{db} also has a certain amount of discontinuity, it is not as obvious as the other two capacitances. The discontinuities in C_{gd} (as well as other capacitances such as C_{bd}, C_{ds} etc) can cause numerical problem in circuit simulation if V_{DS} ever crosses the zero value, besides giving inaccurate simulation results.

There is no need for inverse modeling in the EKV model because the drain current equation, as well as other charge equations, remains valid whether V_{DS} is greater or smaller than zero. The swapping of terminals does not take place in the EKV model. Although we specifically mentioned only the drain current development, the bulk-referencing approach should have a good chance of symmetry in capacitance equations as well. We caution that bulk-referencing does not automa-

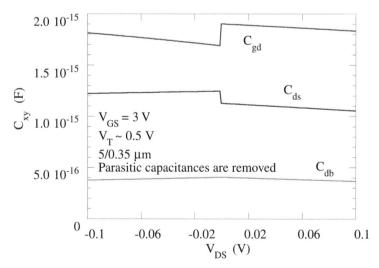

Fig. 1-43 Calculated C_{gd}, C_{ds}, and C_{db} as V_{DS} is scanned from negative to positive values. The discontinuity in C_{gd}, for example, is due to the source-referencing used in BSIM3.

tically guarantee symmetry. When short-channel effects are added, the model often gets complicated enough such that the symmetry is lost if no special effort was spent. One example is an introductory version of EKV [17], in which the symmetry is lost when the narrow-width effects are incorporated. However, in the newer versions of EKV model (such as version 2.6), efforts have been taken to ensure the capacitance symmetry even after accounting for the short-channel effects.

The accuracy of C_{gs} and C_{gd} in bulk-referencing models is more forgiving. When $C_{gs} \neq C_{gd}$ in a bulk-referencing model, it is merely a model accuracy issue. This will not become a numerical convergence issue, because no interchanging of drain/source terminals takes place in bulk-referencing models. This is to be contrasted with source-referencing models, in which the capacitance asymmetry becomes a numerical nuisance, in addition to model inaccuracy.

We examine more carefully the issue of C_{gs} and C_{gd}'s equivalence at $V_{DS} = 0$. We mentioned that the device symmetry in BSIM3 was long lost even in the I_D equation. We naturally doubt that C_{gs} calculated in BSIM3 will be equal to the calculated C_{gd} at $V_{DS} = 0$. Indeed Fig. 1-39 shows that they are not equal in BSIM3. Well, surprisingly, it is really possible (although somewhat fortuitously) that C_{gd} be equal to C_{gs} in source-referencing models. For example, suppose we concentrate only on long-channel transistors with a constant threshold voltage. The gate charge was given in Eq. (1-49). When V_T is a constant, $\partial \alpha / \partial V_D = -1/(V_G - V_S - V_T)$, $\partial \alpha / \partial V_S = \alpha/(V_G - V_S - V_T)$, and $\partial \delta / \partial V_D = \partial \delta / \partial V_S = 0$. Therefore,

$$C_{gd} = -\frac{\partial Q_G}{\partial V_D} = WLC'_{ox} \times \frac{2}{3}\frac{2\alpha + \alpha^2}{(1+\alpha)^2}; \quad (1\text{-}106)$$

$$C_{gs} = -\frac{\partial Q_G}{\partial V_s} = WLC'_{ox} \times \frac{2}{3}\frac{1+2\alpha}{(1+\alpha)^2}. \quad (1\text{-}107)$$

$V_{DS} = 0$ corresponds to an α of unity, independent of V_{GS}. Under this condition, we see that C_{gd} is identical to C_{gs}. This is a rather coincidental result. We started out without special care for device symmetry, which in fact, results in a asymmetrical drain current equation of Eq. (1-5). But, as long as V_T is treated as a constant (or, only depending on V_{BS}, as in Eq. 1-6), then the capacitance symmetry is preserved. We verify that all other capacitance pairs, such as C_{db} and C_{sb}, also obey the symmetry test when V_T is a constant.

Suppose that we now include a DIBL (drain-induced barrier-lowering) effect parameter into the threshold voltage. That is, instead of having a V_T of Eq. (1-6), we write $V_T = V_{T0} + \gamma$ [sqrt($2\phi_f - V_{BS}$) − sqrt($2\phi_f$)] − $\xi \cdot V_{DS}$ (V_T is less at higher drain biases). Then, $\partial \alpha / \partial V_D = -(1 + \xi + \xi\alpha)/(V_G - V_S - V_T)$, and $\partial \alpha / \partial V_S = (\alpha + \alpha\xi - \xi)/(V_G - V_S - V_T)$. Therefore,

$$C_{gd} = WLC'_{ox} \times \left(\frac{2}{3}\frac{2\alpha + \alpha^2}{(1+\alpha)^2} - \frac{2}{3}\xi\frac{1 + 4\alpha + \alpha^2}{(1+\alpha)^2}\right); \quad (1\text{-}108)$$

$$C_{gs} = WLC'_{ox} \times \left(\frac{2}{3}\frac{1+2\alpha}{(1+\alpha)^2} + \frac{2}{3}\xi\frac{1 + 4\alpha + \alpha^2}{(1+\alpha)^2}\right). \quad (1\text{-}109)$$

1.11 PHYSICAL MODEL AND TABLE-LOOKUP MODEL

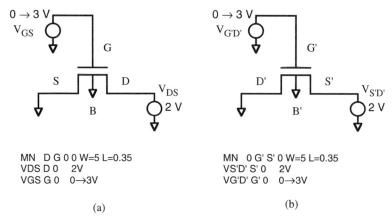

Fig. 1-44 Bias conditions and transistor definition used to compare small-signal conductances calculated when $V_{DS} > 0$ and when $V_{DS} < 0$.

It is easy to verify that $C_{gd} \neq C_{gs}$ at $\alpha = 1$ ($V_{DS} = 0$), unless $\xi = 0$. The capacitance equality is preserved in source-referencing models as long as V_T is a constant, but fails when V_T incorporates some short-channel effects. This is one reason why in some source-referencing models, the threshold voltage for the I-V and C-V calculations are intentionally made to differ. In the I-V characteristics, V_T includes the short-channel effects so that accurate I-V characteristics can be simulated. However, in the calculation of the C-V characteristics, a constant V_T is employed to preserve the capacitance symmetry. Of course, the C-V characteristics cannot be exactly correct since the threshold voltage is made to be a constant. However, this kind of modification has the advantage that the numerical convergence problems associated with a discontinuous C_{gd} disappears.

When a device's $V_{DS} < 0$, the device's conductances are redefined as

$$g_d = \frac{\partial I_S}{\partial V_{SD}}; \quad g_m = \frac{\partial I_S}{\partial V_{GD}}; \quad g_{mb} = \frac{\partial I_S}{\partial V_{GD}} \quad \text{(for } V_{DS} < 0\text{).} \quad (1\text{-}110)$$

Therefore, g_d, g_m, and g_{mb} calculated from Fig. 1-44a (normal mode operation) and Fig. 1-44b (inverse mode operation) are identical, both magnitude and sign. Incidentally, based on our previous discussion, we see that, for example, C_{gd}, C_{ds}, C_{db} calculated from Fig. 1-44a are identical to C_{gs}, C_{sd}, and C_{ds} calculated from Fig. 1-44b. (They are the same even in BSIM3 whose C_{gd} and C_{gs} would differ at $V_{DS} = 0$. The identicalness stems from the inverse modeling.)

1.11 PHYSICAL MODEL AND TABLE-LOOKUP MODEL

Have you seen a journal paper entitled "A *physics-based* model of a semiconductor device?" Sometimes we wonder, isn't everything pretty much physics-based, so why does the author stress that the model he develops is physics-based? Is the author attempting to make a distinction from, say, astrology-based model?

Well, it turns out that a model needs not be based on physics. It can be based on mathematical methods. A *table-lookup model* is one such method. In a table-lookup model, the measured device current and capacitances are stored for different bias points and device geometry in a tabular form. Whenever we want to find out, for example, the current under a particular bias condition, we then need to look up the table entry corresponding to the bias. Limited by the available computer memory, the entries cannot be infinite in number. The recorded values must be those measured at a certain increment of V_{GS}, V_{DS}, and V_{BS}. If the bias voltages do not coincide with those in the table entries, then some kind of interpolation of the stored data is necessary [33]. The interpolation scheme must be carefully chosen to avoid the discontinuities in a model element or the first derivatives of the element. A variation of the table-lookup model eliminates this potential problem. Instead of directly storing the device current, for example, the coefficients of some mathematical functions like the cubic splines are precalculated from the original data for different bias and geometry. Cubic spline, the type of piecewise polynomial approximation using cubic polynomials between each successive pair of data points, ensures not only that the interpolant is continuously differentiable on the interval, but that it has a continuous second derivative on the interval as well. (In our problem, the drain current is a function of V_{GS}, V_{DS} and V_{BS}. The more-than-one-dimension nature of the problem requires us to use the so-called bi-cubic spline, to be exact.) However, the bicubic spline model does not preserve monotonicity and convexity. It is found that the interpolated data may exhibit small local maxima, especially near the saturation-linear and saturation-subthreshold boundaries [34]. It seems that, bicubic bell-shaped spline (B-spine), with its variation diminishing property, preserves the monotonicity and convexity [35].

There are clear advantages in using a table-lookup model. The device characteristics are measured, and directly translated into mathematical functions which will reproduce the data with minimized error. There is no need for parameter extraction, since all measured data are stored in tabular form (or in spline coefficients). The development of a table-lookup model takes much less time than a physical model, at least conceptually. Furthermore, a table-lookup model is independent of technology. The same procedure used to generate a table-lookup model is readily usable in future technologies. This is opposed to a physical model which may need to revise its model equations if some effects unaccounted for in the existing model become significant in future generations of technologies.

Moreover, because a table-lookup model is directly based on measured data, the model will naturally model the non-quasi-static effects if the data are taken at a condition inducing pronounced NQS effects. We caution, however, a non-quasi-static table-lookup model does not necessarily remove the zero-drain-current artifact mentioned in Section 1.6 [35]. A roughly correct explanation is that, in the context of Eq. (1-96), Ref. [35]'s approach only includes some terms in the series expansion. That is, quasi-static approach drops out the higher-order terms in the square brackets of Eq. (1-96). Reference [35] makes an improvement by including the ($j\omega$) terms

1.11 PHYSICAL MODEL AND TABLE-LOOKUP MODEL

inside the square bracket, but still does not include all of the other terms. A complete removal the zero-drain-current problem likely requires the inclusion of a majority of terms. This is clearly an impossible task since the computer memory is limited.

There are quite a number of disadvantages associated with a table-lookup model. The main objection is that it is not scalable. Suppose an initial circuit design calls for the use of $L = 0.5$ and 1.2 μm devices. Two independent table-lookup models are generated for them. However, unforeseen events can eventually lead the designer to replace the 1.2 μm device with a 1 μm. If the transistor model were a physical model such as BSIM3, this change in design poses no major problem, since the scalable nature of the BSIM3 model works equally well for the 1 μm device. However, if the table-lookup model were used, there would be no 1 μm model for the designer to use. The table-lookup model for the 1 μm device can be made only after experimental data of the device are taken, or effectively, after the transistor is made on silicon. In a similar rationale, a table-lookup model cannot be utilized to forecast the technology. For example, if the present technology is $L = 0.35$ μm, there is no way to predict the device characteristics in the next generation technology with a L of 0.2 μm. In contrast, BSIM3 can be used for forecasting, given a certain degree of accuracy. Based on the planned process changes from one generation to the next (such as thinning of the oxide), some idea of the device characteristics of the next generation technology can be established.

There are several other disadvantages, which can be understood after we discuss the usual procedure of generating a table-lookup model. To simplify discussion, we make the table-lookup model based on the quasi-static assumption, although such an assumption is not necessary.

1. Measure s-parameters in a common-source configuration. Measure at different sets of V_{DS}, V_{GS}, and V_{BS}. Although only one frequency point is needed, it is desirable to scan the frequency during measurement.
2. Convert s-parameters to y-parameters using standard conversion tables [24].
3. Since we are constructing a quasi-static table-lookup model, we compare the measured y-parameters with those quasi-static y-parameters given in Eqs. (1-98) and (1-99). From the comparison, extract the small-signal conductances g_m, g_d, g_{mb}, as well as the capacitances C_{gg}, C_{gd}, C_{gb}, C_{dg}, C_{dd}, C_{db}, C_{bg}, C_{bd}, and C_{bb}. All of these parameters are functions of V_{DS}, V_{GS}, and V_{BS}.
4. Determine the dc currents and quasi-static charges: $I_G = I_B = 0$, and

$$I_D(V_{GS}, V_{DS}, V_{BS}) = \int_0^{V_{GS}, V_{DS}, V_{BS}} (g_m \widehat{v_{gs}} + g_d \widehat{v_{ds}} + g_{mb} \widehat{v_{bs}}) \cdot d\vec{v}; \quad (1\text{-}111)$$

$$Q_k(V_{GS}, V_{DS}, V_{BS}) = \int_0^{V_{GS}, V_{DS}, V_{BS}} (C_{kg} \widehat{v_{gs}} + C_{kd} \widehat{v_{ds}} + C_{kb} \widehat{v_{bs}}) \cdot d\vec{v}; \quad (1\text{-}112)$$

where, k denotes either gate (g), drain (d) or bulk (b) node, the caret denotes a unit vector, and

$$d\vec{v} = dv_{gs} \widehat{v_{gs}} + dv_{ds} \widehat{v_{ds}} + dv_{bs} \widehat{v_{bs}}. \quad (1\text{-}113)$$

Due to experimental error inherent in the measurement of the capacitances, we expect the exact values of Q_k obtained with various integration paths to differ [36]. That is, Q_k (1 V, 1 V, 0 V) obtained by integrating from $v_{gs} = 0$ to 1 V, then $v_{ds} = 0$ to 1 V, should differ from the Q_k (1 V, 1 V, 0 V) obtained by integrating from $v_{gs} = 0$ to 0.5 V, then $v_{ds} = 0$ to 1 V, and finally $v_{gs} = 0.5$ to 1 V. In order to maintain charge conservation, we want $Q_k(V_{GS}, V_{DS}, V_{BS})$ to be *path-independent*, that is, having the same value regardless of the integration path.

Mathematically, this means that $\vec{\nabla}Q_k$ is conservative, such that the curl of the vector is zero. However, measurement error and other factors prevent the $Q_k(V_{GS}, V_{DS}, V_{BS})$ matrix from being truly path independent. A good compromise is to use a predetermined integration path known to yield the most accurate result. Although other paths yield other matrices of $Q_k(V_{GS}, V_{DS}, V_{BS})$, we shall deem them less accurate and discard them. It is important that a chosen $Q_k(V_{GS}, V_{DS}, V_{BS})$ are stored as a part of the model, rather than being computed at run time.

5. During circuit simulation, the transient currents are found from

$$i_G(t) = \frac{dQ_G(V_{GS}, V_{DS}, V_{BS})}{dt}; \quad i_B(t) = \frac{dQ_B(V_{GS}, V_{DS}, V_{BS})}{dt};$$
$$i_D(t) = I_D(V_{GS}, V_{DS}, V_{BS}) + \frac{dQ_D(V_{GS}, V_{DS}, V_{BS})}{dt}. \tag{1-114}$$

The above description makes clear that the success of a table-lookup model hinges upon the accuracy of capacitance measurement. In step 3, we basically measure the capacitance with an indirect method, that is, through the use of s-parameters. Because MOSFET is a four-terminal device, measuring the common-source y-parameters means that three-port measurements need to be performed. This presents a problem, if not a major inconvenience, since most high-frequency measurement setups and equipment are designed to make two-port measurement. One typical solution is to tie the bulk to the source, grounding both the bulk and the source terminals. The measured two-port y-parameters will be (see Eq. 1-92)

$$\begin{bmatrix} \tilde{i}_g \\ \tilde{i}_d \end{bmatrix} = \begin{bmatrix} y_{gg} & y_{gd} \\ y_{dg} & y_{dd} \end{bmatrix} \begin{bmatrix} \tilde{v}_g \\ \tilde{v}_d \end{bmatrix}. \tag{1-115}$$

Only four y-parameters are determinable from a measurement. The other y-parameters such as $y_{gb}, y_{db}, y_{bg}, y_{bd}$, and y_{bb} are not identified. Hence, the charge expression given by Eq. (1-112) cannot be fully integrated. For general MOSFET circuit wherein nonzero V_{BS} exists, missing those parameters cannot guarantee accurate circuit simulation. Therefore, for MOSFET, the test structure for 3-port s-parameter measurement needs to be designed with extra care.

1.11 PHYSICAL MODEL AND TABLE-LOOKUP MODEL

III–V field effect transistors such as MESFET, HFET, and HEMT are three-terminal devices [24]. Without the bulk node, the two-port characterization of the device can be easily performed. This is one reason that the *Root model* [36], a proprietary table-lookup model built in a commercial parameter extraction program, is quite popular in III–V industries, although it is less prominent for MOSFET modeling. (Incidentally, the Root model generation follows a similar process as discussed above, although the Root model allows a first-order correction of the non-quasi-static effects. The Root model measures *s*-parameters, which are then converted to *y*-parameters. However, the real parts of the *y*-parameters are not used. Therefore, I_D is not obtained from integration. Instead, I_D is directly measured, from which small-signal conductances can be obtained and, hopefully, these conductances should agree well with those measured from *s*-parameters. This process allows a more accurate modeling I_D because it is measured directly.)

Besides the three-port issues, a question can also be raised about the accuracy of extracted capacitance values from measured *s*-parameters. The following high-frequency data are from a $W/L = 256/0.29\,\mu m$ NMOS, with bulk internally shorted to the source. Figure 1-45 illustrates *y*-parameters converted from measured *s*-parameters. The pad capacitances inherent in the test structure [37], as well as series gate resistance of $6\,\Omega$, drain resistance of $1.2\,\Omega$, and source resistance of $1.2\,\Omega$ have been de-embedded out [37]. At low enough frequencies, the measured *y*-parameters indeed behave as those formulated in Eq. (1-98), indicating that the non-quasi-static effects are negligible. For example, the measured y_{11} is roughly purely imaginary, with a positive magnitude that varies roughly linearly with frequency, just as y_{gg}

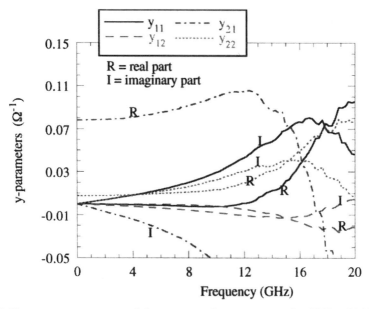

Fig. 1-45 *y*-parameters converted from measured *s*-parameters of a $W/L = 256/0.29\,\mu m$ NMOS. $V_{GS} = 1\,V$, $V_{BS} = 0$, and $V_{DS} = 1\,V$.

given in Eq. (1-98). C_{gg} can then be estimated to be the slope of measured imaginary part of y_{11}. Further, $y_{21} = y_{dg}$ has a relatively constant real part and an imaginary part which varies linearly with frequency. This real part corresponds to g_m, while the negative imaginary part's slope corresponds to C_{dg}. The above measurement result is obtained at $V_{GS} = V_{DS} = 1$ V. We wonder if the theoretical y-parameters of Eq. (1-98) remain equally valid in the subthreshold region. A measurement at $V_{GS} = 0.2$ V and $V_{DS} = 1.5$ V, shown in Fig. 1-46, confirms that the equations still apply. Of course, the values of the conductances and the capacitances change between the strong inversion and the subthreshold region; hence, the magnitudes of the real parts and the slopes of the imaginary parts differ from those of Fig. 1-45. In Fig. 1-46, y_{12} is approximately equal to y_{21}. This is because when the transistor is nearly off, the transistor can be treated as a passive component. A passive two-port network possesses the property that the forward transmission is equal to its reverse transmission.

Although superficially Figs. 1-45 and 1-46 demonstrate that the measurement of capacitances from s-parameters is straightforward, there are problems. First, in order that the device's gate capacitance be much larger than the parasitic pad capacitance, the transistor width has to be on the order of 256 μm. This means the table-lookup model cannot be easily implemented in a transistor with pronounced narrow-width effects. For a width on the order of 1 μm, the device capacitance is just too small, and the s-parameter measurement picks up most of the signals from the parasitic pad capacitance. One method to make a narrow-width device with a larger total gate area is to make the device's gate length long, say, on the order of 256 μm when the gate

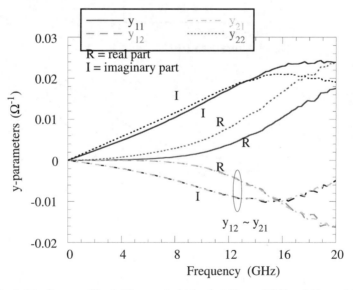

Fig. 1-46 Same as Fig. 1-45, except obtained at $V_{GS} = 0.2$ V and $V_{DS} = 1.5$ V.

width is 1 µm. This device, according to Eq. (1-35), will have a very long transit time. Equivalently, this device will run into non-quasi-static effects at fairly low frequencies. Therefore, modeling the narrow-width effects from such a long device may encounter NQS effects while the actual device with shorter length does not. Second, Figs. 1-45 and 1-46 are the de-embedded results. Without de-embedding, the y-parameters converted from the raw s-parameters will contain parallel and serial parasitics such that they will not agree well with Eq. (1-98), which applies only to the intrinsic transistor. A careful de-embedding procedure appropriate for rf measurement must then be available. The procedure needs to include the measurement of test patterns, such as that to find the polygate sheet resistance. The added measurement and calibration process slow down the overall measurement process, and the time to complete a table-lookup model. Lastly, the measured s-parameters incorporate the substrate effects. This point will be elaborated more in Section 4.6, but Fig. 1-47 also demonstrates this point somewhat. When we measure the s-parameters from a device, we may be tempted into thinking that we are measuring an intrinsic device whose substrate resistance plays no role. After all, the substrate

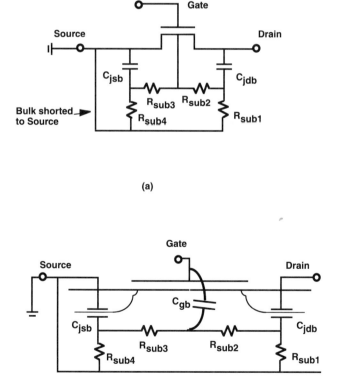

Fig. 1-47 (a) A schematic subcircuit that incorporates the substrate resistances; (b) the physical locations of the substrate resistances.

resistance takes no part in Eq. (1-98). However, in practice, the coupling between the drain-bulk capacitances becomes important at high frequencies. We are really measuring a device whose distributed resistances becomes part of the measurement, as shown in Fig. 1-47a. The substrate resistances cannot be separated from the intrinsic device. The physical origin of the substrate resistance is illustrated in Fig. 1-47b. Therefore, the y-parameters converted from the measured s-parameters are those of the device plus the substrate resistive network. Equating the raw data directly to Eq. (1-98) results in a false determination of the capacitance. Because the value of distributed substrate resistance cannot be easily established, it is difficult to perform de-embedding to ascertain the device capacitance values.

Besides s-parameters, another popular way to measure the device capacitance is to use a multi-frequency LCR meter [38]. An exemplar LCR meter setup is shown in Fig. 1-48, which is used to measure C_{sg} of the device. The "HI" side of the LCR meter consists of a dc voltage source in series with an ac small-signal voltage source at a programmed frequency. The dc voltage source can be ramped between two values, negative or positive. The "LO" side of the LCR meter is connected to a current meter. The LCR meter is designed such that the current meter is always grounded at one end, and at a virtual ground on the other end. The virtual ground is necessary. For example, in the configuration of Fig. 1-48, all of the specified dc bias and small-signal ac voltages at the gate node are supposed to drop across the device. If the source of the transistor is not at virtual ground but is at some other value, then v_{GS} of the device is unknown. Without knowing the exact voltage dropping across the capacitor, the capacitance cannot be ascertained even though we may still read the current off the current meter.

Because the setup is with $V_{DS} = 0$, the measured C_{sg} is equal to C_{gs}. In particular, if the magnitude of V_{GS} is <0, the channel is not inverted, and the measured C_{sg} represents the overlap capacitance between the gate and the source. To see why the setup of Fig. 1-48 measures C_{sg}, we write down the source current equation based on the quasi-static analysis. We are interested in the source current in particular because

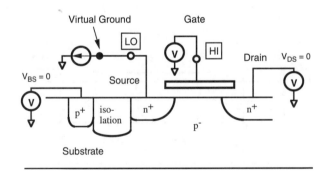

The bottom of the wafer should be isolated (with paper) from the chuck of the measurement assembly.

Fig. 1-48 An exemplar LCR meter setup used to measure C_{sg} of the device.

1.11 PHYSICAL MODEL AND TABLE-LOOKUP MODEL

the current meter of the LCR meter is placed in the source. According to Eq. (1-66c), we have

$$i_S = -I_D + \frac{\partial Q_S}{\partial v_{GS}} \cdot \frac{\partial v_{GS}}{\partial t} + \frac{\partial Q_S}{\partial v_{DS}} \cdot \frac{\partial v_{DS}}{\partial t} + \frac{\partial Q_S}{\partial v_{BS}} \cdot \frac{\partial v_{BS}}{\partial t};$$
$$= -I_D + C_{sg}\frac{\partial v_{GS}}{\partial t} + C_{sd}\frac{\partial v_{DS}}{\partial t} + C_{sb}\frac{\partial v_{BS}}{\partial t}. \quad (1\text{-}116)$$

v_{DS} is identically zero because the applied $V_{DS} = 0$, and because the applied small-signal voltage source is not placed in the drain. Similarly for the bulk side; v_{BS} is identically zero. Further, since $V_{DS} = 0$, the dc current I_D is identically zero. We are then left with $i_S = C_{sg} \cdot \partial v_{GS}/\partial t$. Moreover, $\partial v_{GS}/\partial t = \partial (V_{GS} + v_{gs})/\partial t = \partial v_{gs}/\partial t = j\omega v_{gs}$. Therefore, from the measurement of i_S and the knowledge of the applied v_{gs}, C_{sg} can be established.

The table-lookup model requires measurement of nine independent capacitances, at various V_{GS}, V_{DS} and V_{BS}. The measurement of C_{sg} was shown to be easily done when $V_{DS} = 0$, as specified in Fig. 1-48. Suppose now that V_{DS} is increased to the power-supply value. Now, during part of V_{GS} scan, the device is in strong inversion, and I_D is at some large value. ($\partial v_{DS}/\partial t$ and $\partial v_{BS}/\partial t$ are still zero.) This time, the current meter picks up a large real current I_D, in addition to the imaginary current of $-j\omega C_{sg}$. The typical operating frequency range of a LCR meter is 10 kHz to 1 MHz. At higher frequencies, series resistances may cause inaccuracy in measurement since the now larger ac current will result in voltage drop at the polysilicon gate [24]. At 10 kHz to 1 MHz, the magnitude of ωC_{gs} is too small in comparison to I_D. When the imaginary part sensed by the current meter is much smaller than the real part, the measurement becomes inaccurate. Therefore, it is difficult to use LCR meter to measure all of the required capacitance across all bias ranges. This is the reason why in a table-lookup model [34], only C_{gd}, and C_{gs} are measured and recorded in a lookup table. The missing capacitances such as C_{bd} are calculated from a model equation found in physical models! Further, the drain and source charges cannot be independently identified. Hence the reference finds the inversion charge, equal to $Q_S + Q_D$, and then assumes an arbitrary 50/50 partition for the drain and source charges.

LCR meter measurement is time consuming. Automated measurements are difficult to do because of stray capacitance in probecards, cabling, switching matrices which change if the pin configuration changes. It is preferred that the measurement be done manually when the LCR meter is used.

So, we see that there are many practical issues associated with the measurement of the capacitance, and hence the extraction the table-lookup models. Many publications on table-lookup models in fact bypass these practical issues, taking the capacitance values directly from device simulator such as MEDICI. They are not based on the measured data.

BSIM3, derived from fundamental device physics, is a physical model. From the derivation of Section 1.7 and Section 1.9, we can establish the device capacitance once the device I–V characteristics are measured. There is no need to directly measure the device capacitance. This is in contrast to the table-lookup models, which require precision capacitance measurement in order to establish the charge as a function of biases.

There are some informative table-lookup papers which nonetheless appear quite complicated to the untrained eyes, filled with spline coefficients and succinct matrix notations. Some say this is an inevitable outcome the moment a mathematician takes on a physics problem. It is easy to loose physical insights in a mathematical abstraction. Many of the urgent physical questions, unfortunately, are usually not addressed in these papers. It is difficult to establish whether these methods conserve charge, preserve device symmetry, have discontinuity in g_m/I_D, for example. It seems that, since the current and capacitance are measured at $V_{DS} > 0$ and with source as the reference terminal, most of these table-lookup models would have the same source-referencing problems as BSIM3, that the device symmetry is inherently lost. Further, because charges are integrated from capacitances, which are interpolated, it is possible to have charges which have local minimum or maximum at a particular combination of bias condition. The charge conservation has not been explicitly demonstrated in circuits such as the charge pump of Fig. 1-32.

1.12 SCALABLE MODEL AND DEVICE BINNING

A *scalable model* is one whose equations comprehend the length and width dependencies. Therefore, the same set of model parameters apply to transistors of various lengths and widths. A scalable model is generally a physical model, that is, a model based on device physics. This makes sense. Only after we understand the physics involved in the device operation can we predict correctly what the geometrical dependencies of various parameters are. However, a scalable model needs not be entirely based on device physics. There are occasions when the physics is not well understood, or just too complicated. Some kind of simplification in lieu of physical equation is then necessary for use in a compact SPICE model.

Conversely, a physical model does not necessarily guarantee scalability. It is possible that the model equations may incorrectly formulate geometrical dependencies of some parameters. For example, suppose that the threshold voltage of a unique MOSFET has a quadratic dependence on V_{DS}. If the model equation allows only linear dependence, then the model cannot hope to be scalable. BSIM3 rarely, if ever, runs into this type of problem. After all, BSIM3 is a culmination of the sustained model improvement since the early days of SPICE. For many technologies with channel length down to 0.18 μm, we have found BSIM3v3 to be scalable in long- and short-channel devices, as well as in wide and narrow devices. BSIM3v3 can therefore claim to be both a physical and a scalable model. Figure 1-49 displays the I–V and drain conductance characteristics of several geometries. All of the simulated characteristics are generated with the same set of BSIM3v3 parameters. Incidentally, the I_D vs V_{GS} plot for the $W/L = 4.13/0.49$ μm device is a classic example wherein the simulation is more accurate than the measurement! The measured leakage is apparently due to some mishap, which the simulation should ignore.

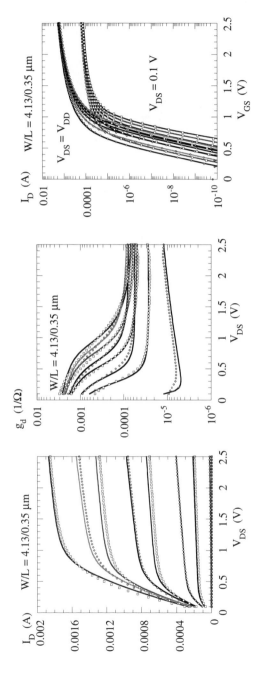

Fig. 1-49 Comparison of measured and calculated I_D-vs-V_{DS}, g_d, and I_D-vs-V_{GS} for various geometries. The simulated results are from one single BSIM3 model, without binning. Because the BSIM3 parameters were extracted by the author in one afternoon, the readers can rest assured that the slight discrepancy in the long/wide device can be eliminated when a more qualified engineer performs the extraction.

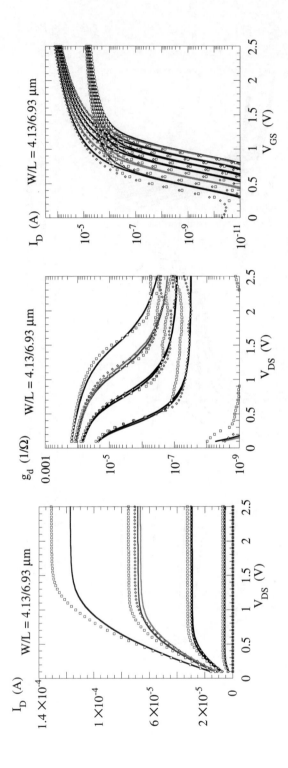

Fig. 1-49 (*continued*)

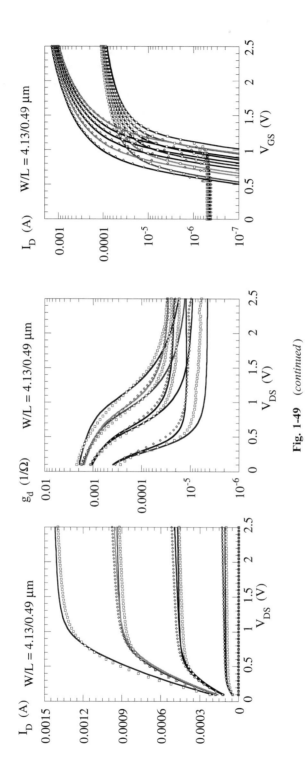

Fig. 1-49 (*continued*)

The opposite of a scalable model is, obviously, a nonscalable model. It is applicable for just one particular channel length and for widths which are wide enough to avoid narrow-width effects. In the past when modeling was geared toward digital circuits, a nonscalable model is considered alright. After all, the majority of a digital circuit usually employed the minimum-geometry transistor. The need to model different channel lengths was often just due to the desire to model process variations.

Then, a time came when it was required to model transistors of several lengths and widths. In the earlier generations of BSIM3, such as BSIM and BSIM2, there was no geometrical dependencies in a lot of variables used to calculate the device characteristics. As a comparison, we write out the equations calculating the threshold voltage in BSIM2 and BSIM3:

$$V_{T,\text{BSIM3}} = \textbf{VTHO} + \textbf{K1}\sqrt{2\phi_f - V_{BS}}$$
$$- \left\{ \exp\left(-\text{DSUB}\frac{L_{\text{eff}}}{2L_{to}}\right) + 2\exp\left(-\text{DSUB}\frac{L_{\text{eff}}}{L_{to}}\right) \right\}(\text{ETA0} + \text{ETAB} \cdot V_{BS})$$
$$\times V_{DS}; \tag{1-117a}$$

$$V_{T,\text{BSIM2}} = \textbf{VTHO} + \textbf{K1} \cdot \sqrt{2\phi_f - V_{BS}}. \tag{1-117b}$$

(*Note*: The threshold voltage equations are simplified here to illustrate the difference between the models. The above are not the exact equations used in BSIM2 or BSIM3. For example, the BSIM3's threshold equation contains several other terms. We included only the DIBL's modification to the long-channel threshold voltage in Eq. (1-117a). In addition, to aid the comparison, we replace the original BSIM2's parameters by the equivalent BSIM3 parameters.) It is clear that, while the length dependence of the threshold voltage is built in BSIM3, the BSIM2 equation does not comprehend the dependence. In fact, the only significant place where there is a geometrical dependence of the variables in BSIM2 is the drain current, which was correctly made to be proportional to W/L.

Even though a nonscalable model may lack geometrical dependence in its model equations, it is still possible to model devices with different sizes. The solution is called *binning*. Suppose that we need to model nine transistor sizes, denoted as $(L_1, W_1), (L_2, W_1), (L_3, W_1), (L_1, W_2)$, for example. We then create nine models for each transistor size as illustrated in Fig. 1-50. The model extracted from the (L_1, W_1) device is said to be valid for all L's such that $L_1 \leq L < L_2$, and all W's such that $W_1 \leq W < W_2$. Likewise, the model extracted from the (L_2, W_3) device is said to be valid for L's such that $L_2 \leq L < L_3$, and all W's exceeding or equal to W_3. There are nine models, each to cover a subregion in the geometrical space of interest. These model subregions are referred to as *bins*, and the practice of creating separate models is known as *binning*.

Binning, though necessary in a nonscalable model, is usually frowned upon. It complicates the process of parameter extraction, since independent efforts need to be

1.12 SCALABLE MODEL AND DEVICE BINNING   91

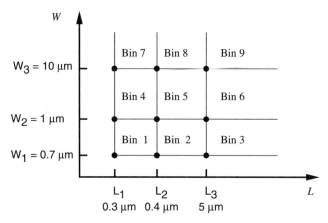

Fig. 1-50 An exemplar binning strategy that breaks up the model's width-length space into nine bins (regions).

spent on each of the model subregions. Some overhead in tracking is necessary to ensure that the proper submodel is selected for a device of a particular set of W and L. Most importantly, binning can result in discontinuity in device characteristics as the device geometry is varied. This is shown in Fig. 1-51, which plots the drive current (the drain current at maximum V_{DS} and V_{GS}) as a function of the channel length for a wide transistor. A discontinuity in the drive current is seen at $L = 0.4\,\mu\text{m}$, which is the boundary between bins 7 and 8. This erroneous simulation

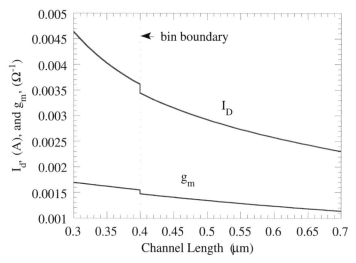

Fig. 1-51 Calculated drive current from a binned model. A discontinuity is observed when the device length is right at the boundary of two bins.

is a possibility inherent in binning since each submodel is built within its own subregion, independent of other submodel development at neighboring subregions. This discontinuity is a problem if the designed channel length or the optimized channel length is close to a bin boundary.

We mentioned that we can perform binning to amend the nonscalability problem of a model such as BSIM or BSIM2. We basically provide several submodels to encompass all geometrical sizes of interest. BSIM and BSIM2 offers a bin-combining strategy to lump all the bin models into one overall model, thereby alleviating the discontinuity problem. A slightly modified strategy is also available in BSIM3, although as we stated, BSIM3 is a scalable model and does not really need binning. Nonetheless, perhaps due to historical reasons or some fabrication-related issues, it is not uncommon to see binned BSIM3 models.

To understand the bin-combining strategy in BSIM3, we classify the BSIM3 parameters into two categories. The first kind are those that make up the backbone of the model, such as VTHO (long-channel threshold voltage). In the absence of binning, these parameters are directly used in the calculation of current, charges, and transconductances. We shall refer them as *primary parameters*. The second kind of parameters are those whose only purpose is to support the bin-combining process. These parameters, referred to as the *bin-combining parameters*, are all defaulted to zero. When they are assigned nonzero values, they are used to modify the values of the primary parameters specified in the SPICE input deck. For example, for a primary parameter VTHO, there are correspondingly three bin-combining parameters, LVTHO, WVTHO, and PVTHO, which are used to calculate the true long-channel threshold voltage in BSIM3. The relationship is given as follows:

$$\mathbf{VTHO} = \text{VTHO} + \frac{\text{LVTHO}}{L_{\mathit{eff}}} + \frac{\text{WVTHO}}{W_{\mathit{eff}}} + \frac{\text{PVTHO}}{L_{\mathit{eff}} \cdot W_{\mathit{eff}}}. \tag{1-118}$$

The **VTHO** on the left-hand side, called a *composite parameter*, is calculated from its right-hand-side terms. It is this value which is fed into the model equations to calculate currents, charges, and transconductances. The second VTHO, that on the right-hand-side, is a device parameter specified in the SPICE model parameter deck (.MODEL). As consistently used in this book, all of the SPICE parameters mentioned in this book are in the planar Courier font, which differs from the bold font chosen for the composite parameter.

Although Eq. (1-118) is just an example, the pattern of the bin-combining equation is exactly the same for other parameters. In general, for a primary parameter P and its corresponding three binning parameters LP, WP, and PP, BSIM3 writes

$$\mathbf{P} = \text{P} + \frac{\text{LP}}{L_{\mathit{eff}}} + \frac{\text{WP}}{W_{\mathit{eff}}} + \frac{\text{PP}}{L_{\mathit{eff}} \cdot W_{\mathit{eff}}}. \tag{1-119}$$

P, in bold font, is that actually used in SPICE calculation. Of course, when binning is not employed, LP, WP, and PP are defaulted to zero and **P** is identical to P, that listed in the model deck.

Just about all of the primary parameters have their corresponding binning parameters. However, it does not make sense to bin some of the primary parameters.

1.12 SCALABLE MODEL AND DEVICE BINNING

For example, TNOM is the temperature at which the SPICE parameters are extracted. This temperature should be identical for all of submodels in the bins. It does not make sense to extract models for one bin at a temperature and another bin at another temperature. Therefore, TNOM is not binnable, and the parameters LTNOM, WTNOM, and PTNOM do not exist.

We are now ready to examine BSIM3's bin-combining strategy. Suppose we have nine bins as shown in Fig. 1-50. Each model in a bin is independently extracted, and the bin-combining parameters (PL, PW, PP) in each model are zero. To combine all these nine bin models, we consider bins 7, 8, and 9 first. Since W_{eff} at these bins are large, we can approximate Eq. (1-118) as

$$\textbf{VTHO} = \text{VTHO} + \frac{\text{LVTHO}}{L_{eff}}. \quad (1\text{-}120)$$

Let VTHO's extracted in these bins be denoted as VTHO_7, VTHO_8, and VTHO_9, respectively, and VTHO without a subscript is that of the final bin-combining model. If we plot these values as a function of $1/L_{eff}$, we obtain three points as shown in Fig. 1-52.

Drawing a line with a least-square error fit, we find that if Eq. (1-120) is used to describe the line, then the y-axis intercept is VTHO and the slope is LVTHO. Because of the short-channel effects, the threshold voltage decreases as the channel length decreases. Hence, LVTHO is negative.

Now we consider bins 3, 6, and 9. Because L_{eff} at these bins are large, we can approximate Eq. (1-118) as

$$\textbf{VTHO} = \text{VTHO} + \frac{\text{WVTHO}}{W_{eff}}. \quad (1\text{-}121)$$

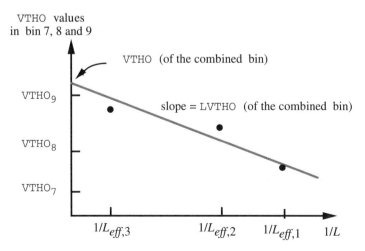

Fig. 1-52 Schematic showing how the parameters from bins 7, 8, and 9 of Fig. 1-50 can be combined.

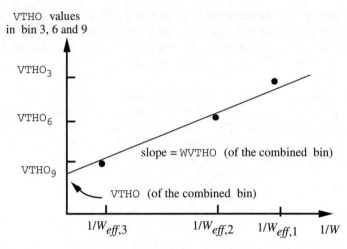

Fig. 1-53 Schematic showing how the parameters from bins 3, 6, and 9 of Fig. 1-50 can be combined.

We plot the VTHO's extracted in these bins, denoted as $VTHO_3$, $VTHO_6$, and $VTHO_9$, in Fig. 1-53. Again, we draw a line with a least-square error fit through the three points. We find that when Eq. (1-121) is used to describe the line, then the y-axis intercept is VTHO and the slope is WVTHO. Note that $WVTHO > 0$. Narrow-width effects usually cause the threshold voltage to increase as the width decreases.

VTHO extracted from the length variation (Eq. 1-120) and the width variation (Eq. 1-121) may not necessarily be the same. Clearly this is not acceptable if we want to combine all nine bins into one overall model. The inconsistency results from the fact that, in our attempt to simplify the discussion, we consider part of the overall equation with either L or W dimension, one at a time. In practice, $VTHO_1$, $VTHO_2$, ... and $VTHO_9$ are simultaneously fed into a program which determines the optimized values of VTHO, LVTHO, WVTHO, and PVTHO to minimize the error in fitting Eq. (1-118). This procedure is repeated for the determination of other parameters such as K1, LK1, WK1, PK1, and so on. One important point is that, the found **VTHO** value of the bin-combining model may not be the same as $VTHO_1$, $VTHO_2$, and $VTHO_9$ at the bin boundaries. The nine VTHO values in the individual bins are used as data to determine, VTHO, LVTHO, WVTHO, and PVTHO. An inherent danger is that the final bin-combining model can predict device characteristics quite different from the individually binned model. The disparity results from the fact that BSIM3 forces the geometrical dependence to be that of Eq. (1-118). At times, such a geometrical dependence can be sufficiently accurate for one particular parameter. However, the equation is made without much physical basis. It is conceivable that Eq. (1-118) can be quite erroneous for other parameters. Generally, the closeness of the dots shown in Figs. 1-52 and 1-53 gives an indication as to how well the overall bin-combining model matches the individual bin models at their respective valid region.

We have all seen some outdated test structures in a test die. The arcane test structure is not deleted from the mask set because we are always afraid that someone else may still have some use for it. Well, binning is a lot like that. In the old time when the models were not scalable, binning started out as a necessity. It was the only way in those models to account for the geometrical dependence of the model parameters. As the model equations in BSIM3 are made scalable, the geometrical dependencies become built-in. The application of binning to a scalable model becomes rather absurd. Let us take on BSIM3' threshold voltage expression given by Eq. (1-117a). As is, the equation already spells out the geometrical dependence of V_T, using parameters such as ETA0, ETAB and DSUB to model ΔV_T brought about by the DIBL effects. The equation shows that the dependence is somewhat exponential in form. If binning is used, with nonzero values of LETA0, LETAB, LDSUB, and so on, then we are adding a roughly inversely linear dependence of L_{eff} (see Eq. 1-118) on top of the exponential dependence. Let us suppose that the previously extracted ETA0, ETAB and DSUB give minimized error for devices of various lengths in a particular bin. Then after the bin-combining process which adds another layer of dependence on L_{eff}, the original fit is then lost. If binning is to be used, it seems that the parameters ETA0, ETAB, and DSUB (among others) should all be made zero, essentially making the BSIM3's threshold voltage to be lose its geometrical dependence, as the BSIM and BSIM2 equations. Then the bin-combining equation of Eq. (1-118) can be given to inject a geometrical dependence into the threshold voltage.

The following contains more details about binning. We describe a way to improve the binning in BSIM3, should binning be used. They are not interesting from a device-physics standpoint, but may interest those who insist on using binning. We can skip the rest of this section without the loss of continuity.

The bin-combining equation in BSIM3 was shown in Eq. (1-119). It has a drawback that, for example, VTHO$_1$, VTHO$_2$ ... and VTHO$_9$ extracted for each of the nine bins do not appear in the final bin-combining model. The bin-combining model may yield device characteristics differing from those predicted by the original bin models in their respective valid region. HSPICE implements Eq. (1-119) in a somewhat different manner, thereby removing the said problems. The HSPICE's bin-combining equation is believed to be

$$P_{HSPICE} = P_{HSPICE} + LP_{HSPICE}\left(\frac{1}{L_{eff}} - \frac{1}{L_{eff,ref}}\right) + WP_{HSPICE}\left(\frac{1}{W_{eff}} - \frac{1}{W_{eff,ref}}\right)$$
$$+ PP_{HSPICE}\left(\frac{1}{L_{eff}} - \frac{1}{L_{eff,ref}}\right)\left(\frac{1}{W_{eff}} - \frac{1}{W_{eff,ref}}\right). \quad (1-122)$$

Again, let us suppose that we have extracted nine bin models as shown in Fig. 1-50. In each of the bin model, LP, WP and PP are zero. Without the bin-combining processes, all these nine models are extracted separately and the problem of current discontinuity across the bin boundary can occur. We attempt bin-combining to remove the discontinuity. The HSPICE's bin-combining strategy will still result in nine bin models (unlike just an overall one as done previously under BSIM3's formulation.) However, P, LP, WP and PP in each of the bin will be assigned with proper values such that the device characteristics across the bin boundaries are continuous. Because there will still be nine bin models after the bin-combining process, we shall denote P, LP, WP, and PP in n-th bin as $P(n)$, $LP(n)$, $WP(n)$, and $PP(n)$, respectively.

Again, for clarity, we shall use VTHO as an example. When Eq. (1-122) is applied, we have

$$\mathbf{VTHO} = \text{VTHO} + \text{LVTHO}\left(\frac{1}{L_{\mathit{eff}}} - \frac{1}{L_{\mathit{eff},\text{ref}}}\right) + \text{WVTHO}\left(\frac{1}{W_{\mathit{eff}}} - \frac{1}{W_{\mathit{eff},\text{ref}}}\right)$$

$$+ \text{PVTHO}\left(\frac{1}{L_{\mathit{eff}}} - \frac{1}{L_{\mathit{eff},\text{ref}}}\right)\left(\frac{1}{W_{\mathit{eff}}} - \frac{1}{W_{\mathit{eff},\text{ref}}}\right). \tag{1-123}$$

In HSPICE's formulation, we choose $L_{\mathit{eff},\text{ref}}$ and $W_{\mathit{eff},\text{ref}}$ of a bin to be those at the lower left corner. So $L_{\mathit{eff},\text{ref}} = L_{\mathit{eff},1}$ and $W_{\mathit{eff},\text{ref}} = W_{\mathit{eff},1}$ for bin 1. Well, for bin 1, the model was extracted at $L_{\mathit{eff},1}$ and $W_{\mathit{eff},1}$. According to Eq. (1-123), VTHO(1) of bin 1 after the bin-combining process is just equal to VTHO_1, the VTHO value of bin 1 right after parameter extraction. Hence,

$$\text{VTHO}(1) = \text{VTHO}_1. \tag{1-124}$$

Similarly for bin 2, we have $\text{VTHO}(2) = \text{VTHO}_2$. However, the lower left corner of bin 2 lands right on the lower right corner of bin 1. To ensure continuous device characteristics across a bin boundary, we demand that **VTHO** of Eq. (1-123) calculated in bins 1 and 2 be equal. The **VTHO** value calculated in bin 1 ($L_{\mathit{eff},\text{ref}} = L_{\mathit{eff},1}$; $W_{\mathit{eff},\text{ref}} = W_{\mathit{eff},1}$) is

$$\mathbf{VTHO} = \text{VTHO}(1) + \text{LVTHO}(1)\left(\frac{1}{L_{\mathit{eff},2}} - \frac{1}{L_{\mathit{eff},1}}\right). \tag{1-125}$$

The **VTHO** value calculated in bin 2 ($L_{\mathit{eff},\text{ref}} = L_{\mathit{eff},2}$; $W_{\mathit{eff},\text{ref}} = W_{\mathit{eff},1}$) is

$$\mathbf{VTHO} = \text{VTHO}(2) = \text{VTHO}_2. \tag{1-126}$$

Equating Eqs. (1-125) and (1-126), we have

$$\text{LVTHO}(1) = [\text{VTHO}_2 - \text{VTHO}_1] \div \left(\frac{1}{L_{\mathit{eff},2}} - \frac{1}{L_{\mathit{eff},1}}\right). \tag{1-127}$$

1.12 SCALABLE MODEL AND DEVICE BINNING

Similarly,

$$\text{WVTHO}(1) = [\text{VTHO}_4 - \text{VTHO}_1] \div \left(\frac{1}{W_{eff,2}} - \frac{1}{W_{eff,1}} \right). \quad (1\text{-}128)$$

We now determine what should the value of PVTHO(1) be. From Fig. 1-50, we see that the upper right corner of bin 1 coincides with the lower left corner of bin 5. At the intersecting point, **VTHO** value calculated in bin 1 ($L_{eff,ref} = L_{eff,1}$; $W_{eff,ref} = W_{eff,1}$) is

$$\begin{aligned}\textbf{VTHO} = &\ \text{VTHO}(1) + \text{LVTHO}(1)\left(\frac{1}{L_{eff,2}} - \frac{1}{L_{eff,1}} \right) + \text{WVTHO}(1)\left(\frac{1}{W_{eff,2}} - \frac{1}{W_{eff,1}} \right) \\ &+ \text{PVTHO}(1)\left(\frac{1}{L_{eff,2}} - \frac{1}{L_{eff,1}} \right)\left(\frac{1}{W_{eff,2}} - \frac{1}{W_{eff,1}} \right). \end{aligned} \quad (1\text{-}129)$$

At the same point, **VTHO** value calculated in bin 5 ($L_{eff,ref} = L_{eff,2}$; $W_{eff,ref} = W_{eff,2}$) is:

$$\textbf{VTHO} = \text{VTHO}(5) = \text{VTHO}_5. \quad (1\text{-}130)$$

Equating Eqs. (1-129) and (1-130), we have

$$\text{PVTHO}(1)$$

$$= \frac{\text{VTHO}_5 - \text{VTHO}_1 - \text{LVTHO}(1)\left(\frac{1}{L_{eff,2}} - \frac{1}{L_{eff,1}} \right) - \text{WVTHO}(1)\left(\frac{1}{W_{eff,2}} - \frac{1}{W_{eff,1}} \right)}{\left(\frac{1}{L_{eff,2}} - \frac{1}{L_{eff,1}} \right)\left(\frac{1}{W_{eff,2}} - \frac{1}{W_{eff,1}} \right)}.$$

$$(1\text{-}131)$$

All the other VTHO(n), LVTHO(n), WVTHO(n), and PVTHO(n) can be similarly obtained. For the outer bins such as bin 9, we can imagine that there are exterior bins located at $L_{eff,4}$ and $W_{eff,4}$ equal to ∞. Therefore, the bin-combining parameters at bin 9 can be calculated using equations similar to Eqs. (1-127), (1-128), and (1-131).

The above bin-combining process results in nine bins, which are modified from those originally extracted nine bins. This bin-combining process ensures the device characteristics to be continuous across a bin boundary. Previously in the bin-combining description for BSIM3, we mentioned that BSIM3's strategy results in one overall bin. However, it needs not be this way. The BSIM3's formulation, though inconvenient, can still be made equivalent to the HSPICE's implementation. By comparing Eqs. (1-122) and (1-118), we can choose P, LP, WP and PP in BSIM3 in the following manner:

$$P_{BSIM3}(n) = P_{HSPICE}(n) - \frac{LP_{HSPICE}(n)}{L_{\mathit{eff},\mathrm{ref}}(n)} - \frac{WP_{HSPICE}(n)}{W_{\mathit{eff},\mathrm{ref}}(n)}$$

$$- \frac{PP_{HSPICE}(n)}{L_{\mathit{eff},\mathrm{ref}}(n) \cdot W_{\mathit{eff},\mathrm{ref}}(n)}; \qquad (1\text{-}132\mathrm{a})$$

$$LP_{BSIM3}(n) = LP_{HSPICE}(n) - \frac{PP_{HSPICE}(n)}{W_{\mathit{eff},\mathrm{ref}}(n)}; \qquad (1\text{-}132\mathrm{b})$$

$$WP_{BSIM3}(n) = WP_{HSPICE}(n) - \frac{PP_{HSPICE}(n)}{L_{\mathit{eff},\mathrm{ref}}(n)}; \qquad (1\text{-}132\mathrm{c})$$

$$PP_{BSIM3}(n) = PP_{HSPICE}(n), \qquad (1\text{-}132\mathrm{d})$$

where P_{HSPICE}, LP_{HSPICE}, WP_{HSPICE}, and PP_{HSPICE} are determined by Eqs. (1-127), (1-128), and (1-131).

REFERENCES AND NOTES

[1] P. Yang, B. Epler, and P. Chatterjee, "An investigation of the charge conservation problem for MOSFET circuit simulation," *IEEE J. Solid-State Circuits*, vol. 18, pp. 128–138, 1983.

[2] R. Rohrer, "Circuit simulation—the early years," *IEEE Circuits and Devices*, pp. 32–37, May 1992.

[3] T. S. Perry, "Donald O. Pederson," *IEEE Spectrum*, pp. 22–27, June, 1998.

[4] A trial version of PSPICE for PC can found in http://www.orcad.com/techserv/pspprint_f.htm. For the Macintosh platform, see http://www.repairfaq.org/ELE/F_Free_-Spice.html.

[5] M. R. Pinto, C. S. Rafferty, and R. W. Dutton, "PISCES II: Poisson and continuity equation solver," Stanford Electronics Laboratory Technical Report, 1984.

[6] H. Shichman and D. Hodges, "Modeling and simulation of insulated-gate field-effect transistor switching circuits," *IEEE J. Solid State Circuits*, vol. 3, pp. 285–289, 1968.

[7] J. Meyer, "MOS models and circuit simulation," *RCA Review*, vol. 32, pp. 42–63, 1971. In this original work, the gate-to-bulk capacitance (C_{gb}) is ignored. However, subsequent implementation of this model for SPICE simulation includes a nonzero C_{gb}.

[8] A. Vladimirescu and S. Liu, The Simulation of MOS Integrated Circuits Using SPICE2, Electronics Research Laboratory Memo. ERL M80/7, University of California, Berkeley, 1980.

[9] D. Ward and R. Dutton, "A charge-oriented model for MOS transistor capacitances," *IEEE J. Solid-State Circuits*, vol. 13, pp. 703–708, 1978.

[10] B. Sheu, D. Sharfetter, P. Ko, and M. Jeng, "BSIM: Berkeley Short Channel IGFET Model for MOS transistors," *IEEE J. Solid-State Circuits*, vol. 22, pp. 558–566, 1987.

REFERENCES AND NOTES

[11] HSPICE User's Manual, Meta-Software, Inc., Campbell, CA, 1996.

[12] M. Jeng, "Design and modeling of deep submicrometer MOSFETs," University of California, Berkeley, Electronics Research Laboratory Memorandum No. UCB/ERL M90/90, 1990.

[13] Y. Cheng et al., "BSIM3 Version 3.0 Manual," University of California/Berkeley, Electronics Research Laboratory (1995). See also, J. Huang et al., "BSIM3 Manual (Version 2.0)," University of California/Berkeley, Electronics Research Laboratory (1994).

[14] W. Liu et al, "BSIM3v3.2.2 MOSFET Model, User's Manual," University of California, Berkeley, posted on http://www-device.eecs.berkeley.edu/~bsim3, 1999. A recent book also provides valuable information: Y. Cheng, and C. Hu, *MOSFET Modeling & BSIM3 User's Guide*, Boston: Kluwer, 1999.

[15] As of March 2000, the latest BSIM3 model released to the public is BSIM3v3.2.2. There has been a discussion to release BSIM3v3.3, which would contain an improvement in the thermal noise formulation (see Section 4.11 for details). Because the effort in early 2000 has been in the release of BSIM4, it is not clear at this point when BSIM3v3.3 will be released, if it is to be released at all. See http://www-device.eecs.berkeley. edu/~bsim3 for the latest development.

[16] R. Velghe, D. Klaasen, and F. Klaasen, "Compact MOS modeling for analog circuit simulation," *IEEE International Electron Device Meeting*, pp. 485–488, 1993.

[17] C. Enz, F. Krummenacher, and E. Vittoz, "An analytical MOS transistor model valid in all regions of operation and dedicated to low voltage and low current applications," *Analog International Circuit and Signal Proceeding*, vol. 8, pp. 83–114, 1995. See also, M. Bucher, C. Lallement, C. Enz, F. Krummenacher, "Accurate MOS modleing for analog circuit simulation using the EKV model, *IEEE Int. Symp. Circuits and Systems*, pp.703–706, 1996.

[18] D. Foty, *MOSFET Modeling With SPICE, Principles and Practice*, Upper Saddle River, NJ: Prentice-Hall, 1997.

[19] See, for example, Y. Tsividis, *Operation and Modeling of the MOS Transistor*, New York: McGraw-Hill, 1987.

[20] The algorithm can be found in, W. H. Press, B. P. Flannery, S. A. Teukolsky, and W. T. Vetterling, *Numerical Recipes, the Art of Scientific Computing*. London: Cambridge University Press, 1986. A step-by-step derivation, for a particular electrical problem, of the required coefficients required by the algorithm can be found in, W. Liu, *Handbook of III-V Heterojunction Bipolar Transistors*, New York: Wiley, 1998, Chapter 6.

[21] K. Kundert, *Designer's Guide to SPICE and SPECTRE*, Boston: Kluwer Academic, 1995. See p. 18 for convergence criteria and p. 173 about the GaAs FET charge model.

[22] Y. Tsivids, and K. Suyama, "MOSFET modeling for analog circuit CAD: problems and prospects," *IEEE J. Solid-State Circuits*, vol. 29, pp. 210–216, 1994.

[23] The standardization effort is promoted by the Compact Model Council, which is affiliated with the Electronics Industries Alliance. The Compact Model Council's history, mission and members can be found in http://www.eia.org/eig/CMC/.

[24] W. Liu, *Fundamentals of III-V Devices: HBTs, MESFETs, HFETs/HEMTs*, New York: Wiley, 1999. For conversion from y-parameters to s-parameters, see p. 249. For gate resistance, see Section 6.5. For charge calculation, see Section 6.1.

[25] Liu, W., Bowen, C., and Chang, M. (1996) "A CAD-compatible non-quasi-static MOSFET model." *IEEE Int. Electron Device Meeting*, pp. 151–154.

[26] B. Sheu, D. Sharfetter, C. Hu, and D. Pederson, "A compact IGFET charge model," *IEEE Trans. Circuits Systems*, vol. 31, pp. 745–748, 1984.

[27] M. Cirit, "The Meyer model revisited: why is charge not conserved?," *IEEE Trans. Computer-Aided Design*, vol. 8, pp. 1033–1037, 1989.

[28] H. Statz, P. Newman, I. Smith, R. Pucel, and H. Haus, "GaAs FET device and circuit simulation in SPICE," *IEEE Trans. Electron Devices*, vol. 34, pp. 160–168, 1987. It is believed that this paper describes the "Staz model" mentioned on p. 173 of Ref. [21].

[29] A. Snider, "Three new mathematical techniques for field effect transistor modeling and analysis," *Proc. IEEE Int. Caracas Conf. on Devices, Circuits, and Systems*, pp. 65–68, 1998.

[30] A. Snider, "Charge conservation and the transcapacitance element: an exposition," *IEEE Trans. Education*, vol. 38, pp. 376–379, 1995.

[31] N. Arora, *MOSFET Models for VLSI Circuit Simulation: Theory and Practice*, New York: Springer-Verlag, 1993.

[32] G. Machado, C. Enz, and M. Bucher, "Estimating key parameters in the EKV MOST model for analog design and simulation," *IEEE Int. Symp. Circuits and Systems*, pp. 1588–1591, 1995.

[33] T. Shima, "Table lookup MOSFET capacitance model for short-channel devices," *IEEE Trans. Computer-Aided Design*, vol. 5, pp. 624–632, 1986.

[34] T. Shima, H. Yamada, and R. Dang, "Table look-up MOSFET modeling system using a 2-D device simulator and monotonic piecewise curbic interpolation," *IEEE Trans. Computer-Aided Design*, vol. 2 pp. 121–126, 1983.

[35] R. Daniels, A. Yang, and J. Harrang, "A universal large/small signal 3-terminal FET model using a nonquasi-static charge-based appraoch," *IEEE Trans. Electron Devices*, vol. 40, pp. 1723–1729, 1993.

[36] D. Root, S. Fan, and J. Meyer, "Technology Independent large signal non quasi-static FET models by direct construction from automatically characterized device data," *Proc. 21st European Microwave Conf.*, Stuttgart, Germany, pp. 927–932, 1991.

[37] W. Liu, *Handbook of III-V Heterojunction Bipolar Transistors*, New York: Wiley, 1998. See Chapter 15 for the pad test structure and de-embedding procedure.

[38] The LCR meter gets its name because an inductor, a capacitor and a resistor are essential for the function of the equipment. Examples of a LCR meter include HP4275 or HP4284.

2

BASIC FACTS ABOUT BSIM3

2.1 WHAT IS AND WHAT'S NOT IMPLEMENTED IN BSIM3

BSIM3 is built upon many models before it. Although several physical effects are now included in BSIM3, some still are not. This section intends to serve as a quick reference to find out what BSIM3 can do, as well as what it cannot do.

Here are some characteristics of BSIM3:

- BSIM3 is a MOSFET model, for both n-channel and p-channel devices. Due to similar equations governing the dynamics of the channel, BSIM3 model may be used for GaAs HFET (heterojunction field-effect transistor) or HEMT (high electron mobility transistor) with some modification [1]. However, BSIM3 is not appropriate for GaAs MESFET (metal semiconductor field-effect transistor) because their starting equations governing the channel dynamics differ.
- BSIM3 works equally well for depletion-mode devices, besides conventional enhancement-mode devices.
- The effective channel length (L_{eff}) for the I-V calculation is approximately equal to $L - 2 \cdot$ LINT, where L is the specified drawn length of the device, and LINT is a SPICE parameter. For the C-V calculation, the CV effective channel length ($L_{eff,CV}$) is roughly equal to $L - 2 \cdot$ DLC.

    The effective channel width (W_{eff}) for the I-V calculation is approximately equal to $W - 2 \cdot$ WINT, where W is the specified drawn width of the device, and WINT is a SPICE parameter. For the C-V calculation, the CV effective channel width ($W_{eff,CV}$) is roughly equal to $W - 2 \cdot$ DWC.

    The parameter K1 can be treated as the body-effect coefficient, labeled as γ in typical device textbooks, as well as in Eq. (1-6a).

- BSIM3 is a scalable model that has demonstrated success in $L = 0.18\,\mu m$ technology, without binning. This is possible because BSIM3 accounts for a

variety of short-channel effects, including the charge-sharing effects, drain-induced barrier-lowering, reverse-short-channel effects, and those effects found in short-channel and narrow-width devices.
- BSIM3 is adequate for a majority of analog and digital circuit simulation. However, for fast-speed analog or rf applications, a subcircuit approach to include parasitic elements is essential (Section 4.6).
- BSIM3 is primarily a quasi-static model; that is, based on the quasi-static approximation (Section 1.6). BSIM3 has a non-quasi-static option, but it is significantly less popular than its quasi-static model. If the BSIM3 parameter NQSMOD is 0 or absent, then the quasi-static model is used in the simulation.
- BSIM3 is a charge-conservative model. The sum of the currents flowing into the four terminals of the MOSFET is always identically zero, even in a transient simulation.
- BSIM3 is considered a physical model, with most parameters having strong correlation with the fabrication process and device design. However, some parameters have weak physical significance and are introduced mainly for the purpose of better fitting.
- BSIM3 accounts for the parasitics such as the overlap and (outer) fringing capacitances as well as the bulk-source and bulk-drain diodes. The inner fringing capacitance, important when the channel is not yet inverted, is not included. The parasitic capacitance between the dain and source contacts, important in sub-quarter-micron devices, is neglected as well.

The modeling of the parasitic diodes' behaviors in BSIM3 is adequate when the diodes are reverse-biased. The modeling in the forward-biased region is not complete, although this operating region is hardly encountered in a normal MOSFET operation. BSIM3 accounts for the rapid increase of the drain (source) and bulk currents when the drain (source)-bulk diode turns on, as well as the depletion capacitance. However, the diffusion capacitance associated with the minority carriers is not accounted for. As far as the noise analysis is concerned, the $1/f$ and shot noises of the p-n junction, either forward- or reverse-biased, are neglected.

- The source and drain terminal resistances are included in BSIM3, but not the gate or the substrate resistances. Likewise, the noises associated with the gate and substrate resistances are absent.
- BSIM3 references the terminal voltages to the source, not bulk. As such, BSIM3 has asymmetry problems at $V_{DS} = 0$, that many device parameters (such as C_{gs}) become discontinuous as V_{DS} crosses zero.
- BSIM3 accounts for the first-order effects of temperature on the device characteristics. The degradation of the mobility and the increase of threshold voltage with temperature are modeled.
- BSIM3's C-V model for the intrinsic transistor is based on long-channel approximation. Many of its device capacitances can exhibit incorrect voltage dependencies, although the circuit simulation can still be quite accurate.

- BSIM3's noise model is based on long-channel derivation. Its noise calculation is inaccurate for short-channel devices.
- A conductance equal to GMIN (default $= 10^{-12} \, \Omega^{-1}$) is often employed in a SPICE model to speed up convergence. In BSIM3, a GMIN conductance is placed between the source and bulk terminals, and another conductance is placed between the drain and the bulk terminals. This is unlike other models which place one single GMIN conductance between the drain and source terminals.

Viewing from another perspective, we can also state the following about BSIM3:

- BSIM3 is not a bipolar model.
- BSIM3 does not include self-heating effects. The device temperature, once specified, remains constant during the entire simulation. It is possible to specify the device temperature to any particular value. BSIM3 contains the parameters to calculate the proper device characteristics under various temperatures.
- BSIM3 does not include the parasitic bipolar effects; it cannot properly simulate the "snap-back" phenomenon in the I-V characteristics. The substrate current due to impact ionization is modeled in BSIM3v3.2, but not in BSIM3v3.1. The accuracy of the substrate current model is limited to the region when the substrate current is low enough such that the parasitic bipolar transistor action has not been triggered. BSIM3 does not predict the avalanche breakdown voltage.
- BSIM3 does not include the drain-induced gate noise. Only the thermal noise and the $1/f$ noise of the channel are considered (Section 4.12).
- BSIM3 does not include the gate-oxide leakage current (due to traps in oxide, for example), or tunneling current through oxide (quantum-mechanical effect). In fact, the dc gate current is always identically zero. There is no leakage path between the gate and other terminals, as intuitively thought by some circuit designers.

    The only leakage currents included in BSIM3 are those associated with the drain-bulk, the source-bulk junctions, and the substrate current brought about by the impact ionization. When V_{GS} is zero or negative, the subthreshold current between the drain and the source is modeled in BSIM3. The subthreshold current continues to decrease exponentially toward zero as V_{GS} decreases, even when V_{GS} becomes negative. Hence, BSIM3 can be used to predict the off current (the drain current at zero V_{GS}).

- There is no GIDL (gate-induced drain leakage) modeling in BSIM3 (Section 4.16). If GIDL is significant at $V_{GS} = 0$, then the off current predicted by BSIM3 is incorrect.
- Sometimes BSIM3 assumes the device is symmetrical, that is, the drain and source is completely interchangeable. For example, The bulk-drain junction capacitance cannot be set independently from the bulk-source capacitance. The parameters calculating the drain-bulk and the source-bulk capacitances, such as

CJ, CJSW, and CJSWG, are identical. So, if a device is asymmetrical, it is not straightfoward to use BSIM3 to model these different junction capacitances.

However, at times BSIM3 seems to allow for asymmetry. For example, the overlap capacitance at the gate-drain side is allowed to differ from the capacitance at the gate-source side. The gate-drain overlap capacitance is modeled with the parameters CGD0 and GGD1, while the gate-source overlap capacitance, with CGS0 and GGS1.

- Some devices have two distinct subthreshold slopes. BSIM3 will not be able to model such a phenomenon.
- There is no parameter which specifies the maximum operating voltages for V_{GS}, V_{DS} and V_{BS}. Care must be exercised to see if the voltages fall outside the breakdown values.

The modeling engineers like to use mathematical formulation to describe the transistor. It is no wonder that there is not a drawing of an equivalent circuit in the BSIM3 manual. Circuit designers, in contrast, prefer to visualize a transistor with its equivalent circuit. In the following sections, we construct equivalent circuits based on BSIM3's mathematical descriptions. We present the equivalent circuits (both exact and approximate) for dc, large-signal, small-signal ac, and noise analyses.

2.2 DC EQUIVALENT CIRCUIT MODEL

Even if we know nothing about BSIM3, we still have some idea of how a MOSFET's dc equivalent model should look like. We think there are terminal parasitic resistances at the drain, gate, and source. We understand that besides the main drain current flowing through the channel, there are parasitic diodes in the source-bulk junction and in the drain-bulk junction. We also envision the existence a leakage gate current through the oxide, either by tunneling mechanism or by thermionic emission after the channel hot electron acquires enough energy from the channel electric field. Our conceptual dc equivalent circuit for a MOS looks like Fig. 2-1. The drain current flowing through the channel toward the source, I_{DS}, should incorporate short-channel effects and is generally a complicated function of biases. It should also be expressed in a smoothing function format, consistent with the desire to eliminate discontinuity in device characteristics. A generalized equation for the channel current can be expressed as

$$I_{DS} = f(V_{GS}, V_{DS}, V_{BS}). \qquad (2\text{-}1)$$

The exact functional form f adopted in BSIM3 is complicated. A simplified form for a long-channel transistor can be taken to be Eq. (1-11).

2.2 DC EQUIVALENT CIRCUIT MODEL   105

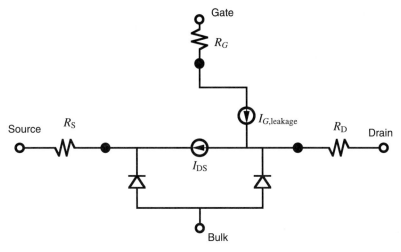

Fig. 2-1 A conceptual dc equivalent circuit for a MOSFET.

Another drain current component originates from the drain-bulk diode. According to elementary device physics, the drain-bulk junction diode current should have the form of

$$I_{j,\text{DB}} = I_{sat.\text{DB}}\left[\exp\left(\frac{qV_{BD}}{kT}\right) - 1\right] \quad \text{(not used in BSIM3, see Eq. 2-3),} \quad (2\text{-}2)$$

where $I_{sat,\text{DB}}$ is some saturation constant and V_{BD} is the terminal voltage across the diode.

It turns out that the BSIM3 model differs from our intuition. To contrast, we show the dc equivalent circuit adopted by BSIM3 in Fig. 2-2. Firstly, BSIM3 does not model any gate leakage current. The current source $I_{G,\text{leakage}}$ in Fig. 2-1 disappears in Fig. 2-2. This means that, in BSIM3, there is a complete dc open between the gate

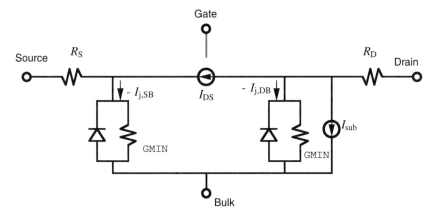

Fig. 2-2 A dc equivalent circuit adopted by BSIM3.

and the other side of the oxide. The dc gate current is always zero. Without any current flowing through the gate resistor, the dc voltage drop there is also zero. Hence, the gate resistance is left out altogether. We believe that the omission of the gate resistance is not a sound modeling practice, especially for low noise/high power applications which employ large-sized transistors. Although the dc gate current is zero when we neglect $I_{G,\text{leakage}}$, the gate current can still be finite during a transient or an ac situation. This current, when flowing through a resistor, produces a voltage drop.

Another difference between the BSIM3 model and our conception is the existence of conductances labeled as GMIN, which are placed in parallel with the drain-bulk and the source-bulk diodes. These conductances do not bear physical significance, at least they are not meant to represent any physical effect. They are added purely to aid the convergence of numerical solution during computer iteration. The value of GMIN is defaulted to $10^{-12}\Omega^{-1}$, independent of the transistor area or the diode area. It can be reset to other values in the .OPTION statement. Because GMIN is placed in parallel with the diode, $I_{j,\text{DB}}$ used in BSIM3, previously given in Eq. (2-2), is modified to

$$I_{j,\text{DB}} = I_{sat,\text{DB}}\left[\exp\left(\frac{qV_{BD}}{kT}\right) - 1\right] + \text{GMIN} \cdot V_{BD}. \qquad (2\text{-}3)$$

The source-bulk current has a similar expression, with the replacement of $I_{sat,\text{DB}}$ by $I_{sat,\text{SB}}$ and V_{BD} by V_{BS}. (We caution that the $I_{j,\text{DB}}$ expression in Eq. (2-3) is still not exactly that used in BSIM3. It is the same only in the reverse-biased condition. In the forward-biased condition, some modification may occur, as detailed in Section 3.2, in the entry of JS). Because GMIN is a conductance whose value can be modified by the user, we use the Courier font for its name, just as we have done for the other BSIM3 SPICE parameters.

Finally, BSIM3 includes a substrate current component labeled as I_{sub}, which results from impact ionization. When an electron travels in the channel, it gains energy from the lateral electric field established by V_{DS}. Usually, the gained energy is quickly dissipated in the lattice as it impacts the lattice. However, at times when the electric field is large enough, the drifting electron can gain an amount of energy exceeding the crystal bonding energy. When the electron finally collides with a crystal atom, it creates an electron–hole pair. The created electron moves toward the drain where a positive voltage of V_{DS} is applied, while the hole moves toward the substrate where the potential is lower. The electron–hole flow constitutes the substrate current, which flows between the drain and the bulk contacts. Just like the channel current, the substrate current can be generalized to the following form:

$$I_{sub} = g(V_{GS}, V_{DS}, V_{BS}). \qquad (2\text{-}4)$$

We use I_{sub} to denote this substrate current generated by impact ionization. While it flows through the bulk terminal, I_{sub} differs from the bulk terminal current, I_B. I_B consists of the drain–bulk and source–bulk junction leakage currents, in addition to

I_{sub}. The substrate current also flows through the drain terminal. The drain terminal current, I_D, thus contains I_{sub} as one component, in addition to the drain–bulk junction leakage and the channel current of the intrinsic MOSFET. The source terminal current, I_S, is not equal to $-I_D$. The relationship would be correct if the channel I_{DS} were the only current in the entire device. Instead, the source current consists of the channel current, as well as the source–bulk junction leakage current. Finally, BSIM3 does not model any dc gate current. So, the gate terminal current, I_G, is identically zero. The relationships between the various terminal currents are summarized in the following:

$$I_G = 0; \quad (2\text{-}5\text{a})$$
$$I_D = I_{DS} + I_{sub} - I_{j,\text{DB}}; \quad (2\text{-}5\text{b})$$
$$I_S = -I_{DS} - I_{j,\text{SB}}; \quad (2\text{-}5\text{c})$$
$$I_B = -I_{sub} + I_{j,\text{SB}} + I_{j,\text{DB}}. \quad (2\text{-}5\text{d})$$

Figure 2-3 shows the measured bulk current of a $L = 0.8\,\mu\text{m}$ NMOS device, plotted as a function of V_{GS} when V_{DS} is fixed at several constant values. If we ignore the junction leakage currents, then the bulk current consists solely of the substrate current. As shown in Fig. 2-3, the substrate current is negative, indicating that the current flows out of the substrate terminal, consistent with the aforementioned physical picture. When V_{DS} is small, the device operates in the linear region independent of the gate-source bias. The channel has not pinched off and the electric field in the channel due to the applied V_{DS} is small. Without a high field region near the drain, the amount of impact ionization is close to nil. The substrate current is practically zero. When V_{DS} is large, the transistor operates in saturation. Near the

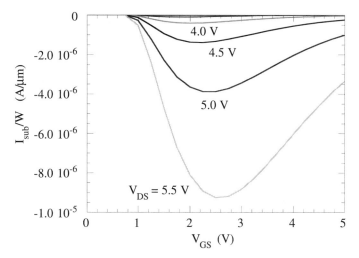

Fig. 2-3 Measured bulk current per width of a $L = 0.8\,\mu\text{m}$ NMOS device.

drain where the channel pinches off, the field between the gate-drain terminals is large and the substrate current due to impact ionization becomes noticeable.

The measured I_{sub}-vs-V_{GS} at a particular V_{DS} is bell-shape-like, indicating the presence of two competing factors that determine the final amount of impact ionization. One factor is the availability of the initiating conducting carriers to start the impact ionization. If the channel current is large (hence there are many electron carriers), then the likelihood of impact ionization increases. The second factor relates to the magnitude of the electric field in the channel region between the gate-drain contacts. If the electric field is large, there will be more impact ionization because an electron can gain a higher amount of energy in a shorter amount of distance. Let us focus on the curve measured at a large V_{DS}. Initially when V_{GS} is fairly small, the channel is not strongly inverted. The number of the conducting electrons is minute, and the substrate current is small. As V_{GS} increases, the channel becomes more strongly inverted, and the amount of current flowing in the channel increases. As there are more carriers to initiate the impact ionization, the substrate current becomes increasingly more negative. However, this trend eventually stops. After V_{GS} increases past a certain value while V_{DS} remains fixed, the voltage drop in the saturated region decreases. This decreases the electric field and the amount of impact ionization dwindles. In the extreme that V_{GS} increases to a large value such that $V_{GS} - V_{DS} > V_T$, then the transistor enters linear region and the substrate current drops back to zero.

The substrate current is a channel hot carrier effect. On the first order, we may say that the electron and the hole impact ionization rates are about the same. However, because the hole has a heavier effective mass than an electron, it takes more time for a hole to acquire the required energy for impact ionization. Consequently, the substrate current in a PMOS is significantly less than NMOS.

We make a cautionary remark here. The physical picture described above is well modeled in BSIM3, with the parameters ALPHA0, ALPHA1 and BETA. However, when the magnitude of I_{sub} is large enough such that there is about a 0.5 V drop across the substrate resistance, another physical effect will be triggered and BSIM3 ceases to be accurate. This new physical effect relates to the parasitic bipolar inherent to the MOS, formed with the *n*-emitter in the source, *p*-base in the substrate, and the *n*-collector in the drain. When there is roughly a 0.5 V drop across the substrate, the base–emitter junction of the bipolar transistor (between the bulk and the source) becomes forward biased. Once the parasitic bipolar transistor action kicks in, the amount of impact ionization can increase rapidly with only a small increment of V_{DS}. Both the drain and the bulk currents will increase substantially. Eventually, the device breakdown associated with the parasitic bipolar occurs and the measured I_D-vs-V_{DS} characteristics exhibit a snap-back behavior, as shown in Fig. 2-4. We qualitatively separate the I-V characteristics into two regions. In the one toward the left where the drain current shows some increase due to the impact ionization, BSIM3 is adequate. However, BSIM3 does not have equations to comprehend the snap-back in the region to the right.

The bulk current (I_B) and gate current (I_G) corresponding to the I-V characteristics of Fig. 2-4 are shown in Fig. 2-5. When the junction leakage is small, I_B is

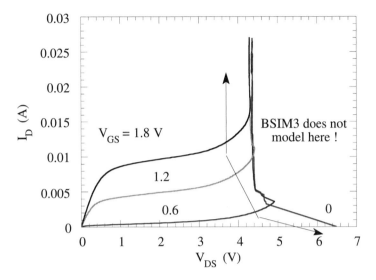

Fig. 2-4 Measured I-V characteristics of a NMOS. Snap-back occurs at the right of the marked boundary.

basically equal to I_{sub}. This figure uses V_{DS} as the x-axis, rather than V_{GS}. Therefore, the shapes of the substrate current differ from the I_{sub} in Fig. 2-3. There are two important points inferable from Fig. 2-5. First, the gate current is nearly zero, independent of bias. The neglect of the gate leakage current in BSIM3 is well justified. (Conceivably, in more advanced technologies, such as those with very thin

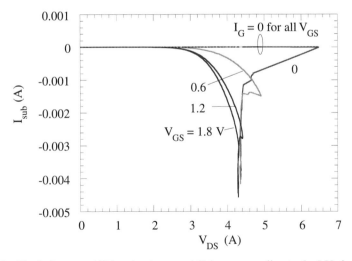

Fig. 2-5 The bulk current (I_B) and gate current (I_G) corresponding to the I-V characteristics of Fig. 2-4.

110 BASIC FACTS ABOUT BSIM3

oxide, gate leakage current through tunneling eventually becomes important.) Secondly, the substrate current can be appreciable. It is at times exceeding 20% of the drain current plotted in Fig. 2-4.

We mentioned that BSIM3 models the substrate current well, at least prior to the parasitic bipolar transistor turns on. However, there is a slight twist. Although BSIM3v3.1 (or prior versions) computes I_{sub}, the calculated result is never outputted to the SPICE simulator. Effectively, the I_{sub} model in BSIM3v3.1 or earlier is absent. This mistake is corrected in BSIM3v3.2. An example where this improper modeling of I_{sub} can lead to wrong circuit design is shown in Fig. 2-6. This is a common charge pump circuit used to charge or discharge the load capacitor of a phase-lock loop, depending on whether the PMOS or the NMOS switch is on. The transistor M1 is intended to set up a current mirror. By design M1 has negligible I_{sub} because its V_{DS} is always small. V_{DS} in M2 is also small, at about 0.5 V when the NMOS switch (M3) is on. Similar to the case of M1, the small V_{DS} means I_{sub} is small in M2. Therefore, the reference current (I_{ref}) is accurately reflected in the drain of M2. The

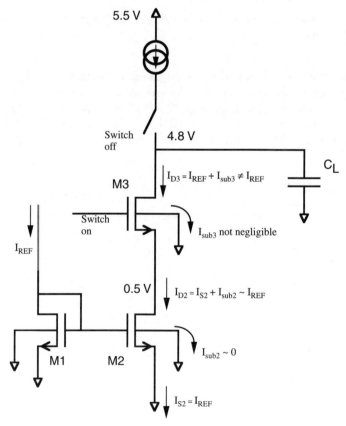

Fig. 2-6 A charge pump circuit for a phase-lock loop. The proper functioning of this circuit depends on the accuracy of the substrate current model.

NMOS switch M3, in contrast, sustains a large V_{DS} drop. The substrate current is a noticeable portion of the drain current. Because the part of its drain current is diverted to substrate before it shows up in the source, its source current is only some fraction of the drain current. The mirrored current at the drain of M3 is therefore larger than the designed I_{ref}. A current mirror that does not mirror current is certainly one of the designer's worst nightmares.

One solution is to lengthen the channel length of M3 (provided that the longer-channel transistor still meets the speed requirement). In long-channel devices, the electric field is smaller, hence the amount of impact ionization is smaller. Figure 2-7 illustrates the measured substrate current for $L = 0.8$, 1, and 4 μm devices at $V_{DS} = 4.5$ V.

After stressing the importance of modeling I_{sub}, we regress to BSIM3 dc equivalent circuit. The last items on Fig. 2-2 are the source and drain parasitic resistances. BSIM3 offers not just one, but two independent ways to account for the parasitic resistances, as schematically illustrated in Fig. 2-8. (It is possible to combine these two independent ways of expressing the resistance. However, often only one of the two methods is used.) For the present discussion, let us suppose, as generally true in practice also, that the drain/source parasitic resistances are constant, independent of the operating biases or current level. In this scenario, the left-hand-side representation is the accurate approach, in which the parasitic resistances are added externally to the transistor as lumped elements. The SPICE simulator treats the overall MOSFET to consist of one intrinsic transistor and two additional resistor elements. Accordingly, the SPICE simulator will solve iteratively for the nodal voltages at the points Source', and Drain', in addition to the Source, Gate and Drain nodes. The drain current flowing in the device, $I_{D,1}$, and the device's

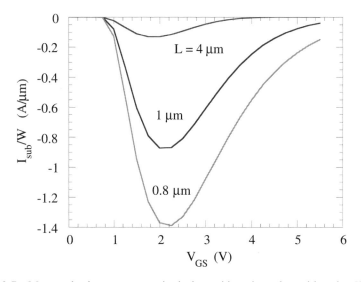

Fig. 2-7 Measured substrate currents in devices with various channel lengths. $V_{DS} = 4.5$ V.

112 BASIC FACTS ABOUT BSIM3

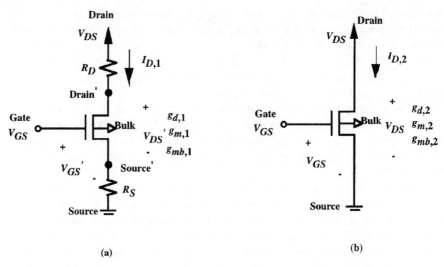

Fig. 2-8 Two BSIM3 approaches to model the parasitic source and drain resistances: (a) lumped-resistance approach; (b) absorbed-resistance approach.

intrinsic transconductances, are direct functions of the terminal voltages of the intrinsic device:

$$I_{D,1} = f_1(V'_{GS}, V'_{DS}) \qquad g_{d,1} = \partial I_{D,1}/\partial V'_{DS}$$
$$g_{m,1} = \partial I_{D,1}/\partial V'_{GS} \qquad g_{mb,1} = \partial I_{D,1}/\partial V'_{BS}. \qquad (2\text{-}6)$$

This *lumped-resistance approach*, while being conceptually simple and, in fact, more accurate than the to-be-discussed second approach, has some drawbacks. First, there are two more nodal voltages which need to be solved for each transistor; namely, the Source' and the Drain' nodes. This translates to lengthened computation time in a SPICE simulation. For simple circuit with few transistors, however, we believe this extra amount of time needed for numerical solution is trivial. Second, generally SPICE parameters are extracted by matching the measured and simulated values of the drain current and the mutual transconductance. In a measurement setting, the measured current is that at the Drain node, and the measured g_m, for example, is equal to $\partial I_D/\partial V_{GS}$ (still referring to Fig. 2-8a). The calculated drain current, which is the current at the Drain' node, will be the same current that flows through the Drain node. Therefore, extracting parameters by fitting the simulated current at the Drain' node against the measured current at the Drain node makes sense. However, the calculated g_m from BSIM3 is really $\partial I'_D/\partial V'_{GS}$, which often differs significantly from measured g_m equaling to $\partial I_D/\partial V_{GS}$. Hence, unless a numerically intensive SPICE subroutines are incorporated to calculate the derivatives at the Drain' and Source' nodes, SPICE parameters should not be extracted by fitting the measured and simulated g'_ms.

Some formula are proposed to correct this problem. With a certain small-signal approximation [2], it is possible to relate the measured conductances to intrinsic conductances:

$$g_d \approx \frac{g_{d,i}}{1 + g_{m,i}R_S + g_{d,i}(R_S + R_D) + g_{mb,i}R_S}; \quad (2\text{-}7a)$$

$$g_m \approx \frac{g_{m,i}}{1 + g_{m,i}R_S + g_{d,i}(R_S + R_D) + g_{mb,i}R_S}; \quad (2\text{-}7b)$$

$$g_{mb} \approx \frac{g_{mb,i}}{1 + g_{m,i}R_S + g_{d,i}(R_S + R_D) + g_{mb,i}R_S}; \quad (2\text{-}7c)$$

where, for example, g_d is $\partial I_D/V_{DS}$, the measured drain conductance, and $g_{d,i}$ is the intrinsic device conductance calculated in BSIM3, equal to $\partial I_D/V'_{DS}$. After the application of Eq. (2-7b), it is possible to extract parameters by fitting BSIM3-generated $g_{d,i}$ to the measured g_d. We caution that, although the approximation of g_d given in Eq. (2-7b) is fairly good, using Eq. (2-7a) to relate g_m and $g_{m,i}$ can run into problems unless the current is fairly low. The equations expressed in Eq. (2-7), after all, are derived with the small-signal assumption, which can sometimes fail at normal operating current levels.

BSIM3 offers another approach to model the parasitic source/drain resistances, basically by absorbing the parasitic resistances' effects on current and conductances into the intrinsic device. The equivalent circuit is shown in the right-hand-side of Fig. 2-8. In this *absorbed-resistance approach*, the nodes Drain' and Source' disappear. The drain current and mutual transconductance are calculated from a totally different set of equations:

$$\begin{array}{ll} I_{D,2} = f_2(V_{GS}, V_{DS}, R_D + R_S) & g_{d,2} = \partial I_{D,2}/\partial V_{DS} \\ g_{m,2} = \partial I_{D,2}/\partial V_{GS} & g_{mb,2} = \partial I_{D,2}/\partial V_{BS}. \end{array} \quad (2\text{-}8)$$

We use f_2 to describe the functional form of $I_{D,2}$, in order to stress that the function differs from f_1 used in the calculation of $I_{D,1}$. Basically, BSIM3 devises some sort of function f_2, which is a modified version of f_1 to account for the effects of the parasitic resistances. The amount of deviation from the original f_1 depends on the magnitude of $R_D + R_S$. The larger the value of $R_D + R_S$, the more significant the deviation. The modification is made in ways such that somehow $I_{D,2}$ is approximately equal to $I_{D,1}$ and more importantly, $g_{d,2}$, $g_{m,2}$, and $g_{mb,2}$ are approximately equal to $g_{d,1}$, $g_{m,1}$, $g_{mb,1}$, respectively. Note the keyword above is *approximately*. No matter how clever the BSIM3 modification may be, there is no way to replace the SPICE simulator's mathematical calculations of the two additional nodal voltages if the exact solution is desired. The approximation is most accurate when the drain current is small such that the voltage drop across R_S is small compared to the applied V_{GS}. However, despite some claims to the contrary, it is believed that the error can be quite high (exceeding 10% at least) when the current flow is large. Setting the issues of dc accuracy aside, the absorbed-resistance approach is also discouraged for rf applications. When the second approach is used, the input resistance of the

114 BASIC FACTS ABOUT BSIM3

MOSFET will be purely imaginary, without seeing the effect of R_S. This is because BSIM3 makes an attempt to equilibrate only the drain current and the conductances at the two sides of Fig. 2-8. The fact that the device y-parameters of the overall device are greatly altered when two extra nodes are added is not considered.

Here are some details of the actual BSIM3 implementation. When the lumped-resistance approach is used, R_S and R_D are specified with the SPICE parameter RSH (sheet resistance of the source and drain contact), together with the fields NRD (number of squares in the drain contact) and NRS (number of squares in the source contact) found in the device statement. They are given by

$$R_S = \text{RSH} \times \text{NRS}; \quad R_D = \text{RSH} \times \text{NRD}. \tag{2-9}$$

RDSW, to be discussed shortly, should be zero in this lumped-resistance approach.

In the absorbed-resistance approach, we see that the function f_2 in Eq. (2-8) depends on the sum of R_S and R_D, rather than each component individually. From this observation alone, we can infer that BSIM3 uses a completely new parameter for this approach, independent from the aforementioned RSH, NRS and NRD. It is RDSW, the sum of source and drain resistances per unit width. When this approach is used, RSH should be set to zero, and RDSW can be approximated from the measured R_S and R_D:

$$R_S + R_D \approx \frac{\text{RDSW}}{W}, \tag{2-10}$$

where W is the device's width. (The actual relationship between the parasitic resistances and RDSW includes other parameters such as WR, PRWG and PRWB to account for secondary effects. See the RDSW entry in Section 3.2 for more details. Equation (2-10) is valid when WR, PRWG, and PRWB assume their default values of 1, 0, and 0, respectively.) For a rf circuit designer who would like to see nonzero real part input resistance due to the source resistance, the lumped-resistance approach is preferred and RDSW should be zero.

Let us examine a typical digital $W/L = 5/0.35\,\mu\text{m}$ MOS transistor which has a parasitic drain resistance of 32.4 Ω and a parasitic source resistance of 32.4 Ω. In the lumped-resistance approach, the SPICE model card's RDSW parameter would be zero. RSH in the model card is nonzero, and can be conveniently set to 1 (Ω/sq). In the MOS device statement, NRD and NRS would be equal to 32.4. Alternatively, if the absorbed-resistance approach is used, RSH is set to zero (and the exact values of NRD and NRS become irrelevant). RDSW is set to be roughly equal to $(32.4 + 32.4) \times 5 = 324$ (Ω-μm). The proper unit used for RDSW in BSIM3 is Ω-μm, not Ω-m, although the majority of other parameters use MKS unit system.

Figure 2-9 illustrates the I-V characteristics simulated with the lumped- (RDSW $= 0$) and absorbed- (RDSW $= 324$) resistance approaches. When the current level is low, the two approaches yield identical I-V characteristics. However, the difference in the models becomes significant at high current levels. This difference is unavoidable, since by nature the absorbed-resistance approach is an approximate alternative to the lumped-resistance approach. Figure 2-10 shows the small-signal

2.2 DC EQUIVALENT CIRCUIT MODEL

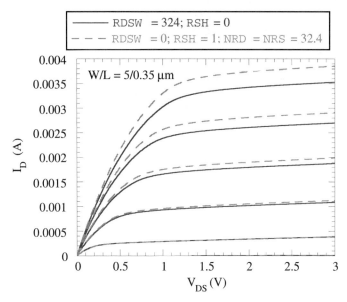

Fig. 2-9 I-V characteristics simulated with the lumped- (RDSW = 0) and absorbed- (RDSW = 324) resistance approaches.

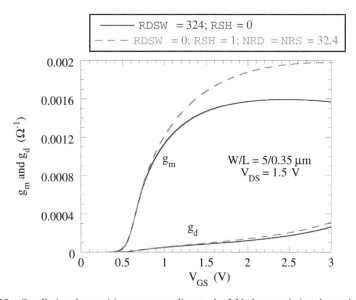

Fig. 2-10 Small-signal quantities corresponding to the I-V characteristics shown in Fig. 2-9.

116   BASIC FACTS ABOUT BSIM3

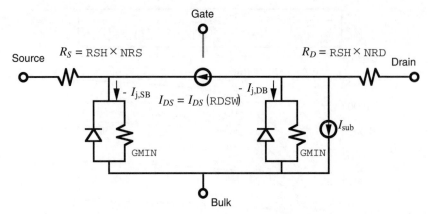

Fig. 2-11   Revised BSIM3 dc equivalent circuit based on Fig. 2-2. The accommodation of the source/drain resistances, with either the lumped- or the absorbed-resistance approach, is specifically shown.

quantities produced by the two approaches. Again, they are quite similar when the bias level is small, but becomes increasingly different at larger bias levels.

BSIM3 allows the mixing of these two approaches. Again for the $R_S = R_D = 32.4\ \Omega$ case, we could set RSH to be 1, with NRD = NRS = 30. The remaining unaccounted 2.4 Ω can be taken up by RDSW, which would then be roughly equal to 24 Ω-μm. One can see that the number of combinations to achieve the desired resistance is infinite.

The dc equivalent model is now redrawn as Fig. 2-11, to emphasize the different modeling approaches in BSIM3 for the source/drain resistances.

2.3   BSIM3'S y-PARAMETERS

Before proceeding to the discussion of BSIM3's large-signal and small-signal models used for transient and ac analyses, it is instructive to examine the y-parameters of BSIM3. These y-parameters include contribution from the parasitic capacitances, as well as those elements associated with the bulk-drain and bulk-source diodes.

There are several parasitic capacitances: drain-bulk junction capacitance, $C_{j,\mathrm{DB}}$; source-bulk junction capacitance, $C_{j,\mathrm{SB}}$; gate-source overlap capacitance (taken to be identical to the gate-drain capacitance), C_{ov}; gate-source fringing capacitance (taken to be identical to the gate-drain fringing capacitance), C_f; and the gate-bulk wiring capacitance, $C_{gb,0}$. Unlike the intrinsic device capacitances, all of these parasitic capacitances are determined exclusively by the terminal voltage appearing between the two ends of the capacitor. These parasitic capacitances are reciprocal. Let us denote $C_{gs,p}$ as the parasitic gate-source capacitance, defined as $\partial Q_{G,p}/\partial V_S$ where $Q_{G,p}$ is the portion of the gate charge associated with the parasitic capacitance. The reciprocal nature of the parasitic capacitance means that $C_{gs,p}$ is the same as $C_{sg,p}$, which is $\partial Q_{S,p}/\partial V_G$. We write all of the parasitic capacitances explicitly:

$$C_{gd,p} = C_{dg,p} = C_{ov} + C_f. \tag{2-11a}$$

$$C_{gs,p} = C_{sg,p} = C_{ov} + C_f. \tag{2-11b}$$

$$C_{bd,p} = C_{db,p} = C_{j,\text{DB}}. \tag{2-11c}$$

$$C_{bs,p} = C_{sb,p} = C_{j,\text{SB}}. \tag{2-11d}$$

$$C_{gb,p} = C_{bg,p} = C_{gb,0}. \tag{2-11e}$$

C_{ov}, C_f, $C_{j,\text{DB}}$, $C_{j,\text{SB}}$, and $C_{gb,0}$ are the parasitic capacitances modeled in BSIM3. The locations of these parasitic capacitances are shown in Fig. 2-12. Specifically, C_{ov}, the overlap capacitance, is the capacitance associated with the overlapped area between the gate terminal and the source junction (or the drain junction). When an

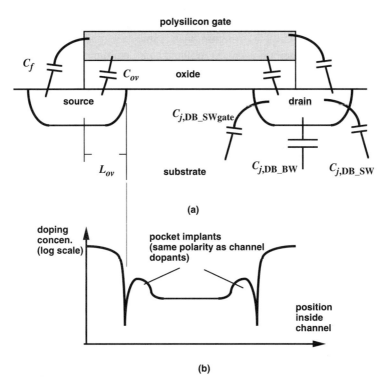

Fig. 2-12 (a) Locations of the parasitic capacitances modeled in BSIM3; (b) doping profile in a transistor with pocket implants.

inversion channel exists, the overlap capacitance is calculated by the standard formula $C_{ov} = \epsilon_{ox}/\text{TOX} \cdot W \cdot L_{ov}$, where W is the width and L_{ov} is the overlap distance marked on Fig. 2-12, and TOX is the oxide thickness. However, when V_{GS} is negative and the channel is not yet inverted, the gate-to-source overlap capacitance gradually decreases from the value of $\epsilon_{ox}/\text{TOX} \cdot W \cdot L_{ov}$ as V_{GS} becomes increasingly more negative. This is because, even though the source is heavily doped, the doping level is finite. As V_{GS} becomes more negative, a thin layer at the top of the source, right next to the oxide, becomes depleted. As the distance separating the gate metal and the undepleted region of the source widens, the capacitance decreases.

The overlap capacitance is not the only parasitic capacitance between the gate and the source. C_f is used to model the capacitance associated with the fringing field at the end of the overlap area. Several expressions of C_f have been derived based on conformal mapping techniques, and will be discussed in Section 3.2 (under the entry of CF). From a measurement perspective, there is no straightforward method to isolate C_f and C_{ov}. One tends to measure the parasitic capacitances as a whole. In BSIM3, C_{ov} is determined by the parameter CGD0, CGD1, and CKAPPA for the drain side, and CGS0, CGS1, and CKAPPA for the source side. (Although for simplicity we employ the same C_{ov} notation for the overlap capacitances at the drain and the source side, BSIM3 allows them to differ.) C_f is modeled by a single parameter CF.

The junction capacitance in BSIM3 is broken down into three components. Let us take the drain side as an example. From Fig. 2-12a, we see the $C_{j,\text{DB}}$ calculated in BSIM3 is a sum of $C_{j,\text{DB\_BW}}$ (bottom-wall capacitance), $C_{j,\text{DB\_SW}}$ (side wall capacitance for the sidewall away from the gate, i.e., the isolation side), and $C_{j,\text{DB\_SWgate}}$ (side wall capacitance adjacent to the gate). The classification into three components is necessary because the substrate doping concentrations at the bottom wall, isolation-side sidewall, and gate-side sidewall are fairly different. BSIM3 calculates $C_{j,\text{DB\_BW}}$ with the parameters CJ, MJ, and PB. $C_{j,\text{DB\_SW}}$ is calculated with CJSW, MJSW, and PBSW. $C_{j,\text{DB\_SWgate}}$ is calculated with CJSWG, MJSWG, and PBSWG. All these three capacitance components are calculated with similar equations. We write out $C_{j,\text{DB\_BW}}$ as an example:

$$C_{j,\text{DB\_BW}} = \text{Bottom\_Wall Area} \times \frac{\text{CJ}}{\left(1 - \frac{V_{BD}}{\text{PB}}\right)^{\text{MJ}}}. \qquad (2\text{-}12)$$

For the side-wall capacitances $C_{j,\text{DB\_SW}}$ and $C_{j,\text{DB\_SWgate}}$, the appropriate sidewall perimeter is used instead of the bottom-wall area, as well as the proper replacement of CJ, PB, and MJ. More details about the BSIM3 equations are found in Section 3.2, under the entry of CJ.

In the past technologies, there is not an acute need to distinguish between $C_{j,\text{DB\_SW}}$ and $C_{j,\text{DB\_SWgate}}$. The substrate doping concentrations at the sidewalls, either adjacent or away from the gate side, were similar. However, modern CMOS technologies often place pocket implants near the gate side and use shallow trench isolation at the isolation side. The pocket implant is a way to increase the effective

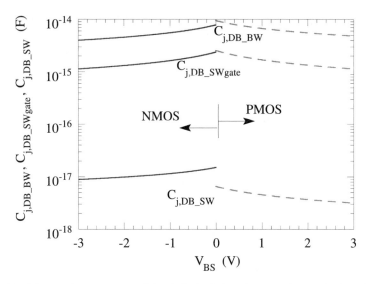

Fig. 2-13 Measured components of the drain–bulk junction capacitances in the NMOS and PMOS of a $L = 0.25\,\mu\text{m}$ technology.

channel doping in short-channel devices, used to mitigate the short-channel V_T-rolloff. The doping profile in a transistor with pocket implants is shown in Fig. 2-12b. Because of the extra doping, the sidewall capacitance adjacent to the gate is much larger than the sidewall capacitance away from the gate. Figure 2-13 are the measured components of the drain-bulk junction capacitances of both the NMOS and PMOS of a $L = 0.25\,\mu\text{m}$ technology. Reverse-biasing the junctions means applying negative V_{BS} in the NMOS and positive V_{BS} in the PMOS. Hence, the NMOS curves are in the negative axis, while the PMOS curves are in the positive axis. The figure demonstrates that the bottom-wall capacitance is the largest component, with the gate-sidewall capacitance being a close second. Because of the use of trench isolation, the sidewall capacitance (at the opposite end of the gate side) is negligible. To reduce the junction capacitance, it is imperative to minimize the size of bottom wall through lithographic advances (so that $C_{j,\text{DB\_BW}}$ is reduced), and to decrease the depth of the drain junction through the use of rapid thermal annealing (so that $C_{j,\text{DB\_SWgate}}$ is reduced).

There are other unmodeled parasitic capacitances, which, depending on one's perspective, can be regarded as significant. An example is the so-called inner fringing capacitance. The C_f above, as illustrated in Fig. 2-12, is the fringing capacitance between the gate and source metal through the external dielectric. The inner fringing capacitance, in contrast, is through the channel region. It can be significant when the channel is yet inverted. Other parasitic capacitances include $C_{ds,p}$, the capacitance between the drain and the source contacts. This capacitance becomes progressively more important as the channel length decreases. These capacitances not accounted for in BSIM3 will be discussed in more detail in Section 4.10.

The sixteen device capacitances modeled in BSIM3, including the parasitic capacitances, are given in the following:

$$C_{gg,t} = C_{gg} + C_{gd,p} + C_{gs,p} + C_{gb,p} = C_{gg} + 2(C_{ov} + C_f) + C_{gb,0}. \quad (2\text{-}13\text{a})$$

$$C_{gd,t} = C_{gd} + C_{gd,p} = C_{gd} + (C_{ov} + C_f). \quad (2\text{-}13\text{b})$$

$$C_{gs,t} = C_{gs} + C_{gs,p} = C_{gs} + (C_{ov} + C_f). \quad (2\text{-}13\text{c})$$

$$C_{gb,t} = C_{gb} + C_{gb,p} = C_{gb} + C_{gb,0}. \quad (2.13\text{d})$$

$$C_{dg,t} = C_{dg} + C_{dg,p} = C_{dg} + (C_{ov} + C_f). \quad (2.13\text{e})$$

$$C_{dd,t} = C_{dd} + C_{dg,p} + C_{db,p} = C_{dd} + (C_{ov} + C_f) + C_{j,\text{DB}}. \quad (2\text{-}13\text{f})$$

$$C_{ds,t} = C_{ds}. \quad (2\text{-}13\text{g})$$

$$C_{db,t} = C_{db} + C_{db,p} = C_{db} + C_{j,\text{DB}}. \quad (2\text{-}13\text{h})$$

$$C_{sg,t} = C_{sg} + C_{sg,p} = C_{sg} + (C_{ov} + C_f). \quad (2\text{-}13\text{i})$$

$$C_{sd,t} = C_{sd}. \quad (2\text{-}13\text{j})$$

$$C_{ss,t} = C_{ss} + C_{sg,p} + C_{sb,p} = C_{ss} + (C_{ov} + C_f) + C_{j,\text{SB}}. \quad (2\text{-}13\text{k})$$

$$C_{sb,t} = C_{sb} + C_{sb,p} = C_{sb} + C_{j,\text{SB}}. \quad (2\text{-}13\text{l})$$

$$C_{bg,t} = C_{bg} + C_{bg,p} = C_{bg} + C_{gb,0}. \quad (2\text{-}13\text{m})$$

$$C_{bd,t} = C_{bd} + C_{bd,p} = C_{bd} + C_{j,\text{DB}}. \quad (2\text{-}13\text{n})$$

$$C_{bs,t} = C_{bs} + C_{bs,p} = C_{bs} + C_{j,\text{SB}}. \quad (2\text{-}13\text{o})$$

$$C_{bb,t} = C_{bb} + C_{bd,p} + C_{bs,p} + C_{bg,p} = C_{bb} + C_{j,\text{DB}} + C_{j,\text{SB}} + C_{bg,0}. \quad (2\text{-}13\text{p})$$

MOSFET, being a four-terminal device, can be characterized with sixteen y-parameters. We discussed the y-parameters of an *intrinsic* MOFETS in Section 1.9. The parasitic capacitances were not included. Further, the effects of the substrate current and the source-bulk and drain-bulk diodes on the y-parameters were not discussed. Here, we present the y-parameters used in the BSIM3 model. They are similar to those given in Section 1.9, but with modifications to reflect the incorporation of the parasitic elements:

$$y_{gg} = j\omega C_{gg,t}. \quad (2\text{-}14\text{a})$$

$$y_{gs} = -j\omega C_{gs,t}. \quad (2\text{-}14\text{b})$$

$$y_{gd} = -j\omega C_{gd,t}. \quad (2\text{-}14\text{c})$$

$$y_{gb} = -j\omega C_{gb,t}. \quad (2\text{-}14\text{d})$$

$$y_{dg} = g_m - j\omega C_{dg,t} + G_{bg}. \quad (2\text{-}14\text{e})$$

$$y_{dd} = g_d + j\omega C_{dd,t} + g_{j,\text{BD}} + G_{bd}. \quad (2\text{-}14\text{f})$$

$$y_{ds} = -g_d - g_m - g_{mb} - j\omega C_{ds,t} - G_{bg} - G_{bd} - G_{bb}. \tag{2-14g}$$

$$y_{db} = g_{mb} - j\omega C_{db,t} - g_{j,BD} + G_{bb}. \tag{2-14h}$$

$$y_{sg} = -g_m - j\omega C_{sg,t}. \tag{2-14i}$$

$$y_{sd} = -g_d - j\omega C_{sd,t}. \tag{2-14j}$$

$$y_{ss} = g_d + g_m + g_{mb} + j\omega C_{ss,t} + g_{j,SB}. \tag{2-14k}$$

$$y_{sb} = -g_{mb} - j\omega C_{ss,t} - g_{j,SB}. \tag{2-14l}$$

$$y_{bg} = -j\omega C_{bg,t} - G_{bg}. \tag{2-14m}$$

$$y_{bd} = -j\omega C_{bd,t} - g_{j,BD} - G_{bd}. \tag{2-14n}$$

$$y_{bs} = -j\omega C_{bs,t} - g_{j,BS} + G_{bg} + G_{bd} + G_{bb}. \tag{2-14o}$$

$$y_{bb} = j\omega C_{bb,t} + g_{j,BS} + g_{j,BD} - G_{bb}, \tag{2-14p}$$

where,

$$G_{bg} = \frac{\partial I_{sub}}{\partial V_{GS}}; \quad G_{bd} = \frac{\partial I_{sub}}{\partial V_{DS}}; \quad G_{bb} = \frac{\partial I_{sub}}{\partial V_{BS}}; \tag{2-15}$$

$$g_{j,DB} = \frac{\partial I_{j,DB}}{\partial V_{BD}}; \quad g_{j,SB} = \frac{\partial I_{j,SB}}{\partial V_{BS}}. \tag{2-16}$$

$I_{j,DB}$ given in Eq. (1-3) includes a current flow through GMIN. Hence, the differential drain-bulk diode conductance, $g_{j,DB}$, contains a CMIN component, in addition to a component associated with the exponential increase of the diode current. Similar statement applies to the differential source-bulk diode conductance, $g_{j,SB}$.

As V_{DS} increases, I_{sub} increases. Hence, $G_{bd} > 0$. The sign of G_{bg} depends on the value of V_{GS}, as shown by the bell curve of Fig. 2-3. Hence, G_{bg} can be positive or negative. A similar situation exists for V_{BS}. As V_{BS} increases (becoming less negative), the threshold voltage decreases. The amount of impact ionization can increase because there are more channel carriers to initiate the impact ionization. On the other hand, the decrease of V_T can also cause the impact ionization to decrease because the device moves away from saturation to linear regions. Therefore, G_{bb} can be either positive or negative. The differential diode conductances, $g_{j,DB}$ and $g_{j,SB}$, are always positive.

The textbook-like expressions for the intrinsic device capacitances for all bias conditions are found in Appendix B. The BSIM3 equations, as well as the detailed implementations of the parasitic capacitances, are found in Appendix A.

Only nine of the sixteen y-parameters are linearly independent. Once a set of nine linearly independent y-parameters is specified, the other seven y-parameters can be established from Eq. (1-87). For simplicity, the nine linearly independent y-parameters can be chosen to be those y-parameters in the common-source configuration: y_{gg}, y_{gd}, y_{gb}, y_{dg}, y_{dd}, y_{db}, y_{bg}, y_{bd}, and y_{bb}. Alternatively, the y-parameters in the common-bulk configuration can be used: y_{gg}, y_{gd}, y_{gs}, y_{dg}, y_{dd}, y_{ds}, y_{sg}, y_{sd}, and y_{ss}. In the numerical analysis routine, BSIM3 uses common-bulk y-parameters to solve iteratively the transient solution. We stress that, this fact has nothing to do with whether BSIM3 is a source-referenced or a bulk-referenced model. As mentioned in Section 1.10, the moment BSIM3 writes its drain current equation in a form similar to Eq. (1-5), BSIM3 is a source-referenced model. After all, the common-bulk y-parameters can be obtained from the common-source y-parameters by comparing Eqs. (1-86) and (1-91) and then using Eq. (1-87).

These 16 y-parameters discussed above are for a four-terminal MOSFET. If Fig. 2-8b is used to model the parasitic source/drain resistances, then the y-parameters are those for the Drain, Source, Gate and Bulk terminals, as expected. If Fig. 2-8a is instead used, then the y-parameters listed above apply to Drain′, Source′, Gate and Bulk terminals. To obtain y-parameters for the overall transistor at the Drain, Source, Gate, and Bulk terminals, we need to first transform the 4-port y-parameters to equivalent 4-port z-parameters. We then add the source and drain resistances to the z-parameter entries related to the source and drain. Finally, these new z-parameters are transformed back to y-parameters. This process is straightforward conceptually, but is quite algebraically intensive [3].

2.4 LARGE-SIGNAL EQUIVALENT CIRCUIT

We briefly revisit our notation for current and voltage. A dc quantity, such as I_D, is expressed with all letters capitalized. A small-signal quantity is expressed in small letters, such as i_d. Their sum represents the instantaneous total value, and is given by a small letter with a capitalized subscript. Therefore, i_D is an instantaneous value.

In Section 1.5 where we discussed the meaning of quasi-static analysis, we related the terminal currents to the terminal charges as

$$i_{D,\text{FET}}(t) = I_{D,\text{FET}} + \frac{dQ_{D,\text{FET}}(t)}{dt}; \qquad (2\text{-}17\text{a})$$

$$i_{S,\text{FET}}(t) = -I_{D,\text{FET}} + \frac{dQ_{S,\text{FET}}(t)}{dt}; \qquad (2\text{-}17\text{b})$$

$$i_{G,\text{FET}}(t) = \frac{dQ_{G,\text{FET}}(t)}{dt}; \qquad (2\text{-}17\text{c})$$

$$i_{B,\text{FET}}(t) = \frac{dQ_{B,\text{FET}}(t)}{dt}. \qquad (2\text{-}17\text{d})$$

We have added the subscript FET to emphasize that these charges, derived from solving the charge distribution in the inverted channel, are the terminal charges associated with the MOSFET transistor action (as opposed to the parasitic diodes, for example, to be discussed shortly).

BSIM3, as a practical model usable for industrial applications, includes current conduction outside the intrinsic transistor. As discussed in the previous section, BSIM3 considers the drain–bulk and source–bulk diode currents, as well as the substrate current due to impact ionization. Let us for convenience consider a semiconductor diode whose p-side is labeled as the bulk and the n-side is labeled as the drain. In dc, the diode current flowing from the positive terminal (p-side) to the negative terminal (n-side) is $I_{j,\text{DB}}$, as shown in Eq. (2-3).

During a transient situation (but still under the quasi-static assumption), we state without proof that the diode transient current entering the drain terminal is

$$i_{D,\text{diode}}(t) = -I_{j,\text{DB}}(v_{BD}) + C_{j,\text{DB}}(v_{BD})\frac{dv_{DB}}{dt} + C_{d,\text{DB}}(v_{BD})\frac{dv_{DB}}{dt} \quad \text{(not BSIM3).} \tag{2-18}$$

The current flowing through the bulk terminal is exactly the negative of Eq. (2-18). Hence, the drain–bulk diode's contribution to the bulk current is

$$i_{B,\text{diode}}(t) = I_{j,\text{DB}}(v_{BD}) + C_{j,\text{DB}}(v_{BD})\frac{dv_{BD}}{dt} + C_{d,\text{DB}}(v_{BD})\frac{dv_{BD}}{dt} \quad \text{(not BSIM3).} \tag{2-19}$$

In these expressions, $C_{j,\text{DB}}$ is the junction capacitance in the drain–bulk junction, and $C_{d,\text{DB}}$ is the diffusion capacitance in the junction. The junction capacitance is the capacitance due solely to the fact that there is a depletion region separating the two terminals (quasi-neutral regions) of the diode. The diffusion capacitance accounts for the capacitance due to minority carriers in the quasi-neutral regions of the diode. Normally the drain-bulk diode in the MOSFET is reverse-biased to ensure that the transistor current is dominated by that associated with the inverted channel. Under this bias condition, the minority carrier concentration is small and $C_{d,\text{DB}}$ is well approximated by zero. Therefore, the transient model of a diode adopted in BSIM3 is that shown in Fig. 2-14, which leaves out the diffusion capacitance. In order to express Eq. (2-18) in a form consistent with charge-based model, BSIM3 rewrites it as

$$i_{D,\text{diode}}(t) = -I_{j,\text{DB}} - \frac{dQ_{j,\text{DB}}}{dt} \quad \text{(BSIM3),} \tag{2-20}$$

where

$$Q_{j,\text{DB}}(t) = Q_{j,\text{DB}}(v_{BD}(t)) = \int_0^{v_{BD}} C_{j,\text{DB}}(v'_{BD})dv'_{BD}. \tag{2-21}$$

In BSIM3, $C_{j,\text{DB}}$ is determined from the parameters CJ, CJSW, CJSWG, MJ, MJSW, MJSWG, PB, PBSW, PBSWG, (Section 3.2, under the entry of CJ). Once

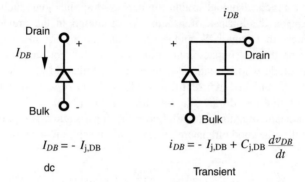

Fig. 2-14 Drain–bulk diode model adopted in BSIM3. The source–bulk diode model is similar.

$C_{j,DB}$'s functional dependence on v_{BD} is known, $Q_{j,DB}$ used in BSIM3 is evaluated from Eq. (2-21).

If forward-biasing the diodes of the MOSFET is somehow a crucial part of the circuit operation, here is a warning. Although $I_{j,DB}$ and the junction capacitance used in BSIM3 are somewhat correctly modeled, the diffusion capacitance is not included at all. During a transient situation, the transient diode current will not be exactly correct if the diode is forward-biased. It is better that the diode areas in the transistor are declared zero. The junctions should instead be represented with semiconductor diodes, declared external to the device. The semiconductor diode model, unlike the BSIM3's diode model, includes the diffusion capacitance.

The accounting of the diffusion capacitance is a difference between BSIM3 diode model and the typical semiconductor diode in SPICE. There is another difference, perhaps more significant. In a semiconductor diode model, the breakdown voltage in the reverse bias condition is an important parameter. However, to improve code efficiency, BSIM3 basically assumes the drain–bulk and source–bulk diodes never break down. Therefore, even if an outrageous drain-bulk bias of $V_{DB} = -1000$ V is mistakenly applied to the MOSFET, BSIM3 will signal no problem. BSIM3 will proceed with the circuit simulation, although the voltage's magnitude far exceeds the diode's reverse breakdown voltage and the circuit would not have functioned.

We now consider BSIM3's modeling on I_{sub}, the substrate current due to impact ionization. Unlike the diode current, the physical mechanism of this current does not involve the charging or discharging of a capacitor associated with a junction depletion layer. There is no minority charge storage, either. Therefore, even in the transient situation, the drain and bulk currents due to the impact ionization current (denoted as I_{sub}) are simply:

$$I_{D,sub}(t) = I_{sub}(v_{GS}(t),\ v_{DS}(t),\ v_{BS}(t)); \quad (2\text{-}22\text{a})$$

$$I_{B,sub}(t) = -I_{sub}(v_{GS}(t),\ v_{DS}(t),\ v_{BS}(t)). \quad (2\text{-}22\text{b})$$

2.4 LARGE-SIGNAL EQUIVALENT CIRCUIT

When Eqs. (2-20) and (2-22) are combined with Eq. (2-17), the instantaneous terminal currents in BSIM3 are given by

$$i_G(t) = \frac{dQ_{G,\text{FET}}(t)}{dt};\tag{2-23a}$$

$$i_D(t) = I_{DS} + I_{sub} - I_{j,\text{DB}} + \frac{dQ_{D,\text{FET}}(t)}{dt} - \frac{dQ_{j,\text{DB}}(t)}{dt};\tag{2-23b}$$

$$i_S(t) = -I_{DS} - I_{j,\text{SB}} + \frac{dQ_{S,\text{FET}}(t)}{dt} - \frac{dQ_{j,\text{SB}}(t)}{dt};\tag{2-23c}$$

$$i_B(t) = -I_{sub} + I_{j,\text{SB}} + I_{j,\text{DB}} + \frac{dQ_{B,\text{FET}}(t)}{dt} + \frac{dQ_{j,\text{SB}}(t)}{dt} + \frac{dQ_{j,\text{DB}}(t)}{dt};\tag{2-23d}$$

where we have changed the notation of $I_{D,\text{FET}}$ to I_{DS}, the drain-to-source current in the channel associated with the intrinsic MOS. We have consistently followed the convention that the current flowing *into* a terminal is counted as positive.

We can rewrite the above expressions, using the chain rule from calculus (Section 1.5):

$$i_G(t) = \frac{\partial Q_{G,\text{FET}}}{\partial v_G} \cdot \frac{dv_G}{dt} + \frac{\partial Q_{G,\text{FET}}}{\partial v_D} \cdot \frac{dv_D}{dt} + \frac{\partial Q_{G,\text{FET}}}{\partial v_S} \cdot \frac{dv_S}{dt} + \frac{\partial Q_{G,\text{FET}}}{\partial v_B} \cdot \frac{dv_B}{dt};\tag{2-24a}$$

$$i_D(t) = I_{DS} + I_{sub} - I_{j,\text{DB}} + \frac{\partial [Q_{D,\text{FET}} - Q_{j,\text{DB}}]}{\partial v_G} \cdot \frac{dv_G}{dt}$$
$$+ \frac{\partial [Q_{D,\text{FET}} - Q_{j,\text{DB}}]}{\partial v_D} \cdot \frac{dv_D}{dt} + \frac{\partial [Q_{D,\text{FET}} - Q_{j,\text{DB}}]}{\partial v_S} \cdot \frac{dv_S}{dt}$$
$$+ \frac{\partial [Q_{D,\text{FET}} - Q_{j,\text{DB}}]}{\partial v_B} \cdot \frac{dv_B}{dt};\tag{2-24b}$$

$$i_S(t) = -I_{DS} - I_{j,\text{DB}} + \frac{\partial [Q_{S,\text{FET}} - Q_{j,\text{SB}}]}{\partial v_G} \cdot \frac{dv_G}{dt}$$
$$+ \frac{\partial [Q_{S,\text{FET}} - Q_{j,\text{SB}}]}{\partial v_D} \cdot \frac{dv_D}{dt} + \frac{\partial [Q_{S,\text{FET}} - Q_{j,\text{SB}}]}{\partial v_S} \cdot \frac{dv_S}{dt}$$
$$+ \frac{\partial [Q_{S,\text{FET}} - Q_{j,\text{SB}}]}{\partial v_B} \cdot \frac{dv_B}{dt};\tag{2-24c}$$

$$i_B(t) = -I_{sub} + I_{j,\text{DB}} + I_{j,\text{SB}} + \frac{\partial [Q_{B,\text{FET}} + Q_{j,\text{DB}} + Q_{j,\text{SB}}]}{\partial v_G} \cdot \frac{dv_G}{dt}$$
$$+ \frac{\partial [Q_{B,\text{FET}} + Q_{j,\text{DB}} + Q_{j,\text{SB}}]}{\partial v_D} \cdot \frac{dv_D}{dt} + \frac{\partial [Q_{B,\text{FET}} + Q_{j,\text{DB}} + Q_{j,\text{SB}}]}{\partial v_S} \cdot \frac{dv_S}{dt}$$
$$+ \frac{\partial [Q_{B,\text{FET}} + Q_{j,\text{DB}} + Q_{j,\text{SB}}]}{\partial v_B} \cdot \frac{dv_B}{dt}.\tag{2-24d}$$

If we choose the source as the reference node and express the terminal voltages in terms of the bias differences between the terminal voltage and the source voltage, then we have $dv_S/dt = 0$, $dv_G/dt = dv_{GS}/dt$, $dv_D/dt = dv_{DS}/dt$, and $dv_B/dt = dv_{BS}/dt$. Therefore,

$$i_G(t) = C_{gg,t}(v_{GS}, v_{DS}, v_{BS}) \cdot \frac{dv_{GS}}{dt} - C_{gd,t}(v_{GS}, v_{DS}, v_{BS}) \cdot \frac{dv_{DS}}{dt}$$
$$- C_{gb,t}(v_{GS}, v_{DS}, v_{BS}) \cdot \frac{dv_{BS}}{dt}; \qquad (2\text{-}25\text{a})$$

$$i_D(t) = I_{DS}(v_{GS}, v_{DS}, v_{BS}) + I_{sub}(v_{GS}, v_{DS}, v_{BS}) - I_{j,\text{DB}}(v_{DB})$$
$$- C_{dg,t}(v_{GS}, v_{DS}, v_{BS}) \cdot \frac{dv_{GS}}{dt} + C_{dd,t}(v_{GS}, v_{DS}, v_{BS}) \cdot \frac{dv_{DS}}{dt}$$
$$- C_{db,t}(v_{GS}, v_{DS}, v_{BS}) \cdot \frac{dv_{BS}}{dt}; \qquad (2\text{-}25\text{b})$$

$$i_B(t) = -I_{sub}(v_{GS}, v_{DS}, v_{BS}) + I_{j,\text{DB}}(v_{DB}) + I_{j,\text{SB}}(v_{SB}) - C_{bg,t}(v_{GS}, v_{DS}, v_{BS}) \cdot \frac{dv_{GS}}{dt}$$
$$- C_{bd,t}(v_{GS}, v_{DS}, v_{BS}) \cdot \frac{dv_{DS}}{dt} + C_{bb,t}(v_{GS}, v_{DS}, v_{BS}) \cdot \frac{dv_{BS}}{dt}; \qquad (2\text{-}25\text{c})$$

where the total capacitances were expressed in Eq. (2-13). In a common-source configuration, the exact expression of the source current is not needed to construct the large-signal model. Just based on the above three equations, we can construct the common-source large-signal model as shown in Fig. 2-15. It is straightforward to verify that the terminal currents in Fig. 2-15 indeed agree with Eq. (2-25).

With some algebra manipulating the various variables in Eq. (2-25), Fig. 2-15 can be shown to be equivalent to Fig. 2-16. The latter figure has the advantage that the various device terminals are connected rather than isolated. This reduces the number of components and makes the equivalent circuit more tractable. (The derivation of Fig. 2-16 follows closely with the derivation leading to the small-signal equivalent circuit of Fig. 2-23, which will be discussed in the next section.) The transcapacitances C_m, C_{mb} and C_{mx} are defined as

$$C_m = C_{dg,t} - C_{gd,t}; \qquad (2\text{-}26)$$

$$C_{mb} = C_{db,t} - C_{bd,t}; \qquad (2\text{-}27)$$

$$C_{mx} = C_{bg,t} - C_{gb,t}. \qquad (2\text{-}28)$$

C_{mx} is almost always zero, while C_m and C_{mb} can be appreciable. The circuits shown in Figs. 2-15 and 2-16 are effectively the same. They are the large-signal equivalent circuits adopted in BSIM3. At times, when we merely want to rationalize how a circuit behaves, we would like to concentrate on the principal components of the equivalent circuit. A good approximation to the exact equivalent circuit is shown in Fig. 2-17. This figure assumes that the substrate current is negligible and the diodes are reverse-biased.

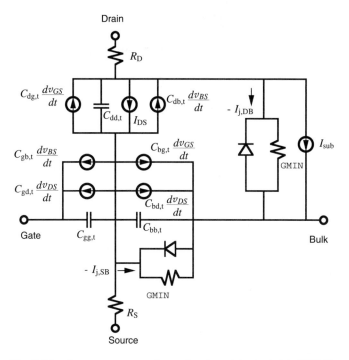

Fig. 2-15 A common-source large-signal model adopted in BSIM3.

Alternatively, we can express the transistor equations in the common-bulk configuration, wherein the reference voltage is made to bulk. If we express the terminal voltages in terms of the bias differences between the terminal voltage and the bulk voltage, then we have $dv_B/dt = 0$, $dv_G/dt = dv_{GB}/dt$, $dv_D/dt = dv_{DB}/dt$, and $dv_S/dt = dv_{SB}/dt$. Therefore,

$$i_G(t) = C_{gg,t}(v_{GB}, v_{DB}, v_{SB}) \cdot \frac{dv_{GB}}{dt} - C_{gd,t}(v_{GB}, v_{DB}, v_{SB}) \cdot \frac{dv_{DB}}{dt}$$
$$- C_{gs,t}(v_{GB}, v_{DB}, v_{SB}) \cdot \frac{dv_{SB}}{dt}; \qquad (2\text{-}29a)$$

$$i_D(t) = I_{DS}(v_{GB}, v_{DB}, v_{SB}) + I_{sub}(v_{GB}, v_{DB}, v_{SB}) - I_{j,DB}(v_{DB})$$
$$- C_{dg,t}(v_{GB}, v_{DB}, v_{SB}) \cdot \frac{dv_{GB}}{dt}$$
$$+ C_{dd,t}(v_{GB}, v_{DB}, v_{SB}) \cdot \frac{dv_{DB}}{dt} - C_{ds,t}(v_{GB}, v_{DB}, v_{SB}) \cdot \frac{dv_{SB}}{dt}; \qquad (2\text{-}29b)$$

$$i_S(t) = -I_{DS}(v_{GB}, v_{DB}, v_{SB}) - I_{j,SB}(v_{SB}) - C_{sg,t}(v_{GB}, v_{DB}, v_{SB}) \cdot \frac{dv_{GB}}{dt}$$
$$- C_{sd,t}(v_{GB}, v_{DB}, v_{SB}) \cdot \frac{dv_{DB}}{dt} + C_{ss,t}(v_{GB}, v_{DB}, v_{SB}) \cdot \frac{dv_{SB}}{dt}. \qquad (2\text{-}29c)$$

128 BASIC FACTS ABOUT BSIM3

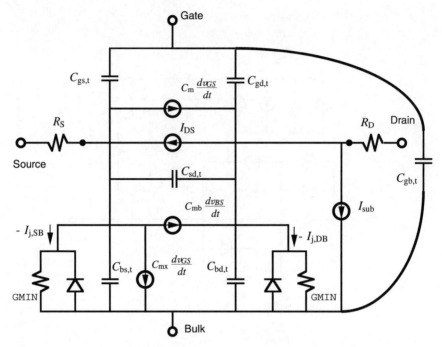

Fig. 2-16 Another common-source large-signal model. Figures 2-15 and 2-16 are equivalent.

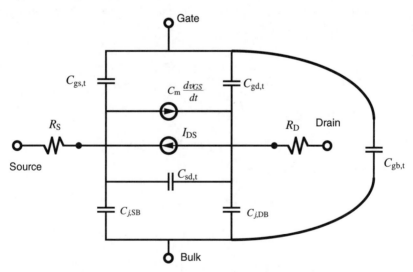

Fig. 2-17 A simplified (but nonexact) common-source large-signal model used in BSIM3.

Based on Eqs. (2-29), we construct the common-bulk large-signal model as shown in Fig. 2-18. At $V_{DS} = 0$, our physical intuition tells us that the transistor's C_{gd} is identical to C_{gs}. The equivalent circuit in the common-bulk configuration makes it clear that the device is symmetrical about the center of the device when $V_{DS} = 0$. In the common-source configuration, such a device symmetry at $V_{DS} = 0$ is not so obvious.

Whether the common-source or the common-bulk equivalent circuit is used, the solutions of the transistor's nodal voltages will come out the same. Choosing one configuration over the other is merely matter of convenience. It is interesting to note that BSIM3 adopts the common-bulk equivalent circuit model to solve for its transient solution. This fact does not imply that BSIM3 is a bulk-referencing model. As emphasized in Section 1.10, BSIM3 is a source-referencing model, using a drain current equation which is based on the implicit assumption that source is grounded. Once that is done, it does not matter whether the equivalent circuit is made common to source, or alternatively and equivalently, made common to bulk. Therefore, although the equivalent circuit of Fig. 2-18 has the appearance of symmetry at $V_{DS} = 0$, it is in reality not symmetrical because BSIM3's source-referencing modeling causes simulated $C_{gd} \neq C_{gs}$ at $V_{DS} = 0$ V.

Occasionally, a circuit designer may want to replace a MOS transistor in a circuit by the equivalent circuit shown in Fig. 2-18. This allows the designer to individually vary any equivalent circuit component such as $C_{gg,t}$, without changing other element's values. The effect of one parameter on the overall circuit performance can then be made clear. In addition, this replacement allows the designer to add extraneous components (such as substrate resistance) to account for some physical effects or processing details inherent only to the designer's chosen technology. Of the

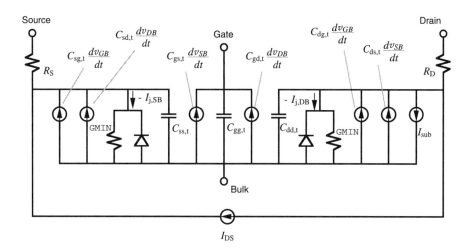

Fig. 2-18 A common-bulk large-signal model adopted in BSIM3. Although this circuit has the appearance of symmetry at $V_{DS} = 0$, it is not symmetrical because BSIM3's source-referencing modeling causes simulated $C_{gd} \neq C_{gs}$ at $V_{DS} = 0$ V.

circuit elements appearing in Fig. 2-18, the diode, the resistor, the capacitor can all be implemented with standard SPICE elements. The only element which is not easily reproduced are the dependent current sources, which depend on the time derivatives of some voltages. We show how we can use regular SPICE elements to reproduce such dependent current sources. We use the circuit of Fig. 2-19 as an example. The goal is to replace $C_{gg} \cdot dv_2/dt$ in Fig. 2-19a with some regular SPICE elements. One implementation is shown in Fig. 2-19b, which uses a combination of the voltage-controlled voltage source and current-controlled current sources to generate $C_{gg} \cdot dv_2/dt$. A SPICE netlist for Fig. 2-19b is listed below for convenience:

```
Vs 1 0 pulse 0v 3v 0s 0.03s 0.03s 0.02s 0.1s
R1 1 2 1e4
R2 2 3 1e4
C 3 0 1e-4
Fcccs 3 0 Vdummy 1
Evcvs 11 0 2 0 1
* this vccs makes v2 to appear the capacitor Cgg
Cgg 12 0 1e-4
Vdummy 11 12 0
.trans tstart=tstop=0.1 tstep=0.001
.print tran v(1) v(2) i(Vdummy)
.option nolist nomod noacct
.end
```

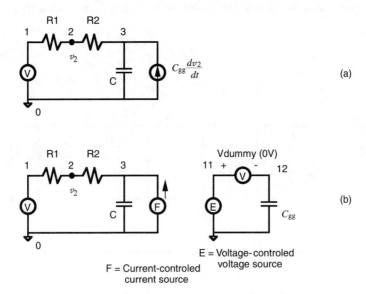

Fig. 2-19 An example showing the use of regular SPICE elements to implement dependent current sources. (a) The dependent current source $C_{gg} \cdot dv_2/dt$ is to be implemented; (b) a combination of SPICE elements to generate $C_{gg} \cdot dv_2/dt$. The SPICE netlist of Fig. 2-19b is given in the text.

2.4 LARGE-SIGNAL EQUIVALENT CIRCUIT 131

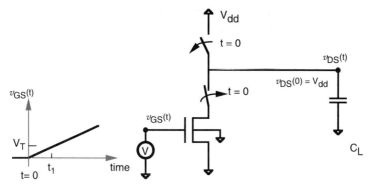

Fig. 2-20 An exemplar circuit showing how BSIM3 calculates current in a transient analysis.

As an instructive illustration of how BSIM3 calculates the transient current in a large-signal situation, let us consider a NMOS whose drain is connected to a loading capacitor initially charged up to the supply voltage, V_{dd}, as shown in Fig. 2-20. At $t = 0$, a gate input voltage ramps up from 0, reaching the threshold voltage at $t = t_1$ and continues its linear increase after $t = t_1$. When the transistor first turns on at $t = t_1$, the transistor is in the saturation region since $v_{GS} - V_T$ is much smaller than $v_{DS}(t)$. As time progresses, the NMOS discharges the load capacitor, lowering $v_{DS}(t)$ at the capacitor. In the meantime, $v_{GS}(t)$ increases in value. Therefore, at a certain time after $t = t_1$, the transistor will enter the linear region. Because of the complexity of the analysis in the linear region, and because our goal here is to describe how the large-signal equations are used in BSIM3, we will analyze the situation when the transistor is either off or in saturation. Even so, we still employ some simplified assumptions to convey the idea. Certainly the accurate solution to the problem requires a full-blown SPICE simulation.

We assume normal device operation, so that the impact ionization current is zero. Other nuisances associated with the diode leakage are also neglected. We can use Fig. 2-15 to guide our thought, and use Eq. (2-25b) directly to write down the equation governing the transistor. With the simplifying assumptions, we write

$$i_D(t) = I_{DS} - C_{dg,t} \cdot \frac{dv_{GS}}{dt} + C_{dd,t} \cdot \frac{dv_{DS}}{dt}. \qquad (2\text{-}30)$$

For the sake of the hand analysis here, we shall take I_{DS} in saturation as $W/(2L) \cdot \mu_n C'_{ox}(v_{GS} - V_T)^2$. (Of course, BSIM3 has a more exact formulation for I_{DS}.) At $t < t_1$, the MOS is off and $I_{DS} = 0$, and the intrinsic device capacitances C_{dd} and C_{dg} are identically zero. The total capacitances $C_{dd,t}$ and $C_{dg,t}$ are made of purely the parasitic components, with $C_{dd,p} = C_{ov} + C_f + C_{j,\text{DB}}$, and $C_{dg,t} = C_{ov} + C_f$. After $t > t_1$, the transistor turns on and enters the saturation region of operation. During saturation, C_{dd} remains zero whereas C_{dg} attains a constant value of $4/15 \cdot C'_{ox} \cdot W \cdot L$ (Appendix B, Eq. B-9), where C'_{ox}, W, and L are the gate capacitance per unit area, the gate width, and the channel length, respectively. Applying the Kirchoff's current law at the top capacitor plate, we have

$$-C_L \frac{dv_{DS}}{dt} = C_{dd,p} \frac{dv_{DS}}{dt} - C_{dg,p} \frac{dv_{GS}}{dt} \qquad t \le t_1. \tag{2-31a}$$

$$-C_L \frac{dv_{DS}}{dt} = \frac{W}{2L}\mu_n C'_{ox}(v_{GS} - V_T)^2 + C_{dd,p}\frac{dv_{DS}}{dt} - (C_{dg} + C_{dg,p})\frac{dv_{GS}}{dt} \qquad t > t_1. \tag{2-31b}$$

The initial condition is

$$v_{DS}(0) = V_{dd}. \tag{2-32}$$

When the SPICE simulator is given with above differential equations, it uses numerical analysis to solve for the solution. For our demonstration here, we shall solve it mathematically. A straightforward integration of Eq. (2-31a) from $t = 0$ to t_1, together with the initial condition of Eq. (2-32), yields

$$v_{DS}(t) = V_{dd} + \frac{C_{dg,p}}{C_L + C_{dd,p}} v_{GS}(t) \qquad t < t_1. \tag{2-33}$$

At $t = t_1$, $V_{DS}(t)$ reaches a peak value of

$$v_{DS}(t_1) = V_{dd} + \frac{C_{dg,p}}{C_L + C_{dd,p}} V_T. \tag{2-34}$$

Equation (2-33) reveals that when $t \le t_1$, the current due to the parasitics flows out of the drain terminal, charging C_L to an output voltage (v_{DS}) larger than the drain supply voltage (V_{dd}). This is because $v_{GS}(t)$ increases linearly with time. After $t > t_1$, we use Eq. (2-31b) to solve for $v_{DS}(t)$:

$$v_{DS}(t) = V_{dd} + \frac{C_{dg,p}}{C_L + C_{dd,p}} V_T + \frac{C_{dg,t}}{C_L + C_{dd,p}} (v_{GS}(t) - V_T)$$

$$- \frac{1}{C_L + C_{dd,p}} \int_{t_0}^{t} \frac{W}{2L} \mu_n C'_{ox}(v_{GS}(t') - V_T)^2 dt'. \tag{2-35}$$

We do not proceed to evaluate the integral. The process thus far demonstrates the key equations involved in a transient analysis.

An examination of Eq. (2-31) shows that, while I_{DS} is always positive, the capacitive components ($C_{dd} \cdot dv_{DS}/dt$ and $-C_{dg} \cdot dv_{GS}/dt$) are always negative. The capacitive components partially cancel the available dc drain current to discharge C_L.

In digital circuits where the transistor terminal voltages toggle between two extreme values, the transistor characteristics change drastically during the transient. However, in some analog circuits, the external bias changes represent only a small perturbation to the prior operating condition. In these small-signal situations, Eq. (2-25) can be simplified.

2.5 SMALL-SIGNAL MODEL

We are used to seeing a small-signal model for a FET in circuit textbooks. A circuit analysis usually begins with a determination of the dc operating point. Then the whole circuit is replaced with a small-signal representation, with the transistor replaced by its small-signal equivalent (such as Fig. 2-21, to be discussed shortly), and with voltage sources shorted and current sources opened. The small-signal nodal voltages and branch currents are established from the small-signal circuit.

That was for the textbook-like analysis of a circuit. In this approach, we need to have a distinct small-signal equivalent circuit for the transistor. The small-signal equivalent circuit of the transistor can be comprehensive or as simple as that of Fig. 2-21, depending on the desired level of complexity. The primary point is that, in this type of analysis, there is a potential disconnect between the large-signal equations governing the transient behavior and the small-signal equivalent used to simulate the device behavior at small-signal. In fact, in some MOSFET models (particularly those for digital applications), the small-signal model is created based on the author's physical understanding of the device. The small-signal model may not have a one-to-one correspondence to the equations which are used to calculate transistor behaviors in a large-signal transient simulation. The problem of having separate large- and small-signal models is that, when a designer simulates small-signal problem using the transient analysis, he gets a different answer from using an ac analysis. The only way to guarantee the same simulation result is to derive the small-signal model based on the large-signal equations used in the transient simulation.

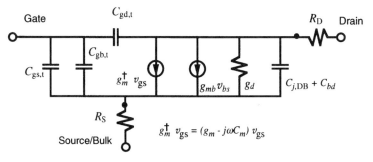

Fig. 2-21 A typical small-signal circuit used to describe MOSFET.

134 BASIC FACTS ABOUT BSIM3

BSIM3 guarantees that the results from ac simulation and transient simulation to be the same because the two simulations use the same set of governing equations. In fact, BSIM3 never seeks a topology for the small-signal equivalent circuit. In contrast to the textbook-like circuit analysis, BSIM3, after the determination of the dc bias point, directly solves for the small-signal node voltages and branch currents using the same set of equations that govern the large-signal transient response of the transistor. There is not a replacement of the transistor by a small-signal equivalent circuit. There is no small-signal model generated independently from the large-signal equations.

We will not see a BSIM3 small-signal equivalent circuit in the BSIM3 manual because an equivalent circuit is never necessary in its solution for nodal voltage and branch currents. Nonetheless, we can still construct one from the large-signal equations used by BSIM3, namely, Eqs. (2-25). We consider small-signal excitations of the terminal voltages and focus on the small-signal terminal currents $i_d = i_D - I_D$, $i_g = i_G - I_G$, $i_s = i_S - I_S$, and $i_b = i_B - I_B$. We temporarily focus on I_{DS}, the component of I_D which flows through the channel toward the source. We know that, in response to an increment in V_{GS}, V_{DS}, and V_{BS},

$$\Delta I_{DS} = I_{DS}(V_{GS} + \Delta V_{GS}, V_{DS} + \Delta V_{DS}, V_{BS} + \Delta V_{BS}) - I_{DS}(V_{GS}, V_{DS}, V_{BS})$$
$$\approx g_m \Delta V_{GS} + g_d \Delta V_{DS} + g_{mb} \Delta V_{BS}, \qquad (2\text{-}36)$$

where g_m, g_d, and g_{mb} were defined in Eq. (1-4).

Similarly, we expect the incremental diode current due to an increment in V_{BD} is

$$\Delta I_{j,\text{DB}} = I_{j,\text{DB}}(V_{BD} + \Delta V_{BD}) - I_{j,\text{DB}}(V_{BD}) \approx g_{j,\text{DB}} \Delta V_{BD}. \qquad (2\text{-}37)$$

The incremental substrate current in response to a increment in V_{GS}, V_{DS}, and V_{BS}, is

$$\Delta I_{sub} = I_{sub}(V_{GS} + \Delta V_{GS}, V_{DS} + \Delta V_{DS}, V_{BS} + \Delta V_{BS}) - I_{sub}(V_{GS}, V_{DS}, V_{BS})$$
$$\approx G_{bg} \Delta V_{GS} + G_{bd} \Delta V_{DS} + G_{bb} \Delta V_{BS}. \qquad (2\text{-}38)$$

Knowing that $I_G = 0$ under the dc condition, we start with Eq. (2-25) for the common-source configuration, and obtain

$$i_g = j\omega C_{gg,t} v_{gs} - j\omega C_{gd,t} v_{ds} - j\omega C_{gb,t} v_{bs}; \qquad (2\text{-}39\text{a})$$

$$i_d = g_m v_{gs} + g_d v_{ds} + g_{mb} v_{bs} + G_{bg} v_{gs} + G_{bd} v_{ds} + G_{bb} v_{bs} + g_{j,\text{DB}} v_{db}$$
$$- j\omega C_{dg,t} v_{gs} + j\omega C_{dd,t} v_{ds} - j\omega C_{db,t} v_{bs}; \qquad (2\text{-}39\text{b})$$

$$i_b = -G_{bg} v_{gs} - G_{bd} v_{ds} - G_{bb} v_{bs} + g_{j,\text{DB}} v_{bd} + g_{j,\text{SB}} v_{bs}$$
$$- j\omega C_{bg,t} v_{gs} - j\omega C_{bd,t} v_{ds} + j\omega C_{bb,t} v_{bs}. \qquad (2\text{-}39\text{c})$$

We have replaced dv_{gs}/dt by $j\omega v_{gs}$. This is consistent with the small-signal sinusoidal excitation under consideration. Equation (2-39) expresses the small-signal terminal currents in terms of the small-signal terminal voltages. The equation is similar in form to the y-parameter equations that $i_g = y_{gg} v_{gs} + y_{gd} v_{ds} + y_{gb} v_{bs}$;

$i_d = y_{dg}v_{gs} + y_{dd}v_{ds} + y_{db}v_{bs}$; and $i_b = y_{bg}v_{gs} + y_{bd}v_{ds} + y_{bb}v_{bs}$. A comparison of Eq. (2-39) and the above equations leads to the y-parameters given in Section 2.3.

Based on Eq. (2-39), a small-signal model of Fig. 2-22 is obtained. (This model differs from Fig. 2-15 in that the latter is a large-signal model.)

We can simplify the small-signal model further, for example, by making the i_g expression in Eq. (2-39a) to depend on v_{gs}, v_{gd}, and v_{gb}. We rewrite Eq. (2-39a) as

$$i_g = j\omega C_{gg,t}v_{gs} - j\omega C_{gd,t}v_{ds} - j\omega C_{gb,t}v_{bs}$$
$$= j\omega(C_{gd,t} + C_{gs,t} + C_{gb,t})v_{gs} - j\omega C_{gd,t}(v_{gs} - v_{gd}) - j\omega C_{gb,t}(v_{gs} - v_{gb})$$
$$= j\omega C_{gs}v_{gs} + j\omega C_{gd}v_{gd} + j\omega C_{gb}v_{gb}. \quad (2\text{-}40a)$$

Similar manipulations of the terminal voltages yield the following:

$$i_d = g_m v_{gs} + g_d v_{ds} + g_{mb}v_{bs} + G_{bg}v_{gs} + G_{bd}v_{ds} + G_{bb}v_{bs} + g_{j,DB}v_{db}$$
$$- j\omega C_{dg,t}v_{gs} + j\omega(C_{gd,t} + C_{sd,t} + C_{bd,t})v_{ds} - j\omega C_{db,t}v_{bs}; \quad (2\text{-}40b)$$
$$i_b = -G_{bg}v_{gs} - G_{bd}v_{ds} - G_{bb}v_{bs} + g_{j,DB}v_{bd} + g_{j,SB}v_{bs}$$
$$- j\omega C_{bg,t}v_{gs} - j\omega C_{bd,t}v_{ds} + j\omega(C_{bg,t} + C_{bd,t} + C_{bs,t})v_{bs}. \quad (2\text{-}40c)$$

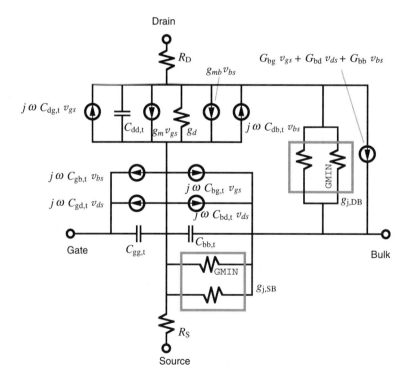

Fig. 2-22 A small-signal equivalent circuit adopted by BSIM3. The transistor is in the common-source configuration.

With the definitions of C_m, C_{mb}, and C_{mx} given in Eqs. (2-26) to (2-28), we further simplify Eq. (2-40) to the following:

$$i_g = j\omega C_{gs}v_{gs} + j\omega C_{gd}v_{gd} + j\omega C_{gb}v_{gb}; \quad (2\text{-}41\text{a})$$

$$i_d = g_m v_{gs} + g_d v_{ds} + g_{mb}v_{bs} + G_{bg}v_{gs} + G_{bd}v_{ds} + G_{bb}v_{bs} + g_{j,DB}v_{db}$$
$$+ j\omega C_{sd,t}v_{ds} + j\omega C_{gd,t}v_{dg} + j\omega C_{bd,t}v_{db} - j\omega C_m v_{gs} - j\omega C_{mb}v_{bs}; \quad (2\text{-}41\text{b})$$

$$i_b = -G_{bg}v_{gs} - G_{bd}v_{ds} - G_{bb}v_{bs} + g_{j,DB}v_{bd} + g_{j,SB}v_{bs}$$
$$+ j\omega C_{bs,t}v_{bs} + j\omega C_{bd,t}v_{bd} + j\omega C_{gb,t}v_{gb} - j\omega C_{mx}v_{bg}. \quad (2\text{-}41\text{c})$$

C_m, C_{mb}, and C_{mx} are always greater or equal to zero, under all operating regions.

These terminal current equations allow us to develop another common-source small-signal model, with a less complicated appearance than Fig. 2-22. This equivalent circuit, shown in Fig. 2-23, is an exact representation of BSIM3's model equations, covering all current components such as those of the diode and the impact ionization current. At times, for the sake of obtaining quick physical understanding of how a circuit operates, even more simplified small-signal equivalent circuit is desired. When the substrate current is negligible, and when the diodes

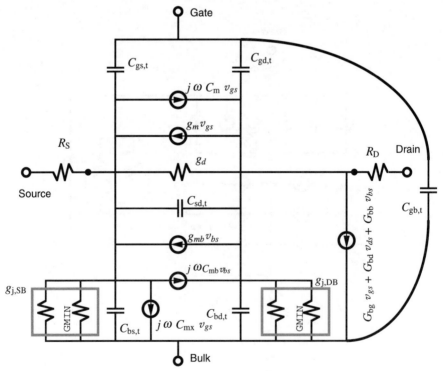

Fig. 2-23 Another common-source small-signal model. Figures 2-22 and 2-23 are equivalent.

are reverse-based, a reduced equivalent circuit of Fig. 2-24 can be used. Moreover, in a typical case where the transistor's source and bulk nodes are tied together, then the equivalent circuit of Fig. 2-21 becomes applicable.

In Fig. 2-24 (or Fig. 2-21), the voltage-controlled current source at the output is written as $g_m^\dagger v_{gs}$, which combines the two current sources in Fig. 2-23: $g_m \cdot v_{gs}$ and $-j\omega C_m \cdot v_{gs}$. C_m, a transcapacitance given in Eq. (2-26), is $C_{dg} - C_{gd}$. From the MOS transistor theory, it can be shown that C_m can be expressed as τ_1/ω_0, such that $g_m^\dagger v_{gs}$ assumes the form of

$$g_m^\dagger v_{gs} = g_m\left(1 - j\frac{\omega}{\omega_0}\tau_1\right)v_{gs}, \qquad (2\text{-}42)$$

where ω_0, a useful frequency scaling variable for MOSFET, was given as Eq. (1-36), and τ_1, a unitless quantity equal to $(C_{gd} - C_{gd}) \cdot \omega_0/g_m$, is given by [1]

$$\tau_1 = \frac{4}{15}\frac{1 + 3\alpha + \alpha^2}{(1+\alpha)^3}. \qquad (2\text{-}43)$$

Although ω_0 is a good scaling variable as far as theoretical modeling is concerned, typical circuit engineers would prefer to scale frequency with something more directly measurable. In this regard, the cutoff frequency, ω_T in rad/s or f_T in Hz, is often used as the scaling variable. Therefore, Eq. (2-42) is sometimes recast to a different form:

$$g_m^\dagger v_{gs} \approx g_m\left(1 - j\frac{\omega}{\omega_T}\tau\right)v_{gs}. \qquad (2\text{-}44)$$

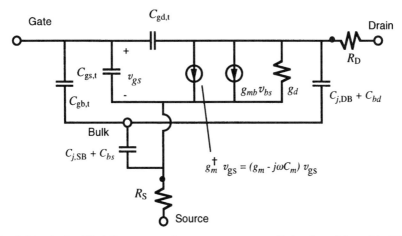

Fig. 2-24 A simplified (but nonexact) common-source small-signal model used in BSIM3. This equivalent circuit is valid when the substrate current is negligible and the diodes, reverse biased.

ω_T turns out to be about 1.5 ω_0 in saturation and is on the same order of magnitude as ω_T even out of saturation. Therefore, approximating Eq. (2-42) by Eq. (2-44) is not problematic, and τ is roughly equal to τ_1.

There is a distinction between g_m and $g_m^\dagger$. The transconductance g_m is constant at a given bias condition. It is by definition equal to $\partial I_D / \partial V_{GS}$ while V_{DS} and V_{BS} are held constant. $g_m^\dagger$, in contrast, denotes the transconductance of the voltage-controlled current source in the small-signal equivalent circuit. $g_m^\dagger$, as demonstrated by Eq. (2-42), is not a constant, but exhibiting some kind of frequency dependence, a dependence that introduces phase lag between the input and output currents.

The above formulation is based on the common-source configuration. We re-derive the above equations with bulk as the reference terminal. In this case, the terminal currents are given by

$$i_g = j\omega C_{gg,t} v_{gb} - j\omega C_{gd,t} v_{db} - j\omega C_{gs,t} v_{sb}; \tag{2-45a}$$

$$i_d = g_m v_{gs} + g_d v_{ds} + g_{mb} v_{bs} + G_{bg} v_{gs} + G_{bd} v_{ds} + G_{bb} v_{bs} + g_{j,\text{DB}} v_{db}$$
$$- j\omega C_{dg,t} v_{gb} + j\omega C_{dd,t} v_{db} - j\omega C_{ds,t} v_{sb}; \tag{2-45b}$$

$$i_s = -g_m v_{gs} - g_d v_{ds} - g_{mb} v_{bs} + g_{j,\text{SB}} v_{sb} - j\omega C_{sg,t} v_{gb}$$
$$- j\omega C_{sd,t} v_{db} + j\omega C_{ss,t} v_{sb}. \tag{2-45c}$$

The common-bulk small-signal model obtained from these equations is shown in Fig. 2-25.

All of the equivalent circuits derived in this chapter are based ultimately on Eq. (2-23), which implicitly employs the quasi-static assumption. In the time domain, the quasi-static assumption breaks down when the time scale of interest is shorter than

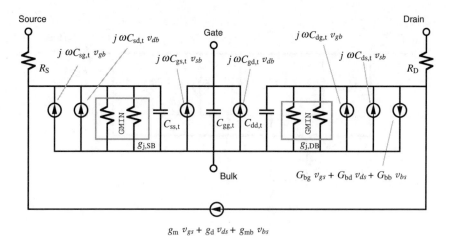

Fig. 2-25 A common-bulk small-signal model adopted in BSIM3. Just as in Fig. 2-18, this equivalent circuit appears symmetrical at $V_{DS} = 0$, but is in reality not because of the source-referencing implicitly assumed in the BSIM3.

2.5 SMALL-SIGNAL MODEL 139

the intrinsic transistor transit time, τ_{tr}. This point was analyzed in Section 1.6, where we illustrated the difference between a non-quasi-static solution and a quasi-static solution. We can infer that, in the frequency domain, the quasi-static assumption becomes problematic when the frequency of the small-signal variation exceeds the inverse of the transit time, or $1/\tau_{tr}$. Although not obvious at this moment, the quasi-static assumption embodied in the equivalent circuit results in two additional deficiencies in the model (besides the limitation to a certain frequency). Firstly, the circuit of Fig. 2-21 (or all other circuits of this section) lacks a circuit element—the channel resistance, which can be important when the transistor is not heavily turned on. Secondly, the form of $g_m^\dagger$ has an incorrect frequency dependence. These problems are addressed in Sections 4.3 and 4.4. In order to improve the model accuracy at high frequencies (or at least to estimate the amount of the error), we have to drop the quasi-static assumption from the very beginning. Although BSIM3 includes non-quasi-static option, it is not derived from first-principle device physics and not expected to fully solve the mentioned problems of the quasi-static model.

Some SPICE simulators permit the various device parameters such as g_m, g_d, g_{mb}, $C_{dg,t}$, $C_{gd,t}$ etc to be outputted. Even if such an option is not available in a given simulator, we can still obtain them through a y-parameter calculation using ac analysis. The following is an example to calculate $C_{gd,t}$, g_d, and $C_{dd,t}$. The netlist of the circuit used for this purpose is shown below:

```
VG gate 0 1.5V
VD drain 0 3.3V AC 1V
VS source 0 0.0V
VB bulk 0 0.0V
mn drain gate source bulk nch w=5 l=0.35
.AC LIN 1 1e9HZ 1e9HZ
.PRINT AC #-iR(VG)# #-iI(VG)# #-iR(VD)# #-iI(VD)#
* #math# denotes a mathematical operation, such as
* taking the negative
.option nolist nomod noacct
.lib dsn=model.bsim3
.end
```

Here are the logics behind the netlist.

1. Construct the netlist. The bulk and source are shorted, with the two-port network's port 1 defined at the gate and port 2, at the drain. An ac voltage (v_2) is set up at the drain, and we print out the real and imaginary parts of ac currents into both ports (i_1 and i_2). In this particular simulation of a device with $L = 0.35\,\mu m$, we find $-iR(V_{GS}) = 0$, $-iI(V_{GS}) = -4.1 \times 10^{-7}\,A$, $-iR(V_{DS}) = 6.2 \times 10^{-5}\,A$, and $-iI(V_{DS}) = -2.2 \times 10^{-7}\,A$. The current $-iR(V_{GS})$ is basically the real part of i_1. A negative sign is needed because i_1 is defined positive when it enters the gate. Likewise, $-iI(V_{GS})$ is the

140   BASIC FACTS ABOUT BSIM3

   imaginary part of i_1. $-\mathrm{iR}(V_{DS})$ and $-\mathrm{iI}(V_{DS})$ are the real and imaginary parts of i_2, respectively.
2. We know $i_1 = y_{11} \cdot v_1 + y_{12} \cdot v_2$ and $i_2 = y_{21} \cdot v_1 + y_{22} \cdot v_2$. Since v_1 (the ac signal at the gate) is zero and v_2 is 1, $\mathrm{Re}(y_{12}) = 0$, $\mathrm{Im}(y_{12}) = -4.1 \times 10^{-7}$ Ω^{-1}, $\mathrm{Re}(y_{22}) = 6.2 \times 10^{-5}$ Ω^{-1}, and $\mathrm{Im}(y_{22}) = -2.2 \times 10^{-7}$ Ω^{-1}.
3. Since we tied the source and the bulk together, the appropriate two-port y-parameters are given by Eq. (1-92). In other words, $y_{11} = y_{gg}$, $y_{12} = y_{gd}$, $y_{21} = y_{dg}$, and $y_{22} = y_{dd}$. From the results of the previous step, we see that $y_{gd} = 0 - j4.1 \times 10^{-7}$ Ω^{-1} and $y_{dd} = 6.2 \times 10^{-5} - j2.2 \times 10^{-7}$ Ω^{-1}.
4. We equate the calculated y_{gd} with the BSIM3 formulation given in Eq. (2-14c). Note that $\mathrm{Re}(y_{gd})$ must be zero; Eq. (2-14c) shows that y_{gd} adopted in BSIM3 does not contain a real part. This is a natural result of the quasi-static assumption used in the BSIM3 formulation. It would be non-zero in a non-quasi-static analysis. By equating of $\mathrm{Im}(y_{12})$ to $-\omega C_{gd,t}$, we find $C_{gd,t}$ to be 6.5×10^{-17} F (the frequency used in the ac analysis, as indicated in the netlist, is 1 GHz).
5. Similarly, equating the calculated y_{dd} to Eq. (2-14f) leads to $g_d = 6.2 \times 10^{-5}$ Ω^{-1} and $C_{dd,t} = -3.5 \times 10^{-17}$ F. (We have assumed the $g_{j,\mathrm{DB}}$ and G_{bd} of Eq. 2-14f to be zero since the transistor is biased in a region where the diode is reverse-biased and the impact ionization current is negligible.) Note that $C_{dd,t}$ in this calculation is negative. This is not a physical result. A calculated negative $C_{dd,t}$ value is indeed one of the BSIM3 model inaccuracies, as elaborated in Section 4.9.
6. Finally, we voice a cautionary remark. In the model card of the device, RSH should be made zero such that $R_D = \mathrm{NRD} \cdot \mathrm{RSH} = 0$ and $R_S = \mathrm{NRS} \cdot \mathrm{RSH} = 0$. Although the actual device may contain parasitic terminal resistances, it is important to know that our goal here is to find the conductances and capacitances of the internal MOS without these parasitic resistances. If the parasitic resistances are included, then the calculated y-parameters will not be exactly given by Eq. (2-14). Equation (2-14) is calculated for the intrinsic transistor free of the terminal resistances. We mentioned this point at the end of Section 2.3.

2.6   NOISE EQUIVALENT CIRCUIT

Noise places the lower limit to the magnitude of the electric signal which can be amplified by a circuit without significant degradation in the signal quality. Even though we may feel comfortable with a large-signal or a small-signal equivalent circuit, we often find the noise equivalent circuit awkward. Somehow we are all more familiar with the voltages and currents themselves, than the noises associated with them. Before we show BSIM3's noise equivalent circuit, we give a brief tutorial as to

2.6 NOISE EQUIVALENT CIRCUIT

how a SPICE simulator calculates noise. We begin with a description of the noise of an ideal resistor, a fundamental circuit element. (Other fundamental elements such as ideal capacitor and inductor are noiseless.)

Suppose we had an extremely sensitive voltage meter which had infinite input resistance. If it were placed to measure the voltage of an otherwise alone resistor, we would observe random voltage variation as a function of time. The random signal variation about its average value is called the (electrical) noise. There are several types of noise. The aforementioned noise found in the resistor is called the *thermal noise*, which is caused by the random thermally excited motion of the charge carriers. The average value of the noise voltage is not a good figure of merit to characterize noise. It is identically zero, due to the fact that the externally applied bias is zero. A good figure of merit is the mean square value, which is given by the following formula:

$$\overline{v_n^2} \triangleq \overline{(v-\bar{v})^2} = \lim_{T \to \infty} \frac{1}{T} \int_0^T (v-\bar{v})^2 dt, \qquad (2\text{-}46)$$

where an overbar indicates an average quantity. Alternatively, the mean square noise current follows a similar equation as the voltage equation given above (just replace v by i). This definition in Eq. (2-46) squares the difference between the instantaneous value at a given time and its average over time, ensuring the noise figure of merit to increase with random variation, rather than averaging it to zero. The thermal noise is due to the Brownian motion of the charge particles, which provides pulses of current even in the absence of an external bias. From intuition we expect the amount of random motion to increase with the temperature. This is indeed reflected in the expression of the mean square value of the noise voltage, obtained from the definition of Eq. (2-46) and given by

$$\overline{v_{n,R}^2} = 4kTR\,\Delta f, \qquad (2\text{-}47)$$

where R is the real part of the resistor's impedance. The reactive component does not generate thermal noise. Δf is the noise bandwidth, which is not the same as the usual 3-dB bandwidth of the circuit being measured. The familiar power gain bandwidth is the frequency corresponding to a 3-dB reduction (half) of the maximum power gain. Or equivalently, it corresponds to the frequency where the voltage gain decreases by a factor of the square root of 2. In contrast, the noise bandwidth is the frequency span, which when multiplied by the low frequency power gain, gives the same area under the actual power gain versus frequency curve. Mathematically, it is given by

$$\Delta f = \frac{1}{G_o} \int_0^\infty g(f) df, \qquad (2\text{-}48)$$

142 BASIC FACTS ABOUT BSIM3

where $G(f)$ is the power gain as a function of frequency and G_o is the constant power gain at low frequencies. Though not equal, the noise bandwidth is often well approximated as the power gain bandwidth. Nevertheless, one important message of Eq. (2-47) is that the amount of the noise depends in some way on the overall circuit bandwidth, which is determined by the circuit as well as the device itself. As the bandwidth is reduced by placing a low pass filter in the circuit, the noise associated with a resistor decreases. As an example, if the transistor circuit has a noise bandwidth of 1 GHz and operates under room temperature (300 K), then the mean square voltage across a 20 Ω gate resistance is calculated to be 3.3×10^{-10} V². This corresponds to a root mean square voltage of 18 μV.

Alternatively, we could characterize the noise of the resistor by measuring the current instead of the voltage. With an ideal current meter with zero input impedance, we would find the resistor current to exhibit a random variation with time, similar to that described for the voltage. We characterize with the mean square current, in an analogous definition as for the voltage. The mean square noise current for the thermal noise is given by

$$\overline{i_{n,R}^2} = \frac{4kT}{R} \Delta f. \qquad (2\text{-}49)$$

The two equivalent representations of a resistor are displayed in Fig. 2-26.

Another quantity characterizing a given noise relates to the noise content in a unit of bandwidth. This is called the spectral density which is denoted as $S_v(f)$ (in V²/Hz) and $S_i(f)$ (in A²/Hz) for the voltage and current, respectively. For now, we focus on the spectral density for the current, which has the following relationship with the mean square current:

$$\overline{i_n^2} = \int_0^{\Delta f} S_i(f) df. \qquad (2\text{-}50)$$

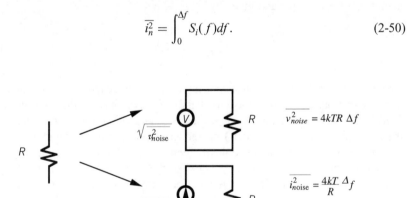

Fig. 2-26 Two equivalent noise representations of a resistor. BSIM3, as well as perhaps all compact models, choose the current noise source representation.

2.6 NOISE EQUIVALENT CIRCUIT

Specifically for the thermal noise of a resistor, we find the current spectral density to be

$$S_{i,R}(f) = \frac{\overline{di^2_{n,R}}}{df} = \frac{\overline{i^2_{n,R}}}{\Delta f} = \frac{4kT}{R}. \tag{2-51}$$

The subscript R is used for the fact that the equation applies to the thermal noise found in a resistor. The spectra density in this case is independent of the frequency; it is thus called the *white noise*. For the MOS transistor, there are other types of noise sources which can be easily identified if the spectral density, rather than the mean square voltage (or current) is known. Sometimes, the square root value of the spectral density is reported. In that case, the root mean square spectral density has a unit of V-Hz$^{-1/2}$ for the voltage and A-Hz$^{-1/2}$ for the current.

We proceed to the noise analysis for a circuit, such as a resistive divider shown in Fig. 2-27a. A sample SPICE netlist calculating the equivalent input and output noises are shown below:

```
Vsource 2 0 1V ac 1V
Rx 1 2 30
Ry 1 0 70
.temp 16.85
.ac LIN 1 100Hz 100Hz
.noise V(1) Vsource 1
.pr noise inoise onoise
.end
```

As defined by the .NOISE statement, the input noise reference is the voltage source, Vsource, and the output noise voltage node is node 1. A sample noise analysis output is shown below:

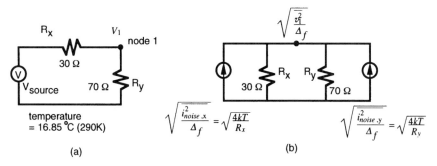

Fig. 2-27 (a) An exemplar circuit showing the noise calculation in a SPICE simulator; (b) the equivalent circuit during the noise analysis.

144 BASIC FACTS ABOUT BSIM3

```
RESISTOR SQUARED NOISE CURRENTS (SQ A/HZ)
NAME RY
TOTAL 2.288E-22 (See Eq. 2-52 for its calculation)
NAME RX
TOTAL 5.338E-22 (See Eq. 2-53 for its calculation)
RESISTOR SQUARED NOISE VOLTAGES (SQ V/HZ)
NAME RY
TOTAL 1.009E-19 (See Eq. 2-54 for its calculation)
NAME RX
TOTAL 2.354E-19 (See Eq. 2-55 for its calculation)
TOTAL NOISE SUMMARY:
INPUT VSOURCE
EQUIVALENT INPUT NOISE 8.285E-10 /RT HZ
OUTPUT V(1)
TOTAL OUTPUT NOISE VOLTAGE 3.363E-19 SQ V/HZ
5.799E-10 V/RT HZ (See Eqs. 2-56 and 2-57)
(OUTPUT) / (INPUT) 0.700
FREQ INOISE_M ONOISE_M
1.000E+02 8.285E-10 5.799E-10 (See Eqs. 2-57 and 2-58)
```

During the noise analysis, the SPICE simulator shorts out all independent voltage sources and opens up the independent current sources. Each element is replaced by the element itself along with its noise source. Both the noise voltage and the noise current representation are equally valid from a theoretical perspective. A SPICE simulator typically uses noise current because adding a branch noise current source is easier in the SPICE frame work. Adding a noise voltage source would have created another node, which would significantly change the matrix originally set up to calculate the circuit voltages and currents in the ac analysis. This would add overhead to the computation time. The equivalent noise circuit analyzed by SPICE is given in Fig. 2-27b.

We first find the noise current sources associated with R_y and R_x. For R_y,

$$\frac{\overline{i^2_{n,Ry}}}{\Delta f} = \frac{4kT}{R_y} = \frac{4 \cdot 1.38 \times 10^{-23} \cdot 290}{70} = 2.29 \times 10^{-22} \frac{A^2}{Hz}. \qquad (2\text{-}52)$$

Similarly for R_x, we have

$$\frac{\overline{i^2_{n,Rx}}}{\Delta f} = \frac{4kT}{R_x} = \frac{4 \cdot 1.38 \times 10^{-23} \cdot 290}{30} = 5.33 \times 10^{-22} \frac{A^2}{Hz}. \qquad (2\text{-}53)$$

2.6 NOISE EQUIVALENT CIRCUIT

We proceed to find the mean noise voltage square per unit bandwidth at node 1 due to the noise from R_y. During this calculation, the noise current source due to R_x is opened up. Therefore, from Fig. 2-27b,

$$\frac{\overline{v_1^2}}{\Delta f} = \left[\sqrt{\frac{4kT}{R_y}} \cdot (R_y // R_x)\right]^2 = 4kT \frac{R_x^2 \cdot R_y}{(R_y + R_x)^2} = 1.01 \times 10^{-19} \frac{V^2}{Hz}. \qquad (2\text{-}54)$$

Similarly, the contribution of the noise from R_x to the mean square noise voltage per bandwidth at node 1 is

$$\frac{\overline{v_1^2}}{\Delta f} = \left[\sqrt{\frac{4kT}{R_x}} \cdot (R_y // R_x)\right]^2 = 4kT \frac{R_y^2 \cdot R_x}{(R_y + R_x)^2} = 2.35 \times 10^{-19} \frac{V^2}{Hz}. \qquad (2\text{-}55)$$

From the superposition principle (applicable when the noise sources are uncorrelated), the total mean square noise voltage is the sum of the components of Eqs. (2-54) and (2-55):

$$\left.\frac{\overline{v_1^2}}{\Delta f}\right|_{total} = 1.01 \times 10^{-19} + 2.35 \times 10^{-19} = 3.36 \times 10^{-19} \frac{V^2}{Hz}. \qquad (2\text{-}56)$$

Equivalently,

$$\sqrt{\left.\frac{\overline{v_1^2}}{\Delta f}\right|_{total}} = \sqrt{3.36 \times 10^{-19}} \sqrt{\frac{V^2}{Hz}} = 5.80 \times 10^{-10} \frac{V}{\sqrt{Hz}}. \qquad (2\text{-}57)$$

Since we specify the output node as node 1, the above quantity is identical to the square root of the output noise spectral density, which is denoted as ONOISE_M in the SPICE printout. To find the input noise spectral density, we determine from Fig. 2-27a that $V_1/V_{source} = R_y/(R_x + R_y) = 0.7$. That is, the voltage gain (V_1/V_{source}) is 0.7. To refer the output noise spectral density to the input, we have

$$\sqrt{\left.\frac{\overline{v_{source}^2}}{\Delta f}\right|_{total}} = \sqrt{\left.\frac{\overline{v_1^2}}{\Delta f}\right|_{total}} / 0.7 = 8.29 \times 10^{-10} \frac{V}{\sqrt{Hz}}. \qquad (2\text{-}58)$$

This is denoted as INOISE_M in the SPICE output.

Noise analysis is a small-signal analysis. The dc bias point of the circuit is first sought for. The SPICE simulator then performs ac analysis, with noise sources appended in the small-signal circuit. The total noise voltage at a given node is equal to the sum of the noises contributed from various noise sources.

It is clear from the above procedure that, as soon as the resistor's noise equivalent circuit is known (such as that shown in Fig. 2-26), the noise analysis of a resistive circuit can proceed. Likewise, to analyze the noise property of a circuit containing a MOS transistor, we must identify its equivalent noise circuit. Let us discuss some of the MOSFET noise mechanisms, which will help us understand BSIM3's noise modeling. In general, a MOSFET has the following noise sources:

1. Thermal noise of the channel current: this noise is generated by the Brownian (thermally excited random) motion of the charge particles in the channel region, which provides pulses of current even in the absence of an external bias. The thermal noise increases with temperature because the amount of random motion increases with the temperature.
2. Flicker ($1/f$) noise of the channel current: its precise origin is has not been unequivocally identified, but its presence is surprisingly universal in all types of semiconductor devices [4]. The magnitude of the noise depends inversely on the frequency; so this noise is important at low frequencies or in an up-converter. The noise may have origin in quantum mechanical effects, occurring when the electrical charges in current are scattered by an arbitrary potential. Or, it can be due to the bulk-related processes, such as the mobility fluctuation. The $1/f$ noise can also be related to the surface effects, resulting from the fluctuations in the occupancy of the surface states.
3. Thermal noise associated with the terminal source resistance R_S: the mean square noise voltage is $4kTR_S \cdot \Delta f$.
4. Thermal noise associated with the terminal drain resistance R_D: the mean square noise voltage for the drain resistance is $4kTR_D \cdot \Delta f$.
5. Thermal noise associated with the distributed gate resistance R_G: the mean square noise voltage is $4kTR_G \cdot \Delta f$. The gate resistance R_G is a distributed resistance, equal to $1/3 \cdot R_{SH,G} \cdot W/L$, where $R_{SH,G}$ is the sheet resistance of the polysilicon gate (see Section 4.5).
6. Thermal noise associated with the distributed substrate resistance: this noise is coupled to the drain current through the substrate transconductance g_{mb}. This noise can be made less significant by decreasing g_{mb}, a result achievable by increasing the source-substrate bias V_{SB} [5].
7. Induced gate noise: a fluctuation in the channel charge induces an equal and opposite fluctuation in the charge on the gate electrode, causing a fluctuation current in the gate. Because this gate current noise and the channel thermal noise both originate from the fluctuation in the channel, the gate and the drain noise current generators are correlated. This induced gate current noise is important at high frequencies where the coupling through the gate capacitance becomes more significant.
8. Shot noise associated with the small but finite gate leakage current: the shot noise relates to the dc current flow across a certain potential barrier. During the thermionic emission from the channel into the gate, the current is a result of carriers with sufficient amount of energy and with velocity at the right direction which pass through a potential barrier. The passage of the carriers is a random event, and the resulting current consists of a large number of independent current pulses.

9. Impact ionization noise: this noise is generated in the impact ionization process. The amount of noise is proportional to I_{sub}. When the impact ionization noise dominates, NMOS devices are expected to have more noise than PMOS devices.
10. The $1/f$ and shot noise associated with the drain-bulk and source-bulk junction diodes: these parasitic junctions contribute to noise the same way a semiconductor p-n junction diode does. In the usual MOS bias condition where the drain–bulk and source–bulk junctions are reverse-biased, both the $1/f$ noise and the shot noise are small.
11. Finally, when the device is biased in the subthreshold region, the drain current noise is a shot noise, not a thermal noise.

More details about these noise sources, a derivation of the equivalent input noise sources, and the MOS noise figure can be found in Ref. [6]. We now turn our attention to BSIM3's noise modeling. The first two noise sources are definitely accounted for in BSIM3, while the incorporation of the third and fourth noise sources depends on the method used to implement R_S and R_D. As discussed in Section 2.2, there are two ways to model these terminal resistances. By using a combination of NRD, NRS, and RSH (the lumped-resistance approach), actual resistances are attached to the MOSFET. During a noise analysis, BSIM3 appends noise current sources associated with the resistances given in Eq. (2-9). An alternative method to implement the terminal resistance employs the parameter RDSW (the absorbed-resistance approach). This parameter at best "effectively" accounts for the effects of the source/drain resistances on the drain current and the conductances, but the resulting input impedance looking into the gate does not include a physical source resistance. The input impedance is in fact purely imaginary. Likewise, looking out through the drain current, there is no drain resistance. Therefore, when the source/drain resistances are modeled with RDSW, no thermal noises sources associated with these resistances will be attached to the transistor. The noise calculation will be incorrect (even if all other mechanisms are accounted for properly.) The proper way to include the terminal resistances, as far as noise analysis is concerned, is to declare the source and drain resistances through the use of NRD, NRS, and RSH, and set RDSW to 0. Finally, the fifth to the eleventh noise sources are ignored in BSIM3 completely.

Of the four noise sources considered in BSIM3, the first two are lumped into one noise source, that associated with the drain current flowing in the intrinsic device. The third and the fourth noise sources are those associated with the terminal resistors. The noise equivalent circuit used by BSIM3 is given in Fig. 2-28a. The added internal small-signal noise sources reflect the location of the corresponding noise sources. For example, $\overline{i_{Rs}^2}/\Delta f$ is the noise source associated with R_S, equal to $4kTR_S$. $\overline{i_{Rd}^2}/\Delta f$ is the noise source associated with R_D, equal to $4kTR_D$. $\overline{i_d^2}/\Delta f$ is the noise source associated with the drain current. It contains the channel thermal noise and the $1/f$ noise components. The exact expressions for these two noise components are to be discussed in Section 3.2 (in the entry of NOIMOD) and Section 4.11.

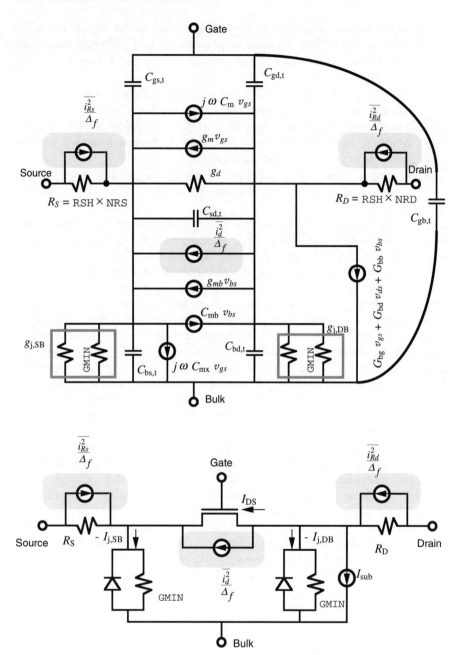

Fig. 2-28 (a) A noise equivalent circuit used by BSIM3 shown at the top; (b) a high-level representation of (a), shown at the bottom. The high-level representation is particularly useful in the discussion of BSIM4 improvement in Chapter 5.

2.7 SPECIAL OPERATING CONDITIONS: $V_{DS} < 0$, $V_{BS} > 0$, $V_{GS} < 0$, OR $V_{BD} > 0$

Because noise is internally generated small signal in the device, Fig. 2-28a resembles the small-signal equivalent circuit of Fig. 2-23. A high-level representation of Fig. 2-28a is shown in Fig. 2-28b.

BSIM3 considers the parasitic junction diodes to be noise free. Therefore, there are no noise sources associated with $g_{j,DB}$ and $g_{j,SB}$, which are the small-signal conductances of the junction diodes. There is no noise source associated with GMIN, either. The resistor GMIN is added to aid numerical convergence. It is not intended to represent any physical process. Lastly, because the noise associated with the impact ionization process is neglected, there is no noise source associated with G_{bg}, G_{bd}, and G_{bb}.

The exclusion of the noise sources associated with the junction diodes, the impact ionization process, and the gate leakage current is reasonable. After all, in a normal MOS operation, the diodes are reverse biased, the amount of impact ionization is insignificant and the gate leakage current is almost nil. The neglect of the noise sources of the gate resistance and the substrate resistance is understandable, simply because BSIM3 does not include these resistances in the model. These resistive noise sources can be easily incorporated into the transistor if we define a MOS subcircuit to include the gate and the substrate resistances. While these noise sources are either truly insignificant or can be incorporated easily into a SPICE simulation, there is one omitted noise source which cannot be easily added. It is the channel-induced gate noise. Without a quantitative analysis, we have no idea whether this noise component is important or not. In Section 4.12 we will examine this omission in BSIM3, assessing the importance of this noise source in various bias conditions.

2.7 SPECIAL OPERATING CONDITIONS: $V_{DS} < 0$, $V_{BS} > 0$, $V_{GS} < 0$, OR $V_{BD} > 0$

Although we have no hesitation using BSIM3 under normal bias conditions, we may wonder if BSIM3 remains valid under some weird bias conditions. In this section, we examine the BSIM3 model when it is used under $V_{DS} < 0$, $V_{GS} < 0$, $V_{BS} > 0$ or $V_{BD} > 0$.

When $V_{DS} < 0$, the drain of the MOSFET acts like the source and vice versa. BSIM3 is designed to handle such a situation. The additional amount of modeling required to handle this inverse mode of operation, the so-called inverse modeling, has been described in Section 1.10. Basically, BSIM3 internally swaps the drain and the source nodes, although externally the drain and source nodes remain to be those specified by the designer. For most circuit simulation, the inverse modeling of swapping nodes works fine and the device symmetry is preserved as far as the current calculation is concerned. However, the drain–source swapping approach has its shortcomings. In analog circuits where the derivative of the drain current with respect to applied voltage is the quantity of interest, then BSIM3 will render unphysical results, as demonstrated in Section 4.7. Another problem is capacitance asymmetry. When the transistor is biased from V_{DS} below zero to above zero, the device capacitances calculated in BSIM3 exhibit discontinuity right at $V_{DS} = 0$. Other problems are elaborated in Section 4.7.

BSIM3 handles the case of $V_{GS} < 0$ without doing anything special. The drain channel current developed in BSIM3 (Eq. 1-12), in the limit that V_{GS} is much smaller than the threshold voltage, is proportional to

$$I_{DS} \propto \left[1 - \exp\left(-\frac{qV_{DS}}{kT}\right)\right] \times \exp\left(\frac{q(V_{GS} - V_T)}{nkT}\right) \quad \text{for } V_{GS} \ll V_T. \quad (2\text{-}59)$$

This expression continues to work correctly even when V_{GS} becomes negative, that is, not giving rise to numerical problems. The current, as given by Eq. (2-59), decreases exponentially toward zero as V_{GS} decreases.

When V_{GS} is smaller than the threshold voltage, the device is said to operate in the subthreshold region, the drain current in the channel is sometimes referred to as the *subthreshold current*. Further, when $V_{GS} = 0$, the small yet finite drain current in the channel, in combination with the leakage current associated with the bulk–drain junction and the substrate current associated with the impact ionization, is referred to as the *leakage current*, or the *off current*. In the ideal case where both the junction leakage and the impact ionization currents can be neglected, then the off current is given by Eq. (2-59) with the substitution of V_{GS} by zero. It is a digital circuit designer's ultimate dream to have a low V_T MOS with a negligible off current. However, as seen from Eq. (2-59), there is a direct relationship between V_T and I_{off}. As the threshold voltage is made less, the off current naturally tends to rise.

We caution that the off current simulated in BSIM3 is not exclusively determined from Eq. (2-59). We first consider the case that the bulk is connected to a large resistance so that it is nearly floating. When the magnitude of the off current is smaller than $V_{DS} \cdot$ GMIN/2, the overall drain current is dominated by the drain current flowing through GMIN to the substrate node, and through another GMIN out to the source node. This current path can be seen in Fig. 2-2, showing that the drain terminal current can either flow through the channel, or, first through the GMIN at the drain side to the bulk, and then from the bulk back up to source through the GMIN at the source side. We mentioned in Section 2.2 that the GMIN conductances, with GMIN defaulted to 10^{-12} Ω^{-1}, are placed intentionally in a BSIM3 model to speed up convergence during numerical analysis. However, in some sense it can be viewed with some physical significance. These GMIN conductances can be used to model the minimum leakage current between the drain and source nodes when V_{GS} is small (or even negative). If the off current is on the order of pA, then a good portion of the off current comes from the conduction through GMIN. In the case where the bulk is tied to the same potential as the source, then the off current is equal to the smaller value of that calculated from Eq. (2-59) and $V_{DS} \cdot$ GMIN. Similar to the previous case, $V_{DS} \cdot$ GMIN is the current that flows through the GMIN between the drain and the bulk, which is at the same potential as the source.

The drain current expression adopted by BSIM3, as given by Eq. (2-59), works well under the dc condition when $V_{GS} < 0$. But how about the transient situation?

2.7 SPECIAL OPERATING CONDITIONS: $V_{DS} < 0$, $V_{BS} > 0$, $V_{GS} < 0$, OR $V_{BD} > 0$

During a transient simulation, Eq. (2-25) shows that the four terminal currents are not determined solely by I_{DS}, but also by the sixteen device capacitances. We need to verify if the expressions of these capacitances continue to work when $V_{GS} < 0$ as well. If somehow BSIM3's model equations for the capacitance did not apply under such a bias condition, we would expect the overall BSIM3 model to fail. The bias condition $V_{GS} < 0$ roughly corresponds to the accumulation region of operation. From Appendix A, where we look into BSIM3's capacitance modeling in more details, or Appendix B, where we discuss simplified BSIM3 capacitance equations, we see that all of the device capacitances have the right behavior, either in the normal inversion region, or the accumulation region under consideration. For example, in the accumulation region where $V_{GS} < 0$, $C_{gg} \sim C_{gb} \sim C_{bg} \sim C_{bb} = WL \cdot C'_{ox}$, while other capacitances are roughly zero. Therefore, we conclude that BSIM3 remains applicable when $V_{GS} < 0$, even in a transient simulation.

Before we leave the discussion about $V_{GS} < 0$, we mention that BSIM3 does not account for gate-induced drain leakage (GIDL) current. Equation (2-59) models only the subthreshold current before the channel inverts. It predicts the drain current to exponentially decrease toward zero as V_{GS} is made negative. Experimentally, this needs not be the case. Figure 2-29 illustrates the measured and BSIM3-simulated I_D-vs-V_{GS} curves for a $L = 0.4\,\mu\text{m}$ device when $V_{DS} = 2.6$ V. The transistor's source and bulk are tied together. The experimental data display an exponential increase as V_{GS} becomes progressively more negative, whereas BSIM3's I_D calculated from Eq. (2-59) decreases. BSIM3's I_D eventually flattens out at 2.6×10^{-12} A because the

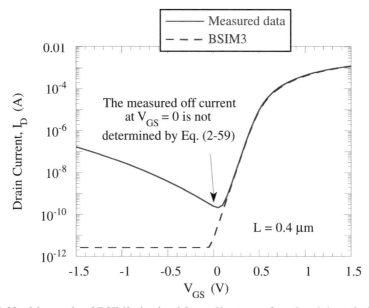

Fig. 2-29 Measured and BSIM3-simulated I_D-vs.-V_{GS} curves for a $L = 0.4\,\mu\text{m}$ device when $V_{DS} = 2.6$ V. The major discrepancy is due to the lack of gate-induced drain leakage (GIDL) current in BSIM3, a shortcoming that is remedied in BSIM4.

drain current then is dominated by $V_{DS} \cdot \mathtt{GMIN}$. This discrepancy between the experimental data and BSIM3 prediction alarms us to two points. One, the off current simulated by BSIM3 can be smaller than the experimentally measured data. Second, when transistor's $V_{GS} < 0$, BSIM3 can significantly underestimate the drain current.

We will discuss the physical origin of GIDL, and the experimental characteristics of GIDL under various bias conditions and in various channel lengths in Section 4.16. A remedy to BSIM3's lack of GIDL modeling will also be discussed.

BSIM3 handles the $V_{DS} < 0$ and $V_{GS} < 0$ situations correctly as far as numerical robustness is concerned. Let us now consider the bias condition $V_{BS} > 0$ or $V_{BD} > 0$. When the bulk–source/bulk–drain diode is heavily forward-biased (e.g., $V_{BS}/V_{BD} = 0.7\,\mathrm{V}$), there are lots of current flow from the bulk to the source/drain. We no longer expect the MOS transistor to behave normally. Because this is not a usual bias condition, there is no need to model the transistor accurately. However, the model equations must still be roughly correct in this extreme operating condition such that a circuit designer will notice some abnormalities in his circuit behavior. BSIM3 achieves this purpose by including an equation to account for the rapid rise of the diode currents in the forward-bias operation. The inclusion of the diodes can be seen in the dc equivalent circuit shown in Fig. 2-2, and the diode current expression was given by Eq. (2-3). A bad MOSFET model would only model the diode with a leakage current at the reverse-bias condition, but not the turn-on at the forward bias condition. This kind of model would place the burden on circuit designers to realize that the bulk-source junction should always be reverse-biased, and hence, to check if the diodes indeed remain reverse-biased at all time.

We all agree that the diodes should not be heavily forward-biased. However, occasionally with tricky circuit design, some designers would like to slightly forward bias the bulk–source diode (e.g., $V_{BS} \sim 0.3$ or even $0.5\,\mathrm{V}$). Because this slight-forward-bias condition is sometimes used intentionally to boost circuit performance, the MOSFET model should try to model this operating condition accurately. Besides the aforementioned modeling of the bulk–source diode current, BSIM3 also accounts for the shift of the threshold voltage brought about by the positive V_{BS}. V_{BS}'s effect on threshold voltage is no stranger to us; it is called the body effect. In the normal operation where V_{BS} is less than zero, the threshold voltage used in BSIM3 (for a large geometry device) is roughly given by

$$V_T = \mathtt{VTH0} + \mathtt{K1} \cdot \left(\sqrt{2\phi_f - V_{BS}} - \sqrt{2\phi_f}\right) \quad \text{(not used in BSIM3 if } V_{BS} > 0\text{)},$$

(2-60)

where K1 can be viewed as the body effect coefficient, commonly denoted as γ, and VTH0 can be viewed as the threshold voltage at zero V_{BS}. (The exact V_T equation used in BSIM3, incorporating the short-channel and narrow-width effects, is quite complicated. See Section 3.2, under the entry of VTH0.) The above equation can be

2.7 SPECIAL OPERATING CONDITIONS: $V_{DS} < 0$, $V_{BS} > 0$, $V_{GS} < 0$, OR $V_{BD} > 0$

compared with Eq. (1-6), which is found in most elementary textbooks. The equation correctly predicts that as V_{BS} becomes increasingly negative, more gate voltage is required to invert the channel and the threshold voltage increases. Conversely, when V_{BS} is slightly positive, V_T decreases. Clearly in the interest of not taking a square root of a negative number, Eq. (2-60) would cease being valid if V_{BS} is positive with a magnitude approaching $2\phi_f$. Nonetheless, before V_{BS} reaches $2\phi_f$, we still believe that this equation works well.

It turns out that BSIM3 abandons Eq. (2-60) as soon as V_{BS} exceeds 0, rather than continuously using the equation until $V_{BS} > 2\phi_f$. When $V_{BS} > 0$, BSIM3 approximates $2\phi_f - V_{BS}$ by

$$2\phi_f - V_{BS} = 2\phi_f \left(1 - \frac{V_{BS}}{2\phi_f}\right) \approx 2\phi_f \left(1 + \frac{V_{BS}}{2\phi_f}\right)^{-1} = \frac{(2\phi_f)^2}{2\phi_f + V_{BS}}. \tag{2-61}$$

The square root of the above quantity, which appears in the threshold voltage expression of Eq. (2-60), is further approximated by the following in BSIM3:

$$\sqrt{2\phi_f - V_{BS}} \approx \frac{2\phi_f}{\sqrt{2\phi_f + V_{BS}}} \approx \frac{\sqrt{2\phi_f}}{\sqrt{1 + V_{BS}/(2\phi_f)}} \approx \frac{(\sqrt{2\phi_f})^3}{2\phi_f + V_{BS}/2}. \tag{2-62}$$

Consequently, at $V_{BS} > 0$, BSIM3 adopts the following threshold voltage expression:

$$V_T = \text{VTH0} + \text{K1} \cdot \left(\frac{(\sqrt{2\phi_f})^3}{2\phi_f + V_{BS}/2} - \sqrt{2\phi_f}\right) \quad \text{(when } V_{BS} > 0\text{)}. \tag{2-63}$$

In this BSIM3 implementation, a positive V_{BS} will never lead to numerical difficulty. This is to be contrasted with Eq. (2-60), wherein a positive V_{BS}, if its magnitude is large enough, can result in imaginary numbers during the V_T calculation.

Figure 2-30 illustrates the calculated threshold voltage using BSIM3's Eqs. (2-63) and (2-60). The calculation uses $\text{VTH0} = 0.54$ V and $\text{K1} = 0.53$, extracted from a long-channel device. The experimental dots are threshold voltages determined from the value at which $I_D = 0.1\ \mu\text{A} \cdot L/W$. The drain bias is 0.01 V, low enough such that the drain-induced barrier-lowering effects do not even come close to being significant. The figure demonstrates that BSIM3 approximation is not good when V_{BS} is slightly positive. When V_{BS} approach $2\phi_f$, we do not have experimental data because the threshold voltage cannot be easily identified when the diode junctions turn on. It is expected neither BSIM3's Eq. (2-63) nor Eq. (2-60) is valid in such extreme bias conditions.

Because BSIM3 overestimates the threshold voltage at $V_{BS} > 0$, we expect the channel current (I_{DS}) to be more than what BSIM3 calculates. Besides the threshold

154 BASIC FACTS ABOUT BSIM3

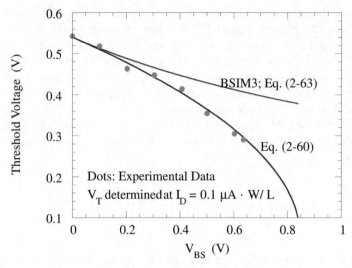

Fig. 2-30 Measured and calculated threshold voltages. The calculated results are from BSIM3's Eqs. (2-63) and (2-60). From numerical-stability consideration, BSIM3 opts for Eq. (2-63) when $V_{BS} > 0$, although the equation is inaccurate. The accurate expression of Eq. (2-60) is used in BSIM3 only at $V_{BS} < 0$.

voltage change, there are other modeling modifications when V_{BS} exceeds 0. One involves the diode current and another, the diode capacitance. We will use the source–bulk diode as the example for discussing these changes at $V_{BS} > 0$. Similar modifications for the drain–bulk diode are made at $V_{BD} > 0$.

When V_{BS} is less than zero, the diode current flowing from the bulk terminal to the source terminal is generally written as (for more exact equations, see the JS entry in Section 3.2)

$$I_{j,\text{SB}} = (\text{JS} \times \text{AS} + \text{JSSW} \times \text{PS}) \cdot \left[\exp\left(\frac{qV_{BS}}{\text{NJ} \cdot kT}\right) - 1 \right] + \text{GMIN} \cdot V_{BS}, \quad (2\text{-}64)$$

where JS, JSSW are diode leakage current density per area and per periphery, respectively. The term $\text{GMIN} \cdot V_{BS}$ is added there to speed up convergence, although it can be interpreted as an extra leakage current which varies linearly with V_{BS}. $I_{j,\text{SB}}$ is negative when V_{BS} is less than zero, meaning that a positive diode leakage current flows from the source to the bulk.

When V_{BS} exceeds zero, BSIM3 provides two similar models to model $I_{j,\text{SB}}$, depending on the value of the parameter IJTH. If IJTH = 0 or negative, then the exact diode equation as that of Eq. (2-64) is used. In this case, when V_{BS} is fairly high such that the diode is heavily turned on, the current calculated from Eq. (2-64) may exceed the largest number allowed in the computer. Although this numerical error gives the circuit designers something to be alerted about and is generally

2.7 SPECIAL OPERATING CONDITIONS: $V_{DS} < 0$, $V_{BS} > 0$, $V_{GS} < 0$, OR $V_{BD} > 0$

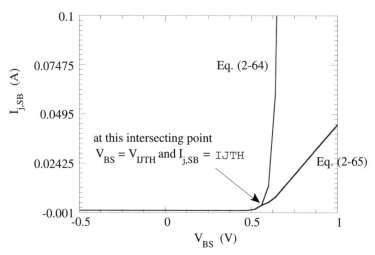

Fig. 2-31 Two diode current models adopted in BSIM3. When IJTH $=0$, the current continues to increase exponentially at forward bias. When IJTH > 0, the current is linearized once the diode current exceeds IJTH. When IJTH < 0, BSIM3 issues a fatal warning.

considered as a good feature, it is sometimes thought that BSIM3 should not allow that to happen. BSIM3 therefore offers another model, whose calculated current when compared with the previous model, is shown in Fig. 2-31. The figure indicates that after $I_{j,SB}$ calculated from Eq. (2-64) reaches IJTH, the second model calculates a current which increases at a much slower pace, at a constant slope equal to that at $I_{j,SB} =$ IJTH. We denote V_{IJTH} as the V_{BS} that results in an $I_{j,SB}$ equal to (IJTH + GMIN $\cdot V_{BS}$). When $V_{BS} > V_{IJTH}$, the second model calculates the diode current as

$$I_{j,SB\_modified} = I_{j,SB}\bigg|_{V_{BS}=V_{IJTH}} + \frac{\partial I_{j,SB}}{\partial V_{BS}}\bigg|_{V_{BS}=V_{IJTH}} \times (V_{BS} - V_{IJTH}). \quad (2\text{-}65)$$

This linear increase prevents $I_{j,SB\_modified}$ from exceeding a computer's largest number. The detailed equations for V_{IJTH} and the derivative $\partial I_{j,SB}/\partial V_{BS}$, are given in Section 3.2, under the entry of JS.

Besides the diode current, the diode capacitance is also affected when V_{BS} exceeds zero. As mentioned in Section 2.3, the overall $C_{j,BS}$ consists of the bottom wall, the isolation-side sidewall, and the gate-side sidewall capacitances. In particular, the bottom-wall capacitance is given by

$$C_{j,SB\_BW} = \text{Bottom\_Wall Area} \times \frac{CJ}{\left(1 - \frac{V_{BS}}{PB}\right)^{MJ}}. \quad (2\text{-}67)$$

156 BASIC FACTS ABOUT BSIM3

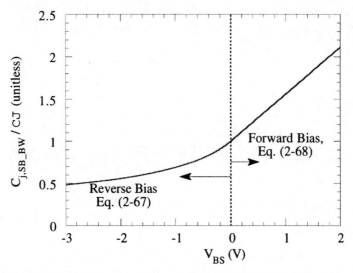

Fig. 2-32 Calculated diode capacitance in BSIM3. When the diode is forward-biased, BSIM3 linearizes the capacitance with the slope obtained at zero bias.

The capacitance is well behaved when the diode is reverse biased, that is, $V_{BS} < 0$. However, as V_{BS} approaches PB, the capacitance of Eq. (2-67) tends infinity. When $V_{BS} > 0$, BSIM3 approximates $C_{j,\text{SB\_BW}}$ as, $C_{j,\text{SB\_BW}}(V_{BS} = 0) + \partial C_{j,\text{SB\_BW}}/\partial V_{BS} \times V_{BS}$:

$$C_{j,\text{SB\_BW}} = \text{Bottom\_Wall Area} \times \text{CJ}\left(1 + \text{MJ}\frac{V_{BS}}{\text{PB}}\right) \quad (V_{BS} > 0). \quad (2\text{-}68)$$

In this equation, $C_{j,\text{SB\_BW}}$ is always finite. The other two components of $C_{j,\text{SB}}$ are also modified accordingly. $C_{j,\text{SB\_BW}}$ as a function of V_{BS} is shown in Fig. 2-32.

REFERENCES AND NOTES

[1] W. Liu, *Fundamentals of III-V Devices: HBTs, MESFETs, HFETs/HEMTs*, New York: Wiley, 1999.

[2] S. Cserveny, "Relationship between measured and intrinsic conductances of MOSFETs," *IEEE Trans. Electron Devices*, vol. 17, pp. 2413–2414, 1990.

[3] A lot of the matrix manipulation can be seen in W. Liu, *Handbook of III-V Heterojunction Bipolar Transistors*, New York: Wiley, 1998. Chapter 8, especially § 8-13 and § 8-14.

[4] D. Harford, "A general model for f^a spectral density random noise with special reference to Flicker noise 1/f," *IEEE Proceeding*, vol. 56, p. 251, March 1968.

[5] R. Jindal, "High frequency noise in fine line NMOS field effect transistors," Proc. IEEE International *Electron Devices Meeting*, pp. 68–71, 1985.
[6] I. Chen, and W. Liu, "High-Speed or Low-Voltage Low-Power Operations," in S. Sze and C. Y. Chang, ed. *ULSI Devices*, New York: Wiley, pp. 547–630, 2000.

3

BSIM3 PARAMETERS

3.1 LIST OF THE PARAMETERS ACCORDING TO FUNCTION

This chapter provides detailed descriptions of the BSIM3 SPICE parameters. Because the word SPICE is used to describe various items, we define some terms to reduce confusion. The circuit in Fig. 1-1 is chosen as an example, for which the input to the SPICE program may appear as

```
IIN 1 0    1MA
RA  1 2 100OHM
RB  2 3 1KOHM
RC  3 0 1KOHM
MN  2 3 0 0 NCH W=5 L=0.35 AS=5 AD=5 PS=9 PD=9 NRD=100
                                              NRS=100 TEMP=85
.OP
.MODEL NCH NMOS LEVEL=8
+ VTH0="0.5"
+ ....
.PRINT OP V(1) V(2) V(3)
.END
```

This is the so-called *circuit netlist*, or sometimes referred to as the SPICE input deck. The circuit netlist provides the information about the circuit, the way the various elements are to be connected. The exact syntax may vary slightly, depending on the SPICE simulator used. The line that begins with "MN 2 3 0 0" is a *MOS device* (or *instance*) *statement*. The statement describes the size of the transistor, its drain–bulk and source–bulk junction areas and perimeters, as well as some information about the source and drain contact resistances. The MOS device statement also renders information about what model characterizes the device. In this case, the device model is called "NCH". The model parameters for the transistor with the "NCH" model, in turn, are given under the .MODEL statement. The model statement makes clear the MOS device is a *n*-channel device (NMOS). The phrase "LEVEL = 8" indicates that this is a BSIM3 model. Depending on the simulator, sometimes other level numbers are used to represent a BSIM3 model. For example, in HSPICE, BSIM3 would be level 49. The entries after the model statement are the BSIM3 *parameter set* (or the SPICE parameter set), which gives the values of the BSIM3 parameters (or sometimes referred to as the *SPICE parameters*.) An argument appearing in the MOS device statement, such as "NRD" for example, is not a SPICE parameter (i.e., not a BSIM3 parameter). However, we use the same Courier font for the arguments in the device statement as for the BSIM3 parameters because their values are modifiable by the user.

Several SPICE simulators, Berkeley SPICE, HSPICE, PSPICE, for example, all support the BSIM3 MOSFET model. However, there is a chance that a BSIM3-look-alike model is not exactly identical to that put on the official BSIM3 web site. For example, a "BSIM3" model adopted in some rf simulation program has a RG parameter (for gate resistance) which does not exist in the official version. We also caution that the arguments specifying the drain junction area (AD), drain junction perimeter (PD), source junction area (AS), and source junction perimeter (PS), in the MOS device statement may also have different default values from that used in BSIM3. For example, in the official BSIM3 code, AD, PD, AS, and PS are defaulted to zero when they are not specified. Some SPICE programs will default AS and AD to 1 µm^2 instead. Nonetheless, we can safely assume that the differences are minor.

We group the BSIM3 parameters according to their functions. We see from the dc equivalent circuit of Fig. 2-2 that there are several distinct elements in a MOSFET. They include: 1) the diodes of the drain–bulk and the source–bulk junctions; 2) the parasitic drain and source terminal resistances; 3) the substrate current due to impact ionization; and finally, and most importantly, 4) the MOS transistor itself, which includes the parasitic overlap and fringing capacitances. BSIM3 utilize several parameters to characterize each of the elements. Some parameters are specifically utilized for one purpose. For example, the parameter CJ is used exclusively for junction capacitance calculation, but not anywhere else. In this case, we say that CJ is a parameter *devoted to* junction capacitance calculation. Other parameters are mainly used for one purpose, but may serve other functions. For example, the parameter TOX is mainly used for calculating the intrinsic transistor current and capacitances. However, TOX also affects the calculation of the substrate current due to impact ionization current, though in a minor way. In this case, we say that TOX is a parameter *mainly used for* intrinsic transistor calculation.

The parameters in parentheses denote temperature-related parameters. In addition, since we are assuming the use of the latest BSIM3 model, LEVEL should be 8 in most SPICE simulators (49 in HSPICE), and VERSION should be 3.2.2 (see the entry of VERSION in Section 3.2). We employ the quasi-static model of BSIM3, not considering its non-quasi-static option (see Section 1.6 for details.) Therefore,

NQSMOD = 0. The exact value for ELM, which is a parameter exclusively used for the non-quasi-static model, is immaterial.

1. Parameters devoted to the junction diodes:
   Current and small-signal conductance calculation: JS, JSSW, NJ, IJTH, (XTI).
   Capacitance and charge calculation: CJ, MJ, PB, CJSW, MJSW, PBSW, CJSWG, MJSWG, PBSWG, (TCJ, TPB, TCJSW, TPBSW, TCJSWG, TPBSWG).
2. Parameters devoted to the parasitic drain source resistances:
   Lumped-resistor approach (see Section 2.2): RDSW, PRWG, PRWB, WR, (PRT).
   Absorbed-resistor approach (see Section 2.2): RSH
3. Parameters devoted to the substrate current due to impact ionization:
   Current and small-signal conductance calculation: ALPHA0, ALPHA1, BETA0
4. MOS transistor:
   Parameters devoted to subthreshold characteristics: VOFF, NFACTOR, CDSC, CDSCD, CDSCB, CIT
   Parameters devoted to threshold voltage calculation: VTH0, NLX, K3, K3B, W0, DVT0, DVT0W, DVT1W, DVT2W, DSUB, ETA0, ETAB
   Parameters mainly used for threshold voltage calculation: K1, K2, DVT1, DVT2, NCH, (KT1, KT1L, KT2)
   Parameter devoted to poly-depletion calculation: NGATE
   Parameters devoted to Early voltage calculation: PCLM, PDIBLC1, PDIBLC2, PDIBLCB, DROUT, PVAG, PSCBE1, PSCBE2
   Parameters devoted to saturation voltage and effective drain–source voltage calculation: A1, A2, DELTA
   Parameters devoted to bulk-charge coefficient (A_{bulk}) calculation: AGS, A0, B0, B1, KETA
   Parameters devoted to effective-channel length and width for I-V calculation: LINT, WINT, LL, LW, LWL, WL, WW, WWL, LLN, LWN, WWN, WLN, DWG, DWB, (Note: LLN, LWN, WWN, WLN are also used for C-V effective-channel length and width calculation.)
   Parameters not devoted to a specific purpose, but is used mainly for channel current calculation: TOX, TOXM, MOBMOD, U0, UA, UB, UC, VSAT, XJ, (UTE, UA1, UB1, UC1, AT, TNOM)
   Parameters which affect the channel current calculation: All of the parameters listed under item 4 (for MOS transistor) thus far.
   Parameters which affect I-V but not C-V calculation: LINT, WINT, A1, A2, DELTA, AGS, MOBMOD, U0, UA, UB, UC, VSAT, PCLM, PDIBLC1, PDIBLC2, PDIBLCB, DROUT, PVAG, PSCBE1, PSCBE2, VOFF, NFACTOR, CDSC, CDSCD, CDSCB, CIT, LL, LW, LWL, WL, WW, WWL (UTE, UA1, UB1, UC1, AT)
   Parameters which affect both I-V and C-V calculation: TOX, TOXM, VTH0, NLX, K3, K3B, W0, DVT0, DVT0W, DVT1W, DVT2W, DSUB, ETA0, ETAB, K1, K2, DVT1, DVT2, NCH, XJ, B0, B1, KETA, LLN, LWN, WWN, WLN (KT1,

3.1 LIST OF THE PARAMETERS ACCORDING TO FUNCTION 161

KT1L, KT2)

Parameters mainly use for intrinsic device capacitance calculation, but may be used for noise calculation: CAPMOD, XPART, DLC, DWC, CLE, CLC, VOFFCV, NOFF, MOIN, ACDE, VFBCV, LLC, LWC, LWLC, WLC, WWC, WWLC

Parameters devoted to gate–drain and gate–source overlap capacitance and charge calculation: CAPMOD, CGD0, CGD1, CGS0, CGS1, CKAPPA, CF

Parameters devoted to gate-bulk overlap capacitance and charge calculation: CGB0

Parameters devoted to noise calculation: NOIMOD, KF, EF, AF, NOIA, NOIB, NOIC, EM

We did not include some processing-related parameters in the above list, such as GAMMA1, GAMMA2, NSUB, VBX, XT and VBM. These parameters are hardly ever used because their effects can be formed with other equivalent parameters. Take GAMMA1 and GAMMA2, for instances: their effects are equivalent to K1 and K2, both to model the body effects. Specifying either set of GAMMA's or K's is sufficient. Usually, K1 and K2 are preferred because they are extracted directly from fitting device characteristics, whereas GAMMA1 and GAMMA2 are thought of as fundamental parameters that can be calculated from doping profiles. The use of GAMMA's takes place primarily when there are no silicon data available.

We did not mention the following five parameters, either: LMIN, LMAX, WMIN, and WMAX, and BINUNIT. The first four are used to restrict the minimum and maximum length and width allowed for the model. They are useful parameters in general, and when binning is used, they are essential. Binning is a reminiscent of non-scalable models and should be discouraged, as discussed in Section 1.12. The parameter BINUNIT determines the unit conversion equation used for binning.

There are a whole bunch of BSIM3 parameters specifically used for binning. We recommend setting all of them to, or leaving them at their default value of, zero, thereby disenabling the binning option in BSIM3.

LA0 LA1, LA2, LAGS, LALPHA0, LAT, LB0, LB1, LBETA0, LCDSC, LCDSCB, LCDSCD, LCF, LCGD1, LCGS1, LCIT, LCKAPPA, LCLC, LCLE, LDELTA, LDROUT, LDSUB, LDVT0, LDVT0W, LDVT1, LDVT1W, LDVT2, LDVT2W, LDWB, LDWG, LETA0, LETAB, LK1, LK2, LK3, LK3B, LKETA, LKT1, LKT2, LNFACTOR, LNLX, LPCLM, LPDIBLC1, LPDIBLC2, LPDIBLC, LPRT, LPRWB, LPRWG, LPSCBE1, LPSCBE2, LPVAG, LRDSW, LU0, LUA, LUA1, LUB, LUB1, LUC, LUC1, LUTE, LVBM, LVBX, LVFBCV, LVOFF, LVSAT, LVTH0, LW0, LWR, LXJ, LXT; and PA0, PA1, PA2, PAGS, PALPHA0, PAT, PB0, PB1, PBETA0, PCDSC, PCDSCB, PCDSCD, PCF, PCGD1, PCGS1, PCIT, PCKAPPA, PCLC, PCLE, PDELTA, PDROUT, PDSUB, PDVT0, PDVT0W, PDVT1, PDVT1W, PDVT2, PDVT2W, PDWB, PDWG, PETA0, PETAB, PK1, PK2, PK3, PK3B, PKETA, PKT1, PKT2, PNFACTOR, PNLX, PPCLM, PPDIBLC1, PPDIBLC2, PPDIBLCB, PPRT, PPRWB, PPRWG, PPSCBE1, PPSCBE2, PPVAG, PRDSW, PU0, PUA, PUA1, PUB1, PUC, PUC1, PUTE, PVBM, PVBX, PVFBCV, PVOFF, PVSAT, PVTH0, PW0, PWR, PXJ, PXT: and at last, WA0 WA1, WA2, WAGS, WALPHA0, WAT, WB0, WB1, WBETA0, WCDSC, WCDSCB, WCDSCD, WCF, WCGD1, WCGS1, WCIT, WCKAPPA, WCLC, WCLE, WDELTA, WDROUT, WDSUB, WDVT0, WDVT0W, WDVT1, WDVT1W,

WDVT2, WDVT2W, WDWB, WDWG, WETA0, WETAB, WK1, WK2, WK3, WK3B, WKETA, WKT1, WKT2, WNFACTOR, WNLX, WPCLM, WPDIBLC1, WPDIBLC2, WPDIBLCB, WPRT WPRWB WPRWG, WPSCBE1, WPSCBE2, WPVAG, WRDSW, WU0, WUA, WUA1, WUB, WUB1, WUC, WUC1, WUTE, WVBM, WVBX, WVFBCV, WVOFF, WVSAT, WVTH0, WW0, WWR, WXJ, WXT.

The list of the binning parameters is excessively long! Inclusion of binning parameters makes BSIM3 a model with more than 300 parameters.

Finally, the above list assumes that we are using the BSIM3v3.2 parameter set (which is identical to the parameter set in BSIM3v3.2.1 and BSIM3v3.2.2). There are some parameters which are introduced in BSIM3v3.2 but are nonetheless absent in BSIM3v3.1. These parameters include: ACDE, ALPHA1, IJTH, MOIN, NOFF, LLC, LWC, LWLC, TCJ, TCJSW, TCJSWG, TOXM, TPB, TPBSW, TPBSWG, VERSION, VFB, VOFFCV, WLC, WWC, and WWLC.

3.2 ALPHABETICAL GLOSSARY OF BSIM3 PARAMETERS

We list all of the BSIM3 parameters in an alphabetical order. Each entry provides quick useful information about a parameter, along with its default value and unit. If a parameter does not appear in the ensuing pages, then the parameter is likely not an official BSIM3 parameter. We have mentioned an example, RG, which is used to denote the gate resistance in an unofficial implementation of BSIM3.

We describe the BSIM3 parameters with the understanding that some models in use are still BSIM3v3.1, even though BSIM3v3.2 (or BSIM3v3.2.1 or BSIM3v3.2.2) are quite common nowadays. New parameters specifically introduced in BSIM3v3.2 and its derivatives are pointed out. The fine details in the differences of these versions, in the context of examining the effects of the parameters, are elaborated.

Just about all the equations here are not exactly the same as those used in BSIM3. We utilize many approximate equations so that the general concept stands out, rather than getting ourselves bogged down by fine details. The exact equations can be found in Appendix A.

A0 (default = 1, unitless) channel-length dependency parameter of the bulk-charge coefficient. In a long-channel MOSFET without the consideration of the bulk charges, the saturation voltage is equal to $V_{GS} - V_T$ and the drain current is proportional to $[(V_{GS} - V_T) \cdot V_{DS} - V_{DS}^2/2]$. (The saturation voltage, $V_{DS,\text{sat}}$, is the drain bias required to pinch off the inversion charge at the drain end.) When the bulk charge is included, $V_{DS,\text{sat}}$ and I_D are modified to the following:

$$V_{DS,\text{sat}} = \frac{V_{GS} - V_T}{1 + \delta}; \tag{3-1}$$

$$I_D = \frac{W_{\text{eff}}}{L_{\text{eff}}} \mu_{\text{eff}} C'_{ox} \left[(V_{GS} - V_T)V_{DS} - \frac{1}{2}(1 + \delta)V_{DS}^2 \right]. \tag{3-2}$$

3.2 ALPHABETICAL GLOSSARY OF BSIM3 PARAMETERS

The bulk-charge factor δ is used to account for the amount of the bulk charge's variation with the channel-to-substrate bias. For long-channel device, δ is given by

$$\delta = \frac{\text{K1}}{2\sqrt{2\phi_f - V_{BS}}}. \tag{3-3}$$

This is basically Eq. (1-7), except that we replaced γ, the commonly used symbol for the body-effect coefficient, by K1, a BSIM3 parameter. Its value as a function of substrate bias and doping level can be found from Fig. 1-18. The strong-inversion surface potential, $2\phi_f$, is given in Eq. (1-6b).

Instead of the bulk-charge *factor* δ typically used in textbooks, BSIM3 uses the bulk-charge *coefficient* A_{bulk}, which is equal to $1 + \delta$ in a long-channel device. When A_{bulk} is used, the long-channel $V_{DS,\text{sat}}$ and I_D in BSIM3 are approximately written as

$$V_{DS,\text{sat}}^{\dagger} = \frac{V_{GS} - V_T}{A_{bulk}}; \tag{3-4}$$

$$I_D^{\dagger} = \frac{W_{eff}}{L_{eff}} \mu_{eff} C_{ox}' \left[(V_{GS} - V_T) V_{DS} - \frac{1}{2} A_{bulk} V_{DS}^2 \right]. \tag{3-5}$$

We denote the saturation voltage with a dagger to emphasize that the voltage given here is a dramatically simplified version of the actual saturation voltage expression used in BSIM3, which accounts for the short-channel and long-channel devices simultaneously. $V_{DS,\text{sat}}^{\dagger}$ given as such, is useful mainly for the discussion of the A0 parameter (as well as other parameters relating to the bulk-charge coefficient). The dagger appearing in the $I_D^{\dagger}$ expression has similar connotation.

In order to account for short-channel effects and narrow-width effects, the actual A_{bulk} used in BSIM3 is more complicated than just $1 + \delta$. When the channel is turned on, we can express A_{bulk} as

$$A_{bulk} = \left\{ 1 + \frac{\text{K1}}{2\sqrt{2\phi_f - V_{BS}}} \left[\frac{\text{A0} \cdot L_{eff}}{L_{eff} + 2\sqrt{\text{XJ} \cdot X_{dep}}} \right. \right.$$
$$\left. \left. \times \left(1 - \text{AGS}(V_{GS} - V_T) \left(\frac{L_{eff}}{L_{eff} + 2\sqrt{\text{XJ} \cdot X_{dep}}} \right)^2 \right) + \frac{\text{B0}}{W_{eff} + \text{B1}} \right] \right\}$$
$$\times \frac{1}{1 + \text{KETA} \cdot V_{BS}}. \tag{3-6}$$

We focus on A0's effects by assuming AGS, B0, and KETA all equal to zero. In this case, A_{bulk} adopted in BSIM3 simplifies to

$$A_{bulk} \approx 1 + \delta \times \frac{\text{A0} \cdot L_{eff}}{L_{eff} + 2\sqrt{\text{XJ} \cdot X_{dep}}}, \tag{3-7}$$

where XJ, the drain/source junction depth, is a SPICE parameter, and X_{dep}, the depletion thickness in the bulk, is given in Appendix A (Eq. A-37). As mentioned, A_{bulk} should be equal to $1 + \delta$ for long-channel devices. If L_{eff} is large, then A_{bulk} given in Eq. (3-7) simplifies to $1 + \delta \times$ A0. A0 can be viewed as a fudging factor used to better fit the device characteristics than if A0 were identically equal to 1. The term $L_{eff}/(L_{eff} + 2\text{sqrt}(\text{XJ} \cdot X_{dep}))$ is attached to A0. This term allows A_{bulk} to become smaller in value as the channel length shortens. This decrease agrees with the empirical observation that the bulk-charge effects are less pronounced in short-channel devices.

While A0 is viewed as a fitting parameter, it should not depart significantly from its default value of 1, so that $A_{bulk} \approx 1 + \delta$ for long-channel devices. A0 should always be greater than 0. Generally it is limited to values between 0.5 and 1.5.

Figure 3-1 illustrates the effects of changing A0 and AGS from their default values of 1 and 0, respectively, for a short-channel device. The details about AGS's effects are discussed separately in the AGS entry. When A0 increases from 1 to 1.5, Eq. (3-6) indicates that A_{bulk} increases. Consequently, I_D according to Eq. (3-5) decreases. By the nature of Eq. (3-7), A0 affects only the short-channel devices. A change of A0 from 1 to 1.5 results in negligible modification of I_D in the long-channel counterpart.

A1 (default = 0, unit = V^{-1}) first non-saturation factor for RDSW $\neq 0$. This parameter bears no relationship with A0. As elaborated in the entry of RDSW, BSIM3 offers two methods to incorporate the effects of the parasitic drain/source resis-

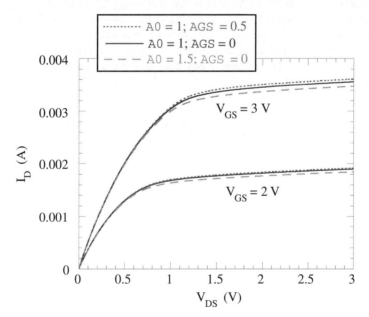

Fig. 3-1 Simulated I-V characteristics with A0 and AGS modified from their default values of 1 and 0, respectively, for a short-channel device.

tances. In one method, wherein RDSW is identical to zero, the parasitic resistances are incorporated as lumped resistances. In this case, the lumped-resistance approach, BSIM3 calculates the saturation voltage according to Eq. (3-4). In another method, RDSW $\neq$ 0. Only the effects of the parasitics resistances on I_D and conductances are reproduced, although the resistances themselves do not appear as part of the device's input impedance. This is the absorbed-resistance approach, in which the saturation voltage is no longer given by Eq. (3-4), but instead by some complicated function of A1 $\cdot (V_{GS} - V_T) +$ A2. Without delving into the equations, we simply state that A1 and A2 are parameters which ultimately determine $V_{DS,\text{sat}}$ when RDSW $\neq$ 0. As RDSW approaches zero, A1 $\cdot (V_{GS} - V_T) +$ A2 should approach unity. Therefore, the defaults of A1 and A2 are 0 and 1, respectively. While we view A1 and A2 as purely fitting parameters, we expect A1 $\cdot (V_{GS} - V_T) +$ A2 to not deviate much from unity.

Because A1 and A2 control the magnitude of saturation voltage, these two parameters are fairly important parameters which affect the drain current in transition between the linear and the saturation regions. However, when RDSW = 0, BSIM3 uses the lumped-resistance approach to determine the device characteristics. A1 and A2 would then produce no effect on the device characteristics.

A2 (default = 0, unit = $\mathbf{V}^{-1}$) second non-saturation factor for RDSW $\neq$ 0. This parameter bears no relationship with A0 but is intimately related to A1. See the entry for A1.

ACDE (default = 1, unit = m/V) exponential coefficient for calculating the centroid of the electron charge distribution in the charge-thickness capacitance model (CAPMOD = 3). The parameter is not used in other capacitance models. It does not affect the I-V characteristics.

CAPMOD = 2, being the most advanced capacitance model at the time of BSIM3v3.1, is considered to be a fairly good model because the C-V curves are smooth and continuous. It is not piecewise linear, as in the CAPMOD = 0 or 1 model. Nonetheless, the CAPMOD = 2 model employs two common assumptions which preclude it from accurately modeling C_{gg}'s voltage dependence [1], particularly when TOX < 100 Å. The first assumption is that the charge layer in the channel is of zero thickness, so that both the accumulation and inversion charges are located right at the semiconductor/oxide interface. The second assumption is the approximation that the surface potential ϕ_s is equal to a constant of $2\phi_f$ in strong inversion, given in Eq. (1-6b). In CAPMOD = 3, both assumptions are removed, and along with the removal, two additional SPICE parameters are introduced: ACDE and MOIN.

According to a detailed quantum simulation [2], the channel charge layer (either in accumulation or inversion) is not a uniform charge layer. It varies in concentration along the normal direction (from the gate toward the bulk). The centroid of the charge concentration, the location where the highest concentration is at, is roughly 16 Å away from the oxide interface. This is for a NMOS with 30 Å oxide thickness, operated in the strong accumulation condition. As V_{GS} is made less negative so that the transistor steps into the depletion operating region, the centroid moves toward the bulk, being roughly 20 Å. As V_{GS} increases to slightly above zero so that the transistor is in the subthreshold region, the centroid becomes only slightly above 20 Å. Further increases of V_{GS} after the transistor enters saturation, the centroid regresses toward the oxide interface. When the transistor is in the heavily inverted operating region, the centroid moves toward 15 Å. For compact modeling, the effective separation between the gate electrode and the channel can be taken to be the sum of TOX and the centroid (X_{DC}). Hence, in CAPMOD = 3, the effective oxide capacitance per unit area ($C'_{ox,\text{eff}}$) is no longer written as ϵ_{ox}/TOX. If the oxide and the silicon had the same dielectric constant, we would write $C'_{ox,\text{eff}}$ as $\epsilon_{ox}/(\text{TOX}+X_{DC})$. However, since they are different, BSIM3 expresses $C'_{ox,\text{eff}}$ as

$$C'_{ox,\text{eff}} = \frac{\epsilon_{ox}}{\text{TOX}} // \frac{\epsilon_s}{X_{DC}}. \tag{3-8}$$

The subscript s denotes silicon, as opposed to oxide. The centroid location is given by two expressions, depending on the bias condition. In the accumulation and depletion regions,

$$X_{DC} = \frac{L_{Debye}}{3} \exp\left[10^{-8}\text{ACDE}\left(\frac{\text{NCH}}{N_0}\right)^{-1/4} \cdot \frac{V_{GS}-V_{BS}-V_{fb,zb}}{\text{TOX}}\right], \tag{3-9}$$

where N_0, an arbitrary reference level, is 2×10^{16} cm^{-3} and L_{Debye}, the Debye length of the channel region, is given by

$$L_{Debye} = \sqrt{\frac{\epsilon_s \cdot kT/q}{q\text{NCH} \cdot 10^6}}. \tag{3-10}$$

The 10^6 factor in the Debye length expression is for unit conversion (NCH is in cm^{-3} whereas L_{Debye} is in m). The 10^{-8} factor in the square brackets of Eq. (3-9) is introduced so that ACDE has values around unity. In the event that ACDE is an arbitrarily large number, the exponent in Eq. (3-9) could assume a rather large number. BSIM3 therefore limits X_{DC} to be $L_{Debye}/3$, which, for a typical doping of NCH = 2×10^{17} cm^{-3}, is about 30 Å. If the oxide thickness greatly exceeds 30 Å, there is no need to use CAPMOD = 3.

ACDE does not affect the capacitance in subthreshold and strong inversion operating regions, for which BSIM3 uses an alternative centroid expression.

In addressing the second assumption, CAPMOD = 3 calculates the surface potential in inversion as $2\phi_f + \phi_\delta$, where ϕ_δ is given by

$$\phi_\delta = \frac{kT}{q} \ln\left[\frac{V_{GST,\text{effCV}}(V_{GST,\text{effCV}}+2\text{K1}\cdot\sqrt{2\phi_f})}{\text{MOIN}\cdot\text{K1}^2\cdot(kT/q)}\right]. \tag{3-11}$$

(A similar equation, found in the BSIM3 manual, is likely erroneous. The above equation is correct, and is that implemented in the codes.) In the equation, $V_{GST,\text{effCV}}$ is given in the entry of NOFF. MOIN is the parameter used to adjust the amount of the delta potential, which lowers the amount of charge inversion. For example, during strong inversion, the textbook formula for the inversion charge density per unit area at a certain location x along the channel is given by, $Q'_{inv}(x) = -C'_{ox} \cdot (V_{GS} - V_T - V_{CS}(x))$, where $V_{CS}(x)$ is the channel potential with respect to the source. Therefore, at the source, $Q'_{inv}(x)$ is $C'_{ox} \cdot (V_{GS} - V_T)$ whereas at the drain end, $Q'_{inv}(x)$ is $C'_{ox} \cdot (V_{GS} - V_T - V_{DS})$ or $C'_{ox} \cdot (V_{GS} - V_T - V_{DS,\text{sat}})$, depending on whether the transistor is in saturation. Now, in CAPMOD = 3, the textbook-like equation is essentially modified to

$$Q'_{inv}(x) = -C'_{ox,\text{eff}} \times (V_{GS} - V_T - \phi_\delta - V_{CS}(x)). \tag{3-12}$$

Note that C'_{ox} is now modified to $C'_{ox,\text{eff}}$ to account for the finite thickness of the charge layer. Further, because the surface potential is no longer fixed at $2\phi_f$, the threshold voltage essentially increases by an amount equal to ϕ_δ. Hence, the capacitance is lower when the non-constant surface potential is considered.

In summary, ACDE is a parameter used to calculate the centroid distance in the accumulation region. Its default is 1. The higher the ACDE value, the thicker the centroid value. Hence, the gate capacitance would drop more rapidly with increasing V_{GS}. MOIN in contrast, is a parameter modeling the incremental of surface potential that needs to be overcome by the gate voltage before the charge is inverted. The smaller the MOIN value, the larger the incremental potential (ϕ_δ), and the more rapidly the gate capacitance decreases with V_{GS} in the inversion region. The default of MOIN is 15. Independent of the PARAMCHK value, BSIM3 will check if ACDE exceeds 1.6 or falls below 0.4. If so, a warning message is issued. Similarly, if MOIN is below 5 or above 25, a warning message appears.

AF (default = 1, unitless) current exponent in calculating the Flicker noise when NOIMOD = 1 or 4. This parameter is not used if NOIMOD = 2 or 3. See the entry NOIMOD for details.

AGS (default = 0, unit = V^{-1}) gate-bias coefficient of the body-charge coefficient, A_{bulk}. A_{bulk} is not a model parameter, but an intermediate variable used by BSIM3 to determine the saturation voltage and current (Eqs. 3-4 and 3-5). When $V_{GS} \gg V_T$, A_{bulk} in BSIM3 is expressed as

$$A_{bulk} \approx \left\{ 1 + \left[\delta \times \frac{A0 \cdot L_{eff}}{L_{eff} + 2\sqrt{XJ \cdot X_{dep}}} \right. \right.$$
$$\left. \times \left(1 - AGS(V_{GS} - V_T) \left(\frac{L_{eff}}{L_{eff} + 2\sqrt{XJ \cdot X_{dep}}}\right)^2 \right) + \frac{B0}{W_{eff} + B1} \right] \right\}$$
$$\times \frac{1}{1 + KETA \cdot V_{BS}}, \tag{3-13}$$

where δ is given by Eq. (3-3). A_{bulk} attains its theoretical value of $1 + \delta$ in an ideal long-channel device. The factor $L_{eff}/(L_{eff} + 2\text{sqrt}(XJ \cdot X_{dep}))$ is used to take care of

A_{bulk}'s dependence on channel length, that the bulk-charge effects diminish in short-channel devices. This was described in the entry of A0.

By a similar token, AGS accounts for the V_{gs} dependence of A_{bulk}. (B0 and B1 are to take care the narrow-width effects and KETA, V_{BS}'s effect on A_{bulk}). The V_{GS}'s dependence of A_{bulk}, the term which multiplies AGS in Eq. (3-13), is derived from some hand-waving argument. Hence, we view AGS as a fitting parameter rather than a purely physical parameter. As such, any value for AGS is deemed acceptable as long as it does not cause anomalies in the I-V characteristics.

Figure 3-1 illustrates the effects of changing A0 and AGS from their default values of 1 and 0, respectively, for a short-channel device. The details about A0's effects are discussed under the entry of A0. When AGS increases from 0 to 0.5, Eq. (3-13) indicates that A_{bulk} decreases, with a larger magnitude of decrement when V_{GS} is large. Consequently, I_D according to Eq. (3-5) increases, and more so for the $V_{GS} = 3$ V curve compared to 2 V.

ALPHA0 (default = 0, unit = m/V)   the first parameter of the substrate current due to impact ionization. BSIM3 attempts to model the impact ionization near the drain with two sets of parameters. One set involves PSCBE1 and PSCBE2, which are discussed in their respective entries. We prefer to turn off the effects of these parameters by setting PSCBE2 = 0 and PSCBE1 to 10^{30} (some large number). These parameters, if used, would yield right drain current behavior but wrong bulk current. (See the entry PSCBE1 for details.) This wrong simulation result can cause the current-mirror circuit discussed in Section 2.2 to malfunction. Instead, we use exclusively the second set of parameters, ALPHA0, ALPHA1, and BETA0, to model the impact ionization current.

When V_{DS} is high, a large voltage is dropped across the depletion region near the drain. The resultant high field accelerates the electrical carriers as they flow through the channel. If these accelerated carriers acquire enough energy, they generate electron-hole pairs upon impacting the lattice. The generated holes move in the direction of the field toward the substrate. The portion of the bulk terminal current (I_B) due to these holes from impact ionization is denoted as I_{sub} in BSIM3, with the subscript *sub* meaning the substrate. The remaining components of I_B are the leakage currents associated with the drain–bulk and source–bulk diodes. I_B in BSIM3 is given as

$$I_B = -(I_{sub} + I_{j,\text{DB}} + I_{j,\text{SB}}). \qquad (3\text{-}14)$$

A minus sign is placed to reflect the convention that the current flowing *into* a terminal is considered positive.

The impact-generated holes move toward the substrate, but the impact-generated electrons go in the opposite direction, toward the drain. The drain current resulting from these electrons has the same magnitude as I_{sub}. The terminal drain current (I_D) consists of I_{sub}, the leakage current from the drain–bulk diode, and most importantly, the channel current flowing through the MOSFET device to the source (I_{DS}):

$$I_D = I_{DS} + I_{sub} + I_{j,\text{DB}}. \qquad (3\text{-}15)$$

Under normal FET operation, the diodes are all reverse biased, so $I_{j,\mathrm{DB}}$ and $I_{j,\mathrm{SB}}$ are negligible. In the absence of impact ionization, I_{sub} is zero altogether and $I_B \sim 0$ and $I_D \sim I_{DS}$, consistent with our conception of a normal MOSFET operation. (*Note*: In BSIM3v3.1 or earlier, I_{sub} is calculated but somehow never gets added to the drain and bulk current expressions. The two terminal currents in BSIM3v3.1 are written as $I_D = I_{DS} + I_{j,\mathrm{DB}}$ and $I_B = -I_{j,\mathrm{DB}} - I_{j,\mathrm{SB}}$. The impact ionization effects are effectively not modeled!)

In BSIM3, I_{sub} is modeled as

$$I_{sub} = \left(\frac{\mathtt{ALPHA0}}{L_{eff}} + \mathtt{ALPHA1}\right)(V_{DS} - V_{DS,\mathrm{eff}})\exp\left[-\frac{\mathtt{BETA0}}{V_{DS} - V_{DS,\mathrm{eff}}}\right] \times I_{DS}.$$

(3-16)

$V_{DS,\mathrm{eff}}$ is the smoothing function defined in BSIM3 to smooth out the transition between the linear and saturation regions (Section 1.4). It is identical to V_{DS} when the externally applied V_{DS} is small, and saturates to $V_{DS,\mathrm{sat}}$ when V_{DS} is large. Hence, at small values of V_{DS}, $V_{DS} - V_{DS,\mathrm{eff}} \to 0$ and I_{sub} is ≈ 0. In contrast, as V_{DS} increases, $V_{DS,\mathrm{eff}}$ approaches $V_{DS,\mathrm{sat}}$, and I_{sub} attains some finite value. Therefore, Eq. (3-16) is able to reproduce the experimentally measured bell-shaped curves of I_{sub} vs. V_{GS}, as shown in Fig. 2-3.

Although Eq. (3-16) is successful in producing the bell-shaped I_{sub}, it fails to simulate the breakdown characteristics of the MOS transistor. Basically, BSIM3 models the impact ionization accurately only when the impact ionization current is small compared to the intrinsic device current. BSIM3 is not able to predict the breakdown voltage nor the bipolar snap-back effects, which are events taking place at large impact ionization currents. The region of accuracy was shown in Fig. 2-4, to the left of the solid line.

The fundamental reason that BSIM3 cannot predict the breakdown voltage is that I_{sub}, as given in Eq. (3-16), approaches infinity only as V_{DS} tends to infinity. However, based on our knowledge of breakdown characteristics, I_{sub} should approach infinity at some finite V_{DS} value, which is experimentally found to be between 6–8 V in modern CMOS digital devices. This discrepancy is illustrated in Fig. 3-2, which plots the drain terminal current equal to the sum of I_{DS} and I_{sub}. There are several papers which discuss the manners at which the BSIM3 model can be modified. One of them [3] uses an I_{sub} expression nearly identical to Eq. (3-16), with the exception that I_{DS} in Eq. (3-16) is replaced by $M \cdot I_{DS}$ where M is the impact multiplication factor. With this replacement, Ref. [3] is able to accurately represent the sharp current increase associated with the breakdown. There are other improvements in the reference to account for the parasitic bipolar transistor action.

When `ALPHA0`, `ALPHA1`, and `BETA0` in a parameter set are at their default values of 0, 0, and 30, respectively, I_{sub} is identically zero. Essentially, BSIM3's I_{sub} model is turned off. This does not necessarily mean that the device lacks the substrate current. It means, more likely, that the substrate current is not adequately modeled.

170 BSIM3 PARAMETERS

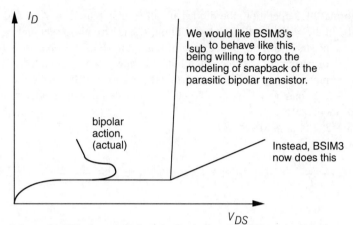

Fig. 3-2 Schematic I-V characteristics obtained from BSIM3 and actual measurement. BSIM3's I_{sub} expression is not capable of producing the rapid current rise associated with breakdown.

NMOS has a higher substrate current than PMOS because the electron has a longer mean free path than the hole. A electron gains more energy before impacting with the lattice, thus is more likely than a hole to generate electron-hole pairs. We expect ALPHA0 to be larger in a NMOS than a PMOS.

Because BSIM3 cannot predict the breakdown characteristics of the device, sometimes a reverse-biased diode can be placed in parallel with the drain and source terminals to implement the breakdown effects. The total drain current is taken to be the sum of the MOS device current and the diode current. This way, when V_{DS} is greater than the reverse breakdown of the diode, the overall drain current increases dramatically. Usually, this drastic change in I-V characteristics is enough to alert a designer that something is wrong and that he should not operate the circuit beyond the specified diode breakdown voltage. This diode-remedy, while representing an improvement over the existing BSIM3, is truly applicable only when the total drain current is small. The moment the total drain current becomes large, parasitic bipolar transistor intrinsic to the MOS turns on and the snap-back behavior of the drain current occurs. At this moment, neither BSIM3 nor the diode-remedy is accurate in reproducing the I-V characteristics. (See Section 2.2 for more details.)

When the term (ALPHA0/L_{eff} + ALPHA1) in Eq. (3-16) is less than or equal to zero, BSIM3 equates I_{sub} to zero.

ALPHA1 (default = 0, unit = V^{-1})   the modified first parameter to account for length variation in the calculation of I_{sub}. As elaborated in the entry of ALPHA0, BSIM3's impact ionization model works fine when the calculated impact ionization current is small. The model does not adequately handle the breakdown behavior or the snap-back I-V characteristics associated with the parasitic bipolar transistor action.

3.2 ALPHABETICAL GLOSSARY OF BSIM3 PARAMETERS

AT (default = 3.3 × 10⁴, unit = m/s) temperature coefficient for carrier saturation velocity. VSAT given in the SPICE parameter set is the carrier saturation velocity *at the nominal temperature*. The nominal temperature, TNOM, is the temperature at which the SPICE parameter set is extracted. It may well differ from the operating device temperature in a particular circuit simulation (see the entry TNOM for details). In the latter case, the actual saturation velocity used in all BSIM3 calculation is modified from the VSAT specified in the SPICE parameter set.

For convenience, we shall denote VSAT specified in the SPICE parameter set as VSAT(TNOM); that is, the saturation velocity at the nominal temperature. We denote VSAT(T_{device}) as the actual saturation velocity used in the BSIM3 calculation when the device temperature of concern is T_{device}. The relationship between them is

$$\text{VSAT}(T_{device}) = \text{VSAT}(\text{TNOM}) - \text{AT} \times \left[\frac{T_{device} + 273.15}{\text{TNOM} + 273.15} - 1\right], \quad (3\text{-}17)$$

where AT is a SPICE parameter characterizing the rate at which the saturation velocity varies as a function of the temperature. AT is a positive number, to reflect the experimental fact that carrier saturation velocity decreases as the device temperature increases. AT has a default value of 3.3×10^4 m/s. When T_{device} turns out to be equal to TNOM, then VSAT specified in the SPICE parameter set is precisely the value used in BSIM3 calculation.

What happens when the device temperature is high enough such that VSAT(T_{device}) given by Eq. (3-17) decreases to a negative value? BSIM3 will catch this unphysical event and issues a fatal error. There is another BSIM3 checking mechanism, which requires the parameter PARAMCHK to be turned on. When PARAMCHK is set to 1, BSIM3 will start issue warning when VSAT(T_{device}) decreases below 1×10^3 m/s.

AT is an important parameter modeling the temperature dependence of the drain current, but has no effect when T_{device} = TNOM. It should always be a positive value (so that the saturation velocity decreases with increasing temperature). Values within 20% of the default value are considered acceptable.

B0 (default = 0, unit = m) channel-width coefficient for the calculation of the body-charge coefficient, A_{bulk}. A_{bulk} is not a model parameter, but an intermediate variable used by BSIM3 to determine the saturation voltage and the current (Eqs. 3-4 and 3-5). When $V_{GS} \gg V_T$, A_{bulk} in BSIM3 is expressed as

$$A_{Bulk} \approx \left\{1 + \left[\delta \times \frac{\text{A0} \cdot L_{eff}}{L_{eff} + 2\sqrt{\text{XJ} \cdot X_{dep}}}\right.\right.$$
$$\left.\left. \times \left(1 - \text{AGS}(V_{GS} - V_T)\left(\frac{L_{eff}}{L_{eff} + 2\sqrt{\text{XJ} \cdot X_{dep}}}\right)^2\right) + \frac{\text{B0}}{W_{eff} + \text{B1}}\right]\right\}$$
$$\times \frac{1}{1 + \text{KETA} \cdot V_{BS}}, \quad (3\text{-}18)$$

where δ is given by Eq. (3-3). A_{bulk} attains its theoretical value of $1 + \delta$ in an ideal long-channel device. The factor $L_{eff}/(L_{eff} + 2\text{sqrt}(\text{XJ} \cdot X_{dep}))$ is used to take care of A_{bulk}'s dependence on channel length, that the bulk-charge effects diminish in short-channel devices. This was described in the entry for A0.

Likewise, B0 and B1 are to take care the narrow-width effects of A_{bulk}. (AGS accounts for the V_{GS} dependence, and KETA, V_{BS}'s effect on A_{bulk}). Both B0 and B1 can be treated as fitting parameters. However, since the bulk-charge effects increase in narrow-width devices, B0 should be greater than zero. In addition, B1, the offset to the effective channel width, should be greater than the negative of the minimum transistor width. Otherwise, the term $\text{B0}/(W_{eff} + \text{B1})$ becomes negative or infinity, where W_{eff} is the effective device width for dc calculations.

B1 (default = 0, unit = m) channel-width offset for the calculation of the body-charge coefficient, A_{bulk}. A_{bulk} is not a model parameter. It is an intermediate variable used by BSIM3 to determine the saturation voltage and the current. See the entry B0 for details.

BETA0 (default = 0, unit = V^{-1}) the second parameter of the substrate current due to impact ionization. As elaborated in the entry ALPHA0, BSIM3's impact ionization model is accurate only when the impact ionization current is small. The model does not adequately handle the breakdown behavior or the snap-back I-V characteristics associated with the parasitic bipolar transistor action. When BETA0 is less than zero, BSIM3 equates the substrate current due to impact ionization to zero.

BINUNIT (default = 1, unitless) binning unit selector. Binning is a process through which physical parameters are modified to acquire some geometrical dependence, in order that a single model card can be used for devices of varying W and L's. A good parameter set should not rely on binning. All the binning parameters (listed in Section 3-1) should take on their default value of 0. However, there will be occasions when a good model cannot be generated in a timely fashion, then binning becomes a necessary compromise.

When a parameter (P) is binned, the true parameter value used in the SPICE calculation is given by

$$P|_{binned\ value} = P|_{unbinned} + \frac{\text{L}P}{L_{eff}} + \frac{\text{W}P}{W_{eff}} + \frac{\text{P}P}{L_{eff} \times W_{eff}}, \quad (3\text{-}19)$$

where L_{eff} and W_{eff} are the effective-channel length and width, respectively. They are given in the entries of LINT and WINT). Let us take P to be the body-effect parameter K1, for example. Then the variables $P|_{unbinned}$, LP, WP, and PP in the above equation correspond to the SPICE parameters K1, LK1, WK1, and PK1, respectively. Suppose further that K1, LK1, WK1, and PK1 are 0.41, 0.005, 0.04, and 0.02, respectively, and that effective-channel length is 0.5 µm and the effective channel width is 4 µm. When BINUNIT = 1, BSIM3 will substitute the values for

L_{eff} and W_{eff} assuming that their unit is μm. Therefore, the binned value for the body-effect parameter is

$$\text{K1}|_{binned\ value} = 0.41 + \frac{0.005}{0.5} + \frac{0.04}{4} + \frac{0.02}{0.5 \times 4} = 0.44 \quad \text{(if BINUNIT} = 1\text{).}$$

If instead BINUNIT = 0, BSIM3 will substitute the values for L_{eff} and W_{eff} assuming that their unit is m. Since the effective-channel length is 0.5 μm, $L_{eff} = 5 \times 10^{-7}$ when the unit is in meter. Similarly, $W_{eff} = 1 \times 10^{-6}$. In this case, the actual value for the body-effect parameter is

$$\text{K1}|_{binned\ value} = 0.41 + \frac{0.005}{5 \times 10^{-7}} + \frac{0.04}{4 \times 10^{-6}} + \frac{0.02}{2 \times 10^{-12}} \quad \text{(if BINUNIT} = 0\text{).}$$

We did not carry out the calculation because clearly the values of LK1, WK1, and PK1 are meant to be used with BINUNIT = 1.

CAPMOD (default = 2 in BSIM3v3.1 and 3 in BSIM3v3.2, unitless) CAPMOD can be 0, 1, 2 or 3. Figure 3-3 plots $C_{gg,t}$ to illustrate the differences of the four models. The CAPMOD = 0 and 1 models produce capacitances which are piece-wise continuous, but the slopes of the capacitances are discontinuous. In contrast, CAPMOD = 2 and 3 capacitance models are continuous, either in the value itself or its derivatives. This nice property is obtained at the cost of model complexity. In the interest of avoiding numerical difficulties during the SPICE simulation, we prefer the CAPMOD = 2 and 3 models. The CAPMOD = 0 and 1 models are provided in BSIM3v3.2 mainly to maintain backward compatibility with previous capacitance models. In particular, CAPMOD = 0 corresponds roughly to the capacitance model used in BSIM, and CAPMOD = 1 resembles that used in BSIM2. There are times when CAPMOD = 0 is useful in practice. For example, in a large digital circuit, the simple equations of CAPMOD = 0 would reduce the number of numerical evaluation necessary to achieve convergence. The simulation will then be faster than if CAPMOD = 2 were used.

In the CAPMOD = 0 capacitance model, the transistor operation is classified into four distinct regions, each region containing its own set of equations to calculate the device capacitances. The operating regions include the accumulation, subthreshold, saturation and linear regions. Because the equations in each region are independent from those in other regions, the derivatives of the device capacitances are not continuous when plotted against V_{GS} or V_{DS}, although the capacitance value are piecewise continuous. The C-V flatband voltage ($V_{FB,CV}$) used in this model is equal to the user-specified parameter VFBCV. The threshold voltage for C-V calculation ($V_{T,CV}$) is equal to VFBCV + $2\phi_f$ + K1 · sqrt($2\phi_f$), where $2\phi_f$ is the strong-inversion surface potential given by Eq. (1-6b). There is no channel-length dependence in this equation, suggesting that this threshold voltage equation applies to the ideal long-channel devices. Although the short-channel effects are neglected in the $V_{T,CV}$ of CAPMOD = 0, they are included in the V_T expression used to calculate the dc I-V characteristics. Therefore, for a short-channel device, V_T and

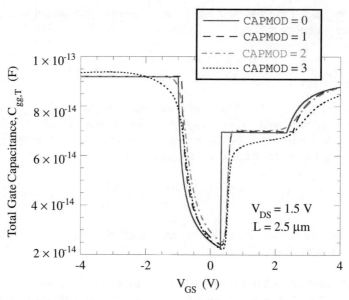

Fig. 3-3 Comparison of $C_{gg,t}$, total gate capacitance including parasitic components, simulated with CAPMOD = 0, 1, 2, and 3.

$V_{T,CV}$ differ in the CAPMOD = 0 capacitance model. The obvious drawback is that the turn-on characteristics of the capacitance (such as C_{gs} or C_{gg}) are not properly modeled. However, the practice of adopting unequal threshold voltage values has the advantage of preserving capacitance symmetry at $V_{DS} = 0$. This was discussed in Section 1.10.

The CAPMOD = 1 model also has the regional characteristics, with discontinuous capacitance derivatives across several operating regions. It improves upon the CAPMOD = 0 model by introducing an "effective V_{GS}" smoothing function (similar to that discussed in Section 1-4), which smoothes out the transition between the subthreshold and the strong inversion regions. However, the overall CAPMOD = 1 model still breaks the device operation into various distinct operation regions—the accumulation and the non-accumulation regions, as well as the linear and saturation regions. Whereas CAPMOD = 1 is in the right step toward removing the capacitance discontinuity, particularly when the device is between the subthreshold and the strong inversion regions, the model still produces a capacitance which changes abruptly at some bias conditions (such as at $V_{GS} \sim -1$ and 2.5 V in Fig. 3-3). In this model, the flatband voltage is not user-specified. It is equated to the zero-bias flatband voltage calculated in the model (for BSIM3v3.2.2), or the bias-dependent flatband voltage (for BSIM3v3.2.1). See the entry under VERSION to clarify the confusion between the two BSIM3 models. The threshold voltage used in the CAPMOD = 1 model is identical to that used for the dc I-V calculation. This is also another fundamental difference between this model and the CAPMOD = 0 model.

The CAPMOD = 1 model employs the effective V_{GS} function to smooth out the device capacitances between the subthreshold and the strong inversion regions. The

3.2 ALPHABETICAL GLOSSARY OF BSIM3 PARAMETERS

CAPMOD = 2 model further introduces effective flatband voltage and effective V_{DS} functions which smooth out the transition between the accumulation and the non-accumulation regions, and between the saturation and linear regions, respectively. The three smoothing functions result in smooth device characteristics at any bias condition. The property of continuous capacitance derivatives is the trademark of CAPMOD = 2 model. This property is one key reason why BSIM3 is well received in the industry. As in CAPMOD = 1 model, the threshold voltage in the CAPMOD = 2 capacitance model is identical to that for the dc I-V calculation. The flatband voltage is equated to the zero-bias flatband voltage calculated in the model (for BSIM3v3.2.2), or the bias-dependent flatband voltage (for BSIM3v3.2.1).

The capacitance models discussed so far all have existed since the introduction of BSIM3v3.1. The CAPMOD = 3 made its debut in the BSIM3v3.2 model. It can be thought to be the CAPMOD = 2 capacitance model, but with two major improvements. The first is the inclusion of the quantum mechanical effects, that the channel carriers do not reside right on the oxide/semiconductor interface, but at a distance into the silicon. As the oxide in modern transistors approaches 30 Å–40 Å range, the channel electrons can no longer be thought to be localized right at the edge of the oxide. Instead, due to the wave nature of the electrons, the centeroid of the electron wavefunctions (where the most probable chance of finding the electrons is located) is at a finite distance away from the oxide/semiconductor interface. The distance between the top metal gate charges and the channel charges is then larger than the physical oxide thickness. We note that generally the quantum mechanical effects are neglected in the I-V modeling, although it is critical in C-V modeling when the oxide thickness is thin. The quantum mechanical effects are not important for dc I-V calculation because the effects are easily absorbed into the effective mobility.

The second improvement of the CAPMOD = 3 model is the abolishment of the constant surface potential. Fixing the surface potential of inverted channel to $2\phi_f$ is an assumption employed in a majority of the compact MOS models. The CAPMOD = 3 model, instead, correctly calculates the surface potential as a function of biases.

Two parameters exclusively created for the CAPMOD = 3 model are ACDE and MOIN. ACDE is used to model the quantum mechanical effects, and MOIN, the varying surface potential. See the entry of ACDE for the detailed equations used in the CAPMOD = 3 capacitance model. The CAPMOD = 3 uses the zero-bias flatband voltage exclusively. There is no option for the user to choose the bias-dependent flatband voltage, which can yield negative C_{gs} anyway. The threshold voltage used in the capacitance model is identical to that used in the dc I-V model.

Although CAPMOD = 3 is more accurate than CAPMOD = 2, the latter is sufficient to result in a trouble-free SPICE simulation (since there is no discontinuity in the capacitance slopes.) Further, in order to use CAPMOD = 3, two additional parameters (ACDE and MOIN) need to be extracted. We consider both CAPMOD = 2 and 3 to be good capacitance models. All four capacitance models account for the poly-depletion effects.

CDSC (default = 2.4×10^{-4}, unit = F/m^2) drain/source to channel coupling capacitance per unit area. In spite of its name, CDSC is purely a dc parameter which does not affect the C-V characteristics of the transistor. CDSC modifies the subthreshold slope of the I_D turn-on characteristics in short-channel devices. A negative CDSC is strongly discouraged, since when carried to an extreme, the overall subthreshold slope can have a negative, unphysical value. The details of CDSC, as well as the ideality factor of the subthreshold turn-on, are described in the entry NFACTOR.

CDSCB (default = 0, unit = F/m^2-V) body-bias sensitivity of CDSC. Although it appears to be a capacitance, CDSCB is purely a dc parameter which does not affect the C-V characteristics of the transistor. CDSCB affects the subthreshold slope of short-channel devices under a nonzero V_{BS}. A positive CDSCB is discouraged, since when carried to an extreme, the subthreshold slope can become negative. The details of CDSCB, as well as the ideality factor of the subthreshold turn-on, are described in the entry NFACTOR.

CDSCD (default = 0, unit = F/m^2-V) drain-bias sensitivity of CDSC. CDSCD is purely a dc parameter without a relationship to the transistor's C-V characteristics. CDSCD affects the subthreshold slope of short-channel devices biased with a nonzero V_{DS}. A negative CDSCD can cause the subthreshold slope to be negative, and thereby should be avoided. The details of CDSCD, as well as the ideality factor of the subthreshold turn-on, are described in the entry NFACTOR.

CF (default = computed, unit = F/m) fringing-field capacitance per side. BSIM3 assumes the fringing capacitances to be the same in the gate-drain and the gate-source sides. The capacitance at each side is denoted as C_f (in the unit of F) which is a constant independent of bias. In BSIM3, the parameter CF (in the unit of F/m) is taken to be the numerical value of C_f divided by the transistor's effective width.

The physical origin of this parasitic capacitance is shown in Fig 2-12. It is basically the extra capacitance due to the fringing fields at the edges of a parallel plate capacitor. C_f (and hence, CF) accounts for the charges on the sidewall and the top edge of the polysilicon gate. C_{ov}, in contrast, accounts for only the charges at the bottom side of the portion of the polysilicon gate that overlaps the junction. In practice, C_f and C_{ov} cannot be separated out from measured capacitances. In BSIM3, the default value for CF is obtained from a conformal mapping technique

$$CF = \frac{2\,\epsilon_{ox}}{\pi} \ln\left(1 + \frac{T_{poly}}{TOX}\right), \qquad (3\text{-}20)$$

where T_{poly} denotes the polysilicon gate thickness. Because T_{poly} is not used anywhere else in the model, it is not economical to make T_{poly} a BSIM3 parameter. In the BSIM3 source code, CF is arbitrarily defaulted to the value calculated with the above formula, except with 4000 Å substituted in the place of T_{poly}. A derivation of Eq. (3-20) can be found in Ref. [4].

Although Eq. (3-20) may be a good approximation, we introduce another formula:

$$\text{CF} = \frac{2\,\epsilon_{ox}}{\pi}\left[\ln\left(1+\frac{T_{poly}}{\text{TOX}}\right) + \ln\frac{\pi}{2} + 0.308\right]. \tag{3-21}$$

This equation has the advantage that its complete derivation (and hence the implicit assumption) can be found in Appendix D. By setting CF of the parameter list to the value calculated from Eq. (3-21), we avoid leaving CF blank, an action which will prompt BSIM3 to use Eq. (3-20) to calculate the defaulted CF at a T_{poly} of 4000 Å.

For completeness, we mention another popular formula for CF (in the dimension of F/m) is

$$\text{CF} = \frac{\epsilon_{ox}}{\pi}\left[\left(\frac{\text{TOX}}{\text{TOX}+T_{poly}} + \frac{\text{TOX}+T_{poly}}{\text{TOX}}\right)\ln\left(1+\frac{2\text{TOX}}{T_{poly}}\right)\right.$$
$$\left. + 2\ln\left(\frac{\text{TOX}+T_{poly}}{4\text{TOX}} - \frac{\text{TOX}}{4(\text{TOX}+T_{poly})}\right)\right]. \tag{3-22}$$

Unfortunately, the above formula's origin is not entirely clear.

CGBO (default = 0, unit = F/m) gate–bulk overlap capacitance per unit gate length. The gate–bulk overlap capacitance ($C_{gb,p}$, in the dimension of F) is due to the overlapping gate metal on top of the field oxide, as shown in Fig. 3-4. Because the oxide thickness is thinner near the device (at the birdsbeak region), CGBO is mostly determined by the overlap nearby the intrinsic device. In BSIM3, CGBO is defaulted to $2\text{DWC}\cdot C'_{ox}$, where $C'_{ox} = \epsilon_{ox}/\text{TOX}$. The factor of 2 appears because DWC is the width offset in one side of the transistor only.

CGBO is a capacitance per unit gate length. The gate–bulk overlap capacitance, $C_{gb,p}$, (in the unit of F rather than F/m) is equal to $\text{CGBO}\cdot L_{\textit{eff},\text{CV}}$. This capacitance is added to the intrinsic gate–bulk capacitance to form the total gate–bulk capacitance.

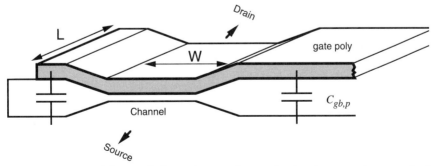

Fig. 3-4 Location of the gate–bulk overlap capacitance, showing that CGBO is appropriately a per unit-length (not unit-width) parameter.

CGDO (default = computed, unit = F/m) voltage-independent gate–drain overlap capacitance per unit gate width. The description about CGDO is identical to that of CGSO, except that CGSO applies to the gate–source capacitance while CGDO concerns with the gate–drain capacitance. See the entry **CGSO** for details.

CGSO (default = computed, unit = F/m) voltage-independent gate–source overlap capacitance per unit gate width. The overall gate–source capacitance in a MOS transistor consists of the intrinsic component, the fringing capacitance (C_f) and the overlap capacitance (C_{ov}). As shown in Fig. 2-12, C_{ov} is the portion of the parasitic capacitance associated purely with the parallel plate capacitor between the gate and the overlapped source region (within the distance L_{ov} marked in the figure). C_f represents the rest of the parasitic capacitance, due to fringing field outside the parallel plate capacitor.

Figure 3-5 illustrates measured gate-to-source capacitance of a NMOS transistor. The device is multi-fingered, so that the overall capacitance is large enough to be measured with a certain precision. In the measurement, the device's source, drain, and bulk are tied to ground while gate bias varies from negative to positive. (However, only the small-signal current flowing out of the source terminal is recorded; hence, the capacitance translated from the small-signal current is the gate-source capacitance.) Because the applied small-signal variation is across the gate and source terminals, the measured capacitance is the total gate-source capacitance ($C_{gs,t}$), equal to the sum of C_{gs} (intrinsic component) and $C_{gs,p}$ (parasitic component). When $V_{GD} = V_{GB} = V_{GS}$ = negative, the channel is depleted of carriers. The measured $C_{gs,t}$ is solely the parasitic component, $C_{gs,p}$, which, as mentioned, is equal to $C_{ov} + C_f$. As discussed in the entry CF, the fringing capacitance is a constant, independent of the bias voltage. The finite voltage dependence seen in Fig. 3-5 is a result of the finite depletion thickness in the

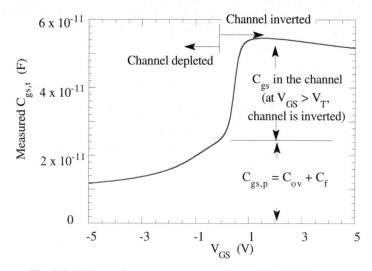

Fig. 3-5 Measured gate-to-source capacitance of a NMOS transistor.

source junction. As V_{GS} becomes progressively more negative, the n^+-doped source region becomes slightly more depleted near the oxide surface. Although the source doping level is high and the depletion thickness is expected to be small, the depletion thickness can still be a noticeable fraction of the gate oxide thickness. Therefore, as more reverse bias is applied, the effective separation between the gate terminal and the undepleted source becomes progressive larger. Consequently, the overlap capacitance decreases. When V_{GS} is positive, there is a sudden dramatic increase in the measured $C_{gs,t}$. This is because a positive V_{GS} causes the intrinsic device's channel region to be inverted. The overall measured $C_{gs,t}$ then includes the additional component originating from the channel.

We now describe how we can estimate the value of CGS0 (and CGS1) from the measured curve shown in Fig. 3-5. For this purpose, the measured capacitance in the inversion region is not interesting as it contains the intrinsic component. We will focus on $V_{GS} < 0$. BSIM3's equation modeling the parasitic capacitance is somewhat complex, but it can be effectively represented by the following equation:

$$C_{gs,p} = W_{\mathit{eff},\mathrm{CV}} \times \left[\mathrm{CF} + \mathrm{CGS0} + \mathrm{CGS1} - \mathrm{CGS1} \left(1 - \frac{1}{\sqrt{1 - \frac{4 \cdot V_{GS,\mathrm{overlap}}}{\mathrm{CKAPPA}}}} \right) \right. $$
$$\left. \times f(V_{GS}) \right]. \qquad (3\text{-}23)$$

Clearly, BSIM3 models C_f as $W_{\mathit{eff},\mathrm{CV}} \cdot \mathrm{CF}$, where $W_{\mathit{eff},\mathrm{CV}}$ is the effective device width for all CV calculations. The rest of the capacitance is simply the overlap capacitance:

$$C_{ov} = W_{\mathit{eff},\mathrm{CV}} \times \left[\mathrm{CGS0} + \mathrm{CGS1} - \mathrm{CGS1} \left(1 - \frac{1}{\sqrt{1 - \frac{4 \cdot V_{GS,\mathrm{overlap}}}{\mathrm{CKAPPA}}}} \right) \times f(V_{GS}) \right], \qquad (3\text{-}24\mathrm{a})$$

where

$$V_{GS,\mathrm{overlap}} = \frac{1}{2}\left[V_{GS} + 0.02 - \sqrt{(V_{GS} + 0.02)^2 + 0.08} \right] \qquad (3\text{-}24\mathrm{b})$$

and $f(V_{GS})$ is a smoothing function which varies from 0 (at $V_{GS} \gg 0$) to 1 (at $V_{GS} \ll 0$). $V_{GS,\mathrm{overlap}}$ has the property that it is equal to 0 when $V_{GS} \gg 0$ and equal to V_{GS} when $V_{GS} \ll 0$. When V_{GS} is very positive (strong inversion), the overlap capacitance is equal to CGS0 + CGS1. When V_{GS} is very negative (gate-source reverse biased), the overlap capacitance asymptotically decrease toward CGS0. The parameter CKAPPA basically adjusts the degree of capacitance variation in these two extreme capacitance values. As indicated in Fig. 3-5, CGS1 can be treated as the amount of capacitance variation between $V_{GS} = 0$ and a very negative V_{GS} value.

Figure 3-6 illustrates the effect of CKAPPA. In typical transistors, CKAPPA is on the order of unity.

The extraction of CGS0, CGS1 and CKAPPA is straightforward from fitting the measured capacitance of Fig. 3-5. Sometimes, if the voltage dependence of the capacitance is weak, we can approximate CGS1 as 0. CGS0 is then equal to the measured capacitance ($C_{gs,t}$) at 0 V subtracted by $CF \cdot W_{\mathit{eff},CV}$.

The default of CGS0 depends on two conditions. If DLC is not given, or if DLC is equal to or less than 0, then CGS0 is defaulted to $0.6 \cdot XJ \cdot C'_{ox}$. In the above expression, XJ is a BSIM3 parameter and C'_{ox} is the gate oxide capacitance per unit area, equal to ϵ_{ox}/TOX. If instead DLC is given and greater than 0, CGS0 is then defaulted to of $DLC \cdot C'_{ox} - CGS1$, but with a caveat. During the error checking stage at the beginning of a SPICE run, BSIM3 checks whether the default of $DLC \cdot C'_{ox} - CGS1$ is greater than 0. If not, CGS0 will be redefaulted to 0, and a warning message is issued to alert the users that a negative CGS0 makes no physical sense.

CGD1 (default = 0, unit = F/m) voltage-dependent gate–drain overlap capacitance per unit gate width. CGD1 is similar to the CGS1 discussed in the entry CGS0, except that CGD1 applies to the gate–drain capacitance while CGS1, the gate–source capacitance. CGD1 represents the amount of capacitance difference between $V_{GD} = 0$ and a heavily reverse-biased V_{GD} value.

CGS1 (default = 0, unit = F/m) voltage-dependent gate-source overlap capacitance per unit gate width. As discussed in the entry CGS0, CGS1 represents the amount of capacitance difference between $V_{GS} = 0$ and a heavily reverse-biased V_{GS} value.

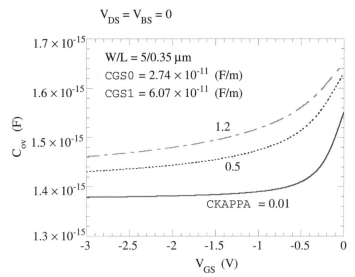

Fig. 3-6 Effects of CKAPPA on BSIM3-calculated overlap capacitance.

CIT (default = 0, unit = F/m²) interface trap capacitance per unit area. In contrast to what its formal name suggests, CIT is purely a dc parameter which does not affect the C-V characteristics of the transistor. As elaborated in the entry NFACTOR, CIT is used mainly to increase (if CIT > 0) or decrease (if CIT < 0) the subthreshold slope of the turn-on characteristics. A negative CIT can cause the subthreshold slope to have a negative, unphysical value. Generally, the term CIT/C'_{ox} should be at most 0.1. C'_{ox} is the gate oxide capacitance per unit area, equal to ϵ_{ox}/TOX. For a technology with TOX of 60 Å, $C'_{ox} = 5.8 \times 10^{-3}$ F/m², CIT should then be restricted between 0 and 5.8×10^{-4} (F/m²).

It is perhaps better to treat CIT just as a fitting parameter which results in the best match in subthreshold characteristics by fixing the ideality factor. It would not be practical to extract this parameter by actually measuring the interface trap capacitance of a device.

CJ (default = 5×10^{-4}, unit = F/m²) source/drain bottom junction capacitance per unit area when the device temperature is equal to TNOM. This parameter is used for the calculation of both the source-bulk and the drain–bulk bottom–wall junction capacitances. BSIM3 assumes the source and drain junctions to be symmetrical, and the same set of parameters are used to describe the source-bulk and drain-bulk capacitances. For example, the parameter CJSW, whose "S" does not mean source but instead denote the sidewall, is a sidewall junction capacitance parameter used for both the drain–bulk and the source–bulk junctions. (When it comes to the overlap capacitance between the gate and source, or between the gate and drain, somehow BSIM3 starts to view MOS as an asymmetrical device and utilizes distinct overlap capacitance parameters for the drain and the source sides. For example, CGD0 and CGD1 are used for the gate–drain overlap capacitance, while CGS0 and CGS1 are used for the gate-source overlap capacitance.)

We describe various parameters using the drain–bulk junction as the example. The same description applies to source–bulk capacitances, except for the appropriate change of $C_{j,\mathrm{DB}}$ to $C_{j,\mathrm{SB}}$, AD to AS, and PD to PS. The total drain-bulk capacitance is given by

$$C_{j,\mathrm{DB}} = \frac{\mathrm{CJ}}{\left(1 - \frac{V_{BD}}{\mathrm{PB}}\right)^{\mathrm{MJ}}} \mathrm{AD} + \frac{\mathrm{CJSW}}{\left(1 - \frac{V_{BD}}{\mathrm{PBSW}}\right)^{\mathrm{MJSW}}} (\mathrm{PD} - W_{e\!f\!f,\mathrm{CV}})$$

$$+ \frac{\mathrm{CJSWG}}{\left(1 - \frac{V_{BD}}{\mathrm{PBSWG}}\right)^{\mathrm{MJSWG}}} W_{e\!f\!f,\mathrm{CV}}. \qquad (3\text{-}25)$$

This equation is applicable when the junction is reverse-biased ($V_{BD} < 0$). In a forward-biased condition, BSIM3 modifies the equation somewhat, as detailed in Section 2.7.

The first component of $C_{j,\mathrm{DB}}$ is the bottom–wall junction capacitance, which is labeled as $C_{j,\mathrm{DB\_BW}}$ in Section 2.3. The second and the third components refer to the sidewall capacitances at the isolation side and at the gate side. They are denoted as $C_{j,\text{DB\_SW}}$ and $C_{j,\text{DB\_SWG}}$, respectively, in Section 2.3. The physical locations of the three capacitances are shown in Fig. 2-12. There is a need to distinguish between the isolation-side and the gate-side sidewall capacitances because the substrate doping is usually lighter in the isolation side than the gate side. This is especially true in modern CMOS devices, in which the channel engineering has resulted high substrate doping at the gate-side sidewall, and the shallow trench isolation has resulted low substrate doping at the device isolation side. The relative magnitudes of these three capacitance components in a modern CMOS technology are shown in Fig. 2-13.

The parameters PB, PBSW, and PBSWG are the built-in potentials of the junctions at the bottom-wall, isolation sidewall, and the gate sidewall, respectively. The built-in potential is a logarithmic function of the substrate doping at the appropriate region; thus, it stays relatively constant even if the substrate doping varies by orders of magnitude. In practice, PB, PBSW, and PBSWG are fixed at about 0.8 V. We rely on other parameters (CJ, CJSW, CJSWG, MJ, MJSW, and MJSWG) to capture the voltage dependencies of the three capacitance components.

The exponents MJ, MJSW, and MJSWG characterize the capacitances' variation with the voltage drop across the junction. In an abrupt junction, where the drain region is considered infinitely doped and the substrate doping is constant, then the exponent has a theoretical value of 1/2. In a linearly graded junction, where the substrate doping increases linearly with the distance away from the junction, the exponent has a theoretical value of 1/3. In a hyper-abrupt junction, where the substrate doping decreases inversely with the distance away from the junction, the exponent is expected to be 1. Generally, the substrate doping is usually higher when it is close to the junction, and becomes lower as we move away from the junction.

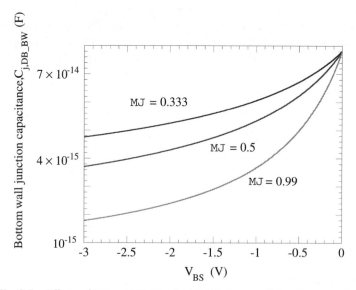

Fig. 3-7 Effects of MJ on BSIM3-calculated bottom–wall junction capacitance.

Therefore, MJ, MJSW, and MJSWG have values between 0.5 and 1. (*Caution*: In some simulators, placing a value of 1 for MJ, for example, can lead to numerical error because there may be a singularity during the derivative evaluation.) Figure 3-7 illustrates the bottom–wall junction capacitance as a function of MJ. When MJ is high, the capacitance varies rapidly with bias voltage.

In the relatively rare situation where the junctions are heavily forward-biased, the junction capacitance can become infinite as V_{BS} (or V_{BD}) approaches PB, PBSW, or PBSWG if Eq. (3-25) is strictly enforced. To avoid this numerical difficulty, BSIM3 modifies the equation to ensure that the calculated capacitance is finite, even at forward-biased conditions. The capacitance expression under positive biases was discussed in Section 2.7.

Prior discussion applies when the device temperature T_{device} is equal to TNOM, the temperature at which the SPICE parameters are extracted. If not, some variables need to be modified. For clarification, we rewrite CJ, CJSW, CJSWG, PB, PBSW, and PBSWG of the SPICE parameter set as CJ(TNOM), CJSW(TNOM), CJSWG(TNOM), PB(TNOM), PBSW(TNOM) and PBSWG(TNOM), respectively. That is, for example, CJ as listed in the SPICE parameters is the junction capacitance per unit area only when T_{device} = TNOM. When $T_{device} \neq$ TNOM, BSIM3 modifies Eq. (3-25) to the following:

$$C_{j,DB}(T_{device}) = \frac{\text{CJ}(T_{device})}{\left(1 - \frac{V_{BS}}{\text{PB}(T_{device})}\right)^{\text{MJ}}} \text{AD} + \frac{\text{CJSW}(T_{device})}{\left(1 - \frac{V_{BS}}{\text{PBSW}(T_{device})}\right)^{\text{MJSW}}} (\text{PD} - W_{\text{eff,CV}})$$

$$+ \frac{\text{CJSWG}(T_{device})}{\left(1 - \frac{V_{BS}}{\text{PBSWG}(T_{device})}\right)^{\text{MJSWG}}} W_{\text{eff,CV}}, \qquad (3\text{-}26)$$

where,

$$\text{CJ}(T_{device}) = \text{CJ}(\text{TNOM}) \times [1 + \text{TCJ} \cdot (T_{device} - \text{TNOM})]; \qquad (3\text{-}27\text{a})$$
$$\text{PB}(T_{device}) = \text{PB}(\text{TNOM}) - \text{TPB} \cdot (T_{device} - \text{TNOM}); \qquad (3\text{-}27\text{b})$$
$$\text{CJSW}(T_{device}) = \text{CJSW}(\text{TNOM}) \times [1 + \text{TCJSW} \cdot (T_{device} - \text{TNOM})]; \qquad (3\text{-}27\text{c})$$
$$\text{PBSW}(T_{device}) = \text{PBSW}(\text{TNOM}) - \text{TPBSW} \cdot (T_{device} - \text{TNOM}); \qquad (3\text{-}27\text{d})$$
$$\text{CJSWG}(T_{device}) = \text{CJSWG}(\text{TNOM}) \times [1 + \text{TCJSW} \cdot (T_{device} - \text{TNOM})]; \qquad (3\text{-}27\text{e})$$
$$\text{PBSWG}(T_{device}) = \text{PBSWG}(\text{TNOM}) - \text{TPBSWG} \cdot (T_{device} - \text{TNOM}). \qquad (3\text{-}27\text{f})$$

If the temperature is high enough, it is possible that CJ(T_{device}), CJSW(T_{device}) or CJSW(T_{device}) becomes negative. When that happens, BSIM3 issues an warning and fix the capacitance values at 0. Likewise, it is possible that PB(T_{device}), PBSW(T_{device}), or PBSWG(T_{device}) goes below 0.01 (V). Under this scenario, BSIM3 issues a warning and fix the built-in voltages at 0.01 (V).

CJSW (default $= 5 \times 10^{-10}$, unit $=$ F/m)   source/drain sidewall junction capacitance per unit length at the isolation sidewall, when the device temperature is equal to TNOM. This parameter is used for the calculation of both the source–bulk and the drain–bulk sidewall junction capacitances at the isolation side. The equation of the total source–bulk capacitance and the discussion about the origin of the capacitance components are given in the entry CJ. The total capacitance includes two sidewall capacitance terms. CJSW characterizes the per length capacitance along the isolation side of the sidewall. CJSWG, in contrast, characterizes that along the gate side of the sidewall. CJSW generally differs from CJSWG because the substrate doping levels on the two sides of the junction are different.

CJSWG (default = CJSW, unit = F/m) source/drain sidewall junction capacitance per unit length at the gate sidewall, when the device temperature is equal to TNOM. This parameter is used for the calculation of both the source-bulk and the drain–bulk sidewall junction capacitances at the gate side. The equation of the total source–bulk capacitance and the discussion about the origin of the capacitance components are given in the entry CJ. The total capacitance includes two sidewall capacitance terms. CJSWG characterizes the per length capacitance along the gate side of the sidewall. CJSW, in contrast, characterizes that along the isolation side of the sidewall. CJSWG generally differs from CJSW because the substrate doping levels on the two sides of the junction are different.

CKAPPA (default = 0.6, unit = V) coefficient of the overlap capacitance variation. It models the dependence of the overlap capacitance on the gate-to-drain or gate-to-source voltage. The overlap capacitance is not a constant, especially when V_{GS} or V_{GD} is negative. This is because the source and drain, though heavily doped, are still finite in doping concentration. The reverse bias between the gate and the source, for example, can partially deplete the source junction. As the effective separation between the gate and source charges increases, the overlap capacitance decreases. See the entry CGS0 for details. The overlap capacitance as a function of CKAPPA was shown in Fig. 3-6.

CLC (default = 1×10^{-7}, unit = m) L_{eff}-dependence coefficient for A_{bulk} in the capacitance calculation. This parameter, along with CLE, is used to calculate the capacitances in short-channel devices. They do not affect the capacitance of long-channel devices.

The channel in a short-channel device can be classified into two regions, depending on the speed at which the electrons travel. At the region near the source, the charges emitted from the source move at a speed equal to the mobility times the channel electric field. As the charges move toward the drain, they experience a higher and higher magnitude of electric field. However, the electron velocity cannot increase indefinitely as the field increases. Once the field exceeds a certain value, they travel at a constant velocity. (BSIM3 uses the parameter VSAT to denote such a saturation velocity). One key outcome of the velocity saturation is that the saturation voltage is smaller in a short-channel device than a long-channel device.

3.2 ALPHABETICAL GLOSSARY OF BSIM3 PARAMETERS

It turns out that I-V modeling accounting for the different physics involved in the two regions can be easily developed. Indeed BSIM3's I-V characteristics fully accounts for the effects associated with the saturation velocity. However, the C-V modeling becomes fairly complicated if the same two-region model were used. Instead of plodding through the equations to derive a C-V model based on the short-channel device physics, BSIM3 adopts a heuristic approach. It develops the C-V model based on the long-channel device physics, assuming that electron velocity is equal to mobility times electric field, throughout the entire channel. But then, knowing that the saturation voltage decreases as the channel length decreases, BSIM3 arbitrarily modifies the saturation voltage in the long-channel C-V model equations to inject some characteristics of short-channel devices. The saturation voltage used in the C-V model, $V_{DS,\text{satCV}}$, if without the modification, would have been written in a manner similar to Eq. (3-4):

$$V_{DS,\text{satCV}} = \frac{V_{GS} - V_T}{A_{bulk,0}} \quad \text{(not used in BSIM3)}, \tag{3-28}$$

where $A_{bulk,0}$ is the bulk-charge coefficient used in BSIM3's C-V modeling. Note, BSIM3 intends to separate the I-V modeling and C-V modeling for the short-channel devices. The parameter A_{bulk} given in Eq. (3-6) is used to calculate I-V characteristics, but a slightly different expression is used for $A_{bulk,0}$ of the C-V model. Despite the difference, they are both roughly equivalent, and both degenerate to $1 + \delta$ in long-channel devices.

$V_{DS,\text{satCV}}$, as is given in Eq. (3-28), lacks any channel length dependence. To reduce the value of $V_{DS,\text{satCV}}$ as channel length decreases, BSIM3 modifies it to

$$V_{DS,\text{satCV}} = \frac{V_{GS} - V_T}{A_{bulk,\text{CV}}} = \frac{V_{GS} - V_T}{A_{bulk,0}\left[1 + \left(\frac{\text{CLC}}{L_{eff,\text{CV}}}\right)^{\text{CLE}}\right]}, \tag{3-29}$$

where $L_{eff,\text{CV}}$ is the effective-channel length for C-V calculation (see the entry DLC for more details.) Basically, BSIM3 replaces A_{bulk} of the I-V model by $A_{bulk,\text{CV}}$ in all of the capacitance expressions originally derived for long-channel devices. The dependence of the capacitances on the channel length is therefore modeled with CLC and CLE, through the use of Eq. (3-29).

This approach is not theoretically robust, but it is better than doing nothing to account for the short-channel effects. While we expect the capacitance to be incorrect in short-channel devices, the simulation results at the circuit level are often surprisingly accurate. This is likely because as channel length gets shorter, the parasitic capacitances play a more important role in the overall circuit performance. In this regard, changing CLC and CLE's values often does not result in a noticeable difference in the simulated circuit performance. Besides, it is not easy to perform experiments or design test structures to establish their values. Therefore, CLC is usually taken to be 0 and CLE is taken to be 1, thereby eliminating $A_{bulk,\text{CV}}$'s channel length dependence. BSIM3 uses a default value of 10^{-7} (m) for CLC and 0.6 for CLE. CLC should be greater or equal to zero to avoid a possible negative value for $A_{bulk,CV}$.

The intrinsic device capacitances are greatly modified by CLC and CLE. Figure 3-8a illustrates the C_{gg} plotted against V_{GS} at $V_{GS} = 1.5$ V, and Fig. 3-8b, C_{gd} and C_{gs} plotted against V_{DS} at $V_{GS} = 2.5$ V. The channel length is 0.35 μm. When CLC = 0 and CLE = 1, the simulated capacitances for the short-channel device would be identical to those in a long-channel device. We can therefore visualize the CLC = 0, CLE = 1 result as a long-channel device result. It is obvious from Fig. 3-8b that

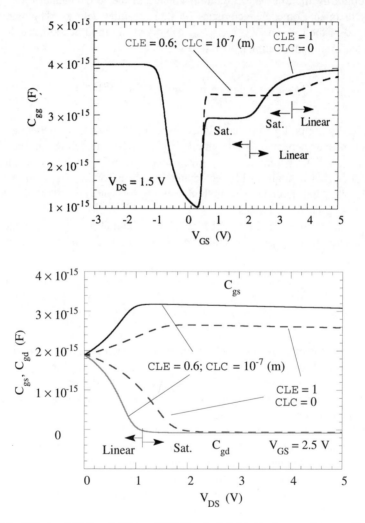

Fig. 3-8 Effects of CLC and CLE on BSIM3-calculated intrinsic capacitances of a 0.35 μm NMOS. (a) C_{gg} plotted against V_{GS} at $V_{DS} = 1.5$ V, (b) C_{gd} and C_{gs} plotted against V_{DS} at $V_{GS} = 2.5$ V.

3.2 ALPHABETICAL GLOSSARY OF BSIM3 PARAMETERS

when CLC = 10^{-7} and CLE = 0.6 causes $V_{DS,\text{satCV}}$ to be smaller than the long-channel device. (The fact that $V_{DS,\text{satCV}}$ is smaller in the short-channel device can be deciphered from Fig. 3-8a also, although more difficult than from Fig. 3-8b. If Fig. 3-8a is used, we can consider the capacitances at $V_{GS} = 2.5$ V, while noting that V_{DS} for the entire figure is at 1.5 V. At that voltage, $V_{DS,\text{satCV}}$ must be smaller than 1.5 V for the short-channel device because the transistor is in the saturation region. Conversely, $V_{DS,\text{sat}}$ must be larger than 1.5 V for the long-channel device because the transistor is in the linear region. From these two statements, we can deduce that $V_{DS,\text{satCV}}$ is smaller in the short-channel device, consistent with Fig. 3-8b.)

Besides the capacitances, CLC and CLE may affect another quantity calculated by BSIM3: the thermal noise. (See Fig. 2-28 for the equivalent noise model adopted in BSIM3.) When NOIMOD is set to be 2 or 4, the terminal noise of the MOSFET is given by

$$\frac{\overline{i_d^2}}{\Delta f} = \frac{4kT\mu_{\text{eff}}}{L_{\text{eff}}^2} |Q_{\text{inv}}|, \tag{3-30}$$

where Q_{inv}, the total channel inversion charge, is equal to the sum of source and drain charges (see Section 1.7 for Q_S and Q_D). Because Q_S and Q_D are used to calculate capacitances, CLC and CLE have a direct impact on the calculated $\overline{i_d^2}$.

CLE (default = 0.6, unitless) L_{eff}-dependence exponent for $A_{\text{bulk,CV}}$ in the capacitance calculation. See the entry CLC for details.

DELTA (default = 0.01, unit = V) effective V_{DS} smoothing parameter. In a regional model, we analyze MOSFET by separating the device operation into several operating regions, such as subthreshold, linear, and saturation regions. In each region, an equation relating to the drain current and the bias voltages is developed. When a bias condition falls into the saturation region, for example, the equation appropriate for the saturation region is used. When the bias condition corresponds to the linear region of operation, then the linear-region equation is used. This kind of model is a *regional model*, which is known to cause numerical difficulties because of the discontinuity in the calculated device characteristics. For instance, while the drain current can somehow be made continuous across different operating regions, its derivative often cannot.

BSIM3 abandons the regional model, which has been the basic framework in textbook and the early SPICE models. BSIM3 ensures that the current, as well as its derivatives, is continuous across various operating regions by defining several smoothing functions. $V_{DS,\text{eff}}$, the smoothing function used to smooth the transition between the linear and saturation regions, is equal V_{DS} when V_{DS} approaches zero, but asymptotically tends toward $V_{DS,\text{sat}}$ (the saturation voltage; Eq. 3-4) when V_{DS} is

188 BSIM3 PARAMETERS

high. DELTA is the parameter used to control the curvature of the transition between $V_{DS} = 0$ and $V_{DS} \gg V_{DS,\text{sat}}$. The $V_{DS,\text{eff}}$ function is given by

$$V_{DS,\text{eff}} = V_{DS,\text{sat}} - \frac{1}{2}\bigg[(V_{DS,\text{sat}} - V_{DS} - \text{DELTA}) + \sqrt{(V_{DS,\text{sat}} - V_{DS} - \text{DELTA})^2 + 4\text{DELTA} \cdot V_{DS,\text{sat}}}\bigg]. \quad (3\text{-}31)$$

(DELTA is the same as the Δ of Eq. 1-10.) The $V_{DS,\text{eff}}$ as a function of V_{DS} at several values of DELTA is shown in Fig. 1-20. The larger the value of DELTA, the faster $V_{DS,\text{eff}}$ starts to make the transition from 0 to $V_{DS,\text{sat}}$. Generally, DELTA can vary from 0.001 to 0.1 to result in a good fit between simulation and measured data. Figure 3-9 illustrates the I-V characteristics when DELTA varies from 0.001, 0.01, to 0.1, while all other parameters remain unchanged.

More details of the $V_{DS,\text{eff}}$ function is found in Section 1.4.

DLC (default = LINT, unit = m) effective channel-length offset for capacitance calculations. When the MOSFET is in strong inversion and V_{DS} is nearly zero, the total gate capacitance is equal to C_{ox}, which is written as $L_{\textit{eff},\text{CV}} \times W_{\textit{eff},\text{CV}} \times \epsilon_{ox}/\text{TOX}$. $L_{\textit{eff},\text{CV}}$, the effective channel length for all CV calculations, is not equal to

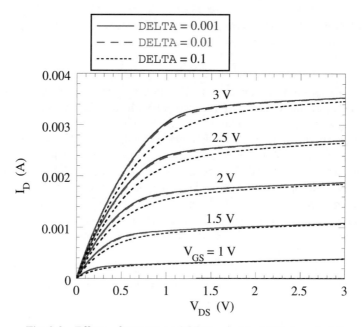

Fig. 3-9 Effects of DELTA on BSIM3-calculated I-V characteristics.

the L specified in the device statement due to some offset resulted from device processing. It can be written as

$$L_{\mathit{eff},\mathrm{CV}} = L - 2 \cdot \mathrm{DLC} - 2 \cdot \Delta L_{\mathrm{geometry},\mathrm{CV}}, \tag{3-32}$$

where L is the drawn length. This equation indicates that most of the channel-length offset is accounted for by DLC, a constant value independent of the drawn L and W of the device. This makes sense, since we tend to think whatever processing steps that give rise to a channel-length offset in one device also cause the same amount of change in other devices. However, during the photolithography, there are potentially some optical reasons which can make the channel-length offset to depend on the device's geometry. These factors are considered of secondary importance, and are lumped into $\Delta L_{\mathrm{geometry},\mathrm{CV}}$, whose exact expression used in BSIM3 is given in the entry LLC. Most often $\Delta L_{\mathrm{geometry},\mathrm{CV}}$ is either zero or close to zero. We then can approximate $L_{\mathit{eff},\mathrm{CV}}$ as $L - 2 \cdot \mathrm{DLC}$.

$L_{\mathit{eff},\mathrm{CV}}$ differs from L_{eff}, the latter being the effective-channel length used in the I-V calculation. Similar to $L_{\mathit{eff},\mathrm{CV}}$ given in Eq. (3-32), L_{eff} is expressed as $L_{\mathit{eff}} = L - 2 \cdot \mathrm{LINT} - \Delta L_{\mathrm{geometry}}$, where $\Delta L_{\mathrm{geometry}}$ is typically zero, barring any photolithographical nonuniformities. We tend to think that $L_{\mathit{eff},\mathrm{CV}}$ for the C-V calculation should be equal to the L_{eff} for the I-V calculation. Therefore, DLC, the length offset for C-V calculation, should be identical to LINT, the length offset for I-V calculation. However, experiences have shown that it is best to not equate DLC and LINT. The values of DLC and LINT should be optimized separately to improve the fit for C-V and I-V characteristics. When DLC is not specified, it is defaulted to LINT.

DLC can be either positive or negative, although it tends to be positive so that the effective-channel length is shorter than the drawn length (L). DLC should not be large enough such that $L_{\mathit{eff},\mathrm{CV}}$ becomes negative.

For details about $W_{\mathit{eff},\mathrm{CV}}$, the effective channel width for all CV calculations, see the entry DWC.

DROUT (default = 0.56, unitless) L_{eff}-dependence exponent in the drain-induced barrier-lowering's (DIBL) correction on the Early voltage. See the entry PCLM for details.

DSUB (default = DROUT, unitless) L_{eff}-dependence exponent of the drain-induced barrier-lowering (DIBL) effects on the threshold voltage. The default of DSUB is equal to the value of DROUT specified in the SPICE parameter set. If DROUT is not specified, then DSUB (as well as DROUT) is defaulted to 0.56. See the entry ETA0 for other details.

DVT0 (default = 2.2, unitless) first coefficient of the charge-sharing's effect on the threshold voltage in short-channel and wide-width devices. The threshold voltage in BSIM3 can be expressed as

$$V_T = \text{VTH0} + \delta_{NP}(\Delta V_{T,\text{body\_effect}} - \Delta V_{T,\text{charge\_sharing}} - \Delta V_{T,\text{DIBL}} \\ + \Delta V_{T,\text{reverse\_short\_channel}} + \Delta V_{T,\text{narrow\_width}} + \Delta V_{T,\text{small\_size}}), \quad (3\text{-}33)$$

where δ_{NP} is +1 for NMOS and −1 for PMOS. A general description of each of the components is found in the entry VTH0. Here, we are specifically interested in the charge-sharing component. In long-channel devices, the threshold voltage is calculated as the voltage required to first deplete the bulk charge enclosed in the channel region and subsequently invert the channel. The channel region can be thought of as the area rigidly defined by the effective-channel length, as marked by the lightly shaded region in Fig. 3-10a. The threshold voltage, so calculated for NMOS, is equal to VTH0 plus the body-effect component. (The exact equation is to be shown shortly.) In short-channel devices, however, some of the bulk charge in the channel region is no longer controlled directly by the gate voltage. The charges in the depletion regions of the source-bulk and drain-bulk junctions, for example, are already depleted by the source and drain voltages, as shown in Fig. 3-10b. The amount of the bulk charge to be depleted by the gate voltage is only that enclosed in the marked trapezoidal region. Because there are fewer bulk charges to be depleted, the threshold voltage decreases in short-channel devices. The charge-sharing component in Eq. (3-33) should always be a positive value. There is a minus sign in front of $\Delta V_{T,\text{charge\_sharing}}$, so that V_T calculated from Eq. (3-33) is smaller when $L_{\textit{eff}}$ decreases. The charge-sharing component in BSIM3 is given by

$$\Delta V_{T,\text{charge\_sharing}} = \text{DVT0}\left[\exp\left(-\text{DVT1}\frac{L_{\textit{eff}}}{2L_t}\right) + 2\exp\left(-\text{DVT1}\frac{L_{\textit{eff}}}{L_t}\right)\right](V_{bi} - 2\phi_f), \quad (3\text{-}34)$$

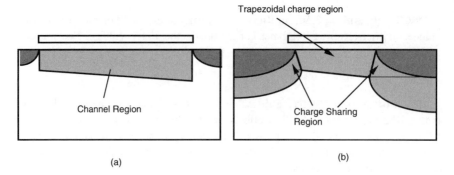

Fig. 3-10 Channel charge regions: (a) in long-channel device; (b) in short-channel device.

where $2\phi_f$ is the strong-inversion surface potential given in Eq. (1-6b) and

$$L_t = \sqrt{\frac{\epsilon_s X_{dep}}{C'_{ox}} (1 + \text{DVT2} \cdot V_{BS})} \qquad (3\text{-}35)$$

and V_{bi} is the built-in voltage of the source-bulk (or, equivalently, the drain–bulk) junction given in Appendix A (Eq. A-33).

The charge-sharing component modeled in BSIM3 is a product of several terms. The term enclosed in the rightmost parentheses, $V_{bi} - 2\phi_f$, does not depend much on any SPICE parameter. It is therefore not interesting to us. We only need to know that $V_{bi} - 2\phi_f$ is positive in practical devices. We shall concentrate on the terms containing DVT0, DVT1 and DVT2, through which the amount of charge-sharing's effect to threshold voltage is modeled. As mentioned, the charge sharing between the drain/source and the bulk nodes is more prominent in short-channel devices. $\Delta V_{T,\text{charge\_sharing}}$ should be zero for long-channel devices, but increasing in magnitude as the channel length becomes shorter. Indeed $\Delta V_{T,\text{charge\_sharing}}$ given by Eq. (3-34) reflects this trend. When L_{eff} is large, the exponents are fairly negative numbers, and the exponential functions in Eq. (3-34) are effectively 0. When L_{eff} is small, in contrast, the exponents become small negative numbers, resulting in finite values in the square brackets of Eq. (3-34). When multiplied by DVT0 and $V_{bi} - 2\phi_f$, the charge-sharing term is of some finite value, consistent with the fact that the charge-sharing correction to the threshold voltage is increasingly large for the short-channel devices. The above description makes clear that the main purpose of DVT1 is to model the charge-sharing component's dependence on L_{eff}. DVT0 is used as a proportional constant to scale the value of the charge-sharing component for various technologies. Because of the minus sign in front of the charge-sharing component, DVT0 is positive. DVT1 should always have positive values, too. Negative DVT1 can cause the exponents of the exponential functions to be numerically too large (to be handled by computer).

The default values for DVT0 and DVT1 are 2.2 and 0.53, respectively. These default values do not have much physical significance. Often, particularly the value of DVT0, can differ from the default value by several orders of magnitude. This is because DVT0 and DVT1 are so-called *paired* parameters which tend to cancel each other's effect when their values are varied in a certain fashion. For example, let us suppose one parameter set is used to model a particular short-channel device. The threshold voltage produced by a certain pair of DVT0 and DVT1 values can be nearly the same if DVT1 of the pair is increased and DVT0 is also made to increase to offset the change in DVT1. Since the two pairs of DVT0 and DVT1 yield the same threshold voltage, either pair can be used. The most effective method to establish the absolute values of DVT0 and DVT1 is to simultaneously extract their values from a set of devices of varying L_{eff}. By fitting the functional dependence of the threshold voltage on L_{eff}, DVT1 can be extracted. Subsequently, DVT0 can be extracted by fitting the required magnitude of the charge-sharing correction to the threshold voltage.

DVT2 is used to model the bulk-bias' effect on the charge-sharing component and, hence, on the threshold voltage. Here are some experimental facts which can guide us to establish the proper sign for DVT2. First, in a long-channel device, the threshold voltage increases as V_{BS} becomes more negative. This can be understood by noting that, as V_{BS} becomes more negative, the lightly shaded region of Fig. 3-10 enlarges. As more bulk charges need to be inverted, the threshold voltage increases. The effect on V_T by V_{BS} is called the body effect. In a long-channel device with wide width, all of the ΔV_T terms in Eq. (3-33) approach 0, except $\Delta V_{T,body\_effect}$. For a NMOS device, the threshold voltage can be written as VTH0 $+ \Delta V_{T,body\_effect}$ or

$$V_T = \text{VTH0} + \text{K1} \cdot \left(\sqrt{2\phi_f - V_{BS}} - \sqrt{2\phi_f}\right) - \text{K2} \cdot V_{BS}, \quad (3\text{-}36)$$

where $2\phi_f$ is the strong-inversion surface potential given in Eq. (1-6b). The second relevant experimental fact is that, in a short-channel device, the threshold voltage change due to the body effect is less significant than the long-channel counterpart. This is because, in a short-channel device, a great portion of the charge in the channel responds to either the drain or source voltage, but not the bulk bias. Since we are discussing DVT2 which relates to the charge-sharing effect, we modify Eq. (3-36) for a short-channel device to account for ΔV_T due to charge-sharing:

$$V_T = \text{VTH0} + \text{K1} \cdot \left(\sqrt{2\phi_f - V_{BS}} - \sqrt{2\phi_f}\right) - \text{K2} \cdot V_{BS} - \Delta V_{T,\text{charge\_sharing}}(V_{BS}).$$
$$(3\text{-}37)$$

$\Delta V_{T,\text{charge\_sharing}}$ is a function of V_{BS} itself. In fact, DVT2 is the parameter controlling the functional dependence of $\Delta V_{T,\text{charge\_sharing}}$ on V_{BS}. $\Delta V_{T,\text{charge\_sharing}}$ is zero in a long-channel device, for which the V_T given by Eq. (3-37) then reduces to Eq. (3-36). As we mentioned, the body effect becomes less dramatic in short-channel devices when V_{BS} becomes more negative. This requires that

$$\Delta V_{T,\text{charge\_sharing}}(V_{BS} = \text{negative}) > \Delta V_{T,\text{charge\_sharing}}(V_{BS} = 0). \quad (3\text{-}38)$$

By examining the exact expression of $\Delta V_{T,\text{charge\_sharing}}$ from Eq. (3-34), we conclude that DVT2 must be negative such that DVT2 $\cdot V_{BS}$ is positive.

DVT2 has a default value of -0.032. We mentioned that DVT0 and DVT1 can be very different from their respective default values. Especially for DVT0, the value in a given SPICE parameter set can be off by several orders of magnitude. However, DVT2 is not a paired parameter. Its value in a SPICE parameter set should not deviate much from the default value.

We comment some more on the choice of the default value for DVT0. As mentioned, BSIM3 chooses a default value of 2.2. However, sometimes it is thought that DVT0's default should be set at zero. After all, the circuit designers generally

3.2 ALPHABETICAL GLOSSARY OF BSIM3 PARAMETERS

assume that the short-channel effects are made negligible when the parameters relating to short-channel effects are not specified.

DVT1 (default = 0.53, unitless) L_{eff}-dependence exponent of the charge-sharing's effect on the threshold voltage. Together with DVT0 and DVT2, DVT1 models the charge-sharing's correction to the threshold voltage in short-channel and wide-width devices. The details of DVT1 are found in the entry DVT0.

The default of DVT1 is 0.53, a value that does not bear much physical significance. As mentioned in the entry of DVT0, DVT0 and DVT1 are considered as paired parameters. As such, the value of DVT1 varies with the choice for DVT0, and both can have fairly different values from their respective default values. DVT1 should always be positive.

DVT2 (default = −0.032, unit = V^{-1}) body-effect coefficient of the charge-sharing's effect on the threshold voltage. Together with DVT0 and DVT1, DVT2 models the charge-sharing's correction to the threshold voltage in short-channel and wide-width devices. The details of DVT2 are found in the entry DVT0.

DVT2 should always be a negative number, and it should be close to its default of −0.032. At times a positive value may be encountered. Although it does not agree with physical intuition, the wrongful effects associated with a positive DVT2 can be offset by other body-effect-related parameters. Therefore, as far as I-V characteristics are concerned, having a nonphysical positive value of DVT2 may still allow for a good fitting. However, an unphysical value of DVT2 is not desirable because it may lead to unphysical (negative) values of the bulk conductance (g_{mb}). It is important to check whether g_{mb} values are always greater than zero under all practical bias conditions. Negative bulk conductances are known to cause numerical problem in SPICE simulation.

DVT0W (default = 0, unitless) first coefficient of the small-size's effect on the threshold voltage. The threshold voltage in BSIM3 can be expressed as

$$V_T = \text{VTH0} + \delta_{NP}(\Delta V_{T,\text{body\_effect}} - \Delta V_{T,\text{charge\_sharing}} - \Delta V_{T,\text{DIBL}}$$
$$+ \Delta V_{T,\text{reverse\_short\_channel}} + \Delta V_{T,\text{narrow\_width}} + \Delta V_{T,\text{small\_size}}) \quad \text{(same as 3-33)}$$

where δ_{NP} is +1 for NMOS and −1 for PMOS. A general description of each of the components is found in the entry of VTH0. Here, we are specifically interested in the small-size effect component, which is given by

$$\Delta V_{T,\text{small\_size}} = \text{DVT0W}\left[\exp\left(-\text{DVT1W}\frac{W_{eff}L_{eff}}{2L_{tw}}\right)\right.$$
$$\left. + 2\exp\left(-\text{DVT1W}\frac{W_{eff}L_{eff}}{L_{tw}}\right)\right](V_{bi} - 2\phi_f), \quad (3\text{-}39)$$

where $2\phi_f$ is the strong-inversion surface potential given in Eq. (1-6b), and

$$L_{tw} = \sqrt{\frac{\epsilon_s X_{dep}}{C'_{ox}}} (1 + \text{DVT2W} \cdot V_{BS}) \qquad (3\text{-}40)$$

and V_{bi} is the built-in voltage of the source-bulk (or, equivalently, the drain–bulk) junction given in Appendix A (Eq. A-33).

Equation (3-34) describes the charge-sharing's effect on the threshold voltage in short-channel devices with wide widths, using DVT0, DVT1, and DVT2 to adjust the amount of the correction. Though appearing nearly the same as Eq. (3-34), Eq. (3-39) is meant to correct the threshold voltage in short-channel devices with *narrow* widths, using instead the parameters DVT0W, DVT1W, and DVT2W.

The small-size component of Eq. (3-39) is a product of several terms. The term enclosed in the parentheses, $V_{bi} - 2\phi_f$, does not depend much on the input SPICE parameters. It is therefore not interesting to us. We only need to know that $V_{bi} - 2\phi_f$ is positive in practical devices. We shall concentrate on the terms containing DVT0W, DVT1W, and DVT2W, through which the amount of charge-sharing's effect in the narrow-width and short-channel devices is modeled. When the product $L_{eff} \cdot W_{eff}$ is large, the exponents in $\Delta V_{T,\text{small\_size}}$ are fairly negative. The exponential functions are basically equal to 0. When $L_{eff} \cdot W_{eff}$ is small, the exponents become small negative numbers, resulting in finite values in the square brackets of Eq. (3-39). When multiplied by DVT0W and $(V_{bi} - 2\phi_f)$, $\Delta V_{T,\text{small\_size}}$ attains some finite value, consistent with the fact that there is a threshold voltage shift in short-channel narrow-width devices. In brief, DVT1W is used to model $\Delta V_{T,\text{small\_size}}$'s dependence on $L_{eff} \cdot W_{eff}$ and DVT0W serves as a proportional constant to scale the amount of ΔV_T correction.

DVT1W should be positive. Negative DVT1W gives rise to positive exponent values in Eq. (3-39), a fact that may cause numerical problems. Establishing the sign of DVT0W is not so straightforward, unfortunately. From elementary device physics, we know that short-channel effects cause V_T to decreases, while the narrow-width effects increase V_T. When a device is both short-channel and narrow-width, it is not clear whether the threshold voltage should increase or decrease. Therefore, DVT0W could be either positive or negative.

DVT2W is used to model the bulk bias' effect on $\Delta V_{T,\text{small\_size}}$. As V_{BS} becomes increasingly negative, the threshold voltage increases. The amount of the increase is not as significant in a short-channel device compared to the long-channel device. Viewed from another perspective, ΔV_T as a result of the body effect is more significant in a narrow-width device compared to a wide-width device. Based on these two conflicting considerations, it is not clear what the sign should be for DVT2W. If the correction for the short-channel effects is more significant than the correction for the narrow-width effects, then DVT2W should be negative, and vice versa. In practice, DVT2W can be viewed as a fitting parameter and is free to assume any value that fits the data well.

DVT0W and DVT1W are *paired* parameters. That is, DVT0W and DVT1W's values tend to cancel each other's effect when their values are varied in a certain fashion.

For example, let us suppose one parameter set is used to model a particular small-size (short-channel and narrow-width) device. The threshold voltage determined by a certain pair of DVT0W and DVT1W values can be nearly the same if DVT1W of the pair is increased and DVT0W is also made to increase to offset the change in DVT1W. Because both pairs of DVT0W and DVT1W produce the same threshold voltage, either pair can be used. The method to establish the absolute values of DVT0W and DVT1W requires a set of devices with varying $L_{eff} \cdot W_{eff}$. DVT1W is extracted by fitting the functional dependence of the threshold voltage on $L_{eff} \cdot W_{eff}$. Subsequently, DVT0W is extracted by fitting the required magnitude of the small-size effect's correction to the threshold voltage.

The default values for DVT0W, DVT1W, and DVT2W are 0 and 5.3×10^6, and -0.032, respectively. When the exact modeling of the narrow-width devices is not critical, we do not need to spend a great deal of time extracting these parameters. We can rely on only K3, K3B and W0 to model the narrow-width effects, although these parameters are meant for the modeling of long-channel devices only. DVT0W, DVT1W, and DVT2W are considered to produce only secondary effects, and we can leave DVT0W at its default value of zero.

DVT1W (default $= 5.3 \times 10^6$, m^{-1}) L_{eff}-dependence exponent of the small-size effect on the threshold voltage. Together with DVT0W and DVT2W, DVT1W models the correction to the threshold voltage in small-size transistors, that is, those with narrow widths and short channel lengths. The details of DVT1W are found in the entry DVT0W. DVT1W should always be positive to avoid numerical difficulty associated with exponential functions in Eq. (3-39).

DVT2W (default $= -0.032$, unit $=$ V^{-1}) body-effect coefficient of the charge-sharing's effect on the threshold voltage. Together with DVT0W and DVT1W, DVT2W models the charge-sharing's correction to the threshold voltage in narrow-width, short-channel devices. The details of DVT2W are found in the entry DVT0W. DVT2W can be either positive or negative.

DWB (default $= 0$, unit $=$ m/V$^{1/2}$) coefficient of the substrate bias' dependence of the width offset. In BSIM3, W_{eff} is given by,

$$W_{eff} = W - 2 \cdot \text{WINT} - 2 \cdot \Delta W_{\text{geometry}} - 2 \cdot \Delta W_{\text{bias\_dependency}}, \quad (3\text{-}41)$$

where W is the drawn width, $\Delta W_{\text{geometry}}$ is detailed in the entry WL, and $\Delta W_{\text{bias\_dependency}}$ is given by

$$\Delta W_{\text{bias\_dependency}} = \text{DWG} \cdot [V_{GS} - V_T] + \text{DWB} \cdot \left[\sqrt{2\phi_f - V_{BS}} - \sqrt{2\phi_f} \right], \quad (3\text{-}42)$$

$2\phi_f$ is the strong-inversion surface potential given in Eq. (1-6b). Equation (3-41) indicates that the majority of the channel-width offset is modeled with the parameter WINT, a constant value specified in the parameter set. However, W_{eff} can depend on the transistor's own drawn W and L, as well as due to biases. The geometrical dependence, accounted for by $\Delta W_{\text{geometry}}$, is due to the processing nonuniformity associated with photolithography. It is generally zero. The bias dependence of W_{eff} is modeled with $\Delta W_{\text{bias\_dependency}}$. From Eq. (3-42), we see that DWB serves to decrease the effective width in response to an increase in the magnitude of V_{BS}. Let us suppose that a transistor's $2\phi_f$ is 1 V and that the most negative V_{BS} is -2.5 V. If at most a 0.1 μm width reduction is caused by the bulk modulation, then DWB should be at most 1.15×10^{-7} m/V$^{1/2}$.

DWC (default = WINT, unit = m) effective channel-width offset for capacitance calculations. When the MOSFET is in inversion and V_{DS} is nearly zero, the total gate capacitance is equal to $C_{ox} = L_{\textit{eff},\text{CV}} \times W_{\textit{eff},\text{CV}} \times \epsilon_{ox} /\text{TOX}$. $W_{\textit{eff},\text{CV}}$, the effective channel width for CV calculation, is not equal to the W specified in the device statement due to some offset resulted from device processing. It is given by

$$W_{\textit{eff},\text{CV}} = W - 2 \cdot \text{DWC} - 2 \cdot \Delta W_{\text{geometry},\text{CV}}. \qquad (3\text{-}43)$$

This equation indicates that most of the channel-width offset is accounted for by DWC, a constant value independent of the drawn L and W of the device. This makes sense. We tend to think whatever processing steps that change the channel-width offset in one device also cause the same amount of change in other devices. However, during the photolithography step, there are potentially some optical reasons which can make the channel-width offset to depend on the device's geometry. These factors are considered to be of secondary importance, and are lumped into $\Delta W_{\text{geometry},\text{CV}}$, whose exact expression used in BSIM3 is given in the entry WLC. Most often $\Delta W_{\text{geometry},\text{CV}}$ is either zero or close to zero. We then can approximate $W_{\textit{eff},\text{CV}}$ as $W - 2 \cdot \text{DWC}$.

$W_{\textit{eff},\text{CV}}$ differs from $W_{\textit{eff}}$, the latter being the effective-channel width used in the I-V calculation. Similar to $W_{\textit{eff},\text{CV}}$ given in Eq. (3-43), $W_{\textit{eff}}$ is expressed as $W_{\textit{eff}} = W - 2 \cdot \text{WINT} - \Delta W_{\text{geometry}} - \Delta W_{\text{bias\_dependency}}$, where $\Delta W_{\text{geometry}}$ is typically zero, barring any photolithographical nonuniformities, and $\Delta W_{\text{bias\_dependency}}$ is also typically zero. Conceptually, $W_{\textit{eff},\text{CV}}$ for the C-V calculation should be equal to the $W_{\textit{eff}}$ for the I-V calculation. Therefore, DWC, the width offset for C-V calculation, should be identical to WINT, the width offset for I-V calculation. However, particularly in devices with narrow widths, experiences have shown that it is best to not equate DWC and WINT. The values of DWC and WINT should be optimized separately to improve the fit for C-V and I-V characteristics. When DWC is not specified, it is defaulted to WINT.

DWC can be either positive or negative, although it tends to be positive so that the effective-channel width is narrower than the drawn width (W). DWC should not be large enough such that $W_{\textit{eff},\text{CV}}$ becomes negative.

For details about $L_{\textit{eff},\text{CV}}$, the effective channel length for all CV calculations, see the entry DLC.

DWG (default = 0, unit = m/V) coefficient of the gate-voltage dependence of the width offset. In BSIM3, the effective width for dc I-V calculation is given by

$$W_{\textit{eff}} = W - 2 \cdot \text{WINT} - 2 \cdot \Delta W_{\text{geometry}} - 2 \cdot \Delta W_{\text{bias\_dependency}}, \qquad (3\text{-}44)$$

3.2 ALPHABETICAL GLOSSARY OF BSIM3 PARAMETERS

where W is the drawn transistor width; WINT is a constant width offset; $\Delta W_{\text{geometry}}$ is the offset which depends on transistor's geometry; and finally, $\Delta W_{\text{bias\_dependency}}$ is the portion of the width offset that varies with V_{GS} and V_{BS}. $\Delta W_{\text{geometry}}$, whose expression is given in the entry WL, is generally zero. The term of interest here is $\Delta W_{\text{bias\_dependency}}$, which is given by

$$\Delta W_{\text{bias\_dependency}} = \text{DWG} \cdot (V_{GS} - V_T) + \text{DWB} \cdot (\sqrt{2\phi_f - V_{BS}} - \sqrt{2\phi_f}), \quad (3\text{-}45)$$

where $2\phi_f$ is the strong-inversion surface potential given in Eq. (1-6b). DWG is the parameter which decreases the effective width in response to an increase in the gate bias. Suppose that a device of a 2.5 V technology with a $V_T \sim 0.5$ V has a channel width of 1 µm. If at most a 0.1 µm width reduction is caused by the gate modulation, then DWG should be at most 5×10^{-8} m/V.

In most cases, the width offset is accounted for by WINT, which represents the constant channel width reduction per side. $\Delta W_{\text{geometry}}$ and $\Delta W_{\text{bias\_dependency}}$, are small and made zero by setting WL, WW, WWL, DWB, and DWG to their default values of 0.

EF (default = 1, unitless) frequency exponent of the Flicker noise. This parameter should be between 0.9 and 1.2. See the entry NOIMOD for details.

ELM (default = 5, unitless) Elmore constant of the channel, used in BSIM3's non-quasi-static (NQS) model. A great majority of simulation using the BSIM3 model is performed with its quasi-static (QS) model. NQS model is important for the accurate modeling of transient effects when the time of interest is comparable to the transit time of charges across the channel. This can happen in an analog circuit in which a short-channel device drives a long-channel device. In digital circuits in which roughly all transistors are of the same minimum geometry, the NQS effects are less significant.

Even in the relatively rare events that NQS is important, a special technique based on BSIM3's QS model can be applied to simulate the NQS effects (Section 4.1). Besides, there is some question about the accuracy of BSIM3's NQS model, that its bulk current is equated to 0 during the entire transient, and the waveform of other terminal currents are not exactly correct (Section 4.1). Due to the concern of the model accuracy and the rarity of significant NQS events, it is better to not invoke the BSIM3 NQS model. This is achieved by assigning the parameter NQSMOD to its default value of 0. In this case, the exact value of ELM is irrelevant.

EM (default = 4.1×10^7, unit = V/m) saturation field used in the Flicker noise calculation when NOIMOD = 2 or 3. This parameter is not used if NOIMOD = 1 or 4. In any event, EM does not affect the calculation of the I-V and C-V characteristics. See the entry NOIMOD for details.

198 BSIM3 PARAMETERS

ETA0 (default = 0.08, unitless)   drain-induced barrier-lowering (DIBL) coefficient for threshold voltage calculation. The threshold voltage in BSIM3 can be expressed as

$$V_T = \text{VTHO} + \delta_{NP}(\Delta V_{T,\text{body\_effect}} - \Delta V_{T,\text{charge\_sharing}} - \Delta V_{T,\text{DIBL}}$$
$$+ \Delta V_{T,\text{reverse\_short\_channel}} + \Delta V_{T,\text{narrow\_width}} + \Delta V_{T,\text{small\_size}}) \quad \text{(same as 3-33)}$$

where δ_{NP} is $+1$ for NMOS and -1 for PMOS. A general description of each of the components is found in the entry of VTH0. Here, we are specifically interested in the DIBL component, which is given by

$$\Delta V_{T,\text{DIBL}} = \left[\exp\left(-\text{DSUB}\frac{L_{\text{eff}}}{2L_{t0}}\right) + 2\exp\left(-\text{DSUB}\frac{L_{\text{eff}}}{L_{t0}}\right) \right] (\text{ETA0} + \text{ETAB} \cdot V_{BS})$$
$$\times V_{DS}, \tag{3-46}$$

where

$$L_{t0} = \sqrt{\frac{\epsilon_s X_{\text{dep},0}}{C'_{ox}}}. \tag{3-47}$$

$X_{\text{dep},0}$, the depletion thickness in the substrate at zero bulk bias, is given in Appendix A (Eq. A-38).

DIBL is a short-channel effect. Figure 3-11 schematically shows the surface potential as a function of x/L, where x is the position away from the source and L is the channel length. The source junction corresponds to $x/L = 0$, and the drain junction, $x/L = 1$. In the long-channel device, the surface potential is fairly flat in the entire channel, independent of V_{DS}. In this case, DIBL is absent and V_{DS} yields little change on the threshold voltage. In the short-channel device, in contrast, there is a noticeable change in the surface potential as V_{DS} changes. When V_{DS} increases, the barrier blocking the carriers in the drain from entering the channel diminishes, and the device turns on sooner. This barrier lowering is induced by V_{DS}. Hence, it is referred as the *drain-induced barrier-lowering effect*. $\Delta V_{T,\text{DIBL}}$, based on this physical picture, should be zero for long-channel devices, but increase in magnitude as the channel length becomes shorter. We verify whether $\Delta V_{T,\text{DIBL}}$ given by Eq. (3-46) behaves as expected. When L_{eff} is large, the exponents in Eq. (3-46) are fairly

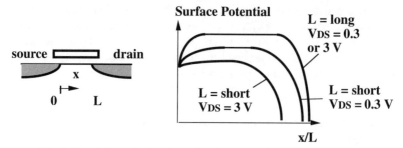

Fig. 3-11 Schematic drawing of surface potential across the channel.

negative. The exponential functions are basically equal to 0, as desired. When L_{eff} is small, the exponents become small negative numbers, resulting in finite values in the square brackets of Eq. (3-46). ETA0 is used as a proportional constant to scale the DIBL value for the threshold correction in the absence of V_{BS}. ETAB serves as a parameter to correct for the V_{BS}'s dependence.

DSUB is a parameter used to control the amount of $\Delta V_{T,\text{DIBL}}$ as a function of L_{eff}. Negative DSUB can cause the exponential functions in Eq. (3-46) to approach infinity. DSUB should always be positive. In addition, based on the fact that the short-channel devices have smaller threshold voltage compared to the long-channel devices, the sign of ETA0 is positive. In this way, the DIBL term given by Eq. (3-46) is positive for short-channel devices and nearly zero for long-channel devices. The negative sign in front of the DIBL term in the overall threshold voltage expression of Eq. (3-33) makes the threshold voltage of the long-channel to be larger than that of the short-channel.

The default value for ETA0 is 0.08. The default of DSUB is equal to the value of DROUT specified in the SPICE parameter set. If DROUT is not specified, then DSUB (as well as DROUT) is defaulted to 0.56. However, these default values do not have much physical significance. Often, particularly the value of ETA0, can differ from the default value by several orders of magnitude. This is because ETA0 and DSUB are so-called *paired* parameters, which tend to cancel each other's effect when their values are varied in a certain fashion. For example, suppose one parameter set is used to model a particular short-channel device. The threshold voltage produced by a certain pair of ETA0 and DSUB values can be nearly the same if DSUB of the pair is increased and ETA0 is made to increase to offset the change in DSUB. Since these two pairs of ETA0 and DSUB yield nearly the same threshold voltage, either pair can be used. Establishing the absolute values of ETA0 and DSUB requires a set of devices with varying L_{eff}. DSUB is extracted by fitting the functional dependence of the threshold voltage on L_{eff}. ETA0 is extracted by fitting the required magnitude of the DIBL's correction to the threshold voltage.

ETAB is used to model the bulk bias' effect on $\Delta V_{T,\text{DIBL}}$. We intend to establish the proper sign for ETAB. In a long-channel device, we know that the threshold voltage increases as V_{BS} becomes more negative. This effect, the body effect, is modeled by the parameters K1 and K2. The threshold voltage can be expressed as

$$V_T = \text{VTH0} + \text{K1} \cdot \left(\sqrt{2\phi_f - V_{BS}} - \sqrt{2\phi_f}\right) - \text{K2} \cdot V_{BS} \quad \text{(same as Eq. 3-36)}$$

where $2\phi_f$ is the strong-inversion surface potential given in Eq. (1-6b). Further, we know that ΔV_T due to the body effect is less significant as the channel length decreases. The channel region in a short-channel device becomes progressively influenced by the drain and source biases. As there are less bulk charges to be depleted by the gate voltage before the inversion, V_T decreases. Since we are discussing ETAB which relates to the DIBL effect, we rewrite the above threshold voltage to account for the change in threshold voltage due to DIBL in short-channel devices:

$$V_T = \text{VTH0} + \text{K1} \cdot (\sqrt{2\phi_f - V_{BS}} - \sqrt{2\phi_f}) - \text{K2} \cdot V_{BS} - \Delta V_{T,\text{DIBL}}(V_{BS}). \quad (3\text{-}48)$$

We have emphasized that $\Delta V_{T,\text{DIBL}}$ is a function of V_{BS} itself, and as mentioned, ETAB is the parameter controlling the functional dependence of $\Delta V_{T,\text{DIBL}}$ on V_{BS}. $\Delta V_{T,\text{DIBL}}$ is a short-channel effect. It is zero in a long-channel device, for which V_T given by Eq. (3-48) then reduces to Eq. (3-36). As we mentioned, the body effect becomes less dramatic in short-channel devices. This requires that

$$\Delta V_{T,\text{DIBL}}(V_{BS} = \text{negative}) > \Delta V_{T,\text{DIBL}}(V_{BS} = 0). \quad (3\text{-}49)$$

By examining the exact expression of $\Delta V_{T,\text{DIBL}}$ from Eq. (3-46), we conclude that ETAB must be negative such that ETAB $\cdot V_{BS}$ is positive.

That was a theoretical determination of ETAB. From a practical standpoint, ETAB can be treated as merely a fitting parameter meant to improve accuracy of the threshold voltage with respect to the applied V_{BS}. Hence, we do not restrict ETAB to be negative; positive values are acceptable. The only requirement is that the magnitude of ETAB $\cdot V_{BS}$ should be much smaller than ETA0; otherwise, we may have inadvertently given V_{BS} too much influence on the DIBL effects. The default value of ETAB is -0.07.

The default value for ETA0 is nonzero. Some believe it should be set at zero. After all, a circuit designer generally assumes that the short-channel effects are made negligible when the parameters relating to short-channel effects are not specified.

ETAB (default = -0.07, unit = V^{-1}) bulk-bias coefficient of the drain-induced barrier-lowering (DIBL) effects. See the entry of ETA0 for details.

GAMMA1 (default = calculated, unit = $V^{1/2}$) first-order textbook-like body-effect factor. It is named the textbook-like body-effect factor because: 1) we want to avoid the confusion with the body-effect factors K1 and K2, which are used in BSIM3 to directly calculate the threshold voltage; and 2) the equation used to calculate the default value of GAMMA1 (when it is not specified in the parameter set) is indeed found in typical textbooks. The default value of GAMMA1 is given by

$$\text{GAMMA1} = \frac{\sqrt{2q\,\epsilon_s\,\text{NCH}}}{C'_{ox}}. \quad (3\text{-}50)$$

When either K1 or K2 is specified in the parameter set, GAMMA1 becomes a useless parameter, independent of whether GAMMA1 is specified or not. For a detailed description of GAMMA1 (as well as all the parameters relating to the modeling of the body-effects in BSIM3), see the entry NCH.

3.2 ALPHABETICAL GLOSSARY OF BSIM3 PARAMETERS

GAMMA2 (default = calculated, unit = $V^{1/2}$) second-order textbook-like body-effect factor. It is named the textbook-like body-effect factor because: 1) we want to avoid the confusion with the body-effect factors K1 and K2, which are used in BSIM3 to directly calculate the threshold voltage; and 2) the equation used to calculate the default value of GAMMA2 (when it is not specified in the parameter set) is found in standard textbooks. The default value of GAMMA2 is given by

$$\text{GAMMA2} = \frac{\sqrt{2q \, \epsilon_s \, \text{NSUB}}}{C'_{ox}}. \tag{3-51}$$

When either K1 or K2 is specified in the parameter set, GAMMA2 becomes a useless parameter, independent of whether GAMMA2 is specified or not. For a detailed description of GAMMA2 (as well as all the parameters relating to the modeling of the body-effects in BSIM3), see the entry NCH.

IJTH (default = 0.1, unit = A) limiting current for the source-bulk and drain-bulk diode turn-on. When IJTH is finite, the diode current does not go exponentially toward infinity with the applied forward bias. When IJTH = 0, the diode is modeled to have the usual exponential turn-on characteristics. A negative value does not make physical sense and BSIM3 issues a fatal error as a consequence. See the entry JS for details. We recommend this parameter be set to 0 explicitly.

JS (default = 10^{-4}, unit = A/m^2) source-bulk and drain-bulk junction saturation current per unit area when the device temperature (T_{device}) is equal TNOM. TNOM is the SPICE parameter which specifies the temperature of the measured device data from which the parameter set is extracted. We will discuss the modification when $T_{device} \neq$ TNOM shortly. But for now, we will concentrate on the situation when $T_{device} =$ TNOM.

We use the drain-bulk junction as an illustration in the following discussion. The models for the source-bulk junction and the drain-bulk junction are identical, except that the appropriate bias voltage used in the source-bulk junction is V_{BS}, while in the drain-bulk junction, V_{BD}.

In BSIM3, the drain current is written as

$$I_D = I_{DS} + I_{sub} - I_{j,\text{DB}}, \tag{3-52}$$

where I_{DS} is the channel current flowing from the drain to source in a normal MOSFET operation, I_{sub} is the current due to impact ionization (described in the entry ALPHA0) that flows between the drain and the bulk, and $I_{j,\text{DB}}$ is the drain-bulk junction current of interest. A minus sign is placed in front of $I_{j,\text{DB}}$ to reflect the fact that the drain terminal is the negative terminal in a *p-n* junction.

BSIM3 provides two similar models to model $I_{j,\text{DB}}$. In the reverse-biased condition, these two models are identical. When the junction is forward-biased (an unlikely event in a normal MOSFET operation), then the calculated current depends on the value of IJTH. When IJTH = 0, $I_{j,\text{DB}}$ for either forward-biased or reverse-biased junction is written as

$$I_{j,\text{DB}} = I_{sat,\text{DB}} \left[\exp\left(\frac{qV_{BD}}{\text{NJ} \cdot kT}\right) - 1 \right] + \text{GMIN} \cdot V_{BD} \quad (\text{IJTH} = 0), \qquad (3\text{-}53)$$

where, the subscript "*sat*" in $I_{sat,\text{DB}}$ means saturation. $I_{sat,\text{DB}}$ is given by

$$I_{sat,\text{DB}} = \begin{cases} \text{JS} \times \text{AD} + \text{JSSW} \times \text{PD} & \text{the result is} > 0 \\ 10^{-14}\text{A} & \text{if AD and PD are both 0} \\ 0 & \text{JS} \times \text{AD} + \text{JSSW} \times \text{PD} < 0. \end{cases} \qquad (3\text{-}54)$$

When $V_{BD} < 0$, $I_{j,\text{DB}}$ calculated from Eq. (3-53) is negative. According to Eq. (3-52), the magnitude of $I_{j,\text{DB}}$ is then added to the overall drain current, hence the minus sign in front of $I_{j,\text{DB}}$ in Eq. (3-52) is correct.

Let us discuss Eq. (3-53) some more. The first component is the usual *p-n* junction current expression, with NJ being the ideality factor of the current's exponential increase with V_{BD}. $I_{sat,\text{DB}}$ is the saturation current of the drain–bulk junction, which is generally the sum of the saturation current in the bulk area, equal to JS (in A/m²) times the drain area AD, and the saturation current in the junction periphery, equal to JSSW (in A/m) times the drain periphery PD. There is a second component in the $I_{j,\text{DB}}$ expression: GMIN · V_{BD}. This component has no physical significance. BSIM3 adds this term to aid the numerical convergence of the dc solution in the SPICE simulation. The variable GMIN is not a SPICE parameter. It has a default value of $10^{-12}(\Omega^{-1})$, but can be reset in the .OPTION control statement.

One problem with Eq. (3-53) is that the diode current increases rapidly with V_{BD} when V_{BD} is positive. If V_{BD} is high enough, the current calculated from Eq. (3-53) can quickly exceed the largest number allowed in the computer. Although this numerical error gives a circuit designer something to be alerted about and is generally considered as a good feature, sometimes it is thought that BSIM3 should have an option to prevent it from happening. BSIM3 therefore offers another model, which is invoked whenever IJTH > 0. The diode currents calculated when IJTH = 0 and when IJTH > 0 are compared in Fig. 3-12.

When IJTH = 0, the calculated diode current increases indefinitely as V_{BD} increases. If IJTH > 0, then after the diode current given by Eq. (3-53) reaches the value of IJTH, the diode current is modified so that it increases at a much slower pace. We denote V_{IJTH} as the V_{BD} that results in an $I_{J,\text{DB}}$ equal to (IJTH + GMIN · V_{BD}). When $V_{BD} > V_{IJTH}$, the second model calculates the diode current as

$$I_{j,\text{DB\_modified}} = I_{j,\text{DB}}\Big|_{V_{BD}=V_{IJTH}} + \frac{\partial I_{j,\text{DB}}}{\partial V_{BD}}\Big|_{V_{BD}=V_{IJTH}} \times (V_{BD} - V_{IJTH}), \qquad (3\text{-}55)$$

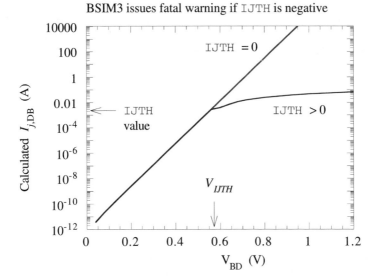

Fig. 3-12 Effects of `IJTH` on BSIM3-calculated diode current.

where $I_{j,\text{DB}}$ is that given in Eq. (3-53). For completeness, we carry out the derivative and the algebraic manipulation, and list the second model $I_{j,\text{DB\_modified}}$ for all bias conditions as

$$I_{j,\text{DB\_modified}} = \begin{cases} I_{sat,\text{DB}}\left[\exp\left(\dfrac{qV_{BD}}{\text{NJ}\cdot kT}\right) - 1\right] + \text{GMIN}\cdot V_{BD} & \text{if } V_{BD} < V_{IJTH}; \\ \text{IJTH} + \dfrac{\text{IJTH} + I_{sat,\text{DB}}}{\text{NJ}} \times \dfrac{q}{kT}(V_{BD} - V_{IJTH}) + \text{GMIN}\cdot V_{BD} & \\ & \text{if otherwise}; \end{cases} \quad (3\text{-}56)$$

where

$$V_{IJTH} = \text{NJ}\frac{kT}{q}\ln\left(\frac{\text{IJTH}}{I_{sat,\text{DB}}} + 1\right). \quad (3\text{-}57)$$

This second model is not recommended. If a designer wants to place a current-limiting resistor in series with the junction, he or she should be the person to do so, rather than letting the model to take care of it.

In a brief summary, BSIM3 uses `JS`, `JSSW`, `NJ` and `IJTH` to model the diode currents. These same parameters are used for the source–bulk and drain–bulk junctions. Depending on the value of `IJTH`, the diode current will be either given by Eq. (3-53) or (3-56). The default values for `JS`, `JSSW`, and `NJ` are 10^{-4} (A/m^2),

0 (A/m), and 1 (unitless), respectively. While the NJ value should not deviate too much from 1 (between 1 and 1.1), JS and JSSW depend on the technology. Their values need not be close to the default values. Because the bulk–drain and bulk–source diodes are usually in reverse bias, the values of JS and JSSW do not affect much the circuit performance. Rather than spending extra effort to extract JS and JSSW separately, sometimes we assume the saturation current to come entirely from the junction area, without a perimeter component. In this simplified extraction methodology, the parameter list specifies JS only, but leaves JSSW defaulted to 0. A JS value on the order of 10^{-3} to 10^{-8} (A/m^2) is common. The IJTH's default value is 0.1 (A). Since its default is not zero, the modified diode current given in Eq. (3-56) will be used in BSIM3 if IJTH is left unspecified. As mentioned, it is preferred that IJTH be set to 0, so that the diode current is calculated from Eq. (3-53), rather than Eq. (3-56). If somehow IJTH is given a negative value, BSIM3 issues a fatal warning, whether the PARAMCHK parameter is 0 or not.

Prior discussion applies when the device temperature T_{device} is equal to TNOM, the temperature at which the SPICE parameters are extracted. If not, some variables need to be modified. In the following discussion, both T_{device} and TNOM are assumed to have been converted from their default unit of Celcius to Kelvin. JS and JSSW appearing in the SPICE parameters are really JS(TNOM) and JSSW(TNOM), respectively. That is, JS and JSSW, as listed in the SPICE parameters, are the area and periphery saturation current densities when T_{device} = TNOM. When $T_{device} \neq$ TNOM, BSIM3's diode current is expressed as

$$I_{j,\text{DB}} = I_{sat,\text{DB}} \left[\exp\left(\frac{qV_{BD}}{\text{NJ} \cdot kT_{device}} \right) - 1 \right] + \text{GMIN} \cdot V_{BD}, \quad (3\text{-}58)$$

where

$$I_{sat,\text{DB}} = \begin{cases} \text{JS}(T_{device}) \times \text{AD} + \text{JSSW}(T_{device}) \times \text{PD} & \text{if the result is} \geq 0 \\ 10^{-14} \text{A} & \text{if AD and PD are both 0} \\ 0 & \text{if JS} \times \text{DA} + \text{JSSW} \times \text{PD} < 0. \end{cases} \quad (3\text{-}59)$$

$$\text{JS}(T_{device}) = \text{JS}(\text{TNOM}) \left[\exp\left(\frac{E_g(T_{device})}{\text{NJ} \cdot kT_{device}} - \frac{E_g(\text{TNOM})}{\text{NJ} \cdot k\text{TNOM}} \right) \right]$$
$$\times \left(\frac{T_{device}}{\text{TNOM}} \right)^{\text{XTI/NJ}}. \quad (3\text{-}60)$$

$$\text{JSSW}(T_{device}) = \text{JSSW}(\text{TNOM}) \left[\exp\left(\frac{E_g(T_{device})}{\text{NJ} \cdot kT_{device}} - \frac{E_g(\text{TNOM})}{\text{NJ} \cdot k\text{TNOM}} \right) \right] \times \left(\frac{T_{device}}{\text{TNOM}} \right)^{\text{XTI/NJ}}.$$
$$(3\text{-}61)$$

In these equation, the silicon bandgap energy as a function of temperature is given by

$$E_g(\text{in eV}) = 1.16 - \frac{7.02 \times 10^{-4} T_{device}^2}{T_{device} + 1108} \quad (T_{device} \text{ is in Kelvin}). \quad (3\text{-}62)$$

(There is a minor typo in the BSIM3 manual.) In the expressions of $\text{JS}(T_{device})$ and $\text{JSSW}(T_{device})$, there are two factors which model the temperature dependence of these saturation current densities. The first one, enclosed in square brackets, relates to the fact that the bandgap energy shrinks as temperature increases, leading to a higher concentration of intrinsic carriers. This is the dominant factor determining JS and JSSW's temperature dependencies. The second factor, relates to the ratio of T_{device} and TNOM, uses a new SPICE parameter, XTI, in the exponent to model the mobility's variation with the temperature. The exponent is to account for 1) the change in the intrinsic carrier concentration due to the temperature dependence of the conduction and valance band density of states; and 2) the change in the carrier diffusion coefficient and recombination lifetime as a function of temperature [5]. If we neglect the contribution from the second effect, then XTI/NJ should be equal to 3. (Since NJ is almost always equal to 1, this means XTI would be 3, which is the default value used in BSIM3.) XTI/NJ may deviate a little bit from 3 due to the aforementioned second effect, however.

Although $E_g(T_{device})$ is the primary contributor to JS and JSSW's temperature dependencies, the exact dependence is hard-coded in BSIM3 and cannot be modified. XTI is the only parameter (besides TNOM) which can affect the temperature dependence of the saturation current densities. When XTI increases, the saturation current densities increase.

JSSW (default = 0, unit = A/m) source–bulk and drain–bulk junction sidewall saturation current per unit periphery at T_{device} = TNOM. See the entry JS for details.

K1 (default = calculated, unit = V$^{1/2}$) first-order body-effect factor. K1, together with K2, is used to model the body effects in BSIM3. These two parameters directly impact the threshold voltage, which is given by

$$V_T = \text{VTH0} + \delta_{NP}(\Delta V_{T,\text{body\_effect}} - \Delta V_{T,\text{charge\_sharing}} - \Delta V_{T,\text{DIBL}}$$
$$+ \Delta V_{T,\text{reverse\_short\_channel}} + \Delta V_{T,\text{narrow\_width}} + \Delta V_{T,\text{small\_size}}) \quad (\text{same as 3-33})$$

where δ_{NP} is +1 for NMOS and −1 for PMOS. A general description of each of the components is found in the entry of VTH0. Here, we are specifically interested in the body effect component, which is given by

$$\Delta V_{T,\text{body\_effect}} = \text{K1} \cdot \left(\sqrt{2\phi_f - V_{BS}} - \sqrt{2\phi_f} \right) - \text{K2} \cdot V_{BS}, \quad (3\text{-}63)$$

where $2\phi_f$ is the strong-inversion surface potential given in Eq. (1-6b). K1 is positive for both NMOS and PMOS and K2 is generally negative for both devices. Sometimes a circuit technique calls for a temporary application of positive V_{BS}. This reduces the threshold voltage and increases the drive current. However, in the normal NMOS operation, V_{BS} is negative so that the threshold voltage becomes larger when the body effects are present. $\Delta V_{T,\text{body\_effect}}$, as given in Eq. (3-63), is independent of the channel length. Empirically, the body effects are less pronounced in short-channel devices due to the charge sharing. The threshold voltage is found to be smaller in short-channel devices, compared to their long-channel counterpart. The reduction in the threshold voltage is taken care by the term $\Delta V_{T,\text{charge\_sharing}}$ in Eq. (3-34), as elaborated in the entry DVT0.

For a typical CMOS process with channel doping on the order of 10^{17} cm^{-3} and an oxide thickness below 100 Å, K1 should be between 0.3 and 0.6. K1 is BSIM3's parameter for the body-effect coefficient, which is generally denoted as γ in textbooks. The value of K1 (or γ) as a function of oxide thickness and substrate doping can be found in Fig. 1-18.

Occasionally, K1 in the parameter set may be very small or very large, such as 0.01 or 1.5. This is an indication that the parameter set was extracted from rf transistors whose bulk is tied to the source. In these transistors, V_{BS} is always zero, and the threshold voltage becomes insensitive to K1 and K2. If these parameters are accidentally optimized during the parameter extraction process, then K1 and K2 will hit their boundary values (such as 0.01 or 1.5) without producing any effect to the I-V characteristics. Hence, the extracted values for K1 and K2 may be wrong. One can argue that, even with the wrong set of K1 and K2, V_T will still be correctly computed since no matter what their values are, $V_{BS} = 0$ will make V_T independent of V_{BS}. However, parameters such as g_{mb} directly depend on K2, for example, rather than K2 · V_{BS}. A wrong value for K1 and K2, while not producing noticeable effect in the I-V characteristics, will impact the calculation of the small-signal parameters.

The higher the value of K1 is, the more significant the body effects are. Therefore, at the same bulk-substrate bias, a transistor with a higher K1 value will have a higher threshold voltage and hence, a lower amount of current flow.

When K2 is specified but K1 is not, K1 is defaulted to 0.53. If neither K1 nor K2 is specified, then K1 and K2 are defaulted in accordance with the following equations:

$$\text{K2} = \frac{(\text{GAMMA1} - \text{GAMMA2})(\sqrt{2\phi_f - \text{VBX}} - \sqrt{2\phi_f})}{2\sqrt{2\phi_f}(\sqrt{2\phi_f - \text{VBM}} - \sqrt{2\phi_f}) + \text{VBM}}. \tag{3-64}$$

$$\text{K1} = \text{GAMMA2} - 2 \cdot \text{K2} \cdot \sqrt{2\phi_f - \text{VBM}}. \tag{3-65}$$

See the entry NCH for more details when K1 and K2 are not specified in the parameter set.

K2 (default = calculated, unitless) second-order body-effect factor. K2, together with K1, is used to model the body effects in BSIM3. These two parameters directly impact the threshold voltage calculation, as discussed in the entry K1. K2 should be between 0 and -0.06, for either NMOS or PMOS. A positive K2 causes the threshold voltage to decrease with increasing more negative V_{BS}, a result which is physically incorrect. A negative K2 with a large magnitude, such as -0.1, is equally troublesome because it tends to cause g_{mb} to be negative. The bulk transconductance is usually not a critical parameter, and a designer rarely checks whether it is negative. From a physical standpoint, g_{mb} should always be positive. More importantly, from a numerical point of view, g_{mb} should stay positive to avoid numerical problems during SPICE simulation (Section 1.2). Sometimes, having positive g_{mb} in all bias conditions is achieved by arbitrarily setting K2 to a very small negative value, such as -1×10^{-10}. In effect, the entire channel region is treated as a uniformly doped area, with the substrate doping equal to NCH. This way, g_{mb} tends to remain positive.

Figure 3-13 illustrates the I-V characteristics with the same parameter set, except that K2 is changed from -0.05 to -1×10^{-10}. Figure 3-14 is the calculated g_{mb} for these two cases. When V_{BS} is small, such as -1 V, the modification in the drain current is small as K2 changes value. However, at the same V_{BS}, the change in K2's value brings a significant change in g_{mb}. Even though K2 is often viewed as a minor parameter because it does not greatly affect I_D, it is fairly important from the viewpoint of fitting g_{mb}.

When K1 is specified but K2 is not, then K2 is defaulted to -0.0186. (*Note*: The BSIM3 manual states that K2 is defaulted to 0. This is inconsistent with the source code and is likely an error.) If neither K1 nor K2 is specified, then K1 and K2 are

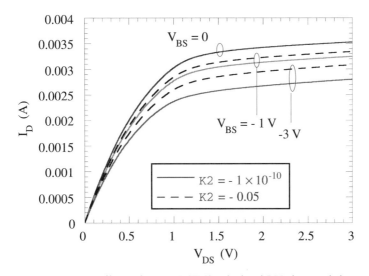

Fig. 3-13 Effects of K2 on BSIM3-calculated I-V characteristics.

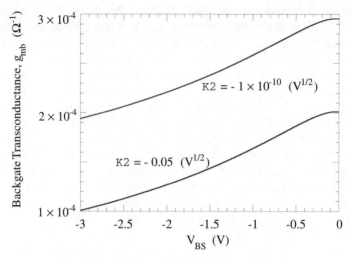

Fig. 3-14 Effects of K2 on BSIM3-calculated g_{mb}.

defaulted in accordance with Eqs. (3-64) and (3-65). See the entry NCH for more details when K1 and K2 are not specified in the parameter set.

K3 (default = 80, unitless) narrow-width coefficient for threshold voltage calculation. When transistor's width is narrow (below 1 μm) and the usual field-oxide isolation is used to define the active region of the device, then the threshold voltage is found to increase compared to the wide devices. For transistors with trench isolation, the narrow-width effects are less pronounced. The distribution of the transistor width in a digital circuit used for a computer CPU is shown in Fig. 3-15. It can be seen that there is a small amount of transistors which fall into the narrow width category. The number of transistors with width between 0 μm and 0.9 μm is roughly 1 percent of the total number of transistors on the circuit.

The threshold voltage in BSIM3 can be expressed as

$$V_T = \text{VTH0} + \delta_{NP}(\Delta V_{T,\text{body\_effect}} - \Delta V_{T,\text{charge\_sharing}} - \Delta V_{T,\text{DIBL}}$$

$$+ \Delta V_{T,\text{reverse\_short\_channel}} + \Delta V_{T,\text{narrow\_width}} + \Delta V_{T,\text{small\_size}}) \quad \text{(same as 3-33)}$$

where δ_{NP} is +1 for NMOS and −1 for PMOS. A general description of each of the components is found in the entry VTH0. Here, we are specifically interested in the narrow-width component, which is given by

$$\Delta V_{T,\text{narrow\_width}} = (\text{K3} + \text{K3B} \cdot V_{BS}) \frac{\text{TOX}}{W_{\text{eff}} + \text{W0}} 2\phi_f. \quad (3\text{-}66)$$

W_{eff} is the effective width of the device, whose expression is given in the entry of WINT. $2\phi_f$, the strong-inversion surface potential given by Eq. (1-6b), is usually about 0.85 V. When the transistor is wide, $\Delta V_{T,\text{narrow\_width}}$ produces a negligible amount of correction to the threshold voltage. It becomes significant only when W_{eff}

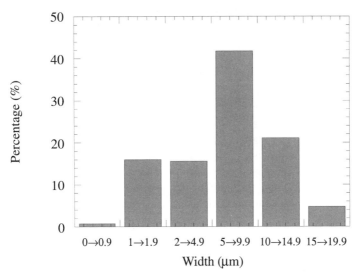

Fig. 3-15 Distribution of the transistor width in a CPU chip. There is a finite small amount of devices which qualify as the narrow-width transistors.

is small, as the name suggests. Physically, the narrow-width effect arises from the fact the fringing field emanating from the gate terminates at bulk charges outside the intrinsic portion of the transistor (as defined by W_{eff}), as illustrated in Fig. 3-16. Therefore, it takes a larger gate voltage to deplete the bulk charges before an inversion layer can be formed. Generally, the ΔV_T due to the narrow-width effect is on the order of 0.1 V when the transistor width is 1 μm.

Of the four parameters appearing in $\Delta V_{T,\text{narrow\_width}}$, TOX is really not an optimizable parameter, since it is fixed for a given process. However, parameters such as K3, K3B and W0 are tunable parameters which can be optimized to improve simulation fit. K3 represents the linear coefficient of the narrow-width effect when

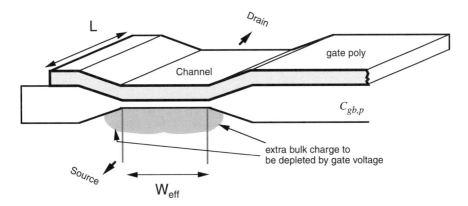

Fig. 3-16 Schematic drawing showing the extra bulk charges responsible for $\Delta V_{T,\text{narrow\_width}}$.

$V_{BS} = 0$. Because the narrow-width effect is known to increase the threshold voltage independent of V_{BS}, K3 should be greater or equal to zero. BSIM3 chooses a default K3 value of 50. The model found in Ref. [6], which has a similar equation for the narrow-width effect as in Eq. (3-66), shows that K3 to be about 9.4. In general, K3 values between 0 and 100 are used. A K3 value exceeding 100 usually indicates that some threshold voltage parameters are incorrectly extracted, so an exceedingly large value of K3 is required to compensate the inadequacy of the threshold voltage.

The default of K3 is not zero. The narrow-width effect is built into BSIM3 even if K3 is left unspecified. If the narrow-width effect is to be disregarded (such as if the transistor widths of the process all exceed 2 µm), K3 must be explicitly set to zero.

K3B quantifies the amount of the narrow-width effect due to V_{BS} variation. In NMOS, V_{BS} is negative in a normal transistor operation. When $|V_{BS}|$ increases, the amount of bulk charge controlled by the body node increases (both inside the intrinsic transistor defined by W_{eff}, or outside the region defined by W_{eff}). Therefore, it takes an even larger gate voltage (relative to when $V_{BS} = 0$) to fully deplete the bulk charges and then invert the channel. Consequently, ΔV_T is larger as $|V_{BS}|$ increases. With V_{BS} being negative in NMOS, K3B is thus a negative number. (For PMOS, an internal sign adjustment is made inside BSIM3 such that K3B should still be negative.)

The default value of K3B is zero. However, according to the model discussed in Ref. [6], K3B should have a default value equal to $-K3/(2\phi_f)$. Because $2\phi_f$ is approximately equal to 0.8 V for practical transistors and K3 was estimated to be 9.42 for the model in Ref. [6], K3B is about $-11.8\,\mathrm{V}^{-1}$. Generally, according to the theoretical relationship that K3B $\sim -K3/(2\phi_f)$, K3B and K3 have values on the same order of magnitude and differing in sign. It is not clear why BSIM3 makes K3 default to a nonzero value while K3B is defaulted to 0.

In the first-order analysis [6], the narrow-width effect on the threshold voltage is found to be inversely proportional to the transistor width. To accommodate some deviation from the theoretical result, BSIM3 introduces a width offset (W0) such that the narrow-width effect becomes inversely proportional to the sum of W_{eff} and W0. W0 should be a positive number. The larger the value, the less significant the narrow-width effect becomes. BSIM3 does not check whether W0 is negative or not. If W0 is negative, some caution needs to be exercised to prevent $W_{\mathit{eff}} + \mathrm{W0}$ from being negative. This is because the narrow-width effect typically gives a positive contribution to V_T, while a negative overall value of $W_{\mathit{eff}} + \mathrm{W0}$ results in a negative narrow-width effect. If W0 in a parameter set is negative, the parameter set is likely not well extracted.

Although BSIM3 does not check whether W0 is negative, BSIM3 checks whether $\mathrm{W0} + W_{\mathit{eff}}$ is identically zero. If so, BSIM3 issues a fatal error since the narrow-width term calculated from the above equation would be infinite. However, due to numerical rounding, this situation almost never arises. Instead, it is believed that BSIM3 should check if the absolute value of $\mathrm{W0} + W_{\mathit{eff}}$ is smaller than a trivially small number (such as 10^{-12}). If the SPICE parameter PARAMCHK is set to be 1 (or any nonzero value), BSIM3 performs one additional check on $\mathrm{W0} + W_{\mathit{eff}}$. If the

3.2 ALPHABETICAL GLOSSARY OF BSIM3 PARAMETERS 211

absolute value of W0 + W_{eff} is smaller than 0.1 µm, BSIM3 issues an warning (not error), since the narrow-width effect's contribution to V_T would be somewhat large and would be inconsistent with the notion that narrow-width effect is a small perturbation.

If it is known that the devices have negligible narrow-width effects, or if the devices are all fairly wide, then it is advisable to set K3 (and K3B also) to zero.

K3B (default = 0, unit = V^{-1}) body-effect coefficient of K3. Together with K3 and W0, K3B models the narrow-width effect's contribution to the threshold voltage. The details of K3B are found in the entry of K3. Although its default value is zero, K3B should generally be a negative number, on the order of −10.

KETA (default = −0.047, unit = V^{-1}) bulk-bias effect coefficient of the bulk-charge coefficient, A_{bulk}. A_{bulk} is not a model parameter. It is an intermediate variable used in BSIM3 to determine the saturation voltage and the current, as elaborated in the entry of A0. A_{bulk} can be expressed as

$$A_{bulk} = \left\{ 1 + \left[\delta \times \frac{A0 \cdot L_{eff}}{L_{eff} + 2\sqrt{XJ \cdot X_{dep}}} \right. \right.$$

$$\left. \times \left(1 - AGS(V_{GS} - V_T)\left(\frac{L_{eff}}{L_{eff} + 2\sqrt{XJ \cdot X_{dep}}}\right)^2 \right) + \frac{B0}{W_{eff} + B1} \right] \right\}$$

$$\times \frac{1}{1 + KETA \cdot V_{BS}} \qquad \text{(same as 3-13)}$$

where δ is given by Eq. (3-3). A_{bulk} attains its theoretical value of $1 + \delta$ in an ideal long-channel device. The factor $L_{eff}/(L_{eff} + 2\text{sqrt}(XJ \cdot X_{dep}))$ accounts for A_{bulk}'s dependence on channel length, that the bulk-charge effects diminish in short-channel devices.

By an examination of Eq. (3-13), it is obvious that the parameter KETA is to take care V_{BS}'s dependence of A_{bulk}. (AGS accounts for V_{GS}'s effect and B0 and B1, the narrow width's effect).

Equation (3-13) appears problematic when KETA · V_{BS} < −1. Under this circumstance, A_{bulk} would become negative, and produce unphysical results. However, in the actual BSIM3 codes, when KETA · V_{BS} starts to be smaller than −0.9, Eq. (3-13) is modified in a manner that A_{bulk} never goes below zero. This coding maneuver avoids numerical problems during a SPICE simulation. However, a good parameter set should be one such that KETA · V_{BS} is never close to −1.

Figure 3-17 illustrates the effects of KETA on the I-V characteristics. When $V_{BS} = 0$, the drain current is not affected by KETA since KETA is multiplied by V_{BS} as shown in Eq. (3-13). (However, g_{mb}, equal to the derivative of the drain current

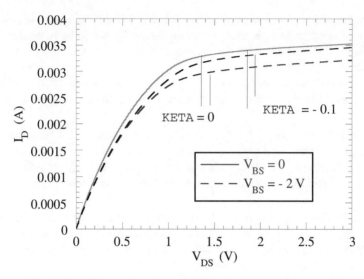

Fig. 3-17 Effects of KETA on BSIM3-calculated I-V characteristics.

with respect to V_{BS}, is affected by KETA even at $V_{BS} = 0$. This is similar to the observation that K2 does not affect the I-V characteristics at $V_{BS} = 0$, but can contribute to negative g_{mb}.) When V_{BS} is negative, KETA influences the drain current level through the modification of the bulk-charge coefficient (A_{bulk}), for both the short-channel and long-channel devices. When KETA changes from 0 to -0.1, the drain current increases because, as indicated in Eq. (3-13), the factor $1/(1 + \text{KETA} \cdot V_{BS})$ becomes smaller. As A_{bulk} is made smaller, the current increases.

KF (default = 1, unitless) proportional constant in the calculation of the Flicker noise when NOIMOD = 1 or 4. This parameter is not used if NOIMOD = 2 or 3. See the entry NOIMOD for details.

KT1 (default = -0.11, unit = V) temperature coefficient of the threshold voltage. All of the SPICE parameters are extracted at the temperature specified by TNOM. It is generally the room temperature. (See TNOM for more details.) If the device temperature is also equal to TNOM, then the threshold voltage is determined exclusively from Eq. (3-33), shown in the entry VTH0. However, the device temperature under simulation needs not be the same as TNOM. For example, due to thermal resistance of the package, the whole circuit may operates at 60 °C above the ambient temperature, often taken to be 27 °C. In that case, we shall denote the threshold voltage calculated in Eq. (3-33) as the threshold voltage at the nominal temperature, V_T(TNOM). The actual threshold voltage at the device temperature,

T_{device}, is denoted as $V_T(T_{device})$. KT1, KT1L and KT2 are used to relate these two quantities:

$$V_T(T_{device}) = V_I(\text{TNOM}) + \left[\text{KT1} + \frac{\text{KT1L}}{L_{eff}} + \text{KT2}\,V_{BS}\right] \times \left(\frac{T_{device}}{\text{TNOM}} - 1\right). \quad (3\text{-}67)$$

KT1 is the main parameter modeling the threshold voltage's temperature dependence. KT1L is used to improve fitting the dependence on the channel length, and KT2 is used to correct the bulk bias' effects. Despite its simplicity, Eq. (3-67) is apparently adequate to result in good fit in the I-V characteristics at various temperatures. Because the threshold voltage decreases with temperature, KT1 should be negative. (Although the threshold voltage decreases with temperature, there is hardly any observation of thermal instability. This is because the mobility's degradation at high temperatures more than offsets the threshold voltage's effects on the drive current. As temperature increases, the drive current decreases.) KT1L and KT2 are viewed as fitting parameters. KT1L is generally positive, whereas KT2 is negative. The default values for KT1, KT1L, and KT2 are -0.11, 0, and 0.22, respectively.

KT1L (default = 0, unit = V · m) channel-length coefficient of the threshold voltage's temperature dependence. See the entry KT1 for details.

KT2 (default = 0.022, unit = V · m) bulk-bias coefficient of the threshold voltage's temperature dependence. See the entry KT1 for details.

The following are binning parameters: LA0, LA1, LA2, LAGS, LALPHA0, LAT, LB0, LB1, LBETA0, LCDSC, LCDSCB, LCDSCD, LCF, LCGD1, LCGS1, LCIT, LCKAPPA, LCLC, LCLE, LDELTA, LDROUT, LDSUB, LDVT0, LDVT0W, LDVT1, LDVT1W, LDVT2, LDVT2W, LDWB, LDWG, LETA0, LETAB, LK1, LK2, LK3, LK3B, LKETA, LKT1, LKT2, LNFACTOR, LNLX, LPCLM, LPDIBLC1, LPDIBLC2, LPDIBLCB, LPRT, LPRWB LPRWG, LPSCBE1, LPSCBE2, LPVAG, LRDSW, LU0, LUA, LUA1, LUB, LUB1, LUC, LUC1, LUTE, LVBM, LVBX, LVFBCV, LVOFF, LVSAT, LVTH0, LW0, LWR, LXJ, LXT;

LEVEL (default = 8, unitless) This is the MOSFET model selector. This parameter is 8 in order to select BSIM3v3 model in most of the SPICE simulators. (It is 49 in the HSPICE simulator.) Other common models of the past are the LEVEL = 1, 2, and 3 models. The predecessor of BSIM3, BSIM, is LEVEL = 4. Choosing LEVEL equal to 8 does not guarantee that we are running the latest BSIM3 version, which is BSIM3v3.2.2. Depending on when the BSIM3v3 model is implemented in the SPICE simulator, LEVEL = 8 could invoke a BSIM3v3.1 (1996), BSIM3v3.2 (1998), BSIM3v3.2.1 (1999) or BSIM3v3.2.2 (1999) model. All these versions of the BSIM3 model are quite similar, except that the latter versions contain some significant improvement over BSIM3v3.1, in the C-V calculation. All of the versions are capable of accurate circuit simulation. For more details, see the entries under VERSION and CAPMOD.

LINT (default = 0, unit = m)   channel-length offset for dc I-V characteristics. In BSIM3, the effective-channel length for dc calculation is given by

$$L_{\mathit{eff}} = L - 2 \cdot \text{LINT} - 2 \cdot \Delta L_{\text{geometry}}, \qquad (3\text{-}68)$$

where L is the channel length specified in the MOSFET device statement. $\Delta L_{\text{geometry}}$ accounts for any geometry-dependent effects which modify the effective channel length. The exact expression of $\Delta L_{\text{geometry}}$ is given in the entry LL, but it can be approximated as zero. Therefore, LINT, representing a the channel length reduction per side, is the major parameter that effects L_{eff}. LINT is generally a positive number, although it may be negative. When it is negative, the effective-channel length becomes larger than the drawn L.

For C-V calculation, the appropriate channel-length offset is DLC. Conceptually, LINT and DLC should be identical. However, from a fitting's point of view, it is better to use these two length reduction parameters to independently model I-V and C-V behaviors. DLC often comes out slightly larger than LINT.

LL (default = 0, unit = m^{LLN})   coefficient of length dependence for channel-length offset in I-V calculation. There are two effective-channel lengths used in BSIM3. One is for dc calculations (including the Flicker noise), denoted as L_{eff}, and another is for C-V calculations (including the thermal noise), denoted as $L_{\mathit{eff},\text{CV}}$. Here, we are concerned with L_{eff}, which is given by

$$L_{\mathit{eff}} = L - 2 \cdot \text{LINT} - 2 \cdot \Delta L_{\text{geometry}}, \qquad \text{(same as 3-68)}$$

where

$$\Delta L_{\text{geometry}} = \frac{\text{LL}}{L^{\text{LLN}}} + \frac{\text{LW}}{W^{\text{LWN}}} + \frac{\text{LWL}}{L^{\text{LLN}} W^{\text{LWN}}}. \qquad (3\text{-}69)$$

L is the drawn channel length. Most of the channel-length offset is modeled with LINT, and typically, $\Delta L_{\text{geometry}} = 0$. However, at times, due to the geometrical effects in the photolithographical development of photoresist, L_{eff} may exhibit dependencies on the exposure area (or, effectively, the transistor's size). $\Delta L_{\text{geometry}}$ is used to characterize such a dependence. LL, in combination of LLN (the exponent of the channel length), specifies the amount of channel-length reduction due to the length dependence.

Disregarding the processing-related issues, LL should be at its default value of zero. LL's counterpart in the $L_{\mathit{eff},\text{CV}}$ expression is LLC.

LLC (default = 0, unit = m^{LLN})   coefficient of length dependence for channel-length offset in the C-V calculation. The effective channel length for I-V calculation (including the Flicker noise), L_{eff}, is given in the entry LL. There is another effective channel length in BSIM3, that used for C-V calculations (including the thermal noise), denoted as $L_{\mathit{eff},CV}$. Here, we are concerned with the latter:

$$L_{\mathit{eff},CV} = L - 2 \cdot \text{DLC} - 2 \cdot \Delta L_{\text{geometry,CV}} \qquad \text{(same as 3-32)}$$

where

$$\Delta L_{\text{geometry,CV}} = \frac{\text{LLC}}{L^{\text{LLN}}} + \frac{\text{LWC}}{W^{\text{LWN}}} + \frac{\text{LWLC}}{L^{\text{LLN}} W^{\text{LWN}}}. \qquad (3\text{-}70)$$

L is the drawn channel length. Most of the channel-length offset is modeled with DLC, and typically, $\Delta L_{\text{geometry,CV}} = 0$. However, at times, due to the geometrical effects in the photolithographical development of photoresist, $L_{\mathit{eff},CV}$ may exhibit dependencies on the exposure area (or, effectively, the transistor's size). $\Delta L_{\text{geometry,CV}}$ is used to characterize such a dependence. LLC, in combination with LNN (the exponent of the channel length), specifies the amount of channel length reduction due to the length dependence.

LLC's counterpart in I-V calculation is LL. LLC is a new parameter introduced in BSIM3v3.2. In BSIM3v3.1, the LLC appearing in the Eq. (3-70) is replaced by LL. Disregarding the processing-related issues, LLC should be at its default value of zero.

LLN (default = 1, unitless) power exponent of the length dependence in the calculation of the I-V and C-V channel-length offsets. BSIM3 calculates two effective channel lengths, L_{eff}, which is used for I-V calculation, and $L_{\mathit{eff},CV}$, which is used for C-V calculation. LLN affects both of them. The effective channel length for I-V calculation can be expressed as: $L_{\mathit{eff}} = L - 2 \cdot \text{LINT} - 2 \cdot \Delta L_{\text{geometry}}$. LLN is the exponent of $1/L$ in the $\Delta L_{\text{geometry}}$ expression, given in the entry LL (Eq. 3-69). $\Delta L_{\text{geometry}}$ is usually zero unless there are processing-related issues which cause L_{eff} to depend on geometry. When LL, LW, and LWL assume their default value of zero such that $\Delta L_{\text{geometry}} = 0$, then the value of LLN is inconsequential.

The effective channel length for C-V calculation can be expressed as: $L_{\mathit{eff},CV} = L - 2 \cdot \text{DLC} - 2 \cdot \Delta L_{\text{geometry,CV}}$. LLN is the exponent of $1/L$ in the $\Delta L_{\text{geometry,CV}}$ expression given in the entry LLC (Eq. 3-70). $\Delta L_{\text{geometry,CV}}$ is usually zero unless there are processing-related issues which cause $L_{\mathit{eff},CV}$ to depend on geometry. When LLC, LWC, and LWLC assume their default value of zero such that $\Delta L_{\text{geometry,CV}} = 0$, then the value of LLN is inconsequential.

LMAX (default = 1, unit = m) maximum channel length. This parameter is meant to specify the maximum channel length allowable for a given MOSFET parameter set. Combined with LMIN, WMIN and WMAX, LMAX appears to be a convenient parameter defining the geometrical space for which a binned model is applicable (Section 1.12). It seems that the intended purpose of this parameter is to specify the maximum channel length permissible for a given MOSFET. For example, if a circuit file uses a $L = 20$ µm MOSFET as part of the circuit element while LMAX in the model card is specified as 5 µm, then the SPICE program should either abort or issue a warning. However, somehow the BSIM3 code does not check whether L given in a MOSFET device statement is smaller than LMAX or not. Consequently, LMAX is a parameter which produces no effect. It is up to the designer to verify that the channel length of a MOSFET is indeed smaller than LMAX.

LMIN (default = 0, unit = m) minimum channel length. This parameter is meant to specify the minimum channel length allowable for a given MOSFET parameter set. Combined with LMAX, WMIN and WMAX, LMIN appears to be a convenient parameter defining the geometrical space for which a binned model is applicable (Section 1.12). It seems that the intended purpose of this parameter is to specify the minimum channel length allowable for a given MOSFET. For example, if a circuit file uses a $L = 0.2$ µm MOSFET as part of the circuit element while LMIN in the model card is specified as 0.5 µm, then the SPICE program should either abort or issue a warning. However, somehow the BSIM3 code does not check whether L given in a MOSFET device statement is larger than LMIN or not. Consequently, LMIN is a parameter which produces no effect. It is up to the designer to verify that the channel length of a MOSFET is longer than LMIN.

LW (default = 0, unit = m^{LWN}) coefficient of width dependence in the calculation of the dc channel-length offset. This parameter is similar to LL; they both modify the effective-channel length for I-V calculation, L_{eff}. LL is a parameter which accounts for the length variation, whereas LW accounts for the width variation. See Eqs. (3-68) and (3-69), listed in the entry LL. Unless there are processing-related issues which cause the effective channel length to depend on the device geometry, LW should be zero. LW's counterpart in C-V calculation is LWC.

LWC (default = 0, unit = m^{LWN}) coefficient of width dependence in the calculation of the C-V channel-length offset. This parameter is similar to LLC; they both modify the effective-channel length for C-V calculation, $L_{eff,CV}$. LLC is a parameter which accounts for the length variation, whereas LWC accounts for the width variation. See Eqs. (3-32) and (3-70), listed in the entry LLC. Unless there are processing-related issues which cause the effective channel length to depend on the device geometry, LWC should be zero.

LWC's counterpart in I-V calculation is LW. LWC is a new parameter introduced in BSIM3v3.2. In BSIM3v3.1, the LWC appearing in Eq. (3-70) is replaced by LW.

LWL (default = 0, unit = m$^{\text{LWN+LLN}}$) coefficient of length and width dependence in the calculation of the dc channel-length offset. This parameter is similar to LL; they both modify the effective-channel length for I-V calculation, L_{eff}. LL is a parameter which accounts for the length variation, whereas LWL accounts for the product of the length and width variation. See Eqs. (3-68) and (3-69), listed in the entry LL. Unless there are processing-related issues which cause the effective channel length to depend on the device geometry, LWL should be zero. LWL's counterpart in C-V calculation is LWLC.

LWLC (default = 0, unit = m$^{\text{LWN+LLN}}$) coefficient of length and width dependence in the calculation of the C-V channel-length offset. This parameter is similar to LLC; they both modify the effective-channel length for C-V calculation, $L_{\textit{eff},\text{CV}}$. LLC is a parameter which accounts for the length variation, whereas LWLC accounts for the length and width variation. See Eqs. (3-32) and (3-70), listed in the entry LLC. Unless there are processing-related issues which cause the effective channel length to depend on the device geometry, LWLC should be zero.

LWLC's counterpart in I-V calculation is LWL. LWLC is a new parameter introduced in BSIM3v3.2. In BSIM3v3.1, the LWLC appearing in Eq. (3-70) is replaced by LWL.

LWN (default = 1, unitless) power exponent of the width dependence in the calculation of the I-V and C-V channel-length offsets. BSIM3 calculates two effective channel lengths, $L_{\textit{eff}}$, which is used for I-V calculation, and $L_{\textit{eff},\text{CV}}$, which is used for C-V calculation. LWN affects both of them. The effective channel length for I-V calculation can be expressed as: $L_{\textit{eff}} = L - 2 \cdot \text{LINT} - 2 \cdot \Delta L_{\text{geometry}}$. LWN is the exponent of $1/W$ in the $\Delta L_{\text{geometry}}$ expression given in the entry of LL (Eq. 3-69). $\Delta L_{\text{geometry}}$ is usually zero unless there are processing-related issues which cause $L_{\textit{eff}}$ to depend on geometry. When LL, LW, and LWL assume their default value of zero such that $\Delta L_{\text{geometry}} = 0$, then the value of LWN is inconsequential.

The effective channel length for C-V calculation can be expressed as: $L_{\textit{eff},\text{CV}} = L - 2 \cdot \text{DLC} - 2 \cdot \Delta L_{\text{geometry}}$. LWN is the exponent of $1/W$ in the $\Delta L_{\text{geometry},\text{CV}}$ expression given in the entry LLC (Eq. 3-70). $\Delta L_{\text{geometry},\text{CV}}$ is usually zero unless there are processing-related issues which cause $L_{\textit{eff},\text{CV}}$ to depend on geometry. When LLC, LWC, and LWLC assume their default value of zero such that $\Delta L_{\text{geometry},\text{CV}} = 0$, then the value of LWN is inconsequential.

MJ (default = 0.5, unitless) grading coefficient of the bottom-wall junction capacitance. It is the bias-voltage's exponent of the bottom-wall capacitance, which is the first term in Eq. (3-25). The bottom-wall capacitance as a function of MJ is shown in Fig. 3-7. See the entry CJ for more details.

MJSW (default = 0.33, unitless) grading coefficient of the isolation-side sidewall junction capacitance. It is the bias-voltage's exponent of the isolation-side sidewall capacitance, which is the second term in Eq. (3-25). See more details in the entry CJ.

MJSWG (default = MJSWG, unitless) grading coefficient of the gate-side sidewall junction capacitance. It is the bias-voltage's exponent of the gate-side sidewall capacitance, which is the last term in Eq. (3-25). See more details in the entry CJ.

MOBMOD (default = 1, unitless) This parameter selects the mobility model to calculate the effective mobility in the channel. It can be either 1, 2, or 3. The equations used in the three mobility models are listed in the entry U0. Either the MOBMOD = 1 or 3 model is preferred for enhancement-mode MOSFETs. These models uses $(V_{GS} + V_T)/(6 \cdot \text{TOX})$ to represent the average normal electric field in the channel [7]. (Normal electric field is the field across the oxide layer, between the gate metal and the channel region.) As electrons travel across the channel, they are constantly pulled toward the gate due to the normal field. The electrons continue to collide with the oxide while they traverse along the channel. The electron mobility in the channel thus decreases as the normal field increases. In MOBMOD = 2, the normal electric field is taken to be $(V_{GS} - V_T)/(6 \cdot \text{TOX})$, a value obtained when the depleted bulk charges are neglected. Therefore, MOBMOD = 2 is preferred for depletion-mode MOSFETs.

The difference between MOBMOD = 1 and MOBMOD = 3 is insignificant, mainly in the ways the body effects are accounted for. Despite the differences in the three models, all three models work equally well, at least as far as the fitted device characteristics are concerned. However, if it is desired to switch the mobility models, such as between MOBMOD = 1 and MOBMOD = 3, the parameters UA, UB, and UC need to be re-extracted.

MOIN (default = 15, unitless) coefficient in calculating the offset surface-potential in the strong inversion region. It is used only in the charge-thickness model to calculate C-V characteristics (CAPMOD = 3). It models the amount of the surface potential in excess of $2\phi_f$ that must be overcome by the gate voltage before the channel layer is inverted. The smaller the MOIN value, the more rapidly the gate capacitance decreases with V_{GS} in the inversion region. MOIN does not affect the I-V characteristics. More details are found in the entry ACDE.

NCH (default = either 1.7×10^{17} or calculated, unit = cm^{-3}) channel doping concentration. This parameter should have positive value, whether the device is NMOS or PMOS. BSIM3 models the complex channel doping profiles as a two-level step profile as shown in Fig. 3-18. NCH is used to represent the higher doping level, nearby the silicon/oxide interface. NSUB is used to denote the doping level toward the substrate. These two doping levels, together with the VBM and VBX parameters, can be used to solve for the two body-effect factors (K1, and K2) that determine the body-effect component of the threshold voltage, $\Delta V_{T,\text{body\_effect}}$ (Eq. 3-63). The relationships between NCH, NSUB and K1, K2 are

$$\text{K2} = \frac{(\text{GAMMA1} - \text{GAMMA2})\left(\sqrt{2\phi_f - \text{VBX}} - \sqrt{2\phi_f}\right)}{2\sqrt{2\phi_f}\left(\sqrt{2\phi_f - \text{VBM}} - \sqrt{2\phi_f}\right) + \text{VBM}} \qquad \text{(same as 3-64)}$$

$$\text{K1} = \text{GAMMA2} - 2 \cdot \text{K2} \cdot \sqrt{2\phi_f - \text{VBM}} \qquad \text{(same as 3-65)}$$

3.2 ALPHABETICAL GLOSSARY OF BSIM3 PARAMETERS 219

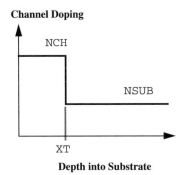

Fig. 3-18 Two-level channel doping profile assumed in BSIM3 to calculate the threshold voltage.

where $2\phi_f$ is the strong-inversion surface potential given in Eq. (1-6b), and

$$\text{GAMMA1} = \sqrt{\frac{2q \, \epsilon_s \, \text{NCH}}{C'_{ox}}} \quad \text{(same as 3-50)}$$

$$\text{GAMMA2} = \sqrt{\frac{2q \, \epsilon_s \, \text{NSUB}}{C'_{ox}}} \quad \text{(same as 3-51)}$$

The body-effect factors used to calculate the threshold voltage in BSIM3 are K1, and K2. They can be extracted once the I-V characteristics are available. So, why does BSIM3 introduce parameters such as NCH, NSUB, GAMMA1, and GAMMA2? There are times when the silicon devices are not available, so a strawman model needs to be created to do a preliminary simulation. In this case, a SUPREME simulation, which outputs the transistor doping profiles, allows us to establish NCH and NSUB directly. (SUPREME is a process simulation program.) The parameters K1 and K2 are not readily available from such a process simulator, and must be inferred from the above equations.

By the fact that NCH, NSUB, GAMMA1, GAMMA2, K1, and K2 are all BSIM3 parameters, it is hinted that there are many ways to calculate $\Delta V_{T,\text{body\_effect}}$. First, when K1 and K2 are given in the parameter set, then NCH, NSUB, GAMMA1, and GAMMA2, whether specified or not, become irrelevant. BSIM3 will directly substitute the specified K1 and K2 parameter values into Eq. (3-63) to find $\Delta V_{T,\text{body\_effect}}$. Second, if neither K1 nor K2 is specified but GAMMA1 and GAMMA2 are, then BSIM3 no longer needs to compute GAMMA1 and GAMMA2 from Eqs. (3-50) and (3-51). Instead, BSIM3 directly substitutes the given GAMMA1 and GAMMA2 parameter values into Eqs. (3-64) and (3-65) to find K1 and K2. From the calculated K1 and K2 values, the $\Delta V_{T,\text{body\_effect}}$ can be determined. Last, when none of GAMMA1, GAMMA2, K1, and K2 are specified in the parameter set, then BSIM3 will take either the specified or defaulted values of NCH and NSUB to calculate GAMMA1 and GAMMA2 from Eqs. (3-50) and (3-51). Subsequently, K1 and K2 are found. These values then enable the calculation of $\Delta V_{T,\text{body\_effect}}$.

In the case K1 and K2 values are provided in the parameter set, BSIM3 will discard GAMMA1, GAMMA2, and NSUB, even though they may be specified with certain values. BSIM3 will print out a warning message, basically stating the fact that once K1 and K2 are specified, GAMMA1 and GAMMA2, NSUB (as well as XT, VBX, which are not discussed here) need not be given. We might think that perhaps the warning message should also apply to NCH. However, NCH is used in other parts of BSIM3. Even if K1 and K2 are specified, NCH is still needed to calculated $2\phi_f$ (strong-inversion surface potential), $X_{dep,0}$ (depletion thickness in the channel region at zero V_{BS}), V_{bi} (drain/source–bulk junction potential), X_{max} (one-third the Debye length of the channel, and X_{DC} (the dc charge thickness). (*Note*: The BSIM3 manual states that X_{DC} depends on NSUB, not NCH. However, the actual BSIM3 code uses NCH instead of NSUB.)

For the curious designers, we pose this question. What happens if either K1 or K2 is given, but not both? Would NCH and NSUB be used to model the body effects at all? The answer is, BSIM3 will first print out a warning message, stressing that one should always give K1 and K2 in pairs, not just one of them. Subsequently, internally in BSIM3, K1 will be defaulted to 0.53 if it is K2 which is specified, or K2 will be defaulted to -0.0186 if it is K1 which gets specified. With the default values, both K1 and K2 values then become available to enable a calculation of $\Delta V_{T,\text{body\_effect}}$. The parameters GAMMA1, GAMMA2, NCH, and NSUB remain unused for the body-effect calculation.

In the event that NCH is not specified in the parameter set, NCH is defaulted to one of two possible values. If GAMMA1 is not given either, then NCH is defaulted to 1.7×10^{17} cm^{-3} (and further, GAMMA1 will then be calculated from this value of NCH by using Eq. 3-50). If GAMMA1 is instead given, then NCH is defaulted to a value given by

$$\text{NCH} = \frac{\text{GAMMA1}^2}{2q \, \epsilon_s} (C'_{ox})^2. \tag{3-71}$$

Note that this default process for NCH takes place regardless of whether K1 or K2 are specified in the SPICE parameter set.

In the event that NSUB is not specified in the parameter set, NSUB is defaulted to 6×10^{16} cm^{-3}. If neither K1, K2, nor GAMMA2 is specified, then NSUB (either the specified value or the default value of 6×10^{16} cm^{-3}) will be used to calculate GAMMA2. However, if GAMMA2 is specified, then NSUB is never used in BSIM3. We have some use for it, though. In the event that either PB, PBSW, or PBSWG is not given, we can use NSUB to estimate their values. See the entry PB for more details.

How about a trick question? What happens if both GAMMA1 and NCH are given? The answer depends on whether K1 and K2 are specified. In the case that either K1 or K2 is given, GAMMA1 will never be used. BSIM3, in fact, issues a warning that GAMMA1 now becomes a redundant parameter. NCH, however, is used to calculate $2\phi_f, X_{dep,0}, V_{bi}, X_{max}$, and X_{DC}. The relationship given in Eq. (3-50) will not be used. If, on the other hand, neither K1 nor K2 is given, then GAMMA1 will be used in Eqs.

(3-64) and (3-65) to calculate K1 and K2. NCH, is still used to calculate $2\phi_f$, $X_{dep,0}$, V_{bi}, X_{max}, and X_{DC}. The relationship given in Eq. (3-50), again, will not be used.

Generally a SPICE model is generated after the silicon devices have been fabricated. From a fit to measured I-V characteristics, the parameters K1 and K2 can be easily extracted. Therefore, a SPICE parameter set often contains values for K1 and K2, obliterating the need to specify NSUB, GAMMA1, GAMMA2, XT, and VBX. (We did not talk about XT and VBX in detail here. These two parameters are used for the calculation of K1 and K2 given in Eqs. 3-64 and 3-65. See the entry XT for more details.) Besides, using GAMMA1, and GAMMA2 (or NCH and NSUB) to model the body effects instead of K1 and K2 has some severe limitation. Sometimes the channel doping profile cannot be well grouped into the two distinct levels shown in Fig. 3-18. Rather, the doping is initially at a low value at the oxide/substrate interface. It gradually increases to a peak value as the position moves away from the oxide and then decreases to a small value at a distance well into the substrate. In this case, it is difficult to specify NCH and NSUB for the structure. Choosing GAMMA1 and GAMMA2 are also difficult since we really do not expect the theoretical equations of Eqs. (3-50) and (3-51) to work well in practice.

NFACTOR (default = 1; unitless) subthreshold turn-on swing factor. The drain current increases exponentially with V_{GS} in the subthreshold region. This empirical result is reflected in BSIM3's I_D expression, which is made to be proportional to $\exp(qV_{GS}/nkT)$ when $V_{GS} < V_T$. Adopting the terminology used to describe the diode current, we shall refer to n in the exponent as the *ideality factor*. Although the ideality factor of a diode often takes on the minimum value of 1, the ideality factor in MOS is always larger than 1 and is bias-dependent. According to Ref. [6],

$$n = 1 + \frac{C'_{dep} + \text{CIT}}{C'_{ox}}, \tag{3-72}$$

where C'_{dep} is the depletion capacitance per unit area in the bulk given in Appendix A (Eq. A-45a), CIT is the interface capacitance per unit area (a BSIM3 parameter) and C'_{ox} is the oxide capacitance per unit area. Figure 3-19 illustrates the turn-on characteristics at $V_{DS} = 0.05$ V for NMOS and -0.05 V for PMOS. The ideality factors for NMOS and PMOS, whether long- or short-channel, are about 1.35 in this modern CMOS technology.

In order to improve the fit between simulation and measured data, as well as to accommodate various device geometries under various bias conditions, BSIM3 modifies the above ideality factor equation to the following:

$$n = 1 + \text{NFACTOR} \cdot \frac{C'_{dep}}{C'_{ox}} + \frac{\text{CIT}}{C'_{ox}} + \frac{\text{CDSC} + \text{CDSCD} \cdot V_{DS} + \text{CDSCB} \cdot V_{BS}}{C'_{ox}}$$
$$\times \left[\exp\left(-\text{DVT1} \frac{L_{eff}}{2L_t}\right) + 2\exp\left(-\text{DVT1} \frac{L_{eff}}{L_t}\right) \right]. \tag{3-73}$$

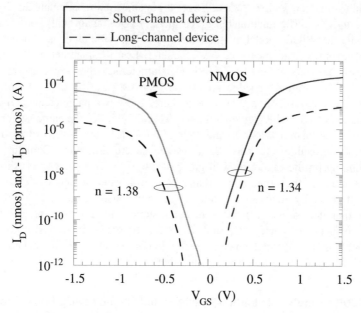

Fig. 3-19 Measured turn-on characteristics at $V_{DS} = 0.05$ V for NMOS and -0.05 V for PMOS. The ideality factor is about 1.35.

Comparing this equation with the idealized equation of Eq. (3-72), we see that NFACTOR should have a value close to 1. However, as long as the ideality factor is matched between the measurement and simulation, we tend not to care exactly how that ideality factor is produced, whether by adjusting CIT or by adjusting NFACTOR. Therefore, despite that NFACTOR should be close to 1, it is not uncommon to see a high value of 5 being used. On the other hand, it is not desirable to have a NFACTOR smaller than unity, or even worse, smaller than zero. A negative NFACTOR may cause unforeseen error in the computation of the drain current, due to a possible negative ideality factor which may result. An NFACTOR smaller than 1 usually indicates other components making up the ideality factor take on excessively high values.

Just like NFACTOR, CIT is treated as a fitting parameter, although physically it is used to denote the interface charge capacitance per unit area. CIT has a default of zero. It should never assume a value smaller than zero, but can be as large as required to result in a good fit in the ideality factor.

The last term in Eq. (3-73) is multiplied by a factor enclosed in square brackets. It is the same factor which appears in the charge-sharing's correction to the threshold voltage in short-channel devices (Eq. 3-34). Therefore, the last term is meant to be important only in short-channel devices. It is used to model the effects from the coupling capacitances between the drain or source to the channel. CDSC, CDSDC, and CDSCB have default values of 2.4×10^{-4} (F/m^2), 0 and 0, respectively. CDSC and CDSCD should be positive numbers, whereas CDSCB should be negative (since

V_{BS} in normal operation is negative.) Generally, when the absolute values of any of the three parameters exceed 0.1, some kinks in simulated drain current may develop under certain bias condition. Because of the danger of unexpected I-V characteristics, and because the last term is meant only for second-order correction to the ideality factor, often CDSC, CDSCB, and CDSCB are simply made to be zero.

NFACTOR is the main BSIM3 parameter that modifies the subthreshold ideality factor. Another parameter, VOFF, is used to fit the transistor's off current (the drain current at $V_{GS} = 0$). Under normal circumstances, VOFF is about 0 or only slightly negative. One can then independently optimize the values of NFACTOR and VOFF, using NFACTOR to fit the subthreshold slope and VOFF to fit the off current. However, when VOFF is fairly negative, these two parameters are no longer independent of each other. The $V_{GST,\text{eff}}$ adopted in BSIM3 is such that NFACTOR and VOFF can both affect the turn-on characteristics when VOFF is fairly negative. In this scenario, I_D no longer increases with V_{GS} with the ideality factor given in Eq. (3-73). A wrongly extracted VOFF value therefore can lead to a unrealistic value for NFACTOR.

Figure 3-20 shows the calculated drain currents, as the values of VOFF and NFACTOR are modified. The lowest curve, produced with VOFF $= -0.11$ (V) and NFACTOR $= 1.12$, fits closely with the experimentally measured data. However, by changing just the VOFF value to -0.5 (dashed curve), we observe some big changes in the turn-on characteristics. The first one, that the off current at $V_{GS} = 0$ increases significantly, is understandable. It is BSIM3's intention to use VOFF to fit to the off current. The second change, unforeseen from the formulation of Eq. (3-73), is that I_D increases with V_{GS} with a fairly large ideality factor. This large ideality factor is not caused by a large NFACTOR, but instead is an artifact of VOFF. Figure 3-20 also shows the effect of changing only the NFACTOR (dotted curve) to 3. As expected,

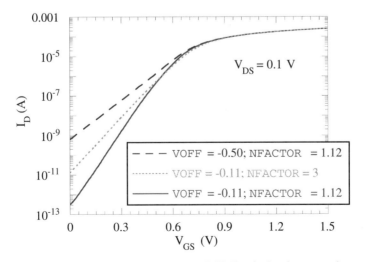

Fig. 3-20 Effects of VOFF and NFACTOR on BSIM3-calculated turn-on characteristics.

the ideality factor of the calculated curve is higher than when NFACTOR = 1.12 (solid curve).

NGATE (default = 0, unit = cm^{-3}) doping concentration in the polysilicon gate. The doping should be inputted as a positive number for both NMOS and PMOS devices. This parameter is used to model the poly-depletion effects, which affect the transistor's I-V and C-V characteristics. When the polysilicon doping is light, the portion of the poly right above the gate oxide is depleted during inversion. Let us consider NMOS whose gate poly is doped n-type. The depletion in the n-type poly is analogous to the depletion in the n-region of a p-n junction. The region is depleted of the free electrons; hence the ionized charges are positive. A positive voltage then drops across the depletion region, between the top of the gate electrode to the polysilicon/oxide interface. This positive voltage drop reduces the available gate voltage to invert the channel. Effectively, poly-depletion reduces the amount of gate voltage seen by the channel. The drive current and device capacitance are lower than if the poly depletion effects were neglected.

NGATE is the parameter used to model the poly-depletion effects. Suppose that we set NGATE to an unrealistic high value of 10^{23} (cm^{-3}), then the poly-depletion model in BSIM3 is essentially bypassed. That is, because the doping is so high, the amount of the depletion is negligible, and the entire applied gate voltage is available to invert the channel. Experiences have shown that, even though poly-depletion effects exist in some devices, we can still obtain good fits to measured I-V characteristics by tweaking parameters other than NGATE. However, the parameters may fall outside their normal physical values, which, in turn, will result in significant overprediction of the device capacitance [8]. Therefore, it is preferred to model the poly-depletion effects properly, by setting the NGATE parameter to its realistic value.

Figure 3-21 illustrates the I-V characteristics of a $L = 0.35\,\mu$m device at three NGATE doping levels: 10^{23}, 10^{20}, and 3×10^{19} (cm^{-3}). The value 10^{23} is not physically possible; but it can be taken to be the case if the poly-depletion effects were totally negligible. As shown in the figure, the poly-depletion effects on the I-V characteristics can be significant at doping levels below 10^{20} cm^{-3}. Figure 3-22 plots C_{gg}'s variation with NGATE. The overall gate capacitance can be thought to consist of a capacitance associated with the poly-depletion layer and the oxide capacitance associated with TOX. Because these two capacitance components are in series, the overall capacitance decreases as NGATE decreases.

Alternatively, we can view the poly-depletion effects as a cause for an incremental threshold voltage. A popular threshold voltage expression accounting for the poly-depletion effects is [8]

$$V_T|_{include\ polydepletion} \approx V_T|_{no\ polydepletion} + \frac{K1^2 (C'_{ox})^2}{2q\,\epsilon_s\,\text{NGATE}} (2\phi_f). \qquad (3\text{-}74)$$

For a typical 0.35 μm process with NGATE $= 3 \times 10^{19}$ (cm^{-3}), K1 $= 0.53$ (V$^{1/2}$), $C'_{ox} = 4.9 \times 10^{-7}$ F/cm^2 (corresponding to TOX $= 70$ Å), and $2\phi_f$ given by Eq.

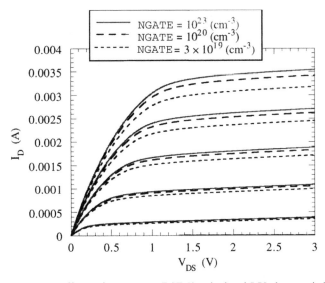

Fig. 3-21 Effects of NGATE on BSIM3-calculated I-V characteristics.

(1-6b) is about 0.85 V, then the additional threshold voltage is 0.0057 V. This appears to be a underestimation, since the I-V characteristics for the NGATE 3×10^{19} cm^{-3} shown in Fig. 3-21 exhibit significant drain current degradation.

The value used for NGATE should be the activated doping concentration, not the chemical doping concentration. For NMOS, the polysilicon is doped n-type to a level at roughly 2×10^{20} cm^{-3}, although the activated portion is only about 6×10^{19} cm^{-3}. Hence, generally NGATE is about 6×10^{19} for NMOS. For PMOS, the polysilicon is doped p-type with boron to roughly 1×10^{20} cm^{-3}.

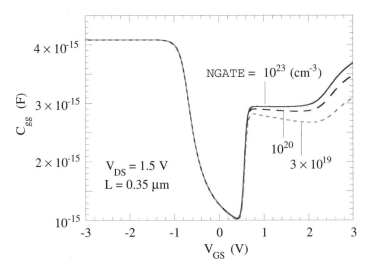

Fig. 3-22 Effects of NGATE on BSIM3-calculated C_{gg}.

However, the activation is only at the low 1×10^{19} cm^{-3} level. (In fact, the activation can be as low as 1×10^{17} cm^{-3} in the area right above the oxide, a result which worsens the poly-depletion effects.) A typical NGATE value for PMOS is 1×10^{19}.

The default of NGATE is 0! This value does not lead to near infinity increase in the threshold voltage, as suggested by Eq. (3-74). The BSIM3 code basically neglects the poly-depletion effects when NGATE is below 10^{18} or above 10^{25} cm^{-3}. Therefore, as we change NGATE from 10^{20} (cm^{-3}) to 10^{19}, and then to 10^{18}, we would see worsened device characteristics with lower drive currents and lower gate capacitance. However, if we continue to decrease NGATE to below 10^{18}, suddenly, the device characteristics improve to the level when NGATE is very large (such as 10^{23}). This behavior is not desirable, because it leads to confusion.

NJ (default = 1, unitless) junction emission coefficient. The current of a *p-n* diode can be expressed as $I_0 \times \exp(qV/nkT)$, where *n* is often referred to as the ideality factor of the diode. BSIM3 employs the same diode expression (with minor modification) for the bulk–source diode and models its junction current as $I_{sat,SB} \times \exp(qV_{BS}/\text{NJ}kT)$. The junction emission coefficient is basically the ideality factor of the bulk–source diode. NJ has an ideal value of unity. In practice, it can deviate slightly from 1, to a value of 1.1 or so. When the recombination current in the diode is excessively large, NJ may approach 2. However, NJ should be very close to unity in most of the diodes. For the bulk–drain diode, BSIM3 uses the same NJ as the junction emission coefficient. There is not a separate junction emission coefficient for the bulk–source and the bulk–drain diodes. See the entry JS for more details.

NLX (default = 1.74×10^{-7}, unit = m) reverse-short-channel-effect coefficient. The threshold voltage in BSIM3 can be expressed as

$$V_T = \text{VTH0} + \delta_{NP}(\Delta V_{T,\text{body\_effect}} - \Delta V_{T,\text{charge\_sharing}} - \Delta V_{T,\text{DIBL}}$$
$$+ \Delta V_{T,\text{reverse\_short\_channel}} + \Delta V_{T,\text{narrow\_width}} + \Delta V_{T,\text{small\_size}}) \quad \text{(same as 3-33)}$$

where δ_{NP} is +1 for NMOS and −1 for PMOS. A general description of each of the components is found in the entry VTH0. Here, we are specifically interested in the reverse-short-channel effect component, given by

$$\Delta V_{T,\text{reverse\_short\_channel}} = \text{K1}\left(\sqrt{1 + \frac{\text{NLX}}{L_{\text{eff}}}} - 1\right)\sqrt{2\phi_f}. \quad (3\text{-}75)$$

L_{eff} is the effective channel length of the device, given by Eq. (3-68), and $2\phi_f$ is the strong-inversion surface potential given by Eq. (1-6b). When the transistor is long (L_{eff} is large), the above term produces a negligible amount of correction to the threshold voltage. In the early CMOS technologies, the threshold voltage is always found to decrease as the channel length decreases. The reduction in threshold voltage, referred to as the short-channel effect, is accounted for by the term,

$-\Delta V_{T,\text{charge\_sharing}}$, in Eq. (3-33). However, in modern technologies, it is often observed that the threshold voltage first increases as L_{eff} decreases, before picking up its expected trend of decrease as L_{eff} decreases. To correctly model the fact that the threshold voltage can temporarily increase as L_{eff} decreases, $\Delta V_{T,\text{reverse\_short\_channel}}$ is a positive term, increasing as L_{eff} decreases. The reverse-short-channel effect is significant only when L_{eff} is small. For a given technology, the measured threshold voltages as a function of channel length is shown in Fig. 3-23. The figure shows that the threshold voltage decreases as L decreases below 0.7 μm, due to the short-channel effects. However, the threshold voltages increases steadily as L decreases from 10 to 0.7 μm.

In the classic short-channel effects, V_T steadily rolls off with decreasing channel length. When the reverse short-channel effects are absent, NLX should be set to zero.

NOFF (default = 1, unitless) smoothing parameter in the $V_{GS} - V_T$ effective function for C-V calculation. BSIM3 relies on smoothing functions to ensure continuous I-V and C-V characteristics across various operation regions. One of the smoothing functions is $V_{GST,\text{eff}}$, which approaches $\exp(V_{GS} - V_T)$ in the subthreshold region and $V_{GS} - V_T$, in the strong inversion region. $V_{GST,\text{eff}}$ is used exclusively for I-V calculation. To achieve a better fitting in the C-V characteristics, BSIM3 defines a similar but different function, $V_{GST,\text{effCV}}$ given by

$$V_{GST,\text{effCV}} = \text{NOFF} \cdot \frac{nkT}{q} \ln\left[1 + \exp\left(\frac{V_{GS} - V_T - \text{VOFFCV}}{\text{NOFF} \cdot nkT/q}\right)\right]. \quad (3\text{-}76)$$

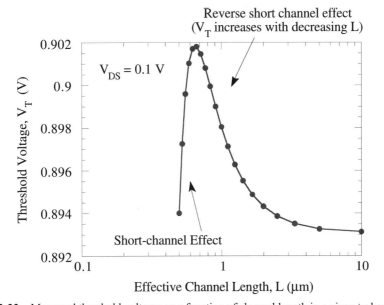

Fig. 3-23 Measured threshold voltages as a function of channel length in a given technology.

When $V_{GS} \ll V_T$, $\exp[q(V_{GS} - V_T - \text{VOFFCV})/(\text{NOFF} \cdot nkT)]$ is a small number. Knowing that $\ln(1+x) \approx x$ when x is small, we find that $V_{GST,\text{effCV}}$ to be proportional $\exp[q(V_{GS} - V_T)/(\text{NOFF} \cdot nkT)]$ in the subthreshold region ($V_{GS} \ll V_T$). It approaches $V_{GS} - V_T - \text{VOFFCV}$ during inversion region ($V_{GS} \gg V_T$). We summarize the differences between $V_{GST,\text{eff}}$ (for I-V calculation) and $V_{GST,\text{effCV}}$ (for C-V calculation) in the following:

	$V_{GST,\text{eff}}$	$V_{GST,\text{effCV}}$
$V_{GS} \ll V_T$	$\propto \exp\left(\dfrac{V_{GS} - V_T - \text{VOFF}}{nkT/q}\right)$	$\propto \exp\left(\dfrac{V_{GS} - V_T - \text{VOFFCV}}{\text{NOFF} \cdot nkT/q}\right)$
$V_{GS} \gg V_T$	$V_{GS} - V_T$	$V_{GS} - V_T - \text{VOFFCV}$

The parameter NOFF enables $V_{GST,\text{effCV}}$ to increase at a different slope from $V_{GST,\text{eff}}$ in the subthreshold region. The parameter VOFFCV represents the offset threshold voltage between the I-V and C-V characteristics. Without these two parameters to independently optimize the fit in the I-V and the C-V characteristics, a good fit in the I-V characteristics may nonetheless result in a poor fit in the C-V characteristics. A pertinent example is Fig. 3-24, which is a comparison of the measured and calculated C_{gg}'s of a long-channel device ($L = 70\,\mu\text{m}$). The simulation is obtained with a BSIM3v3.1 model, while the NOFF parameter is introduced in BSIM3v3.2. Without the availability of NOFF to adjust the turn-on behavior of the C-V characteristics, Fig. 3-24 exhibits different turn-on slopes of the measured and simulated C_{gg}'s. There is, however, a potential problem associated with using a high value of NOFF, as detailed in Section 4.10.

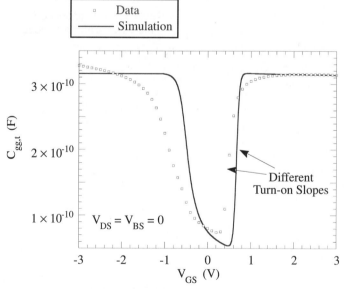

Fig. 3-24 Comparison of the measured and calculated total gate capacitances of a $L = 70\,\mu\text{m}$ device. The total gate capacitance includes the intrinsic as well as the parasitic components.

The default values of NOFF and VOFFCV are 1 and 0, respectively. Although the final optimized values often do not deviate from the default, the freedom to have their values vary is crucial to simultaneously obtain a good fit in the I-V and the C-V characteristics.

Independent of the PARAMCHK value, BSIM3 will check if NOFF exceeds 4 or falls below 0.1. If so, a warning message is issued. Similarly, if VOFFCV is below -0.5 (V) or above 0.5 (V), a warning message appears.

NOIA (default $= 1 \times 10^{20}$ for NMOS and 9.9×10^{18} for PMOS, unitless) noise parameter A for the Flicker noise calculation when NOIMOD $= 2$ or 3. This parameter is not used if NOIMOD $= 1$ or 4. See the entry NOIMOD for details.

NOIB (default $= 5 \times 10^{4}$ for NMOS and 2.4×10^{3} for PMOS, unitless) noise parameter B for the Flicker noise calculation when NOIMOD $= 2$ or 3. This parameter is not used if NOIMOD $= 1$ or 4. See the entry NOIMOD for details.

NOIC (default $= -1.4 \times 10^{-12}$ for NMOS and 1.4×10^{-12} for PMOS, unitless) noise parameter C for the Flicker noise calculation when NOIMOD $= 2$ or 3. This parameter is not used if NOIMOD $= 1$ or 4. See the entry NOIMOD for details.

NOIMOD (default $= 1$, unitless) This is the flag to choose one of the four noise models. BSIM3 does not consider the noise associated with the parasitic gate resistance and substrate resistance. BSIM3 also neglects the noise generated from the parasitic source and drain resistance if the resistances are modeled with the absorbed-resistance approach (i.e., using the parameter RDSW; see the entry RDSW for details). If the source/drain resistances are incorporated through the lumped-resistance approach (i.e., using the RSH parameter), then the usual thermal noise associated with the resistors are included in the BSIM3 calculation.

As far as the intrinsic transistor is concerned, BSIM3 ignores the channel-induced noise at the gate terminal, a noise that can be important at high frequencies. However, both the Flicker ($1/f$) noise and the channel thermal noise are included. The exact model used to calculate the Flicker and the channel thermal noise depends on the value of NOIMOD, which takes on integer values between 1 and 4. Here is a summary:

NOIMOD	Flicker noise model	Channel thermal noise model
1	SPICE2-Flicker	SPICE2-channel
2	BSIM-Flicker	BSIM-channel
3	BSIM-Flicker	SPICE2-channel
4	SPICE2-Flicker	BSIM-channel

NOIMOD $= 4$ is the preferred combination (although NOIMOD $= 2$ can be more accurate, as discussed below). When NOIMOD $= 4$, BSIM3 uses a so-called "SPICE2-Flicker" model to calculate the Flicker noise and a so-called "BSIM-channel" model to calculate the channel thermal noise. The SPICE2-Flicker noise model states that the average value of the square of the drain current at a particular frequency (f) is

$$\overline{i_d^2}|_{Flicker} = KF \times \frac{(I_{DS})^{AF}}{C'_{ox} L_{eff}^2 f^{EF}} \Delta f \quad \text{(SPICE2 - Flicker)}, \quad (3\text{-}77)$$

where KF is the proportional factor; AF characterizes the power dependence of the measured value of $\overline{i_d^2}$ on I_{DS}; and EF characterizes the power dependence of $\overline{i_d^2}$ on the frequency. (I_{DS} is the portion of the drain current which flows through the channel, not including the drain–bulk junction current or the substrate current due to impact ionization. The noises of the latter two drain current components are neglected in BSIM3.) KF depends critically on processing. In fact, $\overline{i_d^2}$ can vary all over the wafer if there is no appropriate monitoring during device processing. The default of KF is 0 in BSIM3, although in real life KF is always greater than zero. KF in NMOS is on the order of 10^{-28} A · F. In PMOS, KF is slightly smaller. AF has a default value of 1, although it may not be surprising if the extracted value for a process differs significantly from it, such as 2. In contrast, EF is usually close to 1, with values generally ranging from 0.9 to 1.2. If EF is much greater than 1, such as 2, then potentially the measured noise is due to bulk traps [9], which are known to give rise to a Lorenztian noise spectrum.

The BSIM3-channel model for channel noise calculation states that $\overline{i_d^2}$ due to channel noise is

$$\overline{i_d^2}|_{channel} = \frac{4kT\mu_{eff}}{L_{eff}^2}|Q_{inv}|\Delta f \quad \text{(BSIM - channel; BSIM3v3.2.2 or before)}.$$

(3-78a)

$$\overline{i_d^2}|_{channel} = \frac{4kT}{R_{DS} + L_{eff}^2/(\mu_{eff}|Q_{inv}|)}\Delta f \quad \text{(BSIM - channel; BSIM3v3.3)}. \quad (3\text{-}78\text{b})$$

At the time of this writing, BSIM3v3.3 has not been formally released (see Ref. [15] in Chapter 1). Equation (3-78b) is more accurate than Eq. (3-78a) because it incorporates the thermal-noise related to R_{DS}, which is nonzero when RDSW $\neq$ 0, as elaborated in Section 4.11. Both equations work well for long-channel devices, but both may underestimate the amount of thermal noise in short-channel transistors (Section 4.11). Equation (3-78) can at times exhibit a kink behavior at the transition between the subthreshold and the inversion regions (Section 4.11). Lastly, some feel that $L_{eff,CV}$ (Eq. 3-32) would have been the more appropriate effective-channel length than L_{eff} (Eq. 3-68) to be used in the denominator. After all, Q_{inv} is the total inversion layer charge integrated from the source to the drain, and is proportional to the product of $L_{eff,CV}$ and $W_{eff,CV}$.

Whereas NOIMOD = 4 is the preferred combination of noise models, at times the BSIM-Flicker and the SPICE2-channel noise models are also used. The BSIM-

3.2 ALPHABETICAL GLOSSARY OF BSIM3 PARAMETERS 231

Flicker noise is considerably more complicated than the SPICE2-Flicker noise of Eq. (3-77). The details can be found in Appendix A. It suffices to represent the model with a simplified equation:

$$\overline{i_d^2}|_{Flicker} \approx f(\text{NOIA}, \text{NOIB}, \text{NOIC}) \times \left[\frac{I_{DS}}{C'_{ox} L^2_{eff} f^{\text{EF}}} + \frac{g(\text{EM})I_{DS}}{W_{eff} L^2_{eff} f^{\text{EF}}}\right]\Delta f$$

(BSIM − Flicker),   (3-79)

where f is some function of the SPICE parameters NOIA, NOIB, NOIC, and g is some function of the SPICE parameter EM. NOIA, NOIB, and NOIC can be treated as fitting parameters, just like KF in the SPICE2-Flicker noise model. EM represents the field at which the carrier velocity saturates. Despite its name, EM is a parameter that affects only the noise calculation, but not the I-V or the C-V characteristics. The biggest problem with using this Flicker noise model lies in the amount of noise measurements needed to extract the NOIA, NOIB, NOIC, and EM parameters. However, if the resources to extract these parameters are available, the BSIM-Flicker noise model provides better accuracy than the SPICE2-Flicker noise model. Figure 3-25a illustrates the measured data and the calculation result based on the SPICE2-Flicker noise model, for a fixed V_{GS} of 2 V and varying V_{DS}. In this model, there is only one primary parameter (KF) which can vary the magnitude of the noise. The figure shows that, although the model captures the trend of varying V_{DS}, the optimizer is not able to fit all three data sets simultaneously. The fit of the SPICE-Flicker noise model is even poorer when we focus on a fixed V_{DS} of 0.8 V and varying V_{GS}, as shown in Fig. 3-25b. Despite that the data are clustered together and independent of V_{GS}, the model predicts some distinct V_{GS} dependency.

In contrast, the BSIM-Flicker noise model fits to those data fairly well. Partly because there are more than one parameter, but more importantly because the physics is well captured by the model, the BSIM-Flicker noise model is accurate for varying V_{DS} and varying V_{GS}. The results are illustrated in Fig. 3-26a for $V_{GS} = 2$ V and varying V_{DS}, and in Fig. 3-26b for $V_{DS} = 0.8$ V and varying V_{GS}.

The SPICE2-channel model states that the $\overline{i_d^2}$ due to channel noise is

$$\overline{i_d^2}|_{channel} = \frac{8kT}{3} \times (g_m + g_d + g_{mb})\Delta f \qquad (\text{SPICE2} - \text{channel}), \qquad (3\text{-}80)$$

where the conductances are defined in Eq. (1-4). This equation is loved by circuit designers who often have no clue about how to calculate Q_{inv}, but have no trouble determining the conductances. Although this equation is good for hand-calculation, it yields wrong result in the linear region (Section 4.11). Therefore, the BSIM3-channel noise model is the preferred model to calculate the thermal noise.

The BSIM3-channel noise model is not problem-free, however. Both Eqs. (3-78) and (3-80) are derived with the long-channel approximation. They will grossly underestimate the noise in short-channel devices. See Section 4.11 for more details.

232 BSIM3 PARAMETERS

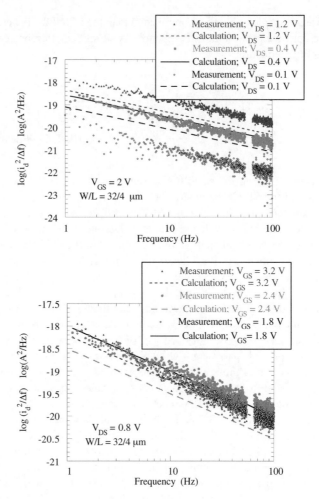

Fig. 3-25 Comparison of measured Flicker noise and the calculation result based on the SPICE2-Flicker noise model. (a) $V_{GS} = 2$ V and V_{DS} varies; (b) $V_{DS} = 0.8$ V and V_{GS} varies. The bias dependence of the SPICE2-Flicker noise model is incorrect. Data are not taken nearby 60 Hz to avoid noises associated with the ac electrical power.

NQSMOD (default = 0, unitless) NQSMOD is a flag for the non-quasi-static (NQS) model. The quasi-static approximation, detailed in Section 1.6, assumes the transistor charges respond instantaneously to any bias voltage variation. This approximation is used in almost all of the compact MOSFET models. The NQS effects are considered insignificant except at high frequencies (higher than the cutoff frequency) or at short transient (at the scale of transistor's transit time.) Fortunately, the NQS effects are negligible in a majority of practical situations (see Section 4.1 for exceptions). Indeed, despite that there is a NQS option, BSIM3 is known for its quasi-static model.

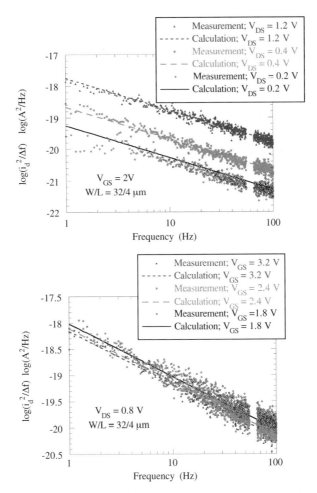

Fig. 3-26 Comparison of measured Flicker noise and the calculation result based on the BSIM-Flicker noise model. (a) $V_{GS} = 2$ V and V_{DS} varies; (b) $V_{DS} = 0.8$ V and V_{GS} varies. The bias dependence of the BSIM-Flicker noise model is fairly good. Data are not taken nearby 60 Hz to avoid noises associated with the ac electrical power.

The NQS and QS models in BSIM3 are not compatible. From a purely theoretical perspective, a NQS solution approaches a QS solution when the time scale of interest far exceeds the intrinsic transistor delay time. However, if we were to simulate a circuit with slow-varying signals with BSIM3, we find that its NQS model solution does not converge to its QS model solution for either a transient or an ac analysis. This is because BSIM3's NQS and QS models are not based on the same set of governing equations.

There are some inherent inaccuracies of the NQS model in BSIM3. For example, the bulk current is always zero, and the terminal currents' waveforms can deviate significantly from the actual (Section 4.1). Because of these inaccuracies, sometimes it is preferred to continue using BSIM3's QS model, but with some modification to the transistor element to "effectively" incorporate the NQS effects. This is detailed in Section 4.1.

Setting NQSMOD = 1 invokes the NQS model, which can be used together with either a transient or an ac analysis. Besides the mentioned problems inherent to BSIM3's NQS model, there is the additional problem that the ac analysis implementation of the model is likely incorrect. Therefore, while NQSMOD = 1 can be used with some advantage in a transient analysis, it does not make sense to use NQSMOD = 1 in an ac analysis. NQSMOD = 0 uses the default quasi-static model for both the transient and ac analyses. Almost always, the quasi-static portion of the BSIM3 should be used, and NQSMOD should be set to 0. More description of BSIM3's NQS model (used in transient analysis) can be found in Appendix E. General guidelines estimating whether a QS model is valid, for both transient and ac analyses, were given by Eqs. (1-35) and (1-36).

NSUB (default = 6×10^{16}, unit = cm^{-3}) substrate doping concentration. It is lower than the channel doping concentration, NCH. The description of NSUB, as well as its default value, is found in the entry NCH.

The following are binning parameters: PA0, PA1, PA2, PAGS, PALPHA0, PAT, PB0, PB1, PBETA0, PCDSC, PCDSCB, PCDSCD, PCF, PCGD1, PCGS1, PCIT, PCKAPPA, PCLC, PCLE, PDELTA, PDROUT, PDSUB, PDVT0, PDVT0W, PDVT1, PDVT1W, PDVT2, PDVT2W, PDWB, PDWG, PETA0, PETAB, PK1, PK2, PK3, PK3B, PKETA, PKT1, PKT2, PNFACTOR, PNLX, PPCLM, PPDIBLC1, PPDIBLC2, PPDIBLCB, PPRT, PPRWB, PPRWG, PPSCBE1, PPSCBE2, PPVAG, PRDSW, PU0, PUA, PUA1, PUB, PUB1, PUC, PUC1, PUTE, PVBM, PVBX, PVFBCV, PVOFF, PVSAT, PVTH0, PW0, PWR, PXJ, PXT.

PARAMCHK: (default = 0, unitless) This parameter, when set to 1, signals BSIM3 to verify whether some parameters and some intermediate variables have physically meaningful values. For example, when the oxide thickness (TOX) in the parameter set is specified to be below 10 Å, we know this is likely an erroneous entry because modern commercial technologies employ oxide thickness thicker than 10 Å. Besides, if the oxide thickness is truly below 10 Å, BSIM3 would not be able to simulate the device characteristics accurately because BSIM3 does not account for the tunneling through thin oxide layer. PARAMCHK, or *parameter check*, is used to check for the value of TOX, and issues a warning message if TOX is less than 10 Å. Other parameters checked by PARAMCHK are: NLX, TOX, NCH, NSUB, NGATE, DVT0, W0, NFACTOR, CDSC, CDSCD, ETA0, B1, A2, RDSW, PDIBLC1, PDIBLC2, CGD0, CGS0, and CGB0. PARAMCHK also verifies whether the following intermediate variables have physically meaningful values: L_{eff} (Eq. 3-68), $L_{eff,CV}$ (Eq. 3-32), W_{eff} (Eq. 3-44), $W_{eff,CV}$ (Eq. 3-43), R_{DS} (Eq. 3-89) and $VSAT(T_{device})$ (Eq.3-17). If not, a warning message is issued.

We caution that the checkings performed when PARAMCHK = 1 represent only a subset of the overall checking routine in BSIM3. PARAMCHK checks whether the value of a given parameter makes physical sense. If not, such as in the case that $0 < \text{TOX} < 10\,\text{Å}$, the SPICE simulation may still proceed without numerical problems. There is another kind of checking in BSIM3, which is performed independent of whether PARAMCHK is 1 or not. This second kind of checking determines whether a given parameter value can cause numerical difficulty during simulation. For example, if TOX is less than zero, then oxide capacitance becomes negative and some terms involving of the square root of the capacitance then become imaginary. This clearly would cause a fatal error during computer calculation. Therefore, BSIM3 always performs an initial checking. When numerically problematic values are identified, BSIM3 prints out a fatal warning message and stops the simulation right away. The following are parameters which are checked for fatal-error-causing values: NLX, TOX, TOXM, NCH, NSUB, NGATE, XJ, DVT1, DVT1W, W0, DSUB, B1, U0, DELTA, NOFF, VOFFCV, IJTH, PCLM, DROUT, PSCBE2, CJSW, CJSWG, CLC, MOIN, and ACDE. It also checks PD and PS in the device statement and the intermediate variable $\text{VSAT}(T_{device})$.

Finally, there is a special checking (again invoked whether PARAMCHK is 1 or not) on the VERSION parameter when either BSIM3v3.2.2 or BSIM3v3.2.1 is used. A warning is issued when the VERSION value specified in the parameter set does not correspond to the version of the BSIM3 model being used. See the entry VERSION for more details.

Although PARAMCHK checks whether $L_{\it eff}$ and $W_{\it eff}$ are of reasonable values, it does not check whether the L specified in the MOSFET device statement is smaller than LMIN, or larger than LMAX. Similar statement can be made for W, with relationship to WMIN and WMAX. For details, see the entries LMIN, LMAX, WMIN, and WMAX.

PB (default = 1.0, unit = V) built-in potential of the bottom-wall junction capacitance, when the device temperature is equal to TNOM. The bottom-wall junction capacitance appears as the first term of the overall junction capacitance given in Eq. (3-25). PB can be estimated as $kT/q \times \ln(N_{SD} \cdot \text{NSUB}/n_i^2)$, where N_{SD} is the source/drain doping level and NSUB is the substrate doping at the bottom wall of the junction. However, it is most often optimized to give the best fit in the junction capacitance. More details of PB, as well as its temperature-dependence formulation, can be found under the entry of CJ.

PBSW (default = 1.0, unit = V) built-in potential of the isolation-side sidewall junction capacitance, when the device temperature is equal to TNOM. This sidewall junction capacitance appears as the second term of the overall junction capacitance given in Eq. (3-25). PBSW can be estimated as $kT/q \times \ln(N_{SD} \cdot \text{NSUB}/n_i^2)$, where N_{SD} is the source/drain doping level and NSUB is the substrate doping at the bottom wall of the junction. However, it is most often optimized to give the best fit in the junction capacitance. More details of PBSW, as well as its temperature-dependence formulation, can be found under the entry CJ.

PBSWG (default = PBSW, unit = V) built-in potential of the gate-side sidewall junction capacitance, when the device temperature is equal to TNOM. This sidewall junction capacitance appears as the third term of the overall junction capacitance given in Eq. (3-25). PBSWG can be estimated as $kT/q \times \ln(N_{SD} \cdot \text{NSUB}/n_i^2)$, where N_{SD} is the source/drain doping level and NSUB is the substrate doping at the bottom wall of the junction. However, it is most often optimized to give the best fit in the junction capacitance. More details of PBSWG, as well as its temperature-dependence formulation, can be found under the entry CJ.

PCLM (default = 1.3, unitless) channel-length modulation parameter in the calculation of drain current. In long-channel MOSFETs, the current remains constant with V_{DS} once V_{DS} exceeds $V_{DS,\text{sat}}$. In short-channel devices, there is some finite of slope to the current due to a combination of the channel-length modulation and DIBL (drain-induced barrier-lowering) effects. (Current may also increase due to impact ionization, an effect not well modeled in BSIM3 if the amount of impact ionization is large. See the entries ALPHA0 and PSCBE1 for details.) The I-V characteristics are schematically shown in Fig. 3-27. Although the following equation is not exactly that used in BSIM3, it describes the essence of BSIM3's modeling of the increase in drain current due to V_{DS}. The drain current in the channel is given by

$$I_{DS} = I_{DS,0}\left(1 + \frac{V_{DS} - V_{DS,\text{sat}}}{V_A}\right), \quad (3\text{-}81)$$

where $I_{DS,0}$ is the ideal long-channel current in the absence of the channel-length modulation and the DIBL effects, and V_A, the *Early voltage*, is a variable that adjusts the slope of the variation of the drain current with respect to V_{DS}. Graphically, V_A can be taken to be value on the $-x$ axis for which the current extrapolates to 0, as shown in Fig. 3-27. Experimentally, V_A is found to increase as V_{GS} increases. BSIM3 writes V_A as

$$V_A = V_{A,\text{sat}} + V_{A,\text{CML-DIBL}}. \quad (3\text{-}82)$$

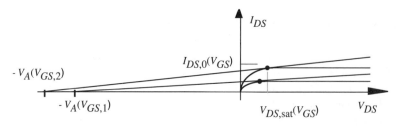

Fig. 3-27 Schematic drawing of I-V characteristics showing the relationship between the Early voltage and the slope of the current in saturation.

$V_{A,\text{sat}}$ is meant to calculate the ideal V_A in absence of short-channel effects. $V_{A,\text{CML-DIBL}}$, in contrast, models V_A's dependence on the channel-length modulation and drain-induced barrier-lowering effects. $V_{A,\text{sat}}$ is given by

$$V_{A,\text{sat}} = f_0(L_{\text{eff}}, \text{VSAT}, R_{DS}, \ldots). \tag{3-83}$$

f_0 denotes a complicated function of a lot of parameters/intermediate variables. We do not intend to write out the exact form of f_0 to reduce complexity (see Appendix A for the exact $V_{A,\text{sat}}$ equations; Eq. A-78). It suffices to know that the main parameters affecting $V_{A,\text{sat}}$ are L_{eff} (Eq. 3-68), VSAT, and R_{DS} (Eq. 3-89). The function f_0 is such that $V_{A,\text{sat}}$ attains a high value when L_{eff} is long, and decreases when L_{eff} is short. The dependence of $V_{A,\text{sat}}$ on L_{eff} is similar to the saturation voltage's ($V_{DS,\text{sat}}$) dependence on L_{eff}. This makes sense. Even in the absence of any short-channel effects, the Early voltage is a function of L_{eff} because the Early voltage is related to the saturation voltage, which, in turn, is a function of L_{eff}. If the channel length is short, or if VSAT is low, or if R_{DS} is high, the carriers quickly accelerate to the saturation velocity at a position not far from the source. In this case, the saturation voltage is small. The saturation voltage would be independent of L_{eff} only if the carrier still travels at a speed smaller than VSAT when it reaches the drain. We emphasize that the fact $V_{A,\text{sat}}$ decreases with decreasing L_{eff} is not due to channel length modulation or DIBL effects, but due to channel pinchoff.

When L_{eff} is long, $V_{A,\text{sat}}$ is large and, further, V_A is independent of the contribution from $V_{A,\text{CML-DIBL}}$. These two characteristics ensure I_{DS} to approach the constant value of $I_{DS,0}$ at a given V_{GS} regardless of V_{DS}. When the channel length is small, however, both $V_{A,\text{sat}}$ and $V_{A,\text{CML-DIBL}}$ contribute to the overall V_A stated in Eq. (3-82). $V_{A,\text{CML-DIBL}}$ is given by

$$V_{A,\text{CML-DIBL}} = \frac{1 + \text{PVAG} \cdot (V_{GS} - V_T) \cdot \mu_{\text{eff}}/(2\text{VSAT} \cdot L_{\text{eff}})}{\text{PCLM} \cdot f_1 + \theta_{\text{rout}}(1 + \text{PDIBLCB} \cdot V_{BS}) \cdot f_2}, \tag{3-84}$$

where

$$\theta_{\text{rout}} = \text{PDIBLC1} \times \left[\exp\left(-\text{DROUT}\frac{L_{\text{eff}}}{2L_{t0}}\right) + 2\exp\left(-\text{DROUT}\frac{L_{\text{eff}}}{L_{t0}}\right)\right] + \text{PDIBLC2}. \tag{3-85}$$

L_{t0} was given in Eq. (3-47), and f_1, and f_2 are rather complicated functions of A_{bulk}, bias conditions, as well as other variables. Again, we do not list out the details of these functions in order to highlight the dependence of the Early voltage on various SPICE parameters.

The parameters appearing in $V_{A,\text{CML-DIBL}}$, including PVAG, PCLM, DROUT, PDIBLC1, PDIBLC2, and PDIBLCB, affect only the short-channel devices. According to Eq. (3-85), PVAG is a parameter used account for $V_{A,\text{CML-DIBL}}$'s variation with V_{GS}. We see the term multiplying PVAG, $(V_{GS} - V_T) \cdot \mu_{\text{eff}}/(2 \cdot \text{VSAT} \cdot L_{\text{eff}})$, which, for a $L_{\text{eff}} = 0.3\,\mu\text{m}$ transistor with a mobility of

$0.067 \, \text{m}^2 \, \text{V}^{-1} \, \text{s}^{-1}$, a saturation velocity of $10^5 \, \text{m/s}$, and a bias of $V_{GS} - V_T = 2 \, \text{V}$, is equal to 2.2. Usually V_A does not vary significantly with V_{GS}. To ensure that the whole term involving PVAG is to be less than 1, then PVAG needs to be smaller than 1/2.2 or about 0.5. In addition, as shown in Fig. 3-27, V_A generally increases with increasing V_{GS}. This empirical observation suggests that PVAG should be a positive number. Having said that, we stress that we really do not assign much physical significance to PVAG. PVAG can be treated merely as a fitting parameter through which we get a match in the I-V characteristics when there is finite slope in the drain current. Hence, negative values of PVAG are also acceptable. BSIM3's default value for PVAG is 0.

Besides PVAG, all the other short-channel parameters such as PCLM, PDIBLC1, PDIBLC2, PDIBLCB, and DROUT appear in the denominator of $V_{A,\text{CML-DIBL}}$. It is obvious that increasing PCLM, PDIBLC1, and PDIBLC2 decreases V_A, an action which will lead to a more rapid increase in the drain current with respect to V_{DS}. Although BSIM3 refers to PCLM as a channel-length modulation parameter and the PDIBL's as the DIBL parameters, there are no need for a circuit designer to separate them. For a designer, these parameters are simply ones to optimize the fit between the measured and simulated I-V characteristics at the saturation region. From this perspective, if better fitting is achieved, it is possible to turn off completely the effects of PCLM, by setting PCLM to a value close to zero, such as 10^{-10}. (Setting PCLM to zero will receive a fatal-error warning from BSIM3 and the simulation will be halted.) We then rely solely PDIBLC1, and PDIBLC2 to obtain the desired fit. Conversely, making PDIBLC1 = PDIBLC2 = 0 but optimizing PCLM is also feasible. The default values for PCLM, PDIBLC1, and PDIBLC2 are 1.3, 0.39, and 0.0086, respectively. There is no particular physical significance to any of these values. They may appear with quite different values in a particular parameter set.

DROUT appears in the exponent of the term multiplying PDIBLC1. If the term associated with PCLM is negligible, then the larger the value for DROUT, the more the dependence of V_A on the channel length. PDIBLC1 and DROUT are *paired* parameters which tend to cancel each other's effect when their values are varied in a certain fashion. For example, let us suppose one parameter set is used to model a particular short-channel device. The Early voltage produced by a certain pair of PDIBLC1 and DROUT values can be nearly the same if DROUT of the pair is increased and PDIBLC1 is made to increase to offset the change in DROUT. Since the two pairs of PDIBLC1 and DROUT produce the same V_A, either pair can be used. A way to establish some absolute values for them is to simultaneously fit a group of devices of varying channel length. DROUT can be determined from a fit of the functional dependence on L_{eff}, and PDIBLC1 is extracted by fitting the required magnitude of the DIBL's correction to the Early voltage. The default of DROUT is 0.56. The value in a particular parameter set can be fairly different. After all, DROUT is best treated as another fitting parameter. However, its value should be greater or equal to 0, since a negative value can cause mathematical difficulty for the exponential function.

PDIBLCB is used to model V_{BS}'s effects on the Early voltage. Typical measurements reveal that V_A decreases as V_{BS} becomes increasingly more negative. This means that PDIBLCB should be negative. However, in the consideration that we are treating PDIBLCB merely as a fitting parameter, whether it is negative or positive is acceptable. The default of PDIBLCB is 0.

Even though the measured I-V characteristics reveal that V_A is modified as V_{BS} changes, the amount of modification is small. Hence, the magnitude of PDIBLCB should be smaller than $1/|V_{BS}|$. When the maximum magnitude of V_{BS} in a technology is 3.3 V, PDIBLCB's magnitude should be less than 0.3.

PDIBLC1 (default = 0.39, unitless) first coefficient of DIBL's correction on the Early voltage. See the entry PCLM for details.

PDIBLC2 (default = 0.0086, unitless) second coefficient of DIBL's correction on the Early voltage. See the entry PCLM for details.

PDIBLCB (default = 0.39, unitless) body-effect coefficient to the DIBL's correction on the Early voltage. See the entry PCLM for details.

PRT (default = 0, unit = $\Omega \cdot \mu m^{WR}$) temperature coefficient of RDSW. As elaborated in the entry RDSW, BSIM3 offers two methods to incorporate the effects of the parasitic drain/source resistances. In one method, wherein RDSW is identically zero, the parasitic resistances are incorporated as lumped resistances. In this lumped-resistance approach, the source/drain resistances are not allowed to vary with temperature. In another method, RDSW $\neq$ 0. Only the effects of the parasitics resistances on I_D and conductances are reproduced, although the resistances themselves do not appear as part of the device's input impedance. This is the absorbed-resistance approach, in which the temperature dependence of the resistance modeled by RDSW can change with temperature. PRT is the parameter used for this purpose. It is meaningful only in the absorbed-resistance approach.

The parameter TNOM in the SPICE parameter set specifies the temperature at which the parameters are extracted. TNOM is generally 27 °C. However, the circuit may operate at an elevated temperature. In this case wherein TNOM in the SPICE parameter set differs from the circuit temperature of the simulation, then R_{DS} to be given in Eq. (3-89) is replaced by

$$R_{DS} = \left[\text{RDSW} + \text{PRT} \times \left(\frac{T_{device}}{\text{TNOM}} - 1 \right) \right]$$
$$\times \frac{[1 + \text{PRWG} \cdot V_{GS} + \text{PRWB}(\sqrt{2\phi_f - V_{BS}} - \sqrt{2\phi_f})]}{(10^6 \times W_{eff})^{WR}}, \quad (3\text{-}86)$$

where $2\phi_f$ is the strong-inversion surface potential given in Eq. (1-6b). The temperatures are in kelvins. PRT is the parameter to correct the temperature dependence of RDSW as the circuit temperature T_{device} deviates from the TNOM specified in the SPICE parameter set. We can treat the term inside the square brackets of Eq. (3-86) as the "effective-RDSW" at various temperatures. The "effective-RDSW" is precisely equal to RDSW of the SPICE parameter at

240 BSIM3 PARAMETERS

T_{device} = TNOM. Figure 3-28a displays extracted "effective-RDSW" at various temperatures for a logic CMOS process. In Fig. 3-28b we replot the figure with the x-axis modified to be (T_{device}/TNOM − 1), so that the parameter PRT as given in Eq. (3-86) can be easily identified. From a least-square-error fit, we see that PRT has the value of 122 (Ω-μm) for this process. In typical digital CMOS processes, PRT can range between 30 to 200 (Ω-μm). In MOSFETs for high-voltage applications (such as 60 V), it is not uncommon to see PRT on the order of 10^4 (Ω-μm; with

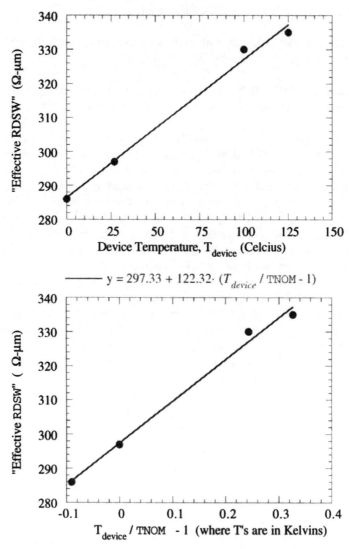

Fig. 3-28 (a) Extracted "effective-RDSW" at various temperatures for a logic CMOS process; (b) same data plotted against (T_{device}/TNOM − 1), so that the parameter PRT as given in Eq. (3-86) can be easily identified.

WR = 1). Sometimes, the temperature dependence of RDSW is not measured and PRT is arbitrarily set to its default value of 0.

PRWG (default = 0, unit = V^{-1}) gate-bias effect coefficient of RDSW. The details of this parameter are described in the entry for RDSW. Generally, PRWG is a secondary parameter and its value should be close to zero, or at least, the product of |PRWG| and V_{GS} should be smaller than 0.1. PRWG, viewed as a fitting parameter without too much physical significance, can take on either a positive or a negative sign.

PRWB (default = 0, unit = V$^{-1/2}$) body-effect coefficient of RDSW. The details of this parameter are described in the entry RDSW. Generally, PRWB is a secondary parameter and its value should be close to zero, or at least, the product of |PRWG| and ($\sqrt{2\phi_f - V_{BS}} - \sqrt{2\phi_f}$) should be smaller than 0.1. PRWB, viewed as a fitting parameter without too much physical significance, can take on either a positive or a negative sign.

PSCBE1 (default = 4.24 × 10^8, unit = V/m) first high-voltage drain-conductance modification factor. In the BSIM3 manual, this parameter's formal name is "the first substrate current body-effect parameter." We change the name because we feel the BSIM3 name is somewhat misleading. Firstly, PSCBE1 does not determine the substrate current at all; I_{sub} is calculated with other parameters, as detailed in the entry of ALPHA0. Secondly, PSCBE1 bears no relationship with the body effects that we typically associate with; when we hear body effects we are generally reminded of the parameters such as K1 which modifies the threshold voltage due to V_{BS}.

So, what exactly does PSCBE1 do? It is meant to model the drain current increase at high V_{DS}, due to the increased current from impact ionization. Neglecting the drain–bulk junction leakage, we can equate the drain terminal current (I_D) as the sum of the MOSFET channel current (I_{DS}) plus the impact ionization current (I_{sub}). With a nonzero I_{sub}, I_D increases while I_{DS}, the channel current in the absence of impact ionization, remains constant. BSIM3 models this increase in drain terminal current by

$$I_D = I_{DS}\left\{1 + (V_{DS} - V_{DS,\text{eff}}) \times \frac{\text{PSCBE2}}{L_{\text{eff}}} \exp\left[-\frac{\text{PSCBE1} \cdot L_{itl}}{V_{DS} - V_{DS,\text{eff}}}\right]\right\}, \quad (3\text{-}87)$$

where the variable L_{itl} is

$$L_{itl} = \sqrt{\frac{\epsilon_s}{\epsilon_{ox}} \text{TOX} \cdot \text{XJ}}. \quad (3\text{-}88)$$

(As always, many equations in this section are not exactly the same as those used in BSIM3. We present the simplified forms to facilitate discussion. For the exact equations, see Appendix A). $V_{DS,\text{eff}}$ is the smoothing function defined in BSIM3 to smooth out the transition between the linear and saturation regions (Section 1.4). It is identical to V_{DS} when the externally applied V_{DS} is small, and saturates to $V_{DS,\text{sat}}$ when V_{DS} is large. Hence, we see that at small values of V_{DS}, $V_{DS} - V_{DS,\text{eff}}$ in Eq. (3-87) approaches 0 and $I_D \approx I_{DS}$. As V_{DS} increases, $V_{DS,\text{eff}}$ approaches $V_{DS,\text{sat}}$, and I_D becomes greater than I_{DS}. The equation makes sense thus far. However, the actual implementation of this equation is where the problem comes. As I_D exceeds I_{DS} when PSCBE1 and PSCBE2 are adjusted to certain values, we would expect to find a nonzero bulk current equal to $I_D - I_{DS}$. However, BSIM3 produces a bulk current of identically zero, even though I_D may greatly exceed I_{DS}. Apparently BSIM3 calculates the effects of the impact ionization on the drain terminal current, but not on the bulk terminal current. This is the reason why we prefer to treat PSCBE1 and PSCBE2 merely as fitting parameters and name them accordingly. They can be used, in addition to PCLM, PDIBLC1, PDIBLC2, PDIBLCB, and DROUT, to modify the Early voltage of the transistor.

The defaults of PSCBE1 and PSCBE2 are 4.24×10^8 (V/m) and 1×10^{-5} (m/V), respectively. The default value of PSCBE1 is roughly equal to the inverse of L_{itl}, which, for a typical technology with TOX = 64 Å and XJ = 2000 Å, is equal to 6.2×10^{-8}. Hence, the exponent enclosed by the square brackets in Eq. (3-87) is on the order of 1 or 10 when V_{DS} is large. A small PSCBE2 default value of 1×10^{-5} ensures $I_D \approx I_{DS}$, a situation wherein the impact ionization can be neglected. A larger value of PSCBE2 leads to increased drain current as V_{DS} increases. The I-V characteristics obtained with two combinations of PSCBE1 and PSCBE2 are shown in Fig. 3-29. In the case where PSCBE1 = 10^9, the drain current behaves normally. When PSCBE1 = 4.6×10^3, the drain current increases rapidly with V_{DS}, even beyond $V_{DS,\text{sat}}$. Despite the significant difference in the drain currents, the bulk currents in these two cases are small, consisting mainly of the drain–bulk junction leakage current.

From a purely physical point of view, PSCBE1 and PSCBE2 should both be positive values. However, we take on the viewpoint that both parameters are fitting parameters. We can tolerate PSCBE2 to be negative, in the rare event that the terminal drain current decreases with V_{DS}. Generally, due to channel modulation, impact ionization, and so on, the drain current increases with V_{DS}. However, if the device's self-heating is significant, it is possible that the drain current decreases with increasing power dissipation brought by an increase in V_{DS}. PSCBE2 can then be used to model this phenomenon. BSIM3 issues a warning when PSCBE2 is negative (whether the parameter check PARAMCHK value is 0 or 1).

As a final note, Eq. (3-87) is similar to the I_{sub} of Eq. (3-16), given in the entry ALPHA0. Therefore, the discussion about the I_{sub} given in Eq. (3-16) applies to the drain current expression under consideration. That is, I_D, as given in Eq. (3-87), will not simulate the rapid rise of drain current associated in the event of avalanche breakdown. I_D approaches infinity only as V_{DS} approaches infinity. However, based on our knowledge of breakdown characteristics, I_D should approach infinite at some finite V_{DS} value, which is experimentally found to be between 6 V to 8 V in modern CMOS digital devices.

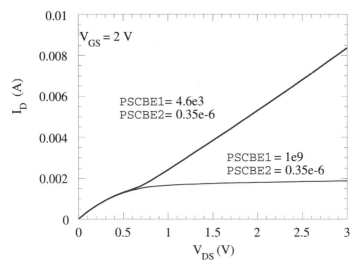

Fig. 3-29 Effects of PSCBE1 and PSCBE2 on BSIM3-calculated I-V characteristics.

PSCBE2 (**default** $= 1 \times 10^{-5}$, **unit** $= $ m/V) second high-voltage drain-conductance modification factor. In the BSIM3 manual, this parameter's formal name is "the second substrate current body-effect parameter." We change the name because the formal name can be misleading. See the entry PSCBE1 for more details.

PVAG (**default** $= 0$, **unitless**) gate-bias dependence of Early voltage. See the entry PCLM for details.

RDSW (**default** $= 0$, **unit** $= \Omega \cdot \mu m^{WR}$) drain-source resistance per μm of gate width. There are terminal resistances associated with the drain and the source in a real device. As discussed in Section 2.2, BSIM3 employs two independent approaches to model such resistances: the lumped-resistance and absorbed-resistance approaches. The lumped-resistance approach relies on using the parameter RSH along with NRD and NRS in the device statement to calculate the values of the drain and source terminal resistances. In this approach, RDSW should be set to zero.

RDSW is a parameter used in the absorbed-resistance approach, wherein BSIM3 assumes $R_D + R_S$ as a whole takes on the following form:

$$R_{DS} \triangleq R_D + R_S = \frac{\text{RDSW}[1 + \text{PRWG} \cdot (V_{GS} - V_T) + \text{PRWB}(\sqrt{2\phi_f - V_{BS}} - \sqrt{2\phi_f})]}{(10^6 \times W_{\mathit{eff}})^{WR}},$$

(3-89)

where W_{eff} is the effective width (see a special note in the entry of WL for the exact definition of W_{eff} in this case), and $2\phi_f$ is the strong-inversion surface potential given in Eq. (1-6b). For simplicity, we did not include the temperature dependence in the above equation. See the entry PRT for more details.

For a transistor in normal operation, PRWG, PRWB, and WR should be at their default values of 0, 0 and 1, respectively. In this case, R_{DS} (the sum of R_D and R_S) is simplified to, $R_{DS} = \text{RDSW}/(10^6 \cdot W_{eff})$. The factor of 10^6 is used to offset the unit imbalance, that W_{eff} is in meters whereas RDSW is in $\Omega \cdot \mu\text{m}^{WR} = \Omega \cdot \mu\text{m}$ (when WR = 1). Basically, R_{DS} of the transistor is taken to be RDSW divided by the width of the transistor. RDSW in typical digital NMOS ranges between 250 and 500 ($\Omega \cdot \mu\text{m}$). For a symmetrical 10 μm NMOS, RDSW = 500 implies that the transistor has a drain resistance of 25 Ω and a source resistance of 25 Ω. Because the hole mobility is lower, RDSW in typical PMOS is several times higher, at about 800–1800 ($\Omega \cdot \mu\text{m}$).

RDSW characterizes the drain-source resistance, which exists right in the current path. A large value lowers the drive current and degrades the device transient performance. In DRAM devices, however, the name of the game is cost reduction. Without a critical need for high speed, the silicided process used to reduce contact resistance is often omitted in a DRAM process. The RDSW value is thus much higher than the quoted values found in typical digital processes.

WR, the exponent in the denominator, should ideally be 1. Generally the drain and the source resistances scale inversely with the device width. To improve fitting, BSIM3 uses WR to allow the exponent to deviate from unity. However, a value of WR exceeding 1.1 or smaller than 0.9 is likely unphysical. Note that WR and RDSW are so-called *paired* parameters, which tend to cancel each other's effect when their values are varied in a certain fashion. For example, if RDSW in the SPICE parameter set is unrealistically high, WR will then need to be large so that the final R_{DS} used in the transistor calculation is adjusted to some reasonable value. (See the entry DVT0 to find another example of paired parameters.)

When we compared the lumped-resistance and the absorbed-resistance approaches in Section 2.2, we assumed that R_D and R_S are constant. This assumption is tantamount to assuming PRWG = PRWB = 0. When the bias dependencies are considered, the absorbed-resistance approach has one additional advantage. Suppose R_S and R_D of Fig. 2-8 are a function of V_{GS} and V_{DS}, rather than having constant values. Because resistors in SPICE generally do not accept dependence on nodal voltages (except with some controlled dependent voltage sources), the lumped-resistance approach in Fig. 2-8a cannot be easily implemented in the SPICE simulator. The absorbed-resistance approach of Fig. 2-8b, however, can easily account for the bias dependencies which are built into Eq. (3-89). PRWG is used to model the incremental resistance due to the application of V_{GS}. In one theoretical view, increasing V_{GS} causes more carriers to be present at the source/drain region, hence reducing R_{DS}. In this regard, PRWG should be a negative number. However, PRWG is really a parameter of secondary importance. In a more pragmatic view, its sign is immaterial, as long as the value of PRWG results in a good fit. Generally, the magnitude of PRWG should be small enough such that PRWG $\cdot V_{GS}$ is much smaller than unity.

Just as PRWG models the gate bias' effect on R_{DS}, PRWB is used to model the incremental resistance due to the application of V_{BS}. In one theoretical view, making

V_{BS} more negative reduces the number of carriers at the source/drain region, hence increasing R_{DS}. In this regard, PRWG should theoretically be a positive number. However, we treat PRWB as a parameter whose main use is to improve fitting. In this practical viewpoint, it does not matter what PRWB's sign is, as long as its value gives good fit. Generally, the magnitude of PRWB is small enough such that PRWB $\cdot (\sqrt{2\phi_f - V_{BS}} - \sqrt{2\phi_f})$ is much smaller than unity.

RSH (default = 0, unit = Ω/square) the sheet resistance in the source diffusion, which is assumed to be identical as the sheet resistance in the drain. This value, when multiplied with NRS (specified in the MOSFET device statement) gives the source resistance appended externally at the source node of the transistor. Similarly, the extrinsic drain resistance is equal to the product of RSH and NRD (the latter is also specified in the MOSFET device statement).

Using this parameter along with NRD and NRS is just one way to establish the values of the source and drain terminal resistances in BSIM3. The second way to account for the source and drain resistances is to use RDSW. When the second way is employed, RDSW is non-zero and RSH is made to have its default value of zero. The details of the two methods to calculate the drain and source resistances were described in Section 2.2. We caution that these two methods are exactly equivalent only when the device current is small, as demonstrated in Figs. 2-9 and 2-10.

Although physically RSH is used to denote the sheet resistance, RSH is often assigned a value of 1 Ω/☐. Suppose through experiment we measure the extrinsic source and drain resistances to be 2.1 Ω and 3.4 Ω, respectively, then assigning NRS to be 2.1 and NRD to be 3.4 will result in the desired extrinsic resistances. This approach, nonetheless, should be used with caution. In some SPICE simulators, both NRS and NRD have a default value of 1, not 0! If RSH takes on the value of 1 (Ω/☐) and NRS and NRD are not specified in the MOSFET device statement, BSIM3 will assign extrinsic source and drain resistances to be 1 Ω.

TCJ (default = 0, unit = K^{-1}) temperature coefficient of CJ. See the entry CJ for details.

TCJSW (default = 0, unit = K^{-1}) temperature coefficient of CJSW. See the entry CJ for details.

TCJSWG (default = 0, unit = K^{-1}) temperature coefficient of CJSWG. See the entry CJ for details.

TNOM (default = device temperature, unit = °C) the nominal temperature; the temperature at which the parameters of a particular model card are extracted. The importance of this parameter is illustrated by the following example. Suppose the entire circuit inside a package operates at a junction temperature $T_{device} = 100\,°C$. However, the SPICE parameters were extracted from measurements performed at the room temperature, that is, with a TNOM = 27 °C. In this scenario, the MOSFET model card will be inaccurate if all of the model parameters (which apply to 27 °C) are used to simulate circuit performance at 100 °C. It is important then to account for the temperature difference by modifying those SPICE parameters and intermediate variables whose values are sensitive to temperature. Although arguably all parameters and intermediate variables depend on the temperature somewhat, BSIM3v3 focuses only on those which show predominant effects, such as the threshold voltage (see KT1). mobility (see UA1 and UTE), saturation velocity (see AT), drain-source resistances (see PRT), junction capacitance (see TCJ), and junction leakage currents (see XTI). Here, we examine the temperature variation of the parameter UA, just to make a point:

$$\text{UA}(T_{device}) = \text{UA}(\text{TNOM}) + \text{UA1} \times \left[\frac{T_{device} + 273.15}{\text{TNOM} + 273.15} - 1 \right]. \quad (3\text{-}90)$$

In the above expression, $\text{UA}(T_{device})$ is the UA at the device temperature, which is the actual UA that gets used for the I-V and C-V calculations. $\text{UA}(\text{TNOM})$ is the UA value specified in the parameter set; it is the UA extracted at TNOM. UA1 is the temperature coefficient of UA; it is also a BSIM3 parameter. If the device temperature in a circuit simulation is identical to the temperature at which the measurement used to extract the parameters is made (i.e., $T_{device} = \text{TNOM}$), then the above equation states that $\text{UA}(T_{device}) = \text{UA}(\text{TNOM})$, as expected. Under this scenario, the value of UA1 becomes irrelevant.

In Eq. (3-90), both T_{device} and TNOM are added with 273.15, the offset used by BSIM3 to convert degree Celsius to Kelvin. Both T_{device} and TNOM have a default unit of °C.

When TNOM is not specified in the parameter list, it is defaulted to the device temperature (T_{device}), which is prescribed by either of the following two ways. First, if the TEMP field of the MOSFET device statement is left blank, then the device temperature is whatever the temperature specified in the ·TEMP control statement in the SPICE circuit deck. The temperature specified in the ·TEMP control statement is often called the circuit temperature. If ·TEMP control statement is absent, then the circuit temperature is defaulted to the TNOM specified in the ·OPTION control statement. (*Note*: The TNOM in the ·OPTION control statement has no relationship with the TNOM in the MOSFET parameter set.) If again TNOM is unspecified in the ·OPTION control statement, then the circuit temperature is defaulted to 27 °C. The aforementioned description of finding the device temperature is nullified if TEMP field of the MOSFET device statement is specified to some value. (Note: This TEMP field is available only in some SPICE simulators.) In this latter case, the device temperature is whatever specified in the TEMP field of the MOSFET device statement. Whatever circuit temperature specified in the ·TEMP or the ·OPTION control statements would have no effect on the default value of TNOM.

TOX (default $= 150 \times 10^{-10}$, unit $=$ m) gate oxide thickness. The TOX value will affect both I-V characters and C-V characteristics, through the intermediate variable of $C'_{ox}(=\epsilon_{ox}/\text{TOX})$. Some SPICE models use TOX as a parameter to calculate C'_{ox}, but use another parameter (such as Kappa) to calculate $\mu_{eff} \cdot C'_{ox}$. In these models, TOX will affect only the C-V calculation, whereas Kappa affects the I-V calculation. In BSIM3, TOX directly affects both I-V and C-V characteristics. This simplifies the statistical modeling.

TOXM (default = TOX, unit = m) gate oxide thickness at which the parameter set was extracted. This is a new parameter in BSIM3v3.2 unavailable in BSIM3v3.1. This parameter is mainly used for next-technology look-ahead or statistical modeling. For example, suppose we currently have a production process based on 80 Å oxide, for which a model extracted from fabricated transistors is available. We are asked to forecast the model for the next technology node based on 60 Å oxide, so that the circuit designers can start designing activities on paper. In a crude approximation, the forecast model can be identical to the existing model, except with TOXM made to be 80×10^{-10} (m) while TOX assigned to be 60×10^{-10} (m). In the SPICE calculation, C'_{ox} will be calculated with the new TOX value of 60 Å. Furthermore, K1 and K2, still being those values of the 80 Å transistor, theoretically depend on the inverse of the oxide thickness. (This fact can be inferred from the first four equations listed in the entry NCH.) Therefore, for the new 60 Å-oxide transistor, K1 and K2 are multiplied by the ratio of TOX/TOXM before the calculation of the I-V and C-V characteristics.

While BSIM3 correctly modifies K1 and K2 in the event of a difference between TOX and TOXM, there is one unfortunate omission. This deficiency will lead to significant simulation error if the device's gate–drain and gate–source overlap capacitances are important factors determining the transient response (or high-frequency) properties. Basically, continuing the above example, we envision the gate–drain and gate–source overlap capacitances to increase by a 33 % as the oxide thickness decreases from 80 Å to 60 Å. However, parameters relating to these overlap capacitances, such as CGD0, CDG1, CGS0, CGS1, and CF are not modified. In fact, the only parameters which are modified are the aforementioned K1 and K2. Since overlap capacitances are known to have profound effects in short-channel devices, some sort of manual modification on the overlap capacitance parameters are necessary. Of the five overlap capacitance parameters, CF is relatively more difficult to be modified than the rest.

TPB (default = 0, unit = V/K) temperature coefficient of PB. See the entry CJ for details.

TPBSW (default = 0, unit = V/K) temperature coefficient of PBSW. See the entry CJ for details.

TPBSWG (default = 0, unit = V/K) temperature coefficient of PBSWG. See the entry CJ for details.

U0 (default = 0.067 for NMOS and 0.025 for PMOS, unit = m^2/V-s) zero-field mobility in the universal mobility formulation, when the device operating temperature is equal to TNOM (the temperature at which the model parameters are

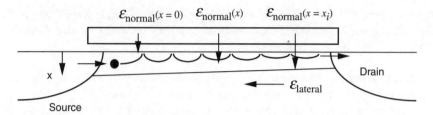

Fig. 3-30 Schematic drawing showing the movement of an electron as it leaves the source toward the drain.

extracted). When an electron leaves the source region, it drifts toward the drain because of the lateral electric field established by V_{DS} along the channel. Under the low-field condition, its velocity has not saturated and is given by the product of the channel mobility and the lateral electric field. The channel mobility is smaller than the bulk mobility because as the electron travels in the channel, it is constantly attracted toward the gate by a normal electric field formed between the gate electrode and the channel region. As shown in Fig. 3-30, the electron makes several collisions with the oxide before it reaches the drain.

The amount of mobility degradation, in this physical picture, depends on the magnitude of the normal electric field, but is independent of the lateral electric field. The normal field, ε_{normal}, is a function of the depth away from the oxide/silicon interface, which is labeled as x in Fig. 3-30. Right on the oxide/silicon interface where $x = 0$, the normal electric field according to the Gauss' law is

$$\varepsilon_{normal}(x = 0) = \frac{Q'_B + Q'_{inv}}{\epsilon_s}, \qquad (3\text{-}91)$$

where Q'_B is the bulk charge per unit area, and Q'_{inv} is the inversion charge per unit area. At a depth equal to the inversion thickness, x_i, the contribution from Q'_{inv} drops out since there is no more inversion charge beyond $x = x_i$. Because the inversion thickness is much smaller than the bulk-charge depletion thickness, the full contribution by Q'_B remains. Therefore, at $x = x_i$,

$$\varepsilon_{normal}(x = x_i) = \frac{Q'_B}{\epsilon_s}. \qquad (3\text{-}92)$$

When the carrier mobility (usually extracted from the measured I_D) is plotted as a function of the normal field right on the interface, the mobility is seen to be a function of bias conditions, substrate doping levels, and so forth [7]. This is schematically shown in Fig. 3-31a. Reference [7] notes that an electron at different depth (x) experiences different amount of normal electric field, with the two extreme

3.2 ALPHABETICAL GLOSSARY OF BSIM3 PARAMETERS

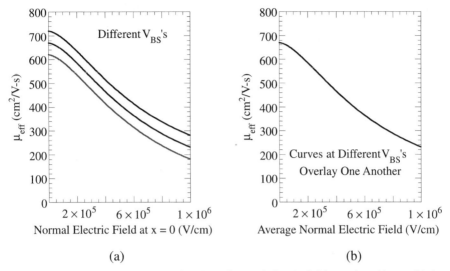

Fig. 3-31 (a) Mobility plotted as a function of normal electric field at various biases; (b) the same set of mobility plotted as a function of the average normal electric field.

values given in Eqs. (3-91) and (3-92). An average normal electric field can be defined, given by

$$\langle \varepsilon_{normal} \rangle = \frac{Q'_B + Q'_{inv}/2}{\epsilon_s}. \tag{3-93}$$

When the data plotted in Fig. 3-31a are replotted with $\langle \varepsilon_{normal} \rangle$ as the x-axis, then all of the mobilities coincide. The curve given in Fig. 3-31b is therefore called the *universal mobility curve*. In real life, we expect the universal mobility curve to be not so universal. After all, the formation of the curve presumes that the mobility is a function of only the normal electric field, but not the lateral. It also assumes that an average effective normal electric field can be used to represent the mobility degradation of all the electrons spread out in the inversion layer. It is not surprising to see measured data to lie outside of the universal mobility curve of Fig. 3-31b.

According to the universal mobility model, the effective mobility at which the carriers drift inside the channel is given by (for NMOS) [10]

$$\mu_{eff} = \frac{670 \text{ cm}^2/\text{V-s}}{1 + \left(\frac{\langle \varepsilon_{normal} \rangle}{6.6 \times 10^5 \text{ V/cm}} \right)^{1.6}}, \quad \text{where } \langle \varepsilon_{normal} \rangle \approx \frac{V_{GS} + V_T}{6 \text{TOX}}. \tag{3-94}$$

The average normal electric field, $\langle \varepsilon_{normal} \rangle$, is the average field emanating from the gate and terminate at the channel charges and the bulk charges. The factor of six in the expression of $\langle \varepsilon_{normal} \rangle$ comes from an averaging of an integral, as well as the fact that $\epsilon_s / \epsilon_{ox} \approx 3$. BSIM3 offers three mobility models, all of which are slight variations of the said universal mobility model. The BSIM3 models explicitly quantify the amount of mobility degradation as a function of V_{GS} and V_{BS}. They do not depend on V_{DS} because BSIM3 assumes mobility to be independent of the lateral electric field. The user chooses the desired model through the parameter MOBMOD, which can have a value of 1, 2 or 3:

$$\mu_{eff} = \frac{U0}{1 + (UA + UC \cdot V_{BS})\left(\frac{V_{GS} + V_T}{TOX}\right) + UB\left(\frac{V_{GS} + V_T}{TOX}\right)} \quad \text{if MOBMOD} = 1;$$

(3-95a)

$$\mu_{eff} = \frac{U0}{1 + (UA + UC \cdot V_{BS})\left(\frac{V_{GS} - V_T}{TOX}\right) + UB\left(\frac{V_{GS} - V_T}{TOX}\right)^2} \quad \text{if MOBMOD} = 2;$$

(3-95b)

$$\mu_{eff} = \frac{U0}{1 + \left[UA\left(\frac{V_{GS} + V_T}{TOX}\right) + UB\left(\frac{V_{GS} + V_T}{TOX}\right)^2\right](1 + UC \cdot V_{BS})} \quad \text{if MOBMOD} = 3;$$

(3-95c)

where U0 is the zero-field mobility of the universal mobility model (with a default value equal to $670\,\text{cm}^2/\text{V-s}$ for NMOS, or in the BSIM3 unit, $0.067\,\text{m}^2/\text{V-s}$; for PMOS, it is $250\,\text{cm}^2/\text{V-s}$). UA, UB, and UC are fitting parameters used to model the amount of mobility degradation due to the average electric field. The mobility models of Eq. (3-95) are somewhat different from the original universal mobility model given in Eq. (3-94). In the original equation, a calculation requires evaluating a power exponent of 1.6. It turns out that evaluating non-integer powers consumes a lot of computation time. BSIM3's mobility models approximate such an exponent with two terms, one being linearly proportional to $\langle\varepsilon_{\text{normal}}\rangle$, as modeled by UA, and another, being parabolically proportional to $\langle\varepsilon_{\text{normal}}\rangle$, as modeled by UB. UC is used in BSIM3 to account for any possible V_{BS} dependence, although the universal mobility model given in Eq. (3-94) does not depend on V_{BS} at all. The MOBMOD = 1 and MOBMOD = 3 models differ only in the modeling of the V_{BS} contribution to the mobility. Choosing between these two models is a matter of preference. The MOBMOD = 2 model differs from the other two models in that the effective electric field is proportional to $(V_{GS} - V_T)/\text{TOX}$, rather than $(V_{GS} + V_T)/\text{TOX}$. This modification is necessary for the modeling of the depletion-mode MOSFET, whose Q'_B can be treated as zero.

The universal mobility model has become a norm against which engineers compare their mobility data. However, as mentioned, the "universal" property of the mobility is based on many assumptions. We need not be alarmed if the optimized U0 which best fits the I-V data of a device is not close to $670\,\text{cm}^2/\text{V-s}$ (for NMOS). It is not uncommon to see values between 350 and $700\,\text{cm}^2/\text{V-s}$. In fact, the U0 in a sample NMOS model card provided in the official BSIM3 website has a value of $388\,\text{cm}^2/\text{V-s}$.

One can raise question about the validity of the MOBMOD = 1 and 3 models. These models correctly predict the mobility degradation with increasing V_{GS}. Unfortunately, the models can be problematic when applied to several devices with several channel lengths. Consider a short-channel device with significant DIBL effects. As V_{DS} increases, its therehsold voltage V_T decreases. This results in a smaller denominator for the short-channel device in comparison to a long-channel device. This means that the short-channel device would have a higher effective mobility than the long-channel device, a result that is at odds with experimental results.

The above discussion is valid only when TNOM (the temperature at which the SPICE parameters are extracted) is identical to T_{device} (the device junction temperature). Since mobility is a function of temperature, U0 extracted at TNOM is identical to the mobility value during device operation only if T_{device} = TNOM. If the device temperature in a circuit simulation is not equal to TNOM, BSIM3 adjusts the U0 parameter by a factor:

$$\text{U0}(T_{device}) = \text{U0}(\text{TNOM}) \times \left(\frac{T_{device} + 273.15}{\text{TNOM} + 273.15}\right)^{\text{UTE}}, \qquad (3\text{-}96)$$

where U0(TNOM) is the U0 specified in the parameter set, whereas U0(T_{device}) is the mobility value actually used for the I-V and C-V calculations. The parameter UTE specifies the amount of temperature dependence of the mobility variation. See UTE for more details about the U0(T_{device}).

UA (default $= 2.25 \times 10^{-9}$, unit $=$ m/V) first-order mobility-degradation coefficient. BSIM3 offers three models of the mobility degradation due to the normal electric field pointing from the gate to the channel. The parameters UA, UB and UC, can be treated as fitting parameters which describe the effective mobility's dependence on biases and the oxide thickness. The equations for the three effective mobility models are given in the entry for U0. Here, we list only the default model (MOBMOD = 1) for illustration:

$$\mu_{eff} = \frac{\text{U0}}{1 + (\text{UA} + \text{UC} \cdot V_{BS})\left(\frac{V_{GS} + V_T}{\text{TOX}}\right) + \text{UB}\left(\frac{V_{GS} + V_T}{\text{TOX}}\right)^2} \quad \text{if MOBMOD} = 1.$$

$$(3\text{-}97)$$

In a first-order approximation, we shall neglect the quadratic effect of the normal electric field (by equating UB to 0). Further, if $V_{BS} = 0$, the effect from UC can also be omitted. The effective mobility is then approximated as $\mu_{eff} = \text{U0}/[1 + \text{UA} \cdot (V_{GS} + V_T)/\text{TOX}]$. Based on a generalization that the effective mobility decreases to roughly 1/3 of the bulk mobility during strong inversion, we can estimate the denominator $1 + \text{UA}(V_{GS} + V_T)/\text{TOX}$ to be equal to 3. When $V_{GS} \sim 2.5$ V, $V_T \sim 0.5$ V, and TOX $= 50$ Å, then UA is roughly equal to 3.3×10^{-9} m/V. Certainly the UA value depends on the technology. The above exercise elucidates how one can estimate UA for various technologies.

Thus far we have neglected the temperature dependence of UA. See UA1 for more details on UA's temperature dependency. The default value of UA, 2.25×10^{-9}, does not have much physical significance. The value in a parameter set may differ significantly from the default.

UA1 (default = 4.31 × 10⁻⁹, unit = m/V) temperature coefficient for UA. UA appearing in the SPICE parameter set represents the first-order mobility degradation coefficient at TNOM (the temperature for which the SPICE parameters are extracted). The device operation temperature, T_{device}, is not necessarily equal to TNOM. Since mobility is a function of temperature, UA extracted at TNOM is identical to the mobility value during device operation only if T_{device} = TNOM. If the device temperature in a circuit simulation differs from TNOM, BSIM3 modifies Eq. (3-97) to the following:

$$\mu_{eff}(T_{device})$$
$$= \frac{UO(T_{device})}{1 + [UA(T_{device}) + UC(T_{device})V_{BS}]\left(\frac{V_{GS}+V_T}{TOX}\right) + UB(T_{device})\left(\frac{V_{GS}+V_T}{TOX}\right)^2}. \tag{3-98}$$

We denote UA(TNOM) as the UA specified in the parameter set. BSIM3 relates UA(TNOM) and UA(T_{device}) according to

$$UA(T_{device}) = UA(TNOM) + UA1 \times \left[\frac{T_{device}+273.15}{TNOM+273.15} - 1\right], \tag{3-99}$$

where UA1 is the parameter quantifying the amount of temperature dependence of UA. For the temperature dependencies of U0, UB, and UC, see the respective entries. See TNOM for more details about the distinction between T_{device} and TNOM.

Carrier mobility is known to decrease with temperature. There are many parameters in Eq. (3-98) that control the mobility's temperature variation, such as UTE, UA1, UB1 and UC1. Let us neglect the temperature dependencies of the latter two parameters so that UB(T_{device}) = UB(TNOM) and UC(T_{device}) = UC(TNOM). Instead, we concentrate on UTE and UA1, which control the temperature variations of U0 and UA, respectively. Suppose that somehow UTE is more negative than it should, then U0(T_{device}) decreases below the correct value. UA1 thus needs to be negative in order to compensate the overdrastic reduction in U0(T_{device}). This mental exercise shows that UTE and UA1 are paired parameters, whose effects tend to cancel the other's. It is quite likely that the temperature-related parameters in Eq. (3-98) are not properly extracted in a particular parameter set. UTE, UA1, UB1, and UC1 can have unphysical values yet still produce a good fit in I-V characteristics. An approach to avoid this problem is to attribute all of the mobility's temperature dependence to U0(T_{device}). By making UTE negative while making all UA1, UB1 and UC1 to zero, we will definitely obtain lowering mobility values at higher temperatures.

The default value of UA1 is 4.31 × 10⁻¹⁹. It is important to realize that the default value of UA1 is some number, not zero. Therefore, a SPICE parameter set can simulate the temperature dependence of I-V characteristics, even though UA1 is not specified. The nonzero default can sometimes be viewed as a bad feature. A designer may falsely believe that, since parameters such as UA1 is not specified, there ought not be any temperature dependence on the I-V characteristics.

UB (default $= 5.87 \times 10^{-19}$, unit $= \text{m}^2/\text{V}^2$) parabolic mobility-degradation coefficient. BSIM3 offers three models of the mobility degradation due to the normal electric field pointing from the gate to the channel. The parameter UA, along with UB and UC, can be treated as fitting parameters which describe the effective mobility's dependence on biases and the oxide thickness. The equations for the three effective mobility models are given in the entry U0. Here, we list only the default model (MOBMOD $= 1$) for illustration:

$$\mu_{\mathit{eff}} = \frac{\text{U0}}{1 + (\text{UA} + \text{UC} \cdot V_{BS})\left(\dfrac{V_{GS} + V_T}{\text{TOX}}\right) + \text{UB}\left(\dfrac{V_{GS} + V_T}{\text{TOX}}\right)^2}$$

if MOBMOD $= 1$ (same as 3-97)

Let us consider $V_{BS} = 0$, so that the effect from UC can be omitted. Based on a generalization that the effective mobility decreases to 1/3 of the bulk mobility during strong inversion, we can estimate the denominator $1 + \text{UA}(V_{GS} + V_T)/\text{TOX} + \text{UB}[(V_{GS} + V_T)/\text{TOX}]^2$ to be equal to 3. To determine an extreme value for UB, we assume that somehow the mobility degradation has a purely parabolic dependence on the normal electric field (see U0 for details). Hence, we suppose that UA $= 0$. If, further, we consider a technology in which $V_{GS} \sim 2.5$ V, $V_T \sim 0.5$ V, and TOX $= 50$ Å, then UB is roughly equal to 5.6×10^{-18} m^2/V^2. Certainly the exact UB value depends on the technology and on the value of UA. The default value of UB in BSIM3, 5.87×10^{-19}, does not have much physical significance. The value in a parameter set may differ significantly from the default.

Thus far we have neglected the temperature dependence of UB. See UB1 for more details on UB's temperature dependency.

UB1 (default $= -7.61 \times 10^{-18}$, m^2/V^2) temperature coefficient for UB. UB appearing in the SPICE parameter set represents the parabolic mobility degradation coefficient at TNOM (the temperature for which the SPICE parameters are extracted). The device operation temperature, T_{device}, is not necessarily equal to TNOM. Since mobility is a function of temperature, UB extracted at TNOM is identical to the mobility value during device operation only if $T_{device} = $ TNOM. If the device temperature in a circuit simulation differs from TNOM, BSIM3 modifies Eq. (3-97) to Eq. (3-98). We denote UB(TNOM) as the UB specified in the parameter set. BSIM3 relates UB(TNOM) and UB(T_{device}) according to

$$\text{UB}(T_{device}) = \text{UB(TNOM)} + \text{UB1} \times \left[\frac{T_{device} + 273.15}{\text{TNOM} + 273.15} - 1\right], \quad (3\text{-}100)$$

where UB1 is the parameter quantifying the amount of temperature dependence of UB. See TNOM for more details about the distinction between T_{device} and TNOM.

The default value of UB1 is -7.61×10^{-18}. See UA1 for a note about the negative value of the temperature coefficient. It is important to realize that the default value of UB1 is some number, not zero. Therefore, a SPICE parameter set can simulate the temperature dependence of I-V characteristics, even though UB1 is not specified. The nonzero default can sometimes be viewed as a bad feature. A designer may falsely believe that, since parameters such as UB1 are not specified, there ought not be any temperature dependence on the I-V characteristics.

UC (default = -4.65×10^{-11} for MOBMOD = 1 or 2, or -0.0465 for MOBMOD = 3, unit = m/V^2 for MOBMOD = 1 or 2, or V^{-1} for MOBMOD = 3) body-effect coefficient of mobility-degradation (due to the application of a finite V_{BS}). As discussed in the entry of U0, BSIM3's mobility model is based on the universal mobility formulation. According to the formulation, the effects of V_{BS} have been absorbed into the average normal electric field; therefore, the universal mobility does not depend on V_{BS}, at least theoretically. However, just in case there is any residual V_{BS} dependence in the measurement results, BSIM3 introduces UC as a parameter to accommodate such a dependence. From this perspective, it seems that UC should be defaulted to 0. However, BSIM3 gives a nonzero default value for UC, reflecting the observation that mobility data measured in various laboratories often do not strictly obey the universal mobility curve. The nonuniversal characteristic of the universal mobility model is discussed in the entry U0.

BSIM3 develops three mobility models out of the universal mobility formulation. The equations for the three models are given in the entry U0. For the convenience of the following discussion, we repeat the equation for MOBMOD = 3 here:

$$\mu_{eff} = \frac{U0}{1 + \left[UA\left(\frac{V_{GS} + V_T}{TOX}\right) + UB\left(\frac{V_{GS} + V_T}{TOX}\right)^2 \right](1 + UC \cdot V_{BS})} \quad \text{if MOBMOD} = 3.$$

(3-101)

We attempt to estimate the value of UC. From the experimental data of [7], we see that when the average normal electric field is high, the effective mobility increases by 5.7 % (from 360 to 380 cm^2/V-s) when V_{BS} changes form 0 V to -20 V. A high effective electric field corresponds to a high value of $(V_{GS} + V_T)/\text{TOX}$ (see the entry U0 for details). Hence, μ_{eff} under high field is proportional to $1/(1 + UC \cdot V_{BS})$. We can therefore equate 1.057 with $1/(1 - UC \cdot 20)$. UC is estimated to be $+ 2.1 \times 10^{-5}$.

The above estimate of UC, when MOBMOD = 3, is based on a set of experimental data. If we were to model the same set of data using MOBMOD = 1 (or 2), the UC parameter would be a different value because the $(1 + UC \cdot V_{BS})$ factor appears at a different location of the equation. Referring to the MOBMOD = 1 equation listed in

Eq. (3-95a), we see that UC is roughly on the order of 2.1×10^{-5} times UA, where the 2.1×10^{-5} value was that estimated for MOBMOD = 3 and according to the entry for UA, UA has an estimated value of 3.3×10^{-9}. Hence, UC for MOBMOD = 1 (or 2) is estimated to be $2.1 \times 10^{-5} \cdot 3.3 \times 10^{-9} = 7 \times 10^{-14}$.

The estimated UC values, 7×10^{-14} for MOBMOD = 1 or 2 and 2.1×10^{-5} for MOBMOD = 3, are obtained from the experimental data of Ref. [7], which fit well to the universal mobility curve. It will not be surprising, however, that experimental data measured out of another laboratory yield a different set of the universal mobility curve. Therefore, it is not uncommon to see UC values differing significantly from the aforementioned estimated values or BSIM3's default values, either in magnitude or sign.

Thus far we have neglected the temperature dependence of UC. See UC1 for more details on UC's temperature dependency.

UC1 (default = -5.6×10^{-11} for MOBMOD = 1 or 2, or -0.056 for MOBMOD = 3, unit = m/V^2 for MOBMOD = 1 or 2, or V^{-1} for MOBMOD = 3) temperature coefficient for UC. UC appearing in the SPICE parameter set is a mobility degradation coefficient at TNOM (the temperature for which the SPICE parameters are extracted). In general, the device operation temperature, T_{device}, is not necessarily equal to TNOM. Since mobility is a function of temperature, UC extracted at TNOM is identical to the mobility value during device operation only if T_{device} = TNOM. If in the device temperature in a circuit simulation differs from TNOM, BSIM3 modifies the Eq. (3-101) to

$$\mu_{eff}(T_{device}) = \frac{U0(T_{device})}{1 + \left[UA(T_{device}) \left(\frac{V_{GS} + V_T}{TOX} \right) + UB(T_{device}) \left(\frac{V_{GS} + V_T}{TOX} \right)^2 \right] (1 + UC(T_{device}) \cdot V_{BS})}.$$

(3-102)

We denote UC(TNOM) as the UC specified in the parameter set. BSIM3 relates UC(TNOM) and UC(T_{device}) according to

$$UC(T_{device}) = UC(TNOM) + UC1 \times \left[\frac{T_{device} + 273.15}{TNOM + 273.15} - 1 \right],$$

(3-103)

where UC1 is the parameter quantifying the amount of temperature dependence of UC. The temperature dependencies of U0, UA and UB are described in their respective entries. As discussed in the entry of UA1, it is sometimes best to drop UA, UB, and UC's temperature dependencies and rely only on U0's temperature dependence to reflect μ_{eff}'s temperature dependency. See TNOM for more details about the distinction between T_{device} and TNOM.

The default value of UC1 is some number, not zero. Therefore, a SPICE parameter set can simulate the temperature dependence of I-V characteristics, even though UC1 is not specified. The nonzero default can sometimes be viewed as a bad feature. A designer may falsely believe that, since parameters such as UC1 are not specified, there ought not be any temperature dependence on the I-V characteristics.

UTE (default = −1.5, unitless) temperature coefficient for the zero-field universal mobility, U0. U0 appearing in the SPICE parameter deck is the value at TNOM, the temperature at which the SPICE parameters are extracted. In general, TNOM differs from T_{device}, the device junction temperature. Since mobility is a function of temperature, U0 extracted at TNOM is identical to the mobility value during device operation only if T_{device} = TNOM. When the device temperature in a circuit simulation is not equal to TNOM, BSIM3 adjusts the U0 parameter by a factor:

$$\text{U0}(T_{device}) = \text{U0}(\text{TNOM}) \times \left(\frac{T_{device} + 273.15}{\text{TNOM} + 273.15}\right)^{\text{UTE}} \quad \text{(same as 3-96)}$$

where T0(TNOM) is the U0 specified in the parameter set and $\text{U0}(T_{device})$ is the mobility value used in the I-V and C-V calculations. The parameter UTE specifies the amount of temperature dependence of the mobility variation.

The default value for UTE is −1.5 in BSIM3, a theoretical value obtained when the mobility is dominated by acoustic phonon scattering [5]. We expect UTE to be negative because the mobility decreases with temperature. In the following, we give an estimate for UTE, based on the experimental data of Ref. [7]. It is found that at a low effective field of 105 V/cm, the mobility decreases from 700 to 350 cm²/V-s as the device temperature increases from 25 °C to 140 °C. Following Eq. (3-96), we can equate UTE with $\log(\text{U0}(T_1)/\text{U0}(T_2))/\log((T_1 + 273.15)/(T_2 + 273.15))$. UTE is calculated to be $\log(700/350)/\log(298.15/413.15) = -2.1$.

It is important to realize that the default value of UTE is a negative number, not zero. Therefore, a SPICE parameter set can simulate the temperature dependence of I-V characteristics, even though UTE is not specified. The nonzero default can sometimes be viewed as a bad feature. A designer may falsely believe that, since parameters such as UTE are not specified, there ought not be any temperature dependence on the I-V characteristics.

VBX (default = calculated, unit = V) the bulk-source bias at which the depletion thickness equals to XT, which is the depth of the channel doping marked in Fig. 3-18. Most often, either K1 or K2 is specified in the SPICE parameter set. In this case, VBX becomes unused, and is often left unspecified.

In the event that neither K1 nor K2 is specified, VBX is used to determine the values for K1 and K2, in accordance with Eq. (3-64). Further, if VBX is not specified in the SPICE parameter set, then its default value is calculated by the following:

$$\text{VBX} = 2\phi_f - \frac{q \cdot \text{NCH} \cdot \text{XT}^2}{2 \, \epsilon_s}, \quad (3\text{-}104)$$

where $2\phi_f$ is the strong-inversion surface potential given in Eq. (1-6b).

If VBX is given in the SPICE parameter set, it should have a negative value, either for NMOS or PMOS. A positive VBX in BSIM3 would mean the bulk–source junction is forward-biased (in either NMOS or PMOS), a condition not normally encountered in MOSFET. Therefore, if VBX is given a positive value, BSIM3 internally reassigns VBX to the negative of its original value.

VBM (default = −3, unit = V) minimum bulk-substrate bias for body-effect calculation. This parameter should have a negative value, for either NMOS or PMOS. A positive VBM in BSIM3 would mean the bulk–source junction is forward-biased (in either NMOS or PMOS), a condition not normally encountered in MOSFET. However, if VBM is specified to have a positive value, BSIM3 internally reassigns VBM to the negative of its original value.

There are two uses of VBM in BSIM3. The first lies in the model of the body effects. In the relatively rare event that neither K1 nor K2 is specified in the SPICE deck, VBM is then used to calculate K1 and K2 according to Eq. (3-64). Often, however, either K1 or K2 is specified. Under such a scenario, VBM is useful only for the determination of a BSIM3 internal variable called the maximum bulk-substrate bias (which is labeled as v_{bc} in the BSIM3 manual and in this book, but as vbsc in the BSIM3 code). To avoid numerical difficulties, BSIM3 does not accept a negative V_{BS} of an arbitrarily large magnitude. To prevent V_{BS} from being more negative than v_{bc}, BSIM3 defines an effective bulk-source bias given by

$$V_{BS,\text{eff}} = v_{bc} + \frac{(V_{BS} - v_{bc} - \delta_1) + \sqrt{(V_{BS} - v_{bc} - \delta_1)^2 - 4\delta_1 v_{bc}}}{2}; \quad \delta_1 = 0.001.$$

(3-105)

The plot of $V_{BS,\text{eff}}$ as a function of V_{BS} is shown in Fig. 3-32. When V_{BS} is in its practical range, $V_{BS,\text{eff}}$ is basically equal to V_{BS}. However, when V_{BS} is more negative than the desired minimum value of v_{bc}, $V_{BS,\text{eff}}$ then becomes equal to v_{bc}. By using

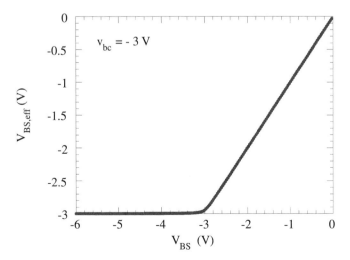

Fig. 3-32 Smoothing function $V_{BS,\text{eff}}$ plotted as a function of V_{BS}.

$V_{BS,\text{eff}}$ in the rest of the program wherever V_{BS} is to be used, BSIM3 avoids using a V_{BS} value which is more negative than v_{bc}.

Now, we are ready to tie VBM with v_{bc}. The actual value chosen for v_{bc} (in the unit of V) is determined through a set of conditions. First, if K2 > 0, v_{bc} is equated to the more negative value of −30 and VBM. As mentioned in the entry for VBM, VBM is defaulted to −3 when it is not specified in the SPICE parameter set. In this case, v_{bc} will then be equal to −30. Note that K2 > 0 is really an unphysical scenario (i.e., K2 should always be smaller than zero). But BSIM3 wants to be able to accommodate unphysical model extraction, so this possibility is included in the source code as well.

In the more likely situation where K2 < 0, v_{bc} is determined as follows. (In this case, the determination of v_{bc} does not require the knowledge of VBM.) First, a temporary voltage, denoted as V_x here, is calculated as

$$V_x = 0.9 \times \left(2\phi_f - \frac{K1^2}{4K2^2}\right), \qquad (3\text{-}106)$$

where $2\phi_f$ is the strong-inversion surface potential given in Eq. (1-6b). If $V_x > -3$ V, v_{bc} is fixed at −3 (V). If $V_x < -30$ (V), v_{bc} is fixed at −30 (V). At $-30 < V_x < -3$, then v_{bc} is equated to V_x.

BSIM3 has four smoothing functions. $V_{FB,\text{effCV}}$ is the smoothing function used to smooth out the transition between the accumulation and the depletion regions; $V_{DS,\text{eff}}$, between the linear and the saturation regions; and $V_{GST,\text{eff}}$, between the subthreshold and the strong inversion regions. In contrast, $V_{BS,\text{eff}}$ is not used to smooth out the device characteristics. It serves only to prevent a very negative V_{BS} value from being applied to the device.

VERSION (default = 3.1 in the BSIM3.1 model, 3.2 in the BSIM3v3.2 model, 3.2.1 in the BSIM3v3.2.1 model and 3.2.2 in the BSIM3v3.2.2 model, unitless) This parameter establishes the version of the BSIM3 model to be used in simulation. It was first created when BSIM3v3.1 was released in December, 1996. However, the parameter VERSION in BSIM3v3.1 does not affect the calculation of device characteristics in any way. It really would not have mattered if VERSION were removed out of the parameter list.

The parameter VERSION begins to be functional in BSIM3v3.2. BSIM3v3.2 contains several important improvements over the 3.1 version (BSIM3v3.1), mostly relating to the C-V modeling, as further discussed in the entry of CAPMOD. Although BSIM3v3.1 remains popular even to date, version 3.2 is preferred. Some circuit simulator vendors may have released their product prior to June 1998. Then, BSIM3v3.1 is the BSIM3 model that is supported in the simulator.

To appreciate the function of VERSION, we need to rave a little about history. When BSIM3v3.1 was released, it was well received because it introduced the CAPMOD = 2 capacitance model which results in smooth C-V characteristics. At that time, the smoothness feature was considered a revolutionary improvement over the previous capacitance models (such as CAPMOD = 1 or 0), whose discontinuity features often cause numerical difficulties during SPICE simulation. However, the CAPMOD = 2 model in BSIM3v3.1 has several problems not fully uncovered at the time of its release. For example, C_{gs} implemented in BSIM3v3.1 can be negative in the accumulation region, as revealed in Fig. 3-33, which are calculated results from BSIM3v3.1 and BSIM3v3.2's CAPMOD = 2 models. The negative C_{gs} in BSIM3v3.1 can be a numerical nuisance. Besides, it indirectly causes C_{gb} to exceed C_{gg} in accumulation.

It was later identified that the bias-dependent flatband voltage ($V_{fb,(v)}$) used in BSIM3v3.1 is responsible for the problem. In BSIM3v3.2, a so-called zero-bias flatband voltage ($V_{fb,zb}$) is formulated to correct the negative C_{gs} problem. It would seem reasonable to simply abolish the bias-dependent flatband voltage of BSIM3v3.1. However, at times, the backward compatibility issue is more important than the negative C_{gs} problem. It is crucial to take the BSIM3v3.1 model parameters extracted previously, run the SPICE simulation with the updated BSIM3v3.2 model, and get the same results as before. It does not matter if a negative C_{gs} is to be reproduced in BSIM3v3.2, but it matters that the identical simulation results be obtained with both BSIM3v3.1 and BSIM3v3.2. The parameter VERSION serves the purpose of backward compatibility. When BSIM3v3.2 is the MOSFET model used in a SPICE simulator, then assigning VERSION with a value of 3.1 instructs the code to utilize the same bias-dependent flatband voltage. The BSIM3v3.2 model will then lead to the same C-V simulation results as BSIM3v3.1. However, if the backward compatibility is not an issue, then VERSION should be set to the default value of 3.2. In this way, the zero-bias flatband voltage is chosen instead, thereby removing the negative C_{gs} problem.

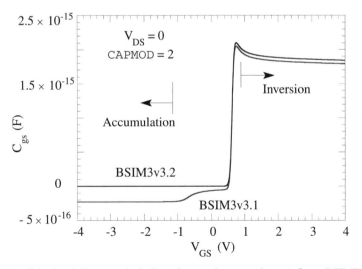

Fig. 3-33 Calculated C_{gs} (not including the overlap capacitance) from BSIM3v3.1 and BSIM3v3.2's CAPMOD = 2 models. The negative C_{gs} in BSIM3v3.1 can be a numerical nuisance.

The expressions for the bias-dependent flatband voltage, $V_{fb,(v)}$, the zero-bias flatband voltage, $V_{fb,zb}$, are given in the entry VFBCV.

In BSIM3v3.2, the parameter VERSION was created as a compromise between the need for accurate results and backward compatibility. However, assigning VERSION to be 3.1 in a BSIM3v3.2 model is awkward. By April 1999 when the next and latest version of BSIM3 was released, it was decided that VERSION should not be used to select between models, (i.e., with the bias-dependent or zero-bias flatband voltages). Two models, BSIM3v3.2.1 and BSIM3v3.2.2, are simultaneously released. BSIM3v3.2.2 uses exclusively the zero-bias flatband voltage which eliminates the negative C_{gs} problem. In this model, VERSION is defaulted to 3.2.2. If the user specifies any other version number, a warning is issued, but the simulation proceeds with the zero-bias flatband voltage regardless the user-input VERSION value. Backward compatibility with BSIM3v3.1 is impossible. In contrast, the BSIM3v3.2.1 model solely uses the bias-dependent flatband voltage which is known to cause the negative C_{gs} problem. In this model, VERSION is defaulted to 3.2.1. If the user specifies some other version numbers, a warning is issued, but the simulation proceeds with the bias-dependent flatband voltage.

Here is a summary:

BSIM3v3.1: The parameter VERSION has no effect. It does not matter whether the user-inputted value is 3.1 or not. The calculation always utilizes the bias-dependent V_{fb}, leading to the negative C_{gs} problem.

BSIM3v3.2: The parameter VERSION has the effect of choosing the desired flatband voltage expression. If VERSION is specified to be 3.1, then the bias-dependent flatband voltage is used. Although this leads to the negative C_{gs} problem, the backward compatibility with BSIM3v3.1 is maintained. If VERSION is set to 3.2, then the zero-bias V_{fb} is used instead.

BSIM3v3.2.1: The parameter VERSION serves only one purpose—for checking. If the user-specified value is not 3.2.1, then a warning is issued. However, independent of the inputted value, the calculation always proceeds with the bias-dependent V_{fb}, as in BSIM3v3.1.

BSIM3v3.2.2: Again the parameter VERSION serves only the checking purpose. If the user-specified value is not 3.2.2, then a warning is issued. However, independent of its value, the calculation always proceeds with the zero-bias V_{fb}, as in BSIM3v3.2.

Here is a method to find out whether a BSIM3 model is BSIM3v3.1, BSIM3v3.2, BSIM3v3.2.1, or BSIM3v3.2.2. First, identify whether IJTH is one of the SPICE parameters. If not, add that parameter in and assign a value of 0. Second, run SPICE simulation. If the simulation terminates normally, then the model is BSIM3v3.2 or above. If the simulation aborts, then the model is BSIM3v3.1. Third, assign the parameter VERSION to a value of 3.2 and then rerun the simulation. If the simulation proceeds without giving a warning, then the model is BSIM3v3.2. If the model is something else, then the warning message would specify whether BSIM3v3.2.2 or BSIM3v3.2.1 is the model being used for simulation.

3.2 ALPHABETICAL GLOSSARY OF BSIM3 PARAMETERS

VFB (default = −1 or calculated, unit = V) flatband voltage for dc I-V calculation. This is a new parameter introduced in BSIM3v3.2. In BSIM3v3.1, the flatband voltage is merely an intermediate variable, without the possibility of user specification. Its value is dictated by the same set of rules that determines the default value of the VFB parameter in BSIM3v3.2, which is described shortly (i.e., Eq. 3-108). There are two places where BSIM3 (both v3.1 and v3.2) utilizes the flatband voltage: a) in the calculation of VTH0 when VTH0 is not specified in the SPICE parameter set; and b) in the calculation of poly-depletion effects. In the case that the poly-depletion effects are negligible, then the dc flatband voltage in BSIM3 is used mainly for the purpose of item (a), that is, the flatband voltage is used to calculate VTH0 (when VTH0 is not specified):

$$\text{VTH0} = \delta_{NP} \cdot \left(\text{VFB} + 2\phi_f + \text{K1}\sqrt{2\phi_f}\right), \tag{3-107}$$

where δ_{NP} is +1 for NMOS and −1 for PMOS, and $2\phi_f$ is the strong-inversion surface potential given in Eq. (1-6b). The above equation of finding VTH0 from VFB is valid only when VTH0 is not given. If VTH0 is specified in the SPICE parameter set, VFB (whether the user-specified value or the default value) is left only with the role of item (b). The details of the dc flatband voltage's effect on the effective V_{GS} due to poly-depletion are seen in Appendix A (Eq. A-41).

The main purpose for adopting VFB as a new parameter in BSIM3v3.2 is to allow accurate modeling for MOS with gate materials other than the conventional polysilicon. When VFB is not specified, its default value depends on whether the VTH0 is specified in the SPICE parameter set or not. If VTH0 is not specified, VFB assumes a value of −1. (The predominant MOS processes uses polysilicon gate, for which the flatband voltage is fairly close to −1 V [6, p. 42].) Otherwise, VFB is defaulted to an equation based on Eq. (3-107):

$$\text{VFB} = \delta_{NP} \cdot \text{VTH0} - 2\phi_f - \text{K1}\sqrt{2\phi_f}. \tag{3-108}$$

In BSIM3, the flatband voltage used for dc calculation (discussed above) generally differs from the flatband voltage used for c.v. calculation. For the latter, the flatband voltage is taken to be VFBCV if CAPMOD = 0, and $V_{fb,zb}$ if CAPMOD = 1, 2, or 3. $V_{fb,zb}$, the zero-bias flatband voltage, is the flatband voltage at $V_{GS} = V_{DS} = V_{BS} = 0$. See the entry VFBCV for more details.

VFBCV (default = −1, unit = V) flatband voltage used to calculate $V_{T,CV}$ when CAPMOD = 0. ($V_{T,CV}$ is the threshold voltage used for C-V calculation.) At other CAPMOD values, this parameter is not used at all. When CAPMOD = 0, the relationship between VFBCV and $V_{T,CV}$ is given by

$$V_{T,CV} = \text{VFBCV} + 2\phi_f + \text{K1} \cdot \sqrt{2\phi_f} \quad (\text{CAPMOD} = 0), \tag{3-109}$$

where $2\phi_f$ is the strong-inversion surface potential given in Eq. (1-6b). $V_{T,CV}$, differs from the threshold voltage for dc I-V calculation, V_T (Eq. 3-33, see the entry VTH0). While V_T accounts for various short-channel effects, only the body effects have been incorporated in the $V_{T,CV}$ in Eq. (3-109). There are at least two advantages to decouple the I-V and C-V threshold voltages, as discussed in Section 1.3. The decoupling enables capacitance symmetry at $V_{DS} = 0$, and C_{gb} to be identically zero in strong inversion, instead of some finite values. Of course, CAPMOD = 0 suffers from the fact that C_{gs}, among others, turns on at a different threshold voltage from the drain current (Fig. 1-17). (The notable problem associated with CAPMOD = 0, that the capacitance's slopes are discontinuous as shown in Fig. 3-3, is a problem unrelated to the use of VFBCV.)

In CAPMOD = 1, 2, or 3, the threshold voltages for I-V and C-V calculations are identical. With $V_{T,CV}$ equated to V_T of Eq. (3-33), the capacitance asymmetry and the finite C_{gb} problem cannot be easily avoided. Therefore, despite its primitiveness, CAPMOD = 0 can be more accurate than the other models in some regards.

The flatband voltage for C-V calculations depends on the values of CAPMOD, as well as the version of the BSIM3 model being used to run the simulation. To find out whether a BSIM3 model is BSIM3v3.2.2, BSIM3v3.2.1, BSIM3v3.2, or BSIM3v3.1, see the entry VERSION. In the following, we list the C-V flatband voltages ($V_{FB,CV}$) used under various scenarios:

BSIM3v3.2.2, (which forces VERSION to be 3.2.2)
    CAPMOD = 0 $V_{FB,CV}$ = VFBCV
    CAPMOD = 1, 2, 3 $V_{FB,CV} = V_{fb,zb}$
BSIM3v3.2.1, (which forces VERSION to be 3.2.1)
    CAPMOD = 0 $V_{FB,CV}$ = VFBCV
    CAPMOD = 1, 2 $V_{FB,CV} = V_{fb,(v)}$
    CAPMOD = 3 $V_{FB,CV} = V_{fb,zb}$
BSIM3v3.2, with VERSION specified as 3.2
    CAPMOD = 0 $V_{FB,CV}$ = VFBCV
    CAPMOD = 1, 2, 3 $V_{FB,CV} = V_{fb,zb}$
BSIM3v3.2, with VERSION specified as 3.1
    CAPMOD = 0 $V_{FB,CV}$ = VFBCV
    CAPMOD = 1, 2 $V_{FB,CV} = V_{fb,(v)}$
    CAPMOD = 3 $V_{FB,CV} = V_{fb,zb}$
BSIM3v3.1 (which ignores VERSION parameter in the codes)
    CAPMOD = 0 $V_{FB,CV}$ = VFBCV
    CAPMOD = 1, 2, (3 is unavailable) $V_{FB,CV} = V_{fb,(v)}$

As shown, there are three kinds of flatband voltages for C-V calculations in BSIM3. VFBCV was discussed above; it is used exclusively for CAPMOD = 0. When CAPMOD = 1 was introduced, it was felt that the I-V and C-V characteristics should share the same turn-on behavior at the threshold voltage. This desire leads to the development of the voltage-dependent flatband voltage expression, which is given by

$$V_{fb,(v)} = V_T - 2\phi_f - K1\sqrt{2\phi_f}. \qquad (3\text{-}110)$$

$V_{fb,(v)}$ is voltage-dependent because V_T (Eq. 3-33, see the entry VTH0) is generally a function of V_{GS}, V_{DS}, and V_{BS}. The fact that $V_{fb,(v)}$ ensures the same threshold voltages for C-V and I-V calculations can inadvertently cause the asymmetrical device capacitances at $V_{DS} = 0$, and unphysical nonzero C_{gb} under strong inversion. These problems are sometimes deemed unimportant. However, the fact that $V_{fb,(v)}$ is bias dependent means the derivatives of $V_{fb,(v)}$ with respect to the bias voltages are nonzero. These nonzero derivatives ultimately cause severe problems in the capacitance calculation, that C_{gg} is smaller than C_{gb} and that C_{gs} can be negative in accumulation. Both problems can be overcome in BSIM3v3.2, by the use of the zero-bias flatband voltage: $V_{fb,zb}$. It is given by

$$V_{fb,zb} = V_T(V_{GS} = V_{DS} = V_{BS} = 0) - 2\phi_f - \text{K1}\sqrt{2\phi_f}. \quad (3\text{-}111)$$

That is, $V_{fb,zb}$ is calculated under the condition that all external bias voltages are zero. This definition makes $V_{fb,zb}$ to be bias-independent, a property which removes both the $C_{gg} < C_{gb}$ and negative C_{gs} problems. However, the asymmetrical capacitance and finite C_{gb} problems still remain (Sections 4.7 and 4.8). (As mentioned, these last two problems can be removed by using different expressions for $V_{T,CV}$ and V_T.)

Once the flatband voltage for C-V calculation ($V_{FB,CV}$) is known, its value is used in Eq. (1-23) to find the effective flatband voltage. This, in turn, enables the calculation of various terminal charges.

VOFF (default = −0.08, unit = V) offset voltage in the subthreshold region. When $V_{GS} = 0$, the drain current is referred to as the off current, I_{off}. It is the leakage current in the off state, and should be minimized. Let us focus on the situation when V_{DS} is large, According to Eqs. (1-12) and (1-21),

$$I_{off}|_{V_{DS}\to\text{large}} = \frac{W}{L}\mu_n C'_{ox}\left(\frac{kT}{q}\right)^2 \exp\left(-\frac{qV_T}{nkT}\right) \times \exp\left(-\frac{q\text{VOFF}}{nkT}\right). \quad (3\text{-}112)$$

If VOFF were absent, we would need to fit I_{off} by optimizing V_T, μ_n, or C'_{ox}. However, all these variables' values are fixed after fitting the I-V characteristics in the strong inversion region. Optimizing them to improve the fit in the off current ultimately creates problem in fitting current in other regions. VOFF renders an extra degree of freedom, in the simultaneous optimization of the drain current in both the subthreshold and the strong-inversion regions.

VOFF is a fitting parameter which can be either positive or negative. The −0.08 V value used as the default does not bear much physical significance. As demonstrated in Fig. 3-20, VOFF can also affect the calculated ideality factor of the subthreshold characteristics, in addition to the off current. When NFACTOR deviates significantly from 1 (such as below 0.6 or above 2), usually it is because VOFF takes on an excessively negative value.

VOFFCV (default = 0, unit = V) offset threshold voltage in the $V_{GS} - V_T$ effective function for C-V calculation. See the entry of NOFF for details.

264   BSIM3 PARAMETERS

VSAT (default = 8×10^4, unit = m/s)   carrier saturation velocity at the nominal temperature, TNOM. This parameter, when used in NMOS, denotes the electron saturation velocity. When used in PMOS, VSAT denotes the hole saturation velocity. Therefore, VSAT in NMOS should be slightly larger than the VSAT in PMOS.

In BSIM3, VSAT is a critical parameter determining the device current in short-channel devices, but has negligible impact on long-channel devices. This agrees with the physical intuition that in long-channel devices, the electric field parallel to the current conduction is small and the electron drift velocity is roughly equal to the mobility times the field. The drift velocity never reaches a magnitude comparable to VSAT. In short-channel devices, in contrast, the field is large enough such that the carriers travel at the saturation velocity in a sizable portion of the channel.

It is desirable that VSAT has a physically meaningful value. The saturation velocity for silicon has been measured to be about 8×10^4 m/s at the room temperature, which is the default value for VSAT. However, because of the particular fashion that the mobility's dependence on electric field is modeled in BSIM3, it is possible to extract a VSAT value twice as large. VSAT values generally range between 8×10^4 and 2×10^5 m/s. A VSAT value outside this range is a sign that other parameters relating to the effective mobility (U0, UA, UB, and UC) have incorrect values.

Figure 3-34 illustrates the effects of VSAT on the I-V characteristics of a short-channel device. When VSAT is at its typical value of 10^5 (m/s), the drain current is

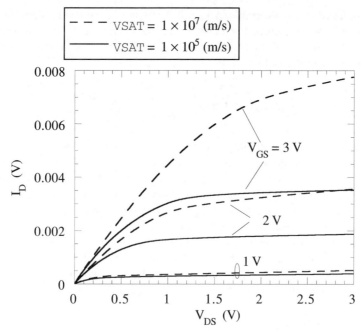

Fig. 3-34   Effects of VSAT on BSIM3-calculated I-V characteristics of a short-channel device.

seen to saturate at normal voltage ranges. However, when VSAT is replaced by an unphysically high value of 10^7 (m/s), the drain current shoots up, without displaying a noticeable saturation behavior. The figure demonstrates the importance of extracting a proper value for VSAT; the drain current can differ by 100 %. When the channel length continues to decrease as the processing technologies evolve, it becomes possible for an electron emitted from the source to not experience a scattering event before reaching the drain. In this non-equilibrium situation, the electron velocity can exceed the saturation velocity, resulting in the so-called *velocity overshoot*. Depending on the extent of the velocity overshoot, we see from Fig. 3-34 that the phenomenon can dramatically increase the drain current.

We said at the beginning that VSAT is the carrier saturation velocity at the nominal temperature, TNOM, the temperature at which the SPICE parameter set is extracted. It may well differ from the operating device temperature (see the entry TNOM for details). When $T_{device} \neq$ TNOM, the actual saturation velocity used in the I-V and C-V calculation is different from VSAT. See the entry AT for details.

VTH0 (default = calculated, unit = V) threshold voltage of a long-channel device at zero V_{BS}. This is one of the most important parameters in the whole parameter set. VTH0 should be a positive number for NMOS, and negative for PMOS. (In BSIM3, VTH0 is the only parameter whose sign should differ for NMOS and PMOS.) The threshold voltage of a transistor is given by

$$V_T = \text{VTH0} + \delta_{NP}(\Delta V_{T,\text{body\_effect}} - \Delta V_{T,\text{charge\_sharing}} - \Delta V_{T,\text{DIBL}}$$
$$+ \Delta V_{T,\text{reverse\_short\_channel}} + \Delta V_{T,\text{narrow\_width}} + \Delta V_{T,\text{small\_size}}) \quad \text{(same as 3-33)}$$

$$\Delta V_{T,\text{body\_effect}} = \text{K1} \cdot \left(\sqrt{2\phi_f - V_{BS}} - \sqrt{2\phi_f}\right) - \text{K2} \cdot V_{BS} \quad \text{(same as 3-63)}$$

$$\Delta V_{T,\text{charge\_sharing}} = \text{DVT0}\left[\exp\left(-\text{DVT1}\frac{L_{eff}}{2L_t}\right) + 2\exp\left(-\text{DVT1}\frac{L_{eff}}{L_t}\right)\right]$$
$$(V_{bi} - 2\phi_f) \quad \text{(same as 3-34)}$$

$$\Delta V_{T,\text{DIBL}} = \left[\exp\left(-\text{DSUB}\frac{L_{eff}}{2L_{t0}}\right) + 2\exp\left(-\text{DSUB}\frac{L_{eff}}{L_{t0}}\right)\right]$$
$$(\text{ETA0} + \text{ETAB} \cdot V_{BS}) \times V_{DS} \quad \text{(same as 3-46)}$$

$$\Delta V_{T,\text{reverse\_short\_channel}} = \text{K1}\left(\sqrt{1 + \frac{\text{NLX}}{L_{eff}}} - 1\right)\sqrt{2\phi_f} \quad \text{(same as 3-75)}$$

$$\Delta V_{T,\text{narrow\_width}} = (\text{K3} + \text{K3B} \cdot V_{BS})\frac{\text{TOX}}{W_{eff} + \text{W0}}2\phi_f \quad \text{(same as 3-66)}$$

$$\Delta V_{T,\text{small\_size}} = \text{DVT0W}\left[\exp\left(-\text{DVT1W}\frac{W_{eff}L_{eff}}{2L_{tw}}\right) + 2\exp\left(-\text{DVT1W}\frac{W_{eff}L_{eff}}{L_{tw}}\right)\right]$$
$$(V_{bi} - 2\phi_f) \quad \text{(same as 3-39)}$$

where $2\phi_f$ is the strong-inversion surface potential given in Eq. (1-6b) and

$$L_{t0} = \sqrt{\frac{\epsilon_s X_{dep,0}}{C'_{ox}}} \qquad \text{(same as 3-47)}$$

$$L_t = \sqrt{\frac{\epsilon_s X_{dep}}{C'_{ox}}} (1 + \text{DVT2} \cdot V_{BS}) \qquad \text{(same as 3-35)}$$

$$L_{tw} = \sqrt{\frac{\epsilon_s X_{dep}}{C'_{ox}}} (1 + \text{DVT2W} \cdot V_{BS}) \qquad \text{(same as 3-40)}$$

δ_{NP} is +1 for NMOS and −1 for PMOS. At least theoretically, all the terms inside the brackets are positive (except $\Delta V_{T,\text{small\_size}}$, which can be either positive or negative.). Therefore, the term which contributes to body effect will add on VTH0 to make V_T larger. Conversely, the term representing the charge-sharing effects will be subtracted from VTH0, leading to a smaller value of V_T. Figure 3-23 illustrates measured threshold voltage as a function of channel length, in a digital CMOS process. The general trend at short L's is that, as L decreases, V_T also decreases. This is the so-called short-channel effects. Because V_{DS} is small (0.1 V) in this figure, the drain-induced barrier-lowering is not yet significant, and most of the short-channel effects are due to charge-sharing effect. Figure 3-23 also reveals the reverse short-channel effect. At intermediate channel length values, the devices' threshold voltage exhibits an increasing trend as L decreases. Eventually, the short-channel effects take control and the threshold voltage drops off.

We highlight the situation under which each term inside the parentheses of Eq. (3-33) becomes important:

Body effect:	Large reverse bulk-source bias (V_{BS})
	Heavy channel doping (NCH)
Charge sharing:	Short-channel transistors
Drain-induced barrier-lowering:	Short-channel devices under large V_{DS}
Reverse short-channel effect:	Short-channel transistors with large NLX
Narrow-width effect:	Narrow-width devices
Small-size effect:	Narrow-width *and* short-channel transistors

It is clear from the above table that, in order to minimize the ΔV_T terms, the device must be wide and long, biased at zero V_{BS} and a small V_{DS}. VTH0 therefore represents the threshold voltage of a large-geometry device biased with zero V_{BS} and a small V_{DS} value of, for example, 0.05 V. As a rough estimate, VTH0 can be approximated to be the V_{GS} value when the drain current per gate width reaches a value of 10^{-8} A/μm. VTH0 generally falls between 0.3 and 0.8 V for NMOS, and −0.2 and −0.8 V for PMOS.

Extracting VTH0 for a transistor has always been the most important process of a model generation. If VTH0 is not extracted and left unspecified in the parameter set,

3.2 ALPHABETICAL GLOSSARY OF BSIM3 PARAMETERS 267

BSIM3 will calculate the default of VTH0 based on a textbook-like equation. The equation makes grossly simplified assumptions about the transistor process. With the complexity of modern transistor structure, VTH0 calculated from equations just would not be good enough to fit the I-V characteristics. If in the rare event that VTH0 is left unspecified, BSIM3v3.1 defaults VTH0 to a value given by

$$\text{VTH0} = \delta_{NP} \cdot \left(-1 + 2\phi_f + \text{K1}\sqrt{2\phi_f} \right) \quad \text{(BSIM3v3.1)}. \quad (3\text{-}113)$$

The flatband voltage is hard-coded to be -1 V. (In the case that VTH0 is given in the parameter set in BSIM3v3.1, the flatband voltage is then not -1 V, but calculated from an equation. See the entry VFB for more details.)

In BSIM3v3.2, the flatband voltage is a user-specifiable parameter (VFB), which can be used to replace as VTH0 as the parameter to model the threshold voltage of the long-channel device under zero V_{BS} and small V_{DS}. If VTH0 is unspecified in BSIM3v3.2, then VTH0 is defaulted to the following:

$$\text{VTH0} = \delta_{NP} \cdot \left(\text{VFB} + 2\phi_f + \text{K1}\sqrt{2\phi_f} \right) \quad \text{(BSIM3v3.2)}. \quad (3\text{-}114)$$

(*Note*: The BSIM3 manual states that VTH0, when unspecified, defaults to 0.7 for NMOS and -0.7 for PMOS. The codes reflecting such a statement are found in b3set.c. However, in b3temp.c, the $\pm$ 0.7 values are erased, replaced with the value given by the above equation. This note applies to BSIM3v3.1 also, that the BSIM3v3.1's description about VTH0's default value of ± 0.7 is also likely incorrect.) If both VFB and VTH0 are unspecified in BSIM3v3.2, then VFB is assigned a value of -1 V, and VTH0 is calculated from Eq. (3-114).

By declaring a value for VFB, we effectively specify the parameter VTH0 through the above equation. When VFB and VTH0 are both given, BSIM3v3.2 simply uses the specified VTH0 value to calculate the threshold voltage in accordance with Eq. (3-33), just like BSIM3v3.1. Equation (3-114) will be ignored. VFB is used only in a very remote part of the codes, concerning poly depletion. Even there, it is used only in a conditional statement, and VFB's effect is usually insignificant. We think that BSIM3 could issue a warning that, when both VFB and VTH0 are specified in the SPICE parameter set, VFB is not much used in the code. Otherwise, users may have a false impression that the threshold voltage is somehow related to the specified VFB.

There is a reason why we would specify VFB (but not VTH0), and thereby using VFB to calculate the threshold voltage. Imagine that the transistor uses polysilicon as the gate material and a model has been extracted. We are now developing a metal-gate process (such as for use in rf power amplifier applications). If our current process relies on VFB to specify the threshold voltage (with VTH0 unspecified), then we can readily predict the new process' I-V characteristics by modifying the VFB value from that of the polysilicon to that of the metal.

Thus far we have neglected the temperature dependence of the threshold voltage. See the entry KT1 for more details.

The following are binning parameters: WA0, WA1, WA2, WAGS, WALPHA0, WAT, WB0, WB1, WBETA0, WCDSC, WCDSCB, WCDSCD, WCF, WCGD1, WCGS1, WCIT, WCKAPPA, WCLC, WCLE, WDELTA, WDROUT, WDSUB, WDVT0, WDVT0W, WDVT1, WDVT1W, WDVT2, WDVT2W, WDWB, WDWG, WETA0, WETAB, WK1, WK2, WK3, WK3B, WKAET, WKT1, WKT2, WNFACTOR, WNLX, WPCLM, WPDIBLC1, WPDIBLC2, WPDIBLCB, WPRT WPRWB WPRWG, WPSCBE1, WPSCBE2, WPVAG, WRDSW, WU0, WUA, WUA1, WUB, WUB1, WUC, WUC1, WUTE, WVBM, WVBX, WVFBCV, WVOFF, WVSAT, WVTH0, WW0, WWR, WXJ, WXT.

W0 (default = 2.5×10^{-6}, unit = m) channel-width offset to calculate narrow-width's effect on the threshold voltage. The details of W0 is found in the entry K3. Although W0 can assume negative values, effort should be made so that W0 is greater than or equal to zero. The default of W0 is 2.5µm. It is not clear how this default value is chosen; it is likely a value that results the best fit for a particular process. The theoretical value for the offset width should be zero.

WINT (default = 0, unit = m) channel-width offset for dc I-V characteristics. In BSIM3, the effective-channel width for dc calculation is given by

$$W_{\mathit{eff}} = W - 2 \cdot \mathtt{WINT} - 2 \cdot \Delta W_{\mathrm{geometry}} - 2 \cdot \Delta W_{\mathrm{bias\_dependency}} \qquad \text{(same as 3-41)}$$

where W is the drawn transistor width; WINT is a constant width offset; $\Delta W_{\mathrm{geometry}}$ is the offset which depends on transistor's geometry; and finally, $\Delta W_{\mathrm{bias\_dependency}}$ is the width offset which varies with V_{GS} and V_{BS}. The detailed expressions for $\Delta W_{\mathrm{geometry}}$ and $\Delta W_{\mathrm{bias\_dependency}}$ are shown in the entries of WL and DWB, respectively. Generally $\Delta W_{\mathrm{geometry}}$ is zero, as WL, W and WWL take their default values of 0. Similarly, $\Delta W_{\mathrm{bias\_dependency}}$ is zero, as DWB, and DWG take their default values of 0. Therefore, WINT, representing a constant channel width reduction per side, is the major parameter that effects W_{eff}. WINT is generally a positive number, although it may be negative. When it is negative, the effective channel width becomes larger than the drawn W.

For C-V calculation, the appropriate channel-width offset is DWC. Conceptually, WINT and DWC should be identical. However, from a fitting's point of view, it is better to use two width-reduction parameters to independently model I-V and C-V behaviors.

WL (default = 0, unit = $m^{\mathtt{WLN}}$) coefficient of width dependence for channel-width offset in I-V calculation. There are two effective-channel widths used in BSIM3. One is for dc calculations (including the Flicker noise), denoted as W_{eff}, and another is for C-V calculations (including the thermal noise), denoted as $W_{\mathit{eff},CV}$. Here, we are concerned with W_{eff}, which is given by

$$W_{\mathit{eff}} = W - 2 \cdot \mathtt{WINT} - 2 \cdot \Delta W_{\mathrm{geometry}} - 2 \cdot \Delta W_{\mathrm{bias\_dependency}} \qquad \text{(same as 3-41)}$$

where

$$\Delta W_{\text{geometry}} = \frac{\text{WL}}{L^{\text{WLN}}} + \frac{\text{WW}}{W^{\text{WWN}}} + \frac{\text{WWL}}{L^{\text{WLN}} W^{\text{WWN}}}; \quad (3\text{-}115)$$

$$\Delta W_{\text{bias\_dependency}} = \text{DWG} \cdot [V_{GS} - V_T] + \text{DWB} \cdot \left[\sqrt{2\phi_f - V_{BS}} - \sqrt{2\phi_f}\right]$$

(same as 3-42)

W is the drawn channel width. Most of the channel-width offset is modeled with WINT, and typically, both $\Delta W_{\text{geometry}}$ and $\Delta W_{\text{bias\_dependency}}$ are 0. However, at times, due to the geometrical effects in the photolithographical development of photoresist, W_{eff} may exhibit dependencies on the exposure area (or, effectively, the transistor's size). $\Delta W_{\text{geometry}}$ is used to characterize such a dependence. WL, in combination with WLN (the exponent of the channel length), specifies the amount of channel width reduction due to some length dependence, as shown in Eq. (3-115).

Disregarding the processing-related issues, WL should be at its default value of zero. WL's counterpart in the $W_{\text{eff,CV}}$ expression is WLC.

Here is a special note. The W_{eff} expression given in Eq. (3-41) are used in just about all of the dc I-V calculation, but not quite. During the Flicker noise calculation (if NOIMOD = 2 or 3) as well as the source-drain resistance (R_{DS}) calculation from the parameter RDSW, somehow the W_{eff} term used there is equal to the W_{eff} of Eq. (3-41) with the substitution of DWG = DWG = 0. This description is true for Eqs. 2.5.11, 8.2, and 8.7 of the BSIM3 manual. We believe this is a very minor mistake, likely caused by the somewhat confusing notation of W_{eff} used in the BSIM3 source code (or manual).

WLC (default = 0, unit = m^{WLN}) coefficient of length dependence for channel-width offset in the C-V calculation. The effective channel width for I-V calculation (including the Flicker noise), W_{eff}, is given in the entry WL. This is another effective channel length in BSIM3, that used for C-V calculations (including the thermal noise), denoted as $W_{\text{eff,CV}}$. Here we are concerned with the latter:

$$W_{\text{eff,CV}} = W - 2 \cdot \text{DWC} - 2 \cdot \Delta W_{\text{geometry,CV}} \quad \text{(same as 3-43)}$$

where

$$\Delta W_{\text{geometry,CV}} = \frac{\text{WLC}}{L^{\text{WLN}}} + \frac{\text{WWC}}{W^{\text{WWN}}} + \frac{\text{WWLC}}{L^{\text{WLN}} W^{\text{WWN}}}. \quad (3\text{-}116)$$

W is the drawn channel width. Most of the channel-width offset is modeled with DWC, and typically, $\Delta W_{\text{geometry,CV}} = 0$. However, at times, due to the geometrical effects in the photolithographical development of photoresist, $W_{\text{eff,CV}}$ may exhibit dependencies on the exposure area (or, effectively, the transistor's size). $\Delta W_{\text{geometry,CV}}$ is used to characterize such a dependence. WLC, in combination WLN (the exponent of the channel length), specifies the amount of channel width reduction due to some length dependence, as shown in Eq. (3-116).

270   BSIM3 PARAMETERS

WLC's counterpart in I-V calculation is WL. WLC is a new parameter introduced in BSIM3v3.2. In BSIM3v3.1, the WLC appearing in the Eq. (3-116) is replaced by WL. Disregarding the processing-related issues, WLC should be at its default value of zero.

WLN (default = 1, unitless)   power exponent of the length dependence in the calculation of the I-V and C-V channel-width offsets. BSIM3 calculates two effective channel widths, $W_{\it eff}$, which is used for I-V calculation, and $W_{\it eff,CV}$, which is used for C-V calculation. WLN affects both of them. The effective channel width for I-V calculation can be expressed as: $W_{\it eff} = W - 2 \cdot \text{WINT} - 2 \cdot \Delta W_{\text{geometry}} - 2 \cdot \Delta W_{\text{bias\_dependency}}$. WLN is the exponent of $1/L$ in the $\Delta W_{\text{geometry}}$ expression given in the entry of WL (Eq. 3-115). $\Delta W_{\text{geometry}}$ is usually zero unless there are processing-related issues which cause $W_{\it eff}$ to depend on geometry. When WL, WW, and WWL assume their default value of zero such that $\Delta W_{\text{geometry}} = 0$, then the value of WLN is inconsequential.

The effective channel length for C-V calculation can be expressed as: $W_{\it eff,CV} = W - 2 \cdot \text{DWC} - 2 \cdot \Delta W_{\text{geometry,CV}}$. WLN is also the exponent of $1/L$ in the $\Delta W_{\text{geometry,CV}}$ expression given in the entry of WLC (Eq. 3-116). $\Delta W_{\text{geometry,CV}}$ is usually zero unless there are processing-related issues which cause $W_{\it eff,CV}$ to depend on geometry. When WLC, WWC, and WWLC assume their default value of zero such that $\Delta W_{\text{geometry,CV}} = 0$, then the value of WLN is inconsequential.

WMIN (default = 0, unit = m)   minimum channel width. This parameter is meant to specify the minimum channel width allowable for a given MOSFET parameter set. Combined with WMAX, LMIN and LMAX, WMIN is a convenient parameter to define the binning region for which the model is applicable (Section 1.12). It seems that this parameter is meant to specify the minimum channel width allowable for a given MOSFET model card. For example, if the circuit file used a $W = 0.5\,\mu\text{m}$ MOSFET as part of the circuit element while WMIN in the model card was specified as $1\,\mu\text{m}$, then the SPICE program would either abort or issue a warning. However, somehow the BSIM3 code does not actually check whether W given in a MOSFET device statement is larger than WMIN or not. Basically, WMIN is a parameter which produces no effect. It does not matter what the value for this parameter is.

WMAX (default = 1, unit = m)   maximum channel width. This parameter is meant to specify the maximum channel width allowable for a given MOSFET parameter set. Combined with WMIN, LMIN and LMAX, WMAX, is a convenient parameter to define the binning region for which the model is applicable (Section 1.12). It seems that this parameter is meant to specify the maximum channel width allowable for a given MOSFET model card. For example, if the circuit file used a $W = 30\,\mu\text{m}$ MOSFET as part of the circuit element while WMAX in the model card was specified as $20\,\mu\text{m}$, then the SPICE program would either abort or issue a warning. However, somehow the BSIM3 code does not actually check whether W given in a MOSFET device statement is smaller than WMAX or not. Basically, WMAX

3.2 ALPHABETICAL GLOSSARY OF BSIM3 PARAMETERS 271

is a parameter which produces no effect. It does not matter what the value for this parameter is.

WR (default = 1, unitless) exponent of the effective device width for the calculation of RDSW. It should be close to unity, between 0.9 and 1.1. The details of this parameter are described in the entry RDSW.

WW (default = 0, unit = m^{WWN}) coefficient of width dependence in the calculation of the dc channel-width offset. This parameter is similar to WL; they both modify the effective-channel width for I-V calculation, W_{eff}. However, WL is a parameter that accounts for the length variation, whereas WW accounts for the width variation. See Eqs. (3-41) and (3-115), listed in the entry WL. Unless there is processing-related issues which cause the effective channel width to depend on the device geometry, WW should be zero. WW's counterpart in C-V calculation is WWC.

WWC (default = 0, unit = m^{WWN}) coefficient of width dependence in the calculation of the C-V channel-width offset. This parameter is similar to WLC; they both modify the effective-channel width for C-V calculation, $W_{\text{eff,CV}}$. However, WLC is a parameter that accounts for the length variation, whereas WWC accounts for the width variation. See Eqs. (3-43) and (3-116), listed in the entry WLC. Unless there are processing-related issues which cause the effective channel width to depend on the device geometry, WWC should be zero.

WWC's counterpart in I-V calculation is WW. WWC is a new parameter introduced in BSIM3v3.2. In BSIM3v3.1, the WWC appearing in Eq. (3-116) is replaced by WW.

WWL (default = 0, unit = m$^{\text{WWN+WLN}}$) coefficient of length and width dependence in the calculation of the dc channel-width offset. This parameter is similar to WL; they both modify the effective-channel width for I-V calculation, W_{eff}. However, WL is a parameter that accounts for the length variation, whereas WWL accounts for the product of the length and width variation. See Eqs. (3-41) and (3-115), listed in the entry WL. Unless there are processing-related issues which cause the effective channel width to depend on the device geometry, WWL should be zero. WWL's counterpart in C-V calculation is WWLC.

WWLC (default = 0, unit = m$^{\text{WWN+WLN}}$) coefficient of length and width dependence in the calculation of the C-V channel-width offset. This parameter is similar to WLC; they both modify the effective-channel width for C-V calculation, $W_{\text{eff,CV}}$. However, WLC is a parameter that accounts for the length variation, whereas WWLC accounts for the product of the length and width variation. See Eqs. (3-43) and (3-116), listed in the entry WLC. Unless there are processing-related issues which cause the effective channel width to depend on the device geometry, WWLC should be zero.

WWLC's counterpart in I-V calculation is WWL. WWLC is a new parameter introduced in BSIM3v3.2. In BSIM3v3.1, the WWLC appearing in Eq. (3-116) is replaced by WWL.

WWN (default = 1, unitless) power exponent of the width dependence in the calculation of the I-V and C-V channel-width offsets. BSIM3 calculates two effective channel widths, W_{eff}, which is used for I-V calculation, and $W_{\mathit{eff},CV}$, which is used for C-V calculation. WWN affects both of them. The effective channel width for I-V calculation can be expressed as: $W_{\mathit{eff}} = W - 2 \cdot \mathtt{WINT} - 2 \cdot \Delta W_{\mathrm{geometry}} - 2 \cdot \Delta W_{\mathrm{bias\_dependency}}$. WWN is the exponent of $1/W$ in the $\Delta W_{\mathrm{geometry}}$ expression given in the entry WL (Eq. 3-115). $\Delta W_{\mathrm{geometry}}$ is usually zero unless there are processing-related issues which cause W_{eff} to depend on geometry. When WL, WW, and WWL assume their default value of zero such that $\Delta W_{\mathrm{geometry}} = 0$, then the value of WWN is inconsequential.

The effective channel length for C-V calculation can be expressed as: $W_{\mathit{eff},CV} = W - 2 \cdot \mathtt{DWC} - 2 \cdot \Delta W_{\mathrm{geometry},CV}$. WWN is the exponent of $1/W$ in the $\Delta W_{\mathrm{geometry},CV}$ expression given in the entry WLC (Eq. 3-116). $\Delta W_{\mathrm{geometry},CV}$ is usually zero unless there are processing-related issues which cause $W_{\mathit{eff},CV}$ to depend on geometry. When WLC, WWC, and WWLC assume their default value of zero such that $\Delta W_{\mathrm{geometry},CV} = 0$, then the value of WLN is inconsequential.

XJ (default = 1.5 × 10⁻⁷, unit = m) source/drain junction depth. This parameter is used in the calculation of the default values of CGS0 and CGD0, as well as the following intermediate variables: A_{bulk} (which quantifies the amount of bulk-charge effects), V_A (the Early voltage). Even though XJ is a very important parameter for process control, XJ is a weak BSIM3 parameter. Most often, CGD0 and CGS0 are specified; hence, the value of XJ does not affect the overlap capacitance. The impact of XJ on the aforementioned intermediate variables is also small. Take A_{bulk} of Eq. (3-6), for example. Even if XJ's value were input incorrectly, other parameters such as A0 and AGS can be optimized (varied) to absorb the effects of XJ.

XPART (default = 0, unitless) charge partition flag. This parameter chooses the desired charge partitioning scheme through which the channel charge is divided into the drain and the source charges. As detailed in Section 1.7, the transient terminal currents in the MOS transistor are related to the time derivatives of the terminal charges. It turns out that, for MOS, it is rather easy to determine the gate and bulk charges, as well as the channel charge which is the sum of the drain charge (Q_D) and the source charge (Q_S). However, it is not straightforward to establish the individual drain charge and source charge from the overall channel charge. XPART is the parameter used to specify the desired manner to partition the channel charge. Although producing absolutely no effect as far as dc characteristics are concerned, XPART has major impact on the frequency dependencies of the device transconductance and input-referred noise (see Sections 4.4 and 4.14).

When XPART = 0 (or any value below 0.5 but not less than 0, such as 0.4), BSIM3 uses the so-called 40/60 charge partition scheme, in which, when the transistor is in saturation, 40 % of the channel charge is assigned to be Q_D while 60 %, Q_S. The Q_D/Q_S ratio is strictly 40/60 only during the saturation operation region. In linear region, the charge ratio is not 40/60, and depends on the bias condition. When XPART = 1 (or any value above 0.5), the 0/100 partitioning scheme applies. Just as in the 40/60 scheme, the charge ratio in the 0/100 scheme is 0 only in saturation. Finally, when XPART = 0.5, the 50/50 partitioning scheme applies. Unlike the previous two schemes, the 50/50 scheme truly divides the channel charge equally between the source and drain charges, regardless of whether the device is in linear or saturation region. If XPART is mistakenly given a negative value, there will be no charge calculation and the transient (or high-frequency) analysis will be incorrect. For PMOS device, XPART should still be between 0 and 1.

The availability of the various partitioning schemes often leads people to believe that the amount of channel charge partitioning is arbitrary. However, as detailed in Section 1.7, the moment we make a quasi-static assumption in the device operation, the proper partitioning scheme is 40/60. Other partition schemes, such as the 0/100 and 50/50 schemes, lack strong physical basis (i.e., they are based from hand-waving arguments and not derivable from device physics). A question naturally arises. If the 40/60 partition is the only one with physical basis, why do we ever need the 0/100 and the 50/50 partitioning schemes? It turns out that the quasi-static approximation embodied in all the charge partition schemes fails when the time scale is comparable to the transit delay time of the transistor. When a transistor is given a step ramp in the gate voltage, then the output voltages simulated from all three quasi-static models are incorrect at the initial part of the transient. Particularly for the case of the 40/60 partition scheme, there is an unphysical initial rise in the output voltage, resulting in a bump feature (Section 4.1). Unfortunately, sometimes this bump can cause numerical difficulty during a SPICE simulation. A case and point is the Killer-NOR gate, discussed in Section 4.2. Although lacking a solid physical foundation, the 0/100 partition scheme becomes popular because it removes the bump problem associated with the 40/60 partition scheme. However, because 0/100 is a quasi-static model, its result is still inaccurate at the initial part of transient. In fact, because it lacks strong physical foundation, it gives a wrong estimation of the rise time even during the latter part of the transient (Section 4.1). Nonetheless, the avoidance of numerical nuisances is a great feature of the 0/100 model.

Figure 3-35 shows the calculated C_{dg} of a $L = 0.35\,\mu m$ NMOS, with XPART equal to 0, 1, and 0.5. The correct C_{dg} behavior is obtained with the 40/60 partition (XPART = 0), with a finite capacitance value at saturation, and increases in the linear region. The 0/100 partition scheme (XPART = 1) yields a zero C_{dg} when the transistor is in saturation. We will show in Section 4.2 that, because the C_{dg} in saturation is zero for XPART = 1, the 0/100 partition scheme is able to avoid the bump problem associated with the 40/60 scheme. It is rather ironic that while the 40/60 partition is physically based and calculates the right C_{dg} behavior, the partition scheme can lead to quite unphysical overall circuit results (at the initial part of the transient). Conversely, with wrong C_{dg} characteristics, the 0/100 partition somehow becomes more numerically robust during a circuit simulation.

The general guideline for XPART is the following. For rf modeling at the frequency domain, XPART should be set to 0 (40/60 partition) unless the frequency of concern is nearby the cutoff frequency of the device. Above the cutoff frequency, it is better to switch to XPART = 1, to avoid the incorrect voltage gain's dependence

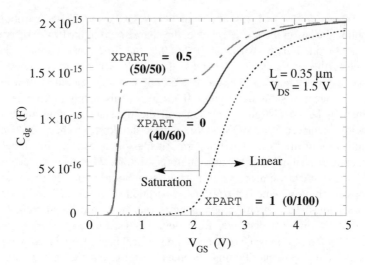

Fig. 3-35 Effects of XPART on BSIM3-calculated C_{dg}.

on frequency (see Sections 4.4 and 4.14). For transient studies, however, XPART should be set to 1 (0/100 partition) so that the unphysical bump feature found in 40/60 will never surface. The 50/50 partition scheme has no fundamental benefits over the other two schemes. It should not be used, other than for the debugging purpose.

XT (default = 1.55 × 10⁻⁷, unit = m) doping depth. It is the depth, from the oxide interface into the substrate, where the channel doping can be approximated to be NCH. Figure 3-18 indicates the position of XT. If either K1, K2, or VBX are specified in the SPICE parameter set, then XT becomes unused. Because very often K1 and K2 in the SPICE parameter set are specified, XT is usually left out of the parameter set.

In the event that neither K1, K2, nor VBX is specified, XT is used (and only used) to calculate the value of VBX, as shown in Eq. (3-104). VBX is then substituted into Eqs. (3-64) and (3-65) to determine the values for K1 and K2. VBX itself is a parameter. Therefore, when the parameter value for VBX is given, the value is directly used in Eqs. (3-64) and (3-65). XT, whether specified or defaulted to the value of 1.55 × 10⁻⁷ (m), then does not produce any effect.

XTI (default = 3.0, unitless) junction saturation current densities' temperature exponent. This parameter is used to model the temperature dependence of JS and JSSW, which are the saturation current densities used to calculated the diode current in the drain–bulk and the source–bulk junctions. For details, see the entry JS.

3.3 FLOW DIAGRAM OF SPICE SIMULATION

The BSIM3 program takes v_{BS}, v_{GS}, v_{DS}, and T_{device} as the input. After some calculation, the BSIM3 model outputs all the I-V and C-V characteristics. The I-V characteristics include the four terminal currents, as well as the small-signal quantities associated with the terminal currents. The C-V characteristics include the four terminal charges and the 16 device capacitances. The noise quantities are also calculated. We present the flow of the BSIM3 calculation in the following diagrams.

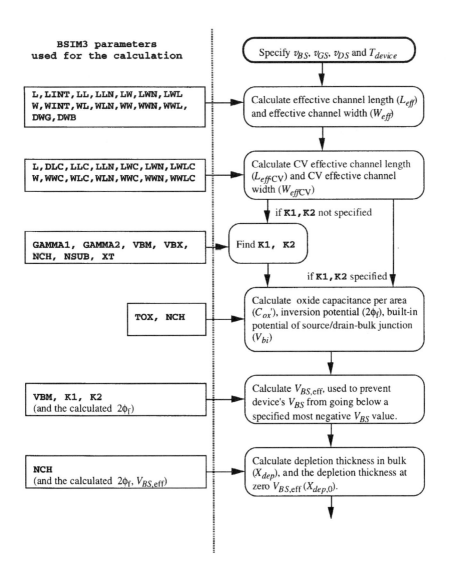

276 BSIM3 PARAMETERS

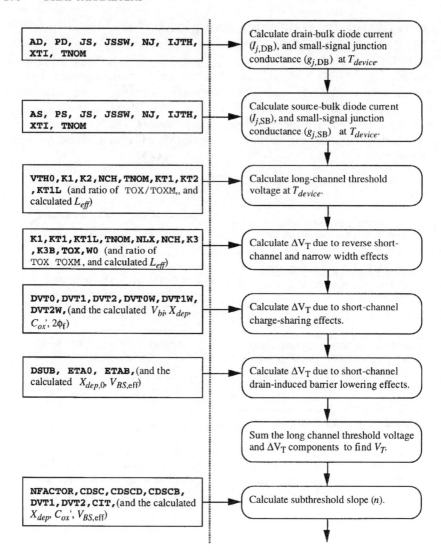

3.3 FLOW DIAGRAM OF SPICE SIMULATION

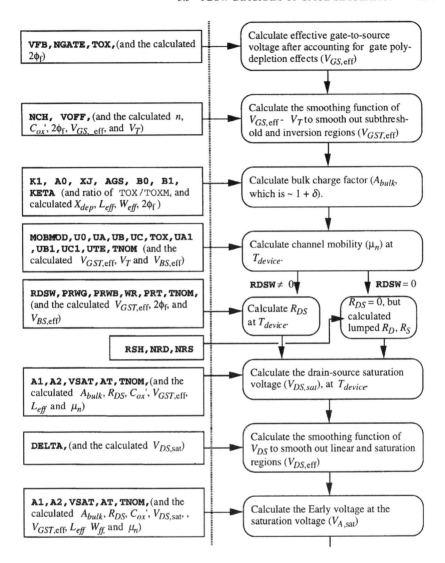

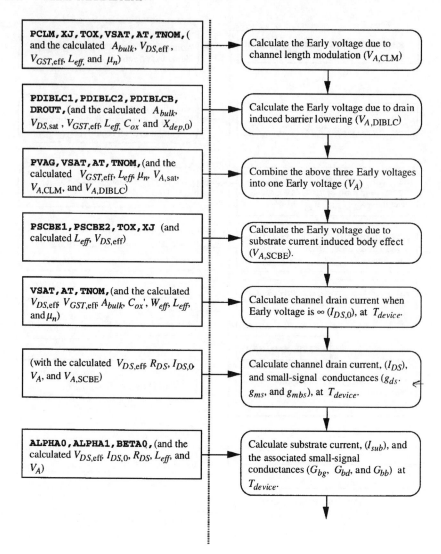

3.3 FLOW DIAGRAM OF SPICE SIMULATION

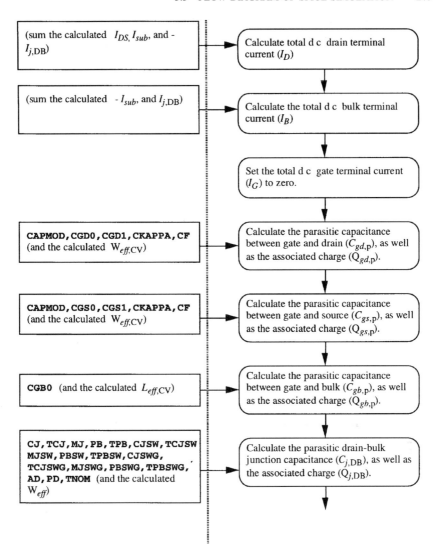

280 BSIM3 PARAMETERS

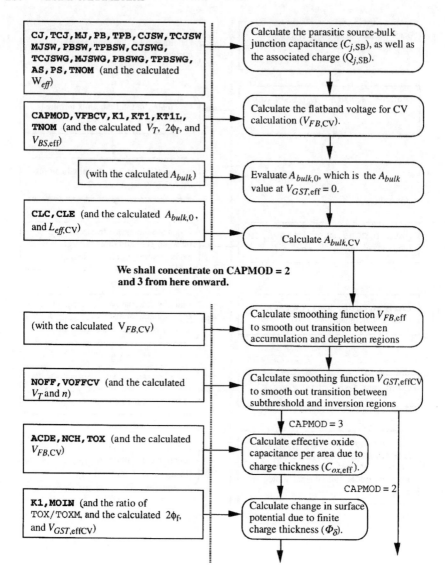

3.3 FLOW DIAGRAM OF SPICE SIMULATION

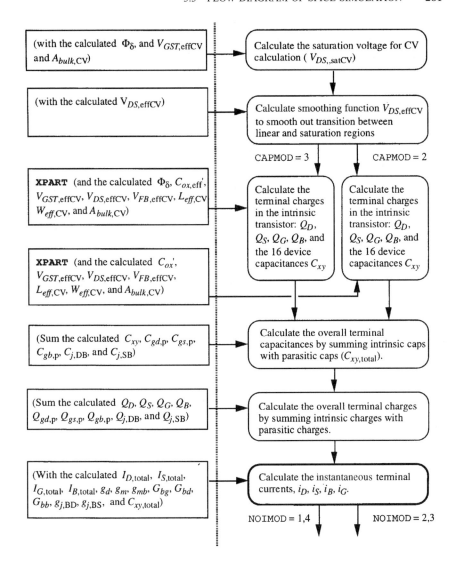

282 BSIM3 PARAMETERS

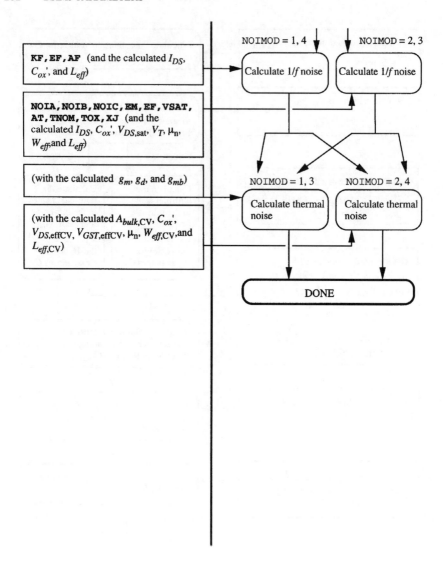

REFERENCES AND NOTES

[1] R. Rios, N. Arora, C. Huang, N. Khalil, J. Faricelli, and L. Gruber, "A physical compact MOSFET model, including quantum mechanical effects, for statistical circuit design applications," *IEEE International Electron Device Meeting*, pp. 937–940, 1995.

[2] W. Liu, X. Jin, Y. King and C. Hu, "An accurate MOSFET intrinsic capacitance model considering quantum mechanical effect for BSIM3v3.2," Memorandum No. UCB/ERL M98/4, University of California, Berkeley, July, 1998.

[3] S. Ramaswamy, A. Amerasekera, and M. Chang, "A unified substrate current model for weak and strong impact ionizatin in sub-0.25 µµ NMOS devices," *IEEE International Electron Device Meeting*, pp. 885–888, 1997.

[4] K. Suzuki, "Parasitic capacitance of submicrometer MOSFET's, *IEEE Trans. Electron Devices*, vol. 46, pp. 1895–1900, 1999.

[5] S. Sze, *Physics of Semiconductor Devices*, 2nd ed., New York: Wiley, 1981. For the discussion on carrier lifetime's temperature dependence, see p. 88. For the discussion on mobility's temperature dependence, see p. 28.

[6] Y. Tsividis, *Operation and Modeling of the MOS Transistor*, New York: McGraw-Hill, 1987. For the discussion on narrow-width effects, see Eq. (5.4.22).

[7] A. Sabnis, and J. Clemens, "Characterization of electron velocity in the inverted ⟨100⟩ Si surface," *IEEE International Electron Devices Meeting*, pp. 18–21, 1979.

[8] R. Rios, N. Arora, and C. Huang, "An analytic polysilicon depletion effect model for MOSFETs," *IEEE Electron Device Lett.*, vol. 15, pp. 129–131, 1994.

[9] D. Costa, and J. Harris, Low-frequency noise properties of Npn AlGaAs/GaAs heterojunction bipolar transistors," *IEEE Trans. Electron Devices*, vol. 39, pp. 2383–2394, 1992.

[10] M. Liang, J. Choi, P. Ko, and C. Hu, "Inversion-layer capacitance and mobility of very thin gate-oxide MOSFETs," *IEEE Trans. Electron Devices*, vol. 33, p. 409, 1986.

4

IMPROVABLE AREAS OF BSIM3

When his wife cooks dinner, Johnny will never criticize the taste of the recipes. Criticism is almost always irritating to his wife after she toils in the kitchen for so long, and for free. In the discussion about BSIM3, likewise, we want to set aside some space for praise. BSIM3 is a fine dish, a noncommercial product of the many graduate students (as well as professors) who have spent years perfecting. In fact, it is probably the best public-domain MOSFET model. According to the official BSIM3 website [1], companies such as Texas Instruments, AMD, Analog Devices, Chartered, Hyundai, IBM, Intel, Lucent, NEC, Samsung, Siemens, and TSMC are either internally using or offering BSIM3 models. The company endorsement is the most convincing testimony of the quality of BSIM3. The whole BSIM3 team ought to be commended for their dedicated works. The entire source code of BSIM3 can be downloaded from the Internet [1].

We have no intention to spoil the dinner here, either. We humbly point out some of the more problematic areas of BSIM3, so that we, as BSIM3 users, are better informed in assessing the simulation result. In each of the following sections of this chapter, we describe a particular area of improvement for BSIM3.

Finally, it is really not our intention to single out BSIM3. Almost all of the problems discussed in this chapter are common to other compact models. We are confident that the BSIM development team is well aware of these problems. With the available resources, they had dealt with the previously more serious problems, and will continue to resolve issues which are pointed out this chapter. Some of these problems are, in fact, addressed in BSIM4, the next generation of the BSIM model beyond BSIM3. We will highlight the modifications of BSIM4 in Chapter 5.

4.1 LACK OF ROBUST NON-QUASI-STATIC MODELS: TRANSIENT ANALYSIS

Non-quasi-static (NQS) modeling is significantly more difficult than quasi-static (QS) modeling. Since quasi-static approximation is good enough for most situations, BSIM3 spends less effort on the NQS model development, in comparison to its QS model. As a consequence, BSIM3's NQS model is less robust than its QS counterpart. A great majority of simulation which uses BSIM3 is performed with its QS model. In this section, we discuss mostly about BSIM3's QS model. We will touch upon BSIM3's NQS model toward the end of this section.

A good rule-of-thumb in deciding whether NQS effects are important is to compare the time scale of interest with the transistor's transit time. The quasi-static approximation assumes that, as the external bias voltages vary with time, the channel charges arrange themselves instantaneously to reach their *steady-state* profile (Section 1.6). When the time scale is short (as compared to the transit time of charges from source to drain, or roughly the inverse of the cutoff frequency), the quasi-static approximation would be violated because the charge definitely has not reached their steady state profile in such a short amount of time. A classic example, discussed in Section 1.6, involves applying a step voltage to the gate to invert the channel. Realistically, as the gate voltage switches to high, electron charges from the source start drifting toward the drain. We estimated the transit time τ_{tr} of the carrier movement from source to drain to be that given in Eq. (1-35). In a QS approximation, however, the steady-state charge profile is assumed to be established instantaneously, at the very moment the gate voltage changes from low to high. A QS model therefore predicts a finite drain current even at $t < \tau_{tr}$. Realistically, the drain current should remain 0 in that duration, and becomes nonzero only after the charges from the source reach the drain.

Equation (1-35) states that the transit time is at the minimum when the transistor operates in saturation. We will thus calculate the transit time when $\alpha = 0$. Considering as an example a $L = 5\,\mu\text{m}$ device in a 3 V technology with a V_T of 0.5 V, ω_0 is found from Eq. (1-36) to be 5×10^9 rad/s (assuming $\mu_n = 500$ cm^2/V-s). τ_{tr}, equal to the inverse of ω_0, is about 2×10^{-10} s. This means that if there is a ramp voltage changing values between two extrema with a time shorter than 2×10^{-10} s, then this $L = 5\,\mu\text{m}$ device would exhibit severe NQS effects. If the ramp time is longer than the transit time, the QS approximation becomes progressively more valid as the ramp time increases. Figure 4-1 illustrates the inverter transient behaviors simulated with NQS and QS simulators. The NMOS inverter is connected to a load capacitor initially charged to the supply voltage, equal to 3 V in this example. The load capacitance, which is basically the capacitance seen in the next inverter stage, is taken to be equal to WLC'_{ox} of the NMOS inverter. The NQS results of Fig. 4-1 is provided by MEDICI, a device simulator that solves the current continuity and the Poisson equation simultaneously, and exactly (within the numerical tolerance). We use BSIM3's QS model to represent the QS results in Fig. 4-1. The BSIM3 QS model allows user to specify the channel charge partition to the drain and the source. If the parameter XPART is set to 0.4, then the so-called

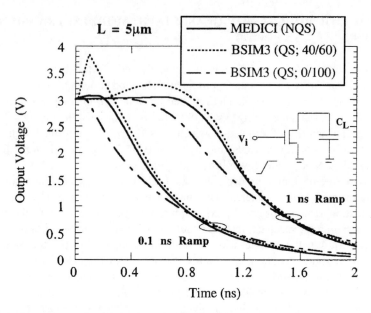

Fig. 4-1 Inverter transient behaviors simulated with NQS (MEDICI) and QS (BSIM3) analyses. The circuit connection is shown as an inset. $L = 5\,\mu m$.

40/60 partition is used in the QS model. If XPART is taken to be 0, then the partition is 0/100. We discussed in Section 1.7 that once the quasi-static approximation is assumed, then the 40/60 partition is the physically correct partition scheme. The 0/100 partition does not make much physical sense, but it has a unique advantage, to be discussed shortly.

The transit time for the $L = 5\,\mu m$ NMOS was calculated to be 0.2 ns. Figure 4-1 shows simulation under two voltage ramps; one increases from 0 to 3 V in 1 ns, and another, 0.1 ns. Compared to the NQS (the exact) MEDICI solution, the QS 40/60 solution gives fairly accurate fall time in the 1 ns ramp, but the accuracy degrades slightly as the ramp time decrease to 0.1 ns. The good agreement between the QS 40/60 model and the NQS results indicates that 40/60 is the physical partition scheme. This partition scheme, nonetheless, suffers from gross discrepancy with the NQS result at the initial portion of the transient. This is understanable because the very premise of the quasi-static approximation is invalid during the initial part of the transient. The 40/60 QS simulation displays a bump, whose magnitude becomes increasingly larger as the ramp time falls shorter than the transit time. This bump in the capacitor voltage is nonphysical, and has been shown to be inconsistent with the energy conservation principle. We examine the fundamental cause of the bump by first writing out the drain current used in BSIM3:

$$i_D(t) = I_D - C_{dg}\frac{dv_{GS}}{dt} + C_{dd}\frac{dv_{DS}}{dt} - C_{db}\frac{dv_{BS}}{dt}. \quad (4\text{-}1)$$

We neglected second-order effects such as the impact ionization; the exact equation was previously given in Eq. (2-25b). We refer I_D as the dc current and the remaining

4.1 LACK OF ROBUST NON-QUASI-STATIC MODELS: TRANSIENT ANALYSIS

three terms as the capacitive currents. We use intrinsic capacitances rather than $C_{dg,t}$, $C_{dd,t}$ and $C_{db,t}$ because we want to focus on the origin of the unphysical bump, which is due to the QS approximation used in the intrinsic transistor. The parasitic capacitances will contribute slightly to the magnitude of the bump, but their contribution are physical and will be addressed separately after the present discussion. Initially when $v_{GS} < V_T$, the transistor is not yet turned on. All of I_D and C_{xy}'s in Eq. (4-1) are zero. With drain current being zero, the output voltage then maintains relatively constant. The moment v_{GS} exceeds V_T, the transistor is considered on, although $v_{GS} - V_T$ is still small. Since $v_{GS} - V_T$ is negligible in the initial part of transient, we can drop I_D out of the equation. Moreover, $C_{db} \cdot dv_{BS}/dt = 0$ because v_{BS} does not change with time, and C_{dd} is 0 because the transistor is in saturation. Equation (4-1) therefore degenerates into: $i_D(t) \approx -C_{dg} \cdot dv_{GS}/dt$. In the 40/60 partition, C_{dg} in saturation region is equal to $4/15 \cdot WLC'_{ox}$, a finite positive number (see Appendix B, Eq. B-9). Because $v_{GS}(t)$ increases with time, $i_D(t)$ is negative, meaning that the drain current exits the device and goes externally toward the load capacitor. This current charges up the capacitor to a value larger than its initial value, resulting in the bump. After a while, v_{GS} eventually exceeds V_T by a certain amount such that the dc component I_D becomes appreciable. This I_D is a positive drain component, which soon exceeds the negative capacitive current. Once this happens, the overall drain current becomes positive. The MOSFET then acts to discharge the capacitor, as it should in an inverter. The output voltage decreases after a certain time.

The above description points out the inherent reason for the bump in a 40/60 QS model. In many inverter-type simulations, the input voltage switches from L (low) to H (high) while the NMOS' drain is initially at H. Because the transistor is initially in saturation, C_{dg} is positive and the capacitive component of the QS drain current (Eq. 4-1) is negative. During the portion of time when $v_{GS} - V_T$ is still small so that the dc component I_D is nearly zero, the overall drain current is then negative. This negative current charges up the output node capacitance, resulting in the bump. The bump disappears only after some time elapses so that $v_{GS} - V_T$ is appreciable and I_D is large.

A comparison of C_{dg} in 0/100 (XPART = 0) and 40/60 (XPART = 0.4) was shown in Fig. 3-35. In the 0/100 partition scheme, C_{dg} is zero during saturation rather than being finite. The bump problem caused by the negative drain current in the 40/60 model therefore disappears. The paper which introduced the 0/100 partition scheme, which is also the foundation of all of the charge models, noted that the 0/100 partition eliminates spikes in transient simulation, although the model is not necessarily physical (see Ref. [1] of Chapter 1). In a 50/50 partition scheme, C_{dg} is positive in saturation just like the 40/60 scheme. Hence, the bump problem is also present in the 50/50 partition.

The large bump due to the 40/60 partitioning in some SPICE simulator can cause numerical problems, or at least cause unphysical voltage values in a particular time interval during which the external voltage sources switch levels rapidly. Therefore, sometimes the nonphysical 0/100 partition scheme is used for simulation. In Section 4.2 we will discuss a so-called Killer NOR-Gate circuit which accentuates the bump problem of the 40/60 partition. As shown in Fig. 4-1, although the 0/100 scheme does not have the bump problem during the initial portion of the transient, it gives rise to an erroneous waveform.

If we examine carefully, and particularly at the 0.1 ns ramp time, we see that even the NQS MEDICI simulation exhibits an initial increase of voltage across the capacitor to values above 3 V. This relatively small increase, in contrast to the bump of the 40/60 QS model, is physical. This increase in voltage is associated with the parasitic overlap and fringing capacitances between the gate and drain terminal, $C_{ov} + C_f$. Figure 4-2 captures the situation at the initial portion of the transient. When $t \sim 0$, the transistor is not turned on yet, and both the gate and the drain current flows in the transistor are insignificant. As v_{GS} ramps up, it sends out a transient current as indicated by the arrow nearby the input voltage source. The current flows toward the drain node of the transistor. (In the BSIM3 QS model, the transistor is assumed to reach its QS state the moment transistor turns on. Although I_D is still nearly zero initially, there is a negative capacitive drain current proportional to C_{dg} which flows out of the transistor. This additional and unphysical current does not exist in the MEDICI simulation because MEDICI is a NQS simulator.) But since the transistor is off, the current originating from the voltage source has no place to go but to the loading capacitor. Consequently, the output voltage increases. This is understandable by writing the Kirchoff's current law at the output node:

$$(C_{ov} + C_f) \frac{d(v_{GS} - v_{DS})}{dt} = C_L \frac{d\, v_{DS}}{dt}. \qquad (4\text{-}2)$$

We rearrange the equation, obtaining

$$\frac{d\, v_{DS}}{dt} = \frac{C_{ov} + C_f}{C_{ov} + C_f + C_L} \times \frac{d\, v_{GS}}{dt} \qquad (v_{GS} < V_T). \qquad (4\text{-}3)$$

Since v_{GS} increases with time, v_{DS} also increases with time (when the transistor is not yet turned on). Once the transistor turns on, however, the above equations neglecting the MOSFET's transistor current are no longer valid. As the transistor turns on and sinks down the charges accumulated in C_L, the output voltage decreases.

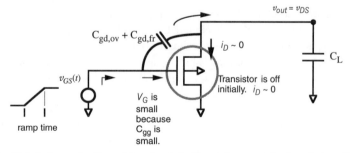

Fig. 4-2 Detailed examination of the circuit in Fig. 4-1 to explain the relationship between the bump in voltage and the parasitic capacitances between the gate and drain terminal.

4.1 LACK OF ROBUST NON-QUASI-STATIC MODELS: TRANSIENT ANALYSIS

We have made similar simulations at other ramp times and with a different value of the loading capacitor. We define the error of the QS 40/60 model as $(V_{peak,QS} - V_{peak,NQS})/V_{peak,NQS}$, where V_{peak} denotes the peak of the bump. Figure 4-3 shows the error as a function of the ramp time for the $L = 5\,\mu m$ device.

Incidentally, meticulous care has been taken so that the comparison of MEDICI and BSIM3 simulations are meaningful. (This is also true for other similar comparisons discussed in this section.) To ensure this, we extract the SPICE parameters using the following method. A MEDICI grid for a particular device length is created using realistic doping profiles and dielectric structures obtained from process simulators. The dc current–voltage characteristics and parasitic capacitances are generated at various drain-to-source, gate-to-source, and bulk-to-source biases. An optimization routine is then used to extract SPICE parameters so that the simulated device characteristics are well reproduced by BSIM3. We basically followed the same procedure to extract SPICE model parameters from the data, except that this time the data were not measured from silicon but simulated from MEDICI.

Figures 4-1 and 4-3 reveal that, the shorter the ramp time, the worse the quasi-static approximation becomes. So, exactly what is the duration of the ramp time beneath which we need to worry about the non-quasi-static effects? A while ago we used the guideline provided by the transit time given in Eq. (1-35) with α set to 0. We caution, however, that the equation was derived without the consideration of velocity saturation, a phenomenon quite frequently encountered in short-channel devices. Because of the excruciating difficulty in a proper derivation of the transit time in velocity saturated devices, we take on a practical approach. We simply use the MOS model (which includes all the parasitic capacitances and resistances) from a given

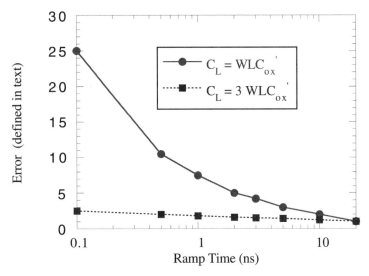

Fig. 4-3 Error of the QS 40/60 model, defined as $(V_{peak,QS} - V_{peak,NQS})/V_{peak,NQS}$, as a function of the ramp time for a $L = 5\,\mu m$ device.

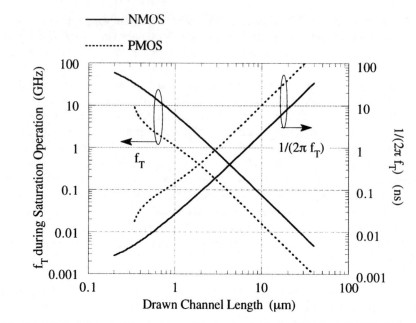

Fig. 4-4 Calculated cutoff frequency and its inverse for both NMOS and PMOS. If the ramp time is longer than $1/(2\pi \cdot f_T)$, then the QS approximation is considered valid.

node of CMOS technology and use BSIM3 to calculate the cutoff frequency as a function of channel length. For all of the calculation, the device is operated in saturation, with $V_{GS} = 1.5$ V and $V_{DS} = 3$ V. Figure 4-4 illustrates the calculated results for both NMOS and PMOS. If the ramp time is longer than $1/(2\pi \cdot f_T)$, then the QS approximation is considered valid, and vice versa.

According to Fig. 4-1, the $L = 5\,\mu m$ NMOS exhibits NQS effects if the ramp time is 0.1 ns. By modern technology standard, a 0.1 ns seems fairly long. We are tempted to believe that NQS effects are severe only in the past technologies, such as in the early 1980s when, in fact, the minimum geometry was 5 µm. However, in that particular technology node, the fastest transistor is also a 5 µm MOS. There is no component which could possibly produce a signal with a ramp time shorter than 0.1 ns. With a 5 µm transistor feeding signals into 5 µm transistors, the ramp time of the bias voltage at any given transistor is longer than the 5 µm transistors' transit time. Therefore, QS approximation is quite accurate for the past technologies, even though the channel length is long by the present standard. Similarly, in modern-day digital circuits, most of the devices are the minimum geometry device. With a 0.18 µm transistor feeding signals into 0.18 µm transistors, the ramp time of the bias voltage at any given transistor is longer than the 0.18 µm transistors' transit time. Hence, QS approximation is valid.

QS approximation would likely fail in analog circuits which use devices of a mixture of lengths. Suppose somehow a $L = 0.18\,\mu m$ transistor is feeding the signals to a $L = 5\,\mu m$ device. If the 0.18 µm is designed to operate at its top speed, then its transit time from Fig. 4-4 is about 2.5 ps. The ramp time of the signal provided to the

4.1 LACK OF ROBUST NON-QUASI-STATIC MODELS: TRANSIENT ANALYSIS

5 μm device can then be on the order of 2.5 ps. In this case, we would expect significant NQS effects in the 5 μm device.

We repeat the inverter simulation with $L = 0.35$ μm, instead of 5 μm as in Fig. 4-1. The simulation results are shown in Fig. 4-5. From Fig. 4-4, the 0.35 μm NMOS has a transit time of 5 ps. Figure 4-5 shows that, even with a ramp time as short as 10 ps, both the 0/100 and 40/60 BSIM3 results agree well with the MEDICI simulation, although the agreement is better for the 40/60 partition scheme. One reason that NQS effects become less noticeable as channel length decreases is that, in shorter channel devices, the parasitic capacitances are a bigger portion of the overall device capacitances (sometimes up to 50%). The transient behavior can be dictated by the parasitic capacitances, and to a lesser degree, by the intrinsic device capacitances. Therefore, the accurate modeling of the intrinsic device itself becomes less critical.

A quick fix to model the NQS effects with BSIM3's QS models is to break to the long-channel device into several shorter channel devices. This is shown in the inset of Fig. 4-6. The figure itself is a comparison between the QS simulation of a 5 μm transistor by itself and the QS simulation of 160 instances of $L = 0.03125$ μm transistors connected in series. The inner subtransistors are assumed to have zero source and drain areas. Each of the 160 subtransistors, now with a channel length of only 0.03125 μm, has a transit time much smaller than the input ramp time of 1 ns. Therefore, the NQS effects disappear in the 160 serially connected subtransistors, and the bump at the initial portion of the transient is gone.

The simulation results of Fig. 4-6 (whether with one whole transistor or 160 subtransistors) are both from BSIM3 40/60 QS models. Being able to account the NQS effects by simply sectioning the transistors into serially connected subtransistors, we might be tempted to conclude that NQS modeling is not needed. This is not

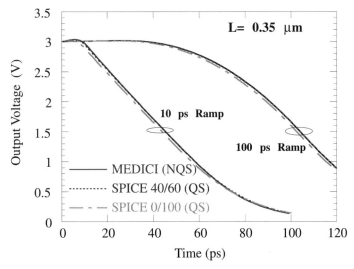

Fig. 4-5 Inverter simulation of Fig. 4-1, except the transistor has a $L = 0.35$ μm, instead of 5 μm.

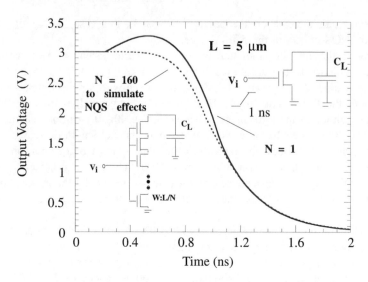

Fig. 4-6 Comparison between the QS simulation of a 5 μm transistor by itself and the QS simulation of 160 instances of $L = 0.03125$ μm transistors connected in series.

a correct attitude. There are several drawbacks associated with breaking up the transistor. First, the short-channel effects must be removed from each of the subtransistors. Otherwise, with its ultrashort channel length, the threshold voltage correction for the subtransistor would be enormous. Next, ΔL used to correct the actual channel length should be set to 0. Otherwise, the subtransistor may likely have a negative effective channel length. Finally, the parameters relating to junction areas, body effect, and subthreshold current need to be properly adjusted so that the subtransistor behaves normally as a transistor. A lot of care needs to be exercised when the transistor-breakup approach is used. It is easy to conceive cases where the breakup approach would be difficult to be implemented in practice.

For completeness, the transistor-breakup technique is applied to the $L = 0.35$ μm transistor. We see from Fig. 4-7 that, whether $N = 1$ or 160, the results simulated with BSIM3's QS models almost overlay each other. This is an indication that, with the ramp time of 10 ps used in the simulation, the NQS effects are insignificant. This finding is consistent with the results of Fig. 4-5, which shows that BSIM3's 40/60 QS results match fairly closely to the NQS MEDICI simulation.

So far we have concentrated on the NQS effects observed in inverter circuits. Another type of circuit commonly cited to have significant NQS effects is the passgate circuit, shown in Fig. 4-8. When the gate voltage is high (usually one threshold voltage above the highest drain/source voltage), the fact that the drain current is practically nil requires v_{DS} be approximately zero. Hence, v_{OUT} is equal to v_{IN}. Let us suppose v_{IN} is tied to a 0 V source voltage. When the gate voltage is high, v_{OUT} also reaches the value of 0. As the gate voltage ramps down, we perceive v_{OUT} to continue to be equal to v_{IN}, which is still held at 0 V. However, in practice, v_{OUT} tends to be slightly negative. The amount of difference between v_{OUT} and v_{IN} is

4.1 LACK OF ROBUST NON-QUASI-STATIC MODELS: TRANSIENT ANALYSIS   293

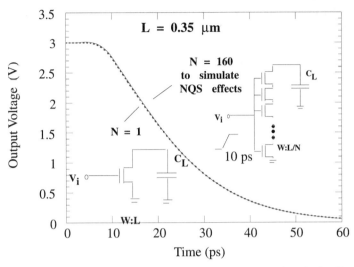

Fig. 4-7 Comparison between the QS simulation of a 0.35 μm transistor by itself and the QS simulation of 160 instances of $L = 0.0021875$ μm transistors connected in series.

termed the *error voltage*. The error voltage is highest in our example in which v_{IN} is tied to 0 V. When v_{IN} is tied to higher voltage values, the error voltage is smaller [2]. We caution that the error voltage exists even if there is no leakage current which dissipates the charges in the capacitor. Indeed we neglect the leakage current in our analysis by properly adjusting the device parameters.

Let us think through what should happen to the circuit as the gate voltage ramps down. When the gate voltage is high (H) in the steady-state condition, v_{IN} drives the v_{OUT}; that is, v_{OUT} approaches the value of v_{IN}, but never exceeds it. Since $v_{IN} > v_{OUT}$, we label v_{IN} as the drain node and v_{OUT} as the source node. As we shall illustrate, the transient values of v_{OUT} during the gate voltage ramp down are also below v_{IN}. Hence v_{IN} is still appropriately named the drain node during the voltage ramp down as well. We base our analysis on the equivalent large-signal model of Fig. 2-16. There are several simplifying assumptions which are valid here. First, while v_{OUT} is not identical to v_{IN}, their difference is small, on the order of 1 mV to 100 mV. With $V_{DS} \sim 0$, we can approximate I_D to be zero. Likewise, C_m, equal to $C_{dg} - C_{gd}$, is also about zero. (These are crude approximations, but we will

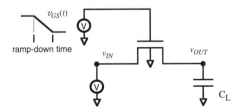

Fig. 4-8 An exemplar passgate circuit which shows significant NQS effects.

do so in order to get a glimpse of the circuit operation.) Second, none of the drain–bulk and the source–bulk junctions are forward-biased. The diode currents are small enough to be neglected. Third, the substrate current associated with impact ionization near the drain is insignificant. Fourth, we make the parasitic terminal source/drain resistances zero. Fifth, we neglect the parasitic gate-to-bulk capacitance due to the crossing of the gate and metal over the field oxide. Lastly, we focus on the voltage transient at the gate, and consider dv_{DB}/dt and $dv_{SB}/dt = 0$. The resulting equivalent circuit pertaining to the passgate circuit is given in Fig. 4-9.

C_{sd} is negative in linear operating region. (It is possible to have negative capacitance in a transistor.) However, because the difference between v_{IN} and v_{OUT} is small, we can obliterate C_{sd} from Fig. 4-9. After this is done, v_{OUT} can be determined from summing the branch currents at the output node:

$$(C_L + C_{j,SB}) \frac{d\,v_{OUT}}{dt} = C_{gd,t} \frac{d(v_{GS} - v_{OUT})}{dt}, \tag{4-4}$$

from which we obtain

$$\frac{dv_{OUT}}{dt} = \frac{C_{gd,t}}{(C_L + C_{j,SB} + C_{gd,t})} \times \frac{d\,v_{GS}}{dt}. \tag{4-5}$$

Since $dv_{GS}/dt < 0$, $dv_{OUT}/dt < 0$. Further, with $v_{OUT}(t=0) = 0$, v_{OUT} is negative as time progresses. Although Fig. 4-9 is based on a QS equivalent circuit and Eq. (4-5) thereby embodies the QS approximation, we expect the conclusion that v_{OUT} is negative to be correct. However, when the v_{GS} ramp-down time is short, NQS effects can occur and the magnitude of v_{OUT} may not be ascertained from our analyses.

Figure 4-10 shows the simulation results for the passgate circuit when the NMOS is 5 µm long. v_{IN} is held constant at zero and C_L is three times the device's oxide capacitance. From the previous studies on the inverter transient, we expect the ramp-

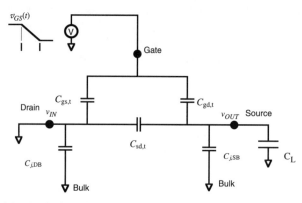

Fig. 4-9 Equivalent circuit of the passgate circuit shown in Fig. 4-8.

4.1 LACK OF ROBUST NON-QUASI-STATIC MODELS: TRANSIENT ANALYSIS

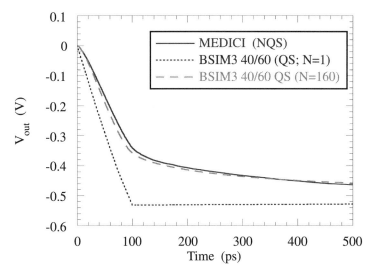

Fig. 4-10 Comparison of the QS (with and without dissecting the transistor into 160 elements) and NQS simulations for the passgate circuit. $L = 5\,\mu m$.

down time of the gate voltage, 0.1 ns, is short enough to dramatize the NQS effects. We see that when the QS analysis (BSIM3) is applied to the transistor as a whole, the resulting error voltage (v_{OUT}) deviates significantly from the NQS simulation (MEDICI). As we apply the transistor-breakup technique and divide the 5 μm into 160 equal sections, then the same QS analysis now approaches the NQS results. These findings all point to the importance of NQS effects in this circuit.

Figure 4-11 is another passgate circuit simulation, except that this time the transistor length is 0.35 μm and the gate voltage ramp down time is 10 ps. The figure again shows that the transistor-breakup technique with $N = 160$ gives nearly the same results as the NQS simulation. This time, due to the fact that parasitic capacitances are a significant portion of a shorter-channel device, the QS analysis with $N = 1$ does not produce significant error compared to the NQS results.

We have described the transistor-breakup technique as a quick fix to account for NQS effects. We also mentioned that, while the technique is beneficial in illustrating the degree of NQS effects, the technique is often difficult to be implemented in practice. The ultimate cure to properly model the NQS effects is not by breaking up the transistor, but by devising a NQS model. BSIM3 has a NQS model, allowing a circuit designer to perform NQS simulation when the NQSMOD parameter is set to 1.

We compare BSIM3's NQS simulation with the exact NQS solution obtained from numerical analysis. We consider a NMOS whose drain is held constant at V_o. The gate voltage increases from V_T at t (time) equal zero to $V_o + V_T$ at the end, where V_T is the threshold voltage. The rate of gate voltage ramp is $\theta V/s$. The

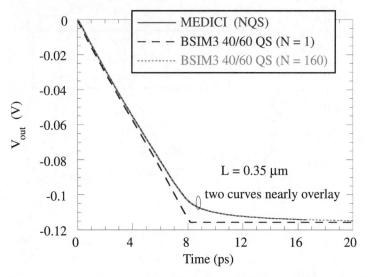

Fig. 4-11 Comparison of the QS (with and without dissecting the transistor into 160 elements) and NQS simulations for the passgate circuit. $L = 0.35\,\mu\text{m}$.

substrate is grounded. In order to allow the graphs to be applicable to all channel lengths and various scales of time, we report the results in normalized variables [3]:

$$t_{normalized} = \frac{\mu_n V_o}{(1+\delta)L^2} t; \qquad (4\text{-}6)$$

$$\theta_{normalized} = \frac{(1+\delta)L^2}{\mu_n V_o^2} \theta; \qquad (4\text{-}7)$$

$$i_{D,normalized} = \left[\frac{W}{L}\mu_n C'_{ox}\frac{V_o^2}{(1+\delta)}\right]^{-1} \times i_D; \qquad (4\text{-}8)$$

where δ is the bulk-charge factor whose value can be identified from Fig. 1-18; and L is the effective channel length. The last equation for the normalized drain current can be similarly applied to other terminal currents. Figure 4-12 plots out the terminal currents with $\theta_{normalized} = 0.25$. For a $L = 5\,\mu\text{m}$ technology and a $\delta = 0.2$, a $\mu_n = 400\,\text{cm}^2/\text{V-s}$ and a $V_o = 2.5\,\text{V}$, we see from Eq. (4-7) that, in the real time domain, $\theta = 0.25 \cdot 400 \cdot (2.5)^2/1.2/(5\times 10^{-4})^2 = 2\times 10^9\,\text{V/s}$. Equivalently, $\theta = 2\,\text{V/ns}$.

Figure 4-12 illustrates the simulation results from numerical analysis and BSIM3's NQS model. The normalized ramp rate, 0.25, is equivalently to a θ of $2\,\text{V/ns}$. We found from the inverter transient study that the transit time is about $0.2\,\text{ns}$ for this transistor. We therefore expect both results to be quite close, and indeed they appear to be so. However, there are signs which signal some flaws in the BSIM3 NQS model. The foremost is that BSIM3 NQS model always produces a

4.1 LACK OF ROBUST NON-QUASI-STATIC MODELS: TRANSIENT ANALYSIS

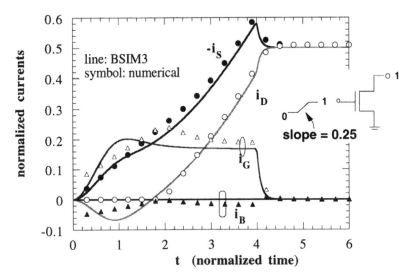

Fig. 4-12 Comparison of terminal currents simulated with BSIM3's NQS model and the exact NQS analysis (MEDICI). $L = 5\,\mu\text{m}$; $\delta = 0.2$; $\mu_n = 400\,\text{cm}^2/\text{V-s}$; $V_o = 2.5\,\text{V}$; and $\theta_{\text{normalized}} = 0.25$. (From Ref. [3], © IEEE 1996, reprinted with permission)

zero bulk current, no matter what the bias conditions are. In this rather slow transient, setting i_B to zero in BSIM3 is not too alarming because the bulk current simulated from the numerical analysis happens to be relatively small anyway. However, as the ramp slope increases, the improper modeling of the bulk current becomes progressively troublesome. This zero bulk current results from the fact that the BSIM3 NQS model is not as robust a physical model as its QS counterpart. Some arbitrary channel partition is needed (we used the NQS default of $\text{XPART} = 0.5$) to separate the channel charges into the source and drain charges. If the bulk charge is not assumed to be zero, then BSIM3 will need to create another parameter just to distribute a portion of the channel charge to the bulk. The second problem with BSIM3's NQS model is that the drain current can be negative at the initial part of the transient. The numerical solution, in contrast, correctly points out that the drain current is identically zero until the charge originating from the source reaches the drain. Although BSIM3 calculates a negative drain current, BSIM3 authors understand this result is unphysical. Therefore, a slight change of coding is made to BSIM3 such that the outputted drain current is made zero if it is calculated to be a negative value.

Many designers, particularly those who never worked on NQS MOSFET modeling, are unforgiving about arbitrarily setting the drain current to zero. There is certainly a good reason for feeling this way. After all, with this kind of arbitrariness, there will be many scenarios, such as dynamic boundary conditions in which drain voltage oscillates between positive and negative values, wherein the actual solution is unknown. However, it seems that the non-zero drain current at the initial part of a transient is not an easily solvable problem. Reference [4] uses table-lookup model (Section 1.11) that considers the first-order correction to the NQS effects. However, the drain current is still found negative. They had to arbitrarily set i_D to zero in the initial part of the transient, too.

The last problem of BSIM3 NQS model, evident from Fig. 4-12, is that the gate current does not agree well with the numerical solution. The gate current simulated by BSIM3's NQS model always starts out being zero. The numerical solution indicates that the gate current is not zero initially. This problem also worsens at faster transient.

Figures 4-13 and 4-14 show the results of faster transients. In these figures, we followed BSIM3 NQS model's somewhat arbitrary fix in setting the drain current to zero when it is negative. Therefore, the drain current looks fairly close to the numerical simulation. However, the bulk and the gate currents reveal significant discrepancy. In addition, in these faster transients, the turn-on characteristics of the drain current is not too well modeled by BSIM3's NQS model.

Besides some of the inaccuracies pointed out previously, there is a bigger (and perhaps a bit academic) problem associated with the BSIM3 NQS model. There is a disconnect between BSIM3's NQS and QS models. The two models are developed from fairly different frameworks. Conceptually, we would expect that, when the time transient is slow enough, the solution obtained from BSIM3's NQS model should converge to that obtained from BSIM3's QS model. However, we doubt that this expectation will be fulfilled. After all, the fact that i_B is set to zero in the NQS model precludes a good equivalence in the QS and NQS models since $i_B \neq 0$ is still quite possible in the QS model.

The author does not have access to the BSIM3's NQS model through a SPICE simulator, so he had to reproduce the BSIM3 NQS model with his own Fortran

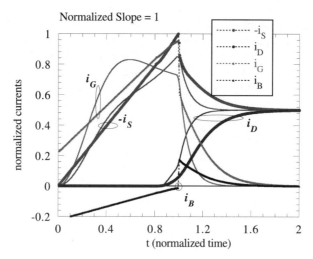

Fig. 4-13 Comparison of terminal currents simulated with BSIM3's NQS model and the exact NQS analysis (MEDICI). $L = 5\,\mu\text{m}$; $\delta = 0.2$; $\mu_n = 400\,\text{cm}^2/\text{v-s}$; $V_o = 2.5\,\text{V}$; and $\theta_{\text{normalized}} = 1$. (From Ref. [3], © IEEE 1996, reprinted with permission).

4.2 PROBLEMS WITH THE 40/60 PARTITION: THE "KILLER NOR GATE"

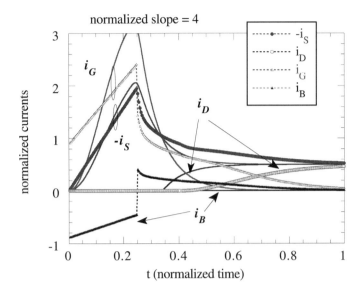

Fig. 4-14 Comparison of terminal currents simulated with BSIM3's NQS model and the exact NQS analysis (MEDICI). $L = 5\,\mu m$; $\delta = 0.2$; $\mu_n = 400\,cm^2/V\text{-s}$; $V_o = 2.5\,V$; and $\theta_{normalized} = 4$. (From Ref. [3], © IEEE 1996, reprinted with permission)

programs. (C programming, like the reverse Polish notation, is too encryptic!) The NQS BSIM3 simulation performed for this section has been simulated from the author's programs, not directly from BSIM3's NQS model. However, the author believes that the implementation is correct and that his program results should be quite close to that simulated with the actual BSIM3 NQS model. For clarity, the details of how BSIM3 NQS model would treat a ramp transient are described in Appendix E. If the content in Appendix E is not exactly that of BSIM3 NQS model, then the claimed BSIM3 NQS results here are perhaps questionable. Nonetheless, the author believes that the basic conclusions of this section are still valid. They are: 1) BSIM3's NQS model results in zero bulk current, for all the time; 2) BSIM3's NQS model does not accurately predict zero current at the initial part of a turn-on transient. Instead, the drain current has to be arbitrarily set to zero to avoid negative drain current.

4.2 PROBLEMS WITH THE 40/60 PARTITION: THE "KILLER NOR GATE"

The bump problem mentioned in the previous section is not a BSIM3 problem *per se*; it is a common to all models employing the quasi-static approximation with the 40/60 charge partitioning scheme. The 50/50 charge partitioning does not eliminate the problem, either. Only the 0/100 partitioning is free of the problem. However, since the 0/100 partitioning scheme is unphysical, using the 0/100 model is subject to the inaccuracies described in the previous section.

In this section, we present a circuit, the Killer NOR-Gate [5], which accentuates the bump problem of the 40/60 model. The Killer NOR-Gate is shown in Fig. 4-15. It is a practical day-to-day circuit, rather than a pathological circuit used for pedantic illustration. BSIM3 simulation results are shown in Figure 4-16. Part a of the figure corresponds to the 40/60 simulation (XPART = 0), and Part b, the 0/100 simulation (XPART = 1). As far as the output voltage is concerned, $V(Q)$, both models give rise to comparable waveforms, in a manner agreeable to the NOR logic function of the two input voltages. The voltage at node X, $V(X)$, however, is drastically different in the two simulations. In the 40/60 simulation, $V(X)$ reaches a -30 V value at $t = 83$ ns, clearly an unphysical value because it is way smaller than the lowest supply voltage of 0. In the 0/100 simulation, $V(X)$ maintains reasonable values throughout the entire transient.

The fundamental reason why $V(X)$ reaches -30 V in the 40/60 simulation is the negative capacitive current discussed in the previous section (that associated with C_{dg}). At $t = 83$ ns, V_{inB} starts to descend in value, thus dv_{GS}/dt in Mp1 is positive. (The source of Mp1 is held at V_{dd}.) Because Mp1 at that time frame is in saturation, C_{dg} is a finite positive value in the 40/60 partition scheme. Moreover, since I_D at that moment is small, the net drain current is negative. The current marked as I_x in Fig. 4-15 is therefore positive, flowing from node X into Mp1. I_x must come from Mp2, which, in turn, comes from either Mn1 or Mn2. Because V_{inA} is low while V_{inB} is high at $t = 83$ ns, Mn1 is off, and a finite current I_y from Mn2 is needed to supply the required I_x at the X node. I_y, in the direction shown in Fig. 4-15, is positive. This means that $V(Q)$ is negative (at this particular instant, the drain of Mn2 is therefore the ground and node Q is the source.) Because V_{inB} is H, the magnitude of $V(Q)$ needs only be small to supply enough current for the required I_x at node Q. Now that I_y flows out of Mn2, it must traverse through Mp2. For this particular transistor at $t = 83$ ns, the bias voltage V_{inA} is zero and $V(Q)$ is slightly negative as mentioned.

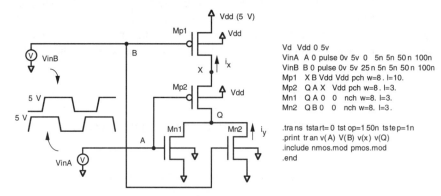

Fig. 4-15 The Killer NOR-Gate [5], which highlights the problems of a QS model with 40/60 partition.

4.2 PROBLEMS WITH THE 40/60 PARTITION: THE "KILLER NOR GATE"

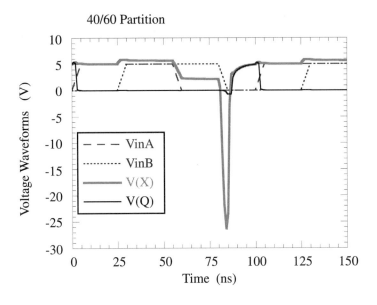

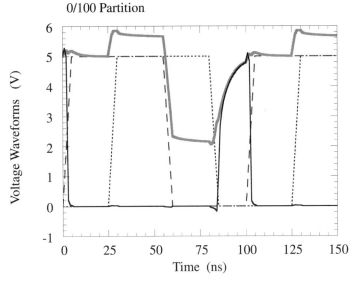

Fig. 4-16 BSIM3 simulation results of the Killer NOR-Gate of Fig. 4-15. (a) With the 40/60 charge partition scheme (XPART = 0); (b) 0/100 partition (XPART = 1).

Further, we want a positive current to flow from node Q to node X. This means that node Q is the source and node X is the drain. V_{GS} in this case is $V_{inA} - V(Q)$, which is slightly negative and Mp2 is in the subthreshold operation. The only manner (at least in the model) that the transistor can generate a current of the size of I_x is to have a fairly negative $V(X)$. The magnitude of the drain-to-source voltage in Mp2,

$V(Q) - V(X)$, can then be fairly large, and subsequently, give rise to an appreciable amount of current. It turns out that $V(X)$ needs to be roughly -30 V to generate an amount equal to I_x.

In real life, the moment V_{inB} switches from H to L at $t = 83$ ns, I_x is roughly zero, instead of equal to $C_{dg} \cdot dv_{GS}/dt$ as calculated in a QS model. The non-quasi-static behavior in Mp1 acts as an inertia element resisting the voltage change. Without a need to supply I_x to Mp1, I_y will be roughly zero and $V(X)$ does not reach -30 V.

The Titanic story tells us that many unfortunate factors need to all come together to forge a great disaster (i.e., fast nautical speed at night, iceberg in ocean, shortage of lifeboats, lack of a sense of emergency, pure bad luck and, according to a 1998 Hollywood film, having fun in an unconventional place.) Similarly, many peculiarities exist in this circuit, and together they give rise to an outrageous simulation result. Although not obvious from the figure, the node X actually connects to several other nodes, besides the source of Mp2. One connection is to the gate of Mp1, through the gate–drain overlap and fringing parasitic capacitances. These capacitances, $C_f + C_{ov}$, form $C_{dg,p}$ which can be lumped with the aforementioned C_{dg} associated with the intrinsic device. When these parasitic capacitances are considered, the magnitude of I_x is even larger than $C_{dg} \cdot dv_{GS}/dt$. Therefore, the problem of making $V(X)$ excessively negative to supply I_x is even more acute.

The second nonobvious connection from node X is to the power supply V_{dd} through $C_{j,\text{DB}}$ of Mp1. This connection would partially supply the needed I_x, in the form of $|C_{j,\text{DB}} \cdot dV(X)/dt|$. Therefore, $V(X)$ would not be as negative as -30 V if $C_{j,\text{DB}}$ is large. However, the Killer NOR-Gate, as given in Fig. 4-15, neglects $C_{j,\text{DB}}$ by not declaring the source and drain junction areas and peripheries in the device statement. This omission cuts off the $C_{j,\text{DB}}$ path, making the Killer NOR-Gate results more unphysical. The third non-obvious connection is again to the power supply V_{dd}, but this time the connection is through $C_{j,\text{SB}}$ of Mp2. Just like $C_{j,\text{DB}}$ of Mp1, this $C_{j,\text{SB}}$ would have supplied part of the required I_x, and would have resulted in a more reasonable value of $V(X)$. Nonetheless, the Killer NOR-Gate circuit given in Fig. 4-15 does not include this junction capacitance.

We note that Mp1 has a L of $10\,\mu\text{m}$ while all other transistors have a much smaller channel length. This abnormally long channel length results in a larger value of C_{dg}, as compared to when $L = 3\,\mu\text{m}$. Hence, I_x is made larger with $L = 10\,\mu\text{m}$, making the problem worse. Finally, the voltage waveforms used in the simulation have a fall/rise time of 5 ns. This is an unrealistically fast transient for the transistors of this technology. A ring-oscillator composed of PMOS and NMOS transistors of $L = 3\,\mu\text{m}$ is found to have a rise/fall time of roughly 25 ns. Therefore, for the $L = 3\,\mu\text{m}$ technology assumed in the circuit, the appropriate rise/fall times used in the simulation should be 5 times slower. The magnitude of $I_x = C_{dg} \cdot dv_{GS}/dt$ would then be 5 times smaller. The magnitude of $V(X)$ then would not be as negative as -30 V.

We resimulate the circuit with following modifications: 1) adding $C_{j,\text{DB}}$ of Mp1, and $C_{j,\text{SB}}$ of Mp1; 2) making Mp2 to have $L = 3\,\mu\text{m}$ as other transistors; 3) revising the input waveforms to $V_{inA} = $ 0V 5V 0 25n 25n 250n 500n and $V_{inB} = $ 0V 5V 125n 25n 25n 250n 500n (basically the rise/fall time is made to be 25 ns instead

4.3 LACK OF CHANNEL RESISTANCE (NQS EFFECT: SMALL-SIGNAL ANALYSIS)

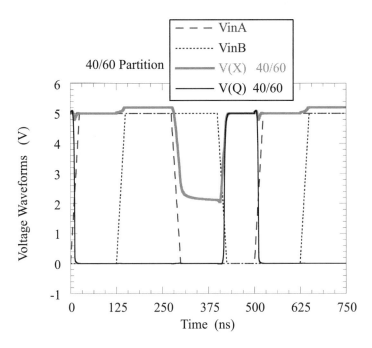

Fig. 4-17 BSIM3 simulation results of the Killer NOR-Gate when the circuit and bias conditions are modified to more realistic values. See text for the modifications. $V(X)$ no longer has absurd values.

of 5 ns). With these more real life scenarios, the Killer NOR-Gate with the 40/60 model is suddenly not as formidable as before. $V(X)$ is found to have the smallest value of 2 V, as shown in Fig. 4-17.

4.3 LACK OF CHANNEL RESISTANCE (NQS EFFECT: SMALL-SIGNAL ANALYSIS)

The last few sections concern primarily with the NQS effects during a transient analysis. We mentioned a simple rule to decide whether NQS effects are significant in a particular event. If the time of interest is long compared to the transistor's delay time, given in Fig. 4-4, then NQS effects are negligible.

We are interested in small-signal ac analysis in this section. The rule establishing the importance of NQS effects follows logically from that proposed for the transient analysis. If the frequency of operation is smaller than the f_T of the device, then NQS effects are negligible. The calculated f_T as a function of the channel length is also found in Fig. 4-4.

What parts of the BSIM3's QS small-signal model become incorrect when the NQS effects are important in the intrinsic device? Although there are several, perhaps only two are considered most important. They are 1) incorrect transconductance frequency dependence; and 2) absence of a channel resistance. Let us examine the first deficiency. In a BSIM3 small-signal equivalent model (such as Figs. 2-21 or 2-24), the transconductance of the output current source due to the input voltage has some frequency dependence. In order to avoid confusion between this frequency-dependent transconductance and the device transconductance ($g_m = \partial I_D/\partial V_{DS}$), we shall denote the former as $g_m^\dagger$ and keep the latter as g_m. Therefore, in BSIM3 QS model,

$$g_m^\dagger = g_m - j\omega(C_{dg} - C_{gd}). \tag{4-9}$$

It is instructive to re-express this equation with the frequency normalization factor, ω_0, which was introduced in Eq. (1-36). After some algebra, it can be shown that Eq. (4-9) is equivalent to [6]

$$g_{m,\text{BSIM3}}^\dagger = g_m\left(1 - j\frac{\omega}{\omega_o}\tau_1\right), \tag{4-10}$$

where τ_1 is given by

$$\tau_1 = \frac{4}{15}\frac{1 + 3\alpha + \alpha^2}{(1+\alpha)^3}. \tag{4-11}$$

We append the subscript BSIM3 to emphasize that $g_m^\dagger$ of Eq. (4-10) is essentially that used in BSIM3. A revisit of Section 2.5 (or Appendix C) shows that $g_{m,\text{BSIM3}}^\dagger$ is simply $y_{dg} - y_{gd}$, where the y-parameters are the quasi-static y-parameters used in BSIM3, and are listed in Eqs. (1-98) and (1-99). If the NQS effects are to be modeled, the NQS y-parameters (denoted $y_{dg,\text{NQS}}$ and $y_{gd,\text{NQS}}$) should be used instead. The derivation of the NQS y-parameters is quite algebraically intensive [6]. According to Appendix C, $g_{m,\text{NQS}}^\dagger$, the $g_m^\dagger$ obtained from $y_{dg,\text{NQS}} - y_{gd,\text{NQS}}$ is

$$g_{m,\text{NQS}}^\dagger = \lim_{k\to\infty}\frac{g_m}{1 + (j\omega)D_1 + (j\omega)^2 D_2 + \cdots + (j\omega)^k D_k}, \tag{4-12}$$

where D_k's are given in Appendix C (Eq. C-4). A good approximation is possible when we just keep the $k = 1$ term. Substituting D_1 from Appendix C (Eq. C-14), we find Eq. (4-12) is simplified to

$$g_{m,\text{NQS}}^\dagger \approx \frac{g_m}{1 + j\omega/\omega_o \cdot \tau_1}. \tag{4-13}$$

At frequencies close to ω_0, we can apply the approximation that $1/(1 + x) \sim 1 - x$. We then see the equivalence between $g_{m,\text{NQS}}^\dagger$ of Eq. (4-13) and $g_{m,\text{BSIM3}}^\dagger$ of Eq. (4-10). This is an important finding. Although BSIM3 is built purely from quasi-static approximation, the resulting transconductance $g_m^\dagger$ does contain a first-order correction of the NQS effects. There has been an assertion that BSIM3's transconductance

4.3 LACK OF CHANNEL RESISTANCE (NQS EFFECT: SMALL-SIGNAL ANALYSIS)

was poorly modeled because it was frequency independent. Such a statement is not entirely correct, and is likely a result of the confusion between the exact meanings of transconductance. As mentioned, g_m given by $\partial I_D/\partial V_{GS}$ is correctly a frequency-independent quantity in BSIM3. $g_m^{\dagger}$ in BSIM3, as given in Eq. (4-10), contains a frequency dependence which is similar to that of $g_{m,\text{NQS}}^{\dagger}$ in the first-order approximation. There is, however, an important distinction between Eqs. (4-10) and (4-13). $|g_{m,\text{BSIM3}}^{\dagger}|$ always increases with frequency, whereas $|g_{m,\text{NQS}}^{\dagger}|$ decreases with frequency. This discrepancy needs to be taken into consideration when we interpret high-frequency simulation results from the BSIM3 model. Figure 4-18 illustrates the calculated $g_{m,\text{BSIM3}}^{\dagger}$ and $g_{m,\text{NQS}}^{\dagger}$ with k as a parameter, for $\alpha = 0.5$ (linear region) and $\alpha = 0$ (saturation region). Even with a k value as small as 1, they are already approximately equal up to certain high frequencies. If a high degree of accuracy is required, then $k = 2$ is desirable.

The discrepancy between $g_{m,\text{BSIM3}}^{\dagger}$ and $g_{m,\text{NQS}}^{\dagger}$ is the first deficiency of the BSIM3 QS small-signal model when NQS effects are important. The deficiency will be elaborated upon in the next section. The next improvable area relates to the absence of a channel resistance. In a BSIM3 small-signal equivalent circuit (such as Fig. 2-21 or 2-24), the branch connecting the intrinsic gate and the intrinsic source nodes consists of a capacitor, $C_{gs,t}$. The impedance of this branch is purely imaginary. However, it is not uncommon to see equivalent circuits reported in the literature whose branch of interest consists of a series connection of the aforementioned capacitance and, in addition, a resistor which we shall denote as the *channel resistance* (r_{ch}).

In the NQS equivalent circuit, the impedance of the branch is no longer purely imaginary. Although the magnitude of the resistive portion can be much smaller than the imaginary part, it is nonetheless finite and needs to be considered during an impedance matching design. To understand the origin of the real part, we notice that

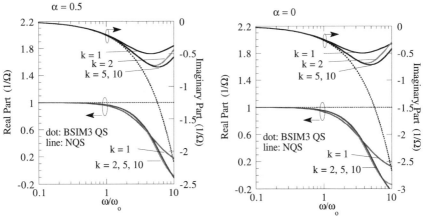

Fig. 4-18 Calculated $g_{m,\text{BSIM3}}^{\dagger}$ (Eq. 4-10) and $g_{m,\text{NQS}}^{\dagger}$ (Eq. 4-13) with k as a parameter, for $\alpha = 0.5$ (linear region) and $\alpha = 0$ (saturation region).

IMPROVABLE AREAS OF BSIM3

the admittance of the branch is basically $-y_{gs}$. (For now, we will concentrate on the intrinsic portion of the transistor only. The parasitic elements will be added later.) BSIM3, a QS model, utilizes the y_{gs} given by Eq. (1-99a), that is, $-y_{gs} = +j\omega C_{gs}$. This y_{gs} is a quasi-static approximation of the more exact non-quasi-static y_{gs}, or, $y_{gs,NQS}$. According to Appendix C, $y_{gs,NQS}$ is quite complicated, equal to a summation of infinite frequency power terms. If we take up to the second frequency power, then $y_{gs,NQS}$ can be written as

$$y_{gs,NQS} = -j\omega C_{gs} \frac{1 + j(\omega/\omega_0)\tau_2}{1 + j(\omega/\omega_0)\tau_1} \tag{4-14a}$$

$$= -j\omega C_{gs} - j\omega C_{gs} \frac{j(\omega/\omega_0)(\tau_2 - \tau_1)}{1 + j(\omega/\omega_0)\tau_1}, \tag{4-14b}$$

where τ_1 was given in Eq. (4-11) and τ_2 is

$$\tau_2 = \frac{1}{15} \frac{2 + 8\alpha + 5\alpha^2}{(1+\alpha)^2(1+2\alpha)}. \tag{4-15}$$

It turns out that working with the impedance between the gate and the source is easier than working with the admittance. We therefore evaluate $-1/y_{gs,NQS}$ as

$$-\frac{1}{y_{gs,NQS}} = \frac{1}{j\omega C_{gs}} \times \frac{1 + j(\omega/\omega_0)\tau_1}{1 + j(\omega/\omega_0)\tau_2} \approx r_{ch} + \frac{1}{j\omega C_{gs}}. \tag{4-15}$$

The channel resistance is the real part of the input impedance, given by

$$r_{ch} = \text{Re}\left(\frac{-1}{y_{gs,NQS}}\right) = \frac{1}{\omega C_{gs}} \times \frac{(\omega/\omega_0)[\tau_1 - \tau_2]}{1 + (\omega/\omega_0)^2 \tau_2^2} = \frac{\tau_1 - \tau_2}{\omega_0 C_{gs}} \times \frac{1}{1 + (\omega/\omega_0)^2 \tau_2^2}. \tag{4-17}$$

Equation (4-17) demonstrates that the channel resistance exists even as the frequency approaches zero.

At frequencies much smaller ω_0, r_{ch} is simplified to

$$r_{ch} = \frac{\tau_1 - \tau_2}{\omega_0 C_{gs}} = \frac{1}{\omega_0 WLC'_{ox}} \frac{3\alpha^3 + 15\alpha^2 + 10\alpha + 2}{15(1+\alpha)^3(1+2\alpha)} \times \frac{3(1+\alpha)^2}{2} \frac{1}{2\alpha+1}$$

$$= \frac{1}{\omega_0 WLC'_{ox}} \frac{3\alpha^3 + 15\alpha^2 + 10\alpha + 2}{10(1+\alpha)(1+2\alpha)^2}. \tag{4-18}$$

4.3 LACK OF CHANNEL RESISTANCE (NQS EFFECT: SMALL-SIGNAL ANALYSIS)

We substitute ω_0 from Eq. (1-36) into Eq. (4-18). The resulting expression is then in terms of the fundamental device parameters:

$$r_{ch} = \left[\frac{W}{L}\mu_n C'_{ox}(V_{GS} - V_T)\right]^{-1} \frac{3\alpha^3 + 15\alpha^2 + 10\alpha + 2}{10(1+\alpha)(1+2\alpha)^2}. \tag{4-19}$$

Sometimes r_{ch} is expressed in terms of g_m, which was given in Eq. (1-70). We write r_{ch} as (neglecting δ in comparison to 1)

$$r_{ch} = \frac{1}{g_m} \frac{(3\alpha^3 + 15\alpha^2 + 10\alpha + 2)(1-\alpha)}{10(1+\alpha)(1+2\alpha)^2}. \tag{4-20}$$

It is common to see a resistive element of $r_{ch} = 1/(5g_m)$ in a circuit model. When the device is in saturation, a substitution of $\alpha = 0$ into Eq. (4-20) reveals the factor of 5 in the denominator.

Because BSIM3 is a QS model, r_{ch} is always zero in its small-signal equivalent circuit and y_{gs} of the intrinsic device is always purely imaginary. Figure 4-19 shows the physical origin of r_{ch}, that the inverted channel lying between the gate and the source terminals is resistive. An equivalent circuit of the branch is shown in Fig. 4-19b. The resistance acts as an inertia element so that the transistor does not turn on and establish its static profile instantaneously. The gate–source parasitic capacitance is in parallel with the branch formed by r_{ch} and C_{gs}. Therefore, a proper NQS small-signal, based on Fig. 2-24, is as shown in Fig. 4-20. The equivalent circuit of Fig. 4-19b appears in the Fig. 4-20, in the branch connecting the intrinsic gate and the source nodes. We have inserted other new elements that are not found in Fig. 2-24. These elements appear when we carry out the non-quasi-static y-parameter analysis for other branches, in a similar fashion that we have done for $y_{gs,NQS}$. The channel

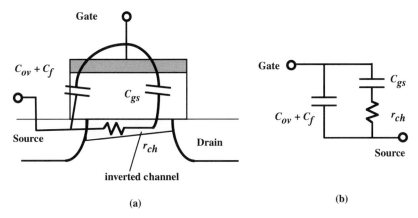

Fig. 4-19 (a) Schematic drawing showing the physical origin of r_{ch}, an element that is absent in a QS model; (b) the location of r_{ch} in relation to intrinsic and extrinsic capacitances. (From Ref. [6], p. 418, © Wiley 1999)

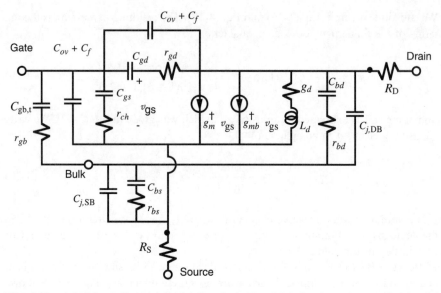

Fig. 4-20 A NQS small-signal model based on Fig. 2-24 that incorporates the channel resistance of Fig. 4-19 as well as other NQS elements.

resistance r_{ch} was given in Eq. (4-20) and $g_m^\dagger$ was given in Eq. (4-13). The other elements associated with the NQS effects in Fig. 4-20 are given by

$$r_{gd} = \left[\frac{W}{L}\mu_n C'_{ox}(V_{GS} - V_T)\right]^{-1} \frac{2\alpha^3 + 10\alpha^2 + 15\alpha + 3}{10\alpha(1+\alpha)(2+\alpha)^2}. \qquad (4\text{-}21)$$

$$r_{gb} = -\left[\frac{W}{L}\mu_n C'_{ox}(V_{GS} - V_T)\right]^{-1} \frac{4\alpha^4 + 34\alpha^3 + 60\alpha^2 + 26\alpha + 4}{5(1+\alpha)(1-\alpha)^4} \times \frac{1+\delta}{\delta}. \qquad (4\text{-}22)$$

$$r_{bs} = \frac{1}{\delta} r_{ch}. \qquad (4\text{-}23)$$

$$r_{bd} = \frac{1}{\delta} r_{gd}. \qquad (4\text{-}24)$$

$$g_{mb}^\dagger = \frac{g_{mb}}{1 + j\omega/\omega_o \cdot \tau_1}. \qquad (4\text{-}25)$$

$$L_d = \frac{\tau_1}{\omega_o} g_d. \qquad (4\text{-}26)$$

Because r_{ch} is the most important effect among all of the above NQS-related elements, most often other elements in Fig. 4-20 do not appear in a so-called NQS equivalent circuit. This is certainly the scenario in typical high-frequency GaAs FET equivalent circuits reported in the literature. Finally, we note that r_{gb} is a negative quantity, albeit small.

4.3 LACK OF CHANNEL RESISTANCE (NQS EFFECT: SMALL-SIGNAL ANALYSIS)

We have defined r_{ch} as $\text{Re}(-1/y_{gs,\text{NQS}})$; it is equal to $1/(5g_m)$ in saturation. Sometimes a similar quantity, r_i, is used to denote $\text{Re}(1/y_{gg,\text{NQS}})$. It is the real part of the impedance looking into the gate terminal when the drain, source and bulk terminals are at ac ground. During saturation, $r_{gd} = \infty$ (so we do not need to consider the branch between gate and drain), and $r_{gb} = -4/(5g_m) \times (1+\delta)/\delta$. For a parallel combination of R_1 in series of C_1 and R_2 in series of C_2, the real part of the impedance of the parallel combination is equal to $(R_1 \cdot C_1^2 + R_2 \cdot C_2^2)/(C_1 + C_2)^2$. Therefore, r_i is given by [7]

$$r_i = \text{Re}\left(\frac{1}{y_{gg,\text{NQS}}}\right) = \frac{r_{ch} \cdot C_{gs}^2 + r_{gb} \cdot C_{gb}^2}{(C_{gs} + C_{gb})^2} = \frac{4(g_m + g_{mb})}{5(2g_m + 3g_{mb})^2} \quad \text{(in saturation)}.$$

(4-27)

Because g_{mb} in modern CMOS can approach 20% of g_m, r_i is roughly $1/(7g_m)$ in saturation.

We would like to evaluate magnitude of r_{ch} in real-life transistors. We focus on two extreme lengths of $L = 35$ and $0.35\,\mu\text{m}$. For the $L = 35\,\mu\text{m}$, we find that f_T is 4.6 MHz. We also run BSIM3 simulation to determine g_m, $V_{DS,\text{sat}}$ (which allows us to find the saturation index α), C_{gs} (i.e., the intrinsic C_{gs}), as well as $C_{ov} + C_f$. From SPICE simulation at three bias conditions, we calculate r_{ch} from Eq. (4-20) and C_{gs} and $C_{ov} + C_f$ in the following:

Bias	V_{GS}	V_{DS}	r_{ch} (Ω)	C_{gs} (F)	$C_{ov} + C_f$ (F)
A	0.02	3.0	2.3×10^{11}	2.58×10^{-16}	1.37×10^{-15}
B	0.5	3.0	4.5×10^{5}	2.23×10^{-13}	1.40×10^{-15}
C	4.0	3.0	4.3×10^{3}	7.57×10^{-13}	1.40×10^{-15}

The data of r_{ch}, C_{gs} and $C_{ov} + C_f$ are then used to simulate the frequency response of the branch equivalent circuit of Fig. 4-19b, as shown in Fig. 4-21. When the transistor is nearly off ($V_{GS} = 0.02$ V; Fig. 4-21a), the channel resistance is huge. The overall frequency response is determined by $C_{ov} + C_f$. Hence, the NQS-related r_{ch} can be safely neglected. In the other extreme when the transistor is heavily turned on ($V_{GS} = 4.0$ V; Fig. 4-21c), the channel resistance is small. The overall frequency response is dominated by the intrinsic transistor this time. However, the frequency response is determined by C_{gs} which is in series with r_{ch}. Hence, the NQS-related r_{ch} is again negligible. The channel resistance is most important when the long-channel transistor nearly turns on ($V_{GS} = 0.5$ V; Fig. 4-21b). In this case, the real part of the overall impedance exceeds the imaginary part at certain high frequencies.

We repeat the above analysis for $L = 0.35\,\mu\text{m}$. We find the cutoff frequency to be 44 GHz and the relevant parameters are:

Bias	V_{GS}	V_{DS}	r_{ch} (Ω)	C_{gs} (F)	$C_{ov} + C_f$ (F)
A	0.02	3.0	6.8×10^{8}	2.82×10^{-16}	1.37×10^{-15}
B	0.5	3.0	1.9×10^{3}	1.21×10^{-15}	1.37×10^{-15}
C	4.0	3.0	1.2×10^{2}	3.06×10^{-15}	1.37×10^{-15}

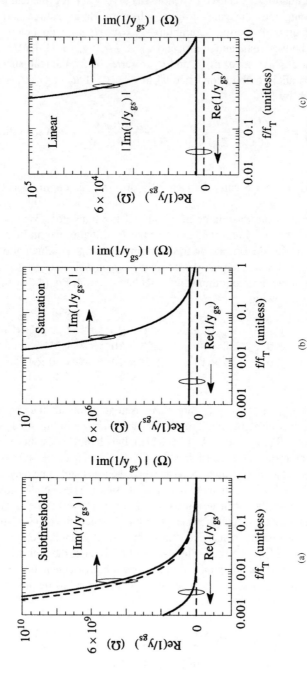

Fig. 4-21 Calculated frequency response of the branch equivalent circuit of Fig. 4-19b, for $L = 35\,\mu m$ (a) Transistor is nearly off ($V_{GS} = 0.02\,V$); (b) transistor nearly turns on ($V_{GS} = 0.5\,V$); (c) transistor heavily turns on ($V_{GS} = 4.0\,V$). The dashed line is obtained with r_{ch} assumed to be zero, and the solid line, obtained with the r_{ch} given in the text. In part b and part c of the figure, $\text{Im}(1/y_{gs})$'s calculated with zero and with non-zero r_{ch} are nearly the same. Therefore, the dashed line nearly coincides with the solid line, and is not clearly visible in the figure.

4.3 LACK OF CHANNEL RESISTANCE (NQS EFFECT: SMALL-SIGNAL ANALYSIS)

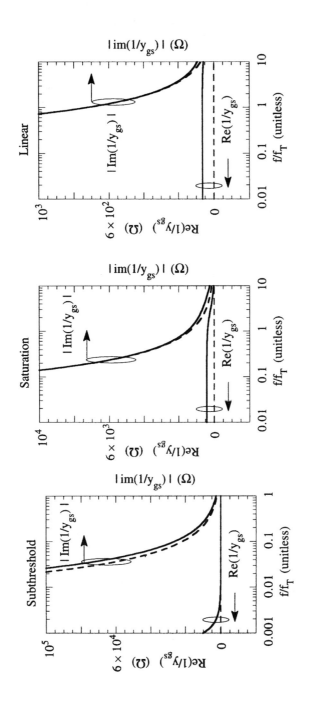

Fig. 4-22 Calculated frequency response of the branch equivalent circuit of Fig. 4-19b, for $L = 0.35$ μm (a) Transistor is nearly off ($V_{GS} = 0.02$ V); (b) transistor nearly turns on ($V_{GS} = 0.5$ V; (c) transistor heavily turns on ($V_{GS} = 4.0$ V). The dashed line is obtained with r_{ch} assumed to be zero, and the solid line, obtained with the r_{ch} given in the text. In part b and part c of the figure, $\text{Im}(1/y_{gs})$'s calculated with zero and with non-zero r_{ch} are nearly the same. Therefore, the dashed line nearly coincides with the solid line, and is not clearly visible in the figure.

In short-channel devices, the parasitic capacitance often exceeds the intrinsic device capacitances (Fig. 4-22). In all three parts of the figure, we see that the real part is smaller than the imaginary part in all useful frequency range. This means that the channel resistance is not an overwhelming factor deciding the transistor's performance.

The channel resistance is most important when the transistor is near the threshold voltage, a condition under which the channel resistance is significantly reduced while the gate-to-source capacitance has not increased much yet. Figure 4-21 demonstrates that r_{ch} of a long-channel device can exceed the impedance of C_{gs}. Therefore, the dominant impedance of the branch connecting the gate and the source is r_{ch}, no longer C_{gs}. The QS approximation of the entire branch by a simple capacitor (of magnitude C_{gs}) is hence grossly in error. For the short-channel device shown in Fig. 4-22, we see that r_{ch}, though becoming comparable to $1/j\omega C_{gs}$, never exceeds the latter. So, it seems that the quasi-static representation by a pure capacitor is sufficient and the presence of r_{ch} can be neglected. However, the calculation of Fig. 4-22 is based on the r_{ch} formula given in Eq. (4-20). It is a formula, we stress, obtained for long-channel devices whose velocity is always given by the mobility times the electric field and does not saturate. The r_{ch} formula therefore does not

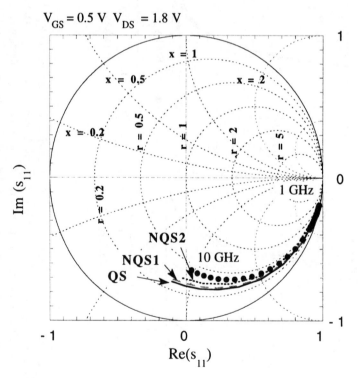

Fig. 4-23 Comparison of measured and calculated s_{11} for a $L = 0.29\,\mu\text{m}$ transistor. The NQS1 and NQS2 calculations are described in the text.

really apply to a short-channel devices wherein the velocity saturation plays a critical role in the transistor dynamics. When the velocity saturation is accounted for the r_{ch} calculation by making a particular adjustment in τ_1 [8], we see that r_{ch} is, in fact, important for short-channel devices near the threshold voltage region. Figure 4-23 illustrates the comparison of measured s_{11} and that calculated from a $L = 0.29\,\mu\text{m}$ transistor model. The QS calculation is when r_{ch} is identically zero. The NQS1 calculation includes r_{ch} in series with C_{gs}, with r_{ch} calculated from the long-channel expression given in Eq. (4-20). The NQS2 calculation is obtained with a larger r_{ch} value calculated with the aforementioned adjustment. The adjustment is necessary in accurately fitting the measured s_{11} (in dots) of the short-channel device.

Although BSIM3 is a QS model and r_{ch} is not accounted for, there is a work-around which can correct this problem. The work-around, to be described in the next section, is applicable only for ac analysis, not large-signal transient analysis.

4.4 INCORRECT TRANSCONDUCTANCE DEPENDENCY ON FREQUENCY

We mentioned that $|g_m^\dagger|$ calculated in BSIM3 increases with frequency. ($g_m^\dagger$ is the transconductance of the voltage-controlled current source at the output, whereas g_m denotes $\partial I_D/\partial V_{DS}$ for the intrinsic device.) The actual $g_m^\dagger$ derived from an NQS analysis, given by Eq. (4-13), has the correct property that its magnitude decreases with frequency. Although $g_{m,\text{BSIM3}}^\dagger$ given by Eq. (4-9), and equivalently in Eq. (4-10), is approximately equal to $g_{m,\text{NQS}}^\dagger$, many unphysical results can surface just because $|g_{m,\text{BSIM3}}^\dagger|$ increases with frequency. Let us demonstrate this problem with a simple amplifier circuit shown in Fig. 4-24. A simplified small-signal equivalent circuit, based on Fig. 2-21, is shown in Fig. 4-24b. We are interested in the magnitude of the ratio of the small-signal output voltage (v_o) to the small-signal input voltage (v_i). Out of intuition, as well as demonstrated in a non-quasi-static

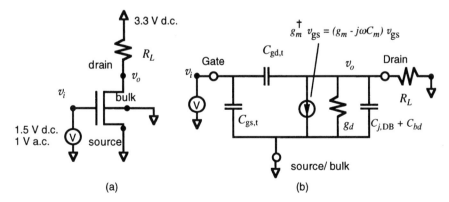

Fig. 4-24 (a) An amplifier circuit that highlights the wrong frequency response of the device transconductance, $g_m^\dagger$; (b) its equivalent circuit.

analysis, we expect $|v_o/v_i|$ to decrease with frequency. At low frequencies, $|v_o/v_i| \sim g_{m,\text{BSIM3}}^\dagger \cdot R_L \sim g_m \cdot R_L$.

We temporarily neglect the junction capacitance $C_{j,\text{DB}}$. Figure 4-25, illustrating results from BSIM3 simulation, shows that $|v_o/v_i|$ increases with frequency. This is a direct consequence of the fact that $g_{m,\text{BSIM3}}^\dagger$ simulated in BSIM3 increases with frequency. This unphysical simulation can overestimate a circuit's frequency response. This problem is somewhat mitigated when $C_{j,\text{DB}}$ is considered. $C_{j,\text{DB}}$, placed in parallel with R_L at the output port, begins to shunt current away from R_L at high frequencies. Therefore, $C_{j,\text{DB}}$'s presence reduces $|v_o/v_i|$. For the circuit of concern, $|v_o/v_i|$'s decrease due to $C_{j,\text{DB}}$ apparently more than offsets $|g_{m,\text{BSIM3}}^\dagger|$'s increase with frequency. The net result is a decrease of $|v_o/v_i|$ with frequency, just as expected. Although the overall simulation result agrees qualitatively with expectation, this result nonetheless is not accurate at high frequencies. This is because $|v_o/v_i|$'s decrease comes from $C_{j,\text{DB}}$, but $|g_{m,\text{BSIM3}}^\dagger|$ still increases with frequency.

For long-channel devices whose $C_{j,\text{DB}}$ is much smaller than the intrinsic device capacitance, $|v_o/v_i|$ will increase with frequency, whether $C_{j,\text{DB}}$ is considered or not. We should also emphasize that Fig. 4-25 is obtained with XPART = 0. When XPART = 1, C_{dg} and C_{gd} appearing in Eq. (4-9) are nearly zero when the device is in saturation. Therefore, $|g_m^\dagger|$'s does not increase with frequency. When XPART = 0, C_{dg} becomes nonzero. $|g_m^\dagger|$ thus incorrectly increases with frequency. The difference in the voltage-gain's dependencies on frequency, for XPART = 0 and 1, will be examined again in Section 4.14.

Although we have demonstrated that the magnitude of v_o/v_i is problematic in BSIM3, we comment that the phase of v_o/v_i is equally wrong.

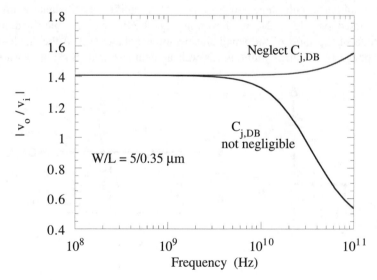

Fig. 4-25 BSIM3-calculated $|v_o/v_i|$ for the circuit of Fig. 4-24a, with XPART = 0. The voltage gain unphysically increases with frequency when $C_{j,\text{DB}} = 0$.

4.4 INCORRECT TRANSCONDUCTANCE DEPENDENCY ON FREQUENCY

It is not easy to correct this problem, since $|g^\dagger_{m,\text{BSIM3}}|$'s frequency dependence builds on top the quasi-static assumption. To remove this assumption would mean a major revision of BSIM3. There is, however, a work-around. This work-around is not really recommended for day-to-day simulation, so the readers may skip the rest of this section. However, this work-around can be useful under special circumstances, such as debugging the simulation results. For the simplicity of illustrating the concept, we will assume that the device's overlap, fringing capacitances, and parasitic terminal resistances are negligible. First, we determine the device's C_{dg}, C_{gd}, and g_m. Second, we notice that, from a comparison of Eqs. (4-9), (4-10), and (4-13), the desired $g^\dagger_{m,\text{NQS}}$ can be rewritten as

$$g^\dagger_{m,\text{NQS}} = \frac{g_m}{1 + j\omega(C_{dg} - C_{gd})}. \quad (4\text{-}28)$$

Third, we append a voltage-controlled current source (VCCS), VCCS1, in the right-hand-side dashed box of Fig. 4-26. This right-hand-side dashed box shows the modification to the default SPICE MOS element to account for the desired $g^\dagger_{m,\text{NQS}}$. (The left-hand-side dashed box, used to modify $1/y_{gs,\text{BSIM3}}$ to $1/y_{gs,\text{NQS}}$, is to be described shortly.) VCCS1, equal to $g_m \cdot v_x$, is basically identical to $g^\dagger_{m,\text{NQS}} \cdot v_{gs,i}$. However, because BSIM3 already produces internally a current source equal to $g^\dagger_{m,\text{BSIM3}} \cdot v_{gs,i}$, we need to subtract out this component. This subtraction is implemented with VCCS2.

The left-hand-side dashed box is used to result in a finite r_{ch}. By choosing $R_1 = \tau_1/(\omega_0 \cdot C_{gs})$, we find the overall gate current of the modified MOS to be

$$i_g = -i_y + j\omega C_{gs} v_{gs,i}, \quad (4\text{-}29)$$

where i_y is

$$i_y = \frac{j\omega C_2}{1 + 1/(j\omega R_1 C_{gs})} v_{gs,i} = \frac{j\omega C_2}{1 + 1/(j\omega \tau_1)} v_{gs,i}. \quad (4\text{-}30)$$

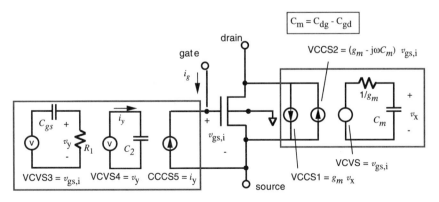

Fig. 4-26 A workaround to correct $|g^\dagger_{m,\text{BSIM3}}|$'s frequency dependence.

The overall y_{gs}, equal to $i_g/v_{gs,i}$, then becomes

$$y_{gs} = j\omega C_{gs} \times \frac{1+j\omega\tau_1(1-C_2/C_{gs})}{1+j\omega\tau_1}. \qquad (4\text{-}31)$$

By choosing $C_2 = C_{gs} \cdot (1 - \tau_2/\tau_1)$, we then yield a y_{gs} equal to $y_{gs,\text{NQS}}$ given in Eq. (4-14a).

In the event that the parasitic capacitances and resistances are significant, the same concept remains applicable. However, the device parameters need to be properly adjusted to move the capacitances and terminal resistances outside the transistor. They are incorporated externally to the transistor, allowing the above work-around to be communicated with the intrinsic nodes of the device.

This work-around is a little bit complicated. A much more straightforward solution to this problem is for the circuit designers to pay more attention in interpreting their simulation results. Because the frequency at which $|g_m^\dagger|$ increases noticeably occurs nearby f_T, the circuit designers need to watch out peculiar result when $|v_o/v_i|$ increases beyond f_T.

4.5 LACK OF GATE RESISTANCE (AND ASSOCIATED NOISE)

The gate of modern MOS technologies is made of polysilicon. Metal gate, often mentioned in idealized examples, is not used because the typical metal work functions are at values which would result in high-threshold voltage. The adoption of polysilicon as the gate material has its drawback, that polysilicon is more resistive than metal. An important parameter characterizing the gate resistance is the gate sheet resistance, $R_{SH,G}$ (in $\Omega/\square$). It is equal to the gate material's resistivity (ρ_G, in Ω-cm) divided by the gate thickness (Λ_G):

$$R_{SH,G} = \frac{\rho_G}{\Lambda_G}. \qquad (4\text{-}32)$$

Generally, the silicided polysilicon gate sheet resistance ranges from 3 $\Omega/\square$ at $L = 1\,\mu\text{m}$ to 6 $\Omega/\square$ at $L = 0.35\,\mu\text{m}$. The unsilicided polysilicon sheet resistance is generally on the order of 40 $\Omega/\square$. The metal gate sheet resistance can retain a low value at 2 $\Omega/\square$ as L shrinks to 0.05 μm [9].

$R_{SH,G}$ is a useful parameter because it is easily measurable from a test structure consisting of a wide metal gate with probe pads at the two ends, as shown in Fig. 4-27. The material underneath the gate structure is designed to be isolated from the active area, so that the current injected from one end of the gate all comes out at the other end. To ensure accurate measurement, two extra voltage-sensing probe pads are attached at the ends of the test structure. The gate sheet resistance is related to the measured gate resistance ($R_{measure}$) by

$$R_{measure} = R_{SH,G} \times \frac{W}{L}. \qquad (4\text{-}33)$$

W and L are the width and the length of the gate metal, respectively.

4.5 LACK OF GATE RESISTANCE (AND ASSOCIATED NOISE) 317

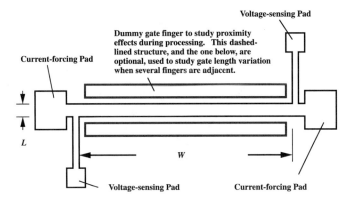

Fig. 4-27 A test structure used to measure the sheet resistance of the gate material. (From Ref. [6], p. 423, © Wiley 1999)

When we wrote Eq. (4-33) for the test structure of Fig. 4-27, the current is restricted in the metal, not flowing into the semiconductor. Whatever charge entering one end comes out at the other end. This kind of signal flow does not occur in a device operation. As shown in Fig. 4-28, the gate charges entering one end of the gate do not all come out at the other end. Instead, the gate current is continuously diverted to the channel through C_{gg} [in the form of $C_{gg} \cdot dv_{GS}/dt$, as seen in Eq. (2-25a)]. At the other end of the gate, there is no more gate current since all of the gate charges have been diverted to the channel. This kind of current flow makes R_G a *distributed resistance*, which differs from that given in Eq. (4-33). The formula for the distributed gate resistance has been determined for small-signal situation [6], as well as large-signal transient situation [10]. In either case, the formula is the same, and is given by

$$R_G = \frac{1}{3} R_{SH,G} \frac{W}{L}. \quad (4\text{-}34)$$

In the gate structure of an actual transistor, some signals remain in the gate for a longer portion of the gate width before exiting to the channel. Some signals depart sooner. The factor 1/3 accounts for the distributed nature of the current conduction.

If the metal is contacted from both ends, then the gate resistance becomes

$$R_G = \frac{1}{12} \frac{W}{L} R_{SH,G}. \quad (4\text{-}35)$$

One might expect the factor to be 1/6. The reason that the resistance decreases by four-fold instead of two-fold is the following. When the gate metal is contacted by two sides, then the location at which the metal current equals zero occurs at $x = W/2$, rather than W. The resistance already decreases by two-fold due to the width reduction. The overall resistance decreases by another two-fold due to the fact

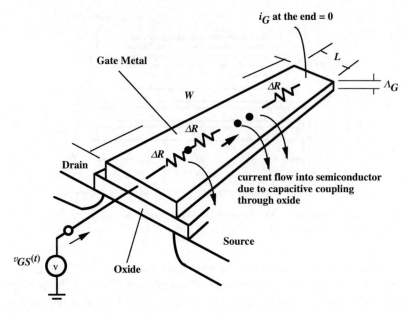

Fig. 4-28 Signal flow in the gate of a transistor. The gate charges enter through one end of the gate but none comes out at the other end. (From Ref. [6], p. 425, © Wiley 1999)

that there are two contacts connected in parallel. For those designers working with bipolar transistors, this factor of 1/12 is similar to that encountered in the intrinsic base resistance of the bipolar transistors [11].

Equation (4-34) works for a single-finger device. If the device consists of several fingers, then the overall device gate resistance is given by

$$R_G = \frac{1}{3}\frac{W}{L}R_{SH,G} \times \frac{1}{N}. \qquad (4\text{-}36)$$

Here, W represents the gate width per finger and N is the number of fingers. If all of the N fingers are connected at both ends, then the factor 1/3 should be replaced by 1/12.

If all we care is the dc analysis, then the neglect of gate resistance in BSIM3 would not cause any harm. After all, the dc gate current is always identical to zero. (As discussed in Section 2.1, BSIM3 does not include any gate leakage current, so the dc gate current in BSIM3 is truly zero all the time.) The voltage drop across the gate resistor would be zero anyway. However, in an ac or transient analysis, a finite amount of transient gate current exists due to the coupling through the gate capacitance. This current flowing through the gate resistor develops a finite voltage drop, reducing the available gate-to-source voltage in the intrinsic device to invert the channel.

Besides the issue associated with reduced effective gate-to-source voltage, there are two additional errors when the gate resistance is neglected. First, the input

4.5 LACK OF GATE RESISTANCE (AND ASSOCIATED NOISE)

resistance seen through the gate becomes dramatically smaller if the gate resistance is omitted. As discussed in Section 4.1, the input resistance in the intrinsic MOSFET is purely imaginary, without any real part. The inclusion of the gate resistance greatly affects the overall device's y-parameters (or s-parameters) and thus have a profound impact in the impedance matching design. Second, the noise associated with the gate resistance automatically drops out of the picture if the gate resistance is neglected. Some people may wonder if the noise spectral density of a distributed resistance is the same as if the resistance were a normal, lumped element. It turns out that the same formula of spectral density of Eq. (2-49) can be used if R in the equation is substituted by a distributed resistance. Therefore, specifically for the distributed gate resistance under consideration, the noise spectral density associated with it is given by

$$\frac{\overline{i_{n,RG}^2}}{\Delta f} = \frac{4kT}{R_G} \quad \left(\text{in } \frac{A^2}{Hz}\right), \tag{4-37}$$

where R_G has the identical expressions as those given above. [That is, depending on whether it is contacted both sides, we should use either Eqs. (4-34), (4-35), or (4-36).] A sample calculation of the relatively magnitudes of the gate resistance noise compared to the channel's thermal noise can be found in Ref. [12].

Figure 4-29a is a physical representation of a transistor with a distributed gate resistance. The BSIM3 model does not account for the gate resistance, a problem which can be solved by two techniques. In the first technique, we replace the transistor by an ideal transistor without any gate resistance and then add a lumped R_G, as shown in Fig. 4-29b. We are basically approximating the distributed nature of the current conduction with a lumped resistor whose value is given Eq. (4-34). However, the transformation from Fig. 4-29a to Fig. 4-29b can fail under some circumstances. A more physically correct technique, yet more laborious, is to dissect the single $W \times L$ transistor of Fig. 4-29a into N sections, each subtransistor then has a width equal to W/N and a length of L, as shown in Fig. 4-29c. Because Fig. 4-29c is meant to reproduce the distributed nature of the transistor, each gate resistance associated with a given section, R, is given by $R_{SH,G} \times W/L$, without the distributed-current factor of $1/3$.

According to a transient analysis of the distributed nature of the gate resistance [10], the approximation of Fig. 4-29a by Fig. 4-29b fails when the time of interest is shorter than the RC time constant τ_{RG}:

$$\tau_{RG} = R_G C_{gg} = \frac{1}{3}\frac{W}{L} R_{SH,G} C_{gg}, \tag{4-38}$$

where C_{gg} in the saturation region is roughly $2/3$ times the oxide capacitance (see Appendix B, Eq. B-5). Generally, the time scale of interest will be shorter than τ_{RG} for typical gate width. Therefore, the approximation with a lumped R_G is valid under almost all cases. For example, let us consider a $W/L = 56/0.35\,\mu m$ NMOS. It is established that, if ideal NMOS and PMOS having zero distributive gate resistance

320 IMPROVABLE AREAS OF BSIM3

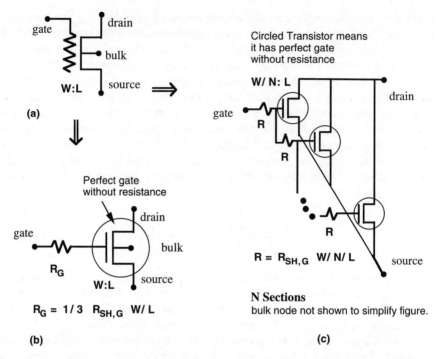

Fig. 4-29 (a) A physical representation of a transistor with a distributed gate resistance; (b) an equivalent circuit with a lumped gate resistance; (c) an equivalent circuit with N elements to more accurately represent the distributed nature of the gate resistance. (From Ref. [10], © IEEE 1998, reprinted with permission)

are concatenated to form a ring oscillator, the natural rise time of the output voltages coming out of n-th stage (n much larger than 1) is about 0.1 ns. This represents the shortest ramp time that is physically possible for this given technology. Then, we simulate an inverter transient analysis, with a 56/0.35 μm NMOS' source and bulk nodes connected to ground, drain connected to a load capacitor of 8 pF (its exact value is not important as long as it is not too small.) The gate sheet resistance is assumed to be 40 Ω/□, a fairly high value by today's standard. The load capacitor is initially charged to 3 V. The gate input voltage increases from 0 V to 3 V in a ramp time of 0.1, 0.2, 0.5, 1 to 2 ns. The simulation results are found in Fig. 4-30. The distributed simulation result, simulated with the circuit of Fig. 4-29c with $N = 160$ sections, is considered the exact solution since the distributed nature of the gate resistance is captured. The lumped resistance result with Fig. 4-29b, although being slightly less accurate, is seen to agree well with the distributed results, even with ramp time as small as 0.5 ns. For the 56/0.35 μm NMOS, $R_G = 1/3 \cdot R_{SH,G} \cdot W/L = 2133$ Ω. $C_{gg} = 2/3 \cdot WLC'_{ox} = 5.94 \times 10^{-14}$ F (the oxide thickness for this technology is 75 Å). τ_{RG}, calculated from Eq. (4-38), is equal to 0.13 ns. Therefore, the results between the distributed and the lumped circuits in Fig. 4-30 do not show much discrepancy, even for the ramp time down to 0.5 ns.

4.5 LACK OF GATE RESISTANCE (AND ASSOCIATED NOISE) 321

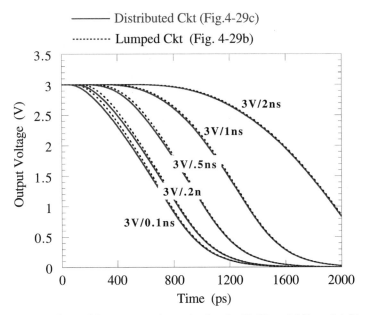

Fig. 4-30 Comparison of inverter transients simulated with Figs. 4-29b and 4-29c, under various gate voltage ramp times. $R_{SH,G} = 40\,\Omega/\square$. The details of the inverter circuit are described in the text. (From Ref. [10], © IEEE 1998, reprinted with permission)

However, at a ramp time of 0.1 ns (the limit of the technology according to the ring-oscillator simulation), some deviation starts to become noticeable.

We repeat the simulation for a ramp time of 1 ns only, in Fig. 4-31, but the $R_{SH,G}$ is varied from $4\,\Omega/\square$ to $400\,\Omega/\square$. As shown, at $R_{SH,G}$ of $400\,\Omega/\square$, the transient behaviors simulated from the lumped transistor approach begin to exhibit some inaccuracy. For this particular case, $R_G = 21333\,\Omega$, whereas C_{gg} remains at 5.94×10^{-14} F. τ_{RG}, calculated from Eq. (4-38), is equal to 1.3 ns. Since the ramp time in the simulation is only 1 ns, the lumped circuit results in a certain amount of error.

For ac analysis, the approximation by a lumped circuit would fail at high frequencies. The radian frequency above which the approximation becomes poor, according to the analysis of Ref. [6], can be given by

$$\omega_{RG} = \frac{1}{R_G C_{gg}} = \left(\frac{1}{3}\frac{W}{L} R_{SH,G} C_{gg}\right)^{-1}. \qquad (4\text{-}39)$$

It is found that ω_{RG} is equal to $1/\tau_{RG}$, although ω_{RG} and τ_{RG} are derived from completely independent analyses (one is ac, and another, transient). As the equation reflects, the wider the width, the smaller the value of the critical frequency. Figure 4-32 plots the input resistance (R_{in}) and input capacitance (C_{in}) as seen into the gate, for a FET with a C_{gg} of 5.22×10^{-13} F and a R_G of $33.2\,\Omega$. R_{in} equals to R_G and C_{in}

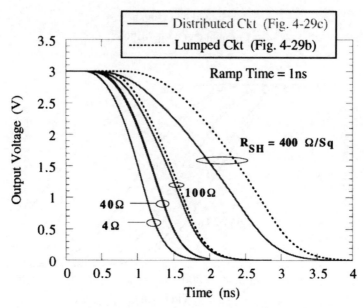

Fig. 4-31 Same as Fig. 4-30, except the ramp time is fixed at 1 ns and $R_{SH,G}$ is varied. (From Ref. [10], ©IEEE 1998, reprinted with permission)

is equal to C_{gg} at low frequencies, but decrease in values as ω reaches $\omega_{RG} = 9$ GHz. A device's performance always degrades over frequency. An intuitive expectation would be that the input resistance increased with frequency, since degradation in device performance was thought to be accompanied by a higher resistance. However, the result shown in Fig. 4-32 suggests that the input resistance decreases with frequency when ω exceeds ω_{RG}. This is because when the frequency gets higher, less and less amount of signal travels through the entire width of the gate before diverting to the channel. When only a certain portion of the width is accessible to the input signal, the effective width decreases and the input resistance decreases. This is similar to the intrinsic base resistance's dependence on frequency in bipolar transistors [11].

The inclusion of R_G is especially important for transistor with wide width, generally found in high-power or low-noise applications. The gate resistance must be included for a proper input impedance matching, as well as noise calculation

We mentioned that, in BSIM3, the branch connecting the source and the gate does not contain a real part. If the parasitic source resistance is zero, then y_{11} is purely imaginary. For some designers, such a result is totally unacceptable. In an attempt to introduce a real part to y_{11}, they intentionally add a gate resistance to the device so that y_{11} is no longer purely real [13]. In this case, the resistance appended at the gate is not meant to be a physical resistance associated with the gate terminal. This is a good work-around of the QS model. Despite that some success can be achieved, we

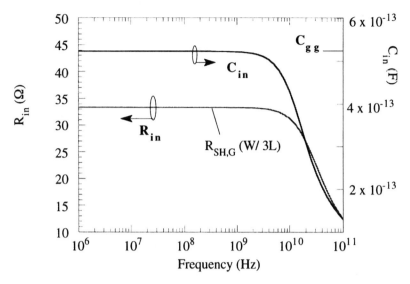

Fig. 4-32 Calculated input resistance (R_{in}) and input capacitance (C_{in}) as seen into the gate. They decrease at high frequencies because a smaller amount of signal reaches the end of the gate finger. (From Ref. [6] p. 428, © Wiley 1999)

will not expect such a quick fix to work for all frequency ranges and the fundamental solution to the modeling of NQS effects requires the change of the model itself.

We compare the geometrical dependency of the physical gate resistance R_G and the intrinsic channel resistance r_{ch}. According to Eq. (4-34), R_G is proportional to W. The wider the transistor, the higher the gate resistance. The r_{ch} given by Eq. (4-19), in contrast, is inversely proportional to W.

4.6 LACK OF SUBSTRATE DISTRIBUTED RESISTANCE (AND ASSOCIATED NOISE)

Unlike bipolar junction transistor (BJT) modeling, fitting the *s*-parameters is not an active part of the parameter extraction for the MOS transistor. For example, in BJTs, the fitting of the *s*-parameters as a function of biases is required to properly extract the base resistance parameters. In MOS modeling, however, *s*-parameters, if measured, are often used only as a check but serves no active role in the model extraction. After a modeling engineer measures a couple of dc I-V characteristics and some parasitic junction and overlap capacitances, he can extract the whole set of SPICE parameters. Most of the parameters, including those with profound high frequency impact, are in fact determined from a fit of the dc I-V characteristics. This practice, at least up to now, has been acceptable because the SPICE model had mainly served the digital circuit designers. For this type of circuits, especially in shorter channel devices, the key to an accurate fit in the delay time is to have an accurate fit in the dc drive current and the parasitic capacitances. The digital circuit is designed with a large I_D, such that it dominates over the transient terms given in Eq. (2-25b), for example. A good modeling of the parasitic capacitances further reduces the error from the capacitive currents.

For rf engineers, the accurate fit to the dc drive current is not as critical. But it is vital for them to have reasonable fit to the s-parameters, with which the impedance matching and the power gain calculation are closely tied. Let us demonstrate a problem confronting the rf circuit designers when a model which accurately simulates the behaviors of digital circuits is used to calculate the s-parameters. The rf NMOS has a cutoff frequency of 46 GHz and a maximum oscillation frequency of 21 GHz. It has a channel length of 0.29 µm, composing of 32 fingers to yield a total width of 256 µm. The threshold voltage is about 0.35 V. Figure 4-33 illustrates the comparison between the measured (de-embedded) and BSIM3-simulated s_{22}, at two bias conditions. Results of other bias conditions can be found in Ref. [8]. The calculated s_{22} from a modified BSIM3 model with added-on substrate resistive network, to be discussed shortly, is also shown. Similarly, Fig. 4-34 exhibits the comparison of the measured (de-embedded) and BSIM3-simulated s_{21}. These figures reveal a clear mismatch between the measured and simulated s-parameters, although the BSIM3 model has been found to work well in digital circuits and accurately predict the delay times in digital gates.

s_{22} characterizes the output impedance when the input port (gate) is terminated with $R_o = 50\,\Omega$. The measured s_{22} data in Fig. 4-33 occupy on a roughly semicircle characterized by $r \approx 0.6$, indicating that the real part of the output impedance (R_{out}) is $\approx 0.6 \times 50 \approx 30\,\Omega$ and is independent of frequency. Figure 4-35a illustrates an intuitive but wrong approach to determine the value of R_{out}. Because the transistor is in saturation, on the first order the impedance path to ground seen from the output port consists of a serial connection of the overlap and fringing gate–drain capacitance ($C_{ov} + C_f$) and the terminating resistance of 50 Ω. R_{out}, in this consideration, is identical to 50 Ω. However, the circuit of Fig. 4-35a neglects the presence of a larger parasitic capacitance, the drain–bulk junction capacitance ($C_{j,\mathrm{DB}}$). Because the default BSIM3 configuration ties the bulk to the source, the path connecting the output to the ground gives rise to a R_{out} of about 0 Ω, as shown in Fig. 4-35b.

An approach to amend the problem is to connect a resistor to the bulk node. As shown in Fig. 4-35c, the output impedance consists of $C_{j,\mathrm{DB}}$ in series with R_{sub}. R_{out} is therefore said to be equal to R_{sub}.

Unfortunately, this reasoning is still incorrect. Fig. 4-35d gives a more accurate representation of the transistor. Besides $C_{j,\mathrm{DB}}$, the source–bulk junction capacitance ($C_{j,\mathrm{SB}}$) also exists at the bulk node. The overall output impedance consists of $C_{j,\mathrm{DB}}$ in series with a parallel combination of R_{sub} and $C_{j,\mathrm{SB}}$. At high frequencies, $C_{j,\mathrm{SB}}$ bypasses the signal which normally flows through R_{sub}, reducing R_{out} from the low frequency value of R_{sub} toward 0 at high frequencies. The measurement, in contrast, reveals that R_{out} remains constant with frequency. Merely adding a substrate resistance to the bulk node of the transistor clearly does not lead to a good fitting of s_{22} across all frequencies. (It may be sufficient at lower frequencies.)

4.6 LACK OF SUBSTRATE DISTRIBUTED RESISTANCE (AND ASSOCIATED NOISE)

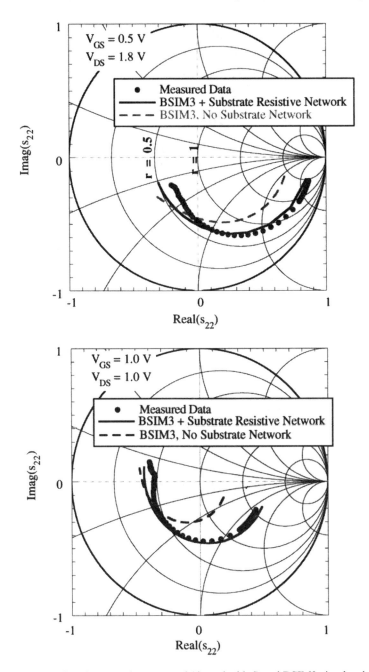

Fig. 4-33 Comparison between the measured (de-embedded) and BSIM3-simulated s_{22} for a $W/L = 256/0.29\,\mu\text{m}$ device. (a) $V_{GS} = 0.5\,\text{V}$, $V_{DS} = 1.8\,\text{V}$. (b) $V_{GS} = V_{DS} = 1.0\,\text{V}$. (From Ref. [8], © IEEE 1997, reprinted with permission)

326 IMPROVABLE AREAS OF BSIM3

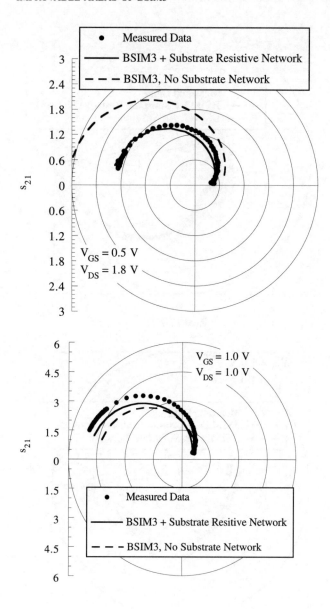

Fig. 4-34 Comparison between the measured (de-embedded) and BSIM3-simulated s_{22}. (a) $V_{GS} = 0.5$ V, $V_{DS} = 1.8$ V. (b) $V_{GS} = V_{DS} = 1$ V. (From Ref. [8], © IEEE 1997, reprinted with permission)

4.6 LACK OF SUBSTRATE DISTRIBUTED RESISTANCE (AND ASSOCIATED NOISE)

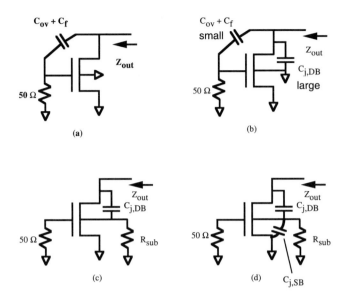

Fig. 4-35 Several intuitive but wrong approaches to determine the value of R_{out}, the real part of the output impedance of a MOS. (a) Considers overlap capacitances only; (b) considers junction capacitance but no substrate resistance; (c) appends a substrate resistance; (d) considers both junction capacitances and a substrate resistance. (From Ref. [8], © IEEE 1997, reprinted with permission)

The above exercise reveals the importance of inserting a resistance between $C_{j,DB}$ and $C_{j,SB}$. Without such a resistance, $C_{j,SB}$ always shunts the signal flow from the output to ground at high frequencies. In the BSIM3 model, $C_{j,DB}$ and $C_{j,SB}$ are hardwired to the bulk node, not allowing for an addition of resistance between the two ends of the capacitors. We propose declaring the transistor's junction areas and junction peripheries to be zero, thereby eliminating $C_{j,DB}$ and $C_{j,SB}$ in the transistor. Instead, semiconductor capacitances tantamount to $C_{j,DB}$ and $C_{j,SB}$ are declared externally. In this manner, resistances ($R_{sub,2} + R_{sub,3}$) can be inserted between the junction capacitances, as shown in Fig. 4-36. In accordance with the symmetry criteria, $R_{sub,2} = R_{sub,3}$ and $R_{sub,1} = R_{sub,4}$. Figure 4-37 is the small-signal equivalent circuit when the substrate resistive network is added to the basic equivalent circuit of Fig. 2-24. From Fig. 4-37, we observe that R_{out} is essentially equal to $R_{sub,1}$. Its magnitude is not affected by $C_{j,SB}$ because of the high-resistance path through $R_{sub,2} + R_{sub,3}$. We placed a channel resistance (r_{ch}) in the model so that it becomes even more accurate for all bias conditions. However, as discussed, BSIM3 is a QS model and r_{ch} does not exist in a BSIM3 equivalent circuit. The rest of the substrate resistive network, however, can be readily implemented in a BSIM3 framework by devising a subcircuit. The equivalent circuit model includes terminal resistances and inductances. Their values are extracted from a shorted test structure. While the majority of the serial impedances is introduced by the feedthroughs of the rf probing structure, R_G is due to the distributed gate resistance discussed in Section 4.5.

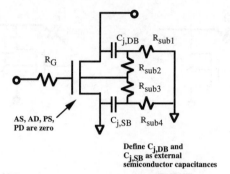

Fig. 4-36 An approach to correctly model the substrate effects at high frequencies. The transistor's junction areas and junction peripheries are deliberately made to be 0. Semiconductor capacitances tantamount to $C_{j,\text{DB}}$ and $C_{j,\text{SB}}$ are declared externally. (From Ref. [8], © IEEE 1997, reprinted with permission)

We have two comments. First, the substrate resistive model of Fig. 4-36 is not the only model that can give desired accuracy. Any model that separates the junction capacitances from the intrinsic device's bulk node and inserts a certain substrate resistance between the two junction capacitances will do the job. Second, the substrate resistances in the network, such as $R_{\text{sub},1}$, and so on, are not easily estimated because the current flow is multidimensional. A good formula to calculate the resistance can be that used to calculate the subcollector resistance in a bipolar structure [6, Fig. 4-19]. Alternatively, it has been published that the sheet resistivity

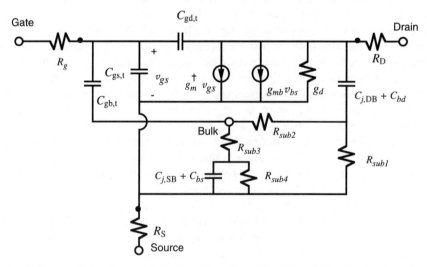

Fig. 4-37 Small-signal equivalent circuit based on Fig. 2-24, with the addition of the substrate resistive network of Fig. 4-36. (From Ref. [8], © IEEE 1997, reprinted with permission)

is 0.76 Ω-cm for a 0.5 µm epitaxial CMOS process, and 0.33 Ω-cm for a 0.25 µm bulk CMOS process [14].

The substrate resistance network is required for an accurate fit in s_{22} and s_{21} for all biases, as demonstrated in Fig. 4-36. Additionally, the resistive network is important for the fitting in s_{11} when the transistor is not yet turned on. This is due to the coupling of $C_{gb,t}$ [8].

4.7 INCORRECT SOURCE/DRAIN ASYMMETRY AT $V_{DS} = 0$

Power MOS structures, such as drain-extended MOS, have an intentionally lower doped region in the drain junction to sustain high voltage drop. However, the source does not have such a region so that the source resistance can be small. The structures wherein the source and the drain differ are called *asymmetrical structures*. Most frequently, MOS transistors are fabricated in a symmetrical manner, with the source and drain terminals indistinguishable from each other. These symmetrical structures is the subject of this section.

Based on the device symmetry, we expect the characteristics associated with the drain and the source terminals to be identical when $V_{DS} = 0$. This means

$$C_{dg} = C_{sg}; \quad C_{gd} = C_{gs}; \quad C_{db} = C_{sb}; \quad C_{bd} = C_{bs}; \quad C_{ds} = C_{sd}$$
$$@ V_{DS} = 0. \quad (4\text{-}40)$$

Further, when $V_{DS} = 0$, we can consider the drain and source are tied together to form a common node. Therefore, the gate and drain/source terminals of the MOSFET constitute a two-terminal device, and we expect $C_{dg} = C_{gd}$ at $V_{DS} = 0$. A similar logic can be applied to the bulk terminal. Therefore, $C_{db} = C_{bd}$ at $V_{DS} = 0$. Expressing these as well as other similar expressions, we write

$$C_{dg} = C_{gd}; \quad C_{sg} = C_{gs}; \quad C_{db} = C_{bd}; \quad C_{sb} = C_{bs} \quad @ V_{DS} = 0.$$
$$(4\text{-}41)$$

In fact, at $V_{DS} = 0$ and under strong inversion, according to Appendix B:

$$C_{dg} = C_{gd} = C_{sg} = C_{gs} = \frac{1}{2} WLC'_{ox}. \quad (4\text{-}42)$$

$$C_{db} = C_{bd} = C_{sb} = C_{bs} = \frac{\delta}{2} WLC'_{ox}. \quad (4\text{-}43)$$

$$C_{ds} = C_{sd} = -\frac{1}{6} WLC'_{ox}. \quad (4\text{-}44)$$

We mentioned in Section 1.10 that BSIM3 is a source-referenced model. That is, its model equations are formulated with the implicit assumption that the source is the terminal to which all the voltages are referenced. Source-referencing, while convenient to use due to historical reasons, nonetheless destroys the device symmetry in the model construction. The device symmetry is more easily preserved in bulk-referencing models, due to the fact that the device is symmetrical about the bulk of the MOSFET.

Because of its use of source-referencing, BSIM3 fails to maintain the equalities specified in Eqs. (4-40) and (4-41). Figure 4-38 shows the plot of device capacitances at $V_{DS} = 0$. The equivalence should be observed for all V_{GS} values. However, as the figure demonstrates, the equivalence in BSIM3's capacitance

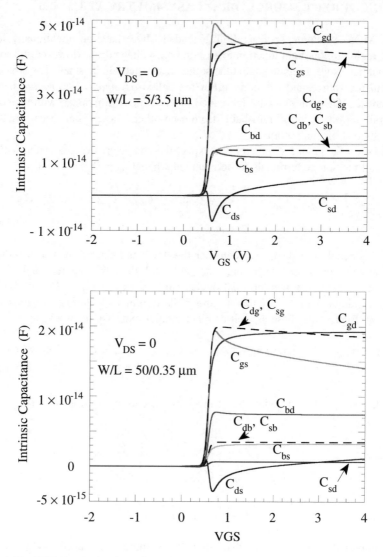

Fig. 4-38 BSIM3-calculated device capacitances at $V_{DS} = 0$. (a) $W/L = 5/3.5\,\mu\text{m}$; (b) $W/L = 50/0.35\,\mu\text{m}$.

4.7 INCORRECT SOURCE/DRAIN ASYMMETRY AT $V_{DS} = 0$

modeling is observed only with $C_{dg} = C_{sg}$ and $C_{db} = C_{sb}$. All the other equivalencies fail.

There are four capacitance models in BSIM3. The parameter CAPMOD can be set to 0, 1, 2, or 3 to select a particular model (see Section 3.2, under the entry of CAPMOD). We mentioned that CAPMOD 2 or 3 should be used, because they result in continuous capacitance as well as its derivatives. This property is important in maintaining numerical robustness during simulation. All of the simulation discussed so far in this section is obtained with CAPMOD = 2. Now, here is a Catch-22. While CAPMOD = 2 ensures continuity and differentiability, it has the asymmetry problem exemplified in Fig. 4-38. CAPMOD = 0, on the other hand, has the differentiability problem, but its C_{gs} and C_{gd} are fairly close for all V_{GS} values at $V_{DS} = 0$. Figure 4-39 compares the simulation results for CAPMOD = 0 and 2. We see that the difference between C_{gs} and C_{gd} is much smaller for CAPMOD = 0, especially at V_{GS} nearby V_T, a region where significant error exists in the CAPMOD = 2 result. This is a reason why, even to this day, CAPMOD = 0 is used (besides the issue of numerical efficiency). The CAPMOD = 1 result is similar to CAPMOD = 0. That is, the difference between C_{gd} and C_{gs} is small, but the capacitance is also piecewise-connected.

In addition to capacitance, there is a practical analog circuit which highlights the symmetry problem in BSIM3. The circuit is shown in Fig. 4-40. As V_{inp} varies, the drain voltage (equal to $V_{inp} + V_{common}$) increases from 1.5 to 3 V, whereas the source voltage (equal to $-V_{inp} + V_{common}$) decreases from 1.5 to 0 V. When $V_{inp} = 0$, $V_{DS} = 0$ and we expect the device characteristics to be inversely symmetrical about $V_{inp} = 0$. That is,

$$I_D(V_{inp}) = -I_D(-V_{inp}). \tag{4-45}$$

Because I_D is an odd function of V_{inp}, the even-order derivatives of I_D with respect to V_{inp} at $V_{inp} = 0$ must be zero. In particular,

$$\left.\frac{\partial^2 I_D}{\partial V_{inp}^2}\right|_{V_{inp}=0} = 0. \tag{4-46}$$

Figure 4-41 illustrates $\partial I_D/\partial V_{DS}$ and g_d for $L = 5$ μm and 0.35 μm devices. We show g_d merely to point out that it is not identical to $\partial I_D/\partial V_{DS}$; rather, g_d is equal to the derivative at the condition that V_{GS} and V_{BS} are constant. In this circuit, V_{GS} and V_{BS} are not constant.

The derivative $\partial I_D/\partial V_{DS}$ of the long-channel device shapes like a parabola, reaching the minimum at $V_{inp} = 0$. This fact implies that Eq. (4-46) is upheld and BSIM3 model behaves as expected. However, for the short-channel device, we find a kink right at $V_{inp} = 0$, and Eq. (4-46) is not fulfilled. The presence of the kink indicates that the BSIM3 model, although properly extracted and found to predict digital circuit performance, suffers from the device symmetry problem. Caution must be exercised when the MOSFETs in the circuits operate at the $V_{DS} = 0$ region.

332 IMPROVABLE AREAS OF BSIM3

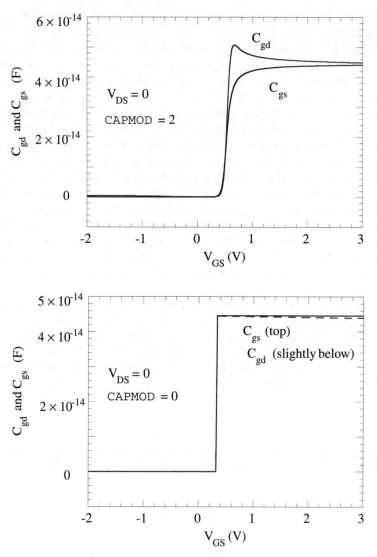

Fig. 4-39 BSIM3-calculated C_{gd} and C_{gs} at $V_{DS} = 0$. (a) CAPMOD = 2; (b) CAPMOD = 0.

Conceivably, a new effective V_{DS} function could be introduced to remove the asymmetry in I_D, through purely numerical means. (For example, $V_{DS,\text{new}}$ could be defined as $\sqrt{[(V_{DS}^2 + \delta^2)] - \delta^2/\sqrt{(V_{DS}^2 + \delta^2)}}$, as once considered in BSIM4 but then later dropped due to some numerical issues.) This type of solution would remove the current problem that $C_{dg} \neq C_{gd}$ at $V_{DS} = 0$. However, the problem that $C_{dg} \neq C_{sg}$ at $V_{DS} = 0$ would likely remain.

4.8 INCORRECT C_{gb} BEHAVIORS 333

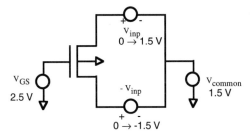

Fig. 4-40 An analog circuit that reveals the symmetry problem of BSIM3.

We have a last item on the subject of symmetry. Because the source and drain terminals are interchangeable, we require y_{ds} be equal to y_{sd} at $V_{DS} = 0$. These two y-parameters, first derived in Section 1.9, are given below:

$$y_{ds} = -g_m - g_{mb} - g_d - j\omega C_{ds}. \qquad (4\text{-}47)$$

$$y_{sd} = -g_d - j\omega C_{sd}. \qquad (4\text{-}48)$$

When $C_{ds} \neq C_{sd}$ at $V_{DS} = 0$, the imaginary parts of y_{ds} and y_{sd} will not agree. Further, it seems that the real part of y_{ds} and y_{sd} would never coincide at $V_{DS} = 0$ since the real parts of their expressions are quite different. It turns out that at $V_{DS} = 0$, g_m and g_{mb} are identically zero as simulated in BSIM3. Therefore, there is no discontinuity in the real part of y_{ds} (or y_{sd}) as V_{DS} switches from positive to negative.

4.8 INCORRECT C_{gb} BEHAVIORS

The most significant enhancement from BSIM3v3.1 to BSIM3v3.2 is the so-called "charge-thickness capacitance model" discussed in Section 3.2 (under the entry ACDE). Another important, yet less prominent, capacitance-model enhancement involves the revision of the flatband voltage used for C-V calculation, $V_{FB,CV}$. The various expressions for $V_{FB,CV}$ used in BSIM3v3.2 and BSIM3v3.1 are clarified in Section 3.2 (under the entry VFBCV). They include, among others, a bias-dependent flatband voltage ($V_{fb,(v)}$), used mostly in BSIM3v3.1 and a zero-bias flatband voltage ($V_{fb,zb}$), used mostly in BSIM3v3.2.

Figure 4-42a displays C_{gg}, C_{gd}, C_{gs}, and C_{gb} of a $L = 0.35\,\mu m$ MOS, while Fig. 4-42b exhibits $C_{gg,t}$, $C_{gd,t}$, $C_{gs,t}$, and $C_{gb,t}$. The figure is simulated with the bias-dependent flatband voltage of BSIM3v3.1. From the plot of the intrinsic capacitances, we immediately spot three problems. First, C_{gg} should always be greater than or at least equal to C_{gb}. Yet in the accumulation, Fig. 4-42a shows a C_{gb} which is larger than C_{gg}. A side effect of this is that C_{gg} in the inversion region is slightly larger than the C_{gg} in the accumulation region. Experimentally, it should be the other way around. The second problem lies in that C_{gd} and C_{gs} are negative, a rather unphysical outcome. The third problem, less troublesome, is that C_{gb} is finite rather than zero during strong inversion. When $V_{DS} = 0$ and V_{GS} is large enough such that

334 IMPROVABLE AREAS OF BSIM3

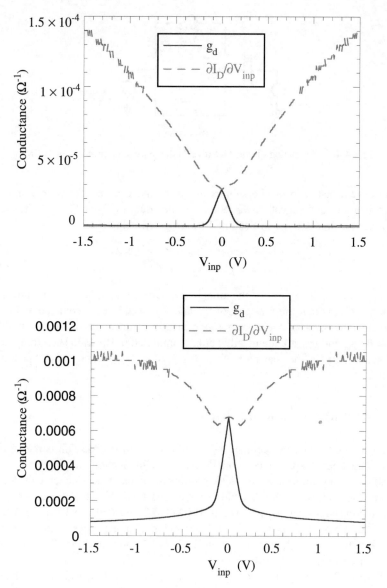

Fig. 4-41 Comparison of $\partial I_D/\partial V_{DS}$ and g_d for the circuit of Fig. 4-40. (a) $L = 5\,\mu\text{m}$; (b) $L = 0.35\,\mu\text{m}$.

the channel inverts, we expect the channel charge to shield off the body from the gate. Therefore, C_{gb} should be zero, an intuitive result which has also been verified with MEDICI simulation. These problems are not as visible when the total capacitances are plotted instead, as in Fig. 4-42b. The most noticeable problem is that $C_{gg,t}$ in the inversion region exceeds that in the accumulation region, by a

4.8 INCORRECT C_{gb} BEHAVIORS   335

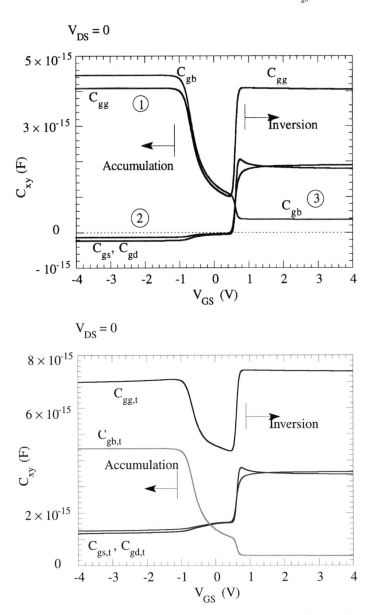

Fig. 4-42 Calculated capacitances of a $L = 0.35\,\mu\text{m}$ NMOS, with the bias-dependent flatband voltage of BSIM3v3.1. (a) C_{gg}, C_{gd}, C_{gs}, and C_{gb}; (b) $C_{gg,t}$, $C_{gd,t}$, $C_{gs,t}$, and $C_{gb,t}$.

336 IMPROVABLE AREAS OF BSIM3

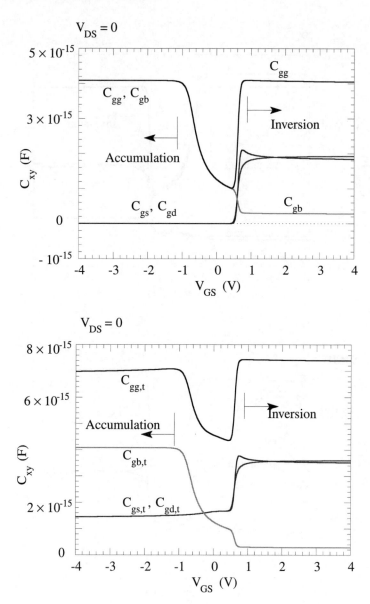

Fig. 4-43 Same calculation as Fig. 4-42, except with the zero-bias flatband voltage of BSIM3v3.2.

noticeable amount. Although $C_{gs,t}$, for example, is positive because the parasitic $C_{ov} + C_f$ masks out the negative intrinsic C_{gs} value, the negative C_{gs} still indicates something is fundamentally incorrect about the model. It turns out that the rudimentary cause for all the mentioned problems is the bias-dependent flatband voltage expression adopted in BSIM3v3.1.

4.8 INCORRECT C_{gb} BEHAVIORS

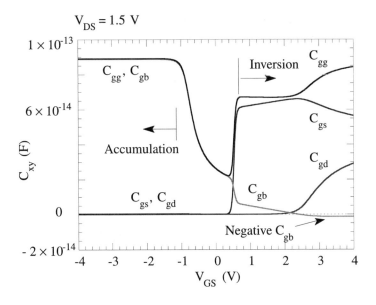

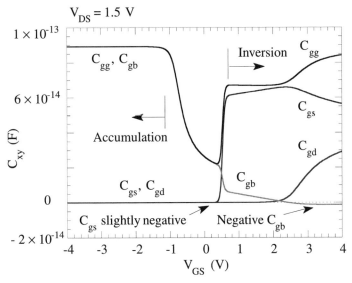

Fig. 4-44 Calculated capacitances of a long-channel ($L = 3.5\,\mu\text{m}$) NMOS. (a) with BSIM3v3.1; (b) with BSIM3v3.2. The two simulation results are nearly identical; they differ only in some fine features.

In BSIM3v3.2, a new, bias-independent, flatband voltage expression is used. Immediately the capacitance behaviors in the accumulation region are corrected. As shown in Fig. 4-43, with the modified flatband expression, C_{gs} and C_{gd} are no longer negative in the accumulation region. C_{gb} no longer exceeds C_{gg} there, either. There

338 IMPROVABLE AREAS OF BSIM3

are only two fairly minor problems that remain. The first is that C_{gb} is still greater than zero in the inversion region. Another problem is that C_{gg} of inversion is slightly larger than the C_{gg} of accumulation.

Figures 4-42 and 4-43 are short-channel results. The long-channel intrinsic capacitances are shown in Fig. 4-44 ($L = 3.5\,\mu\text{m}$). Figure 4-44a, simulated with the BSIM3v3.1 flatband voltage expression, does not have the same negative C_{gd}, C_{gs} problem seen in the short-channel device. However, we observe that C_{gb} can become negative, as shown in Fig. 4-44a. This problem occurs for all V_{DS}

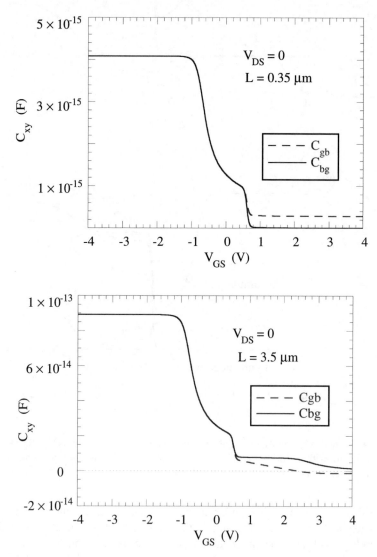

Fig. 4-45 Calculated C_{gb} and C_{bg} with BSIM3v3.2. (a) $L = 0.35\,\mu\text{m}$; (b) $L = 3.5\,\mu\text{m}$.

values, although the figure applies to $V_{DS} = 1.5$ V. The negative C_{gb} is clearly an unphysical result. This problem persists even when BSIM3v3.2 flatband voltage expression is used, as shown in Fig. 4-44b. In fact, Figs. 4-44a and 4-44b appear identical except for some fine features.

The last problem associates with the gate-to-bulk capacitance relates to the fact that C_{gb} should be equal to C_{bg} under all bias conditions. BSIM3, however, fails to maintain this property. Figure 4-45 illustrates the calculated C_{gb} and C_{bg}, for a short-channel transistor ($L = 0.35\,\mu\text{m}$) in part a and for a long-channel transistor ($L = 3.5\,\mu\text{m}$) in part b. In both calculations, the zero-bias flatband voltage adopted in BSIM3v3.2 is used. The figure demonstrates that C_{gb} and C_{bg} calculated by BSIM3 differ, particularly in the inversion region.

4.9 CAPACITANCES WITH WRONG SIGNS

We focus only on the intrinsic capacitances in this section. The parasitic capacitances are made negligible by zeroing BSIM3 parameters such as `CJ`, `CGD0`, and `CF`. The BSIM3 model, prior to the modification on the parasitic capacitance parameters, has been properly extracted and found to fit digital circuit performances well.

C_{gd}, according to Eq. (B-6) in Appendix B, has the minimum value of zero. It is never negative. However, as demonstrated in Fig. 4-46 for an $L = 0.35\,\mu\text{m}$ device biased at $V_{DS} = 2$ V, BSIM3 can produce negative C_{gd} values. This problem is especially serious in short-channel devices operating in saturation. Sometimes it was argued, somewhat qualitatively, that short-channel effects can cause C_{gd} to be

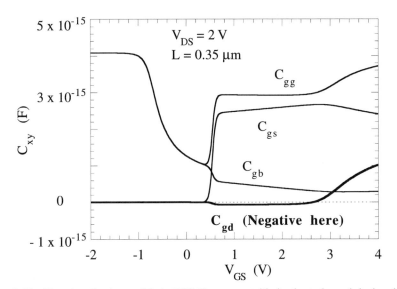

Fig. 4-46 Negative C_{gd} is possible in BSIM3, most notable in short-channel devices biased in saturation.

340 IMPROVABLE AREAS OF BSIM3

negative. However, in light of so many capacitance problems, either mentioned before or to follow, it is likely that the negative C_{gd} is simply caused by improper modeling, perhaps associated with the source-referencing.

Partly because C_{gd} can be negative, C_{dd} calculated by BSIM3 sometimes also takes on negative values. This is again unphysical. According to the formula given by Eq. (B-10), as well as MEDICI simulation, C_{dd} should always be positive. For a change, we display this negative C_{dd} problem by plotting C_{dd} against V_{DS} (rather than V_{GS}) in Fig. 4-47. We see that, just like the negative C_{gd} problem, negative values of C_{dd} occur mostly at high V_{DS} values, in the saturation operation region.

In a two-terminal device, we expect the device capacitances to be positive. In three-terminal or four-terminal devices, it is possible to have negative capacitances. For MOSFET, C_{ds} and C_{sd} are negative, and physical explanation for having negative values are found in Ref. [15]. Figure 4-48 illustrates these two capacitances calculated by BSIM3, for both a long- and a short-channel device. The figure indicates that C_{ds} and C_{sd} calculated by BSIM3 are not always negative, especially in the linear operating region. In fact, C_{sd} is always positive, never reaching zero. Further, the magnitudes of these two capacitances do not match with the theoretical expressions given in Eqs. (B-11) and (B-14). For example, according to the equations, we expect $C_{sd} = C_{ds} \neq 0$ when $V_{DS} = 0$. However, the figure shows that C_{sd} and C_{ds} both start out being nearly zero at $V_{DS} = 0$.

It is fair to say that many capacitances in BSIM3 do not agree with their theoretical expressions, especially the intrinsic components. Nonetheless, as long as the key capacitances (such as C_{gg}, C_{gs}, and C_{gd}) are correctly modeled, BSIM3 has been quite successful in accurately simulating digital and small-signal analog circuits.

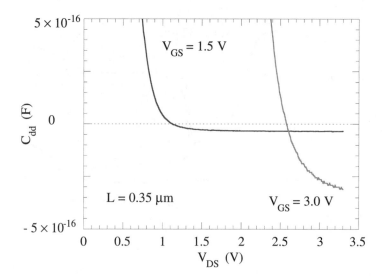

Fig. 4-47 Negative C_{dd} is possible in BSIM3, occurring mostly at high V_{DS} values.

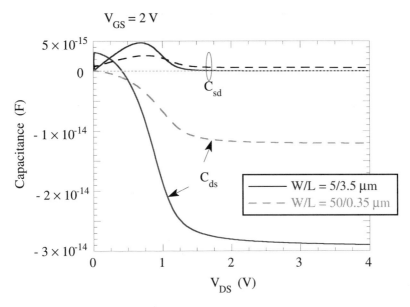

Fig. 4-48 BSIM3-calculated C_{ds} and C_{sd}; their behaviors are inconsistent with a theoretical analysis.

4.10 C_{gg} FIT AND OTHER CAPACITANCE ISSUES

The most up-to-date capacitance model in BSIM3 is the charge-thickness model (CAPMOD = 3). Because it requires more fitting parameters than the simpler CAPMOD = 2 model, the latter is still quite popular. The CAPMOD = 2 (or 1 or 0) model, unfortunately, does not produce accurate fitting in C_{gg}, especially for short-channel devices. Figure 4-49, plotting both simulated and measured C_{gg} as a function of V_{GS} at $V_{DS} = V_{BS} = 0$, illustrates the problem. The figure shows that the measured capacitance is higher at the accumulation operation region than inversion. The transition from accumulation to depletion, and then to inversion, is gradual. However, the CAPMOD = 2 simulation always gives the same maximum capacitance for both the accumulation and the inversion regions, and the transition between operating regions is more abrupt than the measured results. The inaccuracy in C_{gg} can be important in many analog circuits which bias the MOS transistor just above the threshold to maximize the ratio of g_m/I_D. It is therefore critical to accurately model the exact voltage at which C_{gg} turns on, as well as C_{gg}'s magnitude. The temptation to use MOSFET as a capacitor is especially great for analog designers who prefer high-density capacitor to reduce chip size. Typical metal-poly capacitor is about $2\,\text{fF}/\mu m^2$, while a MOS capacitor is about $7\,\text{fF}/\mu m^2$ (for an oxide thickness of 50 Å). However, the MOS capacitor modeling is not as accurate as its I-V modeling. Caution should be exercised when the MOS transistor is used as a capacitor.

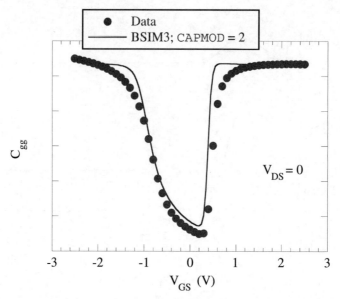

Fig. 4-49 Comparison of measured and calculated $C_{gg,t}$ (total gate capacitance) with CAPMOD = 2. Discrepancy in capacitance values in accumulation can be removed with CAPMOD = 3.

There are numerous occasions when the C_{gg} fit is much worse than that shown in Fig. 4-49. Figure 4-50 illustrates such a fit for a high-voltage asymmetrical MOS. The individual components making up $C_{gg,t}$, such as $C_{gd,t}$, $C_{gs,t}$, and $C_{gb,t}$, are also shown. The simulated C_{gg} is produced from a set of parameters extracted from I-V characteristics and parasitic capacitances. Figure 4-50 exhibits a considerable discrepancy between the simulated and measured curves, particularly in the slope at which $C_{gg,t}$ increases with V_{GS} during inversion.

Although the inaccurate fit of $C_{gg,t}$ in the inversion region is the center of our discussion here, we briefly mention other problems revealed in Fig. 4-50. The simulation, with CAPMOD = 2 and VERSION = 3.2, shows that the simulated $C_{gg,t}$ in accumulation decreases as V_{GS} becomes progressively more negative. This trend is opposite to the measured data, which attain the highest value when V_{GS} is most negative. The decrease of simulated $C_{gg,t}$ in the negative V_{GS} axis is due to a combination of two facts. First, the $C_{gd,p}$ of this high-voltage device shows significant bias dependence, a property also observed experimentally as discussed under the entry CGSO in Section 3.2. Second, the $C_{gd,p}$ decrease spans over a wide negative V_{GS} range. The decrease continues at V_{GS} values well beyond (more negative than) the simulated $V_{FB,CV}$. It is believed that the magnitude of the simulated $V_{FB,CV}$ is much smaller than the realistic value. Therefore, C_{gb} saturates at its peak value earlier than expected experimentally. These two facts work together to cause a decrease in the simulated $C_{gg,t}$. If the simulated $V_{FB,CV}$ were to have a much more negative value (than the roughly 0 value shown in Fig. 4-50), then C_{gb}

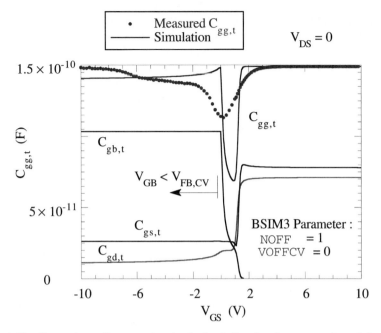

Fig. 4-50 Comparison of measured and calculated $C_{gg,t}$ (total gate capacitance) for a high-voltage MOS. Significant disagreement is found in the slope at which $C_{gg,t}$ increases with V_{GS}.

could continue to increase toward its maximum value even when $C_{gd,p}$ has decreased and approached an asymptotic value. In this manner, the simulated $C_{gg,t}$ would continue to increase with $-V_{GS}$, as observed in the measured data.

Another problem (not displayed in Fig. 4-50) is that C_{gd} is slightly negative in the accumulation region. This problem was discussed in Section 4.8. The observation that the simulated $C_{gs,t} \neq C_{gd,t}$ in inversion should not be construed as an error, however. The high-voltage transistor, such as the one under discussion, typically employs asymmetrical source–drain structure to maximize the breakdown voltage. The structure inherently has different $C_{gs,ov}$ and $C_{gd,ov}$.

Finally, the most serious problem revealed in Fig. 4-50, as far as the circuit design is concerned, is the wrong slope at which the simulated $C_{gg,t}$ increases with V_{GS} in the strong inversion region. The error is especially severe for analog-circuit transistors which are biased near the threshold voltage value. The simulation was performed with BSIM3v3.2, using defaulted NOFF of 1 and VOFFCV of 0. These two parameters are absent in BSIM3v3.1. They are introduced in BSIM3v3.2 to allow a greater flexibility in adjusting the effective V_{GS} value, and thereby improving the accuracy of C_{gs}. Applying default values of NOFF and VOFFCV in BSIM3v3.2 yields essentially the same results as BSIM3v3.1.

We take advantage of the fact that NOFF and VOFFCV are tunable parameters made available to us (since we are using BSIM3v3.2). After experimenting with several combinations of NOFF and VOFFCV values, we find that a NOFF of 8 and VOFFCV of −0.5 yield the best fit to $C_{gg,t}$, as shown in Fig. 4-51. We see that now the simulated $C_{gg,t}$ increases with V_{GS}, with the slope observed experimentally. The minimum $C_{gg,t}$ values, simulated or measured, are roughly the same. The displacement between the two curves in the negative V_{GS} axis can be explained by a discrepancy between the simulated $V_{FB,CV}$ and the realistic value. It seems that, at least from Fig. 4-51, we have made great stride toward accurate fitting in $C_{gg,t}$.

However, there is a suprise. Figure 4-52 illustrates the simulated $C_{gd,t}$, $C_{gs,t}$, and $C_{gb,t}$, which together comprise the simulated $C_{gg,t}$ shown in Fig. 4-51. These three components' dependencies on V_{GS} are totally unexpected, although their sum, $C_{gg,t}$, matches closely with the experimental data. For example, in the depletion region, $C_{gs,t}$ becomes negative while $C_{gd,t}$ continues to have relatively high values. These unexpected behaviors can be eventually attributed to the use of source-referencing in the device model.

Besides the problems associated with the fit to the total gate capacitance, there are other minor capacitance problems. When the transistor is not yet turned on, there is still a finite gate-to-drain capacitance. It is not zero because the gate overlaps with a portion of the drain junction. In a first-order analysis, the overlap capacitance (C_{ov}) can be treated as an ideal parallel plate capacitance whose magnitude is fixed independent of V_{GD}. In practice, because the n^+ drain is not infinitely doped, a small finite depletion region right underneath the oxide expands gradually as V_{GD} becomes

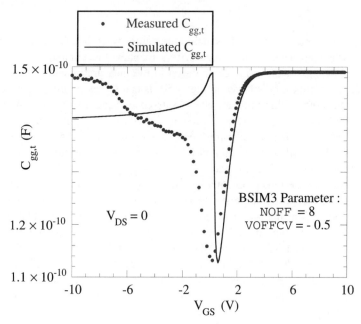

Fig. 4-51 The fit of $C_{gg,t}$ in Fig. 4-50 can be improved after setting NOFF to 8 and VOFFCV to − 0.5.

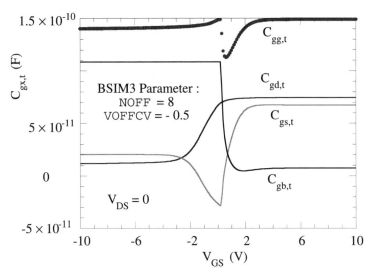

Fig. 4-52 Individual components of the $C_{gg,t}$ shown in Fig. 4-51 exhibit unphysical behaviors. For example, $C_{gs,t}$ can be negative.

more negative. Therefore, C_{ov} decreases gradually with $|V_{GD}|$. The degree of change is a direct function of the doping level in the drain. The doping profiles of a digital CMOS transistor are such that measured C_{ov} vs. V_{GD} exhibits a behavior shown as the solid line in Fig. 4-53. The overlap capacitance varies roughly 20% as V_{GD} changes from 0 V to -2.5 V. However, for analog CMOS, such as drain-extended or lightly-doped drain (LDD) structures, sometimes a deliberately low drain doping is used to increase the breakdown voltage (with a side benefit of lowered overlap capacitance). It is not impossible then to have a C_{ov} shown as dashed line in Fig. 4-53, which decreases by 10-fold as V_{GD} varies. The equation used in BSIM3 to model C_{ov}'s V_{GD} dependence works well with the solid curve of Fig. 4-53. However, the equation will not fit the dashed curve of the same figure. Just due to the way the equation is constructed in BSIM3, the equation likely cannot reproduce a capacitance variation of more than one order of magnitude. Although this shortcoming can be easily eliminated with a change of equation, it is nonetheless a problem now. An analog circuit designer using a MOS with exotic drain structures should always verify whether the overlap capacitance is modeled correctly. Chances are that the voltage-dependency is not well captured.

As the transistor dimension shrinks, the parasitic capacitances become a sizable portion of the overall device capacitance. It is therefore crucial to model the parasitics accurately. For the gate–drain (and the gate–source) parasitics, BSIM3 accounts for both the overlap capacitance and the fringing capacitance. However, BSIM3 misses a so-called *inner fringing capacitance*. The various components of the capacitances at $V_{GS} < V_T$ are shown in Fig. 4-54a. The inner fringing capacitance is present only when the channel is not yet turned on. Once V_{GS}

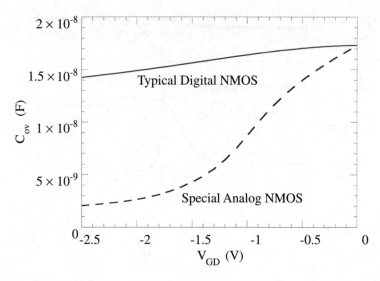

Fig. 4-53 $C_{gd,t}$ measured in a typical CMOS and a high-voltage MOS. They have fairly different voltage dependencies.

exceeds V_T, the inversion layer is formed, screening off the field line emanating from the source and the drain junctions. The inner fringing capacitance then becomes zero. The inner fringing capacitance can be estimated from conformal mapping, using simplified drawing of Fig. 4-54b. In the calculation of $C_{f,\text{inner}}$, the top gate metal is assumed to extend to $-\infty$ and $+\infty$. The bottom plate, of the n-well, extends from $-\infty$ to $x = 0$. All of the capacitances at $x < 0$ are assumed to be known, equal to $C_{ov} + C_f$ (see Fig. 4-54a). The inner fringing capacitance is therefore the

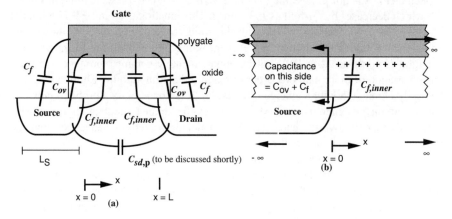

Fig. 4-54 (a) Various parasitic capacitances in a device prior to strong inversion. C_{inner} and $C_{sd,p}$ are not considered in BSIM3; (b) idealized device geometry used to estimate the inner fringing capacitance with conformal mapping.

capacitance between the top plate from $x = 0$ to $+\infty$, and the bottom plate. This capacitance, arising from the fringing field of the parallel capacitor, is given by [6]

$$C_{f,\text{inner}} = \frac{\epsilon_s}{\pi} W \quad (V_{GS} < V_T). \tag{4-49}$$

Let us compare the magnitude of $C_{f,\text{inner}}$ with C_{ov}, which is $\epsilon_{ox}/\text{TOX} \cdot W \cdot L_{ov}$, where L_{ov} is the overlap length. We have

$$\frac{C_{f,\text{inner}}}{C_{ov}} = \left(\frac{\epsilon_s}{\pi \epsilon_{ox}}\right) \frac{\text{TOX}}{L_{ov}}. \tag{4-50}$$

The factor inside the parentheses is roughly unity. In a $L = 0.35\,\mu m$ technology, TOX is $\sim 60\,\text{Å}$ and $L_{ov} \sim 600\,\text{Å}$. Hence, $C_{f,\text{inner}}$ is roughly 10% of C_{ov}. Because the inner fringing capacitance is not accounted for in a BSIM3 model, it needs to be added across the gate/source and the gate/drain terminals. $C_{f,\text{inner}}$ should be replaced by zero once the transistor turns on.

BSIM3 also neglects another parasitic capacitance, the capacitance between the source and the drain junctions ($C_{sd,p}$), shown in Fig. 4-54a. As the channel length is continually scaled down, $C_{sd,p}$ becomes important. Because the drain and source contacts are laid out horizontally on the same level, the usual parallel plate capacitance does not apply. However, through the use of conformal mapping, such a capacitance can still be found [6]

$$C_{sd,p} = \frac{\epsilon_s}{2K(k)} K'(k) W, \tag{4-51}$$

where $k = L/(L + 2 \cdot L_S)$, with L being the drawn length between the source and drain junctions, and L_S (see Fig. 4-54a) is the length of the source junction (assumed to be the same as L_D for the drain junction). $K(k)$, the complete elliptic integral of the first kind, is given by

$$K = \int_0^1 \frac{dw}{(1-w^2)^{1/2}(1-k^2w^2)^{1/2}} \tag{4-52}$$

and $K'(k) = K(\sqrt{(1-k^2)})$. We calculate $C_{sd,p}$ per µm width in Fig. 4-55, assuming that $L_S = L_D = 0.5\,\mu m$. In the same figure, we plot the oxide capacitance per µm width $= \epsilon_{ox}/\text{TOX} \cdot L$ for TOX $= 60\,\text{Å}$. Certainly the TOX value is arbitrary, but it provides a reference against which we can estimate the significance $C_{sd,p}$. The figure demonstrates that $C_{sd,p}$ becomes significant (10% of C_{ox}) as channel length decreases below 0.3 µm. The remedy to BSIM3's lack of modeling of $C_{sd,p}$ is to add a parallel capacitance between the source and drain nodes of the transistor, with a value calculated from Eq. (4-51) or estimated from Fig. 4-55.

BSIM3's capacitance model is based on long-channel device physics. There are some tweaks implemented so that BSIM3 contains a first-order improvement for

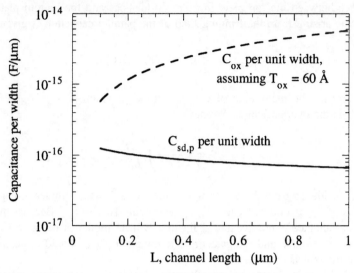

Fig. 4-55 Calculated $C_{sd,p}$ per μm width from Eq. (4-51) and C_{ox} per μm width. $L_S = L_D = 0.5$ μm and TOX = 60 Å.

short-channel devices. We recognize that short-channel capacitance modeling, taking account of the velocity saturation effects, is extremely difficult. In fact, we believe that BSIM3 has taken the right approach in adopting the long-channel framework for the development of device capacitances. This simplifies the model development time, allowing the BSIM3 team to work on other more pressing problems. However, with more and more devices exhibiting pronounced short-channel effects, the issues of short-channel capacitance modeling may need to be revisited.

The modeling of short-channel capacitance is further hampered by the fact that many of the MOS device capacitances cannot be easily measured (see Section 1.11). Lacking experimental data to compare against, we cannot assert that the short-channel C-V characteristics simulated by BSIM3 is incorrect, although we tend to think so because BSIM3 C-V is really a long-channel model. The next best alternative to experimental data is device simulation result. In the following we present the simulation data from MEDICI. The doping profiles used in the MEDICI simulation are generated from SUPREME3 simulation, whose simulation parameters, in turn, have been calibrated with experimental data of the dopant profiles. Some material parameters used by MEDICI, such as mobility and semiconductor work function, are tuned so that the MEDICI simulation agrees with measured I-V characteristics as well as C_{gg}-vs-V_{GS}. Once the simulation parameters have been properly tuned, we run an ac analysis at low frequency to extract the C-V characteristics.

Two devices are subjected to the above procedure. One is a long-channel device of $L = 48.9$ μm (the width is assumed 1 μm in MEDICI). The C-V characteristics,

4.10 C_{gg} FIT AND OTHER CAPACITANCE ISSUES

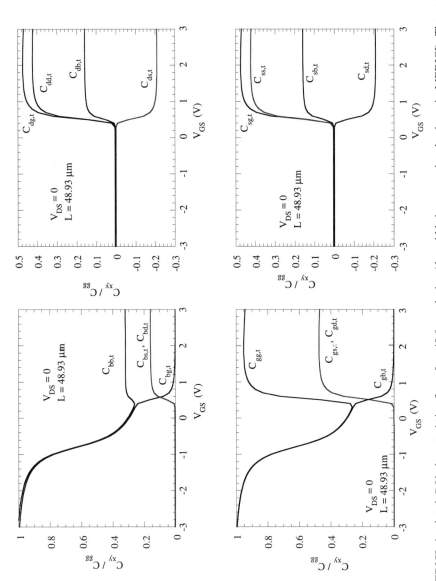

Fig. 4-56 MEDICI-simulated C-V characteristics for a $L = 48.9\,\mu m$ device (the width is assumed to be $1\,\mu m$ in MEDICI). The capacitances are normalized to the $C_{gg,t}$ of the accumulation region. $V_{DS} = 0$.

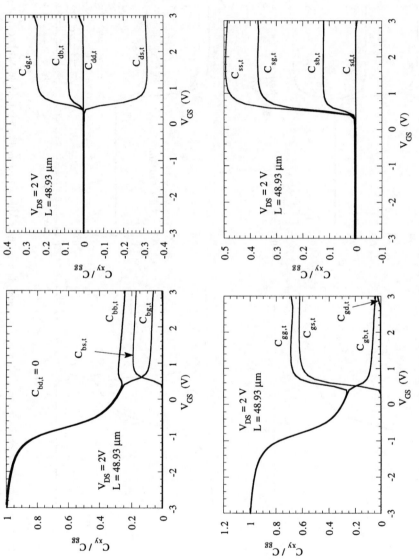

Fig. 4-57 Same as Fig. 4-56, except with $V_{DS} = 2$V.

4.10 C_{gg} FIT AND OTHER CAPACITANCE ISSUES

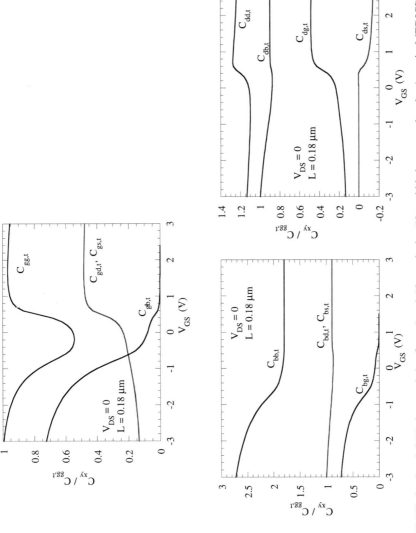

Fig. 4-58 MEDICI-simulated C-V characteristics for a $L = 0.18$ μm device (the width is assumed to be 1 μm in MEDICI). The capacitances are normalized to the $C_{gg,t}$ of the accumulation region. $V_{DS} = 0$.

352　IMPROVABLE AREAS OF BSIM3

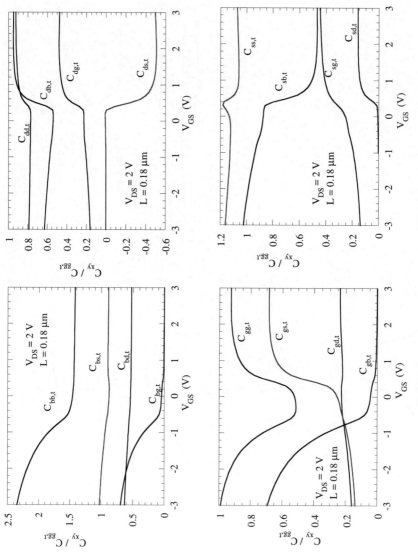

Fig. 4-59　Same as Fig. 4-58, except with $V_{DS} = 2$ V.

4.10 C_{gg} FIT AND OTHER CAPACITANCE ISSUES

normalized to the $C_{gg,t}$ of the accumulation region, are given in Fig. 4-56 for $V_{DS} = 0$ and Fig. 4-57 for $V_{DS} = 2$ V. These plots agree well with the long-channel expressions given in Appendix B. Because the parasitic capacitances are small, the total capacitances are essentially equal to the intrinsic device capacitances.

The other simulated device is a short-channel device with a drawn length of 0.18 µm. The total capacitances, which include C_{ov}, C_f, $C_{j,\text{SB}}$, and $C_{j,\text{DB}}$, are shown in Figs. 4-58 and 4-59, for $V_{DS} = 0$ and 2 V, respectively. Again, the capacitances are normalized by the $C_{gg,t}$ value at $V_{GS} = -3.3$ V. Because $C_{j,\text{SB}}$ and $C_{j,\text{DB}}$ do not add to $C_{gg,t}$, the $C_{bs,t}$ and $C_{bd,t}$ ratios, and so forth, are seen to have values exceeding 1. With some judgment, we estimated the magnitude of the parasitic capacitances and subtract them from the total capacitances. The resulting intrinsic device capacitances are plotted in Figs. 4-60 and 4-61, for $V_{DS} = 0$ and 2 V, respectively. The intrinsic capacitances are normalized with C_{gg} (not $C_{gg,t}$) at $V_{GS} = -3.3$ V. We note that there is a certain degree of arbitrariness in determining the values of the parasitic capacitances. MEDICI does not make a distinction between the intrinsic and the parasitic part of the device.

BSIM3's C-V model is based on long-channel equations. The SPICE results simulated with BSIM3 agree well with the long-channel MEDICI results shown in Figs. 4-56 and 4-57. For example C_{gs} is about 2/3 C_{ox} and C_{ds} is $-1/6\ C_{ox}$ in saturation. However, because these C-V characteristics differ from those short-channel results shown in Figs. 4-60 and 4-61, we expect the BSIM3 model to cease being accurate in short-channel devices. The degree that various capacitances simulated by BSIM3 differ from the actual can be estimated by comparing Figs. 4-56 and 4-60 for the $V_{DS} = 0$ case, and Figs. 4-57 and 4-61 when $V_{DS} = 2$ V.

There are some interesting results in Fig. 4-61, simulated under the saturation operating condition for the short-channel device. First, several publications based on MEDICI simulation have found C_{gs} to exceed its long-channel limit of 2/3 C_{ox}. For example, for a short-channel device exhibiting strong velocity saturation effects, C_{gs} can be as high as $0.9 \cdot C_{ox}$ [16]. Here, our simulation does not reproduce such a claim. We do not know whether it is simply due to the fact that the simulation here has been tuned to experimental results of a production process. With the consideration of poly-depletion, mobility degradation, perhaps C_{gs} of short-channel devices does not deviate much from the long-channel theoretical value.

Another observation is that C_{sd} in short-channel devices is positive in saturation. Its long-channel theoretical value should be zero. We are not sure if this positive capacitance is due to parasitic capacitance between the source and drain contacts, which was assumed zero during the conversion from Fig. 4-59 to Fig. 4-61. Further, C_{sb} is negative. This result is rather counter-intuitive. We do not know the apparent physical reason for this occurrence. The MEDICI results have been checked to make sure the results are not due to convergence error or simply wrongful data entry.

There has not been a detailed study on how BSIM3 C-V models, based on long-channel derivation, affect simulation results in short-channel devices. It seems that the digital circuits are not greatly affected by the capacitance accuracy. For them, the drive current and parasitic capacitances are the primary factors determining the gate delay time. The error is probably more significant in analog circuits.

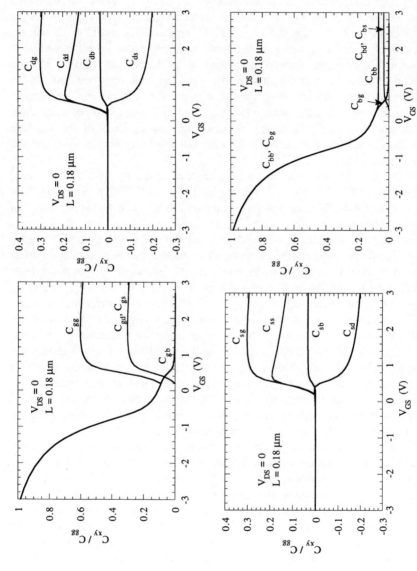

Fig. 4-60 The intrinsic capacitances corresponding to the MEDICI-simulated values in Fig. 4-58, after the parasitic capacitances are estimated and subtracted out. $V_{DS} = 0$.

4.10 C_{gg} FIT AND OTHER CAPACITANCE ISSUES 355

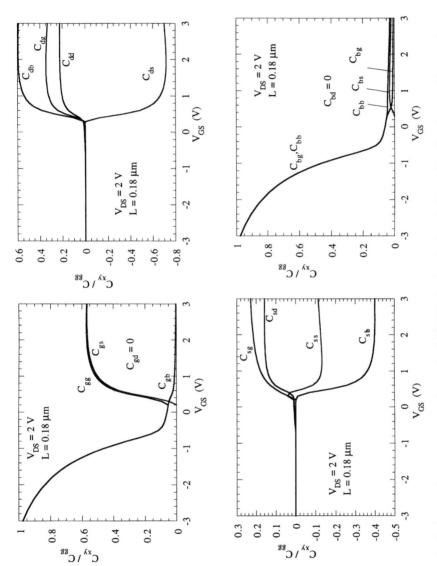

Fig. 4-61 The intrinsic capacitances corresponding to the MEDICI-simulated values in Fig. 4-59, after the parasitic capacitances are estimated and subtracted out. $V_{DS} = 2V$.

4.11 INSUFFICIENT NOISE MODELING (NO EXCESS SHORT-CHANNEL THERMAL NOISE)

BSIM3 develops two distinct models to calculate $1/f$ noise, and another two models to calculate the channel thermal noise. There are a total of four combinations, depending on the particular $1/f$ noise and thermal noise model desired. We are primarily interested in the thermal noise in this section; therefore, we are concerned with NOIMOD = 1 and 4 (see Section 3.2, under the entry NOIMOD, for more details).

When NOIMOD = 1, BSIM3 employs the thermal noise model adopted by circuit designers:

$$\overline{i_d^2} = 4kT\Delta f \times \frac{2}{3}(g_m + g_{mb} + g_d). \qquad (4\text{-}53)$$

There is some rationale behind this heuristic equation. When the transistor is in saturation, $\overline{i_d^2}$ is theoretically found to be $4kT\Delta f \times 2/3\ (g_m + g_{mb})$. This equation, although valid only in the saturation region, was soon applied indiscriminately to all operating regions, including when $V_{DS} \to 0$ (linear region). This practice was bad because when $V_{DS} \to 0$, g_m and g_{mb} both approached zero, leading a prediction of zero $\overline{i_d^2}$. This problem was realized, and another $\overline{i_d^2}$ equation is derived specifically for the linear region. This time, it is found that $\overline{i_d^2} = 4kT\Delta f \times g_d$. There is not a factor of 2/3. These two equations are derived specifically for their respective applicable region of operation. A good compromise accounting for both operating regions would then to combine them, as done in Eq. (4-53). For example, in saturation, $g_d = 0$ and Eq. (4-53) degenerates to $4kT\Delta f \times 2/3\ (g_m + g_{mb})$, as expected. However, at $V_{DS} \to 0$, $g_m = g_{mb} = 0$ and Eq. (4-53) degenerates to $4kT\Delta f \times 2/3\ g_d$. Note that, the theoretical expression in the linear region lacks the 2/3 factor. This suggests that Eq. (4-53) is not exactly correct in all operating regions.

An example circuit exposing this problem is the inset of Fig. 1-11, with the figure itself showing the simulation results. According to the discussion about Fig. 1-11, the circuit operates at $V_{DS} = 0$. By printing out the operating condition, we know that the device's $g_d = 1.4 \times 10^{-2}\ \Omega^{-1}$, whereas $g_m = g_{mb} = 0$. At low frequencies, the MOS transistor behaves like a resistor, with a resistance equal to $1/g_d$. The thermal noise of the drain current is $4kT\Delta f \cdot g_d$ and noise voltage at the drain is expected to be $4kT\Delta f/g_d$. At the circuit temperature of 298 K, we calculate $\overline{v_d^2}/\Delta f$ to be $4 \cdot 1.38 \times 10^{-23} \cdot 298/1.4 \times 10^{-2} = 1.2 \times 10^{-18}$ V^2/Hz. The simulation result with BSIM3, shown in Fig. 1-11, is only 0.78×10^{-18} V^2/Hz at low frequencies. It is clear that BSIM3 underestimates the noise square voltage by a factor of 2/3 when NOIMOD = 1.

Up to BSIM3v3.2.2 (officially released in 1999), the thermal noise model offered in NOIMOD = 4 writes

$$\overline{i_d^2} = 4kT\Delta f \times \frac{\mu_{eff}}{L_{eff}^2}|Q_{inv}|, \qquad (4\text{-}54)$$

where Q_{inv} (in coulombs) is the inversion charge in the channel, and L_{eff} is the effective channel length. (*Note:* In the actual BSIM3 implementation, L_{eff} rather than $L_{eff,CV}$ appears in the denominator of Eq. (4-54), although Q_{inv} is proportional to $L_{eff,CV} \cdot W_{eff,CV}$.) Equation (4-54) is a fundamental equation. When the relationships between Q_{inv} and the transconductances are established, it is found that, for long-channel transistors, Eq. (4-54) degenerates into $4kT\Delta f \times 2/3\ (g_m + g_{mb})$ in saturation, and into $4kT\Delta f \times g_d$, in the linear region, without the extra 2/3 factor. Although Q_{inv} is a difficult quantity to calculate by hand, it is nonetheless a simple quantity to be evaluated in computer. After all, when BSIM3 calculates the device capacitances, Q_{inv} is already a calculated variable. There is thus no overhead associated with calculating the thermal noise from Eq. (4-54). We recommend NOIMOD = 4 over NOIMOD = 1, at least for its accuracy in long-channel devices.

Having made this recommendation, we want to warn the designers that there is a potential drawback of NOIMOD = 4. This is not a concern if the circuit has a stable dc operating point, but can be troublesome if the bias point moves around during circuit operation. Figure 4-62 compares the NOIMOD = 1 and 4 models as a function of V_{GS}. The results are obtained with the circuit of Fig. 1-11, with $V_{DS} = 0.1$ V and V_{GS} scanned between 0 and 3 V. ONOISE refers to the noise voltage (in $V/\sqrt{(Hz)}$) at the drain node. The figure demonstrates that there is a kink behavior of the drain noise voltage at the transition between the subthreshold and the strong inversion region, for NOIMOD = 4. The kink behavior is absent in NOIMOD = 1, in contrast. This is because the noise voltage simulated by the NOIMOD = 1 model depends on the small-signal quantities, which are ensured to be continuous in their derivatives by the use of the smoothing functions.

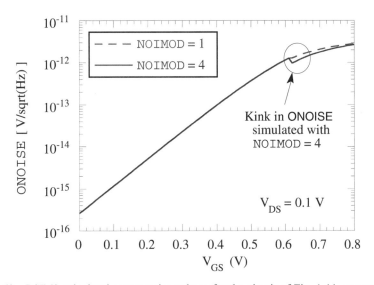

Fig. 4-62 BSIM3-calculated output noise voltage for the circuit of Fig. 1-11. NOIMOD = 1 and 4.

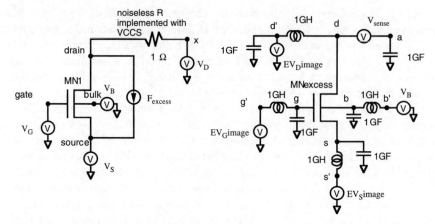

Fig. 4-63 A workaround to add the excess noise in short-channel devices (i.e., that associated with the fact that γ exceeds 2/3).

Let us now discuss the short-channel devices, the main focus of this section. For convenience, we shall concentrate on the saturation region, for which the channel thermal noise can be written as

$$\overline{i_d^2} = 4kT\Delta f \times \gamma g_m. \qquad (4\text{-}55)$$

For a long-channel device, $\gamma = 2/3$. In short-channel devices, γ is found to be larger [17]. It can vary between 1–2 [18], or all the way up to 4–8 [19]. It generally increases with drain current as well as V_{DS}. A possible (but not necessarily unequivocally verified) explanation of the apparent increase of γ in short-channel devices is that the hot electrons, after being accelerated by the channel field, can have hot electron temperatures several times larger than the room temperature [20]. As the electron temperature increases, γ of Eq. (4-55) must increase if T in the equation remains to be the room temperature. Unfortunately, γ is not a BSIM3 parameter; it has a fixed value of 2/3 at saturation region, whether it is long- or short-channel. There is no BSIM3 parameter that allows us to adjust the level of the thermal noise.

Nonetheless, there is a work-around to add this excess noise (i.e., that associated with the fact that γ exceeds 2/3), albeit not too simplistic. This is illustrated in Fig. 4-63 [21]. The transistor on the left-hand side (MN1) is the MOS of interest. The one on the right (MNexcess) has the sole purpose of introducing additional noise of the MOS on the left. The SPICE netlist for this circuit is:

```
* Add Excess Noise to MN1
psqrtEXTRA = 3 /*assign value to the parameter*/

VG gate 0 1.5V AC 1
VD x 0 3.3V
VS source 0 0.0V
VB bulk 0 0.0V
GnoiselessR x drain x drain 1
MN1 drain gate source bulk NCH w=5 l=0.35
```

4.11 INSUFFICIENT NOISE MODELING

```
* Add excess noise here
.option capmax=1e10
mnExcess d g s b nch w=5 l=0.35
EVDimage dprime 0 drain 0 1
CVDbypass dprime 0 1e9
LVDdcfeed dprime d 1e9
EVGimage gprime 0 gate 0 1
CVGbypass g 0 1e9
LVGdcfeed gprime g 1e9
EVBimage bprime 0 bulk 0 1
CVBbypass b 0 1e9
LVBdcfeed bprime b 1e9
EVSimage sprime 0 source 0 1
CVSbypass s 0 1e9
LVSdcfeed sprime s 1e9
vsense d a 0
csense a 0 1e9
Fexcess drain source vsense #psqrtEXTRA#
.OP
.AC LIN 1 1e1HZ 1e1HZ
.NOISE V(drain) VG 1
.PRINT noise inoise onoise
.lib dsn=model.bsim3 (containing NCH MOS model files)
.end
```

In this circuit, the voltages at the drain, gate, source, and bulk of MN1 are reflected to MNexcess through the use of voltage-controlled voltage sources (VCVS). We are only interested in reflecting the dc components of the voltages, so that the dc current, and hence, g_m, g_{mb}, and g_d of MNexcess are identical to those of MN1. The inductors and capacitors placed in series and in parallel to the VCVS serve precisely this purpose, ensuring that the small-signal quantities do not enter MNexcess. Because MNexcess has the same width and length as MN1, the thermal noise given by Eq. (4-54) is identical in the two transistors once the dc bias conditions are identical.

The thermal noise in MNexcess appears at the drain node. We place the sensing capacitor over there to sink in the noise current, allowing it to pass through a sensing voltage source of 0 V, which acts as a current meter. This current meter measures the square root of $\overline{i_d^2}$. A current-controlled current source (CCCS) denoted by Fexcess multiplies this current by the parameter psqrtEXTRA. Suppose that we know that the thermal noise of MN1 is characterized by $\gamma = 2/3 \times 10$. BSIM3, as mentioned, fixes the MN1 thermal noise to be $\gamma = 2/3$ only. In this case, psqrtEXTRA should have a value of 3, such that after it is squared, Fexcess introduces 9 times the thermal noise of MN1. The total thermal noise appearing at the drain node of MN1 is thus 10 times as much, leading to an effective γ of $2/3 \times 10$.

The drain resistor between the drain node and the x node is implemented with a voltage-controlled current source (VCCS), GnoiselessR. This way the resistor is noiseless, simplifying our calculation of noise voltage due to the transistors. There are several ways to implement noiseless resistor. Some SPICE simulators allow the user to simply turn off the noise calculation for a resistor. The noiseless resistor can also be implemented with a current-controlled voltage source.

There is another potential source of error in the thermal-noise calculation of short-channel devices. As first pointed out in Section 2.2 and then emphasized in Section 2.6, when the absorbed-resistance approach is used to model the source/drain resistance, the noises associated with the resistances are not accounted for in Eq. (4-54). In the absorbed-resistances approach, the parameter RDSW is used to model the effects of the source/drain resistance on current and conductances, although there is really no physical resistance and the real part of y_{11} is identically zero. (The resistance R_S in the lumped-resistance approach, in contrast, is a physical resistance which would cause the real part of y_{11} to be nonzero.) For long-channel devices, the channel resistance dominates over the source/drain resistances. Hence, Eq. (4-54) still yields correct results. However, as channel length becomes shorter, a significant part of the device thermal noise originates from the source/drain resistances, in addition to that associated with the intrinsic channel. To accommodate the shortcoming of the absorbed-resistance approach in noise modeling, BSIM3v3.3 has been considered to be introduced in 2000, after the release of BSIM4 (see Ref. [15] of Chapter 1). In this future BSIM3 version, Eq. (4-54) will be modified to

$$\frac{\overline{i_d^2}}{\Delta f} = \frac{4kT}{R_{DS} + L_{eff}^2/(\mu_{eff}|Q_{inv}|)}. \tag{4-56}$$

R_{DS} is given by

$$R_{DS} = \frac{\text{RDSW}[1 + \text{PRWG} \cdot V_{GST,\text{eff}} + \text{PRWB}(\sqrt{2\phi_f - V_{BS,\text{eff}}} - \sqrt{2\phi_f})]}{(10^6 \times W_{eff,\text{special}})^{WR}}, \tag{4-57}$$

where $W_{eff,\text{special}}$ is found in Appendix A (Eq. A-3), although it can be approximated to be $W - 2 \cdot \text{WINT}$. The 10^6 factor is inserted because W_{eff} in BSIM3 is in meters, while RDSW is in Ω-μm. We caution that R_{DS} is still a nonphysical resistance. Hence, despite the modification in the noise expression, Re(y_{11}) remains zero when the absorbed-resistance approach is used.

It is interesting to compare $\overline{i_d^2}/(4kT\Delta f)$ with the device's g_d when $V_{DS} \to 0$ (linear region), particularly when the device's source/drain resistance is accounted for by the absorbed-resistance approach. Figure 4-64 illustrates the ratio of $\overline{i_d^2}/(4kT\Delta f)$ to g_d, as a function of L, for devices with RDSW = 423.5 (Ω-μm), and PRWG = PRWB = 0. (The $1/f$ noise's contribution is deliberately made zero.) The figure shows that when Eq. (4-54) of BSIM3v3.2.2 (or before) is used, the ratio is 1 for the long-channel devices, but deviates significantly from 1 as L becomes shorter. This is an indication that the source/drain resistance becomes comparable to the channel resistance. When Eq. (4-56) of BSIM3v3.3 is used, the ratio remains close to 1 for all channel lengths.

4.11 INSUFFICIENT NOISE MODELING

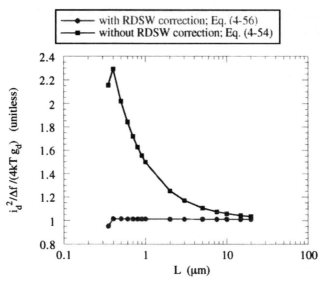

Fig. 4-64 BSIM3-calculated ratio of $i_d^2/(4kT\Delta f)$ to g_d, as a function of L, with Eq. (4-54) of BSIM3v3.2.2 and Eq. (4-56) of the proposed BSIM3v3.3 (see Ref. [15] of Chapter 1). RDSW = 423.5 (Ω-μm); PRWG = PRWB = 0. The $1/f$ noise's contribution is deliberately made to be zero.

We have used the phrase "excess noise" to mean the amount of thermal noise in short-channel devices exceeding the theoretical value found in long-channel devices. Basically, if γ in the device exceeds 2/3, there is an excess noise, which can be accounted for with Fig. 4-63. Sometimes, excess noise refers to another phenomenon, which is not modeled by BSIM3, either. Resistors, depending on the composition of the material, are known to produce more noise than that attributable solely to temperature [22]. In this case, the excess current noise is the $1/f$ noise not accountable by the usual $4kT\Delta f/R$ expressions. As shown in the noise equivalent circuit of Fig. 2-28 there are several resistive (parasitic) components in the MOS transistor. Each of them contributes to noise. In BSIM3, they are modeled with the thermal noise given in Eq. (2-49). The excess $1/f$ noise, potentially important at low frequencies, can be added in with a method similar to that of Fig. 4-63 [22].

The substrate diodes (source–bulk and drain–bulk junctions) appended in the BSIM3 model are noiseless. However, in reality, there are $1/f$ noise and shot noise associated with the diodes. Both noises are proportional to the diode current (or the square of the diode current). Because the source–bulk and drain–bulk junctions in MOSFET are designed to be reverse-biased, the diode currents ($I_{j,\text{SB}}$ and $I_{j,\text{DB}}$) are nearly zero. The omission of these noises is thus reasonable. If it is desirable to consider the $1/f$ noise in the diodes, particularly when the diode is slightly forward biased, then this excess noise can be added with a method similar to that of Fig. 4-63 [22].

4.12 INSUFFICIENT NOISE MODELING (NO CHANNEL-INDUCED GATE NOISE)

BSIM3 models two noise sources in MOSFETs: the thermal noise and the $1/f$ noise, and both are associated with the drain current. Another noise source often described in the noise literature, the drain-induced gate noise, is conspicuously absent. This is likely a result that the induced gate noise is thought to be important only at high frequencies, and that for a long time SPICE (not just BSIM) caters only to the needs of baseband (the low-frequency) designers. Prior to the advent of wireless communications, the concerns of rf engineers in the past had largely been buried in the background noise.

The noise equivalent circuit in BSIM3 was given in Fig. 2-28. We would like to simplify to circuit somewhat to focus on the essential. If we neglect the parasitic terminal resistances, tie the source and bulk to ground, and bias the transistor in saturation so C_{gd} as well as other elements can be approximated as 0, then we have a fairly simplified equivalent circuit shown in Fig. 4-65a. Only the drain current noise appears in the figure; the gate current noise is not modeled in BSIM3. The drain current noise, in the BSIM3 implementation, consists of the thermal noise and the $1/f$ noise. As discussed in Section 3.2 (in the NOIMOD entry), BSIM3 offers two models of thermal noise and two models of $1/f$ noise, thus giving rise to 4 different combinations (NOIMOD = 1, 2, 3, or 4). Although the noises calculated in each of the model will differ, in general, they are on the same order of magnitude. When a transistor is at saturation, the drain current noise is roughly equal to

$$\overline{i_d^2} = 4kT\Delta f \times \frac{2}{3} g_m. \tag{4-58}$$

We dropped the $1/f$ component because we assume the transistor to operate at a certain high frequency.

In a more accurate representation of the noise equivalent circuit (Fig. 4-65b), the gate current noise is included. The gate current noise is equal to the thermal noise associated with the appended conductance in parallel with C_{gs} [23]. It is given by

$$\overline{i_g^2} = 4kT\Delta f \times \frac{4}{15} \frac{\omega^2 C_{gs}^2}{g_m}. \tag{4-59}$$

The MOS transistor can be viewed as an RC distributed network, with the capacitive coupling to the gate representing the distributed capacitance and the channel itself representing the distributed resistance. At high frequency, the local channel voltage fluctuations due to thermal noise couple to the gate through the oxide capacitance, inducing the gate noise current to flow. Therefore, the noise sources given in Eqs.

4.12 INSUFFICIENT NOISE MODELING (NO CHANNEL-INDUCED GATE NOISE)

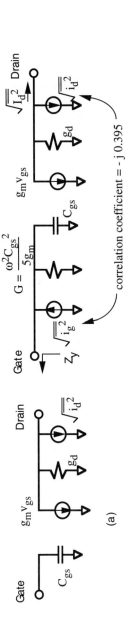

Fig. 4-65 (a) A simplified noise equivalent circuit from Fig. 2-28; (b) a more accurate noise equivalent circuit, the Van der Ziel representation, which incorporates the induced gate noise.

(4-54) and (4-59) are correlated. Their correlation coefficient in saturation region is given by [23]

$$c \equiv \frac{\overline{i_g \cdot i_d^*}}{\sqrt{\overline{i_g^2}}\sqrt{\overline{i_d^2}}} = -j\sqrt{\frac{5}{32}} = -j0.395. \quad (4\text{-}60)$$

The correlation coefficient has a negative sign here because the drain and gate noise current sources are assigned with opposite directions in Fig. 4-65b (one is *into* the gate terminal, and the other is *out of* the drain terminal). Often in other references both the noise currents are assigned with the same direction; then the correlation coefficient would be $+j0.395$. The magnitude of the correlation coefficient is slightly higher (~ 0.445) in the linear operating region [23].

We bias the transistor with a choke inductor placed between the gate of the transistor and the input voltage source. This inductor serves as a dc short between the gate and the input voltage source, but presents an infinite ac impedance looking out of the gate port (i.e., $Z_y = \infty$ in Fig. 4-65b). We are now interested in finding the noise current flowing through the drain node, $\overline{I_d^2}$. Note that $\overline{I_d^2}$ differs from $\overline{i_d^2}$. The latter denotes the individual noise source, whereas the former is the overall effected noise due to both $\overline{i_d^2}$ and $\overline{i_g^2}$. Knowing that the conductance is much smaller than the admittance associated with C_{gs}, we ignore the presence of the conductance G in Fig. 4-65b. The drain node current, I_d, is then given by

$$I_d = i_d + \frac{i_g}{j\omega C_{gs}} \cdot g_m^\dagger, \quad (4\text{-}61)$$

where $g_m^\dagger$ is the output transconductance in the small-signal equivalent circuit, given in Eq. (4-9). For most applications, $g_m^\dagger$ can be approximated as g_m, the device transconductance equaling to $\partial I_D / \partial V_{DS}$.

We show the process of determining the drain-node-current noise step by step, a process that underscores the reason why the correlation coefficient is defined as the way in Eq. (4-60):

$$\overline{I_d^2} = \overline{I_d \cdot I_d^*} = \overline{\left(i_d + i_g \frac{g_m}{j\omega C_{gs}}\right) \cdot \left(i_d^* - i_g^* \frac{g_m}{j\omega C_{gs}}\right)}$$

$$= \overline{i_d \cdot i_d^*} + \overline{i_g \cdot i_g^*} \frac{g_m^2}{\omega^2 C_{gs}^2} + \overline{\left(i_g i_d^* \frac{g_m}{j\omega C_{gs}} - i_g^* i_d \frac{g_m}{j\omega C_{gs}}\right)}$$

$$= \overline{i_d^2} + \overline{i_g^2} \frac{g_m^2}{\omega^2 C_{gs}^2} + 2\mathrm{Re}\left(\overline{i_g i_d^*} \frac{g_m}{j\omega C_{gs}}\right)$$

$$= \overline{i_d^2} + \overline{i_g^2} \frac{g_m^2}{\omega^2 C_{gs}^2} - 2\sqrt{\frac{5}{32}} \frac{g_m}{\omega C_{gs}} \sqrt{\overline{i_g^2}}\sqrt{\overline{i_d^2}} \quad (Z_y = \infty). \quad (4\text{-}62)$$

4.12 INSUFFICIENT NOISE MODELING (NO CHANNEL-INDUCED GATE NOISE)

Substituting $\overline{i_g^2}$ and $\overline{i_d^2}$ from Eqs. (4-58) and (4-59), we get

$$\overline{I_d^2} = 4kT\Delta f \left[\frac{2}{3} g_m + \frac{4}{15} g_m - \frac{1}{3} g_m \right] \quad (Z_y = \infty). \quad (4\text{-}63)$$

The first term is due to $\overline{i_d^2}$, the second term, $\overline{i_g^2}$, and the last term, their correlation. $\overline{i_g^2}$, along with the correlation term, is neglected in BSIM3. It turns out that in this case of $Z_y = \infty$, such an omission does not result in a significant error. This is because the correlation term, being negative, partially cancels out the contribution from $\overline{i_g^2}$ ($4/15 - 1/3 = -1/15$, which is much smaller than $2/3$).

However, there are situations where neglecting the contribution from $\overline{i_g^2}$ (and the correlation) is error-prone. One simplistic example is the circuit shown in Fig. 4-66a, whose netlist is given below:

```
VG y1 0 1.5V AC 1
HCCVS y1 gate VG 1
VD x1 0 3.3V
LD x1 drain 1
VS source 0 0.0V
VB bulk 0 0.0V
MNMOS drain gate source bulk NCH w=5 l=0.35
.AC LIN 1 1e10HZ 1e10HZ
.NOISE V(gate) VG 1
.PRINT noise inoise onoise
.lib dsn=model.bsim3 (containing NCH MOS model files)
.end
```

A current-controlled voltage source (CCVS) is used to create a 1 Ω resistor at the input. This resistor is noiseless, allowing SPICE to compute the noise voltage at the

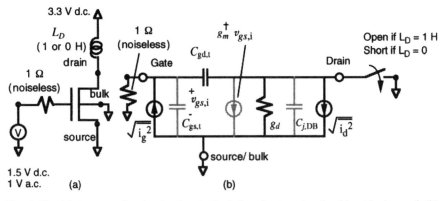

Fig. 4-66 (a) An exemplar circuit wherein the induced gate noise should not be ignored; (b) its equivalent circuit in a noise analysis.

gate node due solely to the transistor. Since this $1\,\Omega$ resistor is in parallel with C_{gs}, whose impedance is much larger, we can drop C_{gs} out of the equivalent circuit. Consequently, the current source $g_m^\dagger v_{gs,i}$ is also negligible. For now, we intentionally eliminate the effects of $C_{j,\mathrm{DB}}$ by properly modifying the relevant SPICE parameters. Under the bias condition $V_D = 3.3$ V and $V_G = 1.5$ V, the transistor has a g_m, g_d, and C_{gd} of $1.53 \times 10^{-3}\,\Omega^{-1}$, $6.2 \times 10^{-5}\,\Omega^{-1}$, and 1.42×10^{-15} F, respectively. We calculate $\overline{i_d^2}/\Delta f$ to be $\sim 4kT \cdot 2/3 \cdot g_m \sim 1.7 \times 10^{-23}$ A^2/Hz.

Let us consider the case wherein the drain inductor is 1 H. Because the inductor is an ac open, all of the drain current noise $\overline{i_d^2}$ moves toward the left. At the frequency of operation (10 GHz), the admittance of C_{gd}, $j\,8.9 \times 10^{-5}\,\Omega^{-1}$, is just slightly larger than g_d. Therefore, slightly more than half of $\overline{i_d^2}$ shows up at the $1\,\Omega$ resistor. The square noise voltage $(\overline{v_g^2})$ at the `gate` node computed in the SPICE simulation is hence close to 1.7×10^{-23} A$^2 \cdot (1\,\Omega)^2/2 = 0.8 \times 10^{-23}$ V^2. Whereas we understand that BSIM3 does not calculate the drain-induced gate noise's contribution to $\overline{v_g^2}$, we know that the combination of the contribution and the correlation term nearly cancels out. We would not expect the final answer to differ from that calculated above.

Now we consider the case when the drain inductance is 0, that the `drain` node of the transistor is tied directly to the voltage source. In the noise analysis, the `drain` node is shorted to the ground, as shown in Fig. 4-66b. All of the drain current noise goes to the nearby ground and never passes through the $1\,\Omega$ resistor. Since BSIM3 considers only the $\overline{i_d^2}$ noise source, BSIM3 predicts a zero $\overline{v_g^2}$. (The SPICE output for $\sqrt{\overline{v_g^2}/\Delta f}$, however, is reported as 10^{-20} V/sqrt(Hz), which is the defaulted minimum value in some SPICE simulator.) Since $\overline{i_d^2}$'s contribution to $\overline{v_g^2}$ is zero in this case, it is critical to consider $\overline{i_g^2}$. BSIM3 therefore underpredicts the noise present in this circuit.

The effect of a large $C_{j,\mathrm{DB}}$ is tantamount to shorting the `drain` node to ground at high frequencies.

It is said that the induced gate noise is not too important in modern CMOS. Often the noises associated with the gate resistance and the distributed substrate resistance are more significant [24].

4.13 INCORRECT NOISE FIGURE BEHAVIOR

We assume that the readers are somewhat familiar with the basic terminology about noise figure. For a detailed review, see Ref. [22]. The noise figure (NF) is a good figure of merit characterizing the device. At a fixed frequency, NF is a function of the impedance at the input seen toward outside the device, but is independent of the impedance at the output seen toward outside the device. Experimental results have shown that the noise figure first decreases with frequency (due to lowered $1/f$ noise) at low frequencies, maintains a relatively constant value at mid-frequencies, and then increases again at high frequencies. Figure 4-67 illustrates the measured NF at high frequencies for two devices, both with a total width of 192 μm. One device has 24 fingers, and hence the per-finger width is 8 μm. The other has 48 fingers, with 4 μm width in each finger. The different layout geometries result in a difference in their

4.13 INCORRECT NOISE FIGURE BEHAVIOR

gate distributed resistances. As mentioned in Section 4.5, the gate resistance introduces thermal noise. Indeed, Fig. 4-67 shows that the two devices have different noise figures. However, in the subsequent discussion, for convenience we will neglect the gate resistance's contribution to noise figure. The important point of Fig. 4-67 is that, for various bias conditions, the NF always increases with frequency at high frequencies.

It has been ambiguously stated that the gate-induced noise is a high-frequency noise. A wrong but plausible corollary is that, since BSIM3 encompasses only the drain current noise, BSIM3 model cannot predict NF's increase with frequency. In fact, it can be shown that the input equivalent noise source due solely to $\overline{i_d^2}$ is [23]

$$\overline{I_n^2} = 4kT\Delta f \times \frac{2}{3}g_m \left(\frac{\omega^2 C_{gs}^2}{g_m^2}\right) \approx 4kT\Delta f \times \frac{2}{3}g_m \left(\frac{\omega}{\omega_T}\right)^2. \quad (4\text{-}64)$$

Since NF is proportional to $\overline{I_n^2}$ times R_P (source resistance of the input power source), NF increases with frequency even though Eq. (4-64) is derived without considering $\overline{i_g^2}$. Another rather elaborate explanation for $\overline{I_n^2}$'s dependence on frequency is found in [22, pp. 143–145].

In this section, we show that the BSIM3 model often correctly predicts the NF to increase with frequency. In some cases where BSIM3 predicts NF to decrease with frequency, we demonstrate that the problem is actually associated with the wrongful

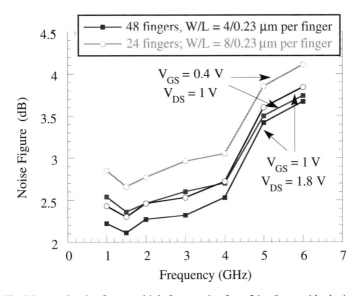

Fig. 4-67 Measured noise figure at high frequencies for a $24 \times 8\,\mu m$ wide device (24 is the number of fingers), and a $48 \times 4\,\mu m$ device. The different layout geometries result in a difference in their gate distributed resistance.

368 IMPROVABLE AREAS OF BSIM3

dependence of $g_m^\dagger$ on frequency (Section 4.4), not the fact that BSIM3 omits $\overline{i_g^2}$. (BSIM3's omission of $\overline{i_g^2}$, however, makes its simulation results less accurate.) Because the NF is not obtainable from SPICE simulation in a straightforward manner, we start by describing a procedure to calculate NF. For simplicity, we consider a two-port consisting of purely resistive elements, as shown inside the dashed box in Fig. 4-68. From an independent simulation tool, we find this 3-resistor network has a NF of 11.76 dB when the power source resistance is 50 Ω. The noise figure is independent of the ambient temperature in this case, and is independent of the load impedance attached to the output of the two-port.

Figure 4-68 demonstrates the proper circuit schematic used to calculate the noise figure. Most noteworthy is the use of a noiseless resistor R_L at the output. Although the noise figure is independent of the value of R_L, we need to place a nonzero R_L there. This is because SPICE does not calculate noise current and we need some resistor at the output to allow us calculate the noise voltage at the output node. The output resistor must be noiseless; otherwise, the noise voltage at the output node will have contribution from the load resistor itself. We implement the noiseless resistor through the use of a current-controlled voltage source (CCVS). The SPICE netlist is provided below:

```
VS 1 0 1V AC 1V
RS 1 2 50 TEMP=16.85
RA 2 3 50
RB 3 0 50
RC 3 4 50
Vdum 4 5 0
HCCVS 5 0 Vdum 50
.temp 27
.AC LIN 10 1e9HZ 1e10HZ
.NOISE V(4) VS 1
.PRINT noise inoise onoise
.end
```

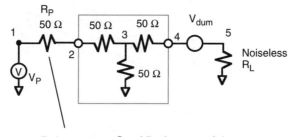

R_p is at 18.65 C, while the rest of the network is at an arbitrary ambient temperature

Fig. 4-68 A bias setup used to determine the noise figure of a 3-resistor two-port network.

4.13 INCORRECT NOISE FIGURE BEHAVIOR

The noise figure is the ratio of the input signal-to-noise ratio to the output signal-to-noise ratio:

$$\text{NF} = \frac{S_i/N_i}{S_o/N_o}. \tag{4-65}$$

(*Note*: In a strict definition, this ratio is the noise factor, and the noise figure should be $10 \times \log_{10}$ of the noise factor. However, we will not distinguish between these two quantities.) In general, both the signal (S) and the noise (N) can be either square-voltage or square-current quantities. However, again, because SPICE does not calculate noise current, we will concentrate on the voltage. In the exemplar circuit of Fig. 4-68, we take N_o to be the noise square-voltage at the output node (node 4). In the SPICE simulation printout, ONOISE denotes the noise voltage at the output node per Hz. Hence, $N_o = (\text{ONOISE})^2 \cdot \Delta f$. N_i is the noise square-voltage introduced by the power-source resistance itself. Therefore,

$$N_i = 4kT_0 \Delta f \, R_P, \tag{4-66}$$

where k, the Boltzmann constant, is 1.38×10^{-23} J/°C, and $T_0 = 290$ K. We caution that $N_i \neq 4kT\Delta f \cdot R_P$, where T is the ambient temperature (27°C in the netlist). According to the IEEE definition, NF is reported with reference to $T_0 = 290$ K. Even if the ambient temperature T differs from T_0, the input noise square voltage should be calculated with $T_0 = 290$ K (or approximately 16.85°C). This is the reason why the power-source resistance R_p in the netlist is specified to be at the temperature of 16.85°C. (Incidentally, T_0 is so chosen because kT_0 is equal to 4.0×10^{-21} J, a convenient round number. It is not because IEEE is headquartered in New Jersey, where the room temperature is generally colder, at 290 K.)

The signal square voltage at the output is $(v(4))^2$. Likewise, $S_i = (v(1))^2$. The noise figure is, therefore,

$$\begin{aligned} \text{NF} &= \frac{(\text{ONOISE})^2 \cdot \Delta f}{4kT_0 \Delta f \, R_P} \times \frac{1}{(\text{voltage gain})^2} \\ &= \frac{(\text{ONOISE})^2 \cdot \Delta f}{4kT_0 \Delta f \, R_P} \times \frac{v(1)^2}{v(4)^2} = \frac{1}{4kT_0 R_P} \times (\text{INOISE})^2, \end{aligned} \tag{4-67}$$

where INOISE is quantity appearing in the SPICE simulation printout. (In some simulators, ONOISE is in the unit of $V^2/$Hz or $A^2/$Hz, rather than in $V/\sqrt{(\text{Hz})}$ or $A/\sqrt{(\text{Hz})}$ as in Eq. 4-67. A similar statement can be made about INOISE.) INOISE is the input-referred noise, equal to ONOISE divided by the voltage gain between the input voltage ($v(1)$) and the output voltage ($v(4)$). SPICE simulation indicates that INOISE $= 3.46 \times 10^{-9}$ V$/\sqrt{(\text{Hz})}$. Because $4kT_0 \cdot R_P = 8.0 \times 10^{-19}$ V$^2/$Hz, Eq. (4-67) calculates NF to be 15, or 11.76 dB, agreeing with a commercial rf simulation tool.

370   IMPROVABLE AREAS OF BSIM3

We now calculate the NF of a MOSFET biased at $V_{GS} = 1.5$ V and $V_{DS} = 3.3$ V. The schematic circuit used for this purpose is shown in Fig. 4-69. The SPICE netlist is shown below for convenience:

```
.option CAPMAX=10
VP 1 0 0V AC 1
RP 1 2 50 Temp = 16.85
Cbias 1 2 3 1
Lbias1 3 5 1
VG 5 0 1.5V
MNMOS 4 3 0 0 NCH w=5 l=0.35
Lbias2 4 6 1
VD 6 0 3.3V
Cbias2 4 7 1
Vdum 7 8 0
HCCVS 8 0 Vdum 50
.AC DEC 10 1e6HZ 1e11HZ
.print ac #|v(4)/v(1)|# #|v(4)/v(3)|#
.NOISE V(4) VP 0
.PRINT noise inoise onoise
.lib dsn=model.bsim3 (containing NCH MOS model files)
.end
```

We eliminate $C_{j,DB}$ to minimize confusion. We plot out $|v(4)/v(1)|$ and $|v(4)/v(3)|$ in Fig. 4-70. The noise figure calculated in accordance with Eq. (4-67) is also shown. NF initially decreases with frequency because of the reduction in $1/f$ noise. It then starts to saturate as the $1/f$ becomes insignificant in comparison to the contribution from the thermal noise, while the gain ratio remains constant. However, as frequency continues to increase, NF decreases again with frequency. This is an

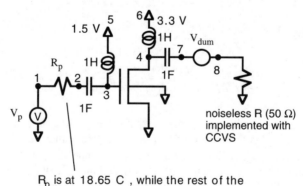

Fig. 4-69 A bias setup used to find the noise figure of a MOSFET biased at $V_{GS} = 1.5$ V and $V_{DS} = 3.3$ V.

4.13 INCORRECT NOISE FIGURE BEHAVIOR

unphysical result. We expect NF to increase with frequency at high frequencies, as observed experimentally. We trace such a NF frequency behavior to the fact that $|v(4)/v(1)|$ increases with frequency. In this circuit, wherein $R_P = 50\,\Omega$ and $C_{gs} = 4 \times 10^{-15}\,\text{F}$, we see that $R_P > (1/\omega C_{gs})$ for frequencies all the way to 800 GHz. At frequencies of interest (below 100 GHz), we therefore see that most of the voltage from V_P drops across the capacitor. This is verified from Fig. 4-70, which shows $|v(4)/v(1)|$ to be indistinguishable from $|v(4)/v(3)|$, which, in turn, is equal to $|g_m^\dagger \cdot R_L|$. Although in reality $g_m^\dagger$ decreases with frequency, we discussed in Section 4.4 that $g_m^\dagger$ implemented in BSIM3 wrongfully increases with frequency. This unphysical behavior causes $|v(4)/v(1)|$ to increase with frequency, resulting in the incorrect NF's dependency on frequency.

Let us now resimulate the circuit with $R_P = 5000\,\Omega$. The results of $|v(4)/v(1)|$, $|v(4)/v(3)|$, and NF are shown in Fig. 4-71. With R_P being $5000\,\Omega$ and $C_{gs} = 4 \times 10^{-15}\,\text{F}$, R_P becomes smaller than $(1/\omega C_{gs})$ at frequencies above 8 GHz. Therefore, above 8 GHz, most of V_P drops across the resistor, not C_{gs}. Consequently, $|v(4)/v(1)|$ decreases, despite that $|v(4)/v(3)|$ still increases due to the wrong behavior of $g_m^\dagger$ in BSIM3. The fact that $|v(4)/v(1)|$ decreases with frequency causes INOISE, and hence NF, to increase with frequency. Although the calculated NF is not too accurate due to the wrong behavior of $g_m^\dagger$, at least the trend that NF increases with frequency at high frequencies agrees with experimental results.

The noise figure is a function of the power-source resistance, or more exactly, source impedance. It is also a function of transistor size. Figure 4-72 plots the calculated noise figures for a single-finger device ($L \times W = 0.35 \times 5\,\mu\text{m}^2$) and a

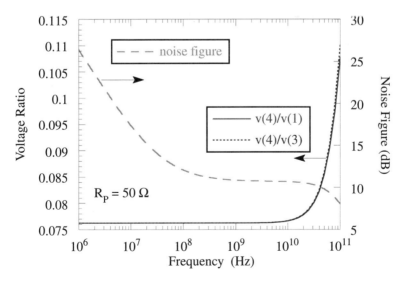

Fig. 4-70 Calculated noise figure, $|v(4)/v(1)|$, and $|v(4)/v(3)|$ of the circuit in Fig. 4-69. $C_{j,\text{DB}}$ is made to be zero by adjusting parameters. $R_P = 50\,\Omega$. The noise figure unphysically decreases with frequency at high frequencies.

372 IMPROVABLE AREAS OF BSIM3

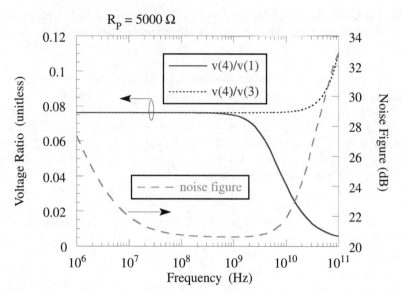

Fig. 4-71 The same results as Fig. 4-70, except with $R_P = 5000\,\Omega$. The noise figure now increases with frequency at high frequencies.

40-finger device ($0.35 \times 5 \times 40\,\mu m^2$), as a function of R_P at 2 GHz. If the transistor properties scale exactly with the transistor area, then the minimum noise figure for the two devices would be the same, although the corresponding power-source resistances to achieve the minimum noise figure differ. In practical circuits where the maximization of the power transfer is a concern, the power-source resistance is maintained at about $50\,\Omega$. An input impedance matching network is inserted between the device and the $50\,\Omega$ R_P to achieve a compromise between maximizing the power transfer and minimizing the noise figure. Figure 4-72 demonstrates that when the transistor consists of 40 fingers, the resistance to achieve the minimum noise figure is not far from $50\,\Omega$. Because the impedance match network can be easily designed when the desired resistance match is close to $50\,\Omega$, we often see multi-finger device used as a low-noise amplifier, although the minimum size device is used for high-speed applications.

It is not convenient to use SPICE to calculate the minimum noise figure (NF_{min}). We present in Appendix F a method to determine NF_{min}, as well as other parameters that characterize the noise of a two-port network.

We end this section with a notorious modeling artifact, that it is possible to simulate 0 dB noise figure if the gate and substrate resistances are zero, and the drain and source resistances are modeled by RDSW instead of parasitic resistances placed externally. Figure 4-73a illustrates the circuit schematic for simulating the noise figure of an input-matched transistor. For the simplicity of analysis, we adjust the device parameters to yield a nearly zero $C_{gd,t}$ (by making the overlap capacitance

4.13 INCORRECT NOISE FIGURE BEHAVIOR

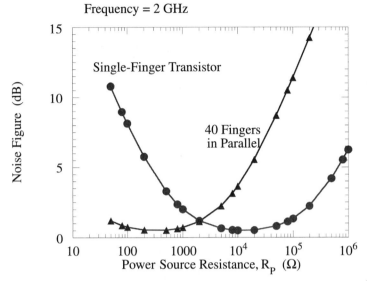

Fig. 4-72 Calculated noise figures for a single finger device ($L \times W = 0.35 \times 5 \, \mu m^2$) and a 40-finger device ($0.35 \times 5 \times 40 \, \mu m^2$), as a function of R_P at 2 GHz.

zero and operating in the saturation region). The input impedance matching is provided by a serial inductor whose value is given by

$$L = \frac{1}{\omega^2 C} = \frac{1}{\omega^2 (C_{gs,t} + C_{gb,t})}. \tag{4-68}$$

The small-signal equivalent circuit, along with the thermal noise of the FET and the thermal noise of R_P, is shown in Fig. 4-73b. Because the inductor is designed to resonate with the input device capacitance at the operating frequency, all of the $\overline{i_{RP}^2}$ noise current flows through the inductor into the capacitor, without going into R_P itself. The noise current is amplified by $g_m^\dagger$ and appears as an output noise. From inspection, we write

$$\sqrt{\frac{\overline{v_o^2}}{\Delta f}}\bigg|_{\text{due to } \overline{i_{RP}^2}} = \frac{|g_m^\dagger|}{\omega C} \left(R_L // \frac{1}{g_d}\right) \sqrt{\frac{\overline{i_{RP}^2}}{\Delta f}}, \tag{4-69}$$

where,

$$\frac{\overline{i_{RP}^2}}{\Delta f} = \frac{4kT_0}{R_P}; \qquad T_0 = 290 \, \text{K}. \tag{4-70}$$

The output noise due to the drain thermal noise of the intrinsic transistor is given as

$$\sqrt{\frac{\overline{v_o^2}}{\Delta f}}\bigg|_{\text{due to } \overline{i_d^2}} = \left(R_L // \frac{1}{g_d}\right) \sqrt{\frac{\overline{i_d^2}}{\Delta f}}. \tag{4-71}$$

374 IMPROVABLE AREAS OF BSIM3

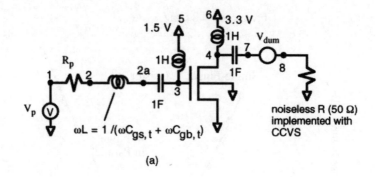

(a)

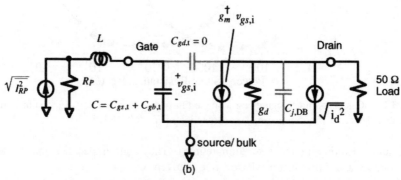

(b)

Fig. 4-73 (a) Circuit schematic of an input-matched transistor; (b) its equivalent circuit. The input matching provided by the inductor L can incorrectly lead to 0 dB noise figure in BSIM3 because BSIM3 ignores the induced gate noise.

For a 0.35 μm transistor biased at $V_{GS} = 1.5$ V and $V_{DS} = 3.3$ V and operated at 2 GHz, $C_{gd,t} \sim 0$, $C_{gs,t} + C_{gb,t} = 4.66 \times 10^{-15}$ F, $g_m^\dagger \sim g_m = 1.53 \times 10^{-3}\ \Omega^{-1}$, $g_d = 6.2 \times 10^{-5}\ \Omega^{-1}$. Further, let us take the common case with $R_P = R_L = 50\ \Omega$. According to Eq. (4-70), $\overline{i_{RP}^2}/\Delta f = 3.2 \times 10^{-22}$ A^2/Hz. When biased in saturation, $\overline{i_d^2}/\Delta f$ is roughly equal to $4kT \cdot 2/3 \cdot g_m = 2.0 \times 10^{-23}$ A^2/Hz. The output square noise due to $\overline{i_{RP}^2}/\Delta f$, calculated from Eq. (4-69), is 5.41×10^{-16} V^2/Hz. In contrast, the output square noise due to $\overline{i_d^2}/\Delta f$, calculated from Eq. (4-71), is 5.0×10^{-20} V^2/Hz. The overall output square noise is the sum of these two components, which is equal to 5.4105×10^{-16} V^2/Hz. The voltage gain of the transistor, $v(8)/v(1)$ of Fig. 4-73a, is equal to

$$A_v = \left|\frac{v(8)}{v(1)}\right| = \frac{1}{R_P} \times \frac{|g_m^\dagger|}{\omega C}\left(R_L // \frac{1}{g_d}\right). \quad (4\text{-}72)$$

The gain is calculated to be 26. The noise figure is computed from Eq. (4-67): $NF = 5.4105 \times 10^{-16}/(4 \cdot 1.38 \times 10^{-23} \cdot 290 \cdot 50 \cdot 26^2) \approx 1$, or effectively, 0 dB.

One may be tempted to conclude that, because all of the extrinsic resistances are removed, the minimum noise figure naturally should be 0 dB. This is a wrong conception. In Chapter 5 we will demonstrate that when the induced gate noise (Section 4.12) is incorporated into the model, the minimum noise figure is no longer close to 0 dB, even after all extrinsic resistances are removed.

4.14 INCONSISTENT INPUT-REFERRED NOISE BEHAVIOR

Input-referred noise (INOISE) is the output noise (ONOISE) divided by the gain of the amplifier. Both INOISE and ONOISE can be either noise in a node voltage or in a branch current, although generally SPICE simulation calculates these quantities only for a node voltage. In this section, we examine the INOISE and ONOISE of a cascaded two-stage amplifier shown in Fig. 4-74a. The device sizes and load resistors are chosen so that both transistors are dc-biased at $V_{GS} = 1.5$ V, where g_m is near the highest value. To avoid complication, the junction capacitances are neglected in the analysis, by properly assigning junction capacitance parameters to zero. The small-signal noise equivalent circuit is shown in Fig. 4-74b, which demonstrates that the noise of the input transistor is amplified by the output transistor and shows up as part of the ONOISE.

It turns out that the simulation results depend on whether XPART is 0 or 1, and on whether the NQS model is invoked. The noise's frequency response in each of the four combinations is drastically different from the other. Figure 4-75a illustrates the ONOISE and INOISE simulated with the QS analysis (NQSMOD = 0), for both the cases XPART = 0 and 1. The corresponding voltage gains at the drains of the two stages are shown in Fig. 4-75b. The figures shows that ONOISE's frequency behaviors are comparable for the two XPART values, but INOISE's exhibits great discrepancies. The difference originates from the fact that when XPART = 0, the voltage gain increases unrealistically with frequency, as discussed in Section 4.4. When XPART = 1, $|g_m^\dagger|$ does not increase with frequency, and the voltage gains are seen to roll off continuously with frequency. The INOISE with XPART = 1 therefore behaves as expected, at least qualitatively.

When NQSMOD is set to 1 so that the NQS analysis is invoked, we expect the NQS effects of the transistor to introduce another pole to $|g_m^\dagger|$. Consequently, we expect the voltage gain, and hence ONOISE, to roll off at a smaller frequency. The NQS simulation results are shown in Fig. 4-76a for ONOISE and INOISE, and in Fig. 4-76b for the voltage gains. The results certainly do not meet our expectation. Instead of rolling off, ONOISE and INOISE in this NQS simulation flattens out at high frequencies, for both XPART = 0 and 1.

Without an experimental verification, it is difficult to judge whether the right result is obtained with XPART = 0 or 1, and with NQSMOD = 0 or 1. Nonetheless, it is noted that drastically different behaviors in the noises' frequency responses can be simulated by merely changing these two parameters.

376 IMPROVABLE AREAS OF BSIM3

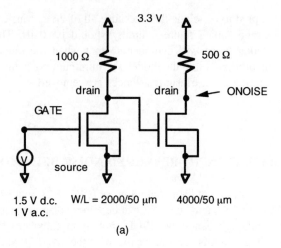

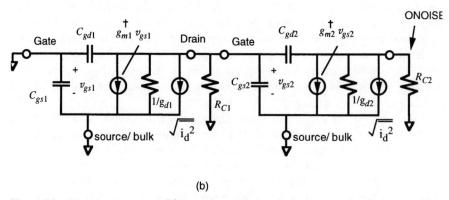

Fig. 4-74 (a) A two-stage amplifier used to examine the noise models; (b) its equivalent circuit. The noise of the input transistor is amplified by the output transistor.

4.15 POSSIBLE NEGATIVE TRANSCONDUCTANCES

Figure 4-77 plots measured g_{mb} for both a short and a long-channel device at various bias conditions. The back-gate transconductance, defined as $\partial I_D / \partial V_{BS}$, is always positive.

In a basic BSIM3 parameter extraction process, the SPICE parameters are obtained by matching the experimental I_D curves under various bias conditions. Sometimes the matching of small-signal quantities, such as g_d or/and g_m, is also part of the extraction to improve accuracy. The g_{mb} data, in contrast, are rarely used in parameter extraction; therefore, we do not expect g_{mb} calculated from a BSIM3 parameter set to match closely with the measured g_{mb}. However, for numerical stability during a circuit simulation, it is desirable to ensure that g_{mb} of the transistors be positive for all values of V_{BS}. Figure 4-78 illustrates calculated g_{mb} from a parameter set which otherwise fits I_D, g_m, and g_d within 7% r.m.s. error in a wide

4.15 POSSIBLE NEGATIVE TRANSCONDUCTANCES 377

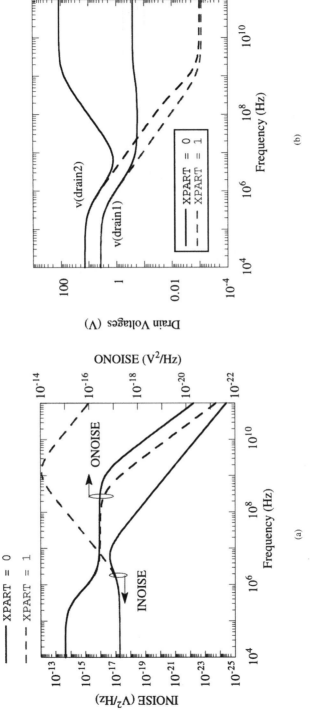

Fig. 4-75 (a) Input and output noise voltages simulated with QS model (NQSMOD=0), for both the cases XPART=0 and 1; (b) corresponding voltage gains at the drains of the two stages.

378 IMPROVABLE AREAS OF BSIM3

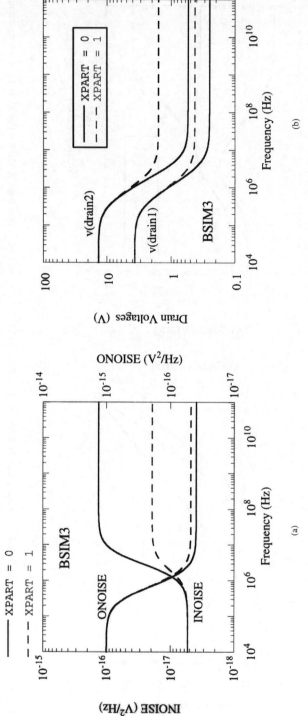

Fig. 4-76 Same as Fig. 4-75, except with BSIM3's NQS model (NQSMOD = 1).

4.15 POSSIBLE NEGATIVE TRANSCONDUCTANCES

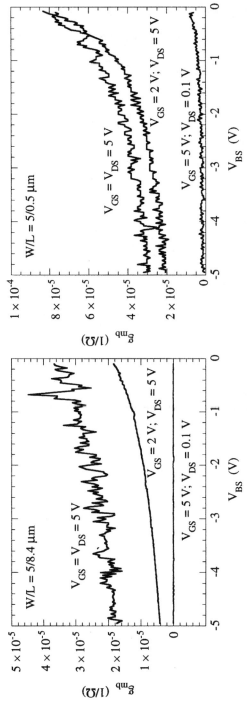

Fig. 4-77 Measured g_{mb} for both a short and a long-channel device at various bias conditions.

380 IMPROVABLE AREAS OF BSIM3

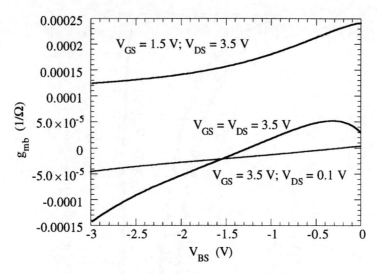

Fig. 4-78 BSIM3-calculated g_{mb}. It can be negative at times.

variety of bias conditions. This parameter set is generally considered accurate for digital circuit simulation and is likely sufficient for general analog circuit simulation. Despite the excellent fit to I_D, g_m and g_d, the calculated g_{mb} exhibits quite different behaviors from the measured results shown in Fig. 4-77. As mentioned, we are really not surprised by the difference. However, we are concerned with the fact that the calculated g_{mb} is negative.

It is not easy to identify why one parameter set gives rises to positive g_{mb}, whereas another seemingly equivalent set produces negative g_{mb}. A systematic procedure to determine the cause for negative g_{mb} requires one to access the BSIM3 codes and print out the numerous intermediate variables (no less than 30 of them!). This procedure is clearly not suitable for the average users. The following is a checklist which helps correct the negative g_{mb} problem: 1) make sure that the parameter UC < 0; 2) make sure that K1 > K2, and K1 should be on the order of 0.4; 3) try to make K2 = 0 (or a very small value such as 10^{-10}); 4) increase the value of KETA (either make it less negative or more positive); 5) increase the values of PDIBLCB, ETAB, and DVT2 (either make them less negative or more positive); and 6) make sure that V_{BS} is 0 or negative; if it has to be positive for specialized circuits, V_{BS} should not approach the value of $2 \cdot \phi_f$. Often after the above correction, the SPICE parameters need to be reoptimized so that I_D, g_d and g_m are once again closely match to experimental results.

It is also possible that calculated g_m becomes negative. From Fig. 1-6, we see that g_m initially increases with V_{GS} as the channel turns on, reaches a maximum value, but then gradually decreases due to mobility degradation with increasing V_{GS}. When carried to an extreme, calculated g_m can eventually become negative, especially when V_{DS} is small, as illustrated in Fig. 4-79. Generally this problem does not deserve much attention. After all, g_m becomes negative at fairly high values of V_{GS}

4.15 POSSIBLE NEGATIVE TRANSCONDUCTANCES

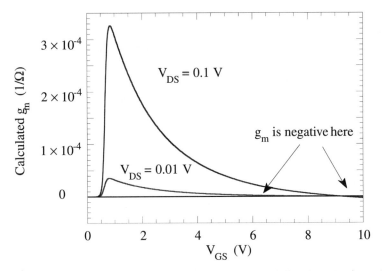

Fig. 4-79 BSIM3-calculated g_m. It can be negative, especially when V_{DS} is small.

(unless V_{DS} is extremely small), usually at values outside of range used in a technology. However, it is still good to bear in mind that g_m can decrease below zero when V_{GS} is high. In real devices, measured g_m is always positive except in advanced processing technologies [25].

A related problem that occasionally surfaces is that calculated g_m can be discontinuous in its derivative. In Chapter 1 we emphasized that BSIM3 does not

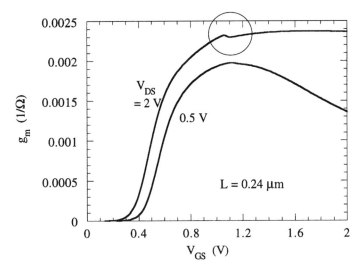

Fig. 4-80 BSIM3-calculated g_m. A bump related to bad choice of A1 and A2 may become apparent in a fine V_{GS} grid.

share many problems of a digital MOS model. One of improvement that BSIM3 brings with its use of smoothing functions is that g_m is smooth, exhibiting no discontinuity. While this is true, in general, a bad parameter set can still cause problems. Figure 4-80 shows the calculated g_m from a parameter set which fits I_D, g_m and g_d well. In fact, when plotted on a coarse V_{GS} scale, the bumpy g_m problem highlighted by a circle does not appear. This problem relates to a variable λ (see Eq. A-71), used to determine the saturation voltage. It is known that this problem often occurs when the value of A1 is high (such as 0.4), and can be eliminated once the A1 parameter is chosen below 0.1. There are also problematic occasions when both A1 and A2 are small, in the range of 0.02. In these cases, increasing A1 and A2 to values close to 1 often removes the g_m kink. However, the exact cause is not well understood. Regardless, the lesson here is that one should always plot out calculated g_m (and other values) to check for kink features or discontinuities.

4.16 LACK OF GIDL (GATE-INDUCED DRAIN LEAKAGE) CURRENT

When we discussed BSIM3's modeling for $V_{GS} < 0$ in Section 2.7, we mentioned that BSIM3 does not account for the gate-induced drain leakage. Consequently, BSIM3 potentially underestimates the off current at $V_{GS} = 0$, as well as the leakage current when $V_{GS} < 0$. A typical figure contrasting the experimentally measured I_D versus that calculated by BSIM3 was shown in Fig. 2-29. In order that we make proper remedy to this situation, we need to understand the physics of GIDL so that a judgment of how to account for the device leakage current can be properly made. Figure 4-81a shows the measured I_D and $-I_B$ for a short-channel device, and Fig. 4-81b, for a long-channel device of the same technology. We make several observations. Afterward, we discuss a physical mechanism of GIDL which agrees with the experimental results.

First, let us have a sanity check. In the short-channel device, we see that the drain current turns on at a smaller V_{GS} as V_{DS} increases from 1.1 V to 2.6 V. This makes sense. The change in the threshold voltage is due to DIBL (drain-induced barrier-lowering). For the long-channel device, DIBL is insignificant and we see that the drain current has the same turn-on characteristics, independent of V_{DS}.

Second, at $V_{GS} < 0$, the amount of the drain current increase is identical to the amount of the bulk current increase. This fact is especially obvious in the $V_{DS} = 2.6$ V curves in both Figs. 4-81a and 4-81b, where the leakage current is significantly larger than the noise floor. We therefore conclude that the GIDL current flows between the drain and the bulk. We have also verified that the source current at $V_{GS} < 0$ remains at the instrument-limitation level. In Section 2.2, we discussed the dc equivalent circuit used in BSIM3, which was shown in Fig. 2-2. The equivalent circuit does not include GIDL. Based on the observation that GIDL flows into the drain and exits through the bulk, we propose an improved dc equivalent circuit which aids our following discussion, as shown in Fig. 4-82. The addition of the GIDL current source reflects the presence of the GIDL current between the drain and

4.16 LACK OF GIDL (GATE-INDUCED DRAIN LEAKAGE) CURRENT

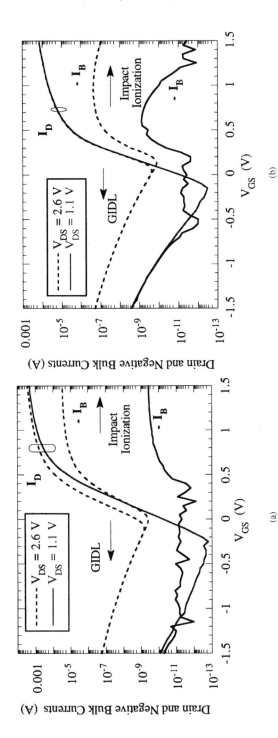

Fig. 4-81 Measured I_D and $-I_B$ at $V_{DS} = 1.1$ and 2.6 V. The gate-induced drain leakage current is prominent at negative V_{GS}. (a) Short-channel device; (b) long-channel device.

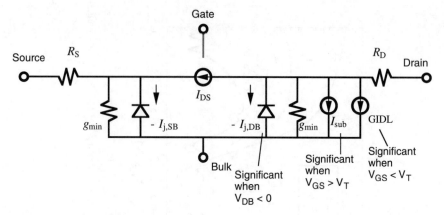

Fig. 4-82 A non-BSIM3 dc equivalent circuit which aids our understanding of the gate-induced drain leakage current.

bulk, in addition to those already accounted for in BSIM3: $I_{j,DB}$ (drain–bulk diode leakage), GMIN (a nonphysical resistor for numerical robustness), and I_{sub} (impact ionization current). GIDL is a drain-to-bulk leakage current induced by the gate bias. The dc gate current remains at zero.

Third, I_B is dominated by GIDL at $V_{GS} < V_T$ and by I_{sub} at $V_{GS} > V_T$. This observation is more clearly seen in Fig. 4-81b for the long-channel device. When V_{GS} is -1.5 V, I_D and $-I_B$ are identical, meaning that $I_D + I_B = 0$. This is the GIDL current which flows between the drain and the bulk. As V_{GS} increases toward zero, the leakage current decreases. Because the leakage current is a function of the gate-to-source bias, this drain leakage current is properly called the *gate-induced* drain leakage current. As V_{GS} increases above zero, GIDL decreases rapidly and I_B is small. I_B picks up in magnitude once again when V_{GS} exceeds the threshold voltage V_T, due to impact ionization. The magnitude of the bulk current first increases with V_{GS} because there are more carriers for impact ionization. However, the impact ionization current is a function of both the availability of the carriers as well as the electric field dropping across the drain depletion region. As V_{GS} increases beyond a certain level, the transistor exits the saturation operation region and enters the linear region. The electric field at the drain then decreases. Hence, $-I_B$ eventually decreases with V_{GS}.

Fourth, we see that the magnitude of GIDL is roughly the same for the long- and the short-channel devices. For example, at $V_{GS} = -1.5$ V and $V_{DS} = 2.6$ V, Figs. 4-81a and 4-81b both demonstrate that $-I_B \sim 2 \times 10^{-7}$ A. This means that GIDL is not a function of the channel length! This is markedly distinct from the impact ionization current, the other component of I_B. The impact ionization current depends on the field in the channel, and hence is a strong function of the channel length. The two figures show that, at $V_{GS} = 1.5$ V, $-I_B$'s of the two devices are dramatically different.

There are many theories explaining the origin of GIDL [26]. They all agree roughly on the big picture, but may differ in details. We present an explanation

4.16 LACK OF GIDL (GATE-INDUCED DRAIN LEAKAGE) CURRENT

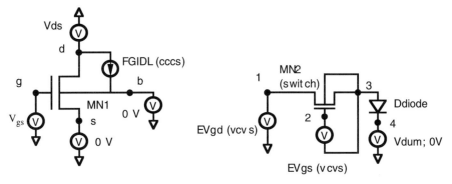

Fig. 4-83 A workaround to incorporate the gate-induced drain leakage current in BSIM3.

which is consistent with the four observations we have made, using NMOS as the exemplary device. The leakage current is related to the generation of carriers in the overlap region between the gate and the drain. From basic device physics, we know that a positive gate bias tends to invert the *p*-type channel. Likewise, a negative gate bias tends to invert the *n*-type drain junction in the overlapped region. The inversion of the drain does not easily take place, though, since the drain is doped much more heavily than the channel. Nonetheless, when V_{GD} is fairly negative, the bias at least causes the overlap region to be depleted of carriers. As the minority carriers, generated either by band-to-band tunneling or trap-assisted tunneling, arrive at the surface to attempt to form the inversion layer, they are immediately swept laterally to the substrate. The current that flows as a result of the carriers being swept from the overlapped region constitutes the GIDL current. In the framework of this explanation, we see that GIDL is not a short-channel effect. The leakage current tends to be significant in LDD devices where the overlapped region is lightly doped. GIDL is generally less a problem in sub-0.25 µm devices whose drain is formed with a single-heavily doped junction.

GIDL needs to be accounted for if the standby current of a circuit is an important specification. Unfortunately, BSIM3 does not model the GIDL effects. A quick fix to incorporate the extra leakage current between the drain and the bulk is shown in Fig. 4-83. The associated SPICE netlist is the following:

```
Vds  d 0  3V
Vgs  g 0  -1.5V
Vss  s 0  0V
Vbs  b 0  0V
MN1  d g s b NCH w=5 l=0.35

EVdg 1 0 d g 1      (a voltage-controlled voltage source)
MN2  1 2 3 3 NCH w=1000 l=0.35
EVgs 2 3 s g 1      (a voltage-controlled voltage source)
Ddiode 3 4 diodeMOD area=1
.model diodeMOD D IS=1e-19 N=12 RS =0 TNOM=27
```

```
Vdum 4 0 0V

FGIDL d b vdum 1     (a current-controlled current source)

.dc vgs -3 3 0.01
.pr dc #-I(Vds)#
.lib dsn=model.bsim3 (containing NCH MOS model file)
.end
```

In this quick fix, the drain-to-gate voltage is referenced through EVdg to drive a diode. The diode parameters are set such that it produces the GIDL current, which increases exponentially with V_{GD} at an ideality factor on the order of 10. However, we realize that GIDL exists only when $V_{GS} < 0$. Therefore, a wide NMOS switch is used such that a negative V_{GS} (referenced through EVgs) can switch on the signal to the diode, but a positive V_{GS} shuts off the NMOS (note the EVgs voltage source is referenced to $-V_{GS}$, not V_{GS}). The NMOS switch has an arbitrarily wide width so that the on resistance of the switch is small and the voltage drop across the drain and source of the switch is nearly zero when the switch is on. The diode is connected to a dummy voltage source of 0 V. It acts as a current meter of the diode, whose current is referenced back to the original NMOS through FGIDL. FGIDL is the excess current associated with GIDL. The I-V characteristics produced with Fig. 4-83, along with the BSIM3 simulation, are shown in Fig. 4-84. For different technologies, the circuit configuration, as well as the diode model parameters, can be changed to better reflect the measured GIDL.

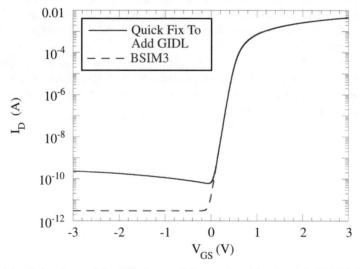

Fig. 4-84 Comparison of the I-V characteristics produced by the default BSIM3 and the workaround circuit of Fig. 4-83.

We end this section with the reminder that, because GIDL is not modeled in BSIM3, the noise due to GIDL is not accounted. Under certain bias conditions, the amount of drain current due to GIDL can be appreciable. Because the current is associated with a certain energy barrier, we believe that the noise associated with GIDL is shot noise in nature. Therefore, suppose that the drain current due to GIDL is I_{gidl}, then the drain noise square current should be

$$\overline{i_d^2}|_{due\ to\ GIDL} = 2qI_{gidl}\Delta f, \qquad (4\text{-}73)$$

where q is 1.6×10^{-19} C. The magnitude of the noise depends on the magnitude of the GIDL current.

4.17 INCORRECT SUBTHRESHOLD BEHAVIORS

We defined the subthreshold-slope ratio in Section 1.3. We calculated the ratio to show the regional nature of a digital model. Figure 1-10 leaves us a clear impression that a digital model can give rise to discontinuity in the device characteristics, while an analog model such as BSIM3 produces smooth characteristics without abrupt transition. However, the smooth characteristic does not necessarily mean the analog model results of Fig. 1-10 are correct. Although qualitatively the BSIM3 behavior exhibited in Fig. 1-10 is right, it does not agree with the theoretical quantitative value.

In a long-channel device, we expect the drain current to be proportional to $1 - \exp(-q \cdot V_{DS}/kT)$ in the subthreshold region as shown in Eq. (1-12). (In short-channel device, the DIBL lowering of the drain barrier may disqualify the above statement.) In a regional model such as that often employed in a digital model, it is easy to make the subthreshold current to exhibit such V_{DS} dependence. However, BSIM3 uses a smoothing function to manipulate its strong inversion models to work in the subthreshold region. Therefore, the subthreshold drain current may not be exactly proportional to $1 - \exp(-q \cdot V_{DS}/kT)$. As demonstrated in Section 1.4, we were able to show that I_D of BSIM3 can have the $1 - \exp(-q \cdot V_{DS}/kT)$ dependence only after making several approximations. Often, the approximations are only grossly approximate, and the subthreshold current's V_{DS} dependence is not exactly correct.

The subthreshold-slope ratio is a benchmark parameter that reveals the frequent failure of the $1 - \exp(-q \cdot V_{DS}/kT)$ dependence in the BSIM3 subthreshold current. The ratio is defined as $(I_{D2} + I_{D1})/(I_{D2} - I_{D1}) \times (V_{DS2} - V_{DS1})/(V_{DS2} + V_{DS1})$. In Fig. 4-85, we plot the ratio of a $L = 10\,\mu\text{m}$ transistor with $V_{DS,1} = 0.001$ V and $V_{DS,2} = 0.002$ V. We have made the JS parameter to be zero and GMIN to be 10^{-26} (Ω^{-1}) so that the leakage current is negligible, even when V_{GS} approaches zero. Let us take the case when the ambient temperature is $-55°$C. Figure 4-85 indicates that the subthreshold slope ratio calculated from BSIM3 approaches 1.027. The theoretical value, after substituting $I = 1 - \exp(-V_{DS}/kT)$ into the ratio

388 IMPROVABLE AREAS OF BSIM3

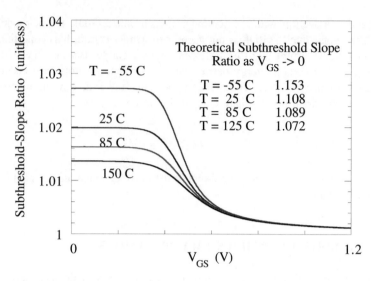

Fig. 4-85 Subthreshold-slope ratio of a $L = 10\,\mu\text{m}$ transistor with $V_{DS,1} = 0.001$ V and $V_{DS,2} = 0.002$ V. The ratio is defined as $(I_{D2} + I_{D1})/(I_{D2} - I_{D1}) \times (V_{DS2} - V_{DS1})/(V_{DS2} + V_{DS1})$.

definition, is 1.153. The discrepancy, as mentioned, is a basic problem when the smoothing function is adopted. The smoothing function necessary converges to the saturation equation when V_{GS} is high. Although the smoothing function attempts to acquire the $1 - \exp(-V_{DS}/kT)$ dependence when V_{GS} is small, it does not quite get there. At other temperatures, we also see that the subthreshold slope ratios calculated from BSIM3 differ from the theoretical values, though with less discrepancy.

Another incorrect subthreshold behavior of BSIM3 is revealed in Fig. 4-86, which plots g_m/I_D vs. V_{GS}, at various V_{BS}. We first showed the g_m/I_D behavior in Fig. 1-8. Back in Section 1.3, we were not interested with the quantitative results of the calculated g_m/I_D. We were mainly concerned with the fact that typical digital model produces discontinuity in the ratio, while BSIM3 is continuous. Here, we are interested in its quantitative value.

In Fig. 1-8, the g_m/I_D ratio is seen to have a peak somewhere near the transition between the subthreshold and the strong inversion regions. The ratio decreases because the mobility degradation associated with high V_{GS} lowers the transconductance and, hence, the ratio. But, what is the reason for the decrement of the ratio as V_{GS} approaches zero? It turns out that BSIM3 predicts the ratio to decrease toward zero because of two leakage currents. The first one is that associated with the drain–bulk junction diode, and the second one is related to GMIN, the "numerical" resistor placed between the drain and the bulk to speed up numerical iteration (see Section 2.2). In order to concentrate on the intrinsic device behaviors, here we make the parameters JS and JSSW to be zero, so that the drain–bulk junction leakage is zero. We further set GMIN to be 10^{-26} (Ω^{-1}) through the .OPTION statement, so that the current through GMIN is negligible in the transistor of interest. After these two steps

4.17 INCORRECT SUBTHRESHOLD BEHAVIORS 389

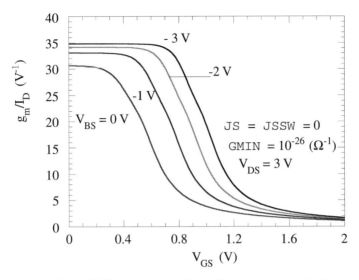

Fig. 4-86 BSIM3-calculated g_m/I_D vs. V_{GS}, at various body biases.

are taken, we then obtain Fig. 4-86, instead of Fig. 1-8. An obvious modification from the characteristic before is the flattening of the g_m/I_D ratio in the subthreshold region. It is now independent of V_{GS}. This is understandable because BSIM3 formulates I_D to be proportional to $\exp(qV_{GS}/nkT)$ when V_{GS} is smaller than V_T, where n is the ideality factor given in Eq. (3-72) or Eq. (3-73). Therefore, g_m/I_D is a constant, independent of V_{GS}. The constant g_m/I_D ratio in the subthreshold region is observed in all V_{BS}'s. As V_{BS} becomes more negative, n becomes smaller. Therefore, the g_m/I_D ratio increases. In fact, as V_{BS} tends $-\infty$, n would approach unity and g_m/I_D would approach q/kT.

Figure 4-86 is simulated with BSIM3. An experimental result showing the measured g_m/I_D for a short-channel device is shown in Fig. 4-87. The device's leakage currents are small. There is a drastic difference between the experimental and the BSIM-simulated curves. In the measured results, g_m/I_D is seen to peak at a certain value, rather than attaining a constant value as calculated by BSIM3. This is an important observation for circuit designers who would like to maximize the g_m/I_D ratio of a particular transistor to obtain high gain at a small dissipation current. The BSIM3 result shown in Fig. 4-85 would mislead the designers that they could bias the transistor all the way to $V_{GS} = 0$ and would still get the maximum g_m/I_D. In practice, the maximum occurs right near the subthreshold-strong-inversion boundary. The idea of a constant subthreshold slope is in reality only an approximate concept, which is not truly correct.

For designers whose circuit functions mostly in either the linear or saturation regions, the subthreshold inaccuracy inherent in BSIM3 is not a big deal. In fact, since most circuits tend to work outside of the subthreshold region, we agree with BSIM3 that we should place priority in ensuring the smoothing function to work in the strong inversion region. The after effects of slight inaccuracy in subthreshold

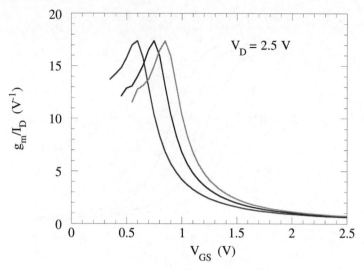

Fig. 4-87 Measured g_m/I_D for a short-channel device. The measured results differ from BSIM3-calculated curves of Fig. 4-86. The leakage current plays a minor role in the decrease below $V_{GS} = 0.5$ V.

characteristics are considered as an unavoidable outcome of the smoothing function. The designers just need to be aware the subthreshold characteristics calculated by BSIM3 may differ from the measured data.

4.18 THRESHOLD VOLTAGE ROLLUP

The threshold voltage adopted in BSIM3 was given under the entry VTH0 in Section 3.2. For the discussion here, let us focus on transistors without narrow-width effects, at low V_{DS} biases and $V_{BS} = 0$. Under this scenario, only two ΔV_T terms remain significant: the charge-sharing term and the reverse short-channel-effect term. The threshold voltage for NMOS can be approximated as

$$V_T = \text{VTH0} - \Delta V_{T,\text{charge\_sharing}} + \Delta V_{T,\text{reverse\_short\_channel}}, \quad (4\text{-}74)$$

where VTH0 is a constant-value parameter and

$$\Delta V_{T,\text{charge\_sharing}} = \text{DVT0}\left[\exp\left(-\text{DVT1}\frac{L_{\text{eff}}}{2L_t}\right) + 2\exp\left(-\text{DVT1}\frac{L_{\text{eff}}}{L_t}\right)\right](V_{bi} - 2\phi_f); \quad (4\text{-}75)$$

$$\Delta V_{T,\text{reverse\_short\_channel}} = \text{K1}\left(\sqrt{1 + \frac{\text{NLX}}{L_{\text{eff}}}} - 1\right)\sqrt{2\phi_f}. \quad (4\text{-}76)$$

The details of various variables such as V_{bi} were discussed in the entry VTH0 in Section 3.2. The details are really not interesting here; just an examination of the functional dependences of the two ΔV_T terms on L_{eff} is sufficient to underscore the point of this section.

Equation (4-75) shows that, as L_{eff} tends to zero, $\Delta V_{T,\text{charge\_sharing}}$ approaches a constant value equal to the product of DVT0 and $3(V_{bi} - 2\phi_f)$. Equation (4-76), in contrast, states that $\Delta V_{T,\text{reverse\_short\_channel}}$ goes to infinity as L_{eff} approaches zero. Therefore, BSIM3 equations predict V_T to roll up as L_{eff} decreases, a phenomenon which is at odds with experimental results. Generally, the measured V_T, shown in Fig. 3-23, first increases slightly as L_{eff} decreases, due to the reverse short-channel effects (modeled by the term Eq. 4-76). As L_{eff} decreases further, the measured V_T decreases rapidly toward a small value, due to the short-channel effects (modeled by Eq. 4-75).

The inconsistency between the modeled and experimental behavior of V_T can be either irrelevant or problematic, depending on whether the model is to be used for predictive purpose. Suppose the model is only to be used for the present technology node, then the inconsistency poses no issue. After all, the parameters of DVT0, DVT1, NLX are extracted such that the BSIM3 equations accurately reflect the measured data. However, if the same model is used to predict the behavior of the next-technology-node transistors, the same set of DVT0, DVT1, NLX may lead to an increase in the threshold voltage because the next-technology-node transistors have a channel length shorter than the minimum channel length of the present technology node.

4.19 PROBLEMS ASSOCIATED WITH A NONZERO RDSW

This section appears just for the completeness of this chapter, even though this problem has been described adequately in Section 2.6. Basically, there are two approaches in modeling the source and drain resistances in BSIM3, as described in Section 2.2. In the lumped-resistance approach, the parameter RSH is used in conjunction with NRS and NRD to calculate the R_S and R_D in Fig. 2-8a. In the absorbed-resistance approach of Fig. 2-8b, $R_S = R_D = 0$, yet the drain current and conductances are modified by a nonzero RDSW such that the final calculated results are approximately equal to those of the lumped-resistance approach. This RDSW parameter at best "effectively" accounts for the effects of the source and drain resistances on the drain current and the conductances, but the resulting input impedance looking into the gate does not include a physical source resistance. The input impedance is in fact purely imaginary. Likewise, looking out through the drain, there is no drain resistance. In other words, only I_D, g_m, g_{mb}, and g_d are correctly calculated when RDSW is used, but y_{11}, s_{11} and so on, will all be incorrect. These are severe limitations as far as the rf characteristics are concerned.

In addition, when the source and drain resistances are modeled with RDSW, no thermal noise sources associated with the resistances will be attached to the transistor. The proper way to include the terminal resistances, as far as noise

392 IMPROVABLE AREAS OF BSIM3

analysis is concerned, is to declare the source and drain resistances through the use of NRD, NRS, and RSH, and set RDSW to zero.

4.20 OTHER NUISANCES

No software can please everyone. You probably know a popular PC/MAC software. Do you find the creepy dancing paper clip (in PC) and the dancing SE/30 (in MAC) of the newer versions of the program to be extremely annoying? Well, according to manuals, they are supposed to be my friends in formatting the document. My advice to those dancing friends—just get out of my sight.

BSIM3, by its nature as a software, cannot please everyone, either. What appear to be nice features ultimately become cumbersome for some. These features are not bugs because they are not necessarily incorrect. They are just nuisances. We list them in no particular order.

1. BSIM3's junction capacitance differs from the SPICE's junction capacitance element in that BSIM3's junction capacitance does not include diffusion capacitance. In the reverse-bias condition which is typically encountered in a normal MOS operation, the BSIM3 model works well. However, in some circuits where V_{BS} is made positive, the neglect of the diffusion capacitance may be problematic, as mentioned in Section 2.4.

2. All diodes have breakdown voltages. When the junction is reverse biased to an extreme, eventually the junction breaks down and a lot of current flows through the diode. In BSIM3, both the source–bulk and the drain–bulk junction diodes are assumed to function indefinitely as the reverse bias increases. Therefore, BSIM3 proceeds normally in calculating the device currents when the drain is connected to 1 V while the bulk is tied to -10 V, for example. In reality, if the reverse breakdown occurs at -7 V, then the reverse bias of -11 V would have caused the drain current (and bulk current) to increase dramatically. The device is useless under the said bias condition. BSIM3, however, would still presume the device to function normally.

   This problem can be corrected by making the junction areas in the device to be zero. The diode behaviors are accounted for by a separate specification of semiconductor junction diodes placed in parallel to the drain–bulk junction and the source–bulk junction. The semiconductor junction diode, as a SPICE device component, allows for the specification of the breakdown voltage.

3. BSIM3 will take a V_{GS} of -30 V and calculates device current as though the device operates under the normal condition. In reality, the oxide has broken down under such a high bias voltage across the oxide. The lack of specification for the maximum (either positive or negative) gate biases is a drawback.

4. The bulk–drain junction capacitance cannot be set independently from the bulk–source capacitance. They both are calculated from the same CJ, CJSW,

CJSWG parameters (among others). Once the value of CJ (junction capacitance per unit area), for example, is assigned a given value for the source–bulk junction, the same value is applied to the drain–bulk junction. The two capacitances can differ only if either the bias or the junction geometry is different.

When the device is asymmetrical, then the junction capacitance per unit areas should be different. Asymmetrical MOS is useful for power applications, because its extended lowly-doped drain region can sustain a large voltage drop. Sometimes for higher voltage capability, the pocket implants placed near the drain is intentionally made to be different from the source to increase the drive current. The solution to model devices with asymmetrical source and drain is to use two independent semiconductor junction diodes to replace the BSIM3 diodes. This approach was also suggested for nuisance #2.

While there is no freedom in independently setting up the junction capacitance parameters, the overlap capacitances (such as CGDO, CGSO) can be made to differ for the gate–source and the gate–drain parasitics. The fringing capacitance of the gate–source and gate–drain parasitics, however, are characterized by the same parameter, CF.

5. When BSIM3 calculates the junction capacitance, it first calculates the built-in potential, which is given by

$$\textit{Built-in potential} = \text{PB} - \text{TPB} \cdot (T_{device} - \text{TNOM}), \qquad (4\text{-}77)$$

where T_{device} is the operating temperature, and TNOM is the temperature at which the SPICE parameters are extracted. PB is the SPICE parameter which characterizes the value of the built-in potential at TNOM. The temperature coefficient, TPB, is another SPICE parameter that user supplies. Sometimes TPB is extracted at a temperature near TNOM. If the value is used for high temperature (large T_{device}), it is possible to obtain unrealistically small values. (BSIM3 will clamp the value at 0.01 V if the built-in potential calculated from above is below 0.01 V.) It seems that the linear temperature dependence of the built-in potential is only a crude approximation that cannot be correct at indefinitely high T_{device}. Perhaps the built-in potential initially decreases linearly with T_{device}, but the rate of decrease may slow down at higher temperatures.

6. BSIM3's calculation of the junction capacitance is based on the transistor layout often used for digital circuits, as shown in Fig. 4-88a. The source is at one side of the gate, and the drain is at the other side. In this case, the capacitance is correctly calculated as

$$C_{j,\text{DB}} = C_{j,\text{DB\_BW}} \cdot \text{AD} + C_{j,\text{DB\_SW}} \cdot (\text{PD} - W_{\textit{eff},\text{CV}}) + C_{j,\text{DB\_SWG}} \cdot W_{\textit{eff},\text{CV}}, \qquad (4\text{-}78)$$

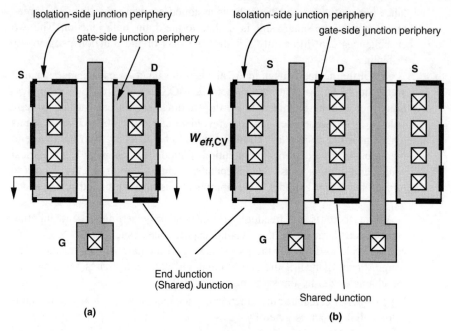

Fig. 4-88 (a) Typical transistor layout adopted for digital circuits, where the source is at one side of the gate, and the drain is at the other side; (b) multi-finger transistor layout used in many analog circuits. In this example, there are two gates with two source junctions but only one drain junction.

where $C_{j,\text{DB\_BW}}$ is the bottom-wall capacitance per area (F/m²), $C_{j,\text{DB\_SW}}$ is the side wall capacitance per length at the isolation side (F/m), and $C_{j,\text{DB\_SWG}}$ is the side wall capacitance per length at the gate side (F/m). The formula works well for the junctions in Fig. 4-88a. It is easy to verify that AD in Eq. (4-78) has the normal meaning of the drain junction area, and PD has the normal meaning of the periphery of the drain junction.

It is not uncommon in analog circuits that a MOS transistor consists of multiple gates, with multiple junction areas. Fig. 4-88b is one example, showing two gates with three junction areas. In this example, the two outer junctions are the source, and the middle acts as the drain. The drain–bulk junction capacitance should be written as

$$C_{j,\text{DB}} = C_{j,\text{DB\_BW}} \cdot \text{AD} + C_{j,\text{DB\_SW}} \cdot (\text{PD} - 2W_{\textit{eff},\text{CV}}) + C_{j,\text{DB\_SWG}} \cdot 2W_{\textit{eff},\text{CV}}. \tag{4-79}$$

This equation differs from Eq. (4-78), which is the equation adopted in BSIM3. In order to properly calculate the drain–bulk capacitance of Fig. 4-

88b with the formula of Eq. (4-78), the meaning of PD or $W_{eff,CV}$ will necessarily be changed. While it is not an unsolvable problem, any tweak to correct the situation adds a certain amount of confusion to the modeling.

Figure 4-88b is just one example wherein the calculation of the junction capacitance requires care. Many other types of layouts have similar problems.

7. We can generalize the source (or drain) parasitic resistance to consist of two components, as shown in Fig. 4-89. The figure is a cross-section view of the transistor region indicated in Fig. 4-88a. The resistance consists of a component associated with the metal-semiconductor contact (denoted R_{cont}), and an access resistance between the n^+ source junction and the inverted channel (denoted R_{access}). Generally, the access resistance tends to dominate. It is a resistance that is inversely proportional to the channel width. The wider the width, the more the periphery to conduct current and the smaller the resistance. The contact resistance, in contrast, does not scale linearly with the width, although it also decreases as the width widens. The nonlinear scalability is shown in Fig. 4-88a, where we see that the contacts are placed with a fixed spacing between them. A stripe contact that would connect a column of the four contacts is not used in practice. The contacts are made with the same size across the wafer to improve the uniformity of the reactive ion etching of the oxide during the formation of the contact.

The source and drain resistance calculated in BSIM3, either using the RDSW parameter or NRD-NRS-RSH combination (see Section 2.2 for details), is assumed to scale linearly with the transistor width. This is indeed the case when the access resistance dominates. However, when contact resistance is a major portion of the overall parasitic source resistance, then BSIM3 equations can become problematic.

8. There is no gate leakage current in BSIM3, nor the gate shot noise associated with the leakage current. As the gate oxide thickness continues to be scaled down, the amount of carrier tunneling through the oxide barrier increases. At present, there does not seem to be an overwhelming need to model the gate

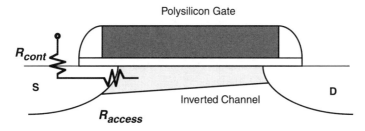

Fig. 4-89 Two components of source (or drain) parasitic resistance: the metal-semiconductor contact (denoted R_{cont}), and an access resistance between the n^+ source junction and the inverted channel (denoted R_{access}).

current, mainly because its magnitude remains small in present production technologies, with the ratio of I_G/I_D often below 10^{-10}. Perhaps because the hole effective mass is larger than the electron, the gate current in a PMOS device is about 1% of the gate current in the NMOS device of the same CMOS technology. More discussion about the gate current can be found in Section 5.18.

BSIM3 does account for the substrate current (I_{sub}) due to impact ionization. However, the noise associated with I_{sub} is not modeled.

9. BSIM3 overestimates the threshold voltage when $V_{BS} > 0$. Somehow BSIM3 switches to a different threshold voltage expression the moment V_{BS} becomes positive. This problem was detailed in Section 2.7.
10. BSIM3 does not support a thermal resistance (neither do popular SPICE simulators). The device self-heating cannot be simulated in BSIM3.
11. Binning should always be used as the last alternative. However, if binning is desired, it is awkward to be used in BSIM3. The way BSIM3 is set up tends to lead to discontinuity in device characteristics across bin boundaries. This problem, as well as a remedy, was detailed in Section 1.12.
12. There is an inconsistency about using TOX and TOXM. See the entry of TOXM in Section 3.2 for more details.
13. We wish BSIM3 to have a reliability model.
14. This item depends on the SPICE simulator which implements the BSIM3 model. Sometimes, funny simulation results are possible when seemingly harmless parameter values are used. Try setting K2 = 0, CKAPPA = 0, MJ = 1, MJSW = 1, or MJSWG = 1. In most simulators, numerical difficulty will result. Perhaps BSIM3 should issue warnings when those values are used.

REFERENCES AND NOTES

[1] The official BSIM3 website is: http://www-device.eecs.berkeley.edu/~bsim3
[2] B. Sheu, and C. Hu, "Switch-induced error voltage on a switched capacitor," *IEEE J. Solid-State Circuits*, vol. 19, pp. 519–525, 1984.
[3] W. Liu, C. Bowen, and M. Chang, "A CAD-compatible non-quasi-static MOSFET model," IEEE International Electron Device Meeting, pp. 151–154, 1996.
[4] R. Daniels, A. Yang, and J. Harrang, "A universal large/small signal 3-terminal FET model using a nonquasi-static charge-based approach," *IEEE Trans. Electron Devices*, vol. 40, pp. 1723–1729, 1993.
[5] The Killer NOR-Gate first came to the author's attention in a broadcast e-mail, dated October 1997, apparently having been copied and forwarded several times. It seems that the original author of the mail, as well as the author who documents this problem, is D. Foty, of Gilgamensh Associates, http://www.sover.net/~dfoty

[6] W. Liu, *Fundamentals of III-V Devices: HBTs, MESFETs, HFETs/HEMTs*, New York: Wiley, 1999.

[7] L. Tiemeijer, L. de Maaijer, R. van Langevelde, A. Scholten, and D. Klaassen, "RF CMOS modeling, Workshop on Advances in Analog Circuit Design, *AACD Proc.*, 1999. (The statement that r_i is $\text{Re}(-1/y_{gs,\text{NQS}})$ in the paper is likely a typo.)

[8] W. Liu, R. Gharpurey, M. C. Chang, U. Erdogan, R. Aggarwal, and J. P. Mattia, "RF MOSFET modeling accounting for distributed substrate and channel resistances with emphasis on the BSIM3v3 SPICE model," *IEEE Int. Electron Device Meeting*, 1997.

[9] J. Hu, H. Yang, R. Kraft, A. Rotondaro, S. Hattangady, W. Lee, A. Chapman, C. Chao, A. Chatterjee, M. Hanratty, M. Rodder, and I. Chen, "Feasibility of using W/TiN as metal gate for conventional 0.13 μm CMOS technology and beyond," *IEEE Int. Electron Device Meeting*, pp. 825–828, 1997.

[10] W. Liu, and M. C. Chang, "Transistor transient studies including transcapacitive current and distributive gate resistance for inverter circuits," *IEEE Trans. Circuits Systems–I: Fundamental Theory Appl.*, vol. 45, pp. 416–422, 1998.

[11] W. Liu, *Handbook of III-V Heterojunction Bipolar Transistors*, New York: Wiley, 1998, Section 4.3.

[12] I. Chen, and W. Liu, "High-speed or Low-voltage low-power operations," in S. Sze and C. Y. Chang, ed. *ULSI Devices*, New York: Wiley, pp. 547–630, 2000.

[13] S. Tin, A. Osman, and K. Mayaram, "BSIM3 MOSFET model accuracy for RF circuit simulation," *IEEE RAWCON Proc.* pp. 351–354, 1998.

[14] T. Kolding, "Test structure for universal estimation of MOSFET substrate effects at gigahertz frequencies," *Proc. Int. Conf. on Microelectronic Test Structures*, pp. 106–111, 2000.

[15] Y. Tsividis, *Operation and Modeling of the MOS Transistor*, New York: McGraw-Hill, 1987. p. 364.

[16] D. Cho, S. Kang, K. Kim, and S. Lee, "An accurate intrinsic capacitance modeling for deep submicrometer MOSFET's," *IEEE Trans. Electron Devices*, vol. 42, pp. 540–548, 1995.

[17] A. Abidi, "High frequency noise measurement on FET's with small dimensions," *IEEE Trans. Electron Devices*, vol. 33, pp. 1801–1805, 1986.

[18] J. Ou, X. Jin, P. Gray, and C. Hu, "Recent developments in BSIM for CMOS RF ac and noise modeling," *Proc. Workshop on Advances in Analog Circuit Design (AACD)*, 1999.

[19] T. Manku, "Microwave noise modeling of CMOS transistors," *Proc. Workshop on Advances in Analog Circuit Design (AACD)*, 1999.

[20] C. Jacobini, "A review of some charge transport properties of silicon," *Solid-State Electronics*, vol. 20, pp. 77–89, 1977.

[21] Private communication with J. Leete, U.C.L.A.

[22] C. Motchenbacher, and J. Connelly, *Low-Noise Electronic System Design*, New York: Wiley, 1993.

[23] A. van der Ziel, *Noise in Solid State Devices and Circuits*, New York: Wiley, 1986, pp. 88–91.

[24] C. Enz, and Y. Cheng, "MOS transistor modeling issues for RF circuit design, *Proc. Workshop on Advances in Analog Circuit Design, AACD*, 1999.

IMPROVABLE AREAS OF BSIM3

[25] R. Versari, and B. Ricco, "MOSFET's negative transconductance at room temperature," *IEEE Trans. Electron Devices*, vol. 46, pp. 1189–1195, 1999.

[26] See, for example, S. Parke, J. Moon, H. Wann, P. Ko, and C. Hu, "Design for suppression of gate-induced drain leakage in LDD MOSFET's using a quasi-two-dimensional analytical model," *IEEE Trans. Electron Devices*, vol. 39, pp. 1694–1703, 1992.

5

IMPROVEMENTS IN BSIM4

5.1 INTRODUCTION

While BSIM3 is being promoted by the Compact Model Council as the industry's standard model [1], the development of BSIM4 is already under way [2,3]. The bulk of BSIM4 is based on BSIM3. However, because there are several significant modifications and additions, the BSIM4 model was not simply made to be another version of BSIM3. Besides, BSIM3v3 currently includes these versions: BSIM3v3.0, BSIM3v3.1, BSIM3v3.2, BSIM3v3.2.1, and BSIM3v3.2.2. (At the time of this writing, BSIM3v3.3 has not been released and may, in fact, never be released. See Ref. [15] of Chapter 1.) By virtue of being a BSIM3v3 model, all these versions are backward compatible, that model cards generated previously remain functional in the updated version. For example, a model card generated with the BSIM3v3.0 model should yield the same simulation characteristics as the BSIM3v3.2 model. This is certainly a valuable feature to the model users. Nonetheless, as the number of versions increases, keeping backward compatibility requires increasingly more overhead in ensuring the codes to be compatible. Perhaps after a certain point, one just has to say enough is enough, and start something fresh, without the burden of maintaining the continuity from past history.

Each of the following sections describes a piece of improvement found in BSIM4. The new model parameters introduced in BSIM4, as well as the BSIM3 parameters they replace, will be discussed and summarized (usually) at the end of each section.

The BSIM4 model described herein is the BSIM4.1.0, released on October 11, 2000. The detailed equations of BSIM4 are given in Appendix G.

5.2 PHYSICAL AND ELECTRICAL OXIDE THICKNESSES

BSIM3 uses TOX as the sole parameter describing the oxide thickness. As the transistor technology advances every year, the oxide becomes progressively thinner. A discernment between the physical and electrical oxide thicknesses has become necessary. The physical oxide thickness is the actual grown thickness of the oxide, such as that measured from a TEM (transmission electron microscope) image. The electrical oxide thickness does not have a strict definition; it is loosely used to mean any oxide thickness that results in a good fit to the measured data. As far as the BSIM4 model is concerned, we can take the electrical thickness to be the oxide thickness inferred from a gate-capacitance measurement. (We left out the detail of specifying whether it is obtained from the accumulation or the inversion region. After all, it is a non-precise thickness. One can argue that from the modeling perspective, the electrical thickness should be that inferred from a capacitance measurement in the inversion region.)

When the oxide thickness is large, distinguishing between the physical and electrical thicknesses produces minor differences. However, when the oxide thickness is 30 Å, for example, a 3 Å discrepancy between the physical and electrical thicknesses can lead to a roughly 10% difference in the I-V and C-V simulation results. BSIM4 introduces several parameters to alleviate the ambiguity encountered in BSIM3:

Parameter Name	Description	Default Value
TOXE	Electrical oxide thickness	4×10^{-9} (m)
TOXP	Physical oxide thickness	TOXE
TOXM	TOXE at which other BSIM4 parameters are extracted	TOXE
DTOX	TOXE − TOXP	0
EPSROX	Oxide's relative dielectric constant	3.9
TOXREF	See section 5.18	3×10^{-9} (m)

Replaced BSIM3 Parameters: TOX, TOXM

EPSROX is the oxide's relative dielectric constant. As technology advances, there tends to be more interest in using higher dielectric-constant material for the gate oxide. We mention in passing that TOXREF is used only in the tunneling current calculation, which will be detailed in Section 5.18.

The parameter TOXM in the above table has the same meaning as that used in BSIM3 (see the TOXM entry in Section 3.2.) We point out this parameter in the above table to show that its default value is TOXE. (TOXM was defaulted to TOX in BSIM3. Since TOX is not used in BSIM4, it is defaulted to TOXE instead.)

The following is a brief algorithm during the parsing of the input SPICE file:

5.2 PHYSICAL AND ELECTRICAL OXIDE THICKNESSES

0. DTOX is defaulted to 0 if it is not specified.
1. Are TOXE and TOXP both given? If no, goto step 2. If yes, then certainly TOXE and TOXP get their respective specified values. BSIM4 also issues warning if somehow the specified values do not obey the relationship DTOX = TOXE − TOXP. (Whether a warning is issued, BSIM4 always proceeds with TOXE and TOXP being their respective specified values.)
2. Is TOXE given? If no, goto step 3. If yes, TOXP is equated with TOXE − DTOX.
3. Is TOXP given? If no, goto step 4. Otherwise, TOXE is equated to TOXP + DTOX.
4. When neither TOXE nor TOXP is given, TOXE and TOXP are both set to 4×10^{-9} (m).

When TOXE is the only specified parameter, then following the above algorithm leads to TOXP = TOXM = TOXE, and DTOX = 0. This resorts back to the BSIM3 scenario wherein TOX is the sole oxide parameter.

TOXE is used to compute the majority of intermediate variables in BSIM4: V_T, A_{bulk}, n (subthreshold ideality factor), μ_{eff}, $V_{DS,sat}$, tunneling currents (examined in Section 5.18), all capacitances and charges in CAPMOD = 0 and 1, and so on. For example, C_{gg} in the strong inversion has the value

$$C_{gg} = C'_{ox,e} \times W_{eff,CV} L_{eff,CV}, \tag{5-1}$$

where $C'_{ox,e}$ is the electrical capacitance per unit area, equal to $\epsilon_{ox}/\text{TOXE}$.

Although TOXP appears less often than TOXE, TOXP is the critical parameter in the charge-thickness capacitance model (CAPMOD = 2 model of BSIM4, which corresponds to CAPMOD = 3 in BSIM3). In this capacitance model, C_{gg} in the strong inversion is given by

$$C_{gg} = C'_{ox,inv} \times W_{eff,CV} L_{eff,CV}, \tag{5-2}$$

where

$$C'_{ox,inv} = \frac{\epsilon_{ox}}{\text{TOXP}} // \frac{\epsilon_s}{X_{DC,inv}}, \tag{5-3}$$

$$X_{DC,inv} = \left[1 + \frac{V_{GST,effCV} + 4(\text{VTH0} - V_{FB} - 2\phi_f)}{2 \times 10^8 \cdot \text{TOXP}}\right]^{0.7}. \tag{5-4}$$

In a classical analysis, the charges are assumed to concentrate right on the oxide–silicon interface. The relevant oxide thickness is TOXP. In reality, the maximum probability of carrier distribution occurs at some distance ($X_{DC,inv}$) away from the interface. The overall gate capacitance can be viewed as a serial combination of two parallel capacitors, one with a thickness of TOXP, and another, of $X_{DC,inv}$. The oxide capacitance for the inversion calculation is thus that given in Eq. (5-3). More details about the charge-thickness model was described under the ACDE entry in Section 3.2.

We have just discussed that the charge-thickness capacitance model is a part of BSIM4 which critically depends on TOXP. There is another area where TOXP plays an important role. It is in the calculation of the drain current, which, in both BSIM3 and BSIM4, has the following general form:

$$I_{DS} = \frac{I_{DS,0}}{1 + \frac{R_{DS}I_{DS,0}}{V_{DS,\text{eff}}}} \times \left(1 + \frac{V_{DS} - V_{DS,\text{eff}}}{V_A}\right), \quad (5\text{-}5)$$

where V_A is the Early voltage. $I_{DS,0}$ in BSIM3 is

$$I_{DS,0} = \frac{W_{\text{eff}} \mu_{\text{eff}} C'_{ox} V_{GST,\text{eff}}}{L_{\text{eff}}[1 + V_{DS,\text{eff}}/(\varepsilon_{sat} L_{\text{eff}})]} \left[1 - \frac{A_{bulk} V_{DS,\text{eff}}}{2(V_{GST,\text{eff}} + 2kT/q)}\right] V_{DS,\text{eff}}, \quad (5\text{-}6)$$

whereas the corresponding $I_{DS,0}$ in BSIM4 is

$$I_{DS,0} = \frac{W_{\text{eff}} \mu_{\text{eff}} C'_{ox,\text{IV}} V_{GST,\text{eff}}}{L_{\text{eff}}[1 + V_{DS,\text{eff}}/(\varepsilon_{sat} L_{\text{eff}})]} \left[1 - \frac{A_{bulk} V_{DS,\text{eff}}}{2(V_{GST,\text{eff}} + 2kT/q)}\right] V_{DS,\text{eff}}. \quad (5\text{-}7)$$

Here is the key difference between the BSIM3 and the BSIM4 expressions. In BSIM3, $I_{DS,0}$ is proportional to C'_{ox}, which is basically $C'_{ox,e}$ in BSIM4. In contrast, $I_{DS,0}$ in BSIM4 is proportional to $C'_{ox,\text{IV}}$, a per-area capacitance approximately equal to the $C'_{ox,\text{inv}}$ of Eq. (5-3). (The fine difference between $C'_{ox,\text{IV}}$ and $C'_{ox,\text{inv}}$ can be found in Appendix G.) We can therefore say that $I_{DS,0}$ is closely related to the electrical oxide thickness in BSIM3, but it is related to the physical oxide thickness in BSIM4.

5.3 STRONG INVERSION POTENTIAL FOR VERTICAL NONUNIFORM DOPING PROFILE

In a standard treatment of MOSFET device operating in the strong inversion, the surface potential is said to be pinned at $2\phi_f$, where ϕ_f is the Fermi level of the bulk material. The typical $2\phi_f$ expression found in textbooks is essentially that adopted in BSIM3:

$$2\phi_f = 2\frac{kT}{q}\ln\left(\frac{\text{NCH}}{n_i}\right). \quad (5\text{-}8)$$

NCH is a BSIM3 parameter, which is the concentration of the substrate doping near the channel region.

In BSIM4, the surface potential under strong inversion, still denoted as $2\phi_f$ throughout this book, assumes a more complicated expression:

$$2\phi_f = \frac{kT}{q} \ln\left(\frac{\text{NDEP}}{n_i}\right) + 0.4 + \text{PHIN}. \quad (5\text{-}9)$$

It is obvious that NDEP now replaces NCH as a BSIM4 parameter. In addition, a new parameter, PHIN, is introduced. It is said that this new formulation is needed to model a vertical nonuniform doping profile in the channel, especially when it is a steep retrograde doping profile. Some limited experiences have shown that the BSIM3 formulation of $2\phi_f$ can do as a good as a job as the BSIM4 formulation. PHIN gives one more degree of freedom in fitting, but does not seem to make significant contribution to the fitting accuracy.

Here are the relevant parameters:

Parameter Name	Description	Default Value
PHIN	Vertical nonuniform doping coefficient	0 (V)
NDEP	Channel Doping Concentration	1.7×10^{17} (cm^{-3})

Replaced BSIM3 Parameters: NCH

5.4 THRESHOLD VOLTAGE MODIFICATIONS

The threshold voltage in BSIM3 is given as

$$V_T = \text{VTH0} + \delta_{NP}(\Delta V_{T,\text{body\_effect}} - \Delta V_{T,\text{charge\_sharing}} - \Delta V_{T,\text{DIBL}} + \Delta V_{T,\text{reverse\_short\_channel}} + \Delta V_{T,\text{narrow\_width}} + \Delta V_{T,\text{small\_size}}). \quad (5\text{-}10)$$

Some of these terms remain unperturbed in BSIM4, but some undergo either a cosmetic modification or a major surgery. We list the BSIM3 terms that get modified in BSIM4:

$$\Delta V_{T,\text{body\_effect}} = \text{K1}\sqrt{2\phi_f - V_{BS}} - \text{K1} \cdot \sqrt{2\phi_f} - \text{K2}V_{BS} \quad \text{(BSIM3)}. \quad (5\text{-}11)$$

$$\Delta V_{T,\text{reverse\_short\_channel}} = \text{K1}\left(\sqrt{1 + \frac{\text{NLX}}{L_{\textit{eff}}}} - 1\right)\sqrt{2\phi_f} \quad \text{(BSIM3)}. \quad (5\text{-}12)$$

$$\Delta V_{T,\text{charge\_sharing}} = \text{DVT0}\left[\exp\left(-\text{DVT1}\frac{L_{\textit{eff}}}{2L_t}\right) + 2\exp\left(-\text{DVT1}\frac{L_{\textit{eff}}}{L_t}\right)\right](V_{bi} - 2\phi_f) \quad \text{(BSIM3)}. \quad (5\text{-}13)$$

$2\phi_f$ in the above expressions is that given in Eq. (5-8) and L_t was given in Eq. (3-35). (The above equations are not exactly correct, to facilitate focusing on the essential changes between BSIM3 and BSIM4. The exact BSIM3 equations are given in Appendix A.)

In BSIM4, the threshold voltage is instead given as

$$V_T = \text{VTH0} + \delta_{NP}(\Delta V_{T,\text{body\_effect}} - \Delta V_{T,\text{charge\_sharing}} - \Delta V_{T,\text{DIBL}}$$
$$+ \Delta V_{T,\text{reverse\_short\_channel}} + \Delta V_{T,\text{narrow\_width}} + \Delta V_{T,\text{small\_size}}$$
$$- \Delta V_{T,\text{pocket\_implant}}). \tag{5-14}$$

In comparison to BSIM3, BSIM4 adds one extra threshold voltage correction, that due to the pocket implants (see Fig. 2-12b). Further, as alluded to, several terms differ from the corresponding terms in BSIM3. These terms as well as $\Delta V_{T,\text{pocket\_implant}}$ are given by

$$\Delta V_{T,\text{body\_effect}} = [\text{K1}\sqrt{2\phi_f - V_{BS}} - \text{K1} \cdot \sqrt{2\phi_f}] \times \sqrt{1 + \frac{\text{LPEB}}{L_{\text{eff}}}} - \text{K2}V_{BS}. \tag{5-15}$$

$$\Delta V_{T,\text{reverse\_short\_channel}} = \text{K1}\left(\sqrt{1 + \frac{\text{LPE0}}{L_{\text{eff}}}} - 1\right)\sqrt{2\phi_f}. \tag{5-16}$$

$$\Delta V_{T,\text{charge\_sharing}} = \text{DVT0}\frac{0.5}{\cosh(\text{DVT1} \times L_{\text{eff}}/L_t) - 1}(V_{bi} - 2\phi_f). \tag{5-17}$$

$$\Delta V_{T,\text{pocket\_implant}} = n\frac{kT}{q}\ln\left[\frac{L_{\text{eff}}}{L_{\text{eff}} + \text{DVTP0}(1 + \exp(-\text{DVTP1} \cdot V_{DS}))}\right]. \tag{5-18}$$

(Again, the above equations are not exactly correct. The exact BSIM4 equations are given in Appendix G.)

Let us examine the rationales behind these changes. The first two ΔV_T's can be considered together; they describe the effects due to nonuniform lateral doping in the channel. In CMOS processing, even without an intentional pocket implants to increase the channel doping nearby the drain and source areas (see Fig. 2-12b), the doping level there is generally higher than that in the mid-channel. The origin of this nonuniform lateral doping has not been unequivocally established, but it is believed to be related to the transient enhanced diffusion of the channel dopant. This effect is particularly significant in NMOS. The heavy source/drain implantation with arsenic creates lattice damages at the two edges of the gate poly. These damages promote the boron channel dopant to pile up there. As the channel length becomes shorter, the portion of the increased-channel-doping region to the overall channel region becomes more significant. This leads to the reverse short-channel effects, that the threshold voltage increases as the channel length decreases. In BSIM3, the sole parameter modeling these effects is NLX, as evidenced in Eq. (5-12). In BSIM4, NLX is renamed LPE0. In addition, BSIM4 considers V_{BS}'s effects on the body effect coefficient K1, which directly impacts $\Delta V_{T,\text{body\_effect}}$. Comparing Eq. (5-11) to Eq. (5-15), we see that BSIM4 uses LPEB to accomplish this goal.

The next modification, on $\Delta V_{T,\text{charge\_sharing}}$, has rather subtle impacts on predictive modeling. (But if the model will not be used for predictive purposes, then the following discussion is not all that relevant.) We mentioned that the reverse short-channel effects cause V_T to increase with decreasing L. However, after L decreases below a certain value, the short-channel effects due to charge sharing eventually overshadows the reverse short-channel effects, and V_T is found to decrease as L approaches zero. $\Delta V_{T,\text{charge\_sharing}}$ is to be subtracted from the overall V_T in Eqs. (5-10) and (5-14), reflecting the short-channel effects that the threshold voltage decreases with decreasing channel length. Its BSIM3 expression, given in Eq. (5-13), approaches a constant value as $L_{\textit{eff}}$ diminishes toward 0. In contrast, $\Delta V_{T,\text{reverse\_short\_channel}}$ (either the BSIM3 or BSIM4 expressions), increases without bound as $L_{\textit{eff}}$ becomes small. As described in Section 4.18, the two ΔV_T's dependencies on L in BSIM3 often results in a modeling artifact, that when the channel length of a future technology is designed to be smaller than L_{min} (minimum channel length) of a given technology, the threshold voltage again increases. In fact, as L tends to zero, BSIM3 predicts an infinite V_T, while measured data suggest V_T continues to decrease, even assume negative values in some cases.

BSIM4's new expression of $\Delta V_{T,\text{charge\_sharing}}$ removes such an artifact. BSIM4 recognizes the origin of the problem resides in that $\Delta V_{T,\text{charge\_sharing}}$ approaches a constant value when L goes to zero. By adopting the $\Delta V_{T,\text{charge\_sharing}}$ given in Eq. (5-17), BSIM4 ensures that $\Delta V_{T,\text{charge\_sharing}}$ goes to infinity at a faster rate than $\Delta V_{T,\text{reverse\_short\_channel}}$. Consequently, the threshold voltage calculated in BSIM4 will be consistent with the experimental results that V_T becomes monotonically smaller as L decreases, even at L's below L_{min}. Although the basic equations for $\Delta V_{T,\text{charge\_sharing}}$ have changed from BSIM3 to BSIM4, BSIM4 continues to use the BSIM3 parameter names of DVT0 and DVT1.

In a conventional CMOS process, the V_{GS} turn-on characteristics for a long-channel device are nearly identical whether V_{DS} is ≈ 0 (such as 0.01 V) or the supply voltage V_{DD}. The ΔV_T shift at the two extreme V_{DS} values is negligible. When pocket implants are used in the process, however, a ΔV_T shift is often observed even in a long-channel ($L = 10\,\mu\text{m}$) device, as shown in Fig. 5-1. BSIM3 is not able to produce such a ΔV_T shift because all of the ΔV_T terms in Eq. (5-10) are nearly zero at $L = 10\,\mu\text{m}$ (and $V_{BS} = 0$). To correct this, BSIM4 introduces the term $\Delta V_{T,\text{pocket\_implant}}$, which induces a V_{DS} dependence in the V_T of a long-channel device. The parameters DVTP0 and DVTP1 controls the amount of the dependence.

Some caution needs to be exercised in extracting the value for DVTP1. It is often observed that a parameter set with DVTP1 = 1000 can fit the I-V characteristics of various-sized transistors well. However, the same parameter set could yield unphysical device capacitances. As shown in Fig. 5-2, a parameter set with DVTP1 = 1000 produces very negative C_{gd} and further, its C_{gs} far exceeds C_{gg} by ten times. Similar C-V behaviors are observed for PMOS as well. This occurs despite a great fit in the I-V characteristics.

Figure 5-3, which plots the $\Delta V_{T,\text{pocket\_implant}}$ term in BSIM4, helps to illustrate the source of the unphysical capacitance values. Independent of the exact value of

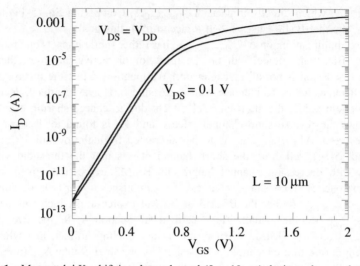

Fig. 5-1 Measured ΔV_T shift in a long-channel ($L = 10\,\mu m$) device, when pocket implants are used.

DVTP1, $\Delta V_{T,\text{pocket\_implant}}$ always varies between two extreme values. The parameter DVTP1 determines the V_{DS} level at which the transition is made. When DVTP1 = 1000, Fig. 5-3 reveals that $\Delta V_{T,\text{pocket\_implant}}$ calculated at $V_{DS} > 0.01$ V is relatively constant. If the to-be-fitted measured data are all at $V_{DS} > 0.01$ V, and if

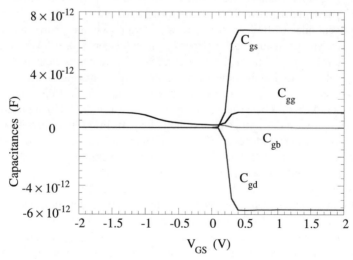

Fig. 5-2 BSIM4-calculated capacitances when DVTP1 = 1000 at $V_{DS} = 0.001$ V.

5.4 THRESHOLD VOLTAGE MODIFICATIONS

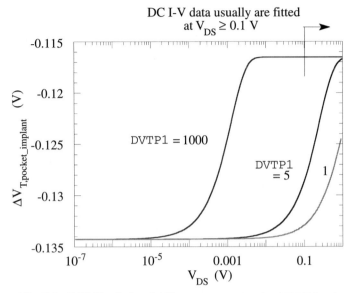

Fig. 5-3 BSIM4-calculated $\Delta V_{T,\text{pocket\_implant}}$ at various DVTP1 values.

$\Delta V_{T,\text{pocket\_implant}}$ is insensitive to V_{DS} at $V_{DS} > 0.01$ V, then it is perfectly fine for an optimizer to extract a DVTP1 of 1000. Although at $V_{DS} < 0.01$ V the derivative of $\Delta V_{T,\text{pocket\_implant}}$ with respect to V_{DS} is high, the extractor never realizes this fact since it is dealing with data taken at $V_{DS} > 0.01$ V. The C-V calculation of Fig. 5-2, however, is carried out at a V_{DS} of 1 mV, which corresponds to the bias condition in which the derivative of $\Delta V_{T,\text{pocket\_implant}}$ is fairly large. This large derivative ultimately causes C_{gd} and C_{gs} to misbehave.

One straightforward way to eliminate the unphysical capacitances is to limit DVTP1 below a certain value, thereby avoiding the sharp derivatives associated with $\Delta V_{T,\text{pocket\_implant}}$. When DVTP1 = 0.2, for example, the C-V values are found to match those produced by a comparable BSIM3 parameter set.

There are two more observations about $\Delta V_{T,\text{pocket\_implant}}$ introduced in BSIM4. First, we mentioned that its purpose is to account for the ΔV_T shift in long-channel devices at $V_{DS} \sim 0$ and V_{DD}. However, $\Delta V_{T,\text{pocket\_implant}}$ itself approaches zero when L_{eff} approaches infinity, rather than asymptotically reaches a nonzero value. Second, $\Delta V_{T,\text{pocket\_implant}}$ increases toward $-\infty$ as L_{eff} goes toward zero, causing the overall V_T to increase toward ∞. This is similar to the increase of $\Delta V_{T,\text{reverse\_short\_channel}}$ with respect to shrinking L_{eff}, but only at a slower pace. We mentioned that $\Delta V_{T,\text{charge\_sharing}}$ will cause V_T to decrease toward $-\infty$ at a faster rate than $\Delta V_{T,\text{reverse\_short\_channel}}$. Hence, $\Delta V_{T,\text{pocket\_implant}}$ will not cause the secondary V_T rollup problem discussed in Section 4.18 and in the earlier portion of this section.

IMPROVEMENTS IN BSIM4

Here are the relevant parameters:

Parameter Name	Description	Default Value
LPE0	Nonuniform lateral doping distance	1.74×10^{-7} (m)
LPEB	Nonuniform lateral doping effect on K1	1.7×10^{17} (cm^{-3})
DVT0†	First coefficient of charge-sharing's correction to V_T	0 (m)
DVT1†	Second coefficient of charge-sharing's correction to V_T	0 (m)
DVTP0	First coefficient of pocket implant correction to V_T	0 (m)
DVTP1	Drain-bias coefficient of pocket implant correction to V_T	0 (V^{-1})

† Same spelling as BSIM3 parameters, but appearing in different equations from BSIM3's. Replaced BSIM3 Parameter: NLX.

5.5 $V_{GST,\text{eff}}$ IN MODERATE INVERSION

The quantity $V_{GST,\text{eff}}$ is used to smooth the transition from subthreshold to strong inversion. It is approximately equal to $V_{GS} - V_T$ in strong inversion, but becomes proportional to $\exp[q(V_{GS} - V_T)/nkT]$ in subthreshold region. In BSIM3, it is given by

$$V_{GST,\text{eff}} = \frac{2nkT/q \ln\left[1 + \exp\left(\frac{V_{GS,\text{eff}} - V_T}{2nkT/q}\right)\right]}{1 + 2n\frac{C'_{ox}}{C'_{dep,0}} \exp\left(-\frac{V_{GS,\text{eff}} - V_T - 2 \cdot \text{VOFF}}{2nkT/q}\right)}, \quad (5\text{-}19)$$

where $V_{GS,\text{eff}}$ is the effective V_{GS} after accounting for the poly-depletion effects; n is the subthreshold slope; and $C'_{dep,0}$ is the capacitance per unit area associated with the channel depletion layer. Their detailed expressions are given in Appendix A.

BSIM4 improves the smoothness of the transition by adding two more parameters. $V_{GST,\text{eff}}$ in BSIM4 is given as

$$V_{GST,\text{eff}} = \frac{n\frac{kT}{q} \ln\left[1 + \exp\left(\frac{m^* V_{GS,\text{eff}} - V_T}{n \, kT/q}\right)\right]}{m^* + n\frac{C'_{ox,e}}{C'_{dep,0}} \exp\left[-\frac{(1 - m^*)(V_{GS,\text{eff}} - V_T) - V_{off}}{nkT/q}\right]}, \quad (5\text{-}20)$$

where

$$m^* = \frac{1}{2} + \frac{\arctan(\text{MINV})}{\pi}; \quad (5\text{-}21)$$

$$V_{off} = \text{VOFF} + \frac{\text{VOFFL}}{L_{eff}}. \quad (5\text{-}22)$$

One obvious change from BSIM3 to BSIM4 is the replacement of the parameter VOFF in Eq. (5-19) by a variable V_{off} in Eq. (5-20). V_{off} in BSIM4, given in Eq. (5-22), is not a constant, but instead shows some dependence on L_{eff}. As discussed in the entry of VOFF in Section 3.2, the constant VOFF in BSIM3 and the variable V_{off} in BSIM4 are used to fit the drain leakage current (I_{off}) at $V_{GS} = 0$. The modification from VOFF to V_{off} is found to be necessary in fitting advanced CMOS technologies with nontrivial channel doping profiles, across an array of channel lengths.

The second change is the introduction of MINV, through the variable m^*. This parameter is particularly useful in minimizing the fitting errors in the moderate inversion regions of the g_m/I_D curve. The dependence of m^* on MINV can be calculated using Eq. (5-21), as plotted in Fig. 5-4. As shown m^* is limited between 0 and 1, a property which improves convergence during the parameter optimization.

Parameter Name	Description	Default Value
VOFF†	Offset voltage in the subthreshold region	−0.08 (V)
VOFFL	Coefficient of offset voltage's length dependence	0 (m)
MINV	Coefficient of moderate inversion	0

† Same spelling as BSIM3 parameters, but appearing in different equations from BSIM3's.

5.6 DRAIN CONDUCTANCE MODEL

The drain terminal current in BSIM4 consists of several components, given by

$$I_D = I_{DS} + I_{sub} + I_{gidl} - I_{j,\text{DB}} - I_{gd,\text{tunnel}} - I_{gcd,\text{tunnel}}, \tag{5-23}$$

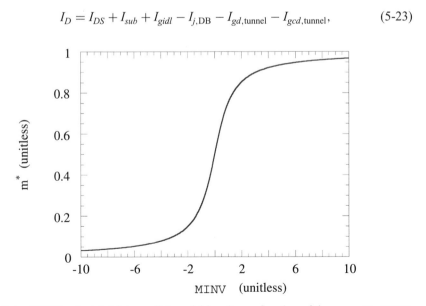

Fig. 5-4 BSIM4-calculated intermediate variable m^* as a function of the parameter MINV.

where I_{DS} is the channel current flowing from the drain to source in a normal MOSFET operation, I_{sub} is the current due to impact ionization (unchanged from the BSIM3 formulation and described in the entry ALPHA0 in Section 3.2), I_{gidl} is the gate-induced drain leakage current to be described in Section 5.10, and $I_{j,DB}$ is the drain–bulk junction current to be described in Section 5.9. A minus sign is placed in front of $I_{j,DB}$ to account for the fact that the drain terminal is the negative terminal in a p-n junction. The last two terms are tunneling currents between the drain and the gate terminals. They will be described in Section 5.18.

We shall concentrate on the portion of drain current which flows through the channel in this section. In BSIM3, it is given by

$$I_{DS} = \frac{I_{DS,0}}{1 + \frac{R_{DS}I_{DS,0}}{V_{DS,\text{eff}}}} \left(1 + \frac{V_{DS} - V_{DS,\text{eff}}}{V_{A,\text{sat}} + V_{A,\text{CLM-DIBL}}}\right) \cdot \left(1 + \frac{V_{DS} - V_{DS,\text{eff}}}{V_{A,\text{SCBE}}}\right), \quad (5\text{-}24)$$

where $I_{DS,0}$ is the constant drain current derived for long-channel devices in typical textbook [4]. The variables $V_{A,\text{sat}}$, $V_{A,\text{CLM-DIBL}}$ and $V_{A,\text{SCBE}}$ have been described under the entries of PCLM and PSCBE1 in Section 3.2. Their exact equations can be found in Appendix A. In BSIM4, we have instead,

$$I_{DS} = \frac{I_{DS,0}}{1 + \frac{R_{DS}I_{DS,0}}{V_{DS,\text{eff}}}} \left(1 + \frac{1}{C_{clm}} \ln \frac{V_{A,\text{sat}} + V_{A,\text{CLM}}}{V_{A,\text{sat}}}\right) \times \left(1 + \frac{V_{DS} - V_{DS,\text{eff}}}{V_{A,\text{DIBL}}}\right)$$

$$\times \left(1 + \frac{V_{DS} - V_{DS,\text{eff}}}{V_{A,\text{DITS}}}\right) \times \left(1 + \frac{V_{DS} - V_{DS,\text{eff}}}{V_{A,\text{SCBE}}}\right) \times \text{NF}. \quad (5\text{-}25)$$

The most obvious change occurring in BSIM4 is the introduction of NF, an instance parameter used to mean the number of fingers. If a transistor has 10 gate fingers, each having a width of 5 μm, then W in the MOSFET instance statement should be 50 μm and NF should be 10 (the number of instances MI is 1). In BSIM3, the same device would be specified with a W of 5 μm, with the number of instances equal to 10. NF and MI together, rather than just MI alone, offers more flexibility in specifying the device geometry in BSIM4. Consider a 2-finger device, with the two outer contacts being the source and only one drain contact located between the fingers. It will be difficult to model the parasitics of this transistor in BSIM3 as discussed in Section 4.20, but it is relatively straightforward in BSIM4, as will be demonstrated in Section 5.19.

Note that if the NF value is so large such that the effective width calculated in Appendix G (Eq. G-2) is negative, then a fatal warning is issued. Furthermore, in BSIM4.0.0., NF is somehow reset to 20 if the user-inputted value exceeds 500.

The second change involves the various Early voltage (V_A) components. There are two Early voltage components which remain the same from BSIM3 to BSIM4. They are $V_{A,\text{SCBE}}$, the Early voltage associated with the impact ionization current, and $V_{A,\text{sat}}$, the Early voltage associated with velocity saturation. Besides these, BSIM4 make several modifications. First, $V_{A,\text{CLM-DIBL}}$ in BSIM3 is separated out into two separate entities, $V_{A,\text{CLM}}$ (Early voltage due to channel-length modulation) and $V_{A,\text{DIBL}}$ (Early voltage due to drain-induced barrier lowering). BSIM4 further introduces $V_{A,\text{DITS}}$, the Early voltage due to drain-induced threshold voltage shift.

Knowing the exact equations of these Early voltages is of no interest to us. They can be found in Appendix G anyway. Here, we concentrate on making some simple observations. First, BSIM4 uses separate $V_{A,\text{CLM}}$ and $V_{A,\text{DIBL}}$ expressions because $V_{A,\text{CLM-DIBL}}$ in BSIM3 was derived with the wrong assumption that the Early voltage due to channel-length modulation is independent of V_{DS}. If we omit the contribution from other Early voltages, then removing this particular BSIM3 assumption results in the following:

$$I_{DS} = \frac{I_{DS,0}}{1 + \frac{R_{DS}I_{DS,0}}{V_{DS,\text{eff}}}} \left(1 + \int_{V_{A,\text{sat}}}^{V_{DS}} \frac{1}{V_{A,\text{sat}} + V_{A,\text{CLM}}} d V_{DS} \right)$$

$$= \frac{I_{DS,0}}{1 + \frac{R_{DS}I_{DS,0}}{V_{DS,\text{eff}}}} \left(1 + \frac{1}{C_{clm}} \ln \frac{V_{A,\text{sat}} + V_{A,\text{CLM}}}{V_{A,\text{sat}}} \right). \quad (5\text{-}26)$$

This explains the origin of the first term in Eq. (5-25).

Another observation is that BSIM4 is intended to work with transistors with significant pocket implants. These implants, placed nearby the drain and source end of the channel as shown in Fig. 2-12b, are designed to increase the drive current in digital circuits. However, they simultaneously cause a noticeable increase in the drain conductance, particularly for the short-channel devices but also somewhat noticeable in long-channel devices. Therefore, many of the Early voltage components in BSIM4 incorporate some corrections due to the pocket implants. The pocket-implant correction factor, F_p, is given by [3]

$$F_p = \left(1 + \text{FPROUT} \cdot \frac{\sqrt{L_{\text{eff}}}}{V_{GS} - V_T + 2kT/q} \right)^{-1}. \quad (5\text{-}27)$$

For example, in the calculation of $V_{A,\text{CML}}$, C_{clm} appearing in Eq. (5-25) is given by

$$C_{clm} \approx \frac{F_p}{\text{PCLM } L_{itl}} \times \left(1 + \frac{R_{DS}I_{DS,0}}{V_{DS,\text{eff}}} \right) \times \left(L_{\text{eff}} + \frac{V_{DS,\text{sat}}}{\varepsilon_{sat}} \right). \quad (5\text{-}28)$$

The parameter FPROUT determines the amount of drain-conductance degradation due to the pocket implants. When the transistor does not have pocket implants, the parameter FPROUT should be set at its default value of 0. FPROUT can be on the order of $500 \, (\text{V/m}^{0.5})$ in devices with significant pocket implants.

The third observation is that BSIM4 introduces $V_{A,\text{DITS}}$, a new BSIM4 term used to model the drain-induced threshold shift [3]. We have seen F_p of Eq. (5-27) affects $V_{A,\text{CML}}$. It turns out that F_p also appears in $V_{A,\text{DITS}}$, which is given by

$$V_{A,\text{DITS}} = \frac{F_p}{\text{PDITS}} [1 + (1 + \text{PDITSL} \cdot L_{\text{eff}}) \times \exp(\text{PDITSD} \cdot V_{DS})]. \quad (5\text{-}29)$$

Basically, the potential barrier at the drain can be lowered in long-channel devices when there is sufficient amount of pocket implants. There are three parameters used to characterize this Early voltage. PDITS affects the magnitude of the voltage, PDITSL describes its dependence on $L_{\it eff}$, and PDITSD, on the drain bias.

We have seen in this section that BSIM4 allows greater flexibility in modeling the output resistance by introducing some more parameters. It is believed that most often BSIM3 can get the job done just as well, as far as fitting the output conductance is concerned. BSIM4, however, offers greater flexibility in fitting the conductances of long- and short-channel devices simultaneously.

Parameter Name	Description	Default Value
NF	Number of fingers (instance)	1
FPROUT	Pocket-implant degradation parameter	$0\,(\text{V}/\text{m}^{0.5})$
PDITS	First coefficient of DITS effects on Early voltage	$0\,(\text{V}^{-1})$
PDITSL	L-dependence of DITS effects on Early voltage	$0\,(\text{m}^{-1})$
PDITSD	V_{DS}-dependence of DITS effects on Early voltage	$0\,(\text{V}^{-1})$

5.7 MOBILITY MODEL

BSIM4 keeps two of the BSIM3 mobility models and introduces one more. The MOBMOD = 0 model in BSIM4 is the MOBMOD = 1 model in BSIM3, and the MOBMOD = 1 model in BSIM4 is the MOBMOD = 3 model in BSIM3. The BSIM3 models were described in the MOBMOD entry in Section 3.2. The new mobility model in BSIM4 is invoked by specifying MOBMOD = 2, which is given by

$$\mu_{\it eff} = \frac{\text{U0}}{1 + (\text{UA} + \text{UC} \cdot V_{BS})\left(\dfrac{V_{GS} - V_T + V_{T\text{-}fb\text{-}\phi}}{\text{TOXE}}\right)^{\text{EU}}}, \quad (5\text{-}30)$$

where

$$V_{T\text{-}fb\text{-}\phi} = \begin{cases} 2 \times (\text{VTH0} - V_{FB} - 2\phi_f) & \text{for NMOS;} \\ 2.5 \times (\text{VTH0} - V_{FB} - 2\phi_f) & \text{for PMOS.} \end{cases} \quad (5\text{-}31)$$

TOXE is the electrical oxide thickness, differing from the physical oxide thickness TOXP by a value DTOX. The factor $(V_{GS} - V_T + V_{T\text{-}fb\text{-}\phi})/\text{TOXE}$ can be roughly treated as the normal electric field between the gate and the channel (see the U0 entry in Section 3.2). As an electron emitted from the source travels through the channel to the drain, it gets attracted to the gate due to the normal electric field established by V_{GS}. As it drifts horizontally in the channel by the field established by V_{DS}, it also tends to collide with the oxide. The effective mobility degrades. This is why the effective mobility given in Eq. (5-30) is some function of this normal field.

To understand the rationale behind this new mobility model, we write out the preferred mobility model in BSIM3 (MOBMOD = 3 in BSIM3, or MOBMOD = 1 in BSIM4):

$$\mu_{eff} = \frac{U0}{1 + \left[UA \left(\frac{V_{GS} + V_T}{TOXE} \right) + UB \left(\frac{V_{GS} + V_T}{TOXE} \right)^2 \right] (1 + UC \cdot V_{BS})}. \quad (5\text{-}32)$$

Equation (5-32) was developed based on the universal-mobility theorem described under the entry U0 in Section 3.2. The model correctly predicts the mobility degradation with increasing V_{GS}. Unfortunately, it can be problematic when it is applied to devices with varying channel length. Consider a short-channel device with significant drain-induced barrier-lowering (DIBL) effects. As V_{DS} increases, its threshold voltage V_T decreases. This results in a smaller denominator value for the short-channel device in comparison to a long-channel device. The short-channel device would have a higher effective mobility than the long-channel device, a result that is at odds with experimental finding.

BSIM4's new mobility model given in Eq. (5-30) is designed to circumvent this unphysical outcome. Because V_T is smaller in short-channel devices due to DIBL, the denominator in Eq. (5-30) is larger, leading to a smaller effective mobility for the short-channel device in comparison to the long-channel device. However, there is still a problem embedded in Eq. (5-30). This problem occurs in CMOS technologies with significant pocket implants and, to a lesser extent, reverse short-channel effects. Both factors increase the average channel doping, with pocket implants being intentional and reverse-short-channel effects coming from boron pileup (Section 5.4). In these devices, the threshold voltage first increases almost linearly as L decreases from, say, 10 μm to 0.3 μm. (Certainly when L decreases below 0.3 μm, for example, the short-channel effects due to charge-sharing dominate and V_T decreases rapidly, as discussed in typical textbooks.) Because V_T for the long-channel device with $L = 10$ μm is smaller than the short-channel device with $L = 0.3$ μm, the denominator in Eq. (5-30) is larger for the long-channel device. This then translates to a smaller effective mobility value for the long-channel device, again contradicting the experimental results. Because of this unphysical artifact, it is not clear whether this newly introduced BSIM4 mobility model offers better predictability than the previous BSIM3 models.

There is only one new parameter introduced in BSIM4 for mobility modeling:

Parameter Name	Description	Default Value
EU	Exponent coefficient for the MOBMOD = 2 model	1.67 (NMOS) 1 (PMOS)
UO†	Low field mobility	0.067 (m²/V-s) (NMOS) 0.025 (m²/V-s) (PMOS)
UA†	First-order mobility-degradation coefficient	10^9 (m/V) (MOBMOD = 0, 1) 10^{-15} (m/V) (MOBMOD = 2)
UB†	Parabolic mobility-degradation coefficient	10^{-19} (m²/V²) (MOBMOD = 0, 1) not used (MOBMOD = 2)
UC†	Body-effect coefficient of mobility-degradation	-0.0465 (V^{-1}) (MOBMOD = 1) -0.0465×10^{-9} (m/V²) (MOBMOD = 0, 2)

† Same spelling as BSIM3 parameters, but appearing in different equations from BSIM3's.

5.8 DIODE CAPACITANCE

BSIM3 distinguishes between the effective device width for the I-V calculation (W_{eff}) and the effective width for the C-V calculation ($W_{eff,CV}$). For simplicity, we can approximate W_{eff} as $W - 2 \cdot$ WINT, and $W_{eff,CV}$ as $W - 2 \cdot$ DWC. In the case of calculating drain current, it is intuitively clear that W_{eff} should be used. In the calculation of the gate oxide capacitance, it is obvious that $W_{eff,CV}$ should be used. How about for the calculation of the drain–bulk junction capacitance? Given a choice between W_{eff} and $W_{eff,CV}$, we naturally would opt for the latter just because a junction capacitance is a capacitance. This is certainly the way adopted in BSIM3.

A more careful examination reveals that the effective width for the junction capacitance needs not be the same as the effective width for the gate oxide capacitance. The effective width in a junction relates to the amount of depletion created by the bias voltage across the junction, at a location away from the intrinsic device. The effective width in a oxide capacitance calculation relates to the amount of inversion created by the gate voltage. BSIM4 takes note of this fine difference, and defines a new effective width. It is the junction effective width ($W_{eff,CJ}$), which can be approximated as $W - 2 \cdot$ DWJ. DWJ is a parameter introduced in BSIM4.

Many of the BSIM3 equations and parameters implicitly assume that the MOSFET device structure is symmetrical with respect to the gate (or substrate). There is no physical difference between the source and the drain. For example, the parameter CJ is the unit-area capacitance parameter used to calculate the areal capacitances of the both source–bulk and the drain–bulk junctions. BSIM4, in contrast, assumes that the device is asymmetrical. BSIM4 provides two unit-area capacitance parameters: CJS for the source–bulk junction, and CJD for the drain–

bulk junction. Other parameters in BSIM3 that apply to both the source and the drain, are made to differ in BSIM4 by appending either S or D. For example, in BSIM4, the gate-side sidewall junction capacitance in the source–bulk diode is CJSWGS, and in the drain–bulk diode, CJSWGD. Similarly, the isolation-side sidewall junction capacitance in the source–bulk diode is CJSWS, and in the drain–bulk diode, CJSWD.

When the diode is reverse biased, the drain–bulk junction capacitance in BSIM4 is given by

$$C_{j,\text{DB}} = \frac{\text{CJD}}{\left(1 - \frac{V_{BD}}{\text{PBD}}\right)^{\text{MJD}}} AD_{\text{eff}} + \frac{\text{CJSWD}}{\left(1 - \frac{V_{BD}}{\text{PBSWD}}\right)^{\text{MJSWD}}} PD_{\text{eff}}$$

$$+ \frac{\text{CJSWGD}}{\left(1 - \frac{V_{BD}}{\text{PBSWGD}}\right)^{\text{MJSWGD}}} W_{\text{eff,CJ}} \times \text{NF}, \qquad (5\text{-}33)$$

where AD_{eff} under most circumstances is just the AD (drain area) specified in the MOSFET instance statement, and PD_{eff} is roughly equal to PD (drain periphery) minus $W_{\text{eff,CJ}} \cdot \text{NF}$. $W_{\text{eff,CJ}}$, the effective width in the junction capacitance calculation, is

$$W_{\text{eff,CJ}} = \frac{W}{\text{NF}} - 2 \cdot \text{DWJ} - 2 \cdot \Delta W_{\text{geometry,CV}}. \qquad (5\text{-}34)$$

W is the width specified in the MOSFET instance statement; it represents the total device width. If the device's number of fingers (NF) is 10 and each finger is 5 μm wide, then the W specified in the instance should be 50 μm. $\Delta W_{\text{geometry,CV}}$ was given in Appendix A; it is usually zero. DWJ is the ΔW for the effective junction width calculation.

Equation (5-33) states that the drain–bulk junction capacitance consists of three components. The first component is the bottom–wall junction capacitance, which is labeled as $C_{j,\text{DB\_BW}}$ in Section 2.3. The second and the third components refer to the sidewall capacitances at the isolation side and at the gate side, respectively. They were denoted as $C_{j,\text{DB\_SW}}$ and $C_{j,\text{DB\_SWG}}$, respectively, in Section 2.3. The physical locations of the three capacitances are shown in Fig. 2-12a. In BSIM3, $C_{j,\text{DB}}$ is given by the following instead of Eq. (5-33):

$$C_{j,\text{DB}} = \frac{\text{CJ}}{\left(1 - \frac{V_{BD}}{\text{PB}}\right)^{\text{MJ}}} AD + \frac{\text{CJSW}}{\left(1 - \frac{V_{BD}}{\text{PBSW}}\right)^{\text{MJSW}}} (\text{PD} - W_{\text{eff,CV}})$$

$$+ \frac{\text{CJSWG}}{\left(1 - \frac{V_{BD}}{\text{PBSWG}}\right)^{\text{MJSWG}}} W_{\text{eff,CV}}, \qquad (\text{BSIM3}) \qquad (5\text{-}35)$$

where $W_{\text{eff,CV}} = W - 2 \cdot \text{DWC} - 2 \cdot \Delta W_{\text{geometry,CV}}$.

A comparison between Eqs. (5-33) and (5-35) reveals a key difference between BSIM4 and BSIM3. In BSIM3, there is no $W_{\text{eff,CJ}}$. As mentioned previously, BSIM3 takes $W_{\text{eff,CV}}$ to be the junction width used for the junction capacitance calculation. In addition, BSIM3 does not use NF as an instance parameter. Therefore, the effective junction width $W_{\text{eff,CV}}$ used in BSIM3 is roughly equal to the W specified in the MOSFET instance statement. On the other hand, $W_{\text{eff,CJ}}$ used in BSIM4 is roughly equal to W/NF. $W_{\text{eff,CJ}}$ is seen to multiply by NF in Eq. (5-33), in order that the total junction width be reproduced. Finally, $(\text{PD} - W_{\text{eff,CV}})$ is used as the multiplication factor in the BSIM3 equation while PD_{eff} is that used in BSIM4. The BSIM3 and BSIM4 expressions are basically equivalent since PD_{eff} of BSIM4 is generally equal to $\text{PD} - W_{\text{eff,CJ}} \cdot \text{NF}$. BSIM4 chooses to use the intermediate variable PD_{eff} to provide more flexibility in specifying the device geometry, as detailed in Section 5.19.

If V_{BD} is positive and approaches PBD, for example, Eq. (5-33) would approach infinity. This is the artifact resulting from the depletion approximation used to derive the capacitance. When $V_{BD} > 0$, BSIM4 linearizes the capacitance, with the slope of the capacitance equal to that at $V_{BD} = 0$. The capacitance at $V_{BD} > 0$ is given by

$$C_{j,\text{DB}} = \text{CJD}\left(1 + \text{MJD}\frac{V_{BD}}{\text{PBD}}\right)AD_{\text{eff}} + \text{CJSWD}\left(1 + \text{MJSWD}\frac{V_{BD}}{\text{PBSWD}}\right)PD_{\text{eff}}$$

$$+ \text{CJSWGD}\left(1 + \text{MJSWGD}\frac{V_{BD}}{\text{PBSWGD}}\right)W_{\text{eff,CJ}} \times \text{NF}. \tag{5-36}$$

The source–bulk junction capacitance has a similar expression as Eqs. (5-33) and (5-36), except with appropriate changes for the source–bulk diode parameters.

Parameter Name	Description	Default Value
DWJ	$\Delta W/2$ in the effective junction width calculation	DWC (m)
CJS	Bottom source–bulk junction capacitance per unit area	5×10^{-4} (F/m^2)
CJD	Bottom drain–bulk junction capacitance per unit area	CJS
PBS	Bottom source–bulk junction built-in potential	1 (V)
PBD	Bottom drain–bulk junction built-in potential	PBS
MJS	Bottom source–bulk junction grading coefficient	0.5
MJD	Bottom drain–bulk junction grading coefficient	MJS
CJSWS	Source–bulk isolation sidewall junction capacitance	5×10^{-10} (F/m^2)
Parameter Name	Description	Default Value

CJSWD	Drain–bulk isolation sidewall junction capacitance	CJSWS
PBSWS	Source–bulk isolation sidewall built-in potential	1 (V)
PBSWD	Drain–bulk isolation sidewall built-in potential	PBSWS
MJSWS	Source–bulk isolation sidewall grading coefficient	0.33
MJSWD	Drain–bulk isolation sidewall grading coefficient	MJSWS
CJSWGS	Source–bulk gate sidewall junction capacitance	CJSWS
CJSWGD	Drain–bulk gate sidewall junction capacitance	CJSWD
PBSWGS	Source–bulk gate sidewall built-in potential	PBSWS
PBSWGD	Drain–bulk gate sidewall built-in potential	PBSWD
MJSWGS	Source–bulk gate sidewall grading coefficient	MJSWS
MJSWGD	Drain–bulk gate sidewall grading coefficient	MJSWD

Replaced BSIM3 parameters: CJ, PB, MJ, CJSW, PBSW, MJSW, CJSWG, PBSWG, MJSWG

5.9 DIODE BREAKDOWN

In both BSIM3 and BSIM4, the diodes' I-V characteristics are totally independent of their C-V characteristics. The CV parameters mentioned in the previous section do not in any way affect the I-V characteristics, and vice versa.

To speed up the dc convergence, BSIM3 treats the source–bulk and the drain–bulk junctions as ideal junctions without avalanche breakdowns. Therefore, the junction leakage current in the reverse-bias region remains relatively constant with bias, never displaying the sharp increase associated with a junction breakdown. Although the lack of breakdown can be beneficial for numerical reasons (less computational time and faster convergence), unphysical values of terminal voltages may become valid solutions. BSIM4 improves upon BSIM3 by offering the option of having breakdown characteristics in the source–bulk and drain–bulk junctions. When the option is turned on (by picking an appropriate value for DIOMOD), the junction current (the $I_{j,\text{DB}}$ component in Eq. 5-23) increases exponentially when the reverse bias approaches the specified breakdown voltage values.

We briefly review how BSIM3 calculates $I_{j,\text{DB}}$. There are actually two $I_{j,\text{DB}}$ models in BSIM3. In the reverse-biased condition, the two models' expressions are identical. When the junction is forward-biased (an unlikely event in a normal

MOSFET operation), then the calculated current depends on the value of IJTH. If IJTH = 0, then

$$I_{j,\text{DB}} = I_{sat,\text{DB}}\left[\exp\left(\frac{qV_{BD}}{\text{NJ}\cdot kT}\right) - 1\right] + \text{GMIN}\cdot V_{BD} \qquad (\text{IJTH} = 0), \qquad (5\text{-}37\text{a})$$

where $I_{sat,\text{DB}}$, the drain–bulk diode junction saturation current, is given by

$$I_{sat,\text{DB}} = \begin{cases} \text{JS}\times\text{AD} + \text{JSSW}\times\text{PD} & \text{the result is } \geq 0 \\ 10^{-14}\text{ A} & \text{if AD and PD are both } \leq 0 \\ 0 & \text{JS}\times\text{AD} + \text{JSSW}\times\text{PD} < 0 \end{cases} \qquad (5\text{-}37\text{b})$$

BSIM3 does not distinguish between the gate-sidewall and the isolation-sidewall components.

That was for IJTH = 0, wherein the diode current in Eq. (5-37) increases indefinitely as V_{BD} increases. If instead IJTH > 0, then after the diode current given by Eq. (5-37) reaches the value of IJTH, the diode current is modified (linearized) so that it increases at a much slower pace. We denote V_{IJTH} as the V_{BD} that results in an $I_{j,\text{DB}}$ equal to (IJTH + GMIN $\cdot V_{BD}$). When $V_{BD} > V_{IJTH}$, the second model calculates the diode current as

$$I_{j,\text{DB\_modified}} = I_{j,DB}\bigg|_{V_{BD}=V_{IJTH}} + \frac{\partial I_{j,\text{DB}}}{\partial V_{BD}}\bigg|_{V_{BD}=V_{IJTH}} \times (V_{BD} - V_{IJTH}), \qquad (5\text{-}38)$$

where $I_{j,\text{DB}}$ is that given in Eq. (5-37). The complete expression which carries out the derivative was given in Section 3.2, under the entry JS. Because $I_{j,\text{DB\_modified}}$ of Eq. (5-38) differs from $I_{j,\text{DB}}$ of Eq. (5-37) only after $V_{BD} > V_{IJTH}$, we said the two BSIM3 models are identical in the reverse-bias region. As mentioned, neither of the two $I_{j,\text{DB}}$ models produces the breakdown characteristics. In the reverse-bias region, $I_{j,\text{DB}}$ in both models is small, roughly equal to $I_{sat,\text{DB}}$, the reverse saturation current.

In BSIM4, the proper $I_{j,\text{DB}}$ expression is selected by DIOMOD, rather than IJTH (or its equivalent in BSIM4, IJTHDFWD for the drain–bulk junction and IJTHSFWD for the source–bulk junction). Because we have just alluded to the different parameters used for the drain–bulk and source–bulk junctions, we formally state another BSIM3/BSIM4 difference besides the different $I_{j,\text{DB}}$ expressions being analyzed shortly. Just as BSIM4 separates out the C-V parameters for the drain–bulk and source–bulk junctions, BSIM4 also separates out the I-V parameters. For example, in Section 5.8 we described that CJS (source–bulk junction) and CJD (drain–bulk junction) used in BSIM4 basically replace CJ (applicable for both junctions) in BSIM3. For the I-V parameters, we therefore expect to have IJTHDFWD, the forward limiting current for the drain–bulk junction, and IJTHSFWD, the forward limiting current for the source–bulk junction. Since we are discussing $I_{j,\text{DB}}$, we will see only the drain-side parameters in the following equations. However, all of the drain-side parameters have their counterpart source-side parameters.

5.9 DIODE BREAKDOWN

Let us continue our discussion of $I_{j,\text{DB}}$ in BSIM4. When DIOMOD is set to 0, the nonlinearized (or resistance-free) expression of $I_{j,\text{DB}}$ is selected. By nonlinearized expression, we mean the junction current is not affected by IJTHDFWD, in the same manner described for Eq. (5-37) given above. Without any linearization, $I_{j,\text{DB}}$ is given by

$$I_{j,\text{DB}} = I_{sat,\text{DB}} \left[\exp\left(\frac{qV_{BD}}{\text{NJD} \cdot kT}\right) - 1 \right] \times f_{breakdown} + \text{GMIN} \cdot V_{BD}, \tag{5-39}$$

where the junction saturation current in BSIM4 is given by

$$I_{sat,\text{DB}} = \begin{cases} \text{JSD} \cdot AD_{eff} + \text{JSWD} \cdot PD_{eff} + \text{JSWGD} \cdot W_{eff,\text{CJ}} \cdot \text{NF} & \text{the result is} \geq 0 \\ 10^{-14} \text{ A} & \text{if } AD_{eff} \text{ and } PD_{eff} \\ & \text{are both} \leq 0 \\ 0 & \text{the result is 0.} \end{cases}$$

(5-40)

and the newly introduced factor $f_{breakdown}$ is given by

$$f_{breakdown} = 1 + \text{XJBVD} \cdot \exp\left(-\frac{q\text{BVD} + qV_{BD}}{\text{NJD} \cdot kT}\right). \tag{5-41}$$

JSD, JSWD, and JSWGD are the drain–bulk junction leakage current per unit area, per unit isolation-side periphery, and per unit gate-side periphery, respectively. As mentioned in Section 5.8, AD_{eff} under most circumstances is just the AD (drain area) specified in the MOSFET instance statement, and PD_{eff} is roughly equal to PD (drain periphery) minus $W_{eff,\text{CJ}} \cdot \text{NF}$.

BVD is the breakdown voltage for the drain–bulk junction. It should be a positive number even though the breakdown occurs in the reverse-bias operation. Whenever a negative BVD is supplied, BSIM4 will reset BVD to 10 V. When V_{BD} is slightly negative, $f_{breakdown} \approx 1$. Hence, $I_{j,\text{DB}}$ calculated in Eq. (5-39) is roughly equal to $I_{sat,\text{DB}} \times f_{breakdown} = I_{sat,\text{DB}}$, just as we expect. As V_{BD} becomes increasingly more negative, with its magnitude approaching BVD, then $f_{breadown}$ becomes a large number dominated by the second component. Consequently, $I_{j,\text{DB}} \approx I_{sat,\text{DB}} \times \text{XJBVD} \cdot \exp[-q(\text{BVD} + V_{BD})/(\text{NJD} \cdot kT)]$. This gives rise to the rapid rise in $I_{j,\text{DB}}$ after $|V_{BD}| > \text{BVD}$. The parameter XJBVD controls the knee voltage at which the junction current increases exponentially with bias. If XJBVD is set to zero, then $f_{breakdown}$ is always 1. The breakdown characteristics then disappear and DIOMOD = 0 model would yield resistance-free diode characteristics in the forward bias region and no breakdown in the reverse bias region. Figure 5-5 displays $I_{j,\text{DB}}$ with various values of XJBVD.

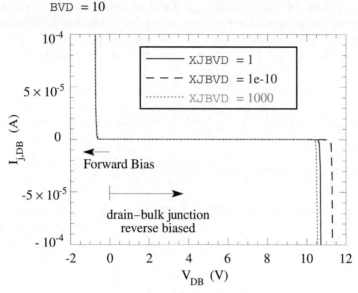

Fig. 5-5 BSIM4-calculated $I_{j,\text{DB}}$ with various values of XJBVD, showing the reverse-breakdown characteristics. DIOMOD = 0.

When DIOMOD = 1, the $f_{breakdown}$ factor in Eq. (5-39) is removed. $I_{j,\text{DB}}$ generally appears as

$$I_{j,\text{DB}} = I_{sat,\text{DB}}\left[\exp\left(\frac{qV_{BD}}{\text{NJD} \cdot kT}\right) - 1\right] + \text{GMIN} \cdot V_{BD} \quad (\text{when } V_{BD} < V_{\text{IJTHD}}). \quad (5\text{-}42)$$

However, $I_{j,\text{DB}}$ gets linearized when the calculated $I_{j,\text{DB}}$ exceeds IJTHDFWD. We denote V_{IJTHD} as the V_{BD} at which $I_{j,\text{DB}}$ in Eq. (5-42) becomes equal to (IJTHDFWD + GMIN $\cdot V_{BD}$). Then, when $V_{BD} > V_{\text{IJTHD}}$, $I_{j,\text{DB}}$ in Eq. (5-42) is replaced by the following:

$$I_{j,\text{DB\_modified}} = I_{j,\text{BD}}\bigg|_{V_{BD}=V_{\text{IJTHD}}} + \frac{\partial I_{j,\text{DB}}}{\partial V_{BD}}\bigg|_{V_{BD}=V_{\text{IJTHD}}} \times (V_{BD} - V_{\text{IJTHD}}). \quad (5\text{-}43)$$

The DIOMOD = 1 model is, in principle, similar to the BSIM3 diode model; it is therefore chosen to be the default model in BSIM4. Both do not allow breakdown characteristics and both have a IJTH-like parameter for linearization. However, in BSIM3, the user can set IJTH to zero to effect a resistance-free diode model in the forward bias region. However, in BSIM4, setting IJTHDFWD ≤ 0 in DIOMOD = 1 will cause IJTHDFWD to be reset to its default value of 0.1. In other words, $I_{j,\text{DB}}$ in the DIOMOD = 1 model of BSIM4 is always linearized at some current level. Certainly, if that current level is set to a very large value, such as 10 A, then $I_{j,\text{DB}}$ is

5.9 DIODE BREAKDOWN

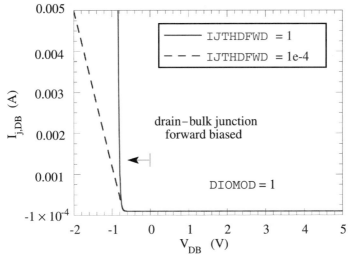

Fig. 5-6 BSIM4-calculated $I_{j,\text{DB}}$ when DIOMOD = 1, for two different values of IJTHDFWD. There is no diode breakdown.

essentially resistance free, at least in the useful practical operating region. Figure 5-6 compares the calculated $I_{j,\text{DB}}$ when DIOMOD = 1, for two different values of IJTHDFWD. For the curve with IJTHDFWD = 10^{-4} (A), the linearization takes place right at 10^{-4} A. In the scale of the figure, the curve with IJTHDFWD = 1 (A) just exponentially increases, in comparison.

The DIOMOD = 2 model is the most comprehensive one; it models the breakdown characteristics just like DIOMOD = 0, and it also allows the linearization of current in both the forward-bias mode (by specifying IJTHDFWD) and in the reverse-bias mode (by specifying IJTHDREV). When the calculated junction current in the forward bias is smaller than IJTHDFWD or when its magnitude in the reverse bias is smaller than IJTHDREV, then it is exactly that given by Eq. (5-39), repeated here for convenience:

$$I_{j,\text{DB}} = I_{sat,\text{DB}} \left[\exp\left(\frac{qV_{BD}}{\text{NJD} \cdot kT}\right) - 1 \right] \times f_{breakdown} + \text{GMIN} \cdot V_{BD} \quad (5\text{-}44)$$

(when $V_{IJTHDREV} < V_{BD} < V_{IJTHDFWD}$).

However, when $V_{BD} > V_{IJTHDFWD}$, which is the V_{BD} at which $I_{j,\text{DB}}$ in Eq. (5-44) becomes equal to (IJTHDFWD + GMIN $\cdot V_{BD}$), then

$$I_{j,\text{BD\_modified}} = I_{j,\text{DB}}\Big|_{V_{BD}=V_{IJTHDFWD}} + \frac{\partial I_{j,\text{DB}}}{\partial V_{BD}}\Big|_{V_{BD}=V_{IJTHDFWD}} \times (V_{BD} - V_{IJTHDFWD}). \quad (5\text{-}45)$$

When $V_{BD} < V_{IJTHDREV}$, which is the V_{BD} at which $I_{j,DB}$ in Eq. (5-44) becomes equal to $(-\text{IJTHDREV} + \text{GMIN} \cdot V_{BD})$, then (*Note:* IJTHDREV is a positive number)

$$I_{j,\text{BD\_modified}} = I_{j,\text{DB}}\bigg|_{V_{BD}=V_{IJTHDREV}} + \frac{\partial I_{j,DB}}{\partial V_{BD}}\bigg|_{V_{BD}=V_{IJTHDREV}} \times (V_{BD} - V_{IJTHDREV}). \quad (5\text{-}46)$$

Figure 5-7 is simulation results with DIOMOD = 2: one with IJTHDFWD = IJTHDREV = 1 (A), and another case with IJTHDFWD = IJTHDREV = 1 × 10⁻⁴ (A). Both cases use XJBVD = 1 and exhibit diode breakdown characteristics. Because the linearization takes place at a smaller current magnitude for the latter case, the calculated $I_{j,DB}$ has a smaller magnitude when the diode is heavily forward-biased or heavily reverse-biased.

We use Fig. 5-8 to illustrate the different diode models in the reverse-bias region. In all three simulations with DIOMOD = 0, 1, and 2, XJBVD is fixed at 1 and IJTHDREV = 10^{-4} (A). When DIOMOD = 0, breakdown characteristics are modeled, but the model does not allow linearization. The IJTHDREV parameter is ignored during the calculation, and the magnitude of $I_{j,DB}$ continues to increase exponentially. The DIOMOD = 1 model does not model the breakdown at all, ignoring the values of both XJBVD and IJTHDREV. Hence, $I_{j,DB}$ remains at a slight negative value in the entire V_{DB} scan. In contrast, the DIOMOD = 2 model captures both the breakdown and the reverse linearization. The magnitude of simulated $I_{j,DB}$ hence does not increase as rapidly as the DIOMOD = 0 model.

We have mentioned about resetting parameter values here and there. Let us summarize, using the drain-side parameters as examples, although the resets apply to the source-side parameters as well. For all three diode models, if BVD ≤ 0, then BVD

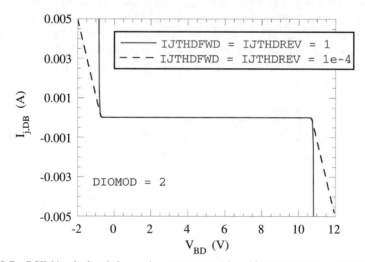

Fig. 5-7 BSIM4-calculated $I_{j,DB}$ when DIOMOD = 2, with IJTHDFWD = IJTHDREV = 1 (A), and with IJTHDFWD = IJTHDREV = 1 × 10⁻⁴ (A). XJBVD = 1. DIOMOD = 2 combines the features of DIOMOD = 0 and 1.

5.9 DIODE BREAKDOWN

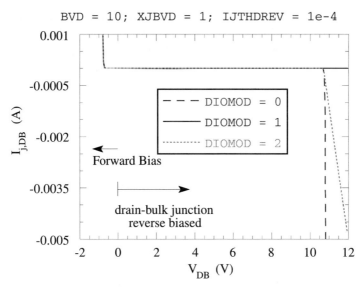

Fig. 5-8 Comparison of BSIM4-calculated $I_{j,\text{DB}}$ with DIOMOD = 0, 1, and 2. XJBVD = 1 and IJTHDREV = 10^{-4} (A).

is set to 10; if IJTHDREV $\leq$ 0, then IJTHDREV is set to 0.1; and if IJTHDFWD $\leq$ 0, IJTHDFWD is set to 0.1. Finally, for DIOMOD = 2 only, XJBVD is reset to 1 if XJBVD $\leq$ 0; and for DIOMOD = 0 only, XJBVD is reset to 1 if XJBVD < 0. From these rules, we see that if we set XJBVD = 0 in the DIOMOD = 0 model, then we remove the breakdown characteristics in the diode so that DIOMOD = 0 becomes comparable to DIOMOD = 1.

For completeness, we mention that JSD, JSWD, and JSWGD appearing in the $I_{sat,\text{DB}}$ expression have temperature dependencies through the parameter XTID. Let's write the temperature variations of JSD and JSWD:

$$\text{JSD}(T_{device}) = \text{JSD}(\text{TNOM}) \left[\exp\left(\frac{E_g(T_{device})}{\text{NJD} \cdot kT_{device}} - \frac{E_g(\text{TNOM})}{\text{NJD} \cdot k\text{TNOM}} \right) \right]$$
$$\times \left(\frac{T_{device}}{\text{TNOM}} \right)^{\text{XTID/NJD}}. \tag{5-47}$$

$$\text{JSWD}(T_{device}) = \text{JSWD}(\text{TNOM}) \left[\exp\left(\frac{E_g(T_{device})}{\text{NJD} \cdot kT_{device}} - \frac{E_g(\text{TNOM})}{\text{NJD} \cdot k\text{TNOM}} \right) \right]$$
$$\times \left(\frac{T_{device}}{\text{TNOM}} \right)^{\text{XTID/NJD}}. \tag{5-48}$$

The bandgap energy's temperature dependence, $E_g(T_{device})$, can be found in Appendix G (Eq. G-398). The temperature dependence of JSWGD is similar to that of JSWD in Eq. (5-48).

When $V_{BD} < 0$, $I_{j,\text{DB}}$ calculated from all of the above diode equations are negative. According to Eq. (5-23), the magnitude of $I_{j,\text{DB}}$ is then added to the overall drain current, hence the minus sign in front of $I_{j,\text{DB}}$ in Eq. (5-23) is correct.

Parameter Name	Description	Default Value
DIOMOD	Diode I–V model selector	1
NJS	Source–bulk junction ideality factor	1
NJD	Drain–bulk junction ideality factor	NJS
JSS	Source–bulk reverse saturation current per unit area	10^{-4} (A/m^2)
JSD	Drain–bulk reverse saturation current per unit area	JSS
JSWS	Source–bulk reverse saturation current per unit isolation-side periphery	0 (A/m)
JSWD	Drain–bulk reverse saturation current per unit isolation-side periphery	JSWS
JSWGS	Source–bulk reverse saturation current per unit gate-side periphery	0 (A/m)
JSWGD	Drain–bulk reverse saturation current per unit gate-side periphery	JSWGS
XTIS	Source–bulk junction current temperature coefficient	1
XTID	Drain–bulk junction current temperature coefficient	XTIS
BVS	Source–bulk junction breakdown voltage	10 (V)
BVD	Drain–bulk junction breakdown voltage	BVS
XJBVS	Source–bulk breakdown voltage fitting parameter	1
XJBVD	Drain–bulk breakdown voltage fitting parameter	XJBVS
IJTHSFWD	Limiting current in forward-biased source–bulk diode	0.1 (A)
IJTHDFWD	Limiting current in forward-biased drain–bulk diode	IJTHSFWD
IJTHSREV	Limiting current in reverse-biased source–bulk diode	0.1 (A)
IJTHDREV	Limiting current in reverse-biased drain–bulk diode	IJTHSREV

Replaced BSIM3 parameters: NJ, JS, JSSW, XTI, IJTH

5.10 GIDL (GATE-INDUCED DRAIN LEAKAGE) CURRENT

We complained in Section 4.16 that the gate-induced drain leakage (GIDL) current is not incorporated in BSIM3 and proposed a BSIM3-compatible subcircuit to account for the GIDL current. This is considered a patch, not a permanent solution, since the subcircuit is somewhat complicated. A more fundamental approach is to include the GIDL effects directly in the model equations. BSIM4 does just that. With a simple equation, BSIM4 calculates the GIDL current as a function of V_{DS}, V_{GS}, and V_{BS}:

$$I_{gidl} = \text{NF} \times \text{AGIDL} \cdot W_{\it{eff},\text{CJ}} \left[\frac{V_{DS} - V_{GS,\text{eff}} - \text{EGIDL}}{3 \cdot \text{TOXE}} \right]$$

$$\times \exp\left(\frac{-3 \cdot \text{TOXE} \times \text{BGIDL}}{V_{DS} - V_{GS,\text{eff}} - \text{EGIDL}} \right) \frac{V_{DB}^3}{\text{CGIDL} + V_{DB}^3}, \qquad (5\text{-}49)$$

where NF, the number of fingers, was described in Section 5.6, and $W_{\it{eff},\text{CJ}}$, the effective channel width for junction-related-quantity calculation, was defined in Section 5.8. Agreeing with GIDL's physical origin outlined in Section 4.16, the I_{gidl} expression adopted in BSIM4 is independent of the channel length, and exhibits weak bulk-bias dependence. The magnitude of I_{gidl} is most affected by V_{DG}, the voltage dropping between the drain and the gate. I_{gidl} can be made zero by assigning AGIDL to its default value of zero.

BSIM3's dc equivalent circuit, shown in Fig. 2-2, does not contain I_{gidl}. In Fig. 4-82 we suggested I_{gidl} should be placed between the drain and the bulk terminals, so that it parallels with the impact ionization current. This fact is observed in a simplified equivalent circuit adopted in BSIM4, shown in Fig. 5-9. (Figure 5-9 is

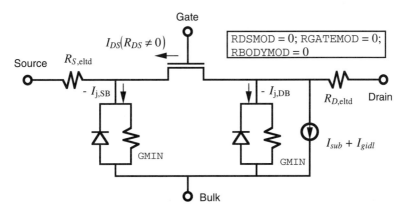

Fig. 5-9 BSIM4's dc equivalent circuit, when the tunneling currents discussed in Section 5.18 are ignored (IGCMOD = IGBMOD = 0). This equivalent circuit, more specifically, is for RDSMOD = 0 (Section 5.11), RGATEMOD = 0 (Section 5.12), and RSUBMOD = 0 (Section 5.13). Other parasitic drain/source, gate, and substrate resistances may need to be added under other conditions.

slightly modified from Fig. 4-82 to reflect the terminology used in BSIM4. Moreover, in order to avoid complexity, Fig. 5-9 is constructed assuming the tunneling currents discussed in Section 5.18 are zero, a result achieved when IGCMOD = IGBMOD = 0.) We caution that Fig. 5-9 applies to the case when RDSMOD = 0 (Section 5.11), RGATEMOD = 0 (Section 5.12), RSUBMOD = 0 (Section 5.13). These three parameters specify whether the parasitic drain/source, gate, and substrate resistances should be added into the model. Although the exact BSIM4 equivalent circuit depends on the values of RDSMOD, RGATEMOD, and RSUBMOD, the circuit will all appear similar to Fig. 5-9 as far as the location of I_{gidl} is concerned.

BSIM4's terminal-current equations can be understood from the equivalent circuit of Fig. 5-9.

$$I_D = I_{DS} + I_{sub} + I_{gidl} - I_{j,\text{DB}}. \tag{5-50}$$

$$I_S = -I_{DS} - I_{j,\text{SB}}. \tag{5-51}$$

$$I_B = -I_{sub} - I_{gidl} + I_{j,\text{SB}} + I_{j,\text{DB}}. \tag{5-52}$$

$$I_G = 0. \tag{5-53}$$

The above equations exclude the tunneling currents, since Fig. 5-9 does not include those components. The complete current equations including the tunneling currents will be discussed in Section 5.18 and we found in Appendix G (Eqs. G-264 to G-267).

BSIM4's GIDL equation is physics-based. When the junction leakage currents are negligible, the magnitude of the bulk current becomes a sum of the GIDL current and the substrate impact ionization current. As mentioned in Section 4.16, I_B is dominated by I_{gidl} when $V_{GS} < V_T$ and by I_{sub} at $V_{GS} > V_T$. Figure 5-10 illustrates the drain terminal current as a function of V_{GS} at $V_{DS} = 1.5$ V. In the simulation, GMIN is set to be 10^{-20} to prevent the calculated I_D from being affected by the GMIN resistors (Fig. 5-9). The junction leakage currents are made negligible by assigning JSS, JSD, JSWS, JSWD, JSWGS and JSWGD to small values. The impact ionization current is made zero by making ALPHA0 = ALPHA1 = 0. The GIDL parameters are: AGIDL = 1; BGIDL = 6.33 × 10^9; and CGIDL = 10^{-3}. Figure 5-10 shows that the gate-induced drain leakage current is prominent when V_{GS} is negative, and I_{gidl} increases exponentially with V_{GS} throughout the entire negative V_{GS} axis. The drain current in the strong inversion region (positive V_{GS}), in contrast, is not perturbed by the GIDL equations.

Parameter Name	Description	Default Value
AGIDL	Pre-exponential coefficient of I_{gidl}	0 (Ω^{-1})
BGIDL	Exponential coefficient of I_{gidl}	2.3 × 10^9 (V/m)
CGIDL	Bulk-bias parameter of I_{gidl}	0.5 (V^3)
EGIDL	Minimum bend bending	0.8 (V)

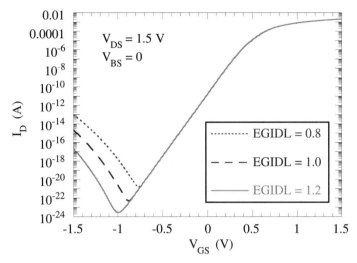

Fig. 5-10 BSIM4-calculated I_D as a function of V_{GS} at $V_{DS} = 1.5$ V, showing the BSIM4's capability to calculate the gate-induced drain leakage current. GMIN = 10^{-20}; JSS, and JSD are set to small values; and JSWS = JSWD = JSWGS = JSWGD = ALPHA0 = ALPHA1 = 0. The GIDL parameters are: AGIDL = 1; BGIDL = 6.33×10^9; CGIDL = 10^{-3}; and IGCMOD = IGBMOD = 0.

5.11 BIAS-DEPENDENT DRAIN–SOURCE RESISTANCE

BSIM3 offers two methods to represent the source and the drain resistances, as detailed in Section 2.2. In the lumped-resistance approach, the parameter RSH is used in conjunction with NRS and NRD to calculate the R_S and R_D in Fig. 2-8a, and RDSW = 0. In the absorbed-resistance approach of Fig. 2-8b, $R_S = R_D = 0$, yet the drain current and conductances are modified by a nonzero RDSW, such that the final calculated I_D, g_d, g_m and g_{mb} are approximately equal to those of the lumped-resistance approach.

BSIM4 has a similar arrangement, except that it introduces a parameter RDSMOD to select the desired resistance approach, rather than relying on RDSW's magnitude to choose between the lumped-resistance and the absorbed-resistance approaches. Furthermore, BSIM4 allows greater flexibility in defining the resistance. A RDSMOD = 1 equivalent circuit is given Fig. 5-11. This figure assumes RGATEMOD = RSUBMOD = 0. Nonzero RGATEMOD and RSUBMOD only complicates the circuit and our discussion here; these two BSIM4 parameters will be detailed in the subsequent two sections. They key point in Fig. 5-11 is that the internal device resistances are modeled by bias-dependent resistances $R_{S,\text{bias-dep}}$ and $R_{D,\text{bias-dep}}$ and the parasitic electrode resistances are modeled by $R_{S,\text{eltd}}$ and $R_{D,\text{eltd}}$. When RDSMOD is equated with its default value of zero, the equivalent circuit becomes Fig. 5-9, in which $R_{S,\text{eltd}}$ and $R_{D,\text{eltd}}$ remain but the bias-dependent resistances become zero. Although $R_{S,\text{bias-dep}} = R_{D,\text{bias-dep}} = 0$, the channel drain

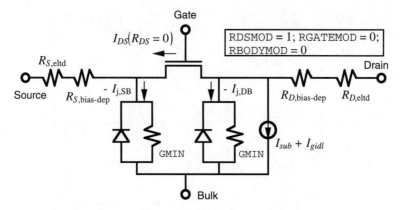

Fig. 5-11 A high-level equivalent circuit of BSIM4 when RDSMOD = 1, showing the bias-dependence resistances and the parasitic electrode resistances at the drain and the source. RGATEMOD = RSUBMOD = 0. IGCMOD = IGBMOD = 0.

current I_{DS} and small-signal conductances are modified by an effective internal source–drain resistance (R_{DS}) such that the final calculated current and conductances are approximately the same as those in RDSMOD = 1. R_{DS} is not a real resistance, similar to the fact that the absorbed-resistance approach of BSIM3 results in no real part in the input impedance (Section 4.19). R_{DS} only ensures I_{DS}, g_m, g_{mb}, and g_d are calculated correctly, but y_{11}, s_{11}, and so forth, will be incorrect. In this sense, RDSMOD = 0 is comparable to the absorbed-resistance approach in BSIM3, and RDSMOD = 1, the lumped-resistance approach.

The effective internal source–drain resistance is calculated as

$$R_{DS} = \begin{cases} \dfrac{\text{RDSWMIN} + \text{RDSW} \times \frac{1}{2}[Tmp + \sqrt{Tmp^2 + 0.01}]}{(10^6 \times W_{\textit{eff},CJ})^{WR}} & \text{if RDSMOD} \neq 1; \\ 0 & \text{if RDSMOD} = 1; \end{cases}$$

(5-54)

where

$$Tmp = \frac{1}{1 + \text{PRWG} \cdot (V_{GS} - V_T)} + \text{PRWB}\left(\sqrt{2\phi_f - V_{BS}} - \sqrt{2\phi_f}\right). \quad (5\text{-}55)$$

The multiplier to RDSW, $[Tmp + \text{sqrt}(Tmp^2 + 0.01)]/2$, has the property that it will never be negative. Even if *Tmp* somehow becomes a very negative number due to a bad choice of parameter values, the multiplier has the minimum value of 0. The multiplier approaches to *Tmp* when *Tmp* is a large positive number. The parameters appearing in *Tmp*, PRWG and PRWB, account for gate bias and substrate bias' effects on the effective resistance, respectively. The factor 10^6 is to convert the unit between the meter used in $W_{\textit{eff},CJ}$ (Section 5.8) and the Ω·μm used for RDSW.

5.11 BIAS-DEPENDENT DRAIN–SOURCE RESISTANCE

R_{DS} is a finite number for RDSMOD = 0. It affects in the calculation for I_D, g_m, g_{mb}, and g_d in a rather complicated fashion, as detailed in Appendix G. R_{DS} is zero in RDSMOD = 1, a model setting that relies on physical resistances $R_{S,\text{bias-dep}}$ and $R_{D,\text{bias-dep}}$ to account for the internal device's source–drain resistances' effects on transistor properties. The source bias-dependent resistance is given by

$$R_{S,\text{bias-dep}} = \begin{cases} \dfrac{\text{RSWMIN} + \text{RSW} \times \frac{1}{2}[Tmp + \sqrt{Tmp^2 + 0.01}]}{\text{NF} \times (10^6 \times W_{\text{eff,CJ}})^{\text{WR}}} & \text{if RDSMOD} = 1; \\ 0 & \text{if RDSMOD} \neq 1; \end{cases} \quad (5\text{-}56)$$

where

$$Tmp = \frac{1}{1 + \text{PRWG}(V_{GS} - V_{FB,\text{SD}})} - \text{PRWB}\, V_{BS}; \quad (5\text{-}57)$$

$$V_{FB,\text{SD}} = \begin{cases} \dfrac{k[\text{TNOM} + 273.15]}{q} \ln\left(\dfrac{\text{NGATE}}{\text{NSD}}\right) & \text{if NGATE} > 0; \\ 0 & \text{if otherwise.} \end{cases} \quad (5\text{-}58)$$

$V_{FB,\text{SD}}$ is the flatband voltage in the source/drain junctions, whose doping level is NSD. Note that NF appears in Eq. (5-56), but not in Eq. (5-54). Conceptually, resistance is inversely proportional to the number of fingers. NF does not appear in Eq. (5-54) because R_{DS} is used to find individual finger current only. Once the individual current is determined, the overall channel current gets multiplied by NF, as seen in Eq. (5-25). Similarly, for the drain side,

$$R_{D,\text{bias-dep}} = \begin{cases} \dfrac{\text{RDWMIN} + \text{RDW} \times \frac{1}{2}[Tmp + \sqrt{Tmp^2 + 0.01}]}{\text{NF} \times (10^6 \times W_{\text{eff,CJ}})^{\text{WR}}} & \text{if RDSMOD} = 1; \\ 0 & \text{if RDSMOD} \neq 1; \end{cases} \quad (5\text{-}59)$$

where

$$Tmp = \frac{1}{1 + \text{PRWG}(V_{GD} - V_{FB,\text{SD}})} - \text{PRWB}\, V_{BD}. \quad (5\text{-}60)$$

For completeness, we also write out the electrode source and drain resistances, $R_{S,\text{eltd}}$ and $R_{D,\text{eltd}}$:

$$R_{S,\text{eltd}} = \begin{cases} 0.001\ (\Omega) & \text{if } R^*_{S,\text{eltd}} = 0 \text{ and (RGEOMOD} \neq 0 \text{ or} \\ & \quad \text{RDSMOD} \neq 0 \text{ or TNOIMOD} \neq 0); \\ R^*_{S,\text{eltd}} & \text{if otherwise;} \end{cases} \quad (5\text{-}61)$$

where $R^*_{S,\text{eltd}}$ is

$$R^*_{S,\text{eltd}} = \begin{cases} 0 & \text{if RGEOMOD} = 0; \\ \text{NRS} \cdot \text{RSH} & \text{if NRS is given and RGEOMOD} \neq 0; \\ \text{see Section 5.19} & \text{if NRS is not given and RGEOMOD} \neq 0. \end{cases} \quad (5\text{-}62)$$

Setting $R_{S,\text{eltd}}$ to an arbitrary 0.001 Ω in Eq. (5-61) is necessary at times to avoid numerical problems (singular matrix) associated with a 0-Ω resistor element. Likewise, for the drain side:

$$R_{D,\text{eltd}} = \begin{cases} 0.001\ (\Omega) & \text{if } R^*_{D,\text{eltd}} = 0 \text{ and (RGEOMOD} \neq 0 \text{ or} \\ & \text{RDSMOD} \neq 0 \text{ or TNOIMOD} \neq 0) \\ R^*_{D,\text{eltd}} & \text{if otherwise} \end{cases} \quad (5\text{-}63)$$

where $R^*_{D,\text{eltd}}$ is

$$R_{D,\text{eltd}}{}^* = \begin{cases} 0 & \text{if RGEOMOD} = 0 \\ \text{NRD} \cdot \text{RSH} & \text{if NRD is given and RGEOMOD} \neq 0; \\ \text{see Section 5.19} & \text{if NRD is not given and RGEOMOD} \neq 0. \end{cases} \quad (5\text{-}64)$$

RSH in BSIM4 is the same parameter in BSIM3. NRD and NRS are both instance parameters declared in a MOSFET instance statement; they are also available for BSIM3.

Parameter Name	Description	Default Value
RDSMOD	Bias-dependent source–drain resistance model selector	0
RDSW†	Zero-bias effective internal device resistance per width	$200\ (\Omega \cdot \mu m^{WR})$
RDSWMIN	Minimum effective internal device resistance per width	$0\ (\Omega \cdot \mu m^{WR})$
RDW	Zero-bias resistance per width for $R_{D,\text{bias\_dep}}$ calculation	$100\ (\Omega \cdot \mu m^{WR})$
RDWMIN	Minimum resistance per width for $R_{D,\text{bias\_dep}}$ calculation	$0\ (\Omega \cdot \mu m^{WR})$
RSW	Zero-bias resistance per width for $R_{S,\text{bias\_dep}}$ calculation	$100\ (\Omega \cdot \mu m^{WR})$
RSWMIN	Minimum resistance per width for $R_{S,\text{bias\_dep}}$ calculation	$0\ (\Omega \cdot \mu m^{WR})$
PRWG†	Gate bias dependence of internal device resistance	$1\ (V^{-1})$
PRWB†	Body bias dependence of internal device resistance	$0\ (V^{-0.5})$
WR†	Width dependence of internal device resistance	1

† Same as BSIMS parameters, repeated here for convenience

5.12 GATE RESISTANCE

BSIM3 does not have a built-in gate resistance. The finite resistance in the polysilicon gate can be accounted for only by adding an external resistor. BSIM4 offers several gate resistance models, allowing the users greater flexibility in modeling the gate resistance. When RGATEMOD = 1, a gate electrode resistance is appended to the intrinsic transistor's gate terminal:

$$R_{G,\text{eltd}} = \frac{\text{RSHG} \cdot \left(\text{XGW} + \dfrac{W_{\textit{eff},\text{CJ}}}{3\text{NGCON}}\right)}{\text{NGCON} \cdot (L - \text{XGL}) \cdot \text{NF}}, \quad (5\text{-}65)$$

where RSHG is the polysilicon gate's sheet resistance in $\Omega/\square$. NGCON is the number of gate contact for each gate finger; it can be either one if the gate finger is contacted from one side, or two if the gate finger is contacted from both sides. (BSIM4 issues a warning if the user-supplied NGCON is neither 1 nor 2.) XGW is distance between the gate contact to the channel edge, and XGL is the difference between the L specified in the MOSFET instance statement and the physical gate length. These geometrical details are illustrated in Fig. 5-12, for both NGCON = 1 and 2. When NGCON = 1, XGW = XGL = 0, and NF = 1, Eq. (5-65) is reduced to

$$R_{G,\text{eltd}} = \frac{1}{3} \times \text{RSHG}\, \frac{W_{\textit{eff},\text{CJ}}}{L}. \quad (5\text{-}66)$$

Where does the 1/3 factor come from? If all of the gate signals that go into one end of the gate finger in Fig. 5-12a were to go out the other end, then the gate resistance would follow our intuitive formula of simple resistance: $R_{G,\text{eltd}} = \text{RSHG} \cdot W_{\textit{eff},\text{CJ}}/L$. However, in the gate structure of an actual transistor, all of the signals go into one end of the gate, but none of them goes out at the other end. These ac signals eventually all leave the gate and enter the channel through C_{gg} (in the form of $C_{gg} \cdot dv_{GS}/dt$). Some signals remain in the gate for a longer portion of the gate width before exiting to the channel. Some signals depart sooner. The factor 1/3 accounts for the distributed nature of the current conduction, as derived in Ref. [4]. Because of the way the current flows in the gate finger, the gate resistance is called a *distributed* resistance.

Figure 5-12b illustrates the layout when NGCON = 2. If XGW = XGL = 0, and NF = 1, Eq. (5-65) then becomes

$$R_{G,\text{eltd}} = \frac{1}{12} \times \text{RSHG}\, \frac{W_{\textit{eff},\text{CJ}}}{L}. \quad (5\text{-}67)$$

When NGCON increases from 1 to 2, one might expect the factor to halve from 1/3 to 1/6. The reason that the resistance decreases by four-fold instead of two-fold is the following: When the gate metal is contacted by two sides, then the location at which the gate current equals zero occurs at $x = W_{\textit{eff},\text{CJ}}/2$, rather than $W_{\textit{eff},\text{CJ}}$. The resistance already decreases by two-fold due to the width reduction. The overall resistance decreases by another two-fold due to the fact that there are two contacts

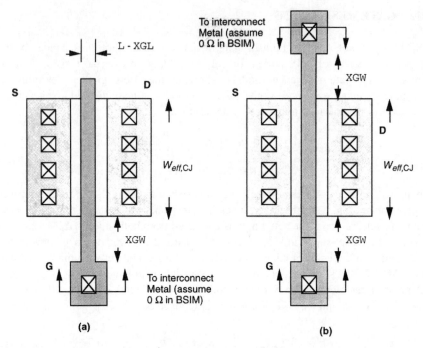

Fig. 5-12 Geometrical details used in calculating the gate resistance in BSIM4. (a) NGCON = 1; (b) NGCON = 2.

connected in parallel. This factor of 1/12 is similar to that encountered in the intrinsic base resistance of bipolar transistors [4,5].

When RGATEMOD = 2, a "channel-reflected gate resistance", $R_{G,\mathrm{crg}}$, is added to the gate terminal in addition to gate electrode resistance, $R_{G,\mathrm{eltd}}$. A description about $R_{G,\mathrm{crg}}$ is given in Ref. [6]. $R_{G,\mathrm{crg}}$ is not a physical resistance; thus the noise equivalent circuits of Section 5.15 do not include a thermal noise source for this resistance. $R_{G,\mathrm{crg}}$ can be thought of as an equivalent resistance that represents the first order non-quasi-static effects in the channel (Section 5.17). Reference [6] establishes that this gate resistance component is critical in matching the noise data. $R_{G,\mathrm{crg}}$ is given by

$$R_{G,\mathrm{crg}} = \frac{1}{\mathrm{XRCRG1} \times \left(\dfrac{I_{DS}}{V_{DS,\mathrm{eff}}} + \mathrm{XRCRG2}\dfrac{kT}{q}\mu_{\mathrm{eff}}\dfrac{W_{\mathrm{eff}}}{L_{\mathrm{eff}}}C'_{ox,\mathrm{IV}} \cdot \mathrm{NF}\right)}, \qquad (5\text{-}68)$$

where $V_{DS,\mathrm{eff}}$ was given by Eq. (1-10) and $C'_{ox,\mathrm{IV}}$ was described in Section 5.2. XRCRG1 and XRCRG2 are two parameters used to correct the value of $R_{G,\mathrm{crg}}$.

RGATEMOD = 3 is nearly identical to RGATEMOD = 2, except that the location of the overlap capacitances differs. The equivalent circuits for various gate resistance

5.12 GATE RESISTANCE   **433**

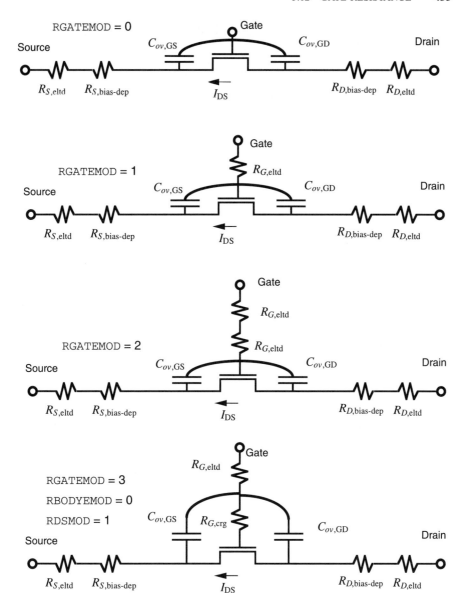

Fig. 5-13 High-level equivalent circuits of BSIM4 at various RGATEMOD values. RDSMOD = 1, RSUBMOD = 0. IGCMOD = IGBMOD = 0.

models are shown in Fig. 5-13. The relative position of the overlap capacitances are indicated in the figure.

Parameter Name	Description	Default Value
RGATEMOD	Gate resistance model selector	0
RSHG	Gate electrode sheet resistance	0 (Ω/square)
XGW	Distance from the gate contact to the channel edge	0 (m)
XGL	Difference between L and the physical gate length	0 (m)
NGCON	Number of gate contacts per finger; either 1 or 2	1
XRCRG1	First parameter of channel-reflected gate resistance	12
XRCRG2	Second parameter of channel-reflected gate resistance	1

5.13 SUBSTRATE RESISTANCE

The MOSFET modeled in BSIM3 does not contain internal substrate resistance. A straightforward remedy to incorporate the substrate's resistive effects is to add an external resistor at the substrate node. If the distributed effects of the substrate resistance are to be included, a more complicated subcircuit with several resistors needs to be used. When the parameter RBODYMOD is changed from its default value of 0 to 1, BSIM4 appends a rather rigid, five-resistor network to the substrate of the intrinsic transistor, as shown in Fig. 5-14. (Figure 5-14 is specifically for the case RDSMOD = RGATEMOD = 0, and IGCMOD = IGBMOD = 0. Appropriate changes for RDSMOD $\neq$ 0 and/or RGATEMOD $\neq$ 0 can be found in Sections 5.11 and 5.12, respectively. The changes for IGCMOD or IGBMOD $\neq$ 0 can be found in Section 5.18.) The BSIM4's resistive network is a good compromise between retaining a

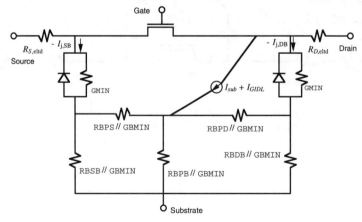

Fig. 5-14 The five-resistor substrate resistance network adopted in BSIM4. RSUBMOD is set to 1. RDSMOD = RGATEMOD = 0. IGCMOD = IGBMOD = 0.

simplistic subcircuit and accounting for the first-order distributed substrate effects, similar to that proposed in Ref. [7].

The key advantage of using the BSIM4 body model is that the source–bulk capacitance does not directly connect to the drain–bulk capacitance. They are separated by the distributed substrate resistances RBPS and RBPD. This feature has been identified as a key to properly model the output impedance of the transistor, especially at high frequencies [7].

The BSIM4 substrate resistance model is not scalable unless the parameters are effectively parameterized with geometrical dependence through the SPICE simulator. As the transistor size becomes twice as large, the substrate resistance network which appends to the device remains unaltered, until the parameter values for RBPS, RBPD, RBPB, RBSB, or RBDB are specifically changed by the user. The substrate resistance model is also non-scalable with respect to NF, the number of fingers. When NF changes from 1 to 2, for example, again the same substrate resistance network is used unless any of the aforementioned substrate resistance parameters is modified. A corollary is the following. The substrate resistance parameters should be extracted for the total device as a whole, not on a per-finger basis.

In order to avoid numerical difficulty (singular matrix) when the substrate resistance is too small, the resistance is replaced by 0.001 Ω if it is smaller than 0.001 Ω. For example, if RBPS//GBMIN of Fig. 5-14 is 10^{-4} Ω, then it is replaced by 0.001 Ω during the numerical calculation.

Several ways to estimate the values of substrate resistances were discussed in Section 4.6.

Parameter Name	Description	Default Value
RBODYMOD	Substrate resistance model selector	0
RBPB	Substrate resistance element, see Fig 5-14	50 (Ω)
RBPD	Substrate resistance element, see Fig 5-14	50 (Ω)
RBPS	Substrate resistance element, see Fig 5-14	50 (Ω)
RBDB	Substrate resistance element, see Fig 5-14	50 (Ω)
RBSB	Substrate resistance element, see Fig 5-14	50 (Ω)
GBMIN	Minimum conductance for the substrate resistance model	10^{-12} (1/Ω)

5.14 OVERLAP CAPACITANCES

The gate-to-source and gate-to-drain overlap capacitances in BSIM4 are given by

$$C_{ov,GS} = NF \times W_{eff,CV} \times \left[CGS0 + CGS1 - CGS1 \left(1 - \frac{1}{\sqrt{1 - \frac{4 \cdot V_{GS,overlap}}{CKAPPAS}}} \right) \right.$$

$$\left. \times \left(\frac{1}{2} - \frac{(V_{GS} + \delta_1)}{2\sqrt{(V_{GS} + \delta_1)^2 + 4\delta_1}} \right) \right]; \qquad (5\text{-}69)$$

$$C_{ov,\text{GD}} = \text{NF} \times W_{\text{eff,CV}} \times \left[\text{CGD0} + \text{CGD1} - \text{CGD1} \left(1 - \frac{1}{\sqrt{1 - \frac{4 \cdot V_{GD,\text{overlap}}}{\text{CKAPPAD}}}} \right) \right.$$

$$\left. \times \left(\frac{1}{2} - \frac{(V_{GD} + \delta_1)}{2\sqrt{(V_{GD} + \delta_1)^2 + 4\delta_1}} \right) \right]; \qquad (5\text{-}70)$$

where NF is the number of fingers, $W_{\text{eff,CV}}$ is the effective width per finger for CV calculations, and

$$V_{GS,\text{overlap}} = \frac{1}{2}\left[V_{GS} + \delta_1 - \sqrt{(V_{GS} + \delta_1)^2 + 4\delta_1} \right]; \qquad \delta_1 = 0.02; \qquad (5\text{-}71)$$

$$V_{GD,\text{overlap}} = \frac{1}{2}\left[V_{GD} + \delta_1 - \sqrt{(V_{GD} + \delta_1)^2 + 4\delta_1} \right]; \qquad \delta_1 = 0.02. \qquad (5\text{-}72)$$

These expressions produce the voltage dependence of the overlap capacitance due to the depletion in the heavily doped source and drain junctions (see CGS0 entry in Section 3.2). The rate at which $C_{ov,\text{GD}}$ decreases with V_{GD} is controlled by CKAPPAD, and the rate at which $C_{ov,\text{GS}}$ decreases with V_{GS}, by CKAPPAS. These expressions are identical to those used in BSIM3, except that BSIM3 does not distinguish between CKAPPAS and CKAPPAD. Instead, BSIM3 uses the same CKAPPA in both Eqs. (5-69) and (5-70).

The replacement of CKAPPA in BSIM3 by CKAPPAS and CKAPPAD in BSIM4 is consistent with BSIM4's attempt to break the symmetry between the source- and the drain-side parasitics. In practice, particularly in high-voltage devices, $C_{ov,\text{GS}}$ and $C_{ov,\text{GD}}$ differ considerably, not just in magnitude but also in the voltage dependence.

Parameter Name	Description	Default Value
CKAPPAS	Voltage dependence coefficient of $C_{ov,\text{GS}}$	0.6 (V)
CKAPPAD	Voltage dependence coefficient of $C_{ov,\text{GD}}$	CKAPPAS

Replaced BSIM3 Parameters: CKAPPA

5.15 THERMAL NOISE MODELS

There are two thermal noise models in BSIM4. TNOIMOD = 0, the default, chooses the charge-based long-channel thermal noise model used in BSIM3 (more specifically, BSIM3v3.3, to be elaborated upon shortly). TNOIMOD = 1 invokes the *holistic thermal noise model*, a new model developed for BSIM4. We shall devote a great deal of discussion to the holistic model, mainly because it models the induced gate noise. The incorporation of this second intrinsic noise source represents a fairly significant advancement over BSIM3's thermal noise models.

5.15 THERMAL NOISE MODELS

For the discussion of the TNOIMOD = 0 model, some historical perspective is in order. Up to BSIM3v3.2.2 (officially released in 1999), the charge-based thermal noise model in BSIM3 is given by

$$\frac{\overline{i_d^2}}{\Delta f} = \frac{4kT}{L_{eff}^2/(\mu_{eff}|Q_{inv}|)}. \tag{5-73}$$

As discussed in Section 4.11, for long-channel devices, this expression correctly degenerates to $4kT\Delta f \times 2/3(g_m + g_{mb})$ in saturation, and into $4kT\Delta f \times g_d$, in the linear region. However, for short-channel devices, Eq. (5-73) becomes problematic in two respects. One, if the absorbed-resistance approach is used to model the source/drain resistance, the noise associated with the resistance is not accounted for. (This point was also discussed in Section 2.6.) In order to accommodate the shortcoming of the absorbed-resistance approach in noise modeling, BSIM3v3.3 has been discussed to be introduced in 2000, after the release of BSIM4 (see Ref. [15] of Chapter 1 for more information about this yet released BSIM3 version). In this particular BSIM3 version, Eq. (5-73) will be modified to

$$\frac{\overline{i_d^2}}{\Delta f} = \frac{4kT}{R_{DS} + L_{eff}^2/(\mu_{eff}|Q_{inv}|)}. \tag{5-74}$$

R_{DS} was given by Eq. (4-57). Equation (5-74) remains valid in the lumped-resistance approach in which physical source and drain resistances are added to the intrinsic transistor (Section 2.2). Under this approach, RDSW and hence R_{DS} is zero. The overall device noise is then the sum of the intrinsic device's thermal noise (given by Eq. 5-74 with $R_{DS} = 0$), and the thermal noises coming from the attached source and drain resistances.

Equation (5-73) is also improvable for the short-channel devices in another aspect. As discussed in Section 4.11, γ appearing in Eq. (4-55) is no longer its long-channel value of 2/3 when L shrinks. γ can be much larger than 2/3. When Eq. (5-73) is used, γ is implicitly made to be fixed at 2/3.

BSIM4's TNOIMOD = 0 model corrects both shortcomings of Eq. (5-73), by writing the device thermal noise as

$$\frac{\overline{i_d^2}}{\Delta f} = \frac{4kT \cdot \text{NTNOI}}{R_{DS} + L_{eff}^2/(\mu_{eff}|Q_{inv}|)} \quad (\text{TNOIMOD} = 0), \tag{5-75}$$

where R_{DS} in BSIM4, given in Eq. (5-54) in Section 5.11, is zero when RDSMOD = 1 and determined by RDSW and RDSMIN when RDSMOD = 0. By keeping R_{DS} in the denominator, BSIM4 properly accounts for the noise due to R_{DS} even though R_{DS} is not a physical resistance. Moreover, by introducing the parameter NTNOI, BSIM4 effectively allows the short-channel devices to have a γ much larger than 2/3. Although having NTNOI as a parameter is an improvement, it still does not allow scaling in the noise modeling. For example, once fixing NTNOI to a value of 10 for a noisy short-channel device, the same value of NTNOI applied to a long-channel device would then overestimate the thermal noise. In addition, as mentioned in Section 4.11, the increase in the γ value in short-channel devices can be viewed as a manifestation of the hot electron temperature in the channel. As the electron temperature depends on the channel field, we expect NTNOI in Eq. (5-75) to be a strong function of V_{DS} as well. However, NTNOI in BSIM4 is a fixed parameter, not exhibiting any bias dependency.

We plotted in Fig. 4-64 the ratio of $i_d^2/(4kT\Delta f)$ to g_d, as a function of L, for devices with RDSW = 423.5 (Ω-µm). That was BSIM3. Figure 5-15 displays the same ratio as a function of the channel length, except with BSIM4's TNOIMOD = 0 and 1 models. (The TNOIMOD = 1 model will be described shortly.) For the TNOIMOD = 0 simulation, NTNOI is set to its default value of 1. For the TNOIMOD = 1 simulation, TNOIA and TNOIB are set to their default values of 1.5 and 3.5, respectively. The transistor is biased at $V_{DS} = 0.001$ V and $V_{GS} = 5$ V, and the $1/f$ noise's contribution is deliberately made zero by setting FNOIMOD = 0 and KF = 0. The ratio is approximately equal to 1 for all devices, consistent with the fact that R_{DS} is included in the thermal noise calculation. It is interesting to note that the TNOIMOD = 1 model, which includes the induced gate noise, has a ratio closer to unity than the TNOIMOD = 0 model.

The noise equivalent circuits for TNOIMOD = 0 are shown in Fig. 5-16. (Thus far we have not considered the tunneling currents, which will be elaborated in Section

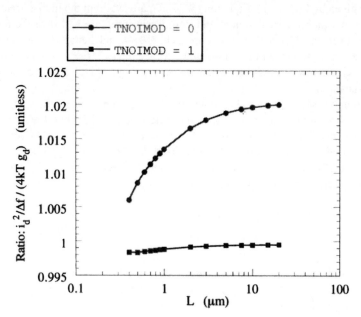

Fig. 5-15 BSIM4-calculated ratio of $i_d^2/(4kT\Delta f)$ to g_d, as a function of L. Just as in Fig. 4-64, RDSW = 423.5 (Ω-µm); PRWG = PRWB = 0. The $1/f$ noise's contribution is deliberately made to be zero.

5.15 THERMAL NOISE MODELS

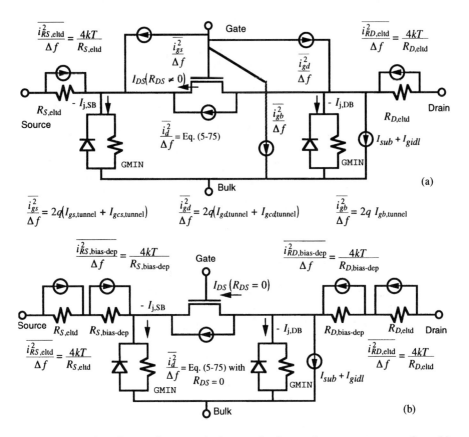

Fig. 5-16 BSIM4's noise equivalent circuits when TNOIMOD = 0. (a) RDSMOD = RGATEMOD = RBODYMOD = 0; (b) RDSMOD = 1, RGATEMOD = RBODYMOD = 0; (c) RDSMOD = 0 RGATEMOD = 3, and RBODYMOD = 1. The $1/f$ noise, given in Section 5.16, should be added in the $i_d^2/\Delta f$ expressions in the figure to complete the noise model. The shot noises associated with the tunneling currents are shown only in Fig. 5-16a to maintain clarity (see Fig. 5-17 for other cases). The tunneling currents are discussed in Section 5.18.

5.18. Although BSIM4 neglects the shot noises associated with I_{sub}, I_{gidl}, $I_{j,DB}$, it incorporates the shot noises associated with the tunneling currents, as shown in Figs. 5-16 and 5-17. The noise equations for the tunneling currents, $i_{gs}^2/\Delta f$, $i_{gd}^2/\Delta f$, and $i_{gb}^2/\Delta f$, are given in the figures.) The equivalent circuits are shown for various model selectors. Note that in Fig. 5-16c, when the comprehensive gate resistance model is turned on, $R_{G,crg}$ does not have an associated thermal noise source. This point was discussed in Section 5.12.

We devote the remainder of this section to TNOIMOD = 1. This model is touted as the *holistic* thermal model, in the sense that the induced gate noise is included and that the drain current noise implicitly incorporates the short-channel effects through its dependence on g_m, g_{mb}, and g_d. The derivation of the model is found in Ref. [8]. We will not repeat the derivation here, nor will we describe the logics behind it. We

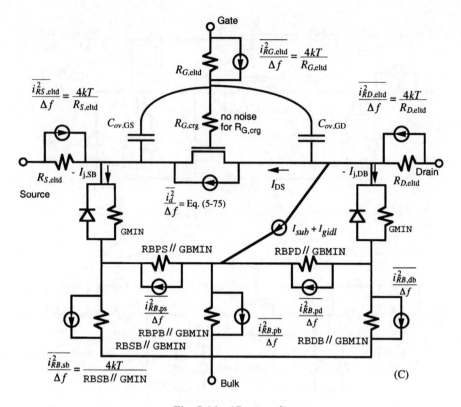

Fig. 5-16 (*Continued*)

will, however, show that the model yields correct simulation results in several test circuits. The equivalent circuits for BSIM4's holistic model are shown in Fig. 5-17, for various combinations of RDSMOD, RGATEMOD, and RBODYMOD. There are two main noise sources: $i_d^2/\Delta f$, and $i_{RS}^2/\Delta f$. (The noise sources $i_{RD}^2/\Delta f$, $i_{RG,\text{eltd}}^2/\Delta f$, $i_{RB,\text{pd}}^2/\Delta f$, e.g., are conventional resistor thermal noise sources, which are not essential to this discussion.) A typical noise-equivalent representation, referred to as the van der Ziel representation [9,10], accounts for the induced gate noise with a $i_g^2/\Delta f$ at the gate, as shown in Fig. 4-65b. A BSIM4's representation, such as Fig. 5-17a, in stark contrast, does not have $i_g^2/\Delta f$. Through equivalent circuit transformation, BSIM4 accounts for the $i_g^2/\Delta f$ noise source of the van der Ziel representation by $i_{RS}^2/\Delta f$. The noise sources in the holistic model are given by

$$\frac{\overline{i_d^2}}{\Delta f} = 4kT\frac{V_{DS,\text{eff}}}{I_{DS}}(\beta g_m + \beta g_{mb} + g_d)^2 - 4kTR_x(g_m + g_{mb} + g_d)^2; \quad (5\text{-}76)$$

$$\frac{\overline{i_{RS}^2}}{\Delta f} = 4kT\frac{R_x + R_{S,\text{eltd}} + R_{S,\text{bias\_dep}}}{(R_{S,\text{eltd}} + R_{S,\text{bias\_dep}})^2}; \quad (5\text{-}77)$$

5.15 THERMAL NOISE MODELS

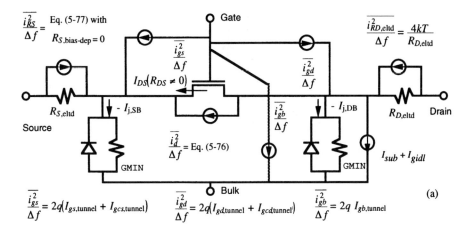

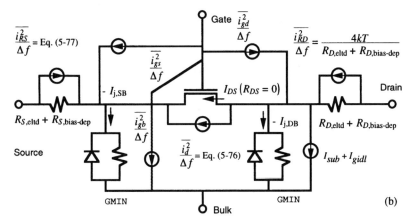

Fig. 5-17 Same as Fig. 5-16, except with `TNOIMOD = 1`, the holistic thermal model. The $1/f$ noise, given in Section 5.16, should be added in the $i_d^2/\Delta f$ expressions in the figure to complete the model. See also Section 5.18 for the tunneling currents.

where

$$R_x = \theta^2 \frac{V_{DS,\text{eff}}}{I_{DS}}; \tag{5-78}$$

$$\beta = 0.577 \times \left[1 + \text{TNOIA} \cdot L_{\text{eff}} \left(\frac{V_{GST,\text{eff}}}{\varepsilon_{sat} L_{\text{eff}}}\right)^2\right]; \tag{5-79}$$

$$\theta = 0.37 \times \left[1 + \text{TNOIB} \cdot L_{\text{eff}} \left(\frac{V_{GST,\text{eff}}}{\varepsilon_{sat} L_{\text{eff}}}\right)^2\right]. \tag{5-80}$$

ε_{sat}, the saturation electric field, is defined to be $2 \cdot \text{VSAT}/\mu_{\text{eff}}$, just as in th dc I-V calculation.

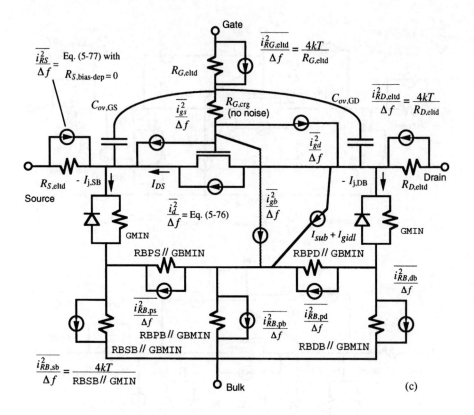

Fig. 5-17 (*Continued*)

Equation (5-77) applies to RDSMOD = 0 or 1 ($R_{S,\text{bias\_dep}}$ becomes zero in RDSMOD = 0). There is no appearance of R_{DS} in the above equations, unlike Eq. (5-75), because the noise sources in the holistic model are made to be functions of the conductances (g_d, g_m, g_{mb}), which have already accounted for R_{DS}.

BSIM4's transformation of $i_g^2/\Delta f$ at the gate terminal into $i_{RS}^2/\Delta f$ at the source terminal has at least two advantages. First, $i_g^2/\Delta f$ of the van der Ziel representation is frequency dependent. On the other hand, $i_{RS}^2/\Delta f$ in the holistic model is independent of frequency. Second, as shown in the derivation of the holistic model [8], the channel noise is easily partitioned in a manner such that the resulting $i_d^2/\Delta f$ and $i_{RS}^2/\Delta f$ are uncorrelated. In contrast, $i_d^2/\Delta f$ and $i_g^2/\Delta f$ of the van der Ziel representation are strongly correlated. These two features facilitate the model implementation into typical SPICE simulators.

We use two examples to demonstrate that the holistic model of BSIM4 and the van der Ziel representation yield comparable results, despite that they appear fairly different. Consider first a circuit shown in Fig. 5-18. In this bias setup with an inductor at the gate terminal, the impedance looking out of the gate port is infinite (i.e., $Z_y = \infty$ in Fig. 4-65b), so all of the gate noise current flows into the gate node of the device. A small drain resistance of 10^{-8} Ω is inserted, mainly to establish drain current noise $(\overline{I_d^2})$ from the drain noise voltage $(\overline{V_{drain}^2})$, with the relationship $\overline{V_{drain}^2} = \overline{I_d^2} \times (10^{-8})^2$. Without this resistance, the output voltage noise at the drain would be zero and those SPICE simulators without the capability to output noise current would not yield information about $\overline{I_d^2}$. A little thought suggests that the noise equivalent circuit of Fig. 5-18 can be represented by Fig. 4-65b, for which Eq. (4-63) applies. We repeat the result here for convenience:

$$\overline{I_d^2} = 4kT\Delta f \left[\tfrac{2}{3} g_m + \tfrac{4}{15} g_m - \tfrac{1}{3} g_m \right] \qquad (Z_y = \infty). \tag{5-81}$$

As discussed in Section 4.12, the first term is due to $\overline{i_d^2}$, the second term, $\overline{i_g^2}$, and the last term, their correlation. When the induced gate noise is neglected, $\overline{I_d^2}$ is proportional to $2/3 \cdot g_m$. When it is included, $\overline{I_d^2}$ is proportional to $3/5 \cdot g_m$. Therefore, and interestingly, $\overline{I_d^2}$ calculated from the van der Ziel representation decreases by 10% the moment the induced gate noise is considered. Let us examine whether BSIM4 predicts the same decrease in $\overline{I_d^2}$, as TNOIMOD is switched from 0 (no induced gate noise) to 1 (accounting for induced gate noise). In the calculation, we make sure that $1/f$ noise is zero so that this noise component does not enter the picture. The small-signal frequency is chosen to be 10 Hz, low enough to prevent capacitances such as C_{gs} and C_{gd} from having any impact. The parameters NTNOI, TNOIA, and TNOIB assume their default values of 1, 1.5 and 3.5, respectively. We find for a particular transistor $\overline{I_d^2} = 1.32 \times 10^{-24}$ A²/Hz when TNOIMOD = 0. As TNOIMOD is switched to 1, indeed $\overline{I_d^2}$ is found to decrease by roughly 10%, to a value of 1.215×10^{-24} A²/Hz. We conclude that the holistic model passes our first test.

The second test circuit is again rather simple, as shown in Fig. 5-19. With $V_{GS} = 1.5$ V and $V_{DD} = 3.3$ V, the transistor is biased in the saturation region. The drain end is deliberately connected to a voltage source, so that all of the drain current thermal noise $(\overline{i_d^2})$ as well as the $1/f$ noise flows to the ground, and does not affect

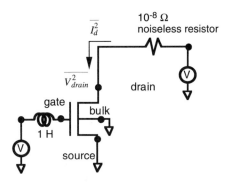

Fig. 5-18 The first circuit used to test the equivalency between BSIM4's TNOIMOD = 1 model (such as Fig. 5-17a) and the van der Ziel representation of Fig. 4-65b.

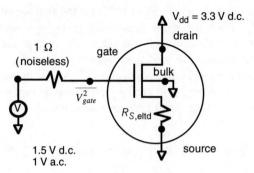

Fig. 5-19 The second circuit used to test the equivalency between BSIM4's TNOIMOD=1 model and the van der Ziel representation.

the noise voltage at the gate, $\overline{V_{gate}^2}$. This is a classic example in which the induced gate noise must be included to produce the correct $\overline{V_{gate}^2}$. If either BSIM3's thermal noise model or BSIM4's TNOIMOD = 0 model is used, $\overline{V_{gate}^2}$ will be identically zero. When BSIM4's holistic model (TNOIMOD = 1) is used, we obtain a nonzero $\overline{V_{gate}^2}$ as shown in Fig. 5-20. Let us see if we can quantitatively reproduce BSIM4 simulation results with a hand analysis based on van der Ziel's representation. For simplicity, we consider the case when $R_{S,\text{eltd}} = 10^{-3}$ Ω, which is ≈ 0, so the equivalent noise circuit is that shown in Fig. 4-66b (in the case L_D is zero and the drain is shorted to ground). It is clear that $\overline{V_{gate}^2} = \overline{i_g^2} \times (1\,\Omega)^2$. The induced gate noise current, in saturation, was given by Eq. (4-59). From a separate SPICE run, we find the device parameters at $R_{S,\text{eltd}} = 10^{-3}$ Ω are: $g_d = 7.7 \times 10^{-8}$ Ω^{-1}; $g_m = 8.75 \times 10^{-5}$ Ω^{-1}; $C_{gs} = 9.4 \times 10^{-13}$ F; $C_{gd} \approx 0$ F; $I_D = 4.23 \times 10^{-5}$ A. At 10^9 Hz,

$$\overline{i_g^2} = 4 \cdot 1.38 \times 10^{-23} \cdot 300 \cdot \frac{4}{15} \frac{(2\pi \cdot 10^9)^2 (9.4 \times 10^{-13})^2}{8.75 \times 10^{-5}} = 1.76 \times 10^{-21} \frac{A^2}{Hz}.$$
(5-82)

Consequently, $\overline{V_{gate}^2} = 1.76 \times 10^{-21}$ V^2/Hz, a value fairly close to the BSIM4 calculation shown in Fig. 5-20. When the frequency is 10^6 Hz, it is straightforward to determine that $\overline{V_{gate}^2} = 1.76 \times 10^{-27}$ V^2/Hz, again close to the BSIM4 result.

It is instructive to derive the above results using BSIM4's holistic equivalent noise circuit shown in Fig. 5-17 (more specifically, Fig. 5-17a since Fig. 5-20 was obtained with RDSMOD = 0). The circuit of Fig. 5-19, with the replacement of the transistor by its equivalent, becomes that shown in Fig. 5-21a. The device is biased in saturation so $C_{gd} \approx 0$ and is omitted from the figure. Further, because the impedance looking into the source of the device is much larger than $R_{S,\text{eltd}}$, all of the noise current $\overline{i_{RS}^2}$ flows through $R_{S,\text{eltd}}$ and the noise voltage at the source node is relatively constant. This is represented in Fig. 5-21b, which replaces the parallel combination of $R_{S,\text{eltd}}$ and $\overline{i_{RS}^2}$ by $\overline{V_{source}^2}$.

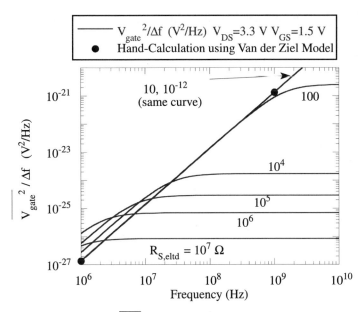

Fig. 5-20 BSIM4-calculated $\overline{V_{gate}^2}$ of Fig. 5-19 when TNOIMOD = 1. Two hand-calculated values (see text) from the van der Ziel representation are shown as dots.

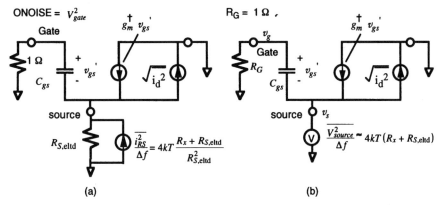

Fig. 5-21 (a) The circuit of Fig. 5-19 after the transistor is replaced by BSIM4's noise equivalent circuit of Fig. 5-17a; (b) redrawing of (a) by replacing the noise current source in the source by a noise voltage source. The noise sources of BSIM4, at the drain and at the source, are uncorrelated.

Summing the current flow at the gate node, we can write $-v_g/R_G = j\omega C_{gs}(v_g - v_s)$. Therefore,

$$\frac{\overline{V_{gate}^2}}{\Delta f} = \frac{\overline{V_{source}^2}}{\Delta f} \times \left| \frac{j\omega C_{gs} R_G}{1 + j\omega C_{gs} R_G} \right|^2. \qquad (5\text{-}83)$$

As mentioned, the device has a C_{gs} of 9.4×10^{-13} F. With R_G being $1\,\Omega$, $j\omega C_{gs}$ is much smaller than 1 for frequencies below 10 GHz. We will therefore approximate $\overline{V_{gate}^2}$ as

$$\frac{\overline{V_{gate}^2}}{\Delta f} = 4kT(R_x + R_{S,\text{eltd}}) \times (\omega C_{gs} R_G)^2. \qquad (5\text{-}84)$$

Let us evaluate R_x. For the long-channel device under consideration, the velocity saturation is not a significant effect. $\varepsilon_{sat} \cdot L_{eff}$ is a large quantity, so Eq. (5-80) reduces to $\theta = 0.37$. Substituting this value into Eq. (5-78), with the knowledge that $V_{DS,\text{eff}} \sim V_{DS,\text{sat}} = 0.664$ V and $I_{DS} = 4.22 \times 10^{-5}$ A, we find R_x to be $2.32 \times 10^3\,\Omega$. This is certainly much larger than $R_{S,\text{eltd}} = 10^{-3}\,\Omega$. At 10^9 Hz, according to Eq. (5-84) is

$$\frac{\overline{V_{gate}^2}}{\Delta f} = 4 \cdot 1.38 \times 10^{-23} \cdot 300 \cdot 2.32 \times 10^3 \cdot (2\pi \cdot 10^9)^2 (9.4 \times 10^{-13})^2.$$

$\overline{V_{gate}^2}$ is calculated to be 1.34×10^{-21} V^2/Hz with the holistic equivalent circuit, a value which compares favorably with the 1.76×10^{-21} V^2/Hz calculated with the van der Ziel representation.

Thus far we have shown that BSIM4's noise equivalent circuit yields the same or similar results as the van der Ziel representation. We now go one step further. We shall show that the moment the induced gate noise is included, the minimum noise figure ceases to be 0 dB in an intrinsic transistor without R_G. We first mentioned the simulation artifact in Section 4.13, that a BSIM3 model for a transistor without extrinsic resistances produces 0 dB minimum noise figure. We further demonstrated that, with the inductor value given in Eq. (4-68), the circuit shown in Fig. 4-73 yields a noise figure (and thus the minimum noise figure) of 0 dB. In the following we shall prove that when the induced gate noise is incorporated, the circuit of Fig. 4-73 will no longer produce a 0 dB noise figure.

We use the van der Ziel representation to illustrate this point. When the induced gate noise enters the picture, we redraw Fig. 4-73b in Fig. 5-22. We use the superposition principle to determine the total noise current flowing through the drain node. We shall approximate $g_m^\dagger$ by g_m, and also keep in mind that the inductor L is

5.15 THERMAL NOISE MODELS

designed such that $j\omega L + 1/(j\omega C) = 0$. When the device noises $\overline{i_d^2}$ and $\overline{i_g^2}$ are omitted, the drain current noise due to $\overline{i_{RP}^2}$ is given by

$$\sqrt{\frac{\overline{I_d^2}}{\Delta f}}\bigg|_{\text{due to } \overline{i_{RP}^2}} = \frac{g_m}{\omega C}\sqrt{\frac{\overline{i_{RP}^2}}{\Delta f}}, \qquad (5\text{-}85)$$

where $\overline{i_{RP}^2}$ was given by Eq. (4-70). In the second step of applying the superposition principle, we focus on the noise contribution from $\overline{i_d^2}$ and $\overline{i_g^2}$ of the transistor, and omit the contribution from $\overline{i_{RP}^2}$. The drain node current in general can be expressed as

$$I_d = i_d + i_g\left[\frac{1}{j\omega C}//(j\omega L + R_P)\right] \times g_m = i_d + i_g\frac{g_m}{j\omega C} + i_g\frac{g_m L}{R_P C}. \qquad (5\text{-}86)$$

Hence, the drain noise current due to $\overline{i_d^2}$ and $\overline{i_g^2}$ is

$$\overline{I_d^2}\bigg|_{\text{due to device}} = \overline{\left(i_d + i_g\frac{g_m}{j\omega C} + i_g\frac{g_m L}{R_P C}\right) \cdot \left(i_d^* - i_g^*\frac{g_m}{j\omega C} + i_g^*\frac{g_m L}{R_P C}\right)}$$

$$= \overline{i_d^2} + \overline{i_g^2}\frac{g_m^2}{\omega^2 C^2} + \overline{i_g^2}\frac{g_m^2 L^2}{\omega^2 R_P^2} + 2\operatorname{Re}\left(\overline{i_g i_d^*}\frac{g_m}{j\omega C}\right) + 2\operatorname{Re}\left(\overline{i_g i_d^*}\frac{g_m L}{R_P C}\right)$$

$$= \overline{i_d^2} + \overline{i_g^2}\frac{g_m^2}{\omega^2 C^2} + \overline{i_g^2}\frac{g_m^2 L^2}{\omega^2 R_P^2} - 2\sqrt{\frac{5}{32}}\frac{g_m}{\omega C}\sqrt{\overline{i_d^2}}\sqrt{\overline{i_g^2}} - 2\sqrt{\frac{5}{32}}\frac{g_m L}{R_P C}\sqrt{\overline{i_d^2}}\sqrt{\overline{i_g^2}}, \qquad (5\text{-}87)$$

where the correlation coefficient of $\overline{i_d^2}$ and $\overline{i_g^2}$, $c = -j(5/32)^{1/2}$, was used. In saturation, $\overline{i_d^2}$ and $\overline{i_g^2}$ are given by Eqs. (4-58) and (4-59), respectively.

Substituting these quantities into Eq. (5-87), we obtain

$$\frac{\overline{I_d^2}}{\Delta f}\bigg|_{\text{due to device}} = 4kT\left[\frac{2}{3}g_m + \frac{4}{15}g_m - \frac{1}{3}g_m + \frac{4}{15}\frac{\omega^2 L^2}{R_P^2}g_m - \frac{1}{3}\frac{\omega L}{R_P}g_m\right]. \qquad (5\text{-}88)$$

The ac voltage gain is similar to that given in Eq. (4-72):

$$A_v = \left|\frac{v(\text{drain})}{v_P}\right| = \frac{R_L}{R_P} \times \frac{g_m}{\omega C}. \qquad (5\text{-}89)$$

448 IMPROVEMENTS IN BSIM4

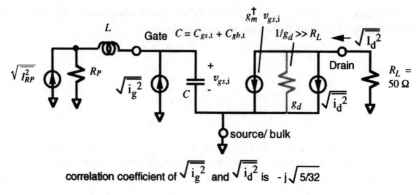

Fig. 5-22 The circuit of Fig. 4-73a after the transistor is replaced by the van der Ziel representation. The original circuit yielded a minimum noise figure of 0 dB in BSIM3 (Section 4.13), but gives rise to reasonable value with BSM4 which includes the induced gate noise source. The correlation coefficient is negative because the currents in the drain and gate noise sources are drawn to be of opposite directions.

Using the noise figure definition of Eq. (4-67), we can write it as

$$NF = \frac{(\overline{I_d^2}|_{\text{due to } i_{RP}^2} + \overline{I_d^2}|_{\text{due to device}})R_L^2}{4kT_0 R_P} \times \frac{1}{A_v^2}$$

$$= \frac{T}{T_0} + \frac{T}{T_0} \frac{\omega^2 C^2 R_P}{g_m} \times \left(\frac{3}{5} + \frac{4 \omega^2 L^2}{15 R_P^2} - \frac{1}{3} \frac{\omega L}{R_P}\right). \quad (5\text{-}90)$$

We perform the calculation on a $W/L = 20/0.5$ μm device, whose device parameters at $V_{GS} = 1.5$ V and $V_{DS} = 3.3$ V are: $C_{gd} \sim 0$, $C = C_{gs} + C_{gb} = 1.7 \times 10^{-14}$ F, $g_m = 3.6 \times 10^{-3}$ Ω^{-1}. At 2 GHz, L is designed to be 3.727×10^{-7} H. Using $R_L = R_P = 50$ Ω, $T = 300$ K and $T_0 = 290$ K, we find NF $= 1.034 + 1.034 \cdot 6.33 \times 10^{-4} \cdot (2244) = 4$ dB.

Figure 5-23 contrasts the noise figures calculated for the said $W/L = 20/0.5$ μm device, with TNOIMOD $= 0$ and 1 models. The source and drain resistances are accounted for by RDSW. When TNOIMOD $= 0$, $R_{S,\text{eltd}} = 0$ and there is no noise source associated with the source and drain resistances. When TNOIMOD $= 1$, $R_{S,\text{eltd}}$ is defaulted to 0.001 Ω, as given by Eqs. (5-61). Although $R_{S,\text{eltd}}$ is small, $i_{RS}^2/\Delta f$ given in Eq. (5-77) is significant since it incorporates transformed contribution from the gate induced noise through R_x. When TNOIMOD $= 0$, the noise figure can be fairly close to 0 dB. However, as the induced gate noise is included in the TNOIMOD $= 1$ model, the noise figure is some finite number.

The equivalent circuits in Fig. 5-17 apply to the normal operating condition of $V_{DS} > 0$. Because of the asymmetry between the noise sources $i_{RS}^2/\Delta f$ and $i_{RD}^2/\Delta f$, it matters to the holistic model whether user-defined source-terminal is physically the source of the transistor. When $V_{DS} < 0$, the two noise sources need to be swapped. Figure 5-24 displays the equivalent circuit for RDSMOD $= 1$ when $V_{DS} < 0$.

5.15 THERMAL NOISE MODELS

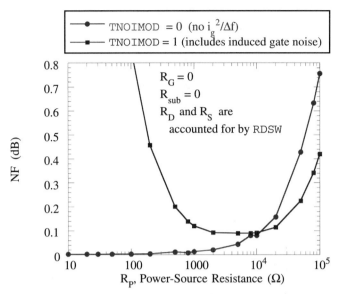

Fig. 5-23 Noise figures calculated for the circuit of Fig. 4-73a, when $W/L = 20/0.5\,\mu\text{m}$. `TNOIMOD` = 0 does not have induced gate noise and suffers from the same 0 dB result as BSIM3. `TNOIMOD` = 1 model of BSIM4 results in correct noise figures.

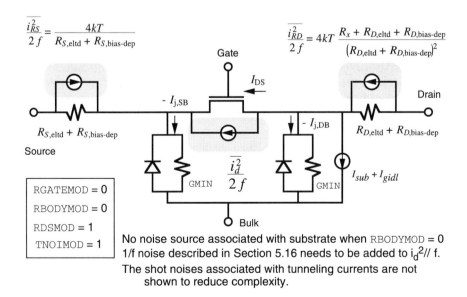

Fig. 5-24 BSIM4's noise equivalent circuit for `TNOIMOD` = 1 when $V_{DS} < 0$. `RDSMOD` = 1.

Although this is the longest section in Chapter 5, the parameters relating to the thermal noise models are surprisingly few.

Parameter Name	Description	Default Value
TNOIMOD	Thermal noise model selector	0
NTNOI	Noise factor for short-channel devices in TNOIMOD = 0	1.0
TNOIA	L-dependence coefficient in TNOIMOD = 1	1.5
TNOIB	Noise partition parameter for L-dependence in TNOIMOD = 1	1.0

Replaced BSIM3 Parameters: NOIMOD

5.16 FLICKER NOISE MODELS

BSIM4 provides the same two Flicker noise models available in BSIM3, the SPICE-Flicker noise model (FNOIMOD = 0) and the BSIM-Flicker noise model (FNOIMOD = 1). The SPICE-Flicker noise expression in BSIM4 is given by

$$\overline{i_d^2}|_{Flicker} = KF \times \frac{(I_{DS})^{AF}}{C'_{ox,e}L_{eff}^2 f^{EF}} \Delta f. \quad \text{(SPICE-Flicker)} \quad (5-91)$$

The SPICE-Flicker noise model in BSIM3 was given by Eq. (3-77). A comparison shows that Eqs. (5-91) and (3-77) are identical, except for the difference in use of C_{ox} in BSIM3 and $C_{ox,e}$ in BSIM4. This is a natural consequence of the fact that BSIM4 adopts the electrical oxide thickness for most of its capacitance calculations, while BSIM3 does not make a concerted effort to distinguish between the physical and the electrical oxide thicknesses (Section 5.2).

The BSIM-Flicker noise model in BSIM4 also follows the basic framework of BSIM3, but with the improvements of using smoothing functions and considering the bulk charge effect. Just as the BSIM3 expression given in Eq. (3-79), the BSIM-Flicker noise in BSIM4 can be represented as

$$\overline{i_d^2}|_{Flicker} \approx f'(\text{NOIA}, \text{NOIB}, \text{NOIC})$$
$$\times \left[\frac{I_{DS}}{C'_{ox,e}L_{eff}^2 f^{EF}} + \frac{g'(\text{EM})I_{DS}}{W_{eff}L_{eff}^2 f^{EF}} \right] \Delta f \quad \text{(BSIM-Flicker)}, \quad (5-92)$$

where f' is some function of the SPICE parameters NOIA, NOIB, NOIC, and g' is some function of the SPICE parameter EM. NOIA, NOIB, and NOIC can be treated as fitting parameters, just like KF in the SPICE2-Flicker noise model given in Eq. (5-91). We use primes to emphasize that BSIM4's f' and g' of Eq. (5-92) differ somewhat from BSIM3's f and g of Eq. (3-79). However, it is not important to find out the details of their difference; the exact expressions for f and g can be found in

Appendix A, and f' and g', in Appendix G. The important fact to realize is that, because of the change in these functions, the values of NOIA, NOIB, and NOIC of a BSIM3 model can not be directly transferred to a BSIM4 model. The definitions of NOIA, NOIB, and NOIC all have been modified. In fact, these parameters are unitless in BSIM3, but have some rather complicated units in BSIM4. EM is a parameter that retains the same meaning in both BSIM3 and BSIM4. It represents the field at which the carrier velocity saturates. Despite its name, EM is a parameter that affects only the noise models, but not the I-V or the C-V characteristics.

In BSIM4, the Flicker noise expression given in Eq. (5-91) or Eq. (5-92) is simply added into $i_d^2/\Delta f$ in the noise equivalent circuits of Figs. 5-16 and 5-17. This addition of Flicker noise to the thermal noise to form the overall intrinsic device noise is similarly done in BSIM3.

Parameter Name	Description	Default Value
FNOIMOD	Flicker noise model selector	1
EF†	Flicker noise frequency exponent in FNOIMOD = 0 model	1
KF†	Flicker noise coefficient in FNOIMOD = 0 model	0 ($A^{2-AF} \cdot s^{1-EF} \cdot F$)
NOIA††	First Flicker noise parameter in FNOIMOD = 1 model default: NMOS = 6.25×10^{41}; PMOS = 6.188×10^{40} ($s^{1-EF}/eV/m^3$)	
NOIB††	Second Flicker noise parameter in FNOIMOD = 1 model default: NMOS = 3.125×10^{26}; PMOS = 1.5×10^{25} ($s^{1-EF}/eV/m$)	
NOIC††	Third Flicker noise parameter in FNOIMOD = 1 model default: 8.75×10^9 ($s^{1-EF} \cdot m/eV$)	
EM†	Saturation field in FNOIMOD = 1 model	4.1×10^7 (V/m)

† Same as BSIM3 parameters, repeated here for convenience.
†† Same spelling as BSIM3 parameters, but appearing in different equations from BSIM3's.
Replaced BSIM3 Parameters: NOIMOD

5.17 NON-QUASI-STATIC AC MODEL

BSIM3 offers a non-quasi-static (NQS) model for transient analysis, which can be invoked by setting the parameter NQSMOD to 1. (Although NQSMOD = 1 can be used together in an ac analysis, the implementation there is likely incorrect. Therefore, the NQSMOD = 1 model in BSIM3 should not be used in an ac analysis.) The transient NQS model equations were detailed in Appendix E. Any attempt to NQS modeling faces the difficulty that there is no general solution to the partial differential equation governing the MOS dynamics (Eq. 1-25). The quasi-static (QS) approximation, discussed in Section 1.6, is an effective method to deliver a good solution without significant computation cost. Most simulations can be handled effectively with the QS approximation, which comprises the bulk of all compact MOS models including BSIM3 and BSIM4. Only occasionally will there be a need to select the NQS option.

A NQS model removes the underlying assumption of the QS approximation, that the channel charge could respond instantaneously to the change in terminal voltages. The degree of the NQS model accuracy is a tradeoff between the computation efficiency and model complexity. The most accurate NQS model would probably require a partial equation solver, which is beyond the capability of standard SPICE simulators. (SPICE simulators can solve ordinary differential equations with one unknown variable, usually time in the transient analysis.) We therefore do not expect BSIM3 NQS model, or any compact NQS model, to be accurate in all regards. We have discussed some of the shortcomings of the BSIM3 NQS model in the NQSMOD entry in Section 3.2, as well as in Section 4.1.

BSIM4 uses two different NQS model selectors to turn on its NQS portion of the code. Setting TRNQSMOD = 1 invokes BSIM4's transient NQS model for transient analysis, and setting ACNQSMOD = 1 invokes BSIM4's ac NQS model for ac analysis. The TRNQSMOD model is nearly identical to, and hence shares the problems with, the transient NQS model of BSIM3. Their only difference relates to the relaxation time expression. According to Eq. (E-9) in Appendix E, the relaxation time in BSIM3 can be viewed to be

$$\tau = \frac{L_{\text{eff,CV}}^2}{\text{ELM}\mu_n(v_{GS} - V_T) + 16\mu_n \times kT/q}. \tag{5-93}$$

The exact expression in the code is more complicated, mainly to take care of smoothing of $v_{GS} - V_T$ among various regions of operation. ELM is the NQS parameter adjusting the relaxation time. In BSIM4, a completely new relaxation time expression is used. It is related to the channel-reflected gate resistance given by Eq. (5-68).

$$\tau = R_{G,\text{crg}} \times W_{\text{eff,CV}} L_{\text{eff,CV}} C'_{ox,e} \times \text{NF} \tag{5-94}$$

Instead of ELM, BSIM4 relies on XRCRG1 and XRCRG2 to adjust the time constant through modifying $R_{G,\text{crg}}$. We shall refer the readers to Appendix E for other details of the transient NQS model in BSIM4. In addition, because the channel-reflected gate resistance has been accounted for in either the transient or the ac NQS model, RGATEMOD of 2 and 3 cannot be selected if either TRNQSMOD or ACNQSMOD is equal to 1.

The ACNQSMOD = 1 model in BSIM4, to be used in an ac analysis, is completely rewritten. The implementation errors thought to be associated with BSIM3's ac NQS model are eliminated. The ac NQS model is developed from the same fundamental equations governing as the transient NQS model. However, the ac NQS model does not require the internal NQS charge node required by the transient model.

The ac NQS model's major advantage over the QS model is the removal of the unphysical increase of device's transconductance as frequency increases. This QS model artifact was detailed in Section 4.4, and is briefly reviewed here. In a BSIM4

5.17 NON-QUASI-STATIC AC MODEL

small-signal equivalent model, which is similar to BSIM3 equivalent model given in Fig. 2-21, the transconductance of the output current source due to the input voltage has some frequency dependence. In order to avoid confusion between this frequency-dependent transconductance and the device transconductance ($g_m = \partial I_D/\partial V_{DS}$), we have denoted the former as $g_m^\dagger$ and kept the latter as g_m. In BSIM4, $g_m^\dagger$ relates to g_m according to (see Eq. 4-9):

$$g_m^\dagger = g_m - j\omega(C_{dg} - C_{gd}). \quad (5\text{-}95)$$

In the QS model, g_m, C_{dg}, and C_{gd} are all independent of frequency. Once a set of bias conditions is known, their values are determined. It is obvious that, as frequency increases, $|g_m^\dagger|$ in the QS model increases with frequency, clearly an unphysical event. In the ac NQS model, in contrast, the above three quantities display some frequency dependence through the relaxation time τ:

$$g_m = \frac{g_{m,0}}{1+\omega^2\tau^2} - \frac{2\omega^2\tau I_{DS,0}}{(1+\omega^2\tau^2)^2}\frac{\partial \tau}{\partial v_{gs}}$$
$$- j\left[\frac{g_{m,0}\omega\tau}{1+\omega^2\tau^2} + \frac{\omega I_{DS,0}(1-\omega^2\tau^2)}{(1+\omega^2\tau^2)^2}\frac{\partial \tau}{\partial v_{gs}}\right]. \quad (5\text{-}96)$$

$$C_{dg} = \frac{C_{dg,0}}{1+\omega^2\tau^2} - \frac{2\omega^2\tau Q_{D,0}}{(1+\omega^2\tau^2)^2}\frac{\partial \tau}{\partial v_{gs}}$$
$$- j\left[\frac{C_{dg,0}\omega\tau}{1+\omega^2\tau^2} + \frac{\omega Q_{D,0}(1-\omega^2\tau^2)}{(1+\omega^2\tau^2)^2}\frac{\partial \tau}{\partial v_{gs}}\right]. \quad (5\text{-}97)$$

$$C_{gd} = \frac{C_{gd,0}}{1+\omega^2\tau^2} - \frac{2\omega^2\tau Q_{G,0}}{(1+\omega^2\tau^2)^2}\frac{\partial \tau}{\partial v_{gs}}$$
$$- j\left[\frac{C_{gd,0}\omega\tau}{1+\omega^2\tau^2} + \frac{\omega Q_{G,0}(1-\omega^2\tau^2)}{(1+\omega^2\tau^2)^2}\frac{\partial \tau}{\partial v_{gs}}\right]. \quad (5\text{-}98)$$

Because all three quantities roll off with frequency, we expect $|g_m^\dagger|$ to decrease with frequency.

Figure 4-24 is a useful circuit to test $|g_m^\dagger|$'s frequency dependence. As demonstrated in Fig. 4-25, a QS model predicts $|g_m^\dagger|$ to increase with frequency when the drain–bulk junction capacitance is negligible. To demonstrate the effectiveness of BSIM4's ac NQS model, we simulate the circuit of Fig. 4-24 with appropriate change in parameters so that $C_{j,DB}$ is zero. Figure 5-25 displays $|v_o/v_i|$ for a $L = 0.5\,\mu\text{m}$ device. With the NQS modeling, $|g_m^\dagger|$ and hence $|v_o/v_i|$ decreases with frequency as expected. Figure 5-26 is the simulation result of a $L = 10\,\mu\text{m}$ device. When a device's channel length increases, the corner frequency at which NQS effects become significant is lowered (see Eq. 1-36). Figure 5-26 is consistent with

454 IMPROVEMENTS IN BSIM4

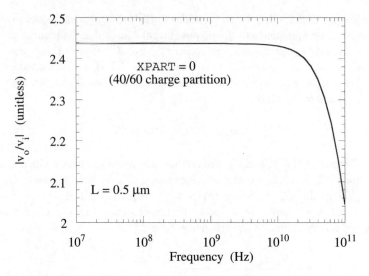

Fig. 5-25 BSIM4-calculated $|v_o/v_i|$ for the circuit of Fig. 4-24, with $L=0.5\,\mu m$.

this fact, showing that the gain rolloff frequency scales roughly with L^{-2}. The low-frequency voltage gain is lower in this $L = 10\,\mu m$ device because g_m is smaller in the longer channel device.

The ac NQS model in BSIM4 works seamlessly with a noise analysis, for both the TNOIMOD $= 0$ and 1 models. Figure 5-27 illustrates BSIM4-simulated voltages and noise quantities for the 2-stage circuit shown in Fig. 4-74. In this simulat-

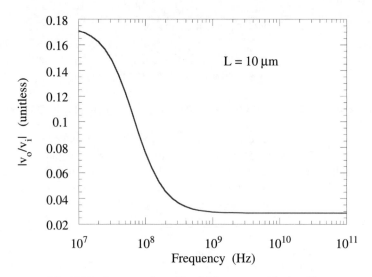

Fig. 5-26 Same result as Fig. 5-25, except with $L = 10\,\mu m$.

5.17 NON-QUASI-STATIC AC MODEL   455

ion, TNOIMOD = 0 and ACNQSMOD = 1. Figure 5-28 is a resimulation with TNOIMOD = 1 (the holistic thermal model) and ACNQSMOD = 1.

A useful test to verify the NQS model's implementation is to compare some voltage magnitudes simulated by an ac analysis and by a transient analysis. In the following discussion, XPART is set to 1 for the 0/100 charge partitioning. The two circuits shown in Figs. 5-29a and 5-29b are basically the same. In one, an ac analysis is performed with v_{GS} equated to a dc component of 1.5 V plus a 1 MHz ac component of 0.01 V. In another, a transient analysis is performed with v_{GS} specified as $1.5 + 0.01 \sin(2\pi \cdot 10^6 t)$. These two analyses should yield the same $|v(\text{drain})|$, for example, after the steady state is achieved in the transient analysis. This statement should remain correct whether a QS or a NQS model is used. As a starting point, we verify that the same $|v(\text{drain})|$ is obtained when ACNQSMOD = TRNQSMOD = 0 (i.e., QS model). The ac analysis yields a $|v(\text{drain})|$ of 0.0154 V. Figure 5-30 displays $v(\text{drain})$ as a function of time, calculated with the transient analysis. As a reference, the voltage waveform of $|v(\text{gate})|$ is also shown. We see that $v(\text{drain})$ oscillates with the same forcing frequency of 1 MHz, with a peak-to-peak value of 0.0308 V, which is exactly twice the 0.0154 V calculated from the ac analysis. Therefore, we conclude that BSIM4's QS model passes this consistency test.

Next, we test the NQS model. We set ACNQSMOD = 1 in the ac analysis, and TRNQSMOD = 1 in the transient analysis. Because 1 MHz is too small a frequency for NQS effects to become significant in the 0.5 μm device, we find again the ac analysis yields a $|v(\text{drain})|$ of 0.0154 V, and the transient analysis yields the same plot as Fig. 5-30. At this point, we conclude that BSIM4's NQS model passes the consistency test when the NQS effects are not expected to be significant.

Finally, we test the NQS model when NQS effects are expected to be important. We change the channel lengths of the devices in Fig. 5-29 from 0.5 μm to 50 μm. Moreover, we increase the operating frequency from 1 MHz to 100 MHz. The ac analysis with ACNQSMOD = 1 determines $|v(\text{drain})|$ to be 0.004573 V. The result from the transient analysis with TRNQSMOD = 1 is shown in Fig. 5-31. The peak-to-peak value of $|v(\text{drain})|$ in the transient analysis is 0.0487 V, which greatly exceeds 2×0.00573 V from the ac analysis. The difference is found to be smaller when XPART is changed to 0. However, it is still alarming to see such a drastic difference at XPART = 1. We therefore conclude that there is some inconsistency between the ac NQS and transient NQS models when the NQS effects are very significant. In our example, the operating frequency of 100 MHz is roughly 30 times the cutoff frequency of the $L = 50$ μm device. It is believed that the transient NQS model is just not accurate enough at this high a frequency. There will not be too much interest at this kind of frequency anyway (for the $L = 50$ μm device).

Parameter Name	Description	Default Value
TRNQSMOD	Transient NQS model selector	0 (off)
ACNQSMOD	ac NQS model selector	0 (off)

Replaced BSIM3 Parameters: ELM

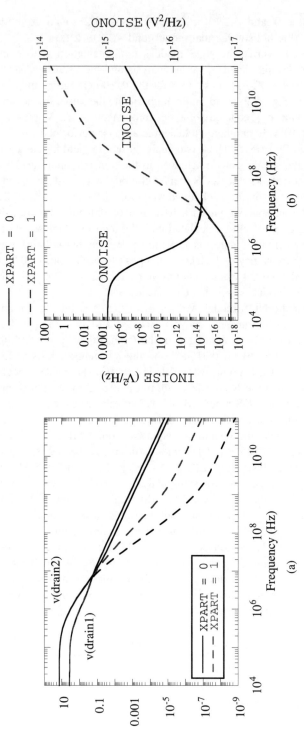

Fig. 5-27 BSIM4-calculated voltages and noise quantities for the two-stage circuit shown in Fig. 4-74. TNOIMOD = 0 and ACNQSMOD = 1.

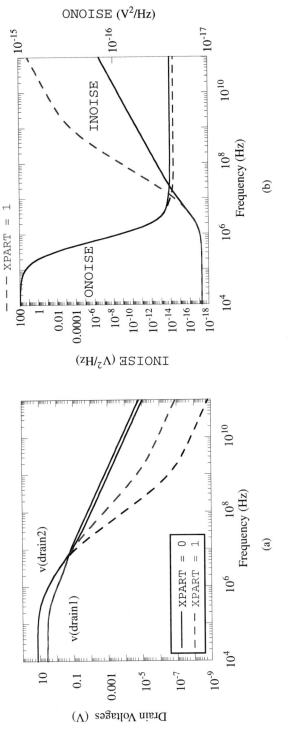

Fig. 5-28 Same result as Fig. 5-27, except with TNOIMOD = 1.

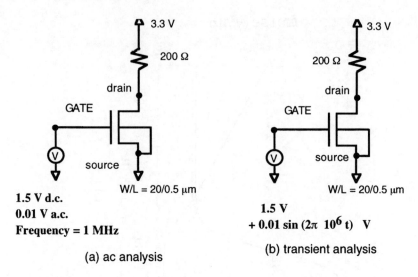

Fig. 5-29 Circuits used to test BSIM4's equivalency of transient and ac small-signal models. (a) ac analysis circuit; (b) transient analysis circuit.

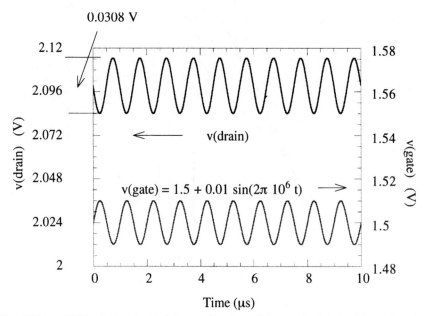

Fig. 5-30 BSIM4-calculated v(drain) as a function of time, calculated with the transient analysis, TRNQSMOD $=1$ (Fig. 5-29b). $L=0.5\,\mu\text{m}$; oscillating frequency is 1 MHz.

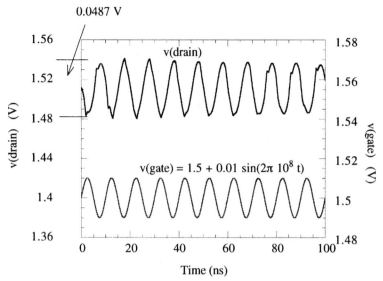

Fig. 5-31 Same result as Fig. 5-30, except with $L = 50\,\mu\text{m}$, and oscillating frequency is 100 MHz.

5.18 GATE TUNNELING CURRENTS

As the oxide becomes progressively thinner in each new generation of technology, the magnitudes of tunneling currents through the oxide become more significant. BSIM4 considers five tunneling components, as shown in Fig. 5-32. $I_{gd,\text{tunnel}}$ is the tunneling current between the gate and the heavily doped drain junction. $I_{gcd,\text{tunnel}}$ denotes the tunnel current from the gate to the channel and then to the drain. Likewise, $I_{gs,\text{tunnel}}$ and $I_{gcs,\text{tunnel}}$ are similar currents, but associated with the source junction. Finally, $I_{gb,\text{tunnel}}$ is the component that tunnels between the gate and the bulk. A thorough analysis of these tunneling currents unavoidably involves quantum mechanics. We point out some references for the physics of tunneling [11,12], but shall concentrate on the BSIM4 formulation of these currents.

One key intermediate variable in the calculation of tunneling currents is the voltage dropping across the oxide. It is given by

$$V_{ox} = V_{FB,\text{tunnel}} - V_{FB,\text{eff\_tunnel}} + \text{K1} \cdot \sqrt{\phi_{s,\text{dep}}} + V_{GST,\text{eff}}. \qquad (5\text{-}99)$$

$V_{FB,\text{tunnel}}$, the flatband voltage for tunneling-current calculation, is chosen to be $V_{fb,\text{zb}}$, defined in Eq. (3-111). $V_{FB,\text{eff\_tunnel}}$ is the flatband smoothing function, made to approach $V_{FB,\text{tunnel}}$ when $V_{GB} \ll V_{FB,\text{tunnel}}$, and V_{GS} when $V_{GB} \gg V_{FB,\text{tunnel}}$. This smoothing function is similar to $V_{FB,\text{effCV}}$, the flatband smoothing function for CV calculation, described in Section 1.4. $\phi_{s,\text{dep}}$ is the surface potential. We add a subscript, dep, to emphasize that it is meant to be applicable in depletion region, as well as strong inversion. If the transistor operates purely in strong inversion, we

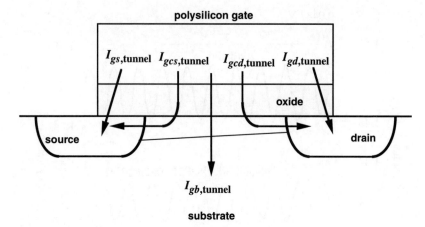

Fig. 5-32 Five dc tunneling current components considered in BSIM4. There are no ac or transient tunneling components.

could have replaced $\phi_{s,\text{dep}}$ by $2\phi_f$, the strong-inversion surface potential used throughout the book. The equation for the surface potential is

$$\sqrt{\phi_{s,\text{dep}}} = \sqrt{\frac{K1^2}{4} + (V_{GB} - V_{FB,\text{eff\_tunnel}} - V_{GST,\text{eff}})} - \frac{K1}{2}. \qquad (5\text{-}100)$$

Finally, $V_{GST,\text{eff}}$ is given by Eq. (1-14). It is the smoothing function used to smooth the device characteristics from the subthreshold region to the strong inversion region. $V_{GST,\text{eff}}$ decreases exponentially toward zero when $V_{GS} < V_T$, but approaches $V_{GS} - V_T$ in the strong inversion region.

It turns out that it is advantageous to separate V_{ox} into two terms, that associated with the accumulation region ($V_{ox,\text{acc}}$), and the depletion/inversion region ($V_{ox,\text{depinv}}$):

$$V_{ox,\text{acc}} = V_{FB,\text{tunnel}} - V_{FB,\text{eff\_tunnel}}. \qquad (5\text{-}101)$$

$$V_{ox,\text{depinv}} = K1 \cdot \sqrt{\phi_{s,\text{dep}}} + V_{GS,\text{eff}}. \qquad (5\text{-}102)$$

Once the oxide potentials are established, all of the tunneling components are written in a similar fashion. In order to maximize flexibility in turning on various components, BSIM4 utilizes two model selectors. When IGBMOD has its default value of 0, $I_{gb,\text{tunnel}}$ is left zero. $I_{gb,\text{tunnel}}$ is calculated only when IGBMOD $\neq 0$. Similarly, when IGCMOD $= 0$, all $I_{gcs,\text{tunnel}}$, $I_{gs,\text{tunnel}}$, $I_{gcd,\text{tunnel}}$ and $I_{gd,\text{tunnel}}$ are set to 0. They are calculated only when IGCMOD $\neq 0$. Let us write down the expression of $I_{gb,\text{tunnel}}$ first. This tunneling current, flowing between the gate and the bulk terminal,

5.18 GATE TUNNELING CURRENTS

is broken down into that associated with the accumulation region ($I_{gb,\text{acc}}$) and that associated with the inversion region ($I_{gb,\text{inv}}$):

$$I_{gb,\text{tunnel}} = \begin{cases} I_{gb,\text{acc}} + I_{gb,\text{inv}} & \text{if } \texttt{IGBMOD} \neq 0; \\ 0 & \text{if } \texttt{IGBMOD} = 0; \end{cases} \quad (5\text{-}103)$$

where

$$I_{gb,\text{acc}} = \text{NF } W_{\text{eff}} L_{\text{eff}} \frac{A1}{\text{TOXE}^2} \left(\frac{\text{TOXREF}}{\text{TOXE}}\right)^{\text{NTOX}} V_{GB,\text{eff\_tunnel}} \times \texttt{NIGBACC} \frac{kT}{q}$$

$$\ln\left[1 + \exp\left(-\frac{q(V_{GB,\text{eff\_tunnel}} - V_{FB,\text{CV}})}{kT \cdot \texttt{NIGBACC}}\right)\right] \times \exp[-B1 \cdot \text{TOXE}(\texttt{AIGBACC}$$

$$- \texttt{BIGBACC} V_{ox,\text{acc}}) \cdot (1 + \texttt{CIGBACC} V_{ox,\text{acc}})] \quad (5\text{-}104)$$

$$I_{gb,\text{inv}} = \text{NF } W_{\text{eff}} L_{\text{eff}} \frac{A2}{\text{TOXE}^2} \left(\frac{\text{TOXREF}}{\text{TOXE}}\right)^{\text{NTOX}} V_{GB,\text{eff\_tunnel}} \times \texttt{NIGBINV} \frac{kT}{q}$$

$$\ln\left[1 + \exp\left(-\frac{q(V_{ox,\text{depinv}} - \texttt{EIGBINV})}{kT \cdot \texttt{NIGBINV}}\right)\right] \times \exp[-B2 \cdot \text{TOXE}(\texttt{AIGBINV}$$

$$- \texttt{BIGBINV} V_{ox,\text{depinv}}) \cdot (1 + \texttt{CIGBINV} V_{ox,\text{depinv}})]. \quad (5\text{-}105)$$

The coefficients $A1$ and $B1$ are 4.97232×10^{-7} A/V^2 and 7.45669×10^{11} g$^{1/2}$/(F$^{1/2} \cdot$ s). The g in the unit of $B1$ is gram. The coefficients $A2$ and $B2$ are 3.75956×10^{-7} A/V^2 and 9.8222×10^{11} g$^{1/2}$/(F$^{1/2} \cdot$ s). TOXREF and NTOX are useful for predictive modeling, for the case that the measured devices have TOXE but the target devices have TOXREF.

The tunneling currents between the gate and source through the channel ($I_{gcs,\text{tunnel}}$), and between the gate and drain through the channel ($I_{gcd,\text{tunnel}}$), have similar expressions. They are different by some factor which is purely a function of V_{DS} through the use of PIGCD. This parameter has a default value of 1 in BSIM4.0.0. In BSIM4.1.0, however, its default somehow has some bias dependencies! The two channel tunneling currents are given by,

$$I_{gcs,\text{tunnel}} = \begin{cases} I_{gc} \times \dfrac{-1 + \texttt{PIGCD} \cdot V_{DS} + \exp(-\texttt{PIGCD} \cdot V_{DS}) + 10^{-4}}{(\texttt{PIGCD} \cdot V_{DS})^2 + 2 \times 10^{-4}} & \text{if } \texttt{IGCMOD} \neq 0 \\ 0 & \text{if } \texttt{IGCMOD} = 0. \end{cases} \quad (5\text{-}106)$$

$$I_{gcd,\text{tunnel}} = \begin{cases} I_{gc} \times \dfrac{1 - (\texttt{PIGCD} \cdot V_{DS} + 1) \cdot \exp(-\texttt{PIGCD} \cdot V_{DS}) + 10^{-4}}{(\texttt{PIGCD} \cdot V_{DS})^2 + 2 \times 10^{-4}} & \text{if } \texttt{IGCMOD} \neq 0 \\ 0 & \text{if } \texttt{IGCMOD} = 0. \end{cases} \quad (5\text{-}107)$$

Generally $I_{gcs,\text{tunnel}}$ and $I_{gcd,\text{tunnel}}$ do not sum up to I_{gc}. These two components, however, are identical when $V_{DS} = 0$, each equal to half of I_{gc}, which is given by

$$I_{gc} = \text{NF } W_{\text{eff}} L_{\text{eff}} \frac{A3}{\text{TOXE}^2} \left(\frac{\text{TOXREF}}{\text{TOXE}}\right)^{\text{NTOX}} V_{GS,\text{eff}}$$

$$\times \text{NIGC} \frac{kT}{q} \ln\left[1 + \exp\left(+\frac{q(V_{GS,\text{eff}} - \text{VTH0})}{kT \cdot \text{NIGC}}\right)\right]$$

$$\times \exp[-B3 \cdot \text{TOXE}(\text{AIGC} - \text{BIGC} V_{ox,\text{depinv}}) \cdot (1 + \text{GIGC} V_{ox,\text{depinv}})]. \quad (5\text{-}108)$$

The coefficients $A3$ and $B3$ are given by

$$A_3 = \begin{cases} 4.97232 \times 10^{-7} & (\text{nmos}) \\ 3.42537 \times 10^{-7} & (\text{pmos}) \end{cases} \quad \frac{A}{V^2};$$

$$B3 = \begin{cases} 7.45669 \times 10^{11} & (\text{nmos}) \\ 1.16645 \times 10^{12} & (\text{pmos}) \end{cases} \quad \frac{g^{1/2}}{F^{1/2} \cdot s}$$

The last two tunneling currents are those associated between the gate and source/drain junctions. These are the regions where the overlap capacitances reside. These tunneling currents depend on the area formed by W_{eff} (not $W_{\text{eff,CV}}$) and DLCIG (not LINT or DLC). DLCIG is the overlap distance of the gate–source/gate–drain junctions. It has a similar meaning as LINT for I-V calculation and DLC for C-V calculation. This parameter is introduced rather than equated to LINT because sometimes the extracted LINT for a best fit is negative and because LINT may not exactly represent the overlapping diffusion distance. It is not equated to DLC because the gate tunnel currents are dc quantities. Finally, a POXEDGE parameter is used to account for the fact that the oxide thickness in the overlap areas may be slightly different from that of the intrinsic device. The currents are given by

$$I_{gs,\text{tunnel}} = \begin{cases} I_{gs} & \text{if IGCMOD} \neq 0; \\ 0 & \text{if IGCMOD} = 0; \end{cases} \quad (5\text{-}109)$$

$$I_{gd,\text{tunnel}} = \begin{cases} I_{gd} & \text{if IGCMOD} \neq 0; \\ 0 & \text{if IGCMOD} = 0; \end{cases} \quad (5\text{-}110)$$

where

$$I_{gs} = \text{NF } W_{\text{eff}} \text{DLCIG} \frac{A3}{(\text{TOXE} \cdot \text{POXEDGE})^2} \left(\frac{\text{TOXREF}}{\text{TOXE} \cdot \text{POXEDGE}}\right)^{\text{NTOX}} V_{GS} \times V'_{GS}$$

$$\times \exp[-B3 \cdot \text{TOXE} \cdot \text{POXEDGE}(\text{AIGSD} - \text{BIGSD } V'_{GS})$$

$$\cdot (1 + \text{CIGSD } V'_{GS})]; \quad (5\text{-}111)$$

$$V'_{GS} = \sqrt{(V_{GS} - V_{FB,SD})^2 + 10^{-4}}; \quad (5\text{-}112)$$

and

$$I_{gd} = \text{NF } W_{\textit{eff}} \text{DLCIG} \frac{A3}{(\text{TOXE} \cdot \text{POXEDGE})^2} \left(\frac{\text{TOXREF}}{\text{TOXE} \cdot \text{POXEDGE}}\right)^{\text{NTOX}} V_{GD} \times V'_{GD}$$
$$\times \exp[-B3 \cdot \text{TOXE} \cdot \text{POXEDGE}(\text{AIGSD} - \text{BIGSD}V'_{GD}) \cdot (1 + \text{CIGSD}V'_{GD})];$$
(5-113)

$$V'_{GD} = \sqrt{(V_{GD} - V_{FB,SD})^2 + 10^{-4}}.$$
(5-114)

$V_{FB,SD}$ is the flatband voltage in the source/drain junctions, whose doping level is NSD, was given in Eq. (5-58).

Figures 5-9 and 5-11 are dc equivalent circuits when the tunneling currents are zero. When IGCMOD and/or IGBMOD are turned on, the equivalent circuits need to incorporate the five tunneling currents. We show the equivalent circuit when RDSMOD = 1, in Fig. 5-33. The equivalent circuit for RDSMOD = 0 is obtained by removing $R_{S,\text{bias-dep}}$ and $R_{D,\text{bias\_dep}}$ in the figure.

Figure 5-34 compares the measured versus simulated gate current for an advanced device. The scale is removed to protect the anonymity of this device. However, there are some qualititive characteristics worth describing. The simulation is performed with IGBMOD = 0; $I_{gb,\text{tunnel}}$ is determined to be negligible. When $V_{DS} = 0$, I_G is always positive. As V_{GS} increases, the tunneling current from the gate to the channel turns on exponentially. The rise of I_{gcd} dominates all other tunneling components. When $V_{DS} = V_{DD}$, I_G can be negative when V_{GS} is low. Under this bias condition, there is significant tunneling current from the drain to the gate at the drain–gate overlap region. I_G has a negative sign because $V_{DS} > V_{GS}$. As V_{GS}

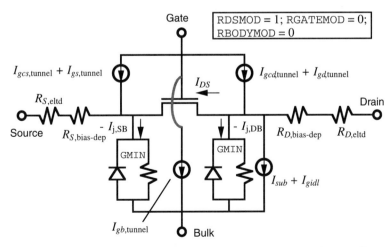

Fig. 5-33 A BSIM4 dc equivalent circuit similar to Fig. 5-11, except the tunneling current components are now included. RDSMOD = 1.

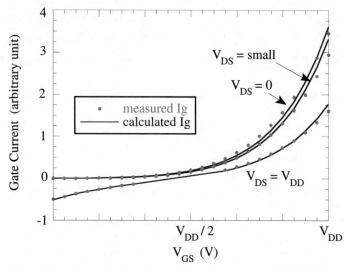

Fig. 5-34 Measured versus BSIM4-calculated gate current for an advanced device.

increases, however, the increase of I_{gcd} eventually overshadows the finite I_{gd}, and I_G becomes positive. Finally, this figure demonstrates that BSIM4's tunneling-current model is accurate. We have also tested the temperature and channel length dependencies, and found the BSIM4 model to produce excellent fit.

Although not advertised in the release note, BSIM4 actually includes shot noises for these tunneling currents. The shot noise expressions are found in the noise equivalent circuits in Figs. 5-16a and 5-17a.

Parameter Name	Description	Default Value
IGCMOD	Model selector for all tunneling current except $I_{gb,\text{tunnel}}$	0 (off)
IGBMOD	Model selector for $I_{gb,\text{tunnel}}$	0 (off)
AIGBACC	First parameter for $I_{gb,\text{acc}}$	$0.43\,(\text{F}\cdot\text{s}^2/\text{g}^{-1})^{1/2}/\text{m}$
BIGBACC	Second parameter for $I_{gb,\text{acc}}$	$0.054\,(\text{F}\cdot\text{s}^2/\text{g})^{1/2}/(\text{V}\cdot\text{m})$
CIGBACC	Third parameter for $I_{gb,\text{acc}}$	$0.075\,\text{V}^{-1}$
NIGBACC	Ideality factor of $I_{gb,\text{acc}}$	1
AIGBINV	First parameter for $I_{gb,\text{inv}}$	$0.35\,(\text{F}\cdot\text{s}^2/\text{g}^{-1})^{1/2}/\text{m}$
BIGBINV	Second parameter for $I_{gb,\text{inv}}$	$0.03\,(\text{F}\cdot\text{s}^2/\text{g})^{1/2}/(\text{V}\cdot\text{m})$
CIGBINV	Third parameter for $I_{gb,\text{inv}}$	$0.006\,\text{V}^{-1}$
EIGBINV	Voltage offset parameter for $I_{gb,\text{inv}}$	1.1 V
NIGBINV	Ideality factor of $I_{gb,\text{inv}}$	3
AIGC	First parameter for I_{gcs}, I_{gcd}	$0.43\,(\text{F}\cdot\text{s}^2/\text{g}^{-1})^{1/2}/\text{m}$ (nmos) $0.31\,(\text{F}\cdot\text{s}^2/\text{g}^{-1})^{1/2}/\text{m}$ (nmos)
BIGC	Second parameter	$0.054\,(\text{F}\cdot\text{s}^2/\text{g})^{1/2}/(\text{V}\cdot\text{m})$ (nmos) $0.024\,(\text{F}\cdot\text{s}^2/\text{g})^{1/2}/(\text{V}\cdot\text{m})$ (pmos)

(*continued*)

Parameter Name	Description	Default Value
CIGC	Third parameter for I_{gcs}, I_{gcd}	0.075 V^{-1} (nmos) 0.03 V^{-1} (pmos)
AIGSD	First parameter for I_{gs}, I_{gd}	0.43 (F·s^2/g^{-1})$^{1/2}$/m (nmos) 0.31 (F·s^2/g^{-1})$^{1/2}$/m (nmos)
BIGSD	Second parameter	0.054 (F·s^2/g)$^{1/2}$/(V·m) (nmos) 0.024 (F·s^2/g)$^{1/2}$/(V·m) (pmos)
CIGSD	Third parameter for I_{gs}, I_{gd}	0.075 V^{-1} (nmos) 0.03 V^{-1} (pmos)
NIGC	Ideality factor of I_{gs}, I_{gcs}, I_{gd}, I_{gcd}	1
DLCIG	Gate–source/gate–drain overlap for calculating I_{gs}, I_{gd}	LINT
POXEDGE	Oxide thickness factor in overlap region	1
PIGCD	V_{DS} dependence of I_{gcs}, I_{gcd}	1 (BSIM4.0.0); calculated (BSIM4.1.0)
NTOX	Exponent of oxide ratio	1
TOXREF	Nominal gate oxide thickness	3×10^{-9} (m)

5.19 LAYOUT-DEPENDENT PARASITICS

BSIM3 concentrates primarily on a single-finger transistor, the dominant type of transistor in digital CMOS circuits. Multi-finger transistor in BSIM3 is treated as a multiple replica of the singe-finger transistor unit, through the use of the "number-of-instance" parameter in the device field (MI). This representation requires a distinct source and drain to be associated with a finger. The whole unit is repeated MI-times. If MI is equal to 2, for example, there will be 2 gate fingers, 2 source contacts, and 2 drain contacts.

This representation is awkward in multi-finger transistors found in analog and large-power circuits, which employs interdigitated layout to save the junction area. In a 2-finger transistor, we may choose to have a layout with two source junctions, one at the leftmost and another at the rightmost sides of the transistor (such as Fig. 4-88b). A single drain contact is placed between the two gate fingers. Essentially, two drain contacts are merged in the center to form just one drain contact. This interdigitated layout cannot be easily represented in a BSIM3 model, because a straightforward assignment of MI = 2 would lead to one extra drain contact area.

This inconvenience is addressed in BSIM4 by introducing several new parameters. Some key parameters include, NF, which denotes the number of fingers, and MIN, GEOMOD and RGEOMOD, which specify the geometrical arrangement of the drain/source contacts in relation to the gate fingers. A whole new subroutine is used by BSIM4 to calculate the drain parasitic resistance, source parasitic resistance, drain–bulk junction capacitance, and source–bulk junction capacitances, depending on the combination of these parameters. The number-of-instance field in the device statement can still be used in BSIM4, but it now means the number of replica of the multi-finger MOS defined by NF, PERMOD, MIN, GEOMOD and RGEOMOD.

Let us back up one step to describe why we are interested in knowing the geometrical details of a device. The main reason is that we would like to calculate the junction capacitances associated with the source–bulk and the drain–bulk diode areas. Let us take the drain–bulk junction capacitance as an example, which was given in Eq. (5-33). The first component of $C_{j,\text{DB}}$ is the bottom-wall junction capacitance, which was labeled as $C_{j,\text{DB\_BW}}$ in Section 2.3. The second and the third components refer to the sidewall capacitances at the isolation side and at the gate side. They were denoted as $C_{j,\text{DB\_SW}}$ and $C_{j,\text{DB\_SWG}}$, respectively, in Section 2.3. The physical locations of the three capacitances are shown in Fig. 2-12a. There is a need to distinguish between the isolation-side and the gate-side sidewall capacitances because the substrate doping is usually lighter in the isolation side than the gate side. In the BSIM4's framework, $PD_{\textit{eff}}$ appearing in Eq. (5-33) is not the entire drain junction periphery. Instead, it should be the total periphery minus the gate-side periphery. However, there is a possible confusion that PD specified by a designer (in the MOS instance statement) is used to mean the entire periphery, without the subtraction of the gate-side periphery. BSIM4 reduces the chance of communication error by defining the effective junction area and periphery as

$$AD_{\textit{eff}} = \begin{cases} AD & \text{if AD is given;} \\ AD_{\textit{eff}}^* & \text{if AD is not given.} \end{cases} \quad (5\text{-}115)$$

$$PD_{\textit{eff}} = \begin{cases} PD & \text{if PD is given and PERMOD} = 0; \\ PD - W_{\textit{eff},\text{CJ}} \cdot NF & \text{if PD is given and PERMOD} \neq 0; \\ PD_{\textit{eff}}^* & \text{if PD is not given,} \end{cases} \quad (5\text{-}116)$$

where $AD_{\textit{eff}}^*$ and $PD_{\textit{eff}}^*$ are discussed shortly. At present, the key message of the above equations is that the new BSIM4 parameter PERMOD clarifies what is meant by PD. When PERMOD = 0, PD given in the instance statement does not include the gate-side periphery. Conversely, when PERMOD $\neq$ 0, PD includes the gate-side periphery. Therefore, prior to the calculation of $C_{j,\text{DB}}$, $PD_{\textit{eff}}$ is determined as the total drain periphery subtracted by $W_{\textit{eff},\text{CJ}} \cdot NF$, where $W_{\textit{eff},\text{CJ}}$ is the effective width for junction-related calculation.

When the device geometry is simple, it is straightforward to calculate AD and PD. However, when the device has more than one finger, computing the correct values for AD and PD can be challenging. Fortunately, BSIM4 can calculate those numbers by specifying NF, MIN, and GEOMOD. Basically, as suggested in Eqs. (5-115) and (5-116), when AD and PD are not specified in the instance statements, BSIM4 go ahead and use NF, MIN and GEOMOD to figure out $AD_{\textit{eff}}^*$ and $PD_{\textit{eff}}^*$. They are then equated to AD and PD, respectively, right before the computation of $C_{j,\text{DB}}$.

It is easy to understand the meaning NF; it is just the number of fingers. When NF is odd, one end junction must be a source, and the other, drain. When NF is even, we only know that the two end junctions are of the same contact. The end contacts can both be drain contacts, or they can both be source contacts. BSIM4 relies on the parameter MIN to remove this ambiguity. When MIN = 1, the number of source contacts is to be minimized. Therefore, both end junctions are drain junctions. When MIN ≠ 1, both end junctions are source junctions. By the way, BSIM4 does not check whether the inputted NF is an integer; doing so possibly degrades the code efficiency. Hence, when NF is assigned with a value of 2.5, BSIM4 will proceed with the code and output some results.

BSIM4 further uses GEOMOD to describe whether the source/drain junctions are isolated, shared, or merged junctions. These terms are illustrated in Fig. 5-35. An *isolated junction* is a junction with a contact, located at the end of a device structure and not connected to any other device. A *merged junction* is a contactless junction. It is a junction shared with another device; therefore, a merged junction must appear at either end of a device. For example, in a NAND-gate circuit, the source of an input transistor is connected to the drain of the second input transistor. There is no need to layout two completely isolated transistors and then make the said connection with an interconnect metal. The two input transistors can be merged into one lumped block instead, with the source of one transistor directly overlay the drain of the second transistor. In this manner, the overall circuit size is reduced and the parasitics associated with a junction are minimized.

BSIM4 assumes all the junctions between gate fingers to be *shared junctions*. A shared junction is a junction with contact, and can be conceptually divided into two equal portions. The right portion can be thought to belong to the finger to the right,

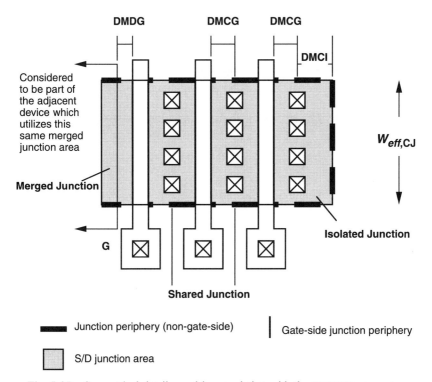

Fig. 5-35 Geometrical details used in association with the GEOMOD parameter.

and the left portion, to the finger to the left. This conceptual division simplifies the equations for resistance, junction area, and junction periphery calculation. A shared junction can also appear as an end contact. It is similar to a merged junction just described, except that a contact is made. The critical dimensions DMCI, DMCG and DMDG are marked in the figure.

BSIM4 assumes the following for a device's structure. The end contacts of the device can either be an isolated, merged, or shared contacts, but all of the intermediate contacts (contacts between two gate fingers) are shared contacts. BSIM4 uses GEOMOD = 0 to denote a device having a source end contact which is isolated and a drain end contact which is isolated. Implicitly in the definition, the NF must be odd so that there can be one source end contact and one drain end contact. When GEOMOD = 1, the device has at least one source end junction, and the source junction(s) is(are) isolated. The drain end contact is either absent (such as NF is even) or a shared contact (when NF is odd). We can keep on going, but there are a total of 11 GEOMOD options. It is best to summarize them in the following table:

GEOMOD	Source End Contact	Drain End Contact	Comment
0	isolated	isolated	NF must be odd
1	isolated	shared or none	NF can be even or odd
2	shared or none	isolated	NF can be even or odd
3	shared or none	shared or none	NF can be even or odd
4	isolated	merged	NF must be odd
5	shared or none	merged	NF can be even or odd
6	merged	isolated	NF must be odd
7	merged	shared or none	NF can be even or odd
8	merged	merged	NF must be odd
9	isolated/shared or none	shared or none	NF must be even
10	shared or none	isolated/shared or none	NF must be even

GEOMOD should be set to 6 for the transistor in Fig. 5-35, if the leftmost contact is a source contact. It should be set to 4 if the leftmost contact is a drain contact. In the case of GEOMOD = 9, NF must be even. If MIN = 0, then the two ends of the device are source contacts. One of the source contacts is shared, and another, isolated. If MIN $\neq$ 0, then the two ends of the device are drain contacts. Both drain contacts are assumed shared in this case. When GEOMOD = 10, NF must also be even. If MIN = 0, then the two ends of the device are source contacts. Both source contacts are assumed shared in this case. If MIN $\neq$ 0, then the two ends of the device are drain contacts. One of the drain contacts is shared, and another, isolated.

The equations used in BSIM4 code to calculate $AD^*_{\it{eff}}$ and $PD^*_{\it{eff}}$ are straightforward but quite convoluted (see Appendix G). This is because the code need to apply

5.19 LAYOUT-DEPENDENT PARASITICS

to all the above 11 GEOMOD cases, for any combination of NF and MIN. Let us carry out a calculation specifically for GEOMOD = 6, which is the layout shown Fig. 5-35 if the leftmost contact is a source contact. For this case,

$$AD_{eff}^* = (\text{DMCG} + \text{DMCI} - 2\text{DMCGT}) \times W_{eff,\text{CJ}} + 2(\text{DMCG} - \text{DMCGT}) \times W_{eff,\text{CJ}}.$$
$$PD_{eff}^* = [2(\text{DMCG} + \text{DMCI} - 2\text{DMCGT}) + W_{eff,\text{CJ}}] + 4(\text{DMCG} - \text{DMCGT}).$$

DMCG, DMCI and DMCG were marked on Fig. 5-35. In both of the above equations, the first term is that associated with the isolated end junction, and the second term, the shared intermediate junction. If DMCGT is 0, then the above equation is straightforward to understand. The reason BSIM4 performs the subtraction of DMCGT is to correct the values of DMCG, DMCI and DMDG of test structures.

If our end goal is only to calculate the junction capacitance, then the above mentioned PERMOD, NF, MIN, GEOMOD, DMCG, DMCI, DMDG, and DMCGT are sufficient. However, BSIM4 also calculates the parasitic epitaxial resistance associated with the junctions. A separate parameter RGEOMOD is introduced to specify the exact type of contacts: point, wide or merged. Figure 5-36 shows the distinction between a point and a wide contact. A point contact may increase packing density, but a wide contact enjoys a smaller associated resistance [13]. There is a need to

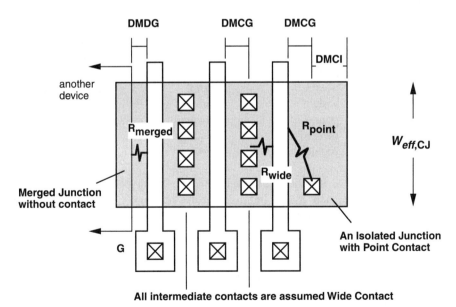

Fig. 5-36 Geometrical details used in association with the RGEOMOD parameter.

distinguish between the two types of contact because the paths of current flows differ. The expressions for R_{point} and R_{wide} shown in the figure are

$$R_{wide} = \frac{\text{RSH} \cdot (\text{DMCG} - \text{DMCGT})}{W_{\textit{eff},\text{CJ}}}. \tag{5-117}$$

$$R_{point} = \frac{\text{RSH} \cdot W_{\textit{eff},\text{CJ}}}{6(\text{DMCG} + \text{DMCI} - 2\text{DMCGT})}. \tag{5-118}$$

Note: R_{point} above is the R_{point} for the isolated end junction. If the end junction is a shared junction, then replace DMCI in the equation with DMCG.

BSIM4 assumes all intermediate junctions to be shared junctions (see the GEOMOD discussion above). Additionally, BSIM4 assumes all these intermediate junction to have wide contacts. The point contact is possible only at an end junction. In the case the end junction is a merged contact, then there is no contact at that end junction. The epitaxial resistance R_{merged} is given by

$$R_{merged} = \frac{\text{RSH} \cdot (\text{DMDG} - \text{DMCGT})}{W_{\textit{eff},\text{CJ}}}. \tag{5-119}$$

RGEOMOD specifies whether an end contact is a wide or a point contact, according to the following table:

RGEOMOD	Source End Contact	Drain End Contact	Intermediate Contact
0	no R_S	no R_D	wide (shared)
1	wide	wide	wide (shared)
2	wide	point	wide (shared)
3	point	wide	wide (shared)
4	point	point	wide (shared)
5	wide	merged	wide (shared)
6	point	merged	wide (shared)
7	merged	wide	wide (shared)
8	merged	point	wide (shared)

These resistance calculations eventually end up being part of $R_{S,\text{eltd}}$ (source electrode resistance) and $R_{D,\text{eltd}}$ (drain electrode resistance), first mentioned in Section 5.11. These electrode resistances are in the BSIM4 equivalent circuit, whether RDSMOD is 0 (Fig. 5-9) or 1 (Fig. 5-11). We repeat the $R_{D,\text{eltd}}$ equations given in Section 5.11 here:

$$R_{D,\text{eltd}} = \begin{cases} 0.001 \ (\Omega) & \text{if } R^*_{D,\text{eltd}} = 0 \text{ and } (\text{RGEOMOD} \neq 0 \text{ or } \text{RDSMOD} \neq 0 \text{ or} \\ & \text{TNOIMOD} \neq 0); \\ R^*_{D,\text{eltd}} & \text{if otherwise}; \end{cases}$$

$$\tag{5-120}$$

where $R^*_{D,\text{eltd}}$ is

$$R^*_{D,\text{eltd}} = \begin{cases} 0 & \text{if RGEOMOD} = 0; \\ \text{NRD} \times \text{RSH} & \text{if NRD is given and RGEOMOD} \neq 0; \\ R_{D,\text{end}}//R_{D,\text{int}} & \text{if NRD is not given and RGEOMOD} \neq 0; \end{cases} \quad (5\text{-}121)$$

where $R_{D,\text{end}}$ is the end drain contact resistance, and $R_{D,\text{int}}$ is the internal drain contact resistance. Let us consider Fig. 5-36 and assume the rightmost contact is a drain contact. The Fig. 5-36 corresponds to RGEOMOD = 8 (and GEOMOD = 6). $R_{D,\text{end}}$ for this particular case is

$$R_{D,\text{end}} = \frac{\text{RSH} \cdot W_{\text{eff,CJ}}}{6(\text{DMCG} + \text{DMCI} - 2\text{DMCGT})};$$

$$R_{D,\text{int}} = \frac{\text{RSH} \cdot (\text{DMCG} - \text{DMCGT})}{2W_{\text{eff,CJ}}}.$$

Note that $R_{D,\text{int}}$ is only half of R_{wide}. This is because there are two equal paths from the internal drain contact to the adjacent gate fingers. As stated in Eq. (5-121), if NRD is not given and if $R_{D,\text{end}}//R_{D,\text{int}}$ is a finite number, then $R_{D,\text{eltd}}$ for this particular case becomes $R_{D,\text{end}}//R_{D,\text{int}}$. This exercise of calculating $R_{D,\text{eltd}}$ has been relatively simple. The actual BSIM4 code is quite involved, since it has to accommodate all GEOMOD and RGEOMOD cases. The details of these equations can be found in Appendix G.

Parameter Name	Description	Default Value
PERMOD	Specifying whether PD and PS includes gate-edge periphery	1 (yes)
GEOMOD	Specify whether end junction is isolated, shared, or merged	0
RGEOMOD	Specify whether end junction contact is wide, point, or merged	0
DMCG	Distance from the contact center to the gate edge	0 (m)
DMCI	Distance from the contact center to the isolation edge	DMCG (m)
DMDG	Same as DMCG but for merged devices only; see Fig. 5-35	0 (m)
DMCGT	DMCG of test structures	0 (m)
NF	Number of fingers	1
MIN	Whether to minimize source contacts when NF = even	0 (no)

REFERENCES AND NOTES

[1] The standardization effort is promoted by the Compact Model Council, which is affiliated with the Electronics Industries Alliance. The Compact Model Council's history, mission, and members can be found in http://www.eia.org/eig/CMC/

[2] See http://www-device.eecs.berkeley.edu/~bsim3 for the latest development.

[3] K. Cao et al., "Modeling of pocket implanted MOSFETs for anomalous analog behavior," Proc. IEEE International Electron Device Meeting, 1999.

[4] W. Liu, *Fundamentals of III-V Devices: HBTs, MESFETs, and HFETs/HEMTs*, New York: Wiley, 1999. See Section 6.5 for the distributed gate resistance. See Section 3.7 for distributed base resistance. See Eq. (5-133) for the expression of $I_{DS,0}$ (written as $I_{DS,sat}$ in the reference).

[5] W. Liu, *Handbook of III-V Heterojunction Bipolar Transistors*, New York: Wiley, 1998.

[6] X. Jin, J. Ou, C. Chen, W. Liu, P. Gray, and C. Hu, "An effective gate resistance model for CMOS RF and noise modeling," *IEEE International Electron Device Meeting*, pp. 961–964, 1998.

[7] W. Liu, R. Gharpurey, M. C. Chang, U. Erdogan, R. Aggarwal, and J. P. Mattia, "RF MOSFET modeling accounting for distributed substrate and channel resistances with emphasis on the BSIM3v3 SPICE model," *IEEE Int. Electron Device Meeting*, 1997.

[8] X. Jin, J. Ou, K. Cao, W. Liu, and C. Hu, "Noise partition—an accurate short-channel MOSFET thermal noise model for RF low noise applications," to be published.

[9] A. van der Ziel, *Noise in Solid State Devices and Circuits*, New York: Wiley, 1986.

[10] A. van der Ziel, "Gate noise in field effect transistors at moderately high frequencies," *IEEE Proc.*, vol. 51, pp. 461–467, 1963.

[11] W. Lee, and C. Hu, "Modeling gate and substrate currents due to conduction- and valence-band electron and hole tunneling,"*IEEE VLSI Tech. Digest*, pp. 198–199, 2000.

[12] N. Yang, W. Henson, J. Hauser, and J. Wortman, "Modeling study of ultrathin gate oxides using direct tunneling current and C-V measurements in MOS devices," *IEEE Trans. Electron Dev.*, pp. 1464–1471, 1999.

[13] I. C. Chen and W. Liu, "High-speed or low-voltage, low-power operations," in C. Y. Chang and S. Sze (eds), *ULSI Devices*, New York: Wiley, pp. 566–567, 2000.

APPENDIX A

BSIM3 EQUATIONS

The equations given in this appendix follows roughly as those used in the actual BSIM3 codes. There are instances where the equations here differ from those in the official BSIM3 manual because the manual equations are not always the actual equations being coded. For example, in the evaluation of charges using the charge-thickness model (CAPMOD = 3), a variable calls for the use of a variable $V_{GB,\text{acc}}$. However, the variable never appears in the BSIM3 code, and thus does not appear in the equations given below. There are also numerous instances where the official manual gives approximate equations. Here, we strive to follow faithfully the exact details given in the codes. A side benefit of using code to formulate our equations here is the following: There are some typos in the manual equations, such as that for ϕ_δ in the charge-thickness capacitance model. However, since we base our equations from the code directly, the likely errors in the official BSIM3 manual are not reproduced.

There are factors such as 10^6 in the following equations. This is because, while the majority of BSIM3 parameters are expressed in the MKS system, some are in cgs. These factors are used to ensure the correctness of unit conversion. For example, in the RDSW equation, the factor of 10^6 is used to offset the unit imbalance, that W_{eff} is in meters, whereas RDSW is in $\Omega \cdot \mu m^{WR} = \Omega \cdot \mu m$ (when WR = 1). The factor 273.15 is to convert the temperature unit from °C to Kelvin.

The following equations apply to BSIM3v3.1, BSIM3v3.2, BSIM3v3.2.1, and BSIM3v3.2.2. The equations relating to binning are not listed here. For binning equations, see Section 1.12.

The Input Variables

    Bias Conditions: V_{GS}, V_{DS}, V_{BS}.
    Temperature: T_{device}.

APPENDIX A

Geometrical Information: L, W (We will consider other geometrical information, such as NRD, PD, AD, and NRS, PS, and AS as SPICE parameters, since they can be modified by the user. Therefore, these parameters, just as those BSIM3 parameters, are in the COURIER font)

Effective Channel Length/Width for I-V Calculation

$$L_{eff} = L - 2 \cdot \text{LINT} - 2 \cdot \Delta L_{\text{geometry}} \tag{A-1}$$

$$W_{eff} = W - 2 \cdot \text{WINT} - 2 \cdot \Delta W_{\text{geometry}} - 2 \cdot \Delta W_{\text{bias\_dependency}}$$

(See more discussion in Eq. A − 7) (A-2)

$$W_{eff,\text{special}} = W - 2 \cdot \text{WINT} - 2 \cdot \Delta W_{\text{geometry}}, \tag{A-3}$$

where

$$\Delta L_{\text{geometry}} = \frac{\text{LL}}{L^{\text{LLN}}} + \frac{\text{LW}}{W^{\text{LWN}}} + \frac{\text{LWL}}{L^{\text{LLN}} W^{\text{LWN}}}, \tag{A-4}$$

$$\Delta W_{\text{geometry}} = \frac{\text{WL}}{L^{\text{WLN}}} + \frac{\text{WW}}{W^{\text{WWN}}} + \frac{\text{WWL}}{L^{\text{WLN}} W^{\text{WWN}}}, \tag{A-5}$$

$$\Delta W_{\text{bias\_dependency}} = \text{DWG} \cdot [V_{GST,\text{eff}}] + \text{DWB} \cdot \left[\sqrt{2\phi_f - V_{BS,\text{eff}}} - \sqrt{2\phi_f}\right]. \tag{A-6}$$

W_{eff} is the true effective width meant in BSIM3. However, under some special circumstances, $W_{eff,\text{special}}$ is used, particularly in the calculation of R_{DS} and noise calculation. See Section 3.2, under the entry WL. (*Note:* W_{eff} defined here is written as pParam->BSIM3weff in the BSIM3 code, and $W_{eff,\text{special}}$ is written as Weff in b3ld.c of the BSIM3 code.) The expression for $V_{GST,\text{eff}}$ appearing in Eq. (A-6) is to be given in Eq. (A-44), and $V_{BS,\text{eff}}$ is given in Eq. (A-22). See also a note in Eq. (A-24).

Here is a special note. BSIM3 is touted as a robust program capable of handling user abuse. As such, BSIM3 needs to accommodate unreasonable SPICE parameter values supplied by the user. For example, what could happen if DWG is a very large positive number? $\Delta W_{\text{bias\_dependency}}$ calculated from Eq. (A-6) can be large, hence leading to a negative W_{eff} as governed by Eq. (A-2). A negative effective channel width is definitely unphysical. In addition, a negative W_{eff} value can cause numerical problem in evaluating many equations that follow. Therefore, BSIM3 performs a check on the value of W_{eff} after it is calculated in Eq. (A-2). If the value is below 0.02 µm (or, in the BSIM3 unit, 2×10^{-8} m), then it is replaced by

$$W_{eff} = 2 \times 10^{-8} \cdot \left(\frac{4 \times 10^{-8} - W_{eff,\text{ from Eq. (A-2)}}}{6 \times 10^{-8} - 2 \cdot W_{eff,\text{ from Eq. (A-2)}}}\right). \tag{A-7}$$

In this manner, even if the effective channel width calculated from Eq. (A-2) is negative, the final W_{eff} used by BSIM3, that calculated from Eq. (A-7), is still positive. The particular choice of factor enclosed in the parentheses of Eq. (A-7) allows W_{eff} and its derivative with respect to $\Delta W_{bias\_dependency}$ to be continuous right at 2×10^{-8} m.

$W_{eff,special}$ and L_{eff} given above, as well as $W_{eff,CV}$ and $L_{eff,CV}$ to be given shortly, do not receive the same special treatment. The moment their values are negative or zero, BSIM3 issues a fatal warning that their values are nonphysical.

Effective Channel Length/Width for C-V Calculation

$$L_{eff,CV} = L - 2 \cdot \text{DLC} - 2 \cdot \Delta L_{geometry,CV}; \tag{A-8}$$

$$W_{eff,CV} = W - 2 \cdot \text{DWC} - 2 \cdot \Delta W_{geometry,CV}; \tag{A-9}$$

where

$$\Delta L_{geometry,CV} = \frac{\text{LLC}}{L^{\text{LLN}}} + \frac{\text{LWC}}{W^{\text{LWN}}} + \frac{\text{LWLC}}{L^{\text{LLN}} W^{\text{LWN}}}; \tag{A-10}$$

$$\Delta W_{geometry,CV} = \frac{\text{WLC}}{L^{\text{WLN}}} + \frac{\text{WWC}}{W^{\text{WWN}}} + \frac{\text{WWLC}}{L^{\text{WLN}} W^{\text{WWN}}}. \tag{A-11}$$

Basic Quantities

$$q = 1.6 \times 10^{-19} \tag{A-12}$$

$$\epsilon_s = 8.85 \times 10^{-12} \cdot 11.7055 \tag{A-13a}$$

$$\epsilon_{ox} = 8.85 \times 10^{-12} \cdot 3.9018 \tag{A-13b}$$

$$k = 1.3806226 \times 10^{-23} \tag{A-14}$$

$$n_i = 1.45 \times 10^{10} \cdot \frac{\text{TNOM}}{300.15} \cdot \sqrt{\frac{\text{TNOM}}{300.15}} \cdot \exp\left[21.5565981 - \frac{E_g(\text{TNOM})}{2k(\text{TNOM} + 273.15)}\right] \tag{A-15}$$

note that n_i is really the intrinsic carrier concentration at TNOM. It is not a function of T_{device}. E_g, the energy bandgap, will be given in Eq. (A-216).

$$2\phi_f = 2 \frac{k[\text{TNOM} + 273.15]}{q} \ln\left(\frac{\text{NCH}}{n_i}\right) \tag{A-16}$$

$$C'_{ox} = \frac{\epsilon_{ox}}{\text{TOX}} \tag{A-17}$$

$$C_{ox} = W_{eff,CV} L_{eff,CV} \times C'_{ox} \tag{A-18}$$

$$T = \begin{cases} \text{TNOM} + 273.15 & \text{if } T_{device} \text{ is not given} \\ T_{device} + 273.15 & \text{if } T_{device} \text{ is given} \end{cases} \tag{A-19}$$

Flatband Voltage for I-V Calculation

If the BSIM3 model is BSIM3v3.2 or above, and if a value for the BSIM3 parameter VFB is given, then the dc flatband voltage (V_{FB}) is given by

$$V_{FB} = \text{VFB}. \tag{A-20}$$

Otherwise, the dc flatband voltage V_{FB} is given by,

$$V_{FB} = \delta_{NP} \cdot \text{VTHO} - 2\phi_f - \text{K1}\sqrt{2\phi_f}, \tag{A-21}$$

where δ_{NP} is +1 for NMOS and −1 for PMOS. The parameter K1 given above is not multiplied by TOX/TOXM, as it often is in the later part of the code.

Note: The flatband voltage for C-V calculation will be described later.

Effective Bulk-to-Source Voltage Function

$$V_{BS,\text{eff}} = v_{bc} + \frac{(V_{BS} - v_{bc} - \delta_1) + \sqrt{(V_{BS} - v_{bc} - \delta_1)^2 - 4\delta_1 v_{bc}}}{2}; \quad \delta_1 = 0.001, \tag{A-22}$$

where v_{bc} is determined by a set of conditions. First, if K2 > 0, v_{bc} is equated to the more negative value of −30 and VBM. As mentioned in the entry for VBM, VBM is defaulted to −3 when it is not specified in the SPICE parameter set. In this case, v_{bc} will then be equal to −30. Note that K2 > 0 is really an unphysical scenario (i.e., K2 should always be smaller than zero). In the more likely situation where K2 < 0, v_{bc} is determined as follows. (In this case, the determination of v_{bc} does not require the knowledge of VBM.) First, a temporary voltage, denoted as V_x here, is calculated as

$$V_x = 0.9 \times \left(2\phi_f - \frac{\text{K1}^2}{4\text{K2}^2}\right), \tag{A-23}$$

where $2\phi_f$ is the strong-inversion surface potential given in Eq. (A-16). If $V_x > -3$ V, v_{bc} is fixed at −3 (V). If $V_x < -30$ (V), v_{bc} is fixed at −30 (V). At $-30 < V_x < -3$, then v_{bc} is equated to V_x.

Special Note about $V_{BS,\text{eff}} > 0$

If $V_{BS,\text{eff}}$ calculated above is greater than zero, then

$$\text{Replace } \sqrt{2\phi_f - V_{BS,\text{eff}}} \text{ by } \frac{\left(\sqrt{2\phi_f}\right)^3}{2\phi_f + V_{BS,\text{eff}}/2} \text{ for Eq. (A-6) and all equations to follow.} \tag{A-24}$$

For example, the $\Delta V_{T,\text{body\_eff}}$ expression, to appear shortly, should be modified to the following when $V_{BS,\text{eff}} > 0$.

$$\Delta V_{T,\text{body\_effect}} = K1 \frac{\text{TOX}}{\text{TOXM}} \frac{\left(\sqrt{2\phi_f}\right)^3}{2\phi_f + \frac{V_{BS,\text{eff}}}{2}} - K1 \cdot \sqrt{2\phi_f} - K2 \frac{\text{TOX}}{\text{TOXM}} V_{BS,\text{eff}}. \quad \text{(A-25)}$$

Threshold Voltage

$$V_T = \text{VTHO} + \delta_{NP}(\Delta V_{T,\text{body\_effect}} - \Delta V_{T,\text{charge\_sharing}} - \Delta V_{T,\text{DIBL}}$$

$$+ \Delta V_{T,\text{reverse\_short\_channel}} + \Delta V_{T,\text{narrow\_width}} + \Delta V_{T,\text{small\_size}}); \quad \text{(A-26)}$$

$$\Delta V_{T,\text{body\_effect}} = K1 \frac{\text{TOX}}{\text{TOXM}} \sqrt{2\phi_f - V_{BS,\text{eff}}} - K1 \cdot \sqrt{2\phi_f} - K2 \frac{\text{TOX}}{\text{TOXM}} V_{BS,\text{eff}}; \quad \text{(A-27)}$$

$$\Delta V_{T,\text{charge\_sharing}} = \text{DVTO}\left[\exp\left(-\text{DVT1}\frac{L_{\text{eff}}}{2L_t}\right)\right.$$

$$\left. + 2\exp\left(-\text{DVT1}\frac{L_{\text{eff}}}{L_t}\right)\right](V_{bi} - 2\phi_f); \quad \text{(A-28)}$$

$$\Delta V_{T,\text{DIBL}} = \left[\exp\left(-\text{DSUB}\frac{L_{\text{eff}}}{2L_{t0}}\right) + 2\exp\left(-\text{DSUB}\frac{L_{\text{eff}}}{L_{t0}}\right)\right]$$

$$\times (\text{ETA0} + \text{ETAB} \cdot V_{BS,\text{eff}}) \times V_{DS};$$

(Not exactly that used in BSIM3; see below) (A-29)

$$\Delta V_{T,\text{reverse\_short\_channel}} = K1 \cdot \frac{\text{TOX}}{\text{TOXM}} \cdot \left(\sqrt{1 + \frac{\text{NLX}}{L_{\text{eff}}}} - 1\right)\sqrt{2\phi_f}; \quad \text{(A-30)}$$

$$\Delta V_{T,\text{narrow\_width}} = (K3 + K3B \cdot V_{BS,\text{eff}}) \frac{\text{TOX}}{W_{\text{eff}} + W0} 2\phi_f; \quad \text{(A-31)}$$

$$\Delta V_{T,\text{small\_size}} = \text{DVT0W}\left[\exp\left(-\text{DVT1W}\frac{W_{\text{eff}}L_{\text{eff}}}{2L_{tw}}\right)\right.$$

$$\left. + 2\exp\left(-\text{DVT1W}\frac{W_{\text{eff}}L_{\text{eff}}}{L_{tw}}\right)\right](V_{bi} - 2\phi_f); \quad \text{(A-32)}$$

where

$$V_{bi} = \frac{k[\text{TNOM} + 273.15]}{q} \ln\left(\frac{\text{NCH} \cdot N_{SD}}{n_i^2}\right); \qquad N_{SD} = 10^{20} \text{ cm}^{-3}. \tag{A-33}$$

$$L_{t0} = \sqrt{\frac{\epsilon_s X_{dep,0}}{C'_{ox}}}. \tag{A-34}$$

$$L_t = \begin{cases} \sqrt{\dfrac{\epsilon_s X_{dep}}{C'_{ox}}} \times (1 + \text{DVT2} \cdot V_{BS,\text{eff}}) & \text{if } \text{DVT2} \cdot V_{BS,\text{eff}} \geq -0.5; \\[2ex] \sqrt{\dfrac{\epsilon_s X_{dep}}{C'_{ox}}} \times \dfrac{1 + 3 \cdot \text{DVT2} \cdot V_{BS,\text{eff}}}{3 + 8 \cdot \text{DVT2} \cdot V_{BS,\text{eff}}} & \text{if } \text{DVT2} \cdot V_{BS,\text{eff}} < -0.5. \end{cases} \tag{A-35}$$

$$L_{tw} = \begin{cases} \sqrt{\dfrac{\epsilon_s X_{dep}}{C'_{ox}}} \times (1 + \text{DVT2W} \cdot V_{BS,\text{eff}}) & \text{if } \text{DVT2W} \cdot V_{BS,\text{eff}} \geq -0.5; \\[2ex] \sqrt{\dfrac{\epsilon_s X_{dep}}{C'_{ox}}} \times \dfrac{1 + 3 \cdot \text{DVT2W} \cdot V_{BS,\text{eff}}}{3 + 8 \cdot \text{DVT2W} \cdot V_{BS,\text{eff}}} & \text{if } \text{DVT2W} \cdot V_{BS,\text{eff}} < -0.5. \end{cases} \tag{A-36}$$

$$X_{dep} = \sqrt{\frac{2 \epsilon_s (2\phi_f - V_{BS,\text{eff}})}{q \text{ NCH}}}. \tag{A-37}$$

$$X_{dep,0} = \sqrt{\frac{2 \epsilon_s (2\phi_f)}{q \text{ NCH}}}. \tag{A-38}$$

The drain-induced barrier-lowering (DIBL) term should always be positive, so that the threshold voltage at large V_{DS} is smaller than when $V_{DS} = 0$. It therefore makes physical sense that $\text{ETA0} + \text{ETAB} \cdot V_{BS,\text{eff}}$ is a positive number. There are times when the values of ETA0 and ETAB supplied by the user yield a negative value for $\text{ETA0} + \text{ETAB} \cdot V_{BS,\text{eff}}$. To prevent the model from producing unexpected results, BSIM3 calculates $\Delta V_{T,\text{DIBL}}$ by first defining a temporary variable (*Tmp*) as

$$Tmp = (\text{ETA0} + \text{ETAB} \cdot V_{BS,\text{eff}}). \tag{A-39}$$

$$\Delta V_{T,\text{DIBL}} = \begin{cases} \left[\exp\left(-\text{DSUB}\dfrac{L_{\text{eff}}}{2L_{t0}}\right) + 2\exp\left(-\text{DSUB}\dfrac{L_{\text{eff}}}{L_{t0}}\right)\right] V_{DS} \times Tmp \\[1ex] \hfill \text{if } Tmp \geq 10^{-4}; \\[2ex] \left[\exp\left(-\text{DSUB}\dfrac{L_{\text{eff}}}{2L_{t0}}\right) + 2\exp\left(-\text{DSUB}\dfrac{L_{\text{eff}}}{L_{t0}}\right)\right] V_{DS} \\[1ex] \qquad \times \dfrac{2 \times 10^{-4} - Tmp}{3 - 2 \times 10^4 \cdot Tmp} \quad \text{if } Tmp < 10^{-4}. \end{cases} \tag{A-40}$$

The $\Delta V_{T,\text{DIBL}}$ calculated from Eq. (A-40) is always a positive term even when Tmp evaluated from Eq. (A-39) is negative.

Effective V_{GS} After Accounting for Poly-Depletion Effects

$$V_{poly} = \frac{q\,\epsilon_s\,\text{NGATE}\,C_{ox}^{\prime\,2}\cdot 10^6}{2}\left[\sqrt{1 + \frac{2(V_{GS} - V_{FB} - 2\phi_f)}{q\,\epsilon_s\,\text{NGATE}\,C_{ox}^{\prime\,2}\cdot 10^6}} - 1\right]^2. \quad \text{(A-41)}$$

$$V_{poly,\text{eff}} = 1.12 - \frac{1}{2}\left(1.12 - V_{poly} - \delta + \sqrt{(1.12 - V_{poly} - \delta)^2 + 4\cdot\delta\cdot 1.12}\right)$$

$$(\delta \text{ fixed at } 0.05). \quad \text{(A-42)}$$

$$V_{GS,\text{eff}} = V_{GS} - V_{poly,\text{eff}}. \quad \text{(A-43)}$$

The main goal of Eq. (A-42) is to prevent V_{poly}, the voltage drop across the polysilicon gate, from exceeding 1.12 V (the silicon bandgap voltage). If it exceeds 1.12 V, then $V_{poly,\text{eff}}$ will be fixed at 1.12 V. The $V_{poly,\text{eff}}$ function is such that $V_{poly,\text{eff}}$ varies between 0 (when $V_{poly} = 0$) and 1.12 (when $V_{poly} \gg 1.12$).

Effective $V_{GS} - V_T$ Smoothing Function

This smoothing function smoothes out the characteristics between the subthreshold and the strong inversion operating regions.

$$V_{GST,\text{eff}} = \frac{2nkT/q\,\ln\left[1 + \exp\left(\dfrac{V_{GS,\text{eff}} - V_T}{2nkT/q}\right)\right]}{1 + 2n\dfrac{C_{ox}^\prime}{C_{dep,0}^\prime}\exp\left(-\dfrac{V_{GS,\text{eff}} - V_T - 2\cdot\text{VOFF}}{2nkT/q}\right)}, \quad \text{(A-44)}$$

where

$$C_{dep,0}^\prime = \frac{\epsilon_s}{X_{dep,0}} \quad \text{(A-45a)}$$

and a similar, soon-to-be used quantity $C_{dep}^\prime$ is defined as

$$C_{dep}^\prime = \frac{\epsilon_s}{X_{dep}}. \quad \text{(A-45b)}$$

APPENDIX A

The ideality n generally appears as

$$n = 1 + \text{NFACTOR} \cdot \frac{C'_{dep}}{C'_{ox}} + \frac{\text{CIT}}{C'_{ox}} + \frac{\text{CDSC} + \text{CDSCD} \cdot V_{DS} + \text{CDSCB} \cdot V_{BS,\text{eff}}}{C'_{ox}}$$

$$\times \left[\exp\left(-\text{DVT1} \frac{L_{\text{eff}}}{2L_t}\right) + 2\exp\left(-\text{DVT1} \frac{L_{\text{eff}}}{L_t}\right) \right]$$

(Not exactly that used in BSIM3). (A-46)

BSIM3 does not exactly use Eq. (A-46) to prevent unreasonable user-specified parameters from producing numerical errors during simulation. For example, if user somehow specifies NFACTOR to be a negative number, then n calculated from Eq. (A-46) would lead to negative ideality factor. BSIM3 calculates the ideality factor in two steps. In the first, a *Tmp* variable is defined:

$$Tmp = \text{NFACTOR} \cdot \frac{C'_{dep}}{C'_{ox}} + \frac{\text{CIT}}{C'_{ox}} + \frac{\text{CDSC} + \text{CDSCD} \cdot V_{DS} + \text{CDSCB} \cdot V_{BS,\text{eff}}}{C'_{ox}}$$

$$\times \left[\exp\left(-\text{DVT1} \frac{L_{\text{eff}}}{2L_t}\right) + 2\exp\left(-\text{DVT1} \frac{L_{\text{eff}}}{L_t}\right) \right]. \quad \text{(A-47)}$$

Then, the ideality factor is given by

$$n = \begin{cases} 1 + Tmp & \text{if } Tmp \geq -0.5; \\ \dfrac{1 + 3 \cdot Tmp}{3 + 8 \cdot Tmp} & \text{if } Tmp < -0.5. \end{cases} \quad \text{(A-48)}$$

With n defined this way, the ideality factor never goes below zero, even though the *Tmp* variable calculated from Eq. (A-47) can be fairly negative.

Mobility Model

There are three mobility models. The user can pick the desired mobility model by assigning the parameter MOBMOD to either 1, 2, or 3. The first step toward the evaluation of the effective mobility (μ_{eff}) is to evaluate a temporary variable *Tmp*:
When MOBMOD = 1, then

$$Tmp = (\text{UA} + \text{UC} \cdot V_{BS,\text{eff}}) \left(\frac{V_{GST,\text{eff}} + 2 \cdot V_T}{\text{TOX}} \right) + \text{UB} \left(\frac{V_{GST,\text{eff}} + 2 \cdot V_T}{\text{TOX}} \right)^2. \quad \text{(A-49)}$$

If MOBMOD = 2, then

$$Tmp = (UA + UC \cdot V_{BS,\text{eff}}) \left(\frac{V_{GST,\text{eff}}}{TOX} \right) + UB \left(\frac{V_{GST,\text{eff}}}{TOX} \right)^2. \quad \text{(A-50)}$$

If MOBMOD = 3, then

$$Tmp = \left[UA \left(\frac{V_{GST,\text{eff}} + 2 \cdot V_T}{TOX} \right) + UB \left(\frac{V_{GST,\text{eff}} + 2 \cdot V_T}{TOX} \right)^2 \right] (1 + UC \cdot V_{BS,\text{eff}}). \quad \text{(A-51)}$$

The effective mobility for all three models is given by the generalized form:

$$\mu_{\text{eff}} = \frac{U0}{\text{Denominator}}, \quad \text{(A-52)}$$

where

$$\text{Denominator} = \begin{cases} 1 + Tmp & \text{if } Tmp \geq -0.8; \\ \dfrac{0.6 + Tmp}{7 + 10 \cdot Tmp} & \text{if } Tmp < -0.8. \end{cases} \quad \text{(A-53)}$$

Drain–Source Resistance

BSIM3 performs a checking prior to the evaluation of the drain–source resistance, to ensure the calculated resistance value makes physical sense. The checking is summarized as

$$\text{If user-inputted RDSW} < 0, \text{ then RDSW is set to 0.} \quad \text{(A-54)}$$

Under most circumstances, the drain–source resistance is calculated as

$$R_{DS} = \frac{RDSW \left[1 + PRWG \cdot V_{GST,\text{eff}} + PRWB \left(\sqrt{2\phi_f - V_{BS,\text{eff}}} - \sqrt{2\phi_f} \right) \right]}{(10^6 \times W_{\text{eff,spercial}})^{WR}}$$

(not exactly that used in BSIM3). (A-55)

BSIM3 is a robust program that needs to account for the fact that sometimes the user-supplied parameter values may lead to unphysical result. For example, if PRWG is a fairly negative value, then the term inside the square parentheses in Eq. (A-55) may become negative. This, in turn, leads to a negative R_{DS}. BSIM3 provides a remedy, defining a temporary variable as

$$Tmp = PRWG \cdot V_{GST,\text{eff}} + PRWB \left(\sqrt{2\phi_f - V_{BS,\text{eff}}} - \sqrt{2\phi_f} \right). \quad \text{(A-56)}$$

Then, R_{DS} is calculated depending on some condition:

$$R_{DS} = \begin{cases} \dfrac{\text{RDSW}}{(10^6 \times W_{\text{eff,special}})^{\text{WR}}} \times [1 + Tmp] & \text{if } Tmp \geq -0.9; \\ \dfrac{\text{RDSW}}{(10^6 \times W_{\text{eff,special}})^{\text{WR}}} \times \left[1 + \dfrac{0.8 \cdot Tmp}{17 + 20 \cdot Tmp}\right] & \text{if } Tmp < -0.9. \end{cases}$$

(A-57)

In this manner, R_{DS} is never negative, even when Tmp defined in Eq. (A-56) decreases to below unity.

Bulk-Charge Coefficient

Under most circumstances, the bulk-charge coefficient is calculated as

$$A_{Bulk} = \left\{ 1 + \frac{\text{K1}}{2\sqrt{2\phi_f - V_{BS,\text{eff}}}} \frac{\text{TOX}}{\text{TOXM}} \left[\frac{\text{A0} \cdot L_{\text{eff}}}{L_{\text{eff}} + 2\sqrt{\text{XJ} \cdot X_{\text{dep}}}} \right. \right.$$
$$\left. \left. \times \left(1 - \text{AGS} \cdot V_{GST,\text{eff}} \left(\frac{L_{\text{eff}}}{L_{\text{eff}} + 2\sqrt{\text{XJ} \cdot X_{\text{dep}}}}\right)^2\right) + \frac{\text{B0}}{W_{\text{eff}} + \text{B1}} \right] \right\}$$
$$\times \frac{1}{1 + \text{KETA} \cdot V_{BS,\text{eff}}}$$

(not exactly that used in BSIM3). (A-58)

A_{Bulk} cannot have arbitrary values. Physically, it should be greater than zero. In fact, in the present technologies, A_{Bulk} should exceed 0.1. BSIM3, being a robust program, ensures that A_{Bulk} retains physical values. Therefore, instead of evaluating A_{Bulk} with Eq. (A-58), BSIM3 evaluates a temporary variable first:

$$A_{Bulk,\text{tmp}} = \left\{ 1 + \frac{\text{K1}}{2\sqrt{2\phi_f - V_{BS,\text{eff}}}} \frac{\text{TOX}}{\text{TOXM}} \left[\frac{\text{A0} \cdot L_{\text{eff}}}{L_{\text{eff}} + 2\sqrt{\text{XJ} \cdot X_{\text{dep}}}} \right. \right.$$
$$\left. \left. \times \left(1 - \text{AGS} \cdot V_{GST,\text{eff}} \left(\frac{L_{\text{eff}}}{L_{\text{eff}} + 2\sqrt{\text{XJ} \cdot X_{\text{dep}}}}\right)^2\right) + \frac{\text{B0}}{W_{\text{eff}} + \text{B1}} \right] \right\}.$$

(A-59)

Then, A_{Bulk} is related to $A_{Bulk,\text{tmp}}$ in accordance with

$$A_{Bulk} = \begin{cases} A_{Bulk,\text{tmp}} \times \dfrac{1}{1 + \text{KETA} \cdot V_{BS,\text{eff}}} & \text{if } A_{Bulk,\text{tmp}} \geq 0.1; \\ \dfrac{0.2 - A_{Bulk,\text{tmp}}}{3 - 20 \cdot A_{Bulk,\text{tmp}}} \times \dfrac{1}{1 + \text{KETA} \cdot V_{BS,\text{eff}}} & \text{if } A_{Bulk,\text{tmp}} < 0.1. \end{cases}$$

(A-60)

In the event that $1 + \text{KETA} \cdot V_{BS,\text{eff}}$ is negative, A_{Bulk} calculated from the above equation becomes negative and thus, unphysical. BSIM3 takes care this situation by a conditional statement. If $\text{KETA} \cdot V_{BS,\text{eff}}$ ever falls below -0.9, then

$$\text{Replace} \frac{1}{1 + \text{KETA} \cdot V_{BS,\text{eff}}} \text{ of Eq. (A-60) by, } \frac{17 + 20\text{KETA} \cdot V_{BS,\text{eff}}}{0.8 + \text{KETA} \cdot V_{BS,\text{eff}}}. \quad \text{(A-61)}$$

This replacement ensures that A_{Bulk} is always positive, even if $1 + \text{KETA} \cdot V_{BS,\text{eff}}$ is negative. The choice of replacement preserves continuity of its value as well as its derivative with respect to $V_{BS,\text{eff}}$ at $\text{KETA} \cdot V_{BS,\text{eff}} = -0.9$.

Drain Saturation Voltage

$$\varepsilon_{sat} = \frac{2 \cdot \text{VSAT}}{\mu_{\text{eff}}}. \quad \text{(A-62)}$$

If $R_{DS} = 0$, then

$$V_{DS,\text{sat}} = \frac{\varepsilon_{sat} L_{\text{eff}} \cdot (V_{GST,\text{eff}} + 2kT/q)}{A_{bulk}\varepsilon_{sat}L_{\text{eff}} + (V_{GST,\text{eff}} + 2kT/q)}. \quad \text{(A-63)}$$

If $R_{DS} \neq 0$, then

$$V_{DS,\text{sat}} = \frac{-b - \sqrt{b^2 - 4ac}}{2a}; \quad \text{(A-64)}$$

$$a = A_{bulk}^2 W_{\text{eff}}\text{VSAT}C'_{ox}R_{DS} + \left(\frac{1}{\lambda} - 1\right)A_{bulk}; \quad \text{(A-65)}$$

$$b = -\left[\left(V_{GST,\text{eff}} + \frac{2kT}{q}\right) \cdot \left(\frac{2}{\lambda} - 1\right) + A_{bulk}\varepsilon_{sat}L_{\text{eff}}\right.$$
$$\left. + 3A_{bulk}\left(V_{GST,\text{eff}} + \frac{2kT}{q}\right)W_{\text{eff}}\text{VSAT}C'_{ox}R_{DS}\right]; \quad \text{(A-66)}$$

$$c = \left(V_{GST,\text{eff}} + \frac{2kT}{q}\right)\varepsilon_{sat}L_{\text{eff}} + 2\left(V_{GST,\text{eff}} + \frac{2kT}{q}\right)^2 W_{\text{eff}}\text{VSAT}C'_{ox}R_{DS}; \quad \text{(A-67)}$$

where λ is determined by the parameters A1 and A2. In general, λ is meant to be equal to $A2 + A1 \cdot V_{GST,\text{eff}}$. However, we need to make sure λ attains only physically meaningful values between 0 and 1. Therefore, λ is obtained according to the following equations/conditions.

Foremost, BSIM3 checks whether the user-inputted values for A2 makes sense. The actions involved in the checking process are summarized as:

If user-inputted A2 < 0.01, then A2 is set to 0.01. (A-68)

If user-inputted A2 > 1, then A2 is set to 1, and A1 is set to 0. (A-69)

Once the checking is done, BSIM3 determines λ depending on the sign of A1. If A1 > 0:

$$Tmp = (1 - A2) - \frac{1}{2}(1 - A2 - A1 \cdot V_{GST,\text{eff}} - \delta$$
$$+ \sqrt{(1 - A2 - A1 \cdot V_{GST,\text{eff}} - \delta)^2 + 4 \cdot \delta \cdot (1 - A2))} \quad \delta = 0.001. \quad \text{(A-70)}$$
$$\lambda = A2 + Tmp. \quad \text{(A-71)}$$

Equation (A-70) follows the same pattern as the smoothing function $V_{DS,\text{eff}}$ given by Eq. (1-10). In that equation, $V_{DS,\text{eff}}$ is made to vary between 0 (when $V_{DS} = 0$) to $V_{DS,\text{sat}}$ (when $V_{DS} \gg V_{DS,\text{sat}}$). Similarly, the temporary variable Tmp varies between 0 (when $A1 \cdot V_{GST,\text{eff}} = 0$) to $1 - A2$ (when $A1 \cdot V_{GST,\text{eff}} \gg 1 - A2$). Consequently, λ varies between its minimum value of A2 and its maximum value of 1.

If instead $A1 \leq 0$, then let $A1'$ be $-A1$ so that $A1'$ is a positive quantity.

$$Tmp = A2 - \frac{1}{2}\left(A2 - A1' \cdot V_{GST,\text{eff}} - \delta + \sqrt{(A2 - A1' \cdot V_{GST,\text{eff}} - \delta)^2 + 4 \cdot \delta \cdot A2}\right)$$
$$\delta = 0.001. \quad \text{(A-72)}$$
$$\lambda = A2 - Tmp. \quad \text{(A-73)}$$

In this case, the Tmp variable varies between 0 (when $A1' \cdot V_{GST,\text{eff}} = 0$) to A2 (when $A1' \cdot V_{GST,\text{eff}} \gg A2$). Consequently, λ varies between its maximum value of A2 and its minimum value of 0.

So, what is going on here? Generally, as long as $A1 \cdot V_{GST,\text{eff}}$ is not too large or too negative, λ is equated to $A2 + A1 \cdot V_{GST,\text{eff}}$. However, when $A1 \cdot V_{GST,\text{eff}}$ is too large, the smoothing function employed in Eq. (A-70) ensures that λ varies only between A2 and 1. Hence, λ cannot exceed 1. (Because of the condition set in Eq. A-68, A2 is never negative. So, λ is below 1, but always greater than 0.) In the other extreme when $A1 \cdot V_{GST,\text{eff}}$ has a very negative value, the smoothing function employed in Eq. (A-72) ensures that λ varies only between 0 and A2. Hence, λ never goes below 0. As $A1 \cdot V_{GST,\text{eff}}$ varies between $-\infty$ and $+\infty$, λ varies between 0 and 1.

Effective Drain-to-Source Voltage

$$V_{DS,\text{eff}} = V_{DS,\text{sat}} - \frac{1}{2}\left(V_{DS,\text{sat}} - V_{DS} - \text{DELTA} \right.$$
$$\left. + \sqrt{(V_{DS,\text{sat}} - V_{DS} - \text{DELTA})^2 + 4 \cdot \text{DELTA} \cdot V_{DS,\text{sat}}}\right). \quad \text{(A-74)}$$

As explained in Section 1.4, $V_{DS,\text{eff}}$ varies between 0 (when $V_{DS} = 0$) and $V_{DS,\text{sat}}$ (when $V_{DS} \gg V_{DS,\text{sat}}$).

APPENDIX A   485

Drain Current in Intrinsic MOS

This is the dominant component of the drain current, which flows through the channel to the source.

$$I_{DS} = \frac{I_{DS,0}}{1 + \frac{R_{DS}I_{DS,0}}{V_{DS,\text{eff}}}} \left(1 + \frac{V_{DS} - V_{DS,\text{eff}}}{V_A}\right) \cdot \left(1 + \frac{V_{DS} - V_{DS,\text{eff}}}{V_{A,\text{SCBE}}}\right), \quad \text{(A-75)}$$

where

$$I_{DS,0} = \frac{W_{\text{eff}}\mu_{\text{eff}}C'_{ox}V_{GST,\text{eff}}}{L_{\text{eff}}[1 + V_{DS,\text{eff}}/(\varepsilon_{sat}L_{\text{eff}})]} \left[1 - \frac{A_{bulk}V_{DS,\text{eff}}}{2(V_{GST,\text{eff}} + 2kT/q)}\right] V_{DS,\text{eff}}. \quad \text{(A-76)}$$

$$V_A = V_{A,\text{sat}} + V_{A,\text{CLM-DIBL}}. \quad \text{(A-77)}$$

$$V_{A,\text{sat}} = \frac{\varepsilon_{sat}L_{\text{eff}} + V_{DS,\text{sat}} + 2R_{DS}\text{VSAT}C'_{ox}W_{\text{eff}}V_{GST,\text{eff}}}{2/\lambda - 1 + R_{DS}\text{VSAT}C'_{ox}W_{\text{eff}}A_{bulk}} \times \left[1 - \frac{A_{bulk}V_{DS,\text{sat}}}{2(V_{GST,\text{eff}} + 2kT/q)}\right].$$

(A-78)

$$V_{A,\text{CLM-DIBL}} = \left(1 + \frac{\text{PVAG } V_{GST,\text{eff}}}{\varepsilon_{sat}L_{\text{eff}}}\right) \times \left(\frac{1}{V_{A,\text{CLM}}} + \frac{1}{V_{A,\text{DIBL}}}\right)^{-1} \quad \text{(see note below.)}$$

(A-79)

$$V_{A,\text{CLM}} = \begin{cases} \dfrac{A_{bulk}\varepsilon_{sat}L_{\text{eff}} + V_{GST,\text{eff}}}{\text{PCLM } A_{bulk}\varepsilon_{sat}L_{itl}}(V_{DS} - V_{DS,\text{eff}}) & \text{if } V_{DS} - V_{Ds,\text{eff}} > 10^{-10}; \\ 5.834617425 \times 10^{14} & \text{if otherwise.} \end{cases}$$

(A-80)

The expression for L_{itl} is given in Eq. (A-84). *Here is a special note for the calculation of $V_{A,\text{CLM}}$.* Before BSIM3 calculates all these Early voltages, BSIM3 performs a check on the value of PCLM. If PCLM is equal to or smaller than zero, then BSIM3 issues a fatal error and the SPICE simulation is halted. The value $5.834617425 \times 10^{14}$ is the maximum number adopted in BSIM3; a computer is not capable of comprehending ∞.

$$\theta_{rout} = \text{PDIBLC1} \times \left[\exp\left(-\text{DROUT}\frac{L_{\text{eff}}}{2L_{t0}}\right) + 2\exp\left(-\text{DROUT}\frac{L_{\text{eff}}}{L_{t0}}\right)\right]$$

$$+ \text{PDIBLC2}. \quad \text{(A-81)}$$

$$V_{A,\text{DIBL}} = \begin{cases} \dfrac{(V_{GST,\text{eff}} + 2kT/q)}{\theta_{rout}(1 + \text{PDIBLCB} \cdot V_{BS,\text{eff}})} & \text{if } \theta_{rout} \geq 0; \\ \times \left[1 - \dfrac{A_{bulk}V_{DS,\text{sat}}}{A_{bulk}V_{DS,\text{sat}} + V_{GST,\text{eff}} + 2kT/q}\right] & \\ 5.834617425 \times 10^{14} & \text{if } \theta_{rout} \leq 0. \end{cases} \quad \text{(A-82)}$$

The above equation for $V_{A,\text{DIBL}}$ represents that used in most circumstances. However, under some special conditions discussed below, the expression is modified slightly to avoid having negative values of $V_{A,\text{DIBL}}$.

$$V_{A,\text{SCBE}} = \begin{cases} \dfrac{L_{\textit{eff}}}{\text{PSCBE2}} \exp\left(\dfrac{\text{PSCBE1}\, L_{\textit{itl}}}{V_{DS} - V_{DS,\text{eff}}}\right) & \textit{if } \text{PSCBE2} > 0; \\ 5.834617425 \times 10^{14} & \textit{if otherwise.} \end{cases} \quad \text{(A-83)}$$

$$L_{\textit{itl}} = \sqrt{\dfrac{\epsilon_s}{\epsilon_{ox}}} \times \text{TOX} \cdot \text{XJ}. \quad \text{(A-84)}$$

In the event that $1 + \text{PVAG} \cdot V_{GST,\text{eff}}/(\varepsilon_{sat} \cdot L_{\textit{eff}})$ is negative, $V_{A,\text{CLM-DIBL}}$ calculated from Eq. (A-79) becomes negative and thus unphysical. BSIM3 takes care this situation by a conditional statement. If $\text{PVAG} \cdot V_{GST,\text{eff}}/(\varepsilon_{sat} \cdot L_{\textit{eff}})$ ever falls below -0.9, then

$$\text{Replace } 1 + \dfrac{\text{PVAG}\, V_{GST,\text{eff}}}{\varepsilon_{sat} L_{\textit{eff}}} \text{ of Eq. (A-79) by } \dfrac{0.8 + \text{PVAG} \cdot V_{GST,\text{eff}}/(\varepsilon_{sat} L_{\textit{eff}})}{17 + 20 \cdot \text{PVAG} \cdot V_{GST,\text{eff}}/(\varepsilon_{sat} L_{\textit{eff}})}. \quad \text{(A-85)}$$

This replacement ensures that $V_{A,\text{CLM-DIBL}}$ is always positive, even if $1 + \text{PVAG} \cdot V_{GST,\text{eff}}/(\varepsilon_{sat} \cdot L_{\textit{eff}})$ is negative.

Similarly, in the event that $1 + \text{PDIBLCB} \cdot V_{BS,\text{eff}}$ is negative, $V_{A,\text{DIBL}}$ calculated from Eq. (A-82) becomes negative and thus unphysical. BSIM3 takes care this situation by a conditional statement. If $\text{PDIBLCB} \cdot V_{BS,\text{eff}}$ ever falls below -0.9, then

$$\text{Replace } \dfrac{1}{1 + \text{PDIBLCB} \cdot V_{BS,\text{eff}}} \text{ of Eq. (A-81) by } \dfrac{17 + 20 \cdot \text{PDIBLCB} \cdot V_{BS,\text{eff}}}{0.8 + \text{PDIBLCB} \cdot V_{BS,\text{eff}}}. \quad \text{(A-86)}$$

This replacement ensures that $V_{A,\text{DIBL}}$ is always positive, even if $1 + \text{PDIBLCB} \cdot V_{BS,\text{eff}}$ is negative. The choice of replacement preserves continuity of its value as well as its derivative with respect to $V_{BS,\text{eff}}$ at $\text{PDIBLCB} \cdot V_{BS,\text{eff}} = -0.9$.

Impact Ionization Current

$$I_{sub} = \left(\dfrac{\text{ALPHA0}}{L_{\textit{eff}}} + \text{ALPHA1}\right)(V_{DS} - V_{DS,\text{eff}}) \exp\left[-\dfrac{\text{BETA0}}{V_{DS} - V_{DS,\text{eff}}}\right] \\ \times \dfrac{I_{DS,0}}{1 + \dfrac{R_{DS} I_{DS,0}}{V_{DS,\text{eff}}}} \left(1 + \dfrac{V_{DS} - V_{DS,\text{eff}}}{V_A}\right). \quad \text{(A-87)}$$

However, either if the term (ALPHA0/L_{eff} + ALPHA1) ≤ 0, or if the parameter BETA0 ≤ 0, then I_{sub} is made to be zero.

Drain–Bulk Junction Current

$$I_{sat,DB} = \begin{cases} JS \times AD + JSSW \times PD & \text{the result is} \geq 0; \\ 10^{-14} A & \text{if AD and PD are both} \leq 0; \\ 0 & JS \times AD + JSSW \times PD < 0. \end{cases} \quad (A\text{-}88)$$

If IJTH = 0, then

$$I_{j,DB} = I_{sat,DB}\left[\exp\left(\frac{qV_{BD}}{NJ \cdot kT}\right) - 1\right] + \text{GMIN} \cdot V_{BD}. \quad (A\text{-}89)$$

If IJTH > 0, then

$$I_{j,DB} = \begin{cases} I_{sat,DB}\left[\exp\left(\dfrac{qV_{BD}}{NJ \cdot kT}\right) - 1\right] + \text{GMIN} \cdot V_{BD} & \text{if } V_{DB} < V_{IJTH}; \\ \text{IJTH} + \dfrac{\text{IJTH} + I_{sat,DB}}{NJ} \times \dfrac{q}{kT}(V_{BD} - V_{IJTH}) + \text{GMIN} \cdot V_{BD} & \\ & \text{if otherwise.} \end{cases}$$
$$(A\text{-}90)$$

$$V_{IJTH} = NJ\frac{kT}{q}\ln\left(\frac{\text{IJTH}}{I_{sat,DB}} + 1\right). \quad (A\text{-}91)$$

Source–Bulk Junction Current

$$I_{sat,SB} = \begin{cases} JS \times AS + JSSW \times PS & \text{the result is} \geq 0; \\ 10^{-14} A & \text{if AS and PS are both} \leq 0, \\ 0 & JS \times AS + JSSW \times PS < 0. \end{cases} \quad (A\text{-}92)$$

If IJTH = 0, then

$$I_{j,SB} = I_{sat,SB}\left[\exp\left(\frac{qV_{BS}}{NJ \cdot kT}\right) - 1\right] + \text{GMIN} \cdot V_{BS}. \quad (A\text{-}93)$$

If IJTH > 0, then

$$I_{j,SB} = \begin{cases} I_{sat,SB}\left[\exp\left(\dfrac{qV_{BS}}{NJ \cdot kT}\right) - 1\right] + \text{GMIN} \cdot V_{BS} & \text{if } V_{BS} < V_{IJTH}; \\ \text{IJTH} + \dfrac{\text{IJTH} + I_{sat,SB}}{NJ} \times \dfrac{q}{kT}(V_{BS} - V_{IJTH}) + \text{GMIN} \cdot V_{BS} & \text{if otherwise} \end{cases}$$
$$(A\text{-}94)$$

$$V_{IJTH} = NJ\frac{kT}{q}\ln\left(\frac{\text{IJTH}}{I_{sat,SB}} + 1\right). \quad (A\text{-}95)$$

Note: V_{BS} is used in the above expressions, not $V_{BS,\text{eff}}$.

APPENDIX A

Dc Current Equations

$$I_D = I_{DS} + I_{sub} - I_{j,\text{DB}}; \qquad \text{(A-96)}$$

$$I_S = -I_{DS} - I_{j,\text{SB}}; \qquad \text{(A-97)}$$

$$I_B = -I_{sub} + I_{j,\text{SB}} + I_{j,\text{DB}}; \qquad \text{(A-98)}$$

$$I_G = 0. \qquad \text{(A-99)}$$

Small-Signal Conductances

We write out definition here. BSIM3 derives the actual equations with the chain rule discussed in Section 1.5. The equations are too complicated to be expressed here.

$$g_m = \frac{\partial I_D}{\partial V_G}; \quad g_d = \frac{\partial I_D}{\partial V_D}; \quad g_{mb} = \frac{\partial I_D}{\partial V_B}. \qquad \text{(A-100a)}$$

$$g_{ms} = \frac{\partial I_{DS}}{\partial V_G}; \quad g_{ds} = \frac{\partial I_{DS}}{\partial V_D}; \quad g_{mbs} = \frac{\partial I_{DS}}{\partial V_B} \qquad \text{(A-100b)}$$

$$G_{bg} = \frac{\partial I_{sub}}{\partial V_G}; \quad G_{bd} = \frac{\partial I_{sub}}{\partial V_D}; \quad G_{bb} = \frac{\partial I_{sub}}{\partial V_B}. \qquad \text{(A-101)}$$

$$g_{j,\text{DB}} = \frac{\partial I_{j,\text{DB}}}{\partial V_{BD}}; \quad g_{j,\text{SB}} = \frac{\partial I_{j,\text{SB}}}{\partial V_{BS}}. \qquad \text{(A-102)}$$

Because of space limitation, we will describe only the CAPMOD = 2 and CAPMOD = 3 capacitance models. CAPMOD = 0 and CAPMOD = 1 models are essentially the same as those given in Appendix B.

Flatband Voltage for C-V Calculation

The flatband voltage expression used for C-V calculations depends on the values of CAPMOD, as well as the version of the BSIM3 model being used to run the simulation. (To find out whether a BSIM3 model is BSIM3v3.2.2, BSIM3v3.2.1, BSIM3v3.2, or BSIM3v3.1, see the entry VERSION in Section 3.2.) The C-V flatband voltage ($V_{FB,\text{CV}}$) is equated to one of the following three variables, VFBCV (a BSIM3 parameter), $V_{fb,(v)}$ (bias-dependent flatband voltage), or $V_{fb,zb}$ (zero-bias flatband voltage), as shown in the following table.

BSIM3v3.2.2, (which forces VERSION to be 3.2.2)
    CAPMOD = 0 $V_{FB,\text{CV}} = \text{VFBCV}$
    CAPMOD = 1,2,3 $V_{FB,\text{CV}} = V_{fb,zb}$
BSIM3v3.2.1, (which forces VERSION to be 3.2.1)
    CAPMOD = 0 $V_{FB,\text{CV}} = \text{VFBCV}$
    CAPMOD = 1,2 $V_{FB,\text{CV}} = V_{fb,(v)}$
    CAPMOD = 3 $V_{FB,\text{CV}} = V_{fb,zb}$

BSIM3v3.2, with VERSION specified as 3.2
 CAPMOD = 0 $V_{FB,CV}$ = VFBCV
 CAPMOD = 1,2,3 $V_{FB,CV} = V_{fb,zb}$
BSIM3v3.2, with VERSION specified as 3.1
 CAPMOD = 0 $V_{FB,CV}$ = VFBCV
 CAPMOD = 1,2 $V_{FB,CV} = V_{fb,(v)}$
 CAPMOD = 3 $V_{FB,CV} = V_{fb,zb}$
BSIM3v3.1 (which really ignores VERSION parameter to the codes)
 CAPMOD = 0 $V_{FB,CV}$ = VFBCV
 CAPMOD = 1,2 (3 is unavailable) $V_{FB,CV} = V_{fb,(v)}$
where

$$V_{fb,(v)} = V_T - 2\phi_f - K1\sqrt{2\phi_f}; \quad \text{(A-103)}$$

$$V_{fb,zb} = V_T(V_{GS} = V_{DS} = V_{BS} = 0) - 2\phi_f - K1\sqrt{2\phi_f}. \quad \text{(A-104)}$$

The $V_{fb,zb}$ equation given here apppears different from the one listed on p. B-25 of the BSIM3 manual.

At the writing of this book, there has been talk of releasing a new version of BSIM3, after the release of BSIM4. This new version has been designated as BSIM3v3.3; see Ref. [15] of Chapter 1. This new BSIM3 version is to be the same as BSIM3v3.2.2, except for a thermal noise expression. Therefore, $V_{FB,CV}$ of this new BSIM3 version can be taken to be that of BSIM3v3.2.2 listed above.

Effective V_{BS} Smoothing Function for C-V Calculation

$$V_{BS,\text{effCV}} = \begin{cases} V_{BS,\text{eff}} & \text{if } V_{BS,\text{eff}} < 0; \\ 2\phi_f - 2\phi_f \dfrac{2\phi_f}{2\phi_f + V_{BS,\text{eff}}} & \text{if } V_{BS,\text{eff}} \geq 0. \end{cases} \quad \text{(A-105)}$$

Effective V_{GB} Smoothing Function for C-V Calculation

$$V_{GB,\text{effCV}} = V_{GS,\text{eff}} - V_{BS,\text{effCV}}. \quad \text{(A-106)}$$

Effective Flatband Smoothing Function for C-V Calculation

If $V_{FB,CV} \geq 0$, then

$$V_{FB,\text{effCV}} = V_{FB,CV} - \frac{1}{2}\left(V_{FB,CV} - V_{GB,\text{effCV}} - \delta_3 \right. \\ \left. + \sqrt{(V_{FB,CV} - V_{GB,\text{effCV}} - \delta_3)^2 + 4 \cdot \delta_3 \cdot V_{FB,CV}}\right), \quad \text{(A-107a)}$$

or, if $V_{FB,CV} < 0$,

$$V_{FB,\text{effCV}} = V_{FB,CV} - \frac{1}{2}\left(V_{FB,CV} - V_{GB,\text{effCV}} - \delta_3\right.$$
$$\left. + \sqrt{(V_{FB,CV} - V_{GB,\text{effCV}} - \delta_3)^2 - 4 \cdot \delta_3 \cdot V_{FB,CV}}\right). \quad \text{(A-107b)}$$

δ_3 is fixed at 0.02. Equation (A-107) is similar to the $V_{DS,\text{eff}}$ function given in Eq. (A-74). $V_{FB,\text{effCV}}$ is made to vary between 0 (when $V_{GB} = 0$) and $V_{FB,CV}$ (when $V_{GB} \gg V_{FB,CV}$). Equation (A-107a) works only when $V_{FB,CV}$ exceeds or equals to zero. For $V_{FB,CV} < 0$, appropriate change in a sign inside the square-root function needs to be made to prevent taking the square root of a negative number as shown in Eq. (A-107b).

Effective $V_{GS} - V_T$ Smoothing Function for C-V Calculation

$$V_{GST,\text{effCV}} = \text{NOFF} \cdot \frac{nkT}{q} \ln\left[1 + \exp\left(\frac{V_{GS,\text{eff}} - V_T - \text{VOFFCV}}{\text{NOFF} \cdot nkT/q}\right)\right]. \quad \text{(A-108)}$$

The ideality factor n was given in Eq. (A-48).

Modified Bulk-Charge Coefficient for C-V Calculation

BSIM3 uses a reduced bulk-charge coefficient ($A_{Bulk,0}$), instead of the A_{Bulk} of Eq. (A-60), to calculate the bulk-charge coefficient of C-V calculation ($A_{Bulk,CV}$). Under most circumstances, the reduced bulk-charge coefficient is calculated as

$$A_{Bulk,0} = \left\{1 + \frac{K1}{2\sqrt{2\phi_f - V_{BS}}} \frac{\text{TOX}}{\text{TOXM}}\left[\frac{A0 \cdot L_{\text{eff}}}{L_{\text{eff}} + 2\sqrt{XJ \cdot X_{dep}}} + \frac{B0}{W_{\text{eff}} + B1}\right]\right\}$$
$$\times \frac{1}{1 + \text{KETA} \cdot V_{BS,\text{eff}}} \quad \text{(Not exactly that used in BSIM3.)} \quad \text{(A-109)}$$

However, $A_{Bulk,0}$ cannot have arbitrary values. Physically, it should be greater than zero. In fact, in the present technologies, $A_{Bulk,0}$ should exceed 0.1. BSIM3, being a robust program, ensures that $A_{Bulk,0}$ retains physical values. Therefore, instead of evaluating $A_{Bulk,0}$ with Eq. (A-109), BSIM3 evaluates a temporary variable first:

$$A_{Bulk,0,\text{tmp}} = \left\{1 + \frac{K1}{2\sqrt{2\phi_f - V_{BS}}} \frac{\text{TOX}}{\text{TOXM}}\left[\frac{A0 \cdot L_{\text{eff}}}{L_{\text{eff}} + 2\sqrt{XJ \cdot X_{dep}}} + \frac{B0}{W_{\text{eff}} + B1}\right]\right\}.$$
$$\text{(A-110)}$$

Then, $A_{Bulk,0}$ is related to $A_{Bulk0,tmp}$ in accordance with

$$A_{Bulk,0} = \begin{cases} A_{Bulk,0,tmp} \times \dfrac{1}{1 + \text{KETA} \cdot V_{BS,\text{eff}}} & \text{if } A_{Bulk,0,tmp} \geq 0.1; \\ \dfrac{0.2 - A_{Bulk,0,tmp}}{3 - 20 \cdot A_{Bulk,0,tmp}} \times \dfrac{1}{1 + \text{KETA} \cdot V_{BS,\text{eff}}} & \text{if } A_{Bulk,0,tmp} < 0.1. \end{cases}$$

(A-111a)

In the event that $1 + \text{KETA} \cdot V_{BS,\text{eff}}$ is negative, then $A_{Bulk,0}$ calculated from the above equation becomes negative and thus, unphysical. BSIM3 takes care this situation by a conditional statement. If $\text{KETA} \cdot V_{BS,\text{eff}}$ ever falls below -0.9, then

$$\text{Replace } \frac{1}{1 + \text{KETA} \cdot V_{BS,\text{eff}}} \text{ of Eq. (A-111a) by } \frac{17 + 20 \, \text{KETA} \cdot V_{BS,\text{eff}}}{0.8 + \text{KETA} \cdot V_{BS,\text{eff}}}.$$

(A-111b)

In the above few equations, $V_{BS,\text{eff}}$ is used (as indicated), not $V_{BS,\text{effCV}}$. Finally, the bulk-charge coefficient for C-V calculation is given by

$$A_{Bulk,CV} = A_{Bulk,0}\left[1 + \left(\frac{\text{CLC}}{L_{\text{eff,CV}}}\right)^{\text{CLE}}\right].$$

(A-112)

Intrinsic Device Charge for CAPMOD = 2 (Which Becomes CAPMOD = 1 in BSIM4)

$$V_{DS,\text{satCV}} = \frac{V_{GST,\text{effCV}}}{A_{Bulk,CV}}.$$

(A-113)

$$V_{DS,\text{effCV}} = V_{DS,\text{satCV}} - (V_{DS,\text{satCV}} - V_{DS} - \delta_4)/2$$
$$+ \sqrt{\frac{(V_{DS,\text{satCV}} - V_{DS} - \delta_4)^2 + 4\delta_4 V_{DS,\text{sat,CV}}}{4}} \qquad \delta_4 \text{ is fixed at } 0.02.$$

(A-114)

A substitution of $V_{DS} = 0$ into the equations involving the evaluation of $V_{DS,\text{effCV}}$ will prove that $V_{DS,\text{effCV}} = 0$ when $V_{DS} = 0$. However, just due to rounding errors, a computer may evaluate $V_{DS,\text{effCV}}$ to be a finite (though very small) value at $V_{DS} = 0$. In order to make sure $V_{DS,\text{effCV}}$ is zero when V_{DS} is zero, BSIM3 specifically adds a if–then statement that sets $V_{DS,\text{effCV}}$ to zero when V_{DS} is precisely zero.

$$Q_{acc} = C_{ox}(V_{FB,\text{effCV}} - V_{FB,CV}).$$

(A-115)

$$Q_{sub,0} = \begin{cases} C_{ox}\left(\text{K1}\dfrac{\text{TOX}}{\text{TOXM}}\right)\left[\sqrt{\dfrac{1}{4}\left(\text{K1}\dfrac{\text{TOX}}{\text{TOXM}}\right)^2 + Tmp} - \dfrac{1}{2}\left(\text{K1}\dfrac{\text{TOX}}{\text{TOXM}}\right)\right] & \text{if } Tmp \geq 0; \\ C_{ox} \times Tmp & \text{if } Tmp < 0; \end{cases}$$

(A-116a)

where C_{ox} and its related C'_{ox} were those used to calculate the I-V characteristics, defined in Eqs. (A-17) and (A-18), and

$$Tmp = V_{GS,\text{eff}} - V_{FB,\text{effCV}} - V_{BS,\text{effCV}} - V_{GST,\text{effCV}}. \quad \text{(A-116b)}$$

The other charge components are

$$\delta Q_{sub} = C_{ox} \left[\frac{1 - A_{bulk,CV}}{2} V_{DS,\text{effCV}} - \frac{(1 - A_{bulk,CV})A_{bulk,CV} V^2_{DS,\text{effCV}}}{12(V_{GST,\text{effCV}} - A_{bulk,CV} V_{DS,\text{effCV}}/2)} \right]. \quad \text{(A-117)}$$

$$Q_{inv} = -C_{ox} \left[\left(V_{GST,\text{effCV}} - \frac{A_{bulk,CV} V_{DS,\text{effCV}}}{2} \right) \right.$$
$$\left. + \frac{A^2_{bulk,CV} V^2_{DS,\text{effCV}}}{12(V_{GST,\text{effCV}} - A_{bulk,CV} V_{DS,\text{effCV}}/2)} \right]. \quad \text{(A-118)}$$

$$Q_G = -Q_{inv} - \delta Q_{sub} + Q_{acc} + Q_{sub,0}. \quad \text{(A-119)}$$

$$Q_B = \delta Q_{sub} - Q_{acc} - Q_{sub,0}. \quad \text{(A-120)}$$

If XPART < 0.5 (40/60 charge partition), then

$$Q_S = -\frac{C_{ox}}{2\left(V_{GST,\text{effCV}} - \frac{A_{bulk,CV} V_{DS,\text{effCV}}}{2}\right)^2}$$
$$\times \left[V^3_{GST,\text{effCV}} - \frac{4}{3} V^2_{GST,\text{effCV}} A_{bulk,CV} V_{DS,\text{effCV}} \right.$$
$$\left. + \frac{2}{3} V_{GST,\text{effCV}} A^2_{bulk,CV} V^2_{DS,\text{effCV}} - \frac{2}{15} A^3_{bulk,CV} V^3_{DS,\text{effCV}} \right]. \quad \text{(A-121)}$$

$$Q_D = -\frac{C_{ox}}{2\left(V_{GST,\text{effCV}} - \frac{A_{bulk,CV} V_{DS,\text{effCV}}}{2}\right)^2}$$
$$\times \left[V^3_{GST,\text{effCV}} - \frac{5}{3} V^2_{GST,\text{effCV}} A_{bulk,CV} V_{DS,\text{effCV}} \right.$$
$$\left. + V_{GST,\text{effCV}} A^2_{bulk,CV} V^2_{DS,\text{effCV}} - \frac{1}{5} A^3_{bulk,CV} V^3_{DS,\text{effCV}} \right]. \quad \text{(A-122)}$$

If XPART = 0.5 (50/50 charge partition), then

$$Q_S = Q_D = \frac{Q_{inv}}{2}. \tag{A-123}$$

If XPART > 0.5 (0/100 charge partition), then

$$Q_S = -C_{ox}\left[\frac{V_{GST,\text{effCV}}}{2} + \frac{A_{bulk,\text{CV}}V_{DS,\text{effCV}}}{4} - \frac{A_{bulk,\text{CV}}^2 V_{DS,\text{effCV}}}{24(V_{GST,\text{effCV}} - A_{bulk,\text{CV}}V_{DS,\text{effCV}}/2)}\right].$$
$$\tag{A-124}$$

$$Q_D = -C_{ox}\left[\frac{V_{GST,\text{effCV}}}{2} + \frac{3A_{bulk,\text{CV}}V_{DS,\text{effCV}}}{4} + \frac{A_{bulk,\text{CV}}^2 V_{DS,\text{effCV}}^2}{8(V_{GST,\text{effCV}} - A_{bulk,\text{CV}}V_{DS,\text{effCV}}/2)}\right].$$
$$\tag{A-125}$$

Intrinsic Device Charge for CAPMOD = 3 (Which Becomes CAPMOD = 2 in BSIM4)

We saw in the CAPMOD = 2 equations that there are several basic components to the gate, bulk, source, and drain charges: Q_{acc}, $Q_{sub,0}$, δQ_{sub}, and Q_{inv}. Q_D and Q_S are basically some kind of partition of Q_{inv}; their sum adds up to Q_{inv}. Q_G and Q_B are some combination of the four basic components. Of these four components, Q_{acc} and $Q_{sub,0}$ are associated with the accumulation and depletion regions. They are most important when V_{GS} is below the threshold voltage value. Due to the smoothing properties, they remain finite (but fairly close to zero) above the threshold. In contrast, δQ_{sub} and Q_{inv} are charges associated with the inversion channel. These latter components are significant only at V_{GS} greater than the threshold voltage, although they are also not quite zero below the threshold.

In CAPMOD = 3, BSIM3 decides that there are really two effective oxide thicknesses to work with. One thickness matters most to the calculation of charges in the accumulation and the depletion region, while another, the strong inversion region. Consequently, the effective oxide thickness for the calculations of Q_{acc}, $Q_{sub,0}$, differs from the effective oxide thickness used to calculate Q_{inv} and δQ_{sub}.

For the calculation of Q_{acc} and $Q_{sub,0}$,

$$C'_{ox,\text{eff}} = \frac{\epsilon_{ox}}{\text{TOX}} // \frac{\epsilon_s}{X_{DC,\text{eff}}}, \tag{A-126}$$

where

$$X_{DC,\text{eff}} = X_{DC,\text{max}} - \frac{(X_{DC,\text{max}} - X_{DC} - \delta_x) + \sqrt{(X_{DC,\text{max}} - X_{DC} - \delta_x)^2 + 4\delta_x X_{DC,\text{max}}}}{2}.$$
$$\tag{A-127}$$

The various terms inside the $X_{DC,\text{eff}}$ are given by

$$X_{DC} = \frac{L_{Debye}}{3} \exp\left[\text{ACDE}\left(\frac{\text{NCH}}{2 \times 10^{16}}\right)^{-1/4} \cdot \frac{V_{GB,\text{eff}} - V_{fb,zb}}{10^8 \times \text{TOX}}\right]. \quad (A\text{-}128)$$

[*Author's note:* The factor involving (NCH/(2×10^{16})) is found in b3temp.c.]

$$L_{Debye} = \sqrt{\frac{\epsilon_s \cdot k[\text{TNOM} + 273.15]/q}{q\text{NCH} \cdot 10^6}}. \quad (A\text{-}129)$$

$$X_{DC,\max} = \frac{L_{Debye}}{3}. \quad (A\text{-}130)$$

$$\delta_x = 10^{-3} \text{TOX}. \quad (A\text{-}131)$$

The $X_{DC,\text{eff}}$ calculated above is used only for the calculation of accumulation charge in the accumulation region.

$$C_{ox,\text{eff}} = L_{\text{eff,CV}} W_{\text{eff,CV}} C'_{ox,\text{eff}}; \quad (A\text{-}132)$$

$$Q_{acc} = C_{ox,\text{eff}}(V_{FB,\text{effCV}} - V_{FB,\text{CV}}); \quad (A\text{-}133)$$

$$Q_{sub,0} = \begin{cases} C_{ox,\text{eff}}\left(K1\frac{\text{TOX}}{\text{TOXM}}\right)\left[\sqrt{\frac{1}{4}\left(K1\frac{\text{TOX}}{\text{TOXM}}\right)^2 + Tmp} - \frac{1}{2}\left(K1\frac{\text{TOX}}{\text{TOXM}}\right)\right] \\ \qquad \qquad \qquad \qquad \qquad \qquad \qquad \qquad \qquad \text{if } Tmp \geq 0; \\ C_{ox} \times Tmp \\ \qquad \qquad \qquad \qquad \qquad \qquad \qquad \qquad \qquad \text{if } Tmp < 0; \end{cases}$$

where

$$Tmp = V_{GS,\text{eff}} - V_{FB,\text{effCV}} - V_{BS,\text{effCV}} - V_{GST,\text{effCV}}. \quad (A\text{-}134)$$

For the calculation of Q_{inv} and δQ_{sub}: The moment the transistor enters subthreshold/depletion or inversion region, another X_{DC} is sought.

$$C'_{ox,\text{inv}} = \frac{\epsilon_{ox}}{\text{TOX}} // \frac{\epsilon_s}{X_{DC,\text{inv}}}, \quad (A\text{-}135)$$

where

$$X_{DC,\text{inv}} = \begin{cases} \dfrac{1.9 \times 10^{-9}}{\left[1 + \dfrac{V_{GST,\text{effCV}} + 4(V_T - V_{fb,zb} - 2\phi_f)}{2 \times 10^8 \cdot \text{TOX}}\right]^{0.7}} & \text{if } (V_T - V_{fb,zb} - 2\phi_f) \geq 0; \\[2em] \dfrac{1.9 \times 10^{-9}}{\left[1 + \dfrac{V_{GST,\text{effCV}} + 10^{-20}}{2 \times 10^8 \cdot \text{TOX}}\right]^{0.7}} & \text{if } (V_T - V_{fb,zb} - 2\phi_f) < 0. \end{cases}$$

(A-136)

$$C_{ox,\text{inv}} = L_{\text{eff,CV}} W_{\text{eff,CV}} C'_{ox,\text{inv}}. \tag{A-137}$$

In addition to the new X_{DC}, BSIM3 also accounts for the fact that the surface potential does not get fixed at $2\phi_f$ in the strong inversion region.

$$\phi_\delta = \begin{cases} \dfrac{kT}{q} \ln\left[1 + \dfrac{V_{GST,\text{effCV}}(V_{GST,\text{effCV}} + 2 \cdot \text{K1} \cdot (\text{TOX}/\text{TOXM}) \cdot \sqrt{2\phi_f})}{\text{MOIN} \cdot \text{K1}^2(\text{TOX}/\text{TOXM})^2 \cdot (kT/q)}\right] & \text{if K1} > 0; \\[2em] \dfrac{kT}{q} \ln\left[1 + \dfrac{V_{GST,\text{effCV}}(V_{GST,\text{effCV}} + \sqrt{2\phi_f})}{0.25 \times \text{MOIN}(kT/q)}\right] & \text{if K1} \leq 0. \end{cases}$$

(A-138)

$$V_{DS,\text{satCV}} = \frac{V_{GST,\text{effCV}} - \phi_\delta}{A_{bulk,\text{CV}}}. \tag{A-139}$$

$$V_{DS,\text{effCV}} = V_{DS,\text{satCV}} - \frac{(V_{DS,\text{satCV}} - V_{DS} - \delta_4) + \sqrt{(V_{DS,\text{satCV}} - V_{DS} - \delta_4)^2 + 4\delta_4 V_{DS,\text{satCV}}}}{2}$$

δ_4 is fixed at 0.02. (A-140)

A substitution of $V_{DS} = 0$ into the equations involving the evaluation of $V_{DS,\text{effCV}}$ will prove that $V_{DS,\text{effCV}} = 0$ when $V_{DS} = 0$. However, just due to rounding errors, a computer may evaluate $V_{DS,\text{effCV}}$ to be a finite (though very small) value at $V_{DS} = 0$. In order to make sure $V_{DS,\text{effCV}}$ is identically zero when V_{DS} is zero, BSIM3 specifically adds a if–then statement that sets $V_{DS,\text{effCV}}$ to zero when V_{DS} is precisely zero.

$$Q_{inv} = -C_{ox,\text{inv}} \left[\left(V_{GST,\text{effCV}} - \phi_\delta - \frac{A_{bulk,\text{CV}} V_{DS,\text{effCV}}}{2} \right) \right.$$
$$\left. + \frac{A_{bulk,\text{CV}}^2 V_{DS,\text{effCV}}^2}{12(V_{GST,\text{effCV}} - \phi_\delta - A_{bulk,\text{CV}} V_{DS,\text{effCV}}/2 + 10^{-20})} \right]. \tag{A-141}$$

APPENDIX A

The factor 10^{-20} is added in the denominator, mainly to prevent the term from going to negative value when the term without the factor is negative and close to 0.

$$\delta Q_{sub} = C_{ox,inv} \left[\frac{1 - A_{bulk,CV}}{2} V_{DS,effCV} - \frac{(1 - A_{bulk,CV})A_{bulk,CV}V_{DS,effCV}^2}{12(V_{GST,effCV} - \phi_\delta - A_{bulk,CV}V_{DS,effCV}/2 + 10^{-20})} \right]. \quad (A\text{-}142)$$

After all four charge components are found (with different effective oxide thicknesses), the terminal charges are given by

$$Q_G = -Q_{inv} - \delta Q_{sub} + Q_{acc} + Q_{sub,0}. \quad (A\text{-}143)$$

$$Q_B = \delta Q_{sub} - Q_{acc} - Q_{sub,0}. \quad (A\text{-}144)$$

If XPART < 0.5 (40/60 charge partition; the 10^{-20} factor is dropped to simplify expression), then

$$Q_S = -\frac{C_{ox,inv}}{2\left(V_{GST,effCV} - \phi_\delta - \frac{A_{bulk,CV}V_{DS,effCV}}{2}\right)^2} \times [(V_{GST,effCV} - \phi_\delta)^3 - \tfrac{4}{3}(V_{GST,effCV} - \phi_\delta)^2 A_{bulk,CV}V_{DS,effCV} + \tfrac{2}{3}(V_{GST,effCV} - \phi_\delta)A_{bulk,CV}^2 V_{DS,effCV}^2 - \tfrac{2}{15}A_{bulk,CV}^3 V_{DS,effCV}^3].$$

$$Q_D = -\frac{C_{ox,inv}}{2\left(V_{GST,effCV} - \phi_\delta - \frac{A_{bulk,CV}V_{DS,effCV}}{2}\right)^2} \times [(V_{GST,effCV} - \phi_\delta)^3 - \tfrac{5}{3}(V_{GST,effCV} - \phi_\delta)^2 A_{bulk,CV}V_{DS,effCV} + (V_{GST,effCV} - \phi_\delta)A_{bulk,CV}^2 V_{DS,effCV}^2 - \tfrac{1}{5}A_{bulk,CV}^3 V_{DS,effCV}^3]. \quad (A\text{-}146)$$

If XPART $= 0.5$ (50/50 charge partition), then

$$Q_S = Q_D = \frac{Q_{inv}}{2}. \quad (A\text{-}147)$$

If XPART > 0.5 (0/100 charge partition), then

$$Q_S = -C_{ox,inv} \left[\frac{V_{GST,effCV} - \phi_\delta}{2} + \frac{A_{bulk,CV}V_{DS,effCV}}{4} - \frac{A_{bulk,CV}^2 V_{DS,effCV}^2}{24(V_{GST,effCV} - \phi_\delta - A_{bulk,CV}V_{DS,effCV}/2 + 10^{-20})} \right]. \quad (A\text{-}148)$$

$$Q_D = -C_{ox,inv} \left[\frac{V_{GST,effCV} - \phi_\delta}{2} - \frac{3A_{bulk,CV}V_{DS,effCV}}{4} + \frac{A_{bulk,CV}^2 V_{DS,effCV}^2}{8(V_{GST,effCV} - \phi_\delta - A_{bulk,CV}V_{DS,effCV}/2 + 10^{-20})} \right]. \quad (A\text{-}149)$$

Intrinsic Device Capacitances

The device capacitances are obtained from chain rule. Although BSIM3 works out the exact expressions, they are quite complicated. We will just express the definition.

$$C_{xy} = \delta_{xy} \frac{\partial Q_x}{\partial V_y}. \tag{A-150}$$

$\delta_{xy} = 1$ if $x = y$, and -1 if $x \neq y$. For example, $C_{gg} = +\partial Q_G/\partial V_G$, and $C_{gd} = -\partial Q_G/\partial V_D$.

Source–Bulk Junction Charge and Capacitance

If $V_{BS} < 0$, then

$$C_{j,\text{SB}} = \frac{\text{CJ}}{\left(1 - \frac{V_{BS}}{\text{PB}}\right)^{\text{MJ}}} \text{AS} + \frac{\text{CJSW}}{\left(1 - \frac{V_{BS}}{\text{PBSW}}\right)^{\text{MJSW}}} (\text{PS} - W_{\text{eff,CV}})$$

$$+ \frac{\text{CJSWG}}{\left(1 - \frac{V_{BS}}{\text{PBSWG}}\right)^{\text{MJSWG}}} W_{\text{eff,CV}}. \tag{A-151}$$

$$Q_{j,\text{SB}} = \int_0^{V_{BS}} C_{j,\text{SB}}(V'_{BS}) dV'_{BS}$$

$$= \frac{\text{CJ} \cdot \text{AS} \cdot \text{PB}}{1 - \text{MJ}} \left[1 - \left(1 - \frac{V_{BS}}{\text{PB}}\right)^{1-\text{MJ}}\right] + \frac{\text{CJSW} \cdot (\text{PS} - W_{\text{eff,CV}}) \cdot \text{PBSW}}{1 - \text{MJSW}}$$

$$\times \left[1 - \left(\frac{V_{BS}}{\text{PBSW}}\right)^{1-\text{MJSW}}\right] + \frac{\text{CJSWG} \cdot W_{\text{eff,CV}} \cdot \text{PBSWG}}{1 - \text{MJSWG}}$$

$$\times \left[1 - \left(1 - \frac{V_{BS}}{\text{PBSWG}}\right)^{1-\text{MJSWG}}\right]. \tag{A-152}$$

If $V_{BS} \geq 0$, then

$$C_{j,\text{SB}} = \text{CJ}\left(1 + \text{MJ}\frac{V_{BS}}{\text{PB}}\right)\text{AS} + \text{CJSW}\left(1 + \text{MJSW}\frac{V_{BS}}{\text{PBSW}}\right)(\text{PS} - W_{\text{eff,CV}})$$

$$+ \text{CJSWG}\left(1 + \text{MJSWG}\frac{V_{BS}}{\text{PBSWG}}\right) W_{\text{eff,CV}}. \tag{A-153}$$

$$Q_{j,\text{SB}} = \text{CJ} \cdot \text{AS}\left(V_{BS} + \frac{\text{MJ} \, V_{BS}^2}{\text{PB} \, 2}\right) + \text{CJSW} \cdot (\text{PS} - W_{\text{eff,CV}})\left(V_{BS} + \frac{\text{MJSW} \, V_{BS}^2}{\text{PBSW} \, 2}\right)$$

$$+ \text{CJSWG} \cdot W_{\text{eff,CV}}\left(V_{BS} + \frac{\text{MJSWG} \, V_{BS}^2}{\text{PBSWG} \, 2}\right). \tag{A-154}$$

Drain–Bulk Junction Charge and Capacitance

If $V_{BD} < 0$, then

$$C_{j,\text{DB}} = \frac{\text{CJ}}{\left(1 - \frac{V_{BD}}{\text{PB}}\right)^{\text{MJ}}} \text{AD} + \frac{\text{CJSW}}{\left(1 - \frac{V_{BD}}{\text{PBSW}}\right)^{\text{MJSW}}} (\text{PD} - W_{\textit{eff},\text{CV}})$$

$$+ \frac{\text{CJSWG}}{\left(1 - \frac{V_{BD}}{\text{PBSWG}}\right)^{\text{MJSWG}}} W_{\textit{eff},\text{CV}}. \quad \text{(A-155)}$$

$$Q_{j,\text{DB}} = \int_0^{V_{BD}} C_{j,\text{DB}}(V'_{BD}) dV'_{BD}$$

$$= \frac{\text{CJ} \cdot \text{AD} \cdot \text{PB}}{1 - \text{MJ}} \left[1 - \left(1 - \frac{V_{BD}}{\text{PB}}\right)^{1-\text{MJ}}\right] + \frac{\text{CJSW} \cdot (\text{PD} - W_{\textit{eff},\text{CV}}) \cdot \text{PBSW}}{1 - \text{MJSW}}$$

$$\times \left[1 - \left(1 - \frac{V_{BD}}{\text{PBSW}}\right)^{1-\text{MJSW}}\right] + \frac{\text{CJSWG} \cdot W_{\textit{eff},\text{CV}} \cdot \text{PBSWG}}{1 - \text{MJSWG}}$$

$$\times \left[1 - \left(1 - \frac{V_{BD}}{\text{PBSWG}}\right)^{1-\text{MJSWG}}\right]. \quad \text{(A-156)}$$

If $V_{BD} \geq 0$, then

$$C_{j,\text{DB}} = \text{CJ}\left(1 + \text{MJ}\frac{V_{BD}}{\text{PB}}\right)\text{AD} + \text{CJSW}\left(1 + \text{MJSW}\frac{V_{BD}}{\text{PBSW}}\right)(\text{PD} - W_{\textit{eff},\text{CV}})$$

$$+ \text{CJSWG}\left(1 + \text{MJSWG}\frac{V_{BD}}{\text{PBSWG}}\right)W_{\textit{eff},\text{CV}}. \quad \text{(A-157)}$$

$$Q_{j,\text{DB}} = \text{CJ} \cdot \text{AD}\left(V_{BD} + \frac{\text{MJ}}{\text{PB}}\frac{V_{BD}^2}{2}\right) + \text{CJSW} \cdot (\text{PD} - W_{\textit{eff},\text{CV}})\left(V_{BD} + \frac{\text{MJSW}}{\text{PBSW}}\frac{V_{BD}^2}{2}\right)$$

$$+ \text{CJSWG} \cdot W_{\textit{eff},\text{CV}}\left(V_{BD} + \frac{\text{MJSWG}}{\text{PBSWG}}\frac{V_{BD}^2}{2}\right). \quad \text{(A-158)}$$

Gate–Source Overlap Capacitance and Charge

$$V_{GS,\text{overlap}} = \frac{1}{2}\left[V_{GS} + \delta_1 - \sqrt{(V_{GS} + \delta_1)^2 + 4\delta_1}\right]; \qquad \delta_1 = 0.02. \tag{A-159}$$

$$Q_{ov,GS} = W_{\text{eff,CV}} \times \left\{\text{CGS0} \cdot V_{GS} + \text{CGS1}\left[V_{GS} - V_{GS,\text{overlap}}\right.\right.$$
$$\left.\left. - \frac{\text{CKAPPA}}{2}\left(-1 + \sqrt{1 - \frac{4 \cdot V_{GS,\text{overlap}}}{\text{CKAPPA}}}\right)\right]\right\}. \tag{A-160}$$

$$C_{ov,GS} = \frac{\partial Q_{ov,GS}}{\partial V_{GS}} = W_{\text{eff,CV}} \times \left[\text{CGS0} + \text{CGS1} - \text{CGS1}\left(1 - \frac{1}{\sqrt{1 - \frac{4 \cdot V_{GS,\text{overlap}}}{\text{CKAPPA}}}}\right)\right.$$
$$\left. \times \left(\frac{1}{2} - \frac{(V_{GS} + \delta_1)}{2\sqrt{(V_{GS} + \delta_1)^2 + 4\delta_1}}\right)\right]. \tag{A-161}$$

Gate–Drain Overlap Capacitance and Charge

$$V_{GD,\text{overlap}} = \frac{1}{2}\left[V_{GD} + \delta_1 - \sqrt{(V_{GD} + \delta_1)^2 + 4\delta_1}\right]; \qquad \delta_1 = 0.02. \tag{A-162}$$

$$Q_{ov,GD} = W_{\text{eff,CV}} \times \left\{\text{CGD0} \cdot V_{GD} + \text{CGD1}\left[V_{GD} - V_{GD,\text{overlap}}\right.\right.$$
$$\left.\left. - \frac{\text{CKAPPA}}{2}\left(-1 + \sqrt{1 - \frac{4 \cdot V_{GD,\text{overlap}}}{\text{CKAPPA}}}\right)\right]\right\}. \tag{A-163}$$

BSIM3 uses the same CKAPPA parameter for the gate–drain and gate–bulk overlap charges.

$$C_{ov,GD} = \frac{\partial Q_{ov,GD}}{\partial V_{GD}} = W_{\text{eff,CV}} \times \left[\text{CGD0} + \text{CGD1}\right.$$
$$\left. -\text{CGD1}\left(1 - \frac{1}{\sqrt{1 - \frac{4 \cdot V_{GD,\text{overlap}}}{\text{CKAPPA}}}}\right)\left(\frac{1}{2} - \frac{(V_{GD} + \delta_1)}{2\sqrt{(V_{GD} + \delta_1)^2 + 4\delta_1}}\right)\right]. \tag{A-164}$$

Fringing Capacitance and Charge

$$C_f = W_{\mathit{eff},\mathrm{CV}} \cdot \mathrm{CF}. \tag{A-165}$$

$$Q_{f,\mathrm{GS}} = \int_0^{V_{GS}} C_f dV'_{GS} = W_{\mathit{eff},\mathrm{CV}} \cdot \mathrm{CF} \cdot V_{GS}. \tag{A-166}$$

$$Q_{f,\mathrm{GD}} = \int_0^{V_{GD}} C_f dV'_{GD} = W_{\mathit{eff},\mathrm{CV}} \cdot \mathrm{CF} \cdot V_{GD}. \tag{A-167}$$

BSIM3 assumes gate–source and gate–drain fringing capacitances are identical.

Parasitic Gate–Bulk Capacitance and Charge

$$C_{gb,0} = L_{\mathit{eff},\mathrm{CV}} \cdot \mathrm{CGB0}. \tag{A-168}$$

$$Q_{GB,\mathrm{p}} = \int_0^{V_{GB}} C_{GB,\mathrm{p}} dV'_{GB} = L_{\mathit{eff},\mathrm{CV}} \cdot \mathrm{CGB0} \cdot V_{GB}. \tag{A-169}$$

Total Terminal Charges (Intrinsic Plus Parasitic Charges)

$$Q_{G,\mathrm{t}} = Q_G + Q_{ov,\mathrm{GS}} + Q_{ov,\mathrm{GD}} + Q_{f,\mathrm{GS}} + Q_{f,\mathrm{GD}} + Q_{GB,\mathrm{p}}. \tag{A-170}$$

$$Q_{D,\mathrm{t}} = Q_D - Q_{ov,\mathrm{GD}} - Q_{f,\mathrm{GD}} - Q_{j,\mathrm{DB}}. \tag{A-171}$$

$$Q_{S,\mathrm{t}} = Q_S - Q_{ov,\mathrm{GS}} - Q_{f,\mathrm{GS}} - Q_{j,\mathrm{SB}}. \tag{A-172}$$

$$Q_{B,\mathrm{t}} = Q_B + Q_{j,\mathrm{SB}} + Q_{j,\mathrm{DB}} - Q_{GB,\mathrm{p}}. \tag{A-173}$$

Total Device Capacitances (Intrinsic Plus Parasitic Capacitances)

In the following equations, $C_{f,\mathrm{GS}}$ and $C_{f,\mathrm{GD}}$ are identical to the C_f given in Eq. (A-165). BSIM3 does not distinguish between the fringing capacitances at the source side and the drain side. We denote them separately to make the equations more easily understandable. The overlap capacitances for the source side and the drain side ($C_{ov,\mathrm{GS}}$ and $G_{ov,\mathrm{GD}}$), in contrast, are different in BSIM3. They were given in Eqs.

(A-161) and (A-164), respectively.

$$C_{gg,t} = C_{gg} + C_{ov,GS} + C_{f,GS} + C_{ov,GD} + C_{f,GD} + C_{gb,0}. \quad \text{(A-174)}$$
$$C_{gd,t} = C_{gd} + C_{ov,GD} + C_{f,GD}. \quad \text{(A-175)}$$
$$C_{gs,t} = C_{gs} + C_{ov,GS} + C_{f,GS}. \quad \text{(A-176)}$$
$$C_{gb,t} = C_{gb} + C_{gb,0}. \quad \text{(A-177)}$$
$$C_{dg,t} = C_{dg} + C_{ov,GD} + C_{f,GD}. \quad \text{(A-178)}$$
$$C_{dd,t} = C_{dd} + C_{ov,GD} + C_{f,GD} + C_{j,DB}. \quad \text{(A-179)}$$
$$C_{ds,t} = C_{ds}. \quad \text{(A-180)}$$
$$C_{db,t} = C_{db} + C_{j,DB}. \quad \text{(A-181)}$$
$$C_{sg,t} = C_{sg} + C_{ov,GS} + C_{f,GS}. \quad \text{(A-182)}$$
$$C_{sd,t} = C_{sd}. \quad \text{(A-183)}$$
$$C_{ss,t} = C_{ss} + C_{ov,GS} + C_{f,GS} + C_{j,SB}. \quad \text{(A-184)}$$
$$C_{sb,t} = C_{sb} + C_{j,SB}. \quad \text{(A-185)}$$
$$C_{bg,t} = C_{bg} + C_{gb,0}. \quad \text{(A-186)}$$
$$C_{bd,t} = C_{bd} + C_{j,DB}. \quad \text{(A-187)}$$
$$C_{bs,t} = C_{bs} + C_{j,SB}. \quad \text{(A-188)}$$
$$C_{bb,t} = C_{bb} + C_{j,SB} + C_{j,DB} + C_{gb,0}. \quad \text{(A-189)}$$

The following transcapacitances are not really used in the code, but it is convenient for us to list them out, anyway.

$$C_m = C_{dg,t} - C_{gd,t}. \quad \text{(A-190)}$$
$$C_{mb} = C_{db,t} - C_{bd,t}. \quad \text{(A-191)}$$
$$C_{mx} = C_{bg,t} - C_{gb,t}. \quad \text{(A-192)}$$

Transient Current Equations

$$i_G(t) = C_{gg,t} \cdot \frac{dv_{GS}}{dt} - C_{gd,t} \cdot \frac{dv_{DS}}{dt} - C_{gb,t} \cdot \frac{dv_{BS}}{dt}. \quad \text{(A-193)}$$
$$i_D(t) = I_{DS} + I_{sub} - I_{j,DB} - C_{dg,t} \cdot \frac{dv_{GS}}{dt} + C_{dd,t} \cdot \frac{dv_{DS}}{dt} - C_{db,t} \cdot \frac{dv_{BS}}{dt}. \quad \text{(A-194)}$$
$$i_B(t) = -I_{sub} + I_{j,DB} + I_{j,SB} - C_{bg,t} \cdot \frac{dv_{GS}}{dt} - C_{bd,t} \cdot \frac{dv_{DS}}{dt} + C_{bb,t} \cdot \frac{dv_{BS}}{dt}. \quad \text{(A-195)}$$
$$i_S(t) = -i_G(t) - i_D(t) - i_B(t). \quad \text{(A-196)}$$

Small-Signal Current Equations

$$i_g = j\omega C_{gs} v_{gs} + j\omega C_{gd} v_{gd} + j\omega C_{gb} v_{gb}. \tag{A-197}$$

$$i_d = g_m v_{gs} + g_d v_{ds} + g_{mb} v_{bs} + G_{bg} v_{gs} + G_{bd} v_{ds} + G_{bb} v_{bs} + g_{j,\text{DB}} v_{db}$$
$$+ j\omega C_{sd,t} v_{ds} + j\omega C_{gd,t} v_{dg} + j\omega C_{bd,t} v_{db} - j\omega C_m V_{gs} - j\omega C_{mb} v_{bs}. \tag{A-198}$$

$$i_b = -G_{bg} v_{gs} - G_{bd} v_{ds} - G_{bb} v_{bs} + g_{j,\text{DB}} v_{bd} + g_{j,\text{SB}} v_{bs}$$
$$+ j\omega C_{bs,t} v_{bs} + j\omega C_{bd,t} v_{bd} + j\omega C_{gb,t} v_{gb} - j\omega C_{mx} v_{bg}. \tag{A-199}$$

$$i_s(t) = -i_g(t) - i_d(t) - i_b(t). \tag{A-200}$$

Temperature Modeling

If the device temperature is not equal to TNOM, then various quantities need to be substituted with the results given below. See the entry TNOM in Section 3.2 for more details.

$$V_T(T_{\text{device}}) = V_T(\text{TNOM}) + \left[\text{KT1} + \frac{\text{KT1L}}{L_{\text{eff}}} + \text{KT2} V_{BS,\text{eff}} \right] \times \left[\frac{T_{\text{device}} + 273.15}{\text{TNOM} + 273.15} - 1 \right]. \tag{A-201}$$

$$\text{VSAT}(T_{\text{device}}) = \text{VSAT}(\text{TNOM}) - \text{AT} \times \left[\frac{T_{\text{device}} + 273.15}{\text{TNOM} + 273.15} - 1 \right]. \tag{A-202}$$

Suppose that the device temperature is high enough such that $\text{VSAT}(T_{\text{device}})$ given above decreases to a negative value. BSIM3 will catch this unphysical event and issues a fatal error. There is another BSIM3 checking mechanism, which requires the parameter PARAMCHK to be turned on. When PARAMCHK is set to 1, then BSIM3 will start to issue a warning when $\text{VSAT}(T_{\text{device}})$ decreases below 1×10^3 m/s.

$$\text{U0}(T_{\text{device}}) = \text{U0}(\text{TNOM}) \times \left(\frac{T_{\text{device}} + 273.15}{\text{TNOM} + 273.15} \right)^{\text{UTE}}. \tag{A-203}$$

BSIM3 checks whether $\text{U0}(T_{\text{device}})$ is less than 0. If so, BSIM3 issues a fatal warning since the mobility must be a positive number. It is clear from the above equation that the only way to make $\text{U0}(T_{\text{device}}) < 0$ is by having nonphysical U0(TNOM).

$$\text{UA}(T_{\text{device}}) = \text{UA}(\text{TNOM}) + \text{UA1} \times \left[\frac{T_{\text{device}} + 273.15}{\text{TNOM} + 273.15} - 1 \right]. \tag{A-204}$$

$$\text{UB}(T_{\text{device}}) = \text{UB}(\text{TNOM}) + \text{UB1} \times \left[\frac{T_{\text{device}} + 273.15}{\text{TNOM} + 273.15} - 1 \right]. \tag{A-205}$$

$$\text{UC}(T_{\text{device}}) = \text{UC}(\text{TNOM}) + \text{UC1} \times \left[\frac{T_{\text{device}} + 273.15}{\text{TNOM} + 273.15} - 1 \right]. \tag{A-206}$$

UA, UB, and UC are coefficients which modify the carrier mobility. They can be negative or positive, as far as calculating a numerical value for the carrier mobility is concerned. BSIM3 does not check whether they are positive or negative.

$$\text{RDSW}(T_{device}) = \text{RDSW}(\text{TNOM}) + \text{PRT} \times \left[\frac{T_{device} + 273.15}{\text{TNOM} + 273.15} - 1\right]. \quad \text{(A-207)}$$

If the calculated $R_{DS} \approx \text{RDSW}(T_{device})/W_{eff}^{WR}$ from Eq. (A-57) is found to be smaller than $0.001\,\Omega$, then BSIM3 issues a warning and sets $\text{RDSW}(T_{device})$ to 0.

$$\text{CJ}(T_{device}) = \text{CJ}(\text{TNOM}) \times [1 + \text{TCJ} \cdot (T_{device} - \text{TNOM})] \quad \text{(A-208)}$$

If $\text{TCJ} \cdot (T_{device} - \text{TNOM})$ is less than -1, then BSIM3 issues a warning and set $\text{CJ}(T_{device})$ to 0.

$$\text{PB}(T_{device}) = \text{PB}(\text{TNOM}) - \text{TPB} \cdot (T_{device} - \text{TNOM}). \quad \text{(A-209)}$$

If the calculated $\text{PB}(T_{device})$ is less than 0.01, then BSIM3 issues a warning and set $\text{PB}(T_{device})$ to 0.01.

$$\text{CJSW}(T_{device}) = \text{CJSW}(\text{TNOM}) \times [1 + \text{TCJSW} \cdot (T_{device} - \text{TNOM})]. \quad \text{(A-210)}$$

If $\text{TCJSW} \cdot (T_{device} - \text{TNOM})$ is less than -1, then BSIM3 issues a warning and set $\text{CJSW}(T_{device})$ to 0.

$$\text{PBSW}(T_{device}) = \text{PBSW}(\text{TNOM}) - \text{TPBSW} \cdot (T_{device} - \text{TNOM}). \quad \text{(A-211)}$$

If the calculated $\text{PBSW}(T_{device})$ is less than 0.01, then BSIM3 issues a warning and set $\text{PBSW}(T_{device})$ to 0.01.

$$\text{CJSWG}(T_{device}) = \text{CJSWG}(\text{TNOM}) \times [1 + \text{TCJSWG} \cdot (T_{device} - \text{TNOM})]. \quad \text{(A-212)}$$

If $\text{TCJSWG} \cdot (T_{device} - \text{TNOM})$ is less than -1, then BSIM3 issues a warning and set $\text{CJSWG}(T_{device})$ to 0.

$$\text{PBSWG}(T_{device}) = \text{PBSWG}(\text{TNOM}) - \text{TPBSWG} \cdot (T_{device} - \text{TNOM}). \quad \text{(A-213)}$$

If the calculated $\text{PBSWG}(T_{device})$ is less than 0.01, then BSIM3 issues a warning and set $\text{PBSWG}(T_{device})$ to 0.01.

Due to space limitation, T_{device} and TNOM in the following three equations are assumed to be in Kelvin.

$$\text{JS}(T_{device}) = \text{JS}(\text{TNOM}) \left[\exp\left(\frac{E_g(T_{device})}{\text{NJ} \cdot kT_{device}} - \frac{E_g(\text{TNOM})}{\text{NJ} \cdot k\text{TNOM}} \right) \right] \times \left(\frac{T_{device}}{\text{TNOM}} \right)^{\text{XTI/NJ}}. \quad \text{(A-214)}$$

$$\text{JSSW}(T_{device}) = \text{JSSW}(\text{TNOM}) \left[\exp\left(\frac{E_g(T_{device})}{\text{NJ} \cdot kT_{device}} - \frac{E_g(\text{TNOM})}{\text{NJ} \cdot k\text{TNOM}} \right) \right]$$
$$\times \left(\frac{T_{device}}{\text{TNOM}} \right)^{\text{XTI/NJ}}, \quad \text{(A-215)}$$

where

$$E_g \text{ (in eV)} = 1.160 - \frac{7.02 \times 10^{-4} T_{device}^2}{T_{device} + 1108} \quad (T_{device} \text{ is in Kelvin}). \quad \text{(A-216)}$$

Noise Modeling

See the special note about the W_{eff} used in the noise models, found in the discussion right after Eq. (A-6).

$$\left. \frac{\overline{i_d^2}}{\Delta f} \right|_{channel, \text{SPICE2}} = \frac{8kT}{3} \times |g_m + g_d + g_{mb}|. \quad \text{(A-217)}$$

$$\left. \frac{\overline{i_d^2}}{\Delta f} \right|_{channel, \text{BSIM3}} = \frac{4kT\mu_{eff}}{L_{eff}^2} |Q_{inv}|. \quad \text{(A-218a)}$$

Equation (A-218a) is applicable for BSIM3v3.3.2 or prior versions. At the writing of this book, there has been talk of releasing a new version of BSIM3, after the release of BSIM4. This new version has been designated as BSIM3v3.3; see Ref. [15] of Chapter 1. This new BSIM3 version is to be the same as

BSIM3v3.2.2, except for the above thermal noise expression, as discussed in Section 4.11. For this new version of BSIM3,

$$\left.\frac{\overline{i_d^2}}{\Delta f}\right|_{\text{channel,BSIM3}} = \frac{4kT}{R_{DS} + L_{\text{eff}}^2/(\mu_{\text{eff}}|Q_{\text{inv}}|)}. \quad \text{(A-218b)}$$

$$\left.\frac{\overline{i_d^2}}{\Delta f}\right|_{\text{Flicker,SPICE2}} = \text{KF} \times \frac{|I_{DS}|^{\text{AF}}}{C'_{ox} L_{\text{eff}}^2 f^{\text{EF}}}. \quad \text{(A-219)}$$

$$\left.\frac{\overline{i_d^2}}{\Delta f}\right|_{\text{Flicker,BSIM3}} = \begin{cases} S_{\text{inversion}} & \text{if } V_{GS} > V_T + 0.1, \\ S_{\text{subthreshold}} & \text{if otherwise.} \end{cases} \quad \text{(A-220)}$$

$$S_{\text{inversion}} = \frac{q^2 kT \mu_{\text{eff}} I_{DS}}{C'_{ox} L_{\text{eff}}^2 f^{\text{EF}} 10^8} \left[\text{NOIA} \ln\left(\frac{N_0 + 2 \times 10^{14}}{N_1 + 2 \times 10^{14}}\right) + \text{NOIB}(N_0 - N_1) \right.$$
$$\left. + \frac{\text{NOIC}}{2}(N_0^2 - N_1^2) \right] + \frac{kT/q I_{DS}^2 \Delta L_{\text{clm}}}{W_{\text{eff,special}} L_{\text{eff}}^2 f^{\text{EF}} 10^8}$$
$$\times \frac{\text{NOIA} + \text{NOIB} \cdot N_1 + \text{NOIC} \cdot N_1^2}{(N_1 + 2 \times 10^{14})^2}. \quad \text{(A-221)}$$

$$S_{\text{subthreshold}} = \frac{S_{\text{limit}} \times S_{\text{weak-inversion}}}{S_{\text{limit}} + S_{\text{weak-inversion}}}, \quad \text{(A-222)}$$

where

$$N_0 = C'_{ox}(V_{GS} - V_T)/q. \quad \text{(A-223)}$$

$$N_1 = \frac{C'_{ox}[V_{GS} - V_T - \min(V_{DS}, V_{DS,\text{sat}})]}{q}. \quad \text{(A-224)}$$

$$\Delta L_{\text{clm}} = \begin{cases} L_{itl} \ln\left[\frac{(V_{DS} - V_{DS,\text{sat}})/L_{ittl} + \text{EM}}{\varepsilon_{\text{sat}}}\right] & \text{if } V_{DS} > V_{DS,\text{sat}}; \\ 0 & \text{if otherwise.} \end{cases} \quad \text{(A-225)}$$

$$S_{\text{limit}} = S_{\text{inversion}} \quad \text{evaluated at } V_{GS} = V_T + 0.1. \quad \text{(A-226)}$$

$$S_{\text{weak-inversion}} = \frac{\text{NOIA } kT/q I_{DS}^2}{W_{\text{eff,special}} L_{\text{eff}} f^{\text{EF}} \cdot 4 \times 10^{36}}. \quad \text{(A-227)}$$

It is L_{eff}, not $L_{\text{eff,CV}}$, in all of the above equations.

External Resistor Calculation (R_D, R_S):

As described in Section 2.2, there are several ways to represent the effects of source/drain resistances. Two fundamental representations were shown in Fig. 2-8. In Fig. 2-8a, the contact resistances are lumped elements connected externally to the intrinsic transistor, which should be characterized with a RDSW of 0. In Fig. 2-8b, the effects of the resistances are absorbed by the MOS transistor, with the parameter RDSW. That is, $R_S = R_D$ found in Fig. 2-8a become zero in Fig. 2-8b. However, a nonzero RDSW modifies the equations in the MOS of Fig. 2-8b, effectively producing nearly identical I-V and small-signal characteristics. The third method to represent the source/drain resistances is a combination of the said two methods, using a combination of calculated lumped elements as well as RDSW. The actual BSIM3 code, as it is written, implicitly uses the third method. Therefore, the overall transistor consists of R_S and R_D plus a MOS whose RDSW can be zero or nonzero. R_S and R_D external to the MOS are calculated as

$$R_S = \text{RSH} \times \text{NRS}; \qquad (A\text{-}228)$$
$$R_D = \text{RSH} \times \text{NRD}; \qquad (A\text{-}229)$$

where NRS and NRD are specified in the MOS device statement, along with the transistor size and junction areas and peripheries. The internal R_{DS} modeled by RDSW has been calculated using the equations starting at Eq. (A-54).

APPENDIX B

CAPACITANCES AND CHARGES FOR ALL BIAS CONDITIONS

We write out the charge and capacitance expressions here. They are not exactly identical to those of BSIM3. However, these simplified expressions are convenient for users to do back-of-envelope type of calculations. They apply to the intrinsic transistor, not including the parasitic capacitances. These expressions are derived with the 40/60 partition scheme, which is the physical partition scheme once the quasi-static approximation is made (see XPART entry in Section 3.2). For the detailed BSIM3 expressions, see Appendix A.

The following expressions apply to the inversion region ($V_{GS} > V_T$, the threshold voltage), whether the transistor is in the linear region or saturation region.

$$Q_G = WLC'_{ox}\left[\frac{V_{GS}-V_T}{1+\delta}\left(\delta + \frac{2}{3}\frac{\alpha^2+\alpha+1}{1+\alpha}\right) + \gamma\sqrt{2\phi_f - V_{BS}}\right]. \quad (B-1)$$

$$Q_B = -WLC'_{ox}\left[\frac{\delta}{1+\delta}\cdot(V_{GS}-V_T)\cdot\left(1 - \frac{2}{3}\frac{\alpha^2+\alpha+1}{1+\alpha}\right) + \gamma\sqrt{2\phi_f - V_{BS}}\right]. \quad (B-2)$$

$$Q_D = -WLC'_{ox}\left[(V_{GS}-V_T)\frac{6\alpha^3+12\alpha^2+8\alpha+4}{15(1+\alpha)^2}\right]. \quad (B-3)$$

$$Q_S = -WLC'_{ox}\left[(V_{GS}-V_T)\frac{4\alpha^3+8\alpha^2+12\alpha+6}{15(1+\alpha)^2}\right]. \quad (B-4)$$

$$C_{gg} = WLC'_{ox}\left[\frac{2}{3}\times\frac{\alpha^2+4\alpha+1}{(1+\alpha)^2} + \frac{\delta}{3(1+\delta)}\times\left(\frac{1-\alpha}{1+\alpha}\right)^2\right]. \quad (B-5)$$

$$C_{gd} = WLC'_{ox}\left[\frac{2}{3}\times\frac{\alpha^2+2\alpha}{(1+\alpha)^2}\right]. \quad (B-6)$$

508 APPENDIX B

$$C_{gs} = WLC'_{ox} \left[\frac{2}{3} \times \frac{2\alpha + 1}{(1+\alpha)^2} \right]. \tag{B-7}$$

$$C_{gb} = WLC'_{ox} \left[\frac{\delta}{3(1+\delta)} \times \left(\frac{1-\alpha}{1+\alpha} \right)^2 \right]. \tag{B-8}$$

$$C_{dg} = WLC'_{ox} \left[\frac{2}{15} \times \frac{3\alpha^2 + 11\alpha^2 + 14\alpha + 2}{(1+\alpha)^3} \right]. \tag{B-9}$$

$$C_{dd} = WLC'_{ox} \left[\frac{2(1+\delta)}{15} \times \frac{3\alpha^3 + 9\alpha^2 + 8\alpha}{(1+\alpha)^3} \right]. \tag{B-10}$$

$$C_{ds} = -WLC'_{ox} \left[\frac{4(1+\delta)}{15} \times \frac{\alpha^2 + 3\alpha + 1}{(1+\alpha)^3} \right]. \tag{B-11}$$

$$C_{db} = \delta \times C_{dg}. \tag{B-12}$$

$$C_{sg} = WLC'_{ox} \left[\frac{2}{15} \times \frac{2\alpha^2 + 14\alpha^2 + 11\alpha + 3}{(1+\alpha)^3} \right]. \tag{B-13}$$

$$C_{sd} = -WLC'_{ox} \left[\frac{4(1+\delta)}{15} \times \frac{\alpha^3 + 3\alpha^2 + \alpha}{(1+\alpha)^3} \right]. \tag{B-14}$$

$$C_{ss} = WLC'_{ox} \left[\frac{2(1+\delta)}{15} \times \frac{8\alpha^2 + 9\alpha + 3}{(1+\alpha)^3} \right]. \tag{B-15}$$

$$C_{sb} = \delta \times C_{sg}. \tag{B-16}$$

$$C_{bg} = C_{gb}. \tag{B-17}$$

$$C_{bd} = \delta \times C_{gd}. \tag{B-18}$$

$$C_{bs} = \delta \times C_{gs}. \tag{B-19}$$

$$C_{bb} = WLC'_{ox} \left[\frac{2\delta}{3} \times \frac{\alpha^2 + 4\alpha + 1}{(1+\alpha)^2} + \frac{\delta}{3(1+\delta)} \times \left(\frac{1-\alpha}{1+\alpha} \right)^2 \right]. \tag{B-20}$$

The saturation index (α) is found in Eq. (1-33). The bulk-charge factor (δ) is found in Eq. (1-7).

In the depletion region ($V_{GS} < V_T$ but $V_{GB} > V_{FB}$, the flatband voltage):

$$Q_G = WLC'_{ox} \left[\gamma \left(-\frac{\gamma}{2} + \sqrt{\frac{\gamma^2}{4} + V_{GB} - V_{FB}} \right) \right]. \tag{B-21}$$

$$Q_B = -Q_G. \tag{B-22}$$

$$Q_D = Q_S = 0. \tag{B-23}$$

$$C_{gg} = C_{gb} = C_{bb} = C_{bg} = WLC'_{ox} \left[\frac{\gamma}{2\sqrt{\frac{\gamma^2}{4} + V_{GB} - V_{FB}}} \right]. \tag{B-24}$$

$$C_{gs} = C_{gd} = C_{dg} = C_{dd} = C_{ds} = C_{db} = C_{sg} = C_{sd} = C_{ss} = C_{sb} = C_{bd} = C_{bs} = 0, \tag{B-25}$$

where γ, the body-effect coefficient, can be identified from Fig. 1-18. Its relationship with δ is described in Eq. (1-7).

In the accumulation region ($V_{GB} < V_{FB}$):

$$Q_G = WLC'_{ox}[V_{GS} - \phi_{ms}]. \tag{B-26}$$
$$Q_B = -Q_G. \tag{B-27}$$
$$Q_D = Q_S = 0. \tag{B-28}$$
$$C_{gg} = C_{gb} = C_{bb} = C_{bg} = WLC'_{ox}. \tag{B-29}$$
$$C_{gs} = C_{gd} = C_{dg} = C_{dd} = C_{ds} = C_{db} = C_{sg} = C_{sd}$$
$$= C_{ss} = C_{sb} = C_{bd} = C_{bs} = 0, \tag{B-30}$$

where ϕ_{ms} is the metal-semiconductor work function. The interface charge is assumed to be zero.

APPENDIX C

NON-QUASI-STATIC y-PARAMETERS

We present the formula for the common-source y-parameters for MOSFETs [1,2]. The remaining seven y-parameters can be obtained from these nine linearly independent y-parameters, according to Eq. (1-87). The common-source y-parameters are expressed in the following forms:

$$y_{gg} = \frac{N_{gg}(j\omega)}{D(j\omega)}; \quad y_{gd} = \frac{N_{gd}(j\omega)}{D(j\omega)}; \quad y_{gb} = \frac{N_{gb}(j\omega)}{D(j\omega)};$$

$$y_{dg} = \frac{N_{dg}(j\omega)}{D(j\omega)}; \quad y_{dd} = \frac{N_{dd}(j\omega)}{D(j\omega)}; \quad y_{db} = \frac{N_{db}(j\omega)}{D(j\omega)}; \quad \text{(C-1)}$$

$$y_{bg} = \frac{N_{bg}(j\omega)}{D(j\omega)}; \quad y_{bd} = \frac{N_{bd}(j\omega)}{D(j\omega)}; \quad y_{bb} = \frac{N_{bb}(j\omega)}{D(j\omega)};$$

where $D(j\omega)$ and $N_{xy}(j\omega)$'s are given by

$$D(j\omega) = D_0 + (j\omega)D_1 + (j\omega)^2 D_2 + \cdots$$

$$N_{gg}(j\omega) = N_{gg,0} + (j\omega)N_{gg,1} + (j\omega)^2 N_{gg,2} + \cdots$$

$$N_{gd}(j\omega) = N_{gd,0} + (j\omega)N_{gd,1} + (j\omega)^2 N_{gd,2} + \cdots \quad \text{(C-2)}$$

$$N_{dg}(j\omega) = N_{dg,0} + (j\omega)N_{dg,1} + (j\omega)^2 N_{dg,2} + \cdots$$

$$N_{dd}(j\omega) = N_{dd,0} + (j\omega)N_{dd,1} + (j\omega)^2 N_{dd,2} + \cdots$$

APPENDIX C 511

For $n = 0$, we have

$$D_0 = 1; \quad N_{gg,0} = 0; \quad N_{gd,0} = 0; \quad N_{gb,0} = 0;$$
$$N_{dg,0} = g_m; \quad N_{dd,0} = g_d; \quad N_{db,0} = \delta g_m = g_{mb};$$
$$N_{bg,0} = 0; \quad N_{bd,0} = 0; \quad N_{bb,0} = 0. \quad \text{(C-3)}$$

The following is for $n \geq 1$:

$$D_n = \omega_0^{-n} \frac{4^{n+1}}{2(1-\alpha^2)^{2n+1}} \times \left\{ \frac{1 - \alpha^{3n+2}}{n! 3^n \prod_{i=1}^{n+1} 3i - 1} \right.$$
$$\left. + \sum_{m=1}^{N} \frac{\alpha^{3n-3m+3} - \alpha^{3m-1}}{(m-1)!(n-m+1)! 3^n \prod_{i=1}^{m} 3i - 1 \prod_{i=1}^{n-m} 3i + 1} \right\} \quad \text{(C-4)}$$

$$\frac{N_{gg,n}}{WLC'_{ox}} = \frac{\delta}{1+\delta} D_{n-1} + \frac{1}{1+\delta} \frac{\omega_0^{1-n} 4^n}{(1-\alpha^2)^{2n-1}(1+\alpha)3^{n-1}} \left\{ \frac{1}{(n-1)! \prod_{i=1}^{n} 3i - 1} \right.$$
$$\times \left[\alpha - \alpha^{3n-1} + \frac{(1-\alpha)(\alpha^{3n-1} + \alpha^{3n-2} + \cdots + 1)}{3n} \right]$$
$$+ \sum_{m=1}^{n-1} \frac{\alpha - \alpha^{3m-1}}{(m-1)!(n-m)! \prod_{i=1}^{m} 3i - 1 \prod_{i=1}^{n-m-1} 3i + 1}$$
$$\times \frac{\alpha^{3n-3m} + \alpha^{3n-3m-1} + \cdots + 1}{3n - 3m + 1}$$
$$\left. + \frac{(\alpha^{3m} - \alpha)(\alpha^{3n-3m-1} + \alpha^{3n-3m-2} + \cdots + 1)}{m!(n-m-1)!(3n-3m) \prod_{i=1}^{m-1} 3i + 1 \prod_{i=1}^{n-m} 3i - 1} \right\}. \quad \text{(C-5)}$$

$$\frac{N_{gd,n}}{WLC'_{ox}} = \frac{\omega_0^{1-n} 4^n \alpha}{(1-\alpha^2)^{2n-1}(1+\alpha)} \left\{ \frac{1}{(n-1)! 3^{n-1} \prod_{i=1}^{n} 3i - 1} \right.$$
$$\times \left(\frac{\alpha^{3n-1} + \alpha^{3n-2} + \cdots + 1}{3n} - 1 \right)$$
$$+ \sum_{m=1}^{n-1} \frac{\alpha^{3m-1} + \alpha^{3m-2} + \cdots + 1}{m!(n-m)! 3^n \prod_{i=1}^{m} 3i - 1 \prod_{i=1}^{n-m-1} 3i + 1}$$
$$\left. - \frac{\alpha^{3m} + \alpha^{3m-1} + \cdots + 1}{m!(n-m-1)! 3^{n-1}(3m+1) \prod_{i=1}^{m-1} 3i + 1 \prod_{i=1}^{n-m} 3i - 1} \right\}. \quad \text{(C-6)}$$

$$\frac{N_{gb,n}}{WLC'_{ox}} = \frac{-\delta}{1+\delta}D_{n-1} + \frac{\delta}{1+\delta}\frac{\omega_0^{1-n}4^n}{(1-\alpha^2)^{2n-1}(1+\alpha)}\left\{\frac{\alpha - \alpha^{3n-1}}{(n-1)!3^{n-1}\prod_{i=1}^{n}3i-1}\right.$$

$$+ \frac{(1-\alpha)(\alpha^{3n-1} + \alpha^{3n-2} + \cdots + 1)}{n!3^n \prod_{i=1}^{n} 3i - 1}$$

$$+ \sum_{m=1}^{n-m} \frac{(\alpha - \alpha^{3m-1})(\alpha^{3n-3m} + \alpha^{3n-3m-1} + \cdots + 1)}{(m-1)!(n-m)!3^{n-1}(3n-3m+1)\prod_{i=1}^{m} 3i - 1 \prod_{i=1}^{n-m-1} 3i + 1}$$

$$\left. \times \frac{(\alpha^{3m} - \alpha)(\alpha^{3n-3m-1} + \alpha^{3n-3m-2} + \cdots + 1)}{m!(n-m)!3^n \prod_{i=1}^{m-1} 3i + 1 \prod_{i=1}^{n-m} 3i - 1}\right\}. \quad \text{(C-7)}$$

$$\frac{N_{dg,n}}{WLC'_{ox}} = \frac{\omega_0^{1-n}4^n}{(1-\alpha^2)^{2n}}\left\{\frac{\alpha^2 - \alpha^{3n}}{(n-1)!3^{n-1}\prod_{i=1}^{n}3i-1}\right.$$

$$+ \frac{\alpha^{3n} - \alpha}{n!3^n \prod_{i=1}^{n-1} 3i + 1} + \frac{(1-\alpha)\alpha^{3n}}{n!3^n \prod_{i=1}^{n} 3i - 1}$$

$$+ \sum_{m=1}^{n-1} \frac{(\alpha^2 - \alpha^{3m}) \cdot \alpha^{3n-3m}}{(m-1)!(n-m)!3^{n-1}\prod_{i=1}^{m} 3i - 1 \prod_{i=1}^{n-m} 3i + 1}$$

$$\left. - \frac{(\alpha^{3m} - \alpha) \cdot \alpha^{3n-3m}}{m!(n-m)!3^n \prod_{i=1}^{m-1} 3i + 1 \prod_{i=1}^{n-m} 3i - 1}\right\}. \quad \text{(C-8)}$$

$$\frac{N_{dd,n}}{WLC'_{ox}} = (1+\delta) \times \frac{\omega_0^{1-n}4^n}{(1-\alpha^2)^{2n}}\left\{\frac{\alpha}{n!3^n \prod_{i=1}^{n-1} 3i + 1} + \frac{\alpha^{3n+1}}{n!3^n \prod_{i=1}^{n} 3i - 1}\right.$$

$$- \frac{\alpha^2}{(n-1)!3^{n-1}\prod_{i=1}^{n} 3i - 1}$$

$$+ \sum_{m=1}^{n-1} \frac{\alpha^{3n-3m+1}}{m!(n-m)!3^n \prod_{i=1}^{m-1} 3i + 1 \prod_{i=1}^{n-m} 3i - 1}$$

$$\left. - \frac{\alpha^{3n-3m+2}}{(m-1)!(n-m)!3^{n-1}\prod_{i=1}^{m} 3i - 1 \prod_{i=1}^{n-m} 3i + 1}\right\}. \quad \text{(C-9)}$$

$$N_{db,n} = \delta \times N_{dg,n}. \quad \text{(C-10)}$$

$$N_{bg,n} = N_{gb,n}. \quad \text{(C-11)}$$

$$N_{bd,n} = \delta \times N_{gd,n}. \quad \text{(C-12)}$$

$$\frac{N_{bb,n}}{WLC'_{ox}} = \frac{\delta}{1+\delta}D_{n-1} + \frac{\delta^2}{1+\delta}\frac{\omega_0^{1-n}4^n}{(1-\alpha^2)^{2n-1}(1+\alpha)}\left\{\frac{\alpha-\alpha^{3n-1}}{(n-1)!3^{n-1}\prod_{i=1}^n 3i-1}\right.$$

$$+ \frac{(1-\alpha)(\alpha^{3n-1}+\alpha^{3n-2}+\cdots+1)}{n!3^n \prod_{i=1}^n 3i-1}$$

$$+ \sum_{m=1}^{n-m} \frac{(\alpha-\alpha^{3m-1})(\alpha^{3n-3m}+\alpha^{3n-3m-1}+\cdots+1)}{(m-1)!(n-m)!3^{n-1}(3n-3m+1)\prod_{i=1}^m 3i-1\prod_{i=1}^{n-m-1} 3i+1}$$

$$\left. + \frac{(\alpha^{3m}-\alpha)(\alpha^{3n-3m-1}+\alpha^{3n-3m-2}+\cdots+1)}{m!(n-m)!3n\prod_{i=1}^{m-1} 3i+1\prod_{i=1}^{n-m} 3i-1}\right\}. \qquad \text{(C-13)}$$

It is convenient to express some of these coefficients explicitly, up to $n=2$. The terms with $n=0$ were described in Eq. (C-3). The following are the terms associated with $n=1$ and 2.

$$D_1 = \frac{4}{15}\frac{1}{\omega_0}\frac{\alpha^2+3\alpha+1}{(1+\alpha)^3}. \qquad \text{(C-14)}$$

$$D_2 = \frac{1}{45}\frac{1}{\omega_0^2}\frac{\alpha^2+4\alpha+1}{(1+\alpha)^4}. \qquad \text{(C-15)}$$

$$N_{gg,1} = WLC'_{ox}\left\{\frac{2}{3(1+\delta)}\frac{\alpha^2+4\alpha+1}{(1+\alpha)^2}+\frac{\delta}{1+\delta}\right\}. \qquad \text{(C-16)}$$

$$N_{gg,2} = \frac{WLC'_{ox}}{\omega_0}\left\{\frac{2}{45(1+\delta)}\frac{2\alpha^2+11\alpha+2}{(1+\alpha)^3}+\frac{4}{15}\frac{\delta}{1+\delta}\frac{\alpha^2+3\alpha+1}{(1+\alpha)^3}\right\}. \qquad \text{(C-17)}$$

$$N_{gd,1} = -WLC'_{ox}\frac{2}{3}\frac{\alpha(\alpha+2)}{(1+\alpha)^2}. \qquad \text{(C-18)}$$

$$N_{gd,2} = -\frac{WLC'_{ox}}{\omega_0}\frac{2}{45}\frac{\alpha(2\alpha^2+8\alpha+5)}{(1+\alpha)^4}. \qquad \text{(C-19)}$$

$$N_{gb,1} = -WLC'_{ox}\left\{\frac{\delta}{3(1+\delta)}\left(\frac{1-\alpha}{1+\alpha}\right)^2\right\}. \qquad \text{(C-20)}$$

$$N_{gb,2} = -\frac{WLC'_{ox}}{\omega_0}\left\{\frac{2}{45}\frac{\delta}{1+\delta}\frac{4\alpha^2+7\alpha+4}{(1+\alpha)^3}\right\}. \qquad \text{(C-21)}$$

$$N_{dg,1} = N_{gd,1}. \qquad \text{(C-22)}$$

$$N_{dg,2} = N_{gd,2}. \qquad \text{(C-23)}$$

$$N_{dd,1} = WLC'_{ox}(1+\delta)\times\frac{2}{3}\frac{\alpha(\alpha+2)}{(1+\alpha)^2}. \qquad \text{(C-24)}$$

$$N_{dd,2} = \frac{WLC'_{ox}}{\omega_0}(1+\delta) \times \frac{2}{45}\frac{\alpha(2\alpha^2+8\alpha+5)}{(1+\alpha)^4}. \tag{C-25}$$

$$N_{db,1} = \delta \times N_{gd,1}. \tag{C-26}$$

$$N_{db,2} = \delta \times N_{gd,2}. \tag{C-27}$$

$$N_{bg,1} = N_{gb,1}. \tag{C-28}$$

$$N_{bg,2} = N_{gb,2}. \tag{C-29}$$

$$N_{bd,1} = \delta \times N_{gd,1}. \tag{C-30}$$

$$N_{bd,2} = \delta \times N_{gd,2}. \tag{C-31}$$

$$N_{bb,1} = WLC_{ox}\left\{\frac{\delta}{1+\delta} + \frac{2\delta^2}{3(1+\delta)}\frac{\alpha^2+4\alpha+1}{(1+\alpha)^2}\right\}. \tag{C-32}$$

$$N_{bb,2} = \frac{WLC'_{ox}}{\omega_0}\left\{\frac{2\delta^2}{45(1+\delta)}\frac{2\alpha^2+11\alpha+2}{(1+\alpha)^3} + \frac{4}{15}\frac{\delta}{1+\delta}\frac{\alpha^2+3\alpha+1}{(1+\alpha)^3}\right\}. \tag{C-33}$$

It can be shown that, for all $n \neq 0$, $N_{dg,n} = N_{gd,n}$, and $N_{gb,n} = N_{bg,n}$.

REFERENCES

[1] Y. Tsividis, *Operation and Modeling of the MOS Transistor*, New York: McGraw-Hill, 1987.

[2] W. Liu, *Fundamentals of III-V Devices: HBTs, MESFETs, and HFET's/HEMTs*, New York: Wiley, 1999. Appendices A and B.

APPENDIX D

FRINGING CAPACITANCE

The fringing capacitance cannot be measured. At best, the sum of the overlap and the fringing capacitance can be measured. The usual practice is to use a theoretical formula to calculate the fringing capacitance based on geometrical information. The rest is then attributed to the overlap capacitance.

The theoretical formula for fringing capacitance requires a knowledge of conformal mapping, a rather archaic branch of mathematics whose importance is dimished with the use of computer calculation. For modeling purpose, where an equation is required to determine the fringing capacitance under all possible geometrical layout, there is no more straightforward way than conformal mapping. Unfortunately, there are several fringing capacitance equations floating in the literature, yet most of them do not give meaningful origin of derivation. It is always bothersome to see an equation without a known source. Therefore, we spend some time developing what we think is a plausible fringing capacitance equation here. The derivation as well as the symbols are based mostly on Ref. [1].

We consider the bottom dark line of Fig. D-1a to be the bottom plate of a capacitor, and the upper dark line, the top plate. The top plate of the capacitor bends at a right angle upward, mimicking the poly gate which does not extend parallely with the bottom plate forever. For convenience, we call the inside top plate as the horizontal portion of the top plate and the outside top plate as the vertical plate. The separation between the bottom plate and the inside top plate is d. Figure D-1a is an ideal representation of the capacitor. In reality, the poly gate is not infinitely thick, so the representation from the corner to infinity is not exactly correct. Although a comformal mapping for a finite polygate thickness is possible, the derivation becomes quite complicated. The idealization is a good approximation of the realistic geometry. The top plate is assumed connected to a bias V_0 and the bottom plate is grounded.

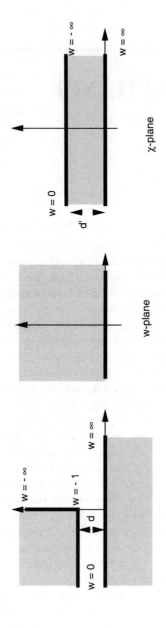

Fig. D-1 (a) an ideal representation of the fringing capacitor to take advantage of the conformal mapping technique. This is the z-plane representation. (b) transformation of z-plane to the w-plane. (c) transformation of w-plane to χ-plane.

The capacitor plates are drawn on the z-plane. As in all of the conformal mapping techniques, the first step is to transform the shaded region in the z-plane to the shaded region in the w-plane (Fig. D-1b). Then, all of the shaded region is transformed into the shaded region in the w-plane (Fig. D-1c). The logic behind ending up in the χ-plane is that, with the two infinitely long capacitor plates as shown in Fig. D-1c, the field distribution are governed by well known formula. Because the electric field is invariant under the conformal transformation, the electric field in the z-plane is known as well. Note that d' is the separation between the two plates in the x-plane; it differs from d in the z-plane, the real-life geometrical plane.

It can be shown that the final transformation equation between the x-plane and the z-plane is given by

$$z = \frac{d}{\pi}\left[2\sqrt{\exp\frac{\pi}{d'}\chi + 1} - 2\ln\left(\sqrt{\exp\frac{\pi}{d'}\chi + 1} + 1\right) + \frac{\pi}{d'}\chi\right]. \quad (D\text{-}1)$$

As a check, we substitute $\chi = (0, jd')$ into Eq. (D-1), and get $z = jd$. Therefore, the point $(0, jd')$ in the χ-plane maps into the point $(0, jd)$ in the z-plane, agreeing with intuition. Substitution of other points demonstrates that the bottom plate in the χ-plane is mapped into the bottom plate of the z-plane. The top plate at the negative x-axis in the χ-plane maps into the top plate at the negative x-axis of the z-plane. The top plate at the positive x-axis of the χ-plane maps into the bent top plate (which runs in the y-axis) in the z-plane.

The electric field at the top plate of the χ-plane, at an arbitrary point $x + jd'$, is given by

$$\varepsilon_x - j\varepsilon_y|_{\chi = x + jd'} = j\frac{V_0}{d}\frac{1 + \sqrt{u}}{u + \sqrt{u}}, \quad (D\text{-}2)$$

where

$$u = 1 - \exp\frac{\pi}{d'}x. \quad (D\text{-}3)$$

Because the electric field is conserved during conformal transformation, the same electric field is found in the z-plane. However, it is much more convenient to work on the χ-plane.

Let us concentrate on the plate between $\chi = (-\infty, jd')$ and $(0, jd')$. In this region, $x < 0$; hence, $u > 0$ (or at least, nonimaginary) and $\varepsilon_x = 0$. The electric field in the y direction is

$$\varepsilon_y = -\frac{V_0}{d}\frac{1 + \sqrt{u}}{u + \sqrt{u}}. \quad (D\text{-}4)$$

The total charge density (in the dimension of C/cm²) associated with that half plane is

$$\sigma = -\epsilon_{ox}\varepsilon_y = \epsilon_{ox}\frac{V_0}{d}\frac{1 + \sqrt{u}}{u + \sqrt{u}} = \sigma_0\frac{1 + \sqrt{u}}{u + \sqrt{u}}, \quad (D\text{-}5)$$

where σ_0 is the ideal charge density in a parallel plate. Part of the fringing charge (that associated with the inside plate) is, therefore,

$$\Delta Q_{inside} = W\sigma_0 \int_{-\infty}^{0} \frac{1+\sqrt{u}}{u+\sqrt{u}} - 1\, dx = W\sigma_0 \frac{d}{\pi} \times 0.615. \tag{D-6}$$

The integral is numerically evaluated.

The other portion of the fringing charge appears in the outside plate. Because $x > 0$ there, $u < 0$ and we shall let $\bar{u}$ be $-u$. The electric field in the y direction is zero, and ε_x is

$$\varepsilon_x = j\frac{V_0}{d}\frac{1+j\sqrt{\bar{u}}}{-\bar{u}+j\sqrt{\bar{u}}} = \frac{V_0}{d}\frac{1}{\sqrt{\bar{u}}}. \tag{D-7}$$

The fringing charge at the outside plate is thus

$$\Delta Q_{inside} = W \int_{d}^{y_0} \epsilon_{ox}\frac{V_0}{d}\frac{1}{\sqrt{\bar{u}}}\, dy = w\sigma_0 \int_{d}^{y_0} \frac{1}{\sqrt{\bar{u}}}\, dy, \tag{D-8}$$

where y_0 is the location in the x-plane that corresponds to the $y = T_{poly}+d$ in the z-plane, and T_{poly} is the polygate thickness. With the help of Eq. (D-1), it can be shown that the $\bar{u}$ corresponding to y_0 is

$$\bar{u} = \frac{\pi}{2}\left(\frac{T_{poly}+d}{d}\right)^2. \tag{D-9}$$

Continuing the evaluation of the $\Delta Q_{outside}$ integral, we find

$$\Delta Q_{outside} = W\sigma_0 \int_{d}^{\frac{\pi}{2}\left(\frac{T_{poly}+d}{d}\right)^2} \frac{1}{\sqrt{\bar{u}}}\, dy = \frac{W\sigma_0 d}{\pi}\left[2\ln\left(\frac{T_{poly}+d}{d}\right) + 2\ln\frac{\pi}{2}\right]. \tag{D-10}$$

The fringing capacitance is equal to $(\Delta Q_{outside} + \Delta Q_{inside})/(W \cdot V_0)$. Letting C'_{ox} be $\epsilon_{ox}/d = \sigma_0/V_0$, we obtain

$$C_f = C'_{ox} \times \frac{d}{\pi}\left[2\ln\left(\frac{T_{poly}+d}{d}\right) + 2\ln\frac{\pi}{2} + 0.615\right]. \tag{D-11}$$

REFERENCE

[1] Gibbs, W. (1958) *Conformal Transformations in Electrical Engineering*, London, Chapman & Hall, Chapter 8, pp. 87–96.

APPENDIX E

BSIM3 NON-QUASI-STATIC MODELING

As mentioned in the NQSMOD entry in Section 3.2, setting NQSMOD = 1 invokes BSIM3's non-quasi-static (NQS) model. Although NQSMOD = 1 can be used in either a transient or an ac analysis, there is likely some code implementation error in the coding of the ac NQS model. Therefore, we shall concentrate describing the transient NQS model.

In BSIM3's quasi-static (Q_S) model, the terminal currents are given by

$$i_D = I_D + \frac{dQ_D}{dt}. \tag{E-1}$$

$$i_S = -I_D + \frac{dQ_S}{dt}. \tag{E-2}$$

$$i_G = \frac{dQ_G}{dt}. \tag{E-3}$$

$$i_B = \frac{dQ_B}{dt}. \tag{E-4}$$

The charges are functions of the instantaneous voltages v_{DS}, v_{GS} and v_{BS}. Take Q_D of Eq. (1-46), for example. Once the instantaneous voltage values are given, Q_D can be found. In its NQS model, the current expressions are totally rewritten, as given by

$$i_D = I_D + \text{XPART} \cdot \frac{Q_{def}}{\tau}. \tag{E-5}$$

$$i_S = -I_D + (1 - \text{XPART}) \cdot \frac{Q_{def}}{\tau}. \tag{E-6}$$

$$i_G = -\frac{Q_{def}}{\tau}. \tag{E-7}$$

$$i_B \equiv 0, \tag{E-8}$$

where XPART is the parameter for partitioning the channel charge to the drain and source, and the relaxation time τ is given by

$$\tau = \frac{L_{eff}^2 \cdot CV}{ELM\mu_n(v_{GS} - V_T) + 16\mu_n \times kT/q}. \quad \text{(E-9)}$$

Note that τ dynamically changes with time because v_{GS} and V_T are functions of time. The SPICE parameter ELM, used to fit the "Elmor's constant," has a default of 5. In the simulation presented in Section 4.1, we set ELM to 10, which has been shown to give the best fit by the BSIM3 authors [1]. Q_{def}, the deficit channel charge (or the surplus charge if its sign is changed), depends not just on the instantaneous voltage values, but also the time revolution of these voltages. This is because Q_{def} is determined from a time-dependent differential equation, rather than a simple algebraic equation as in the case of Q_D in Eq. (1-46). The time derivative of the deficit charge is given by

$$\frac{dQ_{def}}{dt} = \frac{dQ_{ch}}{dt} - \frac{Q_{def}}{\tau}, \quad \text{(E-10)}$$

where Q_{ch}, is the quasi-static channel charge equal to $Q_S + Q_D$. (Q_{ch} was labeled as Q_I, the inversion charge, in Section 1.7. We change the symbol here to conform with the BSIM3 notation.) In the strong inversion, Q_{ch} is then (from Eq. 1-38):

$$Q_{ch} = -WLC'_{ox}(v_{GS} - V_T) \times \frac{2}{3}\frac{1 + \alpha + \alpha^2}{1 + \alpha}. \quad \text{(E-11)}$$

Because Q_{ch} is a function of v_{GS}, v_{DS} (α depends on v_{DS}), and v_{BS} (V_T depends on v_{BS}), $dQ_{ch}/dt = \partial Q_{ch}/\partial v_{GS} \cdot \partial v_{GS}/\partial t + \partial Q_{ch}/\partial v_{GS} \cdot \partial v_{GS}/\partial t + \partial Q_{ch}/\partial v_{GS} \cdot \partial v_{GS}/\partial t$. For the transistor transient considered in Section 4.1, the transistor is always in the saturation region. Hence, the saturation index $\alpha = 0$, and Eq. (E-10) during saturation can be simplified to

$$\frac{dQ_{def}}{dt} = -\frac{2}{3}WLC'_{ox}\frac{d(v_{GS} - V_T)}{dt} - \frac{Q_{def}}{L^2}(ELM\ \mu_n(v_{GS} - V_T) + 16\ \mu_n \times kT/q). \quad \text{(E-12)}$$

For the problem considered in Section 4.1, we therefore substitute $v_{GS}(t) = \theta t$ into Eq. (E-12), and solve numerically the differential equation. Once $Q_{def}(t)$ is established from Eq. (E-12), $Q_{def}(t)$ is substituted into Eqs. (E-5) to (E-8) to find the terminal currents.

We comment that the QS current equations given in Eqs. (E-1) to (E-4) can be derived from fundamental physics principles, as demonstrated in Section 1.7. However, the NQS current equations of Eqs. (E-5)–(E-8) are written down in a less vigorous manner. While the logics behind the equations is not completely hand-waving, it nonetheless does not have a solid physical background. Consequently, waveforms simulated under slow transients using the BSIM3 NQS model can grossly differ from that simulated with the BSIM3 QS model.

REFERENCE

[1] M. Chan, K. Hui, R. Neff, C. Hu, and P. Ko, "A relation time approach to model the non-quasi-static transient effects in MOSFETs," *IEEE Int. Electron Devices Meeting*, pp. 169–172, 1994.

APPENDIX F

NOISE FIGURE

In Section 4.13 we calculated the noise figure (NF) as a function of the source resistance. We see from Fig. 4-72 that NF is high when R_p (the source resistance of the input power source) is small, gradually decreases as R_p keeps on increasing, and finally increases with R_p. The parabolic type of the NF behavior occurs in transistors with different sizes. As the transistor size increases, the minimum point of the parabolic shifts toward lower R_p value.

In the NF discussion in Section 4.13, we varied the source resistance value to minimize the NF. The freedom of changing the source resistance might be possible in low frequency circuits without power transfer considerations. However, in high-frequency amplifier circuits where maximum power transfer is to be delivered, the source resistance of the power source is generally fixed at 50 Ω. Some kind of impedance matching circuit, consisting of lossless inductors and capacitors, is inserted between the input power source and the amplifying transistor. Figure F-1 is an exemplar circuit showing the input impedance matching network, made of a shunting 2 pF capacitor in series with a 1 nH inductor. The circuit is to be operated at 2 GHz. The source resistance of the input power source remains at 50 Ω. If we consider the two-port network to consist of both the input matching circuit as well as the transistor, then the noise figure of the network can still be calculated with Eq. (4-67). A SPICE simulation of Fig. F-1 indicates that INOISE = 4.833×10^{-9} V/sqrt(Hz). Substituting this value and a R_p of 50 Ω into Eq. (4-67), we find the noise figure to be 14.6 dB. This value differs from 10.8 dB, calculated in Section 4.13 (see the 2 GHz point of Fig. 4-70) for the case of no input impedance matching. It is clear that, although the matching circuits capacitor and inductor are noiseless elements, they nonetheless modify the noise figure of the overall two-port network.

APPENDIX F   523

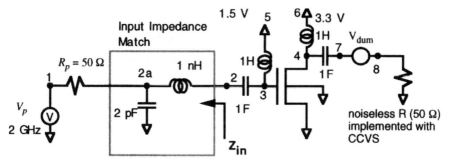

Fig. F-1 An exemplar circuit showing the input impedance matching network, made of a shunting 2pF capacitor in series with a 1nH inductor. For noise figuration calculation, the power-source resistance R_p is at 18.65°C, while the rest is at an arbitrary ambient temperature.

Let us replace the circuit of Fig. F-1 with an equivalent to gain more understanding about the noise figure. Z_{in}, the input impedance indicated in Fig. F-1, is equal to $19.4 + j\,11.8\,\Omega$ at 2 GHz. The input portion of the circuit can be replaced by a series combination of $19.4\,\Omega$ in series with a 6.75 pF capacitor, as shown in Fig. F-2. Because Figs. F-1 and F-2 are equivalent, we expect that the noise figures are identical. Running a SPICE simulation for the circuit of Fig. F-2, we find that INOISE $= 3.016 \times 10^{-9}$ V/sqrt(Hz). Let us continue to use Eq. (4-67), except this time the proper source resistance to be used is $R_0 = 19.4\,\Omega$, instead the previous $50\,\Omega$. We get again a NF of 14.6 dB. This result signifies that Eq. (4-67) can be applied to either circuit configuration. However, for Fig. F-1, a R_p value of $50\,\Omega$ should be used, while for Fig. F-2, R_p should be $19.4\,\Omega$.

We have used voltage source as our input power source, with the power-source resistance R_p connected in series with the source. In the noise network theory, it is often more convenient to represent the power source as a current source in parallel with the power-source conductance (G_s) and the source immitance (B_s), as shown in Fig. F-3. For this type of representation, Eq. (4-67) is no longer applicable. The proper noise figure equation is now developed.

The noise figure definition is still that of Eq. (4-65); $N_o/S_o/(N_i/S_i)$. For the circuit of Fig. F-3, we can think of N_o as the noise square current at node 4, N_i as the noise

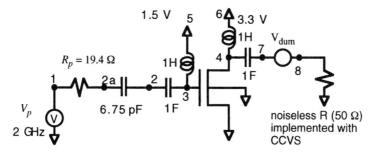

Fig. F-2 An equivalent circuit of Fig. F-1. For noise figure calculation, the power-source resistance R_p is at 18.65°C, while the rest is at an arbitrary ambient temperature.

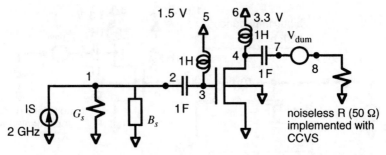

Fig. F-3 A circuit with an input current source in parallel with the power-source conductance and immitance. For noise figure calculation, G_s is at 18.65°C, while the rest is at an arbitrary ambient temperature.

square current associated with the source resistance, S_o as the square of the current at node 4, and S_i as the square of IS, the input source current. Expressed in terms of the SPICE simulation output, $N_o = \text{ONOISE}^2$. Because the noiseless resistor at the output (node 4) is 50 Ω, the output current is simply v(4)/50. Hence, $S_o = v(4)^2/50^2$. N_i, the noise square current associated with the source resistance, is $4kT/R_p$, where R_p is $1/G_s$. Finally, S_i is IS^2. Since in the SPICE circuit files we always make IS to have a unity magnitude, $S_i = 1$.

Substituting N_o, N_i, S_o, and S_i into Eq. (4-65), we write

$$\text{NF} = \frac{\text{ONOISE}^2}{50^2} \frac{R_p}{4\,kT_0} \times \frac{50^2}{v^2(4)} \cdot \text{IS}^2 = \frac{\text{ONOISE}^2}{v^2(4)} \frac{R_p}{4\,kT_0}. \tag{F-1}$$

As an illustration, we apply Eq. (F-1) to calculate the noise figure for a case of $R_p = 1/G_s = 50\,\Omega$ and $B_s = 0$. In this case, we know from basic circuit theory that Fig. F-3 becomes equivalent to Fig. 4-69. A SPICE simulation indicates that $\text{ONOISE} = 2.36 \times 10^{-10}$, and v(4) = 3.811 V. (v(4) is a huge value because we made the ac magnitude of IS to be 1 A.) Substituting these values and a R_p of 50 Ω into Eq. (F-1), we find NF = 10.8 dB, which is exactly the same result we got in Section 4.13.

We further consider the impedance matching's effects on the noise figure calculation. Instead of Fig. F-1, we can redraw an equivalent circuit as shown in Fig. F-4, with the V_p − 50 Ω series branch replaced by an IS in parallel with the 50 Ω. The SPICE simulation result for Fig. F-4 is: $\text{ONOISE} = 2.297 \times 10^{-10}$, and v(4) = 2.371. These values, together with $R_p = 50\,\Omega$, lead to a noise figure of 14.67 dB according to Eq. (F-1). This is again the same result as that calculated for Fig. F-1.

The input impedance seen away from the device, indicated as Y_S in Fig. F-4, is $0.037646 + j\,0.02291\,\Omega^{-1}$ at 2 GHz. (It is equivalent to a 26.56 Ω resistor in parallel with a 1.824 pF capacitor.) Figure F-4 can therefore be represented by Fig. F-3, with $G_s = 1/26.56\,\Omega^{-1}$ and a B_s of $0.02291\,\Omega^{-1}$. The SPICE simulation of Fig. F-3 with the said G_s and B_s values yield an ONOISE of 2.297×10^{-10} V/sqrt(Hz) and a V(4) of 1.728. We substitute these values into

APPENDIX F   525

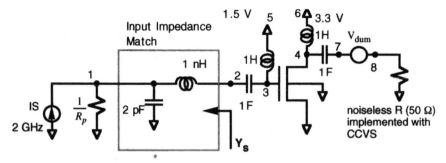

Fig. F-4 Equivalent circuit of Fig. F-1, in the $G_s - B_s$ configuration of Fig. F-3. For noise figure calculation, R_p is at 18.65°C, while the rest of the network is at an arbitrary ambient temperature.

Eq. (F-1), along with $R_p = 26.56\ \Omega$ (not 50 Ω), we obtain a NF of 14.67 dB with Fig. F-3's construct, again identical to that found in Figs. F-1 and F-4. We see that Eq. (F-1) applies to all configurations with current source as the input and a Y_S as the input admittance.

The two-port noise theory states that, there is an optimum pair of G_s and B_s in Fig. F-3 which results in the lowest noise figure for a given transistor. The lowest noise figure value is called the minimum noise figure, denoted by NF_{min}. A conceptually straightforward method to find NF_{min} for a transistor shown in Fig. F-3 is to vary G_s and B_s, calculate the noise figure. After exhausting all combinations of G_s and B_s, we can then find the NF_{min}. This method suffers from the drawback that it is not deterministic, and one always wonder that if the NF_{min} found in the calculated cases is the real NF_{min}. A better alternative relies on network theory, which finds that for all combinations of G_s and B_s, the NF can be expressed as

$$NF = NF_{min} + \frac{R_n}{G_s}[(G_s - G_{opt})^2 + (B_s - B_{opt})^2]. \qquad (F-2)$$

Where G_{opt} and B_{opt} denote the optimum pair of G_s and B_s which leads to the minimum noise figure NF_{min}. R_n, for our purpose, can be treated as a device parameter whose value depends on the magnitudes of the noise sources in the transistor. Once the four noise parameters of a transistor, NF_{min}, G_{opt}, B_{opt}, and R_n, are known, the transistor's noise properties are completely characterized.

Equation (F-2) is an exact equation. It accounts for the fact that the noise sources in a transistor can be correlated. An important revelation of Eq. (F-2) is that the noise figure depends only on the input admittance. NF is independent of the output impedance. There are four unknowns in the equations: NF_{min}, G_{opt}, B_{opt}, and R_n. If there are four linearly independent equations, then we can solve for these unknowns uniquely. To achieve this end, we can first simulate the circuit of Fig. F-3 with a particular G_{s1} and B_{s1}, for which we obtain a noise figure equal to NF_1. We then modify the value of G_s and B_s to G_{s2} and B_{s2} and resimulate. This time, a new noise figure denoted as NF_2 is obtained. This process is repeated two more times, during which we then obtain NF_3 and NF_4. With a numerical routine, we can then obtain

NF$_{min}$, G_{opt}, B_{opt}, and R_n. (For circuit designers with simulation tools such as LIBRA or HP-ADS, the determination of the NF$_{min}$, for example, is just several clicks away. However, for modeling groups, it may not be economical to own those software packages just to determine the NF$_{min}$. Therefore, writing a small numerical routine is worthwhile to some.)

The following notes are made to those with some experiences in numerical analysis. Our common sense states that all we need is four equations to determine the values of the four noise parameters. We could have varied only G_s but kept B_s to be identically zero for all four calculations. That is, we calculate NF$_1$ from G_{s1} and $B_{s1} = 0$, NF$_2$ from G_{s2} and $B_{s2} = 0$, and so on. This seems to simplify the process a bit, especially the coding of the numerical program to solve for four unknowns with four equations. However, setting $B_s = 0$ has a nonobvious deleterious effect. Typically the numerical routine to solve N-equations with N-unknowns is based on the Newton-Raphson iteration technique. A set of nonlinear equations, like what we have here, are first linearized. The linearization involves taking the derivatives of Eq. (F-2) with respect to each of the variables, such as NF$_{min}$, G_{opt}, B_{opt} and R_n. Somehow when $B_s = 0$, these derivatives form a singular matrix when they are arranged in a matrix form. This prevents the numerical routine from proceeding to solve for the unknowns. (In the jargon of numerical analysis—the matrix cannot be inverted when the pivoting element is zero.) It is better to generate the four equations with nonzero G_s and nonzero B_s. For those interested in writing up his own numerical routine to solve for the four unknowns, we recommend the MNEWT subroutine in Ref. [1]. Adopting their use of symbols, we set up the following four equations [2]:

$$g_i = \text{NF}_{min} + \frac{R_n}{G_{s,i}}[(G_{s,i} - G_{opt})^2 + (B_{s,i} - B_{opt})^2] - \text{NF}_i \quad i = 1, 2, 3, 4. \quad \text{(F-3)}$$

Our goal is to minimize $g_i(\text{NF}_{min}, G_{opt}, B_{opt}, R_n)$ during each iteration. To this end, MNEWT requires the evaluation of 16 α and 4 β coefficients (i = 1, 2, 3, 4):

$$\alpha_{i,1} = \frac{\partial g_i}{\partial \text{NF}_{min}} = 1. \quad \text{(F-4)}$$

$$\alpha_{i,2} = \frac{\partial g_i}{\partial G_{opt}} = \frac{R_n}{G_{s,i}}(2G_{opt} - 2G_{s,i}). \quad \text{(F-5)}$$

$$\alpha_{i,3} = \frac{\partial g_i}{\partial B_{opt}} = \frac{R_n}{G_{s,i}}(2B_{opt} - 2B_{s,i}). \quad \text{(F-6)}$$

$$\alpha_{i,4} = \frac{\partial g_i}{\partial R_n} = \frac{1}{G_{s,i}}[(G_{s,i} - G_{opt})^2 + (B_{s,i} - B_{opt})^2]. \quad \text{(F-7)}$$

$$\beta_i = -g_i(\text{NF}_{min}, G_{opt}, B_{opt}, R_n). \quad \text{(F-8)}$$

We demonstrate the procedure to find NF$_{min}$ using the circuit of Fig. F-3. The transistor in the figure is a 5/0.35 μm device with 40 fingers in parallel. We have found in Fig. 4-72 that a fairly high source resistance exceeding 10^4 Ω is needed to achieve NF$_{min}$ for a single finger device of 5/0.35 μm. We would like to make our example more applicable to those used in practice. Therefore, we choose a 40-finger device whose NF$_{min}$ occurs closer to 50 Ω. We run SPICE simulations for the circuit four times, each time with a different combinations of G_s and B_s. The following table displays the simulation results at a frequency of 2 GHz ($\omega = 2\pi \cdot 2 \times 10^9$ rad/s):

$G_{s,1} = (50\,\Omega)^{-1}$;　　$B_{s,1} = \omega C\ \Omega^{-1}$; $C = 2$ pF　　NF$_1 = 1.85156$ (not dB);

$G_{s,2} = (20\,\Omega)^{-1}$;　　$B_{s,2} = \omega C\ \Omega^{-1}$; $C = 3$ pF　　NF$_2 = 2.17347$ (not dB);

$G_{s,3} = (100\,\Omega)^{-1}$;　　$B_{s,3} = \omega C\ \Omega^{-1}$; $C = 1$ pF　　NF$_3 = 1.50644$ (not dB);

$G_{s,4} = (200\,\Omega)^{-1}$;　　$B_{s,4} = \omega C\ \Omega^{-1}$; $C = 0.5$ pF　　NF$_3 = 1.3432$ (not dB);

With these four sets of data points, we are able to establish that NF$_{min} = 1.06$ (0.25 dB); $G_{opt} = 6.85 \times 10^{-4}\ \Omega^{-1}$; $G_B = -2.94 \times 10^{-3}\ \Omega^{-1}$; and $R_n = 13.6\ \Omega$.

Although the previous development provides an easily understood way of calculating NF$_{min}$, this method can run into numerical problems at times. In an attempt to best fit the four data points, for example, NF$_{min}$ so calculated from the described numerical subroutine can become negative. A better approach that provides numerical stability is to find the equivalent noise sources at the input, and then determine NF$_{min}$ based upon these equivalent noise sources. However, this topic is beyond the scope of this book.

REFERENCES AND NOTES

[1] W. Press, B. Flannery, S. Teukolsky, and W. Vetterling, *Numerical Recipes, the Art of Scientific Computing*. London: Cambridge University Press, 1986.

[2] It is possible to reduce four unknowns to three, by developing an equation relating NF$_{min}$ to R_n and G_{opt}. (Communication with J. S. Goo of Stanford University.) See also, F. Danneville, H. Happy, G. Dambrine, J. Belquin, and A. Cappy, "Microscopic noise modeling and macroscopic noise models: how good a connection?" IEEE Trans. Electron Devices, vol. 41, pp. 779–786, 1994.

APPENDIX G

BSIM4 EQUATIONS

There are factors such as 10^6 in the following equations. This is because, while the majority of BSIM4 parameters are expressed in the MKS system, some are in cgs. These factors are used to ensure the correctness of unit conversion. For example, in the RDSW equation, the factor of 10^6 is used to offset the unit imbalance, that $W_{\textit{eff}}$ is in meters whereas RDSW is in $\Omega \cdot \mu m^{WR} = \Omega \cdot \mu m$ (when WR = 1). The factor 273.15 is to convert the temperature unit from °C to Kelvin.

The following equations apply to BSIM4.1.0. The equations relating to binning are not listed here. For binning equations, see Section 1.12.

The Input Variables

Bias Conditions: V_{GS}, V_{DS}, V_{BS}.

Temperature: T_{device}.

Geometrical Information: L, W (We will consider other geometrical information, such as NRD, PD, AD, and NRS, PS, and AS, as SPICE parameters, since they can be modified by the user. Therefore, these parameters, just as those BSIM4 parameters, are in the COURIER font.)

Effective Channel Length/Width for I-V Calculation

$$L_{\textit{eff}} = L - 2 \cdot \mathtt{LINT} - 2 \cdot \Delta L_{\text{geometry}}; \qquad (G\text{-}1)$$

$$W_{\textit{eff}} = \frac{W}{\mathtt{NF}} - 2 \cdot \mathtt{WINT} - 2 \cdot \Delta W_{\text{geometry}} - 2 \cdot \Delta W_{\text{bias\_dependency}}$$

$$\text{(see more discussion about Eq. G-7);} \qquad (G\text{-}2)$$

$$W_{\textit{eff},\text{special}} = \frac{W}{\mathtt{NF}} - 2 \cdot \mathtt{WINT} - 2 \cdot \Delta W_{\text{geometry}}; \qquad (G\text{-}3)$$

where

$$\Delta L_{\text{geometry}} = \frac{\text{LL}}{L^{\text{LLN}}} + \frac{\text{LW}}{W^{\text{LWN}}} + \frac{\text{LWL}}{L^{\text{LLN}} W^{\text{LWN}}}; \quad (G\text{-}4)$$

$$\Delta W_{\text{geometry}} = \frac{\text{WL}}{L^{\text{WLN}}} + \frac{\text{WW}}{W^{\text{WWN}}} + \frac{\text{WWL}}{L^{\text{WLN}} W^{\text{WWN}}}; \quad (G\text{-}5)$$

$$\Delta W_{\text{bias\_dependency}} = \text{DWG} \cdot [V_{GST,\text{eff}}] + \text{DWB} \cdot \left[\sqrt{2\phi_f - V_{BS,\text{eff}}} - \sqrt{2\phi_f} \right]. \quad (G\text{-}6)$$

W_{eff} is the true effective width meant in BSIM4. However, under some special circumstances, W_{eff},special is used, particularly in the calculation of R_{DS} and noise calculation. See Section 3.2, under the entry WL. (*Note:* W_{eff} defined here is written as pParam->BSIM4Weff in the BSIM4 code, and $W_{\text{eff,special}}$ is written as W_{eff} in b4ld.c of the BSIM4 code.) The expression for $V_{GST,\text{eff}}$ appearing in Eq. (G-6) is to be given in Eq. (G-49), and $V_{BS,\text{eff}}$ is given in Eq. (G-26). See also a note in Eq. (G-28).

Here is a special note. BSIM4 is touted as a robust program capable of handling user abuse. As such, BSIM4 needs to accommodate unreasonable SPICE parameter values supplied by the user. For example, what could happen if DWG is a very large positive number? $\Delta W_{\text{bias\_dependency}}$ calculated from Eq. (G-6) can be large, hence leading to a negative W_{eff} as governed by Eq. (G-2). A negative effective channel width is definitely unphysical. In addition, a negative W_{eff} value can cause numerical problem in evaluating many equations that follow. Therefore, BSIM4 performs a check on the value of W_{eff} after it is calculated in Eq. (G-2). If the value is below 0.02 μm (or, in the BSIM4 unit, 2×10^{-8} m), then it is replaced by

$$W_{\text{eff}} = 2 \times 10^{-8} \cdot \left(\frac{4 \times 10^{-8} - W_{\text{eff,from Eq. (G-2)}}}{6 \times 10^{-8} - 2 \cdot W_{\text{eff,from Eq. (G-2)}}} \right). \quad (G\text{-}7)$$

In this manner, even if the effective channel width calculated from Eq. (G-2) is negative, the final W_{eff} used by BSIM4, that calculated from Eq. (G-7), is still positive. The particular choice of factor enclosed in the parentheses of Eq. (G-7) allows W_{eff} and its derivative with respect to $\Delta W_{\text{bias\_dependency}}$ to be continuous right at 2×10^{-8} m.

$W_{\text{eff,special}}$ and L_{eff} given above, as well as $W_{\text{eff,CV}}$, $W_{\text{eff,CJ}}$ and $L_{\text{eff,CV}}$ to be given shortly, do not receive the same special treatment. The moment their values are negative or zero, BSIM4 issues a fatal warning that their values are nonphysical.

Effective Channel Length/Width for C-V, Junction Capacitance and Current Calculation

$$L_{\text{eff,CV}} = L - 2 \cdot \text{DLC} - 2 \cdot \Delta L_{\text{geometry,CV}}; \quad (G\text{-}8)$$

$$W_{\text{eff,CV}} = \frac{W}{\text{NF}} - 2 \cdot \text{DWC} - 2 \cdot \Delta W_{\text{geometry,CV}}; \quad (G\text{-}9)$$

$$W_{\text{eff,CJ}} = \frac{W}{\text{NF}} - 2 \cdot \text{DWJ} - 2 \cdot \Delta W_{\text{geometry,CV}}; \quad (G\text{-}10)$$

where

$$\Delta L_{\text{geometry,CV}} = \frac{\text{LLC}}{L^{\text{LLN}}} + \frac{\text{LWC}}{W^{\text{LWN}}} + \frac{\text{LWLC}}{L^{\text{LLN}}W^{\text{LWN}}}; \quad \text{(G-11)}$$

$$\Delta W_{\text{geometry,CV}} = \frac{\text{WLC}}{L^{\text{WLN}}} + \frac{\text{WWC}}{W^{\text{WWN}}} + \frac{\text{WWLC}}{L^{\text{WLN}}}W^{\text{WWN}}. \quad \text{(G-12)}$$

Basic Quantities

$$q = 1.6 \times 10^{-19}. \quad \text{(G-13)}$$

$$\epsilon_s = 8.85 \times 10^{-12} \cdot 11.7055. \quad \text{(G-14a)}$$

$$\epsilon_{ox} = 8.85 \times 10^{-12} \cdot \text{EPSROX}. \quad \text{(G-14b)}$$

$$n_i = 1.45 \times 10^{10} \cdot \frac{\text{TNOM}}{300.15} \cdot \sqrt{\frac{\text{TNOM}}{300.15}}$$

$$\cdot \exp\left[21.5565981 - \frac{(\text{TNOM})}{2 \cdot k(\text{TNOM} + 273.15)}\right] \quad \text{(G-15)}$$

note that n_i is really the intrinsic carrier concentration at TNOM. It is not a function at T_{device}. E_g is given in Eq. (G-398).

$$k = 1.3806226 \times 10^{-23}. \quad \text{(G-16)}$$

$$2\phi_f = \frac{k[\text{TNOM} + 273.15]}{q} \ln\left(\frac{\text{NDEP}}{n_i}\right) + 0.4 + \text{PHIN}. \quad \text{(G-17)}$$

In BSIM4, there are a total of five oxide-thickness parameters: TOXE, TOXP, DTOX, TOXM, and TOXREF. The last one plays only a minor role, in the tunneling current calculation. It will appear in the latter part of this appendix. The other four parameters, in contrast, affect the whole I-V and C-V characteristics. Their meanings and default values were discussed in Section 5.2. If TOXE is given but not TOXP, then TOXP is set as

$$\text{TOXP} = \text{TOXE} - \text{DTOX}. \quad \text{(G-18)}$$

Conversely, if TOXP is given but not TOXE, then TOXE is set as

$$\text{TOXE} = \text{TOXP} + \text{DTOX}. \quad \text{(G-19)}$$

Most of the time, TOXE is used for the I-V calculation. However, particularly in the charge-thickness capacitance model (CAPMOD = 2), TOXP is the primary oxide parameter.

$$C'_{ox,e} = \frac{\epsilon_{ox}}{\text{TOXE}}. \quad \text{(G-20)}$$

$$C'_{ox,p} = \frac{\epsilon_{ox}}{\text{TOXP}}. \quad \text{(G-21)}$$

$$C_{ox,e} = W_{\text{eff,CV}} L_{\text{eff,CV}} \times C'_{ox,e} \times \text{NF}. \quad \text{(G-22)}$$

$$T = \begin{cases} \text{TNOM} + 273.15 & \text{if } T_{\text{device}} \text{ is not given;} \\ T_{\text{device}} + 273.15 & \text{if } T_{\text{device}} \text{ is given.} \end{cases} \quad \text{(G-23)}$$

Flatband Voltage for I-V Calculation

If a value for the BSIM4 parameter VFB is given, then it is given by

$$V_{FB} = \text{VFB}. \tag{G-24}$$

Otherwise, the dc flatband voltage V_{FB} is given by

$$V_{FB} = \delta_{NP} \cdot \text{VTHO} - 2\phi_f - \text{K1}\sqrt{2\phi_f}, \tag{G-25}$$

where δ_{NP} is $+1$ for NMOS and -1 for PMOS. The parameter K1 given above is not multiplied by TOXE/TOXM, as it often is in the later part of the code. If for some reason VTHO is not specified in the parameter set, then V_{FB} is equated to -1 (V). The flatband voltage for C-V calculation will be described later.

Effective Bulk-to-Source Voltage Function

$$V_{BS,\text{eff}} = v_{bc} + \frac{(V_{BS} - v_{bc} - \delta_1) + \sqrt{(V_{BS} - v_{bc} - \delta_1)^2 - 4\delta_1 v_{bc}}}{2}; \quad \delta_1 = 0.001, \tag{G-26}$$

where v_{bc} is determined by a set of conditions. First, if K2 > 0, v_{bc} is equated to the more negative value of -30 and VBM. As mentioned in the entry for VBM in section 3.2, VBM is defaulted to -3 when it is not specified in the SPICE parameter set. In this case, v_{bc} will then be equal to -30. Note that K2 > 0 is really an unphysical scenario (i.e., K2 should always be smaller than zero). In the more likely situation where K2 < 0, v_{bc} is determined as follows. (In this case, the determination of v_{bc} does not require the knowledge of VBM.) First, a temporary voltage, denoted as V_x here, is calculated as

$$V_x = 0.9 \times \left(2\phi_f - \frac{\text{K1}^2}{4\,\text{K2}^2}\right), \tag{G-27}$$

where $2\phi_f$ is the strong-inversion surface potential given in Eq. (G-16). If $V_x > -3$ V, v_{bc} is fixed at -3 (V). If $V_x < -30$ (V), v_{bc} is fixed at -30 (V). At $-30 < V_x < -3$, then v_{bc} is equated to V_x.

Special Note About $V_{BS,\text{eff}} > 0$

If $V_{BS,\text{eff}}$ calculated above is greater than zero, then

Replace $\sqrt{2\phi_f - V_{BS,\text{eff}}}$ by $\dfrac{(\sqrt{2\phi_f})^3}{2\phi_f + V_{BS,\text{eff}}/2}$ for Eq. (G-6) and all

equations to follow. (G-28)

For example, the $\Delta V_{T,\text{body\_eff}}$ expression, to appear shortly, should be modified to the following when $V_{BS} > 0$.

$$\Delta V_{T,\text{body\_effect}} = K1 \frac{\text{TOXE}}{\text{TOXM}} \frac{(\sqrt{2\phi_f})^3}{2\phi_f + \frac{V_{BS,\text{eff}}}{2}} - K1 \cdot \sqrt{2\phi_f} - K2 \frac{\text{TOXE}}{\text{TOXM}} V_{BS,\text{eff}}. \quad \text{(G-30)}$$

Threshold Voltage

$$V_T = \text{VTHO} + \delta_{NP}(\Delta V_{T,\text{body\_effect}} - \Delta V_{T,\text{charge\_sharing}} - \Delta V_{T,\text{DIBL}}$$
$$+ \Delta V_{T,\text{reverse\_short\_channel}} + \Delta V_{T,\text{narrow\_width}} + \Delta V_{T,\text{small,size}}$$
$$- \Delta V_{T,\text{pocket\_implant}}). \quad \text{(G-31)}$$

$$\Delta V_{T,\text{body\_effect}} = \left[K1 \frac{\text{TOXE}}{\text{TOXM}} \sqrt{2\phi_f - V_{BS,\text{eff}}} - K1 \cdot \sqrt{2\phi_f}\right] \times \sqrt{1 + \frac{\text{LPEB}}{L_{\text{eff}}}}$$
$$- K2 \frac{\text{TOXE}}{\text{TOXM}} V_{BS,\text{eff}}. \quad \text{(G-32)}$$

$$\Delta V_{T,\text{charage\_sharing}} = \text{DVT0} \frac{0.5}{\cosh(\text{DVT1} \times L_{\text{eff}}/L_t) - 1}(V_{bi} - 2\phi_f). \quad \text{(G-33)}$$

$$\Delta V_{T,\text{BIBL}} = \left[\exp\left(-\text{DSUB}\frac{L_{\text{eff}}}{2L_{t0}}\right) + 2\exp\left(-\text{DSUB}\frac{L_{\text{eff}}}{L_{t0}}\right)\right]$$
$$\times (\text{ETA0} + \text{ETAB} \cdot V_{BS,\text{eff}}) \times V_{DS}$$

(not exactly that used in BSIM4; see below). (G-34)

$$\Delta V_{T,\text{reverse\_shortchannel}} = K1 \cdot \frac{\text{TOXE}}{\text{TOXM}} \cdot \left(\sqrt{1 + \frac{\text{LPE0}}{L_{\text{eff}}}} - 1\right) \sqrt{2\phi_f}. \quad \text{(G-35)}$$

$$\Delta V_{T,\text{narrow\_width}} = (K3 + K3B \cdot V_{BS,\text{eff}}) \frac{\text{TOXE}}{W_{\text{eff}} + \text{W0}} 2\phi_f. \quad \text{(G-36)}$$

$$\Delta V_{T,\text{small\_size}} = \text{DVT0W}\left[\exp\left(-\text{DVT1W}\frac{W_{\text{eff}}L_{\text{eff}}}{2L_{tw}}\right)\right.$$
$$\left. + 2\exp\left(-\text{DVT1W}\frac{W_{\text{eff}}L_{\text{eff}}}{L_{tw}}\right)\right](V_{bi} - 2\phi_f). \quad \text{(G-37)}$$

$\Delta V_{T,\text{pocket\_implant}}$ is to be given after the ideality factor n is found. See Eq. (G-57).

The various variables appearing in the above equations are:

$$V_{bi} = \frac{k[\text{TNOM} + 273.15]}{q} \ln\left(\frac{\text{NDEP} \cdot \text{NSD}}{n_i^2}\right). \tag{G-38}$$

$$L_{t0} = \sqrt{\frac{\epsilon_s X_{dep,0}}{C'_{ox,e}}}. \tag{G-39}$$

$$L_t = \begin{cases} \sqrt{\dfrac{\epsilon_s X_{dep}}{C'_{ox,e}}} \times (1 + \text{DVT2} \cdot V_{BS,\text{eff}}) & \text{if } \text{DVT2} \cdot V_{BS,\text{eff}} \geq -0.5; \\[1em] \sqrt{\dfrac{\epsilon_s X_{dep}}{C'_{ox,e}}} \times \dfrac{1 + 3 \cdot \text{DVT2} \cdot V_{BS,\text{eff}}}{3 + 8 \cdot \text{DVT2} \cdot V_{BS,\text{eff}}} & \text{if } \text{DVT2} \cdot V_{BS,\text{eff}} < -0.5. \end{cases} \tag{G-40}$$

$$L_{tw} = \begin{cases} \sqrt{\dfrac{\epsilon_s X_{dep}}{C'_{ox,e}}} \times (1 + \text{DVT2W} \cdot V_{BS,\text{eff}}) & \text{if } \text{DVT2W} \cdot V_{BS,\text{eff}} \geq -0.5; \\[1em] \sqrt{\dfrac{\epsilon_s X_{dep}}{C'_{ox,e}}} \times \dfrac{1 + 3 \cdot \text{DVT2W} \cdot V_{BS,\text{eff}}}{3 + 8 \cdot \text{DVT2W} \cdot V_{BS,\text{eff}}} & \text{if } \text{DVT2W} \cdot V_{BS,\text{eff}} < -0.5. \end{cases} \tag{G-41}$$

$$X_{dep} = \sqrt{\frac{2\epsilon_s (2\phi_f - V_{BS,\text{eff}})}{q\,\text{NDEP}}}. \tag{G-42}$$

$$X_{dep,0} = \sqrt{\frac{2\epsilon_s (2\phi_f)}{q\,\text{NDEP}}}. \tag{G-43}$$

The drain-induced barrier-lowering (DIBL) term should always be positive, so that the threshold voltage at large V_{DS} is smaller than when $V_{DS} = 0$. It therefore makes physical sense that ETA0 + ETAB $V_{BS,\text{eff}}$ is a positive number. There are times when the values of ETA0 and ETAB supplied by the user yield a negative value for ETA0 + ETAB $V_{BS,\text{eff}}$. To prevent the model from producing unexpected results, BSIM4 calculates $\Delta V_{T,\text{DIBL}}$ by first defining a temporary variable (Tmp) as

$$Tmp = (\text{ETA0} + \text{ETAB} \cdot V_{BS,\text{eff}}). \tag{G-44}$$

$$\Delta V_{T,\text{DIBL}} = \begin{cases} \left[\exp\left(-\text{DSUB}\dfrac{L_{eff}}{2L_{t0}}\right) + 2\exp\left(-\text{DSUB}\dfrac{L_{eff}}{L_{t0}}\right)\right] V_{DS} \times Tmp \\[0.5em] \hfill \text{if } Tmp \geq 10^{-4}; \\[1em] \left[\exp\left(-\text{DSUB}\dfrac{L_{eff}}{2L_{t0}}\right) + 2\left(-\text{DSUB}\dfrac{L_{eff}}{L_{t0}}\right)\right] V_{DS} \times \dfrac{2 \times 10^{-4} - Tmp}{3 - 2 \times 10^4 \cdot Tmp} \\[0.5em] \hfill \text{if } Tmp < 10^{-4}. \end{cases}$$

$$\tag{G-45}$$

Effective V_{GS} after Accounting for Poly-Depletion Effects

$$V_{poly} = \frac{q \, \epsilon_s \, \text{NGATE} C_{oxe}^{\prime 2} \cdot 10^6}{2} \left[\sqrt{1 + \frac{2(V_{GS} - V_{FB} - 2\phi_f)}{q \, \epsilon_s \, \text{NGATE} C_{oxe}^{\prime 2} \cdot 10^6}} - 1 \right]^2. \quad (G-46)$$

$$V_{poly,\text{eff}} = 1.12 - \tfrac{1}{2}(1.12 - V_{poly} - \delta + \sqrt{(1.12 - V_{poly} - \delta)^2 + 4 \cdot \delta \cdot 1.12})$$

(δ fixed at 0.05). (G-47)

$$V_{GS,\text{eff}} = V_{GS} - V_{poly,\text{eff}}. \quad (G-48)$$

The main goal of Eq. (G-47) is to prevent V_{poly}, the voltage drop across the polysilicon gate, from exceeding 1.12 V (the silicon bandgap voltage). If it exceeds 1.12 V, then $V_{poly,\text{eff}}$ will be fixed at 1.12 V. The $V_{poly,\text{eff}}$ function is such that $V_{poly,\text{eff}}$ varies between 0 (when $V_{poly} = 0$) and 1.12 (when $V_{poly} \gg 1.12$).

Effective $V_{GS} - V_T$ Smoothing Function

This smoothing function smoothes out the characteristics between the subthreshold and the strong inversion operating regions.

$$V_{GST,\text{eff}} = \frac{n \dfrac{kT}{q} \ln\left[1 + \exp\left(\dfrac{m^* \, V_{GS,\text{eff}} - V_T}{n \, kT/q}\right)\right]}{m^* + n \dfrac{C_{ox,e}'}{C_{dep,0}'} \exp\left[-\dfrac{(1-m^*)(V_{GS,\text{eff}} - V_T) - V_{off}}{n \, kT/q}\right]}, \quad (G-49)$$

where

$$m^* = \frac{1}{2} + \frac{\arctan(\text{MINV})}{\pi}; \quad (G-50)$$

$$V_{off} = \text{VOFF} + \frac{\text{VOFFL}}{L_{eff}}; \quad (G-51)$$

$$C_{dep,0}' = \frac{\epsilon_s}{X_{dep,0}}; \quad (G-52)$$

and a similar, soon-to-be used quantity C_{dep}' is defined as

$$C_{dep}' = \frac{\epsilon_s}{X_{dep}}. \quad (G-53)$$

The ideality n generally appears as

$$n = 1 + \text{FACTOR} \cdot \frac{C'_{dep}}{C'_{ox,e}} + \frac{\text{CIT}}{C'_{ox,e}} + \frac{\text{CDSC} + \text{CDSCD} \cdot V_{DS} + \text{CDSCB} \cdot V_{BS,eff}}{C'_{ox,e}}$$

$$\times \frac{0.5}{\cosh(\text{DVT1} \times L_{eff}/L_t) - 1} \quad \text{(not exactly that used in BSIM4)}. \quad \text{(G-54)}$$

BSIM4 does not exactly use Eq. (G-54), to prevent unreasonable user-specified parameters from producing numerical errors during simulation. For example, if user somehow specifies NFACTOR to be a negative number, then n calculated from Eq. (G-54) would lead to negative ideality factor. BSIM4 calculates the ideality factor in two steps. In the first, a *Tmp* variable is defined:

$$Tmp = \text{NFACTOR} \cdot \frac{C'_{dep}}{C'_{ox,e}} + \frac{\text{CIT}}{C'_{ox,e}} + \frac{\text{CDSC} + \text{CDSCD} \cdot V_{DS} + \text{CDSCB} \cdot V_{BS,eff}}{C'_{ox,e}}$$

$$\times \frac{0.5}{\cosh(\text{DVT1} \times L_{eff}/L_t) - 1}. \quad \text{(G-55)}$$

Then, the ideality factor is given by

$$n = \begin{cases} 1 + Tmp & \text{if } Tmp \geq -0.5; \\ \dfrac{1 + 3 \cdot Tmp}{3 + 8 \cdot Tmp} & \text{if } Tmp < -0.5. \end{cases} \quad \text{(G-56)}$$

With n defined this way, the ideality factor never goes below zero, even though the *Tmp* variable calculated from Eq. (G-55) can be fairly negative.

ΔV_T Correction Due to Pocket Implant

We mentioned that the ΔV_T correction due to pocket implant requires the knowledge of the ideality factor, n. Since n is identified above, we are now ready to calculate $\Delta V_{T,\text{pocket\_implant}}$:

$$\Delta V_{T,\text{pocket\_implant}} = n \frac{kT}{q} \ln\left[\frac{L_{eff}}{L_{eff} + \text{DVTP0}(1 + \exp(-\text{DVTP1} \cdot V_{DS}))}\right]. \quad \text{(G-57)}$$

Mobility Model

There are three mobility models. The user can pick the desired mobility model by assigning the parameter MOBMOD to either 0, 1 or 2. The first step toward the evaluation of the effective mobility (μ_{eff}) is to evaluate a temporary variable *Tmp*:

If MOBMOD = 0 (corresponding to MOBMOD = 1 in BSIM3), then

$$Tmp = (UA + UC \cdot V_{BS,\text{eff}}) \left(\frac{V_{GST,\text{eff}} + 2 \cdot V_T}{TOXE} \right) + UB \left(\frac{V_{GST,\text{eff}} + 2 \cdot V_T}{TOXE} \right)^2. \quad \text{(G-58)}$$

When MOBMOD = 1 (corresponding to MOBMOD = 3 in BSIM3), then

$$Tmp = \left[UA \left(\frac{V_{GST,\text{eff}} + 2 \cdot V_T}{TOXE} \right) + UB \left(\frac{V_{GST,\text{eff}} + 2 \cdot V_T}{TOXE} \right)^2 \right] (1 + UC \cdot V_{BS,\text{eff}}).$$

$$\text{(G-59)}$$

If MOBMOD is any number other than 0 or 1, such as when MOBMOD = 2, then

$$Tmp = (UA + UC \cdot V_{BS,\text{eff}}) \left(\frac{V_{GST,\text{eff}} + V_{T-fb-\phi}}{TOXE} \right)^{EU}, \quad \text{(G-60)}$$

where EU is set to zero if the user-supplied value is negative and

$$V_{T-fb-\phi} = \begin{cases} 2(\text{VTH0} - V_{FB} - 2\phi_f) & \text{for NMOS}; \\ 2.5(\text{VTH0} - V_{FB} - 2\phi_f) & \text{for PMOS}. \end{cases} \quad \text{(G-61)}$$

The effective mobility for all three models is given by the generalized form:

$$\mu_{\text{eff}} = \frac{U0}{\text{Denominator}}, \quad \text{(G-62)}$$

where

$$\text{Denominator} = \begin{cases} 1 + Tmp & \text{if } Tmp \geq -0.8; \\ \dfrac{0.6 + Tmp}{7 + 10 \cdot Tmp} & \text{if } Tmp < -0.8. \end{cases} \quad \text{(G-63)}$$

The MOBMOD = 2 model is newly introduced in BSIM4.

Effective Internal Source–Drain Resistance

BSIM4 performs a checking prior to the evaluation of the drain–source resistance, to ensure the calculated resistance value makes physical sense. The checking is summarized as:

If user-inputted RDSW < 0,   then RDSW is set to 0. (G-64)

If user-inputted RDSWMIN < 0,   then RDSWMIN is set to 0. (G-65)

The effective internal source–drain resistance is calculated as

$$R_{DS} = \begin{cases} \dfrac{\text{RDSWMIN} + \text{RDSW} \times \frac{1}{2}[Tmp + \sqrt{Tmp^2 + 0.01}]}{(10^6 \times W_{\textit{eff},\text{CJ}})^{\text{WR}}} & \textit{if } \text{RDSMOD} \neq 1; \\ 0 & \textit{if } \text{RDSMOD} = 1; \end{cases} \tag{G-66}$$

where

$$Tmp = \frac{1}{1 + \text{PRWG} \cdot V_{GST,\text{eff}}} + \text{PRWB}\left(\sqrt{2\phi_f - V_{BS,\text{eff}}} - \sqrt{2\phi_f}\right). \tag{G-67}$$

The multiplier to RDSW, $[Tmp + \text{sqrt}(Tmp^2 + 0.01)]/2$, has the property that it will never be negative. Even if *Tmp* somehow becomes a very negative number due to a bad choice of parameter values, the multiplier has the minimum value of 0. The multiplier approaches to *Tmp* when *Tmp* is a large positive number.

Bulk-Charge Coefficient

Under most circumstances, the bulk-charge coefficient is calculated as

$$A_{Bulk} = \left\{ 1 - \frac{dV_{T,\text{Long\_L}}}{dV_{BS,\text{eff}}} \times \left[\frac{A0 \cdot L_{\textit{eff}}}{L_{\textit{eff}} + 2\sqrt{XJ \cdot X_{dep}}} \right. \right.$$
$$\left. \left. \times \left(1 - \text{AGS} \cdot V_{GST,\text{eff}} \left(\frac{L_{\textit{eff}}}{L_{\textit{eff}} + 2\sqrt{XJ \cdot X_{dep}}}\right)^2\right) + \frac{B0}{W_{\textit{eff}} + B1} \right] \right\}$$
$$\times \frac{1}{1 + \text{KETA} \cdot V_{BS,\text{eff}}} \quad \text{(not exactly that used in BSIM4)}, \tag{G-68}$$

where

$$\frac{dV_{T,\text{long\_L}}}{dV_{BS,\text{eff}}} = \sqrt{1 + \frac{\text{LPEB}}{L_{\textit{eff}}}} \times \frac{K1}{2\sqrt{2\phi_f - V_{BS,\text{eff}}}} \frac{\text{TOXE}}{\text{TOXM}}$$
$$+ K2 \frac{\text{TOXE}}{\text{TOXM}} - K3 \times \frac{\text{TOXE}}{W_{\textit{eff}} + \text{W0}} 2\phi_f. \tag{G-69}$$

A_{bulk} cannot have arbitrary values. Physically, it should be greater than zero. In fact, in the present technologies, A_{bulk} should exceed 0.1. BSIM4, being a robust program, ensures that A_{bulk} retains physical values. Therefore, instead of evaluating A_{bulk} with Eq. (G-68), BSIM4 calculates a temporary variable first:

$$A_{Bulk,\text{tmp}} = \left\{ 1 - \frac{dV_{T,\text{long\_L}}}{dV_{BS,\text{eff}}} \times \left[\frac{A0 \cdot L_{eff}}{L_{eff} + 2\sqrt{XJ \cdot X_{dep}}} \right. \right.$$
$$\left. \left. \times \left(1 - \text{AGS} \cdot V_{GST,\text{eff}} \left(\frac{L_{eff}}{L_{eff} + 2\sqrt{XJ \cdot X_{dep}}} \right)^2 \right) + \frac{B0}{W_{eff} + B1} \right] \right\}.$$
(G-70)

Then, A_{bulk} is related to $A_{bulk,\text{tmp}}$, in accordance with

$$A_{Bulk} = \begin{cases} A_{Bulk,\text{tmp}} \times \dfrac{1}{1 + \text{KETA} \cdot V_{BS,\text{eff}}} & \text{if } A_{Bulk,\text{tmp}} \geq 0.1; \\ \dfrac{0.2 - A_{Bulk,\text{tmp}}}{3 - 20 \cdot A_{Bulk,\text{tmp}}} \times \dfrac{1}{1 + \text{KETA} \cdot V_{BS,\text{eff}}} & \text{if } A_{Bulk,\text{tmp}} < 0.1. \end{cases}$$
(G-71)

In the event that $1 + \text{KETA} \cdot V_{BS,\text{eff}}$ is negative, A_{bulk} calculated from the above equation becomes negative and thus, unphysical. BSIM4 takes care this situation by a conditional statement. If $\text{KETA} \cdot V_{BS,\text{eff}}$ ever falls below -0.9, then

$$\text{Replace } \frac{1}{1 + \text{KETA} \cdot V_{BS,\text{eff}}} \text{ of Eq. (G-71) by, } \frac{17 + 20\,\text{KETA} \cdot V_{BS,\text{eff}}}{0.8 + \text{KETA} \cdot V_{BS,\text{eff}}}.$$
(G-72)

This replacement ensures that A_{Bulk} is always positive, even if $1 + \text{KETA} \cdot V_{BS,\text{eff}}$ is negative. The choice of replacement preserves continuity of its value as well as its derivative with respect to $V_{BS,\text{eff}}$ at $\text{KETA} \cdot V_{BS,\text{eff}} = -0.9$.

Drain Saturation Voltage

$$\varepsilon_{sat} = \frac{2 \cdot \text{VSAT}}{\mu_{\text{eff}}}.$$
(G-73)

If $R_{DS} = 0$, then

$$V_{DS,\text{sat}} = \frac{\varepsilon_{sat} L_{eff} \cdot (V_{GST,\text{eff}} + 2kT/q)}{A_{bulk}\varepsilon_{sat}L_{eff} + (V_{GST,\text{eff}} + 2kT/q)}.$$
(G-74)

If $R_{DS} \neq 0$, then

$$V_{DS,\text{sat}} = \frac{-b - \sqrt{b^2 - 4ac}}{2a}; \tag{G-75}$$

$$a = A_{bulk}^2 W_{eff} \text{VSAT} C'_{ox,e} R_{DS} + \left(\frac{1}{\lambda} - 1\right) A_{bulk}; \tag{G-76}$$

$$b = -\left[\left(V_{GST,\text{eff}} + \frac{2kT}{q}\right)\cdot\left(\frac{2}{\lambda} - 1\right) + A_{bulk}\varepsilon_{sat}L_{eff}\right.$$
$$\left. + 3A_{bulk}\left(V_{GST,\text{eff}} + \frac{2kT}{q}\right)W_{eff}\text{VSAT}C'_{ox,e}R_{DS}\right]; \tag{G-77}$$

$$c = \left(V_{GST,\text{eff}} + \frac{2kT}{q}\right)\varepsilon_{sat}L_{eff} + 2\left(V_{GST,\text{eff}} + \frac{2kT}{q}\right)^2 W_{eff}\text{VSAT}C'_{ox,e}R_{DS}; \tag{G-78}$$

where λ is determined by the parameters A1 and A2. In general, λ is meant to be equal to $A2 + A1 \cdot V_{GST,\text{eff}}$. However, we need to make sure λ attains only physically meaningful values between 0 and 1. Therefore, λ is obtained according to the following equations/conditions:

If user-inputted A2 < 0.01, then A2 is set to 0.01. (G-79)

If user-inputted A2 > 1, then A2 is set to 1, and A1 is set to 0. (G-80)

Once the checking is done, BSIM4 determines λ depending on the sign of A1. If A1 > 0, then

$$Tmp = (1 - A2) - \frac{1}{2}\left(1 - A2 - A1 \cdot V_{GST,\text{eff}} - \delta\right.$$
$$\left. + \sqrt{(1 - A2 - A1 \cdot V_{GST,\text{eff}} - \delta)^2 + 4 \cdot \delta \cdot (1 - A2)}\right) \quad \delta = 0.001. \tag{G-81}$$

$$\lambda = A2 + Tmp. \tag{G-82}$$

Equation (G-81) follows the same pattern as the smoothing function $V_{DS,\text{eff}}$ given by Eq. (1-10). In that equation, $V_{DS,\text{eff}}$ is made to vary between 0 (when $V_{DS} = 0$) to $V_{DS,\text{sat}}$ (when $V_{DS} \gg V_{DS,\text{sat}}$). Similarly, the temporary variable Tmp varies between 0 (when $A1 \cdot V_{GST,\text{eff}} = 0$) to $1 - A2$ (when $A1 \cdot V_{GST,\text{eff}} \gg 1 - A2$). Consequently, λ varies between its minimum value of A2 and its maximum value of 1.

If instead A1 $\leq$ 0, then let A1' be $-$A1 so that A1' is a positive quantity.

$$Tmp = A2 - \frac{1}{2}\left(A2 - A1' \cdot V_{GST,\text{eff}} - \delta\right.$$
$$\left. + \sqrt{(A2 - A1' \cdot V_{GST,\text{eff}} - \delta)^2 + 4\delta A2}\right) \quad \delta = 0.001, \tag{G-83}$$

$$\lambda = A2 - Tmp. \tag{G-84}$$

In this case, the *Tmp* variable varies between 0 (when $A1' \cdot V_{GST,\text{eff}} = 0$) to A2 (when $A1' \cdot V_{GST,\text{eff}} \gg A2$). Consequently, λ varies between its maximum value of A2 and its minimum value of 0.

So, what is going on here? Generally, as long as $A1 \cdot V_{GST,\text{eff}}$ is not too large or too negative, λ is equated to $A2 + A1 \cdot V_{GST,\text{eff}}$. However, when $A1 \cdot V_{GST,\text{eff}}$ is too large, the smoothing function employed in Eq. (G-81) ensures that λ varies only between A2 and 1. Hence, λ cannot exceed 1. (Because of the condition set in Eq. G-79, A2 is never negative. So, λ is below 1, but always greater than 0.) In the other extreme, when $A1 \cdot V_{GST,\text{eff}}$ has a very negative value, the smoothing function employed in Eq. (G-83) ensures that λ varies only between 0 and A2. Hence, λ never goes below 0. As $A1 \cdot V_{GST,\text{eff}}$ varies between $-\infty$ and $+\infty$, λ varies between 0 and 1.

Effective Drain-to-Source Voltage

$$V_{DS,\text{eff}} = V_{DS,\text{sat}} - \frac{1}{2}\left(V_{Ds,\text{sat}} - V_{DS} - \text{DELTA} + \sqrt{(V_{Ds,\text{sat}} - V_{DS} - \text{DELTA})^2 + 4 \cdot \text{DELTA} \cdot V_{DS,\text{sat}}}\right). \quad \text{(G-85)}$$

As explained in Section 1.4, $V_{DS,\text{eff}}$ varies between 0 (when $V_{DS} = 0$) and $V_{DS,\text{sat}}$). A substitution of $V_{DS} = 0$ into the equation will prove that $V_{DS,\text{eff}} = 0$ when $V_{DS} = 0$. However, just due to rounding errors, a computer may evaluate $V_{DS,\text{eff}}$ to be a finite (though very small) value at $V_{DS} = 0$. In order to make sure $V_{DS,\text{eff}}$ is identically zero when V_{DS} is zero, BSIM4 specifically adds a if–then statement that sets $V_{DS,\text{eff}}$ to zero when V_{DS} is precisely zero.

Effective Oxide Capacitance for I-V Calculation

In the calculation of the drain current, BSIM4 notes that the drain current is significant mainly in the strong inversion region. In classical analysis, the charges are assumed to concentrate right on the oxide–silicon interface. The relevant oxide thickness would then be TOXP. However, in reality, the maximum probability of carrier distribution occurs at some distance ($X_{DC,\text{inv}}$) away from the interface. Therefore, an oxide capacitance for the inversion calculation is sought.

$$C'_{ox,\text{IV}} = \frac{\epsilon_{ox}}{\text{TOXP}} // \frac{\epsilon_s}{X_{DC,\text{IV}}}, \quad \text{(G-86)}$$

where

$$X_{DC,\text{IV}} = \begin{cases} \dfrac{1.9 \times 10^{-9}}{\left[1 + \dfrac{V_{GST,\text{eff}} + 4(\text{VTH0} - V_{FB} - 2\phi_f)}{2 \times 10^8 \cdot \text{TOXP}}\right]^{0.7}} & \text{if } (\text{VTH0} - V_{FB} - 2\phi_f) \geq 0; \\[2em] \dfrac{1.9 \times 10^{-9}}{\left[1 + \dfrac{V_{GST,\text{eff}}}{2 \times 10^8 \cdot \text{TOXP}}\right]^{0.7}} & \text{if } (\text{VTH0} - V_{FB} - 2\phi_f) < 0. \end{cases} \quad \text{(G-87)}$$

APPENDIX G   541

Drain Current in Intrinsic MOS

This is the dominant portion of the drain current, which flows through the channel to the source.

$$I_{DS} = \frac{I_{DS,0}}{1 + \frac{R_{DS}I_{DS,0}}{V_{Ds,\text{eff}}}} \left(1 + \frac{1}{C_{clm}} \ln \frac{V_A}{V_{A,\text{sat}}}\right) \times \left(1 + \frac{V_{DS} - V_{DS,\text{eff}}}{V_{A,\text{DIBL}}}\right)$$

$$\times \left(1 + \frac{V_{DS} - V_{DS,\text{eff}}}{F_{A,\text{DITS}}}\right) \times \left(1 + \frac{V_{DS} - V_{DS,\text{eff}}}{V_{A,\text{SCBE}}}\right) \times NF, \quad \text{(G-88)}$$

where

$$I_{DS,0} = \frac{W_{\text{eff}} \mu_{\text{eff}} C'_{ox,IV} V_{GST,\text{eff}}}{L_{\text{eff}}[1 + V_{DS,\text{eff}}/(\varepsilon_{sat}L_{\text{eff}})]} \left[1 - \frac{A_{bulk} V_{DS,\text{eff}}}{2(V_{GST,\text{eff}} + 2kT/q)}\right] V_{DS,\text{eff}}. \quad \text{(G-89)}$$

(*Note:* The same $I_{DS,0}$ expression above is found in BSIM3. $I_{DS,0}$ is labeled as `Id1` in BSIM3 code. However, `Id1` in the BSIM4 code refers to $I_{DS,0}/V_{DS,\text{eff}}$ to facilitate the implementation of the holistic thermal noise model. Therefore, $V_{DS,\text{eff}}$ needs to be multiplied to `Id1` to effect the overall drain current expression of Eq. G-89).

There are many terms associated with the various Early voltages. They are given by

$$V_A = V_{A,\text{sat}} + V_{A,\text{CLM}}. \quad \text{(G-90)}$$

$$V_{A,\text{sat}} = \frac{\varepsilon_{sat} L_{\text{eff}} + V_{DS,\text{sat}} + 2R_{DS}\text{VSAT}C'_{ox,e}W_{\text{eff}}V_{GST,\text{eff}}}{2/\lambda - 1 + R_{DS}\text{VSAT}C'_{ox,e}W_{\text{eff}}A_{bulk}}$$

$$\times \left[1 - \frac{A_{bulk}V_{DS,\text{sat}}}{2(V_{GST,\text{eff}} + 2kT/q)}\right]. \quad \text{(G-91)}$$

$$F_p = \begin{cases} \left(1 + \text{FPROUT} \cdot \frac{\sqrt{L_{\text{eff}}}}{V_{GST,\text{eff}} + 2kT/q}\right)^{-1} & \textit{if } \text{FPROUT} > 0; \\ 1 & \textit{if } \text{FPROUT} \leq 0. \end{cases} \quad \text{(G-92)}$$

F_p is termed the degradation factor due to pocket implantation, a processing step which affects the characteristics of both the long and short channel devices.

$$F_{VG} = \begin{cases} 1 + \frac{\text{PVAG } V_{GST,\text{eff}}}{\varepsilon_{sat}L_{\text{eff}}} & \textit{if } \frac{\text{PVAG}V_{GST,\text{eff}}}{\varepsilon_{sat}L_{\text{eff}}} > -0.9; \\ \frac{0.8 + \text{PVAG} \cdot V_{GST,\text{eff}}/(\varepsilon_{sat}L_{\text{eff}})}{17 + 20 \cdot \text{PVAG} \cdot V_{GST,\text{eff}}/(\varepsilon_{sat}L_{\text{eff}})} & \textit{if } \text{otherwise.} \end{cases}$$

(G-93)

APPENDIX G

F_{VG} is a factor that accounts for the effects of gate bias on the slope of I_{DS} in the saturation region.

$$Tmp = \frac{F_p}{\text{PCLM } L_{itl}} F_{VG} \times \left(1 + \frac{R_{DS} I_{DS,0}}{V_{DS,\text{eff}}}\right) \times \left(L_{\text{eff}} + \frac{V_{DS,\text{sat}}}{\varepsilon_{sat}}\right). \quad \text{(G-94)}$$

$$C_{clm} = \begin{cases} Tmp & \text{if PCLM} > 0 \text{ and } (V_{DS} - V_{DS,\text{eff}}) > 10^{-10}; \\ 5.834617425 \times 10^{14} & \text{if otherwise.} \end{cases} \quad \text{(G-95)}$$

$$V_{A,\text{CLM}} = \begin{cases} C_{clm}(V_{DS} - V_{DS,\text{eff}}) & \text{if PCLM} > 0 \text{ and } (V_{DS} - V_{DS,\text{eff}}) > 10^{-10} \\ 5.834617425 \times 10^{14} & \text{if otherwise} \end{cases}$$
$$\text{(G-96)}$$

The expression for L_{itl} is given in Eq. (G-102). *Here is a special note for the calculation of $V_{A,\text{CLM}}$.* Before BSIM4 calculates all these Early voltages, BSIM4 performs a check on the value of PCLM. If PCLM is equal to or smaller than zero, then BSIM4 issues a fatal error and the SPICE simulation is halted. The value $5.834617425 \times 10^{14}$ is the maximum number adopted in BSIM4; a computer is not capable of comprehending ∞.

$$\theta_{rout} = \text{PDIBLC1} \times \left[\exp\left(-\text{DROUT} \frac{L_{\text{eff}}}{2L_{t0}}\right) + 2\exp\left(-\text{DROUT} \frac{L_{\text{eff}}}{L_{to}}\right)\right]$$
$$+ \text{PDIBLC2}. \quad \text{(G-97)}$$

$$V_{A,\text{DIBL}} = \begin{cases} \dfrac{(V_{GST,\text{eff}} + 2kT/q)}{\theta_{rout}(1 + \text{PDIBLCB} \cdot V_{BS,\text{eff}})} \times F_{VG} \\ \left[1 - \dfrac{A_{bulk} V_{DS,\text{sat}}}{A_{bulk} V_{DS,\text{sat}} + V_{GST,\text{eff}} + 2kT/q}\right] & \text{if } \theta_{rout} \geq 0; \\ 5.834617425 \times 10^{14} & \text{if } \theta_{rout} \leq 0. \end{cases} \quad \text{(G-98)}$$

In the event that $1 + \text{PDIBLCB} \cdot V_{BS,\text{eff}}$ is negative, $V_{A,\text{DIBL}}$ calculated from Eq. (G-97) becomes negative and thus, unphysical. BSIM4 takes care this situation by a conditional statement. If $\text{PDIBLCB} \cdot V_{BS,\text{eff}}$ ever falls below -0.9, then

Replace $\dfrac{1}{1 + \text{PDIBLCB} \cdot V_{BS,\text{eff}}}$ of Eq. (G-97) by

$$\frac{17 + 20 \cdot \text{PDIBLCB} \cdot V_{BS,\text{eff}}}{0.8 + \text{PDIBLCB} \cdot V_{BS,\text{eff}}}. \quad \text{(G-99)}$$

This replacement ensures that $V_{A,\text{DIBL}}$ is always positive, even if $1 + \text{PDIBLCB} \cdot V_{BS,\text{eff}}$ is negative. The choice of replacement preserves continuity of its value as well as its derivative with respect to $V_{BS,\text{eff}}$ at $\text{PDIBLCB} \cdot V_{BS,\text{eff}} = -0.9$.

APPENDIX G   543

The Early voltage due to drain induced threshold voltage shift is given by

$$V_{A,\text{DITS}} = \begin{cases} \dfrac{F_P}{\text{PDITS}}[1 + (1 + \text{PDITSL} \cdot L_{\text{eff}}) \times \exp(\text{PDITSD} \cdot V_{DS})] & \text{if PDITS} > 0; \\ 5.834617425 \times 10^{14} & \text{if otherwise.} \end{cases} \quad (G\text{-}100)$$

$$V_{A,\text{SCBE}} = \begin{cases} \dfrac{L_{\text{eff}}}{\text{PSCBE2}} \exp\left(\dfrac{\text{PSCBE1} L_{itl}}{V_{DS} - V_{DS,\text{eff}}}\right) & \text{if PSCBE2} > 0; \\ 5.834617425 \times 10^{14} & \text{if otherwise.} \end{cases} \quad (G\text{-}101)$$

$$L_{itl} = \sqrt{\dfrac{\epsilon_s}{\epsilon_{ox}}} \times \text{TOXE} \cdot \text{XJ}. \quad (G\text{-}102)$$

Electrode and Channel-Reflected Gate Resistances ($R_{G,\text{eltd}}$, $R_{G,\text{crg}}$)

BSIM4 allows four options to account for the gate resistance. RGATEMOD = 0 (or any number other than 1, 2, or 3) corresponds to the BSIM3 model, in which no gate resistance was accounted for. When RGATEMOD = 1, a constant gate resistance associated with the electrode is attached to the internal MOS gate. As shown in Fig. 5-13, there is one additional node. When RGATEMOD = 2, a "channel-reflected" gate resistance is added on top of the said electrode resistance. Because these two resistances are in series, BSIM4 uses only one extra node compared to the RGATEMOD – 0 case. When RGATEMOD = 3, two extra nodes are needed. Figure 5-13 shows that one node is devoted to separate out $R_{G,\text{eltd}}$, and $R_{G,\text{crg}}$, so that the overlap capacitance current does not pass through $R_{G,\text{crg}}$.

The formula for $R_{g,\text{eltd}}$, and $R_{G,\text{crg}}$ are given by

$$R_{G,\text{eltd}} = \dfrac{\text{RSHG} \cdot \left(\text{XGW} + \dfrac{W_{\text{eff,CJ}}}{3\,\text{NGCON}}\right)}{\text{NGCON} \cdot (L - \text{XGL}) \cdot \text{NF}}; \quad (G\text{-}103)$$

$$R_{G,\text{crg}} = \text{XRCRG1} \times \left(\dfrac{I_{DS}}{V_{DS,\text{eff}}} + \text{XRCRG2}\dfrac{kT}{q}\mu_{\text{eff}}\dfrac{W_{\text{eff}}}{L_{\text{eff}}}C'_{ox,\text{IV}} \cdot \text{NF}\right). \quad (G\text{-}104)$$

Substrate Resistance Network ($R_{B,\text{pb}}$, $R_{B,\text{pd}}$, $R_{B,\text{ps}}$, $R_{B,\text{db}}$, $R_{B,\text{sb}}$)

BSIM4 allows a substrate resistance network to be attached to the intrinisc transistor, as shown in Fig. 5-14. Previously in BSIM3, if high-frequency accuracy is desired, the users have to create their own subcircuit to add a substrate resistance network. In BSIM4, a fixed network is provided for the users' convenience, when the parameter RBODYMOD is set to 1. When RBODYMOD has its default value of 0, no substrate resistance network is present.

When `RBODYMOD = 1`, BSIM4 calculates the substrate resistances according to the following:

$$R_{B,\text{pb}} = \begin{cases} \text{RBPB}//\text{GBMIN}^{-1} & \text{if RBPB} > 0.001 \ (\Omega); \\ 0.001 & \text{if RBPB} \leq 0.001 \ (\Omega). \end{cases} \quad \text{(G-105)}$$

$$R_{B,\text{pd}} = \begin{cases} \text{RBPD}//\text{GBMIN}^{-1} & \text{if RBPD} > 0.001 \ (\Omega); \\ 0.001 & \text{if RBPD} \leq 0.001 \ (\Omega). \end{cases} \quad \text{(G-106)}$$

$$R_{B,\text{ps}} = \begin{cases} \text{RBPS}//\text{GBMIN}^{-1} & \text{if RBPS} > 0.001 \ (\Omega); \\ 0.001 & \text{if RBPS} \leq 0.001 \ (\Omega). \end{cases} \quad \text{(G-107)}$$

$$R_{B,\text{db}} = \begin{cases} \text{RBDB}//\text{GBMIN}^{-1} & \text{if RBDB} > 0.001 \ (\Omega); \\ 0.001 & \text{if RBDB} \leq 0.001 \ (\Omega). \end{cases} \quad \text{(G-108)}$$

$$R_{B,\text{sb}} = \begin{cases} \text{RBSB}//\text{GBMIN}^{-1} & \text{if RBSB} > 0.001 \ (\Omega); \\ 0.001 & \text{if RBSB} \leq 0.001 \ (\Omega). \end{cases} \quad \text{(G-109)}$$

The purpose of `GBMIN` is to prevent the five resistances from having excessively large resistance values. If, for example, $R_{B,\text{sb}}$ is nearly infinite, then $R_{B,\text{sb}}$ used by BSIM4 will be effectively equal to $(\text{GBMIN})^{-1}$. From this regard, `GBMIN` should not be too small a number. BSIM4 issues a warning if `GBMIN` is below 10^{-20}, although the `GBMIN` value will not be altered.

Effective Source/Drain Area and Perimeter ($AS_{\textit{eff}}$, $PS_{\textit{eff}}$, $AD_{\textit{eff}}$, $PD_{\textit{eff}}$)

As described in Section 5.19, the choice of $AS_{\textit{eff}}$, $PS_{\textit{eff}}$, $AD_{\textit{eff}}$, and $PD_{\textit{eff}}$ is largely governed by the equation used to calculate the source–bulk and drain–bulk junction capacitances and the junctions' epitaxial resistances.

$$AS_{\textit{eff}} = \begin{cases} AS & \text{if AS is given;} \\ AS^*_{\textit{eff}} & \text{if AS is not given.} \end{cases} \quad \text{(G-110)}$$

$$AD_{\textit{eff}} = \begin{cases} AD & \text{if AD is given;} \\ AD^*_{\textit{eff}} & \text{if AD is not given.} \end{cases} \quad \text{(G-111)}$$

$$PS_{\textit{eff}} = \begin{cases} PS & \text{if PS is given and PERMOD} = 0; \\ PS - W_{\textit{eff},\text{CJ}} \cdot NF & \text{if PS is given and PERMOD} \neq 0; \\ PS^*_{\textit{eff}} & \text{if PS is not given.} \end{cases} \quad \text{(G-112)}$$

$$PD_{\textit{eff}} = \begin{cases} PD & \text{if PD is given and PERMOD} = 0; \\ PD - W_{\textit{eff},\text{CJ}} \cdot NF & \text{if PD is given and PERMOD} \neq 0; \\ PD^*_{\textit{eff}} & \text{if PD is not given.} \end{cases} \quad \text{(G-112)}$$

$AS^*_{\textit{eff}}$, $PS^*_{\textit{eff}}$, $AD^*_{\textit{eff}}$ and $PD^*_{\textit{eff}}$ are quite complicated functions of NF (number of fingers), GEOMOD (specifies whether drain/source contacts are shared, isolated or merged), RGEOMOD (specifies whether the drain/source contacts are wide or point contacts). Prior to the determination the junction capacitances and junction epitaxial resistances, we need to establish the number of source and drain contacts in a multi-finger device, at the end as well as in the intermediate locations. ND_{end} denotes the number of drain contacts at the two ends of a transistor. It can be either 0, 1, or 2. ND_{int} denotes the number of drain contacts in the intermediate locations. Likewise, NS_{end} and NS_{int} refer to the number of source contacts at the two ends and in the intermediate locations, respectively.

If the number of finger (NF) is an odd number, then

$$ND_{end} = NS_{end} = 1 \qquad (\textit{if NF odd}); \qquad (\text{G-114})$$

$$ND_{int} = NS_{int} = \frac{\text{NF} - 1}{2} \qquad (\textit{if NF odd}). \qquad (\text{G-115})$$

If, on the other hand, NF is an even number, then the calculation of ND_{end}, NS_{end}, ND_{int}, and NS_{int} depends on whether the number of source or drain contacts is to be minimized. For example, if we intend to minize the number of source contacts, then we would place two drain contacts at the two ends of the transistor. A BSIM4 parameter, MIN, is used to convey this information. If the number of drain contacts is to be minimized, then MIN = 0. Conversely, if the number of the source contacts is to be minimized, then MIN = 1. Therefore,

If the number of finger (NF) is an even number, then

$$ND_{end} = \begin{cases} 2 & \textit{if NF is even and MIN} = 1; \\ 0 & \textit{if NF is even and MIN} \neq 1. \end{cases} \qquad (\text{G-116})$$

$$NS_{end} = \begin{cases} 0 & \textit{if NF is even and MIN} = 1; \\ 2 & \textit{if NF is even and MIN} \neq 1. \end{cases} \qquad (\text{G-117})$$

$$ND_{int} = \begin{cases} \text{NF}/2 - 1 & \textit{if NF is even and MIN} = 1; \\ \text{NF} & \textit{if NF is even and MIN} \neq 1. \end{cases} \qquad (\text{G-118})$$

$$NS_{int} = \begin{cases} 2 & \textit{if NF is even and MIN} = 1; \\ \text{NF}/2 - 1 & \textit{if NF is even and MIN} \neq 1. \end{cases} \qquad (\text{G-119})$$

We can now proceed to find AS^*_{eff}, PS^*_{eff}, AD^*_{eff} and PD^*_{eff}. Their exact value depends on the parameter GEOMOD.

Note: NS_{int} and ND_{int} given in this appendix is half of the "nuEndS" and "nuEndD" values used in the BSIM4 code. In this way, NS_{int} is truly the number of intermediate source fingers and ND_{int} is truly the number of intermediate drain fingers.

If GEOMOD = 0 (both the drain and source end contacts are isolated), then

$$AS^*_{eff} = NS_{end} \cdot (\text{DMCG} + \text{DMCI} - 2\,\text{DMCGT}) \times W_{eff,CJ}$$
$$+ NS_{int} \cdot 2(\text{DMCG} - \text{DMCGT}) \times W_{eff,CJ}; \quad (G\text{-}120)$$

$$PS^*_{eff} = NS_{end} \cdot [2(\text{DMCG} + \text{DMCI} - 2\,\text{DMCGT}) + W_{eff,CJ}]$$
$$+ NS_{int} \cdot 4(\text{DMCG} - \text{DMCGT}); \quad (G\text{-}121)$$

$$AD^*_{eff} = ND_{end} \cdot (\text{DMCG} + \text{DMCI} - 2\,\text{DMCGT}) \times W_{eff,CJ}$$
$$+ ND_{int} \cdot 2(\text{DMCG} - \text{DMCGT}) \times W_{eff,CJ}; \quad (G\text{-}122)$$

$$PD^*_{eff} = ND_{end} \cdot [2(\text{DMCG} + \text{DMCI} - 2\,\text{DMCGT}) + W_{eff,CJ}]$$
$$+ ND_{int} \cdot 4(\text{DMCG} - \text{DMCGT}). \quad (G\text{-}123)$$

If GEOMOD = 1 (source end contact is isolated; drain end contact is either absent or shared), then

$$AS^*_{eff} = NS_{end} \cdot (\text{DMCG} + \text{DMCI} - 2\,\text{DMCGT}) \times W_{eff,CJ}$$
$$+ NS_{int} \cdot 2(\text{DMCG} - \text{DMCGT}) \times W_{eff,CJ}; \quad (G\text{-}124)$$

$$PS^*_{eff} = NS_{end} \cdot [2(\text{DMCG} + \text{DMCI} - 2\,\text{DMCGT}) + W_{eff,CJ}]$$
$$+ NS_{int} \cdot 4(\text{DMCG} - \text{DMCGT}); \quad (G\text{-}125)$$

$$AD^*_{eff} = ND_{end} \cdot (\text{DMCG} - \text{DMCGT})W_{eff,CJ} + 2ND_{int}(\text{DMCG} - \text{DMCGT})W_{eff,CJ}; \quad (G\text{-}126)$$

$$PD^*_{eff} = ND_{end}.2(\text{DMCG} - \text{DMCGT}) + ND_{int} \cdot 4(\text{DMCG} - \text{DMCGT}) \quad (G\text{-}127)$$

If GEOMOD = 2 (source end contact is either absent or shared; drain end contact is isolated), then

$$AS^*_{eff} = NS_{end}(\text{DMCG} - \text{DMCGT})W_{eff,CJ} + 2NS_{int}(\text{DMCG} - \text{DMCGT})W_{eff,CJ}; \quad (G\text{-}128)$$

$$PS^*_{eff} = NS_{end} \cdot 2(\text{DMCG} - \text{DMCGT}) + NS_{int} \cdot 4(\text{DMCG} - \text{DMCGT}); \quad (G\text{-}129)$$

$$AD^*_{eff} = ND_{end} \cdot (\text{DMCG} + \text{DMCI} - 2\,\text{DMCGT}) \times W_{eff,CJ}$$
$$+ ND_{int} \cdot 2(\text{DMCG} - \text{DMCGT}) \times W_{eff,CJ}; \quad (G\text{-}130)$$

$$PD^*_{eff} = ND_{end} \cdot [2(\text{DMCG} + \text{DMCI} - 2\,\text{DMCGT}) + W_{eff,CJ}]$$
$$+ ND_{int} \cdot 4(\text{DMCG} - \text{DMCGT}). \quad (G\text{-}131)$$

If GEOMOD = 3 (both the drain and source end contacts are either absent or shared), then

$$AS^*_{eff} = NS_{end}(\text{DMCG} - \text{DMCGT})W_{eff,\text{CJ}} + 2NS_{int}(\text{DMCG} - \text{DMCGT})W_{eff,\text{CJ}}; \quad \text{(G-132)}$$

$$PS^*_{eff} = NS_{end} \cdot 2(\text{DMCG} - \text{DMCGT}) + NS_{int} \cdot 4(\text{DMCG} - \text{DMCGT}); \quad \text{(G-133)}$$

$$AD^*_{eff} = ND_{end}(\text{DMCG} - \text{DMCGT})W_{eff,\text{CJ}} + 2ND_{int}(\text{DMCG} - \text{DMCGT})W_{eff,\text{CJ}}; \quad \text{(G-134)}$$

$$PD^*_{eff} = ND_{end} \cdot 2(\text{DMCG} - \text{DMCGT}) + ND_{int} \cdot 4(\text{DMCG} - \text{DMCGT}). \quad \text{(G-135)}$$

If GEOMOD = 4 (source end contact is isolated; drain end contact is merged), then

$$AS^*_{eff} = NS_{end} \cdot (\text{DMCG} + \text{DMCI} - 2\,\text{DMCGT}) \times W_{eff,\text{CJ}} + NS_{int} \cdot 2(\text{DMCG} - \text{DMCGT}) \times W_{eff,\text{CJ}}; \quad \text{(G-136)}$$

$$PS^*_{eff} = NS_{end} \cdot [2(\text{DMCG} + \text{DMCI} - 2\,\text{DMCGT}) + W_{eff,\text{CJ}}] + NS_{int} \cdot 4(\text{DMCG} - \text{DMCGT}); \quad \text{(G-137)}$$

$$AD^*_{eff} = ND_{end}(\text{DMDG} - \text{DMCGT})W_{eff,\text{CJ}} + 2ND_{int}(\text{DMCG} - \text{DMCGT})W_{eff,\text{CJ}}; \quad \text{(G-138)}$$

$$PD^*_{eff} = ND_{end} \cdot 2(\text{DMDG} - \text{DMCGT}) + ND_{int} \cdot 4(\text{DMCG} - \text{DMCGT}). \quad \text{(G-139)}$$

If GEOMOD = 5 (source end contact is either absent or shared; drain end contact is merged), then

$$AS^*_{eff} = NS_{end}(\text{DMCG} - \text{DMCGT})W_{eff,\text{CJ}} + 2NS_{int}(\text{DMCG} - \text{DMCGT})W_{eff,\text{CJ}}; \quad \text{(G-140)}$$

$$PS^*_{eff} = NS_{end} \cdot 2(\text{DMCG} - \text{DMCGT}) + NS_{int} \cdot 4(\text{DMCG} - \text{DMCGT}); \quad \text{(G-141)}$$

$$AD^*_{eff} = ND_{end}(\text{DMG} - \text{DMCGT})W_{eff,\text{CJ}} + 2ND_{int}(\text{DMCG} - \text{DMCGT})W_{eff,\text{CJ}}; \quad \text{(G-142)}$$

$$PD^*_{eff} = ND_{end} \cdot 2(\text{DMDG} - \text{DMCGT}) + ND_{int} \cdot 4(\text{DMCG} - \text{DMCGT}). \quad \text{(G-143)}$$

If GEOMOD = 6 (source end contact is merged; drain end contact is isolated), then

$$AS^*_{eff} = NS_{end}(\text{DMDG} - \text{DMCGT})W_{eff,\text{CJ}} + 2NS_{int}(\text{DMCG} - \text{DMCGT})W_{eff,\text{CJ}}; \quad \text{(G-144)}$$

$$PS^*_{eff} = NS_{end} \cdot 2(\text{DMDG} - \text{DMCGT}) + NS_{int} \cdot 4(\text{DMCG} - \text{DMCGT}); \quad \text{(G-145)}$$

$$AD^*_{eff} = ND_{end} \cdot (\text{DMCG} + \text{DMCI} - 2\text{DMCGT}) \times W_{eff,\text{CJ}} + ND_{int} \cdot 2(\text{DMCG} - \text{DMCGT}) \times W_{eff,\text{CJ}}; \quad \text{(G-146)}$$

$$PD^*_{eff} = ND_{end} \cdot [2(\text{DMCG} + \text{DMCI} - 2\text{DMCGT}) + W_{eff,\text{CJ}}] + ND_{int} \cdot 4(\text{DMCG} - \text{DMCGT}). \quad \text{(G-147)}$$

If GEOMOD = 7 (source end contact is merged; drain end contact is either absent or shared), then

$$AS^*_{eff} = NS_{end}(\text{DMDG} - \text{DMCGT})W_{eff,\text{CJ}} + 2NS_{int}(\text{DMCG} - \text{DMCGT})W_{eff,\text{CJ}}; \quad \text{(G-148)}$$

$$PS^*_{eff} = NS_{end} \cdot 2(\text{DMDG} - \text{DMCGT}) + NS_{int} \cdot 4(\text{DMCG} - \text{DMCGT}); \quad \text{(G-149)}$$

$$AD^*_{eff} = ND_{end}(\text{DMCG} - \text{DMCGT})W_{eff,\text{CJ}}$$
$$+ 2ND_{int}(\text{DMCG} - \text{DMCGT})W_{eff,\text{CJ}}; \quad \text{(G-150)}$$

$$PD^*_{eff} = ND_{end} \cdot 2(\text{DMCG} - \text{DMCGT}) + ND_{int} \cdot 4(\text{DMCG} - \text{DMCGT}). \quad \text{(G-151)}$$

If GEOMOD = 8 (both source end and drain end contacts are merged), then

$$AS^*_{eff} = NS_{end}(\text{DMDG} - \text{DMCGT})W_{eff,\text{CJ}}$$
$$+ 2NS_{int}(\text{DMCG} - \text{DMCGT})W_{eff,\text{CJ}}; \quad \text{(G-152)}$$

$$PS^*_{eff} = NS_{end} \cdot 2(\text{DMDG} - \text{DMCGT}) + NS_{int} \cdot 4(\text{DMCG} - \text{DMCGT}); \quad \text{(G-153)}$$

$$AD^*_{eff} = ND_{end}(\text{DMDG} - \text{DMCGT})W_{eff,\text{CJ}}$$
$$+ 2ND_{int}(\text{DMCG} - \text{DMCGT})W_{eff,\text{CJ}}; \quad \text{(G-154)}$$

$$PD^*_{eff} = ND_{end} \cdot 2(\text{DMDG} - \text{DMCGT}) + ND_{int} \cdot 4(\text{DMCG} - \text{DMCGT}). \quad \text{(G-155)}$$

If GEOMOD = 9 (NF must be even in this case. If MIN = 0, then the two ends of the device are source contacts. One of the source contacts is shared, and another, isolated. If MIN ≠ 0, then the two ends of the device are drain contacts. Both drain contacts are assumed to be shared in this case), then

$$AS^*_{eff} = (\text{DMCG} + \text{DMCI} - 2\,\text{DMCGT})W_{eff,\text{CJ}}$$
$$+ (\text{NF} - 1)(\text{DMCG} - \text{DMCGT})W_{eff,\text{CJ}}; \quad \text{(G-156)}$$

$$PS^*_{eff} = [2(\text{DMCG} + \text{DMCI} - 2\,\text{DMCGT}) + W_{eff,\text{CJ}}]$$
$$+ 2(\text{NF} - 1)(\text{DMCG} - \text{DMCGT}); \quad \text{(G-157)}$$

$$AD^*_{eff} = \text{NF} \cdot (\text{DMCG} - \text{DMCGT}) \times W_{eff,\text{CJ}}; \quad \text{(G-158)}$$

$$PD^*_{eff} = \text{NF} \cdot 2(\text{DMCG} - \text{DMCGT}). \quad \text{(G-159)}$$

If GEOMOD = 10 (NF must be even in this case. If MIN = 0, then the two ends of the device are source contacts. Both source contacts are assumed shared in this case.

If MIN $\neq 0$, then the two ends of the device are drain contacts. One of the drain contacts is shared, and the other is isolated), then

$$AS^*_{\mathit{eff}} = \mathrm{NF} \cdot (\mathrm{DMCG} - \mathrm{DMCGT}) \times W_{\mathit{eff},\mathrm{CJ}}; \tag{G-160}$$

$$PS^*_{\mathit{eff}} = \mathrm{NF} \cdot 2(\mathrm{DMCG} - \mathrm{DMCGT}); \tag{G-161}$$

$$AD^*_{\mathit{eff}} = (\mathrm{DMCG} + \mathrm{DMCI} - 2\,\mathrm{DMCGT})W_{\mathit{eff},\mathrm{CJ}}$$
$$+ (\mathrm{NF} - 1)(\mathrm{DMCG} - \mathrm{DMCGT})W_{\mathit{eff},\mathrm{CJ}}; \tag{G-162}$$

$$PD^*_{\mathit{eff}} = [2(\mathrm{DMCG} + \mathrm{DMCI} - 2\,\mathrm{DMCGT}) + W_{\mathit{eff},\mathrm{CJ}}]$$
$$+ 2(\mathrm{NF} - 1)(\mathrm{DMCG} - \mathrm{DMCGT}). \tag{G-163}$$

External Resistor Calculation (R_D, R_S):

As described in Section 5.11, the overall source resistance has two components: the bias-dependent source resistance ($R_{S,\mathrm{bias-dep}}$) associated with the transistor, and the electrode source resistance ($R_{S,\mathrm{eltd}}$) associated with the contacts. The distinction between them is a bit arbitrary. However, $R_{S,\mathrm{bias-dep}}$ can be thought as the resistance between the n^+ source and the inverted channel, while $R_{S,\mathrm{eltd}}$ is purely due to contact formed between the metal interconnect and the n^+ source.

$$R_{S,\mathrm{bias-dep}} = \begin{cases} \dfrac{\mathrm{RSWMIN} + \mathrm{RSW} \times \frac{1}{2}[Tmp + \sqrt{Tmp^2 + 0.01}]}{\mathrm{NF} \times (10^6 \times W_{\mathit{eff},\mathrm{CJ}})^{\mathrm{WR}}} & \textit{if } \mathrm{RDSMOD} = 1; \\ 0 & \textit{if } \mathrm{RDSMOD} \neq 1 \end{cases} \tag{G-164}$$

where

$$Tmp = \frac{1}{1 + \mathrm{PRWG}(V_{GS} - V_{FB,SD})} - \mathrm{PRWB}V_{BS}; \tag{G-165}$$

$$V_{FB,SD} = \begin{cases} \dfrac{k[\mathrm{TNOM} + 273.15]}{q} \ln\left(\dfrac{\mathrm{NGATE}}{\mathrm{NSD}}\right) & \textit{if } \mathrm{NGATE} > 0; \\ 0 & \textit{if otherwise}; \end{cases} \tag{G-166}$$

For the drain side,

$$R_{D,\mathrm{bias-dep}} = \begin{cases} \dfrac{\mathrm{RDWMIN} + \mathrm{RDW} \times \frac{1}{2}[Tmp + \sqrt{Tmp^2 + 0.01}]}{\mathrm{NF} \times (10^6 \times W_{\mathit{eff},\mathrm{CJ}})^{\mathrm{WR}}} & \textit{if } \mathrm{RDSMOD} = 1 \\ 0 & \textit{if } \mathrm{RDSMOD} \neq 1 \end{cases} \tag{G-167}$$

where

$$Tmp = \frac{1}{1 + \text{PRWG}(V_{GD} - V_{FB,SD})} - \text{PRWB}V_{BD}. \quad \text{(G-168)}$$

The multiplier to RSW and RDW, $[Tmp + \text{sqrt}(Tmp^2 + 0.01)]/2$ has the property that it will never be negative. Even if *Tmp* somehow becomes a very negative number due to a bad choice of parameter values, the multiplier has the minimum value of 0. The multiplier approaches to *Tmp* when *Tmp* is a large positive number.

Let us determine an intermediate variable for $R_{S,\text{eltd}}$:

$$R^*_{S,\text{eltd}} = \begin{cases} 0 & \textit{if } \text{RGEOMOD} = 0; \\ \text{NRS} \cdot \text{RSH} & \textit{if } \text{NRS is given and } \text{RGEOMOD} \neq 0; \\ R_{S,\text{end}}//R_{S,\text{int}} & \textit{if } \text{NRS is not given and } \text{RGEOMOD} \neq 0; \end{cases} \quad \text{(G-169)}$$

where $R_{S,\text{end}}$ (end source contact resistance) and $R_{S,\text{int}}$ (internal source contact resistance) are given by a complicated function of RGEOMOD, and so forth. Likewise, an intermediate variable is defined for the drain side:

$$R^*_{D,\text{eltd}} = \begin{cases} 0 & \textit{if } \text{RGEOMOD} = 0; \\ \text{NRD} \cdot \text{RSH} & \textit{if } \text{NRD is given and } \text{RGEOMOD} \neq 0; \\ R_{D,\text{end}}//R_{D,\text{int}} & \textit{if } \text{NRD is not given and } \text{RGEOMOD} \neq 0; \end{cases} \quad \text{(G-170)}$$

where $R_{D,\text{end}}$ is the end drain contact resistance, and $R_{D,\text{int}}$ is the internal drain contact resistance. The $R_{S,\text{eltd}}$ and $R_{D,\text{eltd}}$ expressions given above have an asterisk, indicating that they are not the exact electrode resistances used in BSIM4. In order to avoid numerical problems associated with a singular matrix, BSIM4 will modify these expressions somewhat when they have a value of 0. The modifying equations are to appear shortly. Before that, we present the resistance components $R_{S,\text{end}}$, $R_{S,\text{int}}$, $R_{D,\text{end}}$, and $R_{D,\text{int}}$, which appear in the above expressions. These components depend on GEOMOD, a parameter that specifies the layout geometry of the contacts.

If GEOMOD = 0 (both the drain and source end contacts are isolated, then only NF = odd makes sense. However, no warning message will be given if NF = even, and some meaningless results will be computed by BSIM4.):

$$R_{S,\text{int}} = \text{RSH} \cdot (\text{DMCG} - \text{DMCGT})/(2NS_{int} \cdot W_{\textit{eff},\text{CJ}}); \quad \text{(G-171)}$$

$$R_{S,\text{end}} = \begin{cases} \text{RSH} \cdot (\text{DMCG} - \text{DMCGT})/(2NS_{end} \cdot W_{\textit{eff},\text{CJ}}) \\ \qquad \textit{if } \text{RGEOMOD} = 1, 2, 5; \\ \text{RSH} \cdot W_{\textit{eff},\text{CJ}}/[6NS_{end} \cdot (\text{DMCG} + \text{DMCI} - 2\,\text{DMCGT})] \\ \qquad \textit{if } \text{RGEOMOD} = 3, 4, 6; \end{cases} \quad \text{(G-172)}$$

$$R_{D,\text{int}} = \text{RSH} \cdot (\text{DMCG} - \text{DMCGT})/(2ND_{int} \cdot W_{\textit{eff},\text{CJ}}); \quad \text{(G-173)}$$

$$R_{D,\text{end}} = \begin{cases} \text{RSH} \cdot (\text{DMCG} - \text{DMCGT})/(2ND_{end} \cdot W_{\textit{eff},\text{CJ}}) \\ \qquad \textit{if } \text{RGEOMOD} = 1, 3, 7; \\ \text{RSH} \cdot W_{\textit{eff},\text{CJ}}/[6ND_{end} \cdot (\text{DMCG} + \text{DMCI} = 2\,\text{DMCGT})] \\ \qquad \textit{if } \text{RGEOMOD} = 2, 4, 8. \end{cases} \quad \text{(G-174)}$$

APPENDIX G **551**

If GEOMOD = 1 (source end contact is isolated; drain end contact is either absent or shared), then

$$R_{S,\text{int}} = \text{RSH} \cdot (\text{DMCG} - \text{DMCGT})/(2NS_{\text{int}} \cdot W_{\text{eff,CJ}}); \tag{G-175}$$

$$R_{S,\text{end}} = \begin{cases} \text{RSH} \cdot (\text{DMCG} - \text{DMCGT})/(2NS_{\text{end}} \cdot W_{\text{eff,CJ}}) & \text{if RGEOMOD} = 1, 2, 5; \\ \text{RSH} \cdot W_{\text{eff,CJ}}/[6NS_{\text{end}} \cdot (\text{DMCG} + \text{DMCI} - 2\text{DMCGT})] \\ & \text{if RGEOMOD} = 3, 4, 6; \end{cases} \tag{G-176}$$

$$R_{D,\text{int}} = \text{RSH} \cdot (\text{DMCG} - \text{DMCGT})/(2ND_{\text{int}} \cdot W_{\text{eff,CJ}}); \tag{G-177}$$

$$R_{D,\text{end}} = \begin{cases} \text{RSH} \cdot (\text{DMCG} - \text{DMCGT})/(2ND_{\text{end}} \cdot W_{\text{eff,CJ}}) & \text{if RGEOMOD} = 1, 3, 7; \\ \text{RSH} \cdot W_{\text{eff,CJ}}/[12ND_{\text{end}} \cdot (\text{DMCG} - \text{DMCGT})] & \text{if RGEOMOD} = 2, 4, 8. \end{cases} \tag{G-178}$$

Note: if $ND_{\text{end}} = 0$, then $R_{D,\text{end}} = \infty$. In this scenario, $R_{D,\text{end}}//R_{D,\text{int}}$ of Eq. (G-170) simply reduces to $R_{D,\text{int}}$.

If GEOMOD = 2 (source end contact is either absent or shared; drain end contact is isolated), then

$$R_{S,\text{int}} = \text{RSH} \cdot (\text{DMCG} - \text{DMCGT})/(2NS_{\text{int}} \cdot W_{\text{eff,CJ}}); \tag{G-179}$$

$$R_{S,\text{end}} = \begin{cases} \text{RSH} \cdot (\text{DMCG} - \text{DMCGT})/(2NS_{\text{end}} \cdot W_{\text{eff,CJ}}) & \text{if RGEOMOD} = 1, 2, 5; \\ \text{RSH} \cdot W_{\text{eff,CJ}}/[12NS_{\text{end}} \cdot (\text{DMCG} - \text{DMCGT})] & \text{if RGEOMOD} = 3, 4, 6; \end{cases} \tag{G-180}$$

$$R_{D,\text{int}} = \text{RSH} \cdot (\text{DMCG} - \text{DMCGT})/(2ND_{\text{int}} \cdot W_{\text{eff,CJ}}); \tag{G-181}$$

$$R_{D,\text{end}} = \begin{cases} \text{RSH} \cdot (\text{DMCG} - \text{DMCGT})/(2ND_{\text{end}} \cdot W_{\text{eff,CJ}}) \\ \quad \text{if RGEOMOD} = 1, 3, 7; \\ \text{RSH} \cdot W_{\text{eff,CJ}}/[6ND_{\text{end}} \cdot (\text{DMCG} + \text{DMCI} - 2\,\text{DMCGT})] \\ \quad \text{if RGEOMOD} = 2, 4, 8. \end{cases} \tag{G-182}$$

If GEOMOD = 3 (both the drain and source end contacts are either absent or shared), then

$$R_{S,\text{int}} = \text{RSH} \cdot (\text{DMCG} - \text{DMCGT})/(2NS_{\text{int}} \cdot W_{\text{eff,CJ}}); \tag{G-183}$$

$$R_{S,\text{end}} = \begin{cases} \text{RSH} \cdot (\text{DMCG} - \text{DMCGT})/(2NS_{\text{end}} \cdot W_{\text{eff,CJ}}) & \text{if RGEOMOD} = 1, 2, 5; \\ \text{RSH} \cdot W_{\text{eff,CJ}}/[12NS_{\text{end}} \cdot (\text{DMCG} - \text{DMCGT})] & \text{if RGEOMOD} = 3, 4, 6; \end{cases} \tag{G-184}$$

$$R_{D,\text{int}} = \text{RSH} \cdot (\text{DMCG} - \text{DMCGT})/(2ND_{\text{int}} \cdot W_{\text{eff,CJ}}); \tag{G-185}$$

$$R_{D,\text{end}} = \begin{cases} \text{RSH} \cdot (\text{DMCG} - \text{DMCGT})/(2ND_{\text{end}} \cdot W_{\text{eff,CJ}}) \\ \quad \text{if RGEOMOD} = 1, 3, 7; \\ \text{RSH} \cdot W_{\text{eff,CJ}}/[12ND_{\text{end}} \cdot (\text{DMCG} - \text{DMCGT})] \\ \quad \text{if RGEOMOD} = 2, 4, 8; \end{cases} \tag{G-186}$$

If GEOMOD = 4 (source end contact is isolated; drain end contact is merged. In this GEOMOD, only NF = odd makes sense. However, no warning message will be given if NF = even and some meaningless results will be computed by BSIM4. Because NF should be odd, $ND_{end} = 1$ is implicitly assumed in the BSIM4 code.):

$$R_{S,\text{int}} = \text{RSH} \cdot (\text{DMCG} - \text{DMCGT})/(2NS_{int} \cdot W_{\mathit{eff},\text{CJ}}); \tag{G-187}$$

$$R_{S,\text{end}} = \begin{cases} \text{RSH} \cdot (\text{DMCG} - \text{DMCGT})/(2NS_{end} \cdot W_{\mathit{eff},\text{CJ}}) \\ \qquad \textit{if } \text{RGEOMOD} = 1, 2, 5; \\ \text{RSH} \cdot W_{\mathit{eff},\text{CJ}}/[6NS_{end} \cdot (\text{DMCG} + \text{DMCI} - 2\,\text{DMCGT})] \\ \qquad \textit{if } \text{RGEOMOD} = 3, 4, 6; \end{cases} \tag{G-188}$$

$$R_{D,\text{int}} = \text{RSH} \cdot (\text{DMCG} - \text{DMCGT})/(2ND_{int} \cdot W_{\mathit{eff},\text{CJ}}); \tag{G-189}$$

$$R_{D,\text{end}} = \text{RSH} \cdot (\text{DMDG} - \text{DMCGT})/W_{\mathit{eff},\text{CJ}}. \tag{G-190}$$

If GEOMOD = 5 (source end contact is either absent or shared; drain end contact is merged), then

$$R_{S,\text{int}} = \text{RSH} \cdot (\text{DMCG} - \text{DMCGT})/(2NS_{int} \cdot W_{\mathit{eff},\text{CJ}}); \tag{G-191}$$

$$R_{S,\text{end}} = \begin{cases} \text{RSH} \cdot (\text{DMCG} - \text{DMCGT})/(2NS_{end} \cdot W_{\mathit{eff},\text{CJ}}) \\ \qquad \textit{if } \text{RGEOMOD} = 1, 2, 5; \\ \text{RSH} \cdot W_{\mathit{eff},\text{CJ}}/[12NS_{end} \cdot (\text{DMCG} - \text{DMCGT}) \\ \qquad \textit{if } \text{RGEOMOD} = 3, 4, 6; \end{cases} \tag{G-192}$$

$$R_{D,\text{int}} = \text{RSH} \cdot (\text{DMCG} - \text{DMCGT})/(2ND_{int} \cdot W_{\mathit{eff},\text{CJ}}); \tag{G-193}$$

$$R_{D,\text{end}} = \text{RSH} \cdot (\text{DMDG} - \text{DMCGT})/W_{\mathit{eff},\text{CJ}}. \tag{G-194}$$

If GEOMOD = 6 (source end contact is merged; drain end contact is isolated. In this GEOMOD only NF = odd makes sense. However, no warning message will be given if NF = even and some meaningless results will be computed by BSIM4. Because NF should be odd, $NS_{end} = 1$ is implicitly assumed in the BSIM4 code.):

$$R_{S,\text{int}} = \text{RSH} \cdot (\text{DMCG} - \text{DMCGT})/(2NS_{int} \cdot W_{\mathit{eff},\text{CJ}}); \tag{G-195}$$

$$R_{S,\text{end}} = \text{RSH} \cdot (\text{DMDG} - \text{DMCGT})/W_{\mathit{eff},\text{CJ}}; \tag{G-196}$$

$$R_{D,\text{int}} = \text{RSH} \cdot (\text{DMCG} - \text{DMCGT})/(2ND_{int} \cdot W_{\mathit{eff},\text{CJ}}); \tag{G-197}$$

$$R_{D,\text{end}} = \begin{cases} \text{RSH} \cdot (\text{DMCG} - \text{DMCGT})/(2ND_{end} \cdot W_{\mathit{eff},\text{CJ}}) & \textit{if } \text{RGEOMOD} = 1, 3, 7; \\ \text{RSH} \cdot W_{\mathit{eff},\text{CJ}}/[6ND_{end} \cdot (\text{DMCG} + \text{DMCI} - 2\,\text{DMCGT})] \\ \qquad \textit{if } \text{RGEOMOD} = 2, 4, 8. \end{cases} \tag{G-198}$$

APPENDIX G   553

If GEOMOD = 7 (source end contact is merged; drain end contact is either absent or shared), then

$$R_{S,\text{int}} = \text{RSH} \cdot (\text{DMCG} - \text{DMCGT})/(2NS_{int} \cdot W_{\textit{eff},\text{CJ}}); \quad \text{(G-199)}$$

$$R_{S,\text{end}} = \text{RSH} \cdot (\text{DMDG} - \text{DMCGT})/W_{\textit{eff},\text{CJ}}; \quad \text{(G-200)}$$

$$R_{D,\text{int}} = \text{RSH} \cdot (\text{DMCG} - \text{DMCGT})/(2ND_{int} \cdot W_{\textit{eff},\text{CJ}}); \quad \text{(G-201)}$$

$$R_{D,\text{end}} = \begin{cases} \text{RSH} \cdot (\text{DMCG} - \text{DMCGT})/(2ND_{end} \cdot W_{\textit{eff},\text{CJ}}) \\ \quad \textit{if } \text{RGEOMOD} = 1, 3, 7; \\ \text{RSH} \cdot W_{\textit{eff},\text{CJ}}/[12ND_{end} \cdot (\text{DMCG} - \text{DMCGT})] \\ \quad \textit{if } \text{RGEOMOD} = 2, 4, 8. \end{cases} \quad \text{(G-202)}$$

If GEOMOD = 8 (both source end and drain end contacts are merged. In this GEOMOD, only NF = odd makes sense. However, no warning message will be given if NF = even and some meaningless results will be computed by BSIM4. Because NF should be odd, $NS_{end} = ND_{end} = 1$ is implicitly assumed in the BSIM4 code.):

$$R_{S,\text{int}} = \text{RSH} \cdot (\text{DMCG} - \text{DMCGT})/(2NS_{int} \cdot W_{\textit{eff},\text{CJ}}); \quad \text{(G-203)}$$

$$R_{S,\text{end}} = \text{RSH} \cdot (\text{DMDG} - \text{DMCGT})/W_{\textit{eff},\text{CJ}}; \quad \text{(G-204)}$$

$$R_{D,\text{int}} = \text{RSH} \cdot (\text{DMCG} - \text{DMCGT})/(2ND_{int} \cdot W_{\textit{eff},\text{CJ}}); \quad \text{(G-205)}$$

$$R_{D,\text{end}} = \text{RSH} \cdot (\text{DMDG} - \text{DMCGT})/W_{\textit{eff},\text{CJ}}. \quad \text{(G-206)}$$

If GEOMOD = 9 (NF must be even in this case. If MIN = 0, then the two ends of the device are source contacts. One of the source contacts is shared, and another, isolated. If MIN ≠ 0, then the two ends of the device are drain contacts. Both drain contacts are assumed shared in this case. All contacts are assumed to be wide contacts; there is no point contact.):

$$R_{S,\text{end}}//R_{S,\text{int}} = \text{RSH} \cdot (\text{DMCG} - \text{DMCGT})/(\text{NF} \cdot W_{\textit{eff},\text{CJ}}); \quad \text{(G-207)}$$

$$R_{D,\text{end}}//R_{D,\text{int}} = \text{RSH} \cdot (\text{DMCGT} - \text{DMCGT})/(\text{NF} \cdot W_{\textit{eff},\text{CJ}}). \quad \text{(G-208)}$$

Note: Rather than separating out the internal and end components, we directly calculate the overall resistances in this case. The actual equations used in the BSIM4 codes are somewhat convoluted. We simplify them by taking advantage of the fact that 1/2 in parallel with 1/(NF − 2) is equal to 1/NF.

If GEOMOD = 10 (NF must be even in this case. If MIN = 0, then the two ends of the device are source contacts. Both source contacts are assumed shared in this case. If MIN ≠ 0, then the two ends of the device are drain contacts. One of the drain contacts is shared, and another, isolated. All contacts are assumed to be wide contacts; there is no point contact.):

$$R_{S,\text{end}}//R_{S,\text{int}} = \text{RSH} \cdot (\text{DMCG} - \text{DMCGT})/(\text{NF} \cdot W_{\textit{eff},\text{CJ}}); \quad \text{(G-209)}$$

$$R_{D,\text{end}}//R_{D,\text{int}} = \text{RSH} \cdot (\text{DMCG} - \text{DMCGT})/(\text{NF} \cdot W_{\textit{eff},\text{CJ}}).$$

Note: Rather than separating out the internal and end components, we directly calculate the overall resistances in this case. The actual equations used in the BSIM4 codes are somewhat convoluted. We simplify them by taking advantage of the fact that 1/2 in parallel with 1/(NF − 2) is equal to 1/NF. One would have to examine the BSIM4 codes to decipher what is meant by the above statement.

The above calculations of $R_{S,\text{end}}$, $R_{S,\text{int}}$, $R_{D,\text{end}}$, and $R_{D,\text{int}}$ are used to calculate $R^*_{S,\text{eltd}}$ and $R^*_{D,\text{eltd}}$. As mentioned previously, $R^*_{S,\text{eltd}}$ and $R^*_{D,\text{eltd}}$ are modified to avoid a singular matrix:

$$R_{S,\text{eltd}} = \begin{cases} 0.001(\Omega) & \text{if } R^*_{S,\text{eltd}} = 0 \text{ and } (\text{RGEOMOD} \neq 0 \text{ or } \text{RDSMOD} \neq 0 \\ & \text{or } \text{TNOIMOD} \neq 0); \\ R^*_{S,\text{eltd}} & \text{if otherwise}; \end{cases} \quad \text{(G-211)}$$

$$R_{D,\text{eltd}} = \begin{cases} 0.001(\Omega) & \text{if } R^*_{S,\text{eltd}} = 0 \text{ and } (\text{RGEOMOD} \neq 0 \text{ or } \text{RDSMOD} \neq 0 \\ & \text{or } \text{TNOIMOD} \neq 0); \\ R^*_{D,\text{eltd}} & \text{if otherwise}. \end{cases} \quad \text{(G-212)}$$

Drain–Bulk Junction Current ($I_{j,\text{DB}}$)

$$V_{BD,\text{jct}} = \begin{cases} V_{BD} & \text{if } \text{RBODYMOD} = 0; \\ V_{DBD} & \text{if } \text{RBODYMOD} = 1; \end{cases} \quad \text{(G-213)}$$

$$I_{sat,\text{DB}} = \begin{cases} \text{JSD} \cdot AD_{\text{eff}} + \text{JSWD} \cdot PD_{\text{eff}} + \text{JSWGD} \cdot W_{\text{eff,CJ}} \cdot \text{NF} & \text{if the result is } \geq 0; \\ 10^{-14} \text{A} & \text{if } AD_{\text{eff}} \text{ and } PD_{\text{eff}} \text{ are both } \leq 0; \\ 0 & \text{if the result is } < 0. \end{cases}$$
$$\text{(G-214)}$$

The exact relationship between $I_{j,\text{DB}}$ and $I_{sat,\text{DB}}$ depends on the value of DIOMOD, which can be either 0, 1 or 2.

When DIOMOD = 0, then

$$I_{j,\text{DB}} = I_{sat,\text{DB}} \left[\exp\left(\frac{qV_{BD,\text{jct}}}{\text{NJD} \cdot kT}\right) - 1 \right] \times f_{breakdown,\text{BD}} + \text{GMIN} \cdot V_{BD,\text{jct}}, \quad \text{(G-215)}$$

where

$$f_{breakdown,\text{BD}} = 1 + \text{XJBVD} \cdot \exp\left(-\frac{\text{BVD} + V_{BD,\text{jct}}}{\text{NJD} \cdot kT/q}\right). \quad \text{(G-216)}$$

BVD is the absolute value of the breakdown voltage. $f_{breakdown,\text{BD}}$ remains to be roughly 1 as long as $V_{BD,\text{jct}} > -\text{BVD}$. Only when $V_{BD,\text{jct}}$ becomes much more negative than $-\text{BVD}$ will $f_{breakdown,\text{BD}}$ becomes a large number. This large number, when multiplied with −1 in the square parentheses, becomes a large negative number. This then produces the desired large negative current associated with the reverse breakdown.

When DIOMOD = 1, the model is identical to BSIM3v3.2, without the reverse breakdown voltage. Therefore, at DIOMOD = 1, the value of BVD is irrelevant. The current instead depends on the value of IJTHD. Although this model is, in effect, identical to that of BSIM3v3.2, some intermediate variables are introduced. Therefore, the following equations may appear different from those of Appendix A. The introduction of these interemediate variables are to better extend the equations for DIOMOD = 2.

If IJTHD = 0, then

$$I_{j,\mathrm{DB}} = I_{sat,\mathrm{DB}}\left[\exp\left(\frac{qV_{BD,\mathrm{jct}}}{\mathrm{NJD}\cdot kT}\right) - 1\right] + \mathrm{GMIN}\cdot V_{BD,\mathrm{jct}}. \quad (G\text{-}217)$$

If IJTHD > 0, then

$$I_{j,\mathrm{DB}} = \begin{cases} I_{sat,\mathrm{DB}}\left[\exp\left(\dfrac{qV_{BD,\mathrm{jct}}}{\mathrm{NJD}\cdot kT}\right) - 1\right] + \mathrm{GMIN}\cdot V_{BD,\mathrm{jct}} & \text{if } V_{BD,\mathrm{jct}} < V_{IJTHD}; \\ (I_{Vjdm\_Fwd} - I_{sat,\mathrm{DB}}) + \dfrac{I_{Vjdm\_Fwd}}{\mathrm{NJD}} \\ \times \dfrac{q}{kT}(V_{BD,\mathrm{jct}} - V_{IJTHD}) + \mathrm{GMIN}\cdot V_{BD,\mathrm{jct}} & \text{if otherwise}; \end{cases} \quad (G\text{-}218)$$

$$V_{IJTHD} = \mathrm{NJD}\frac{kT}{q}\ln\left(\frac{\mathrm{IJTHD}}{I_{sat,\mathrm{DB}}} + 1\right); \quad (G\text{-}219)$$

$$I_{Vjdm\_Fwd} = I_{sat,\mathrm{DB}}\cdot \exp\left(\frac{qV_{IJTHD}}{\mathrm{NJD}\,kT}\right); \quad \text{basically equal to } \mathrm{IJTHD} + I_{sat,\mathrm{DB}}. \quad (G\text{-}220)$$

When DIOMOD = 2, then the model is a combination of the DIOMOD = 1 and the DIOMOD = 0 models.

$$Tmp = 1 + \frac{\mathrm{IJTHD}}{I_{sat,\mathrm{DB}}} - \mathrm{XJBVD}\cdot \exp\left(-\frac{q\cdot \mathrm{BVD}}{\mathrm{NJD}\,kT}\right); \quad (G\text{-}221)$$

$$V_{IJTHD} = \mathrm{NJD}\frac{kT}{q}\ln\left[\frac{1}{2}\left(Tmp + \sqrt{Tmp^2 + 4\,\mathrm{XJBVD}\cdot \exp\left(\frac{-q\cdot \mathrm{BVD}}{\mathrm{NJD}\,kT}\right)}\right)\right]. \quad (G\text{-}222)$$

If $V_{BD,\mathrm{jct}} < -\mathrm{BVD}$, then

$$I_{BVD} = I_{sat,\mathrm{DB}} \times (1 + \mathrm{XJBVD})\left[\exp\left(-\frac{q\,\mathrm{BVD}}{\mathrm{NJD}\cdot kT}\right) - 1\right]; \quad (G\text{-}223)$$

$$Slope_{BVD} = \frac{I_{sat,\mathrm{DB}}}{\mathrm{NJD}\cdot kT/q} \times \left[\exp\left(-\frac{q\,\mathrm{BVD}}{\mathrm{NJD}\cdot kT}\right) + \mathrm{XJBVD}\right]; \quad (G\text{-}224)$$

$$I_{j,\mathrm{DB}} = I_{BVD} + Slope_{BVD} \times (\mathrm{BVD} + V_{BD,\mathrm{jct}}) + \mathrm{GMIN}\cdot V_{BD,\mathrm{jct}}. \quad (G\text{-}225)$$

If $V_{BD,\text{jct}} \geq -\text{BVD}$ and $V_{BD,\text{jct}} \leq \text{VIJTHD}$, then

$$I_{j,\text{DB}} = I_{sat,\text{DB}} \times \left[\exp\left(\frac{q \cdot V_{BD,\text{jct}}}{\text{NJD} \cdot kT}\right) - 1\right] \cdot \left[1 + \text{XJBVD} \cdot \exp\left(-\frac{\text{BVD} + V_{BD,\text{jct}}}{\text{NJD} \cdot kT/q}\right)\right]$$
$$+ \text{GMIN} \cdot V_{BD,\text{jct}}. \qquad \text{(G-226)}$$

If $V_{BD,\text{jct}} > \text{VIJTHD}$, then

$$I_{Vjdm\_Fwd} = I_{sat,\text{DB}} \times \left[\exp\left(\frac{q \cdot V_{IJTHD}}{\text{NJD} \cdot kT}\right) - 1\right]$$
$$\cdot \left[1 + \text{XJBVD} \cdot \exp\left(-\frac{\text{BVD} + V_{IJTHD}}{\text{NJD} \cdot kT/q}\right)\right]. \qquad \text{(G-227)}$$

$$Slope_{IJTHD} = \frac{q \cdot I_{sat,\text{DB}}}{\text{NJD}\, kT} \times \left[\exp\left(\frac{qV_{IJTHD}}{\text{NJD}\, kT}\right)\right.$$
$$\left. + \text{XJBVD} \cdot \exp\left(-\frac{\text{BVD} + V_{IJTHD}}{\text{NJD}\, kT/q}\right)\right]. \qquad \text{(G-228)}$$

$$I_{j,\text{DB}} = I_{Vjsm\_Fwd} + Slope_{IJTHD} \times (V_{BD,\text{jct}} - V_{IJTHD}) + \text{GMIN} \cdot V_{BD,\text{jct}}. \qquad \text{(G-229)}$$

Source–Bulk Junction Current ($I_{j,\text{SB}}$)

$$V_{BS,\text{jct}} = \begin{cases} V_{BS} & \text{if } \text{BODYMOD} = 0; \\ V_{SBS} & \text{if } \text{RBODYMOD} = 1. \end{cases} \qquad \text{(G-230)}$$

$$I_{sat,\text{SB}} = \begin{cases} \text{JSS} \cdot AS_{\textit{eff}} + \text{JSWS} \cdot PS_{\textit{eff}} + \text{JSWGS} \cdot W_{\textit{eff},\text{CJ}} \cdot \text{NF} & \text{if the result is } \geq 0; \\ 10^{-14}\,\text{A} & \text{if } AS_{\textit{eff}} \text{ and } PS_{\textit{eff}} \text{ are both } \leq 0; \\ 0 & \text{if the result is } < 0. \end{cases}$$
$$\text{(G-231)}$$

The exact relationship between $I_{j,\text{SB}}$ and $I_{sat,\text{SB}}$ depends on the value of DIOMOD, which can be either 0, 1, or 2.

When DIOMOD = 0, then

$$I_{j,\text{SB}} = I_{sat,\text{SB}}\left[\exp\left(\frac{qV_{BS,\text{jct}}}{\text{NJS} \cdot kT}\right) - 1\right] \times f_{breakdown,\text{BS}} + \text{GMIN} \cdot V_{BS,\text{jct}}, \qquad \text{(G-232)}$$

where

$$f_{breakdown,\text{BS}} = 1 + \text{XJBVS} \cdot \exp\left(-\frac{\text{BVS} + V_{BS,\text{jct}}}{\text{NJS} \cdot kT/q}\right). \qquad \text{(G-233)}$$

BVS is the absolute value of the breakdown voltage. $f_{breakdown,BS}$ remains to be roughly 1 as long as $V_{BSjct} > -\text{BVS}$. Only when $V_{BS,jct}$ becomes much more negative than $-\text{BVS}$ will $f_{breakdown,BS}$ become a large number. This large number, when multiplied with -1 in the square parantheses, becomes a large negative number. This then produces the desired large negative current associated with the reverse breakdown.

When $\text{DIOMOD} = 1$, then the model is the BSIM3v3.2 equivalent, without the reverse breakdown voltage. Therefore, at $\text{DIOMOD} = 1$, the value of BVS is irrelevant. The current instead depends on the value of IJTHS. Although this model is similar to that of BSIM3v3.2, some intermediate variables are introduced. Therefore, the following equations may appear different from those of Appendix A. The introduction of these intermediate variables are to better extend the equations for $\text{DIOMOD} = 2$.

If $\text{IJTHS} = 0$, then

$$I_{j,\text{SB}} = I_{sat,\text{SB}} \left[\exp\left(\frac{qV_{BS,\text{jct}}}{\text{NJS} \cdot kT}\right) - 1 \right] + \text{GMIN} \cdot V_{BS,\text{jct}}. \quad (G\text{-}234)$$

If $\text{IJTHS} > 0$, then

$$I_{j,\text{SB}} = \begin{cases} I_{sat,\text{SB}} \left[\exp\left(\dfrac{qV_{BS,\text{jct}}}{\text{NJS} \cdot kT}\right) - 1 \right] + \text{GMIN} \cdot V_{BS,\text{jct}} & \text{if } V_{BS,\text{jct}} < V_{\text{IJTHS}}; \\ (I_{Vjdm\_Fwd} - I_{sat,\text{SB}}) + \dfrac{I_{Vjdm\_Fwd}}{\text{NJS}} \times \dfrac{q}{kT}(V_{BS,\text{jct}} - V_{\text{IJTHS}}) + \text{GMIN} \cdot V_{BS,\text{jct}} & \text{if otherwise.} \end{cases}$$

$$(G\text{-}235)$$

$$V_{\text{IJTHS}} = \text{NJS}\frac{kT}{q}\ln\left(\frac{\text{IJTHS}}{I_{sat,\text{SB}}} + 1\right); \quad (G\text{-}236)$$

$$I_{Vjdm\_Fwd} = I_{sat,\text{SB}} \cdot \exp\left(\frac{qV_{\text{IJTHS}}}{\text{NJS } kT}\right); \quad \text{basically equal to } \text{IJTHS} + I_{sat,\text{SB}}. \quad (G\text{-}237)$$

When $\text{DIOMOD} = 2$, then the model is a combination of the $\text{DIOMOD} = 1$ and the $\text{DIOMOD} = 0$ models.

$$Tmp = 1 + \frac{\text{IJTHS}}{I_{sat,\text{SB}}} - \text{XJBVS} \cdot \exp\left(-\frac{q \cdot \text{BVS}}{\text{NJS } kT}\right). \quad (G\text{-}238)$$

$$V_{\text{IJTHS}} = \text{NJS}\frac{kT}{q}\ln\left[\frac{1}{2}\left(Tmp + \sqrt{Tmp^2 + 4\,\text{XJBVS} \cdot \exp\left(-\frac{q \cdot \text{BVS}}{\text{NJS } kT}\right)}\right)\right]. \quad (G\text{-}239)$$

If $V_{BS,\text{jct}} < -\text{BVS}$, then

$$I_{BVS} = I_{sat,SB} \times (1 + \text{XJBVS})\left[\exp\left(-\frac{q\,\text{BVS}}{\text{NJS} \cdot kT}\right) - 1\right]. \tag{G-240}$$

$$\text{Slope}_{BVS} = \frac{I_{sat,SB}}{\text{NJS} \cdot kT/q} \times \left[\exp\left(-\frac{q\,\text{BVS}}{\text{NJS} \cdot kT}\right) + \text{XJBVS}\right]. \tag{G-241}$$

$$I_{j,SB} = I_{BVS} + \text{Slope}_{BVS} \times (\text{BVS} + V_{BS,\text{jct}}) + \text{GMIN} \cdot V_{BS,\text{jct}}. \tag{G-242}$$

If $V_{BS,\text{jct}} \geq -\text{BVS}$ and $V_{BS,\text{jct}} \leq \text{VIJTHS}$, then

$$I_{j,SB} = I_{sat,SB} \times \left[\exp\left(\frac{q \cdot V_{BS,\text{jct}}}{\text{NJS} \cdot kT}\right) - 1\right] \cdot \left[1 + \text{XJBVS} \cdot \exp\left(-\frac{\text{BVS} + V_{BS,\text{jct}}}{\text{NJS} \cdot kT/q}\right)\right]$$
$$+ \text{GMIN} \cdot V_{BS,\text{jct}}. \tag{G-243}$$

IF $V_{BS,\text{jct}} > \text{VIJTHS}$, then

$$I_{Vjsm\_Fwd} = I_{sat,SB} \times \left[\exp\left(\frac{q \cdot V_{IJTHS}}{\text{NJS} \cdot kT}\right) - 1\right]$$
$$\cdot \left[1 + \text{XJBVS} \cdot \exp\left(-\frac{\text{BVS} + V_{IJTHS}}{\text{NJS} \cdot kT/q}\right)\right]. \tag{G-244}$$

$$\text{Slope}_{IJTHS} = \frac{q \cdot I_{sat,SB}}{\text{NJS}\,kT}$$
$$\times \left[\exp\left(\frac{qV_{IJTHS}}{\text{NJS}\,kT}\right) + \text{XJBVS} \cdot \exp\left(-\frac{\text{BVS} + V_{IJTHS}}{\text{NJS}\,kT/q}\right)\right]. \tag{G-245}$$

$$I_{j,SB} = I_{Vjsm\_Fwd} + \text{Slope}_{IJTHS} \times (V_{BS,\text{jct}} - V_{IJTHS}) + \text{GMIN} \cdot V_{BS,\text{jct}}. \tag{G-246}$$

Impact Ionization Current

$$I_{sub} = \text{NF} \times \left(\frac{\text{ALPHA0}}{L_{\textit{eff}}} + \text{ALPHA1}\right)(V_{DS} - V_{DS,\text{eff}})\exp\left[-\frac{\text{BETA0}}{V_{DS} - V_{DS,\text{eff}}}\right]$$
$$\times \frac{I_{DS,0}}{1 + \frac{R_{DS}I_{DS,0}}{V_{DS,\text{eff}}}}\left(1 + \frac{1}{C_{clm}}\ln\frac{V_A}{V_{S,\text{sat}}}\right) \times \left(1 + \frac{V_{DS} - V_{DS,\text{eff}}}{V_{A,\text{DIBL}}}\right)$$
$$\times \left(1 + \frac{V_{DS} - V_{DS,\text{eff}}}{V_{A,\text{DITS}}}\right). \tag{G-247}$$

However, if either the term (ALPHA0/L_{eff} + ALPHA1) ≤ 0, or if the parameter BETA0 ≤ 0, then I_{sub} is made to be zero.

Gate-Induced-Drain-Leakage (GIDL) Current

$$I_{gidl} = \text{NF} \times \text{AGIDL} \cdot W_{eff,CJ} \left[\frac{V_{DS} - V_{GS,\text{eff}} - \text{EGIDL}}{3 \cdot \text{TOXE}} \right]$$

$$\times \exp\left(\frac{-3 \cdot \text{TOXE} \times \text{BGIDL}}{V_{DS} - V_{GS,\text{eff}} - \text{EGIDL}} \right)$$

$$\times \frac{V_{DB}^3}{\text{CGIDL} + V_{DB}^3}. \qquad \text{(G-248)}$$

However, if the term in square brackets is ≤ 0, or if the parameter AGIDL ≤ 0, or if BGIDL ≤ 0, or if CGIDL ≤ 0, or if $V_{DB} < 0$, then I_{GIDL} is made to be zero.

Flatband Voltage for C-V and Tunneling Current Calculation

$$V_{FB,CV} = V_{fb,zb} = V_T(V_{GS} = V_{DS} = V_{BS} = 0) - 2\phi_f - \text{K1}\sqrt{2\phi_f}. \qquad \text{(G-249)}$$

Depletion Surface Potential for C-V and Tunneling Current Calculation

$$\sqrt{\phi_s, \text{dep}} = \begin{cases} \left[\sqrt{\frac{1}{4}\left(\text{K1}\frac{\text{TOXE}}{\text{TOXM}}\right)^2 + Tmp} - \frac{1}{2}\left(\text{K1}\frac{\text{TOXE}}{\text{TOXM}}\right) \right] & \text{if } Tmp \geq 0; \\ Tmp/\left(\text{K1}\frac{\text{TOXE}}{\text{TOXM}}\right) & \text{if } Tmp < 0; \end{cases}$$

$$\text{(G-250)}$$

where

$$Tmp = V_{GS,\text{eff}} - V_{FB,\text{effCV}} - V_{BS,\text{eff}} - V_{GST,\text{eff}}. \qquad \text{(G-251)}$$

Effective V_{GB} Smoothing Function for Tunneling Calculation

$$V_{GB,\text{eff\_tunnel}} = V_{GS,\text{eff}} - V_{BS,\text{eff}}. \qquad \text{(G-252)}$$

Effective Flatband Smoothing Function for C-V Calculation

If $V_{FB,CV} \geq 0$, then

$$V_{FB,\text{eff\_tunnel}} = V_{FB,CV} - \frac{V_{FB,CV} - V_{GB,\text{eff\_tunnel}} - \delta_3}{2}$$
$$- \sqrt{\frac{(V_{FB,CV} - V_{GB,\text{eff\_tunnel}} - \delta_3)^2 + 4\delta_3 V_{FB,CV}}{2}}; \quad \text{(G-253a)}$$

or, if $V_{FB,CV} < 0$, then

$$V_{FB,\text{eff\_tunnel}} = V_{FB,CV} - \frac{V_{FB,CV} - V_{GB,\text{eff\_tunnel}} - \delta_3}{2}$$
$$- \sqrt{\frac{(V_{FB,CV} - V_{GB,\text{eff\_tunnel}} - \delta_3)^2 - 4\delta_3 V_{FB,CV}}{2}}. \quad \text{(G-253b)}$$

Accumulation and Inversion Oxide Voltages for Tunneling Current Calculation

$$V_{ox,acc} = V_{FB,CV} - V_{FB,\text{eff\_tunnel}}; \quad \text{(G-254)}$$

$$V_{ox,\text{depinv}} = \left(\text{K1} \frac{\text{TOXE}}{\text{TOXM}}\right)\sqrt{\phi_{s,\text{dep}}} + V_{GST,\text{eff}}. \quad \text{(G-255)}$$

Gate-to-Bulk Tunneling Currents

$$I_{gb,\text{tunnel}} = \begin{cases} I_{gb,acc} + I_{gb,\text{inv}} & \textit{if } \texttt{IGBMOD} \neq 0; \\ 0 & \textit{if } \texttt{IGBMOD} = 0; \end{cases} \quad \text{(G-256)}$$

where

$$I_{gb,acc} = \text{NF} W_{\text{eff}} L_{\text{eff}} \frac{A1}{\text{TOXE}^2} \left(\frac{\text{TOXREF}}{\text{TOXE}}\right)^{\text{NTOX}} V_{GB,\text{eff\_tunnel}}$$
$$\times \texttt{NIGBACC} \frac{kT}{q} \ln\left[1 + \exp\left(-\frac{q(V_{GB,\text{eff\_tunnel}} - V_{FB,CV})}{kT \cdot \texttt{NIGBACC}}\right)\right]$$
$$\times \exp[-B1 \cdot \text{TOXE}(\texttt{AIGBACC} - \texttt{BIGBACC} V_{ox,acc})]$$
$$\cdot (1 + \texttt{CIGBACC}\, V_{ox,acc})]. \quad \text{(G-250)}$$

$$A1 = 4.97232 \times 10^{-7} \frac{A}{V^2}; \quad B1 = 7.45669 \times 10^{11} \frac{g^{1/2}}{F^{1/2} \cdot s}. \quad \text{(G-251)}$$

$$I_{gb,\text{inv}} = \frac{\text{NF } W_{\textit{eff}} L_{\textit{eff}} \text{A2}}{\text{TOXE}^2} \left(\frac{\text{TOXREF}}{\text{TOXE}}\right)^{\text{NTOX}} V_{GB,\text{eff\_tunnel}}$$

$$\times \text{NIGBINV} \frac{kT}{q} \ln\left[1 + \exp\left(-\frac{q(V_{ox,\text{depinv}} - \text{EIGBINV})}{kT \cdot \text{NIGBINV}}\right)\right]$$

$$\times \exp[-\text{B2} \cdot \text{TOXE}(\text{AIGBINV} - \text{BIGBINV} V_{ox,\text{depinv}})$$

$$\cdot (1 + \text{CIGBINV} V_{ox,\text{depinv}})]. \tag{G-252}$$

$$\text{A2} = 3.75956 \times 10^{-7} \frac{\text{A}}{\text{V}^2}; \qquad \text{B2} = 9.82222 \times 10^{11} \frac{\text{g}^{1/2}}{\text{F}^{1/2} \cdot \text{s}}. \tag{G-253}$$

$$I_{gcs,\text{tunnel}} = \begin{cases} I_{gc} \times \dfrac{-1 + \text{PIGCD} \cdot V_{DS} + \exp(-\text{PIGCD} \cdot V_{DS}) + 10^{-4}}{(\text{PIGCD} \cdot V_{DS})^2 + 2 \times 10^{-4}} \\ \qquad\qquad\qquad\qquad\qquad\qquad\qquad \textit{if } \text{IGCMOD} \neq 0; \\ 0 \qquad\qquad\qquad\qquad\qquad\qquad\qquad \textit{if } \text{IGCMOD} = 0. \end{cases} \tag{G-254}$$

$$I_{gcd,\text{tunnel}} = \begin{cases} I_{gc} \times \dfrac{1 - (\text{PIGCD} \cdot V_{DS} + 1) \cdot \exp(-\text{PIGCD} \cdot V_{DS}) + 10^{-4}}{(\text{PIGCD} \cdot V_{DS})^2 + 2 \times 10^{-4}} \\ \qquad\qquad\qquad\qquad\qquad\qquad\qquad \textit{if } \text{IGCMOD} \neq 0; \\ 0 \qquad\qquad\qquad\qquad\qquad\qquad\qquad \textit{if } \text{IGCMOD} = 0. \end{cases} \tag{G-255}$$

$$I_{gs,\text{tunnel}} = \begin{cases} I_{gs} & \textit{if } \text{IGCMOD} \neq 0; \\ 0 & \textit{if } \text{IGCMOD} = 0. \end{cases} \tag{G-256}$$

$$I_{gd,\text{tunnel}} = \begin{cases} I_{gd} & \textit{if } \text{IGCMOD} \neq 0; \\ 0 & \textit{if } \text{IGCMOD} = 0; \end{cases} \tag{G-257}$$

where

$$I_{gc} = \text{NF } W_{\textit{eff}} L_{\textit{eff}} \frac{\text{A3}}{\text{TOXE}^2} \left(\frac{\text{TOXREF}}{\text{TOXE}}\right)^{\text{NTOX}} V_{GS,\text{eff}}$$

$$\times \text{NIGC} \frac{kT}{q} \ln\left[1 + \exp\left(+\frac{q(V_{GS,\text{eff}} - \text{VTH0})}{kT \cdot \text{NIGC}}\right)\right]$$

$$\times \exp[-\text{B3} \cdot \text{TOXE}(\text{AIGC} - \text{BIGC} V_{ox,\text{depinv}}) \cdot (1 + \text{CIGC} V_{ox,\text{depinv}})]. \tag{G-258}$$

$$I_{gs} = \text{NF } W_{\textit{eff}} \text{DLCIG} \frac{\text{A3}}{(\text{TOXE} \cdot \text{POXEDGE})^2} \left(\frac{\text{TOXREF}}{\text{TOXE} \cdot \text{POXEDGE}}\right)^{\text{NTOX}} V_{GS} \times V'_{GS}$$

$$\times \exp[-\text{B3} \cdot \text{TOXE} \cdot \text{POXEDGE}(\text{AIGSD} - \text{BIGSD } V'_{GS})$$

$$\cdot (1 + \text{CIGSD} V'_{GS})]. \tag{G-259}$$

$$V'_{GS} = \sqrt{(V_{GS} - V_{FB,SD})^2 + 10^{-4}}. \tag{G-260}$$

$$I_{gd} = NFW_{\text{eff}} \text{DLCIG} \frac{A3}{(\text{TOXE} \cdot \text{POXEDGE})^2} \left(\frac{\text{TOXREF}}{\text{TOXE} \cdot \text{POXEDGE}}\right)^{\text{NTOX}} V_{GD} \times V'_{GD}$$
$$\times \exp[-B3 \cdot \text{TOXE} \cdot \text{POXEDGE}(\text{AIGSD} - \text{BIGSD}\, V'_{GD})$$
$$\cdot (1 + \text{CIGSD}\, V'_{GD})] \tag{G-261}$$

$$V'_{GD} = \sqrt{(V_{GD} - V_{FB,SD})^2 + 10^{-4}}. \tag{G-262}$$

$$A3 = \begin{cases} 4.97232 \times 10^{-7} & (\text{nmos}) \\ 3.42537 \times 10^{-7} & (\text{pmos}) \end{cases} \frac{A}{V^2};$$

$$B3 = \begin{cases} 7.45669 \times 10^{11} & (\text{nmos}) \\ 1.16645 \times 10^{12} & (\text{pmos}) \end{cases} \frac{g^{1/2}}{F^{1/2} \cdot s}. \tag{G-263}$$

Dc Current Equations

$$I_D = I_{DS} + I_{sub} + I_{gidl} - I_{j,\text{DB}} - I_{gd,\text{tunnel}} - I_{gcd,\text{tunnel}}. \tag{G-264}$$

$$I_S = -I_{DS} - I_{j,\text{SB}} - I_{gs,\text{tunnel}} - I_{gcs,\text{tunnel}}. \tag{G-265}$$

$$I_B = -I_{sub} - I_{gidl} + I_{j,\text{SB}} + I_{j,\text{DB}} - I_{gb,\text{tunnel}}. \tag{G-266}$$

$$I_G = I_{gd,\text{tunnel}} + I_{gcd,\text{tunnel}} + I_{gs,\text{tunnel}} + I_{gcs,\text{tunnel}} + I_{gb,\text{tunnel}}. \tag{G-267}$$

Small-Signal Conductances

We write out definition here. BSIM4 derives the actual equations with the chain rule discussed in Section 1.5. The equations are too complicated to be expressed here.

$$g_m = \frac{\partial I_D}{\partial V_G}; \quad g_d = \frac{\partial I_D}{\partial V_D}; \quad g_{mb} = \frac{\partial I_D}{\partial V_B}. \tag{G-268a}$$

$$g_{ms} = \frac{\partial I_{DS}}{\partial V_G}; \quad g_{ds} = \frac{\partial I_{DS}}{\partial V_D}; \quad g_{mbs} = \frac{\partial I_{DS}}{\partial V_B}. \tag{G-268b}$$

$$G_{bg} = \frac{\partial I_{sub}}{\partial V_G}; \quad G_{bd} = \frac{\partial I_{sub}}{\partial V_D}; \quad G_{bb} = \frac{\partial I_{sub}}{\partial V_B}. \tag{G-269}$$

$$G_{gidl,g} = \frac{\partial I_{gidl}}{\partial V_G}; \quad G_{gidl,d} = \frac{\partial I_{gidl}}{\partial V_D}; \quad G_{gidl,b} = \frac{\partial I_{gidl}}{\partial V_B}. \tag{G-270}$$

$$g_{j,\text{DB}} = \frac{\partial I_{j,\text{DB}}}{\partial V_{BD}}; \quad g_{j,\text{SB}} = \frac{\partial I_{j,\text{SB}}}{\partial V_{BS}}. \tag{G-271}$$

$$g_{gcs,\text{tunnel}\_g} = \frac{\partial I_{gcs,\text{tunnel}}}{\partial V_G}; \qquad g_{gcs,\text{tunnel}\_d} = \frac{\partial I_{gcs,\text{tunnel}}}{\partial V_D};$$

$$g_{gcs,\text{tunnel}\_b} = \frac{\partial I_{gcs,\text{tunnel}}}{\partial V_B}. \tag{G-272}$$

$$g_{gcd,\text{tunnel}\_g} = \frac{\partial I_{gcd,\text{tunnel}}}{\partial V_G}; \qquad g_{gcd,\text{tunnel}\_d} = \frac{\partial I_{gcd,\text{tunnel}}}{\partial V_D};$$

$$g_{gcd,\text{tunnel}\_b} = \frac{\partial I_{gcd,\text{tunnel}}}{\partial V_B}. \tag{G-273}$$

$$g_{gs,\text{tunnel}\_g} = \frac{\partial I_{gs,\text{tunnel}}}{\partial V_G}; \qquad g_{gs,\text{tunnel}\_d} = 0; \qquad g_{gs,\text{tunnel}\_b} = 0. \tag{G-274}$$

$$g_{gd,\text{tunnel}\_g} = \frac{\partial I_{gd,\text{tunnel}}}{\partial V_G}; \qquad g_{gd,\text{tunnel}\_d} = \frac{\partial I_{gd,\text{tunnel}}}{\partial V_D} = -\frac{\partial I_{gd,\text{tunnel}}}{\partial V_G};$$

$$g_{gd,\text{tunnel}\_b} = 0. \tag{G-275}$$

$$g_{gb,\text{tunnel}\_g} = \frac{\partial I_{gb,\text{acc}}}{\partial V_G} + \frac{\partial I_{gb,\text{inv}}}{\partial V_G}; \qquad g_{gb,\text{tunnel}\_d} = \frac{\partial I_{gb,\text{inv}}}{\partial V_D};$$

$$g_{gb,\text{tunnel}\_b} = \frac{\partial I_{gb,\text{acc}}}{\partial V_B} + \frac{\partial I_{gb,\text{inv}}}{\partial V_B}. \tag{G-276}$$

Due to space limitation, we describe only the CAPMOD = 1 and CAPMOD = 2 capacitance models. The CAPMOD = 0 model is essentially the same as those given in Appendix B.

Effective V_{BS} Smoothing Function for C-V Calculation

$$V_{BS,\text{effCV}} = \begin{cases} V_{BS,\text{eff}} & \text{if } V_{BS,\text{eff}} < 0; \\ 2\phi_f - 2\phi_f \dfrac{2\phi_f}{2\phi_f + V_{BS,\text{eff}}} & \text{if } V_{BS,\text{eff}} \geq 0. \end{cases} \tag{G-277}$$

Effective V_{GB} Smoothing Function for C-V Calculation

$$V_{GB,\text{effCV}} = V_{GS,\text{eff}} - V_{BS,\text{effCV}}. \tag{G-278}$$

Effective Flatband Smoothing Function for C-V Calculation

If $V_{FB,CV} \geq 0$, then

$$V_{FB,\text{effCV}} = V_{FB,CV} - \frac{1}{2}\bigg(V_{FB,CV} - V_{GB,\text{effCV}} - \delta_3$$
$$+ \sqrt{(V_{FG,CV} - V_{GB,\text{effCV}} - \delta_3)^2 + 4 \cdot \delta_3 \cdot V_{FB,CV}}\bigg); \quad \text{(G-279a)}$$

or, if $V_{FB,CV} < 0$, then

$$V_{FB,\text{effCV}} = V_{FB,CV} - \frac{1}{2}\bigg(V_{FB,CV} - V_{GB,\text{effCV}} - \delta_3$$
$$+ \sqrt{(V_{FB,CV} - V_{GB,\text{effCV}} - \delta_3)^2 - 4 \cdot \delta_3 \cdot V_{FB,CV}}\bigg). \quad \text{(G-279b)}$$

δ_3 is fixed at 0.02. Equation (G-279) is similar to the $V_{DS,\text{eff}}$ function given in Eq. (A-74). $V_{FB,\text{effCV}}$ is made to vary between 0 (when $V_{GB} = 0$) and $V_{FB,CV}$ (when $V_{GB} \gg V_{FB,CV}$). Equation (G-279a) works only when $V_{FB,CV}$ exceeds or equals to zero. For $V_{FB,CV} < 0$, appropriate change in a sign inside the square-root function needs to be made to prevent from taking the square root of a negative number.

Effective $V_{GS} - V_T$ Smoothing Function for C-V Calculation

$$V_{GST,\text{effCV}} = \text{NOFF} \cdot \frac{nkT}{q} \ln\left[1 + \exp\left(\frac{V_{GS,\text{eff}} - V_T - \text{VOFFCV}}{\text{NOFF} \cdot nkT/q}\right)\right]. \quad \text{(G-280)}$$

The ideality factor n was given in Eq. (G-56).

Modified Bulk-Charge Coefficient for C-V Calculation

BSIM4 uses a reduced bulk-charge coefficient ($A_{bulk,0}$), instead of the A_{bulk} of Eq. (G-71), to calculate the bulk-charge coefficient of C-V calculation ($A_{bulk,CV}$). Under most circumstances, the reduced bulk-charge coefficient is calculated as

$$A_{Bulk,0} = \left\{1 - \frac{dV_{T,\text{long\_L}}}{dV_{BS,\text{eff}}} \times \left[\frac{\text{A0} \cdot L_{\text{eff}}}{L_{\text{eff}} + 2\sqrt{\text{XJ} \cdot X_{\text{dep}}}} + \frac{\text{B0}}{W_{\text{eff}} + \text{B1}}\right]\right\}$$
$$\times \frac{1}{1 + \text{KETA} \cdot V_{BS,\text{eff}}} \quad \text{(not exactly that used in BSIM4)}, \quad \text{(G-281)}$$

where the derivative $dV_{T,\text{long\_L}}/dV_{BS,\text{eff}}$ was given in Eq. (G-69). $A_{bulk,0}$ cannot have arbitrary values. Physically, it should be greater than zero. In fact, in the present technologies, $A_{bulk,0}$ should exceed 0.1. BSIM4, being a robust program, ensures that $A_{bulk,0}$ retains physical values. Therefore, instead of evaluating $A_{bulk,0}$ with Eq. (G-281), BSIM4 evaluates a temporary variable first:

$$A_{Bulk,0,\text{tmp}} = \left\{1 - \frac{dV_{T,\text{long\_L}}}{dV_{BS,\text{eff}}} \times \left[\frac{\text{A0} \cdot L_{\text{eff}}}{L_{\text{eff}} + 2\sqrt{\text{XJ} \cdot X_{dep}}} + \frac{\text{B0}}{W_{\text{eff}} + \text{B1}}\right]\right\}. \quad \text{(G-282)}$$

Then, $A_{bulk,0}$ is related to $A_{bulk,0,\text{tmp}}$ in accordance with

$$A_{Bulk,0} = \begin{cases} A_{Bulk,0,\text{tmp}} \times \dfrac{1}{1 + \text{KETA} \cdot V_{BS,\text{eff}}} & \text{if } A_{Bulk,0,\text{tmp}} \geq 0.1; \\ \dfrac{0.2 - A_{Bulk,0,\text{tmp}}}{3 - 20 \cdot A_{Bulk,0,\text{tmp}}} \times \dfrac{1}{1 + \text{KETA} \cdot V_{BS,\text{eff}}} & \text{if } A_{Bulk,0,\text{tmp}} < 0.1. \end{cases}$$
$$\text{(G-283)}$$

In the event that $1 + \text{KETA} \cdot V_{BS,\text{eff}}$ is negative, $A_{bulk,0}$ calculated from the above equation becomes negative and thus unphysical. BSIM4 takes care this situation by a conditional statement. If $\text{KETA} \cdot V_{BS,\text{eff}}$ ever falls below -0.9, then

$$\text{Replace } \frac{1}{1 + \text{KETA} \cdot V_{BS,\text{eff}}} \text{ of Eq. (G-283) by } \frac{17 + 20\, \text{KETA} \cdot V_{BS,\text{eff}}}{0.8 + \text{KETA} \cdot V_{BS,\text{eff}}}.$$
$$\text{(G-284)}$$

In the above few equations, $V_{BS,\text{eff}}$ is used (as indicated), not $V_{BS,\text{effCV}}$. Finally, the bulk–charge coefficient for C-V calculation is given by

$$A_{Bulk,\text{CV}} = A_{Bulk,0}\left[1 + \left(\frac{\text{CLC}}{L_{\text{eff,CV}}}\right)^{\text{CLE}}\right]. \quad \text{(G-285)}$$

Intrinsic Device Charge for CAPMOD = 1 (Which Was CAPMOD = 2 in BSIM3)

$$V_{DS,\text{satCV}} = \frac{V_{GST,\text{effCV}}}{A_{Bulk,\text{CV}}}. \quad \text{(G-286)}$$

$$V_{DS,\text{effCV}} = V_{DS,\text{satCV}}$$

$$- \frac{(V_{DS,\text{satCV}} - V_{DS} - \delta_4) + \sqrt{(V_{DS,\text{satCV}} - V_{DS} - \delta_4)^2 + 4\delta_4 V_{DS,\text{satCV}}}}{2}$$

$$\delta_4 \text{ is fixed at } 0.02. \quad \text{(G-287)}$$

A substitution of $V_{DS} = 0$ into the equations involving the evaluation of $V_{DS,\text{effCV}}$ will prove that $V_{DS,\text{effCV}} = 0$ when $V_{DS} = 0$. However, just due to rounding errors, a computer may evaluate $V_{DS,\text{effCV}}$ to be a finite (though very small) value at $V_{DS} = 0$.

In order to make sure $V_{DS,\text{effCV}}$ is zero when $V_{DS,\text{effCV}}$ is zero, BSIM4 specifically adds an if–then statement that sets $V_{DS,\text{effCV}}$ to zero when V_{DS} is precisely zero.

$$Q_{acc} = C_{ox,e}(V_{FB,\text{effCV}} - V_{FB,\text{CV}}). \tag{G-288}$$

$$Q_{sub,0} = C_{ox,e}\left(K1 \frac{\text{TOXE}}{\text{TOXM}}\right)\sqrt{\phi_{s,\text{dep}}}. \tag{G-289}$$

$$\delta Q_{sub} = C_{ox,e}\left[\frac{1 - A_{bulk,\text{CV}}}{2}V_{DS,\text{effCV}} - \frac{(1 - A_{Bulk,\text{CV}})A_{bulk,\text{CV}}V_{DS,\text{effCV}}^2}{12(V_{GST,\text{effCV}} - A_{bulk,\text{CV}}V_{Ds,\text{effCV}}/2)}\right]. \tag{G-290}$$

$$Q_{inv} = -C_{ox,e}\left[\left(V_{GST,\text{effCV}} - \frac{A_{bulk,\text{CV}}V_{DS,\text{effCV}}}{2}\right) + \frac{A_{bulk,\text{CV}}^2 V_{DS,\text{effCV}}^2}{12(V_{GST,\text{effCV}} - A_{bulk,\text{CV}}V_{DS,\text{effCV}}/2)}\right]. \tag{G-291}$$

$$Q_G = -Q_{inv} - \delta Q_{sub} + Q_{acc} + Q_{sub,0}. \tag{G-292}$$

$$Q_B = \delta Q_{sub} - Q_{acc} - Q_{sub,0}. \tag{G-293}$$

If XPART $<$ 0.5 (40/60 charge partition), then

$$Q_S = -\frac{C_{ox,e}}{2\left(V_{GST,\text{effCV}} - \frac{A_{bulk,\text{CV}}V_{DS,\text{effCV}}}{2}\right)^2}$$

$$\times [V_{GST,\text{effCV}}^3 - \tfrac{4}{3}V_{GST,\text{effCV}}^2 A_{bulk,\text{CV}}V_{DS,\text{effCV}}$$

$$+ \tfrac{2}{3}V_{GST,\text{effCV}}A_{bulk,\text{CV}}^2 V_{DS,\text{effCV}}^2 - \tfrac{2}{15}A_{bulk,\text{CV}}^3 V_{DS,\text{effCV}}^3]. \tag{G-294}$$

$$Q_D = -\frac{C_{ox,e}}{2\left(\frac{V_{GST,\text{effCV}} - A_{bulk,\text{CV}}V_{Ds,\text{effCV}}}{2}\right)^2}$$

$$\times [V_{GST,\text{effCV}}^3 - \tfrac{5}{3}V_{GST,\text{effCV}}^2 A_{bulk,\text{CV}}V_{DS,\text{effCV}}$$

$$+ V_{GST,\text{effCV}}A_{bulk,\text{CV}}^2 V_{DS,\text{effCV}}^2 - \tfrac{1}{5}A_{bulk,\text{CV}}^3 V_{DS,\text{effCV}}^3]. \tag{G-295}$$

If XPART = 0.5 (50/50 charge partition), then

$$Q_S = Q_D = \frac{Q_{inv}}{2}. \tag{G-296}$$

If XPART > 0.5 (0/100 charge partition), then

$$Q_S = -C_{ox,e}\left[\frac{V_{GST,\text{effCV}}}{2} + \frac{A_{bulk,CV}V_{DS,\text{effCV}}}{4}\right.$$
$$\left. - \frac{A_{bulk,CV}^2 V_{DS,\text{effCV}}^2}{24(V_{GST,\text{effCV}} - A_{bulk,CV}V_{DS,\text{effCV}}/2)}\right]. \tag{G-297}$$

$$Q_D = -C_{ox,e}\left[\frac{V_{GST,\text{effCV}}}{2} + \frac{3A_{bulk,CV}V_{DS,\text{effCV}}}{4}\right.$$
$$\left. + \frac{A_{bulk,CV}^2 V_{DS,\text{effCV}}^2}{8(V_{GST,\text{effCV}} - A_{bulk,CV}V_{DS,\text{effCV}}/2)}\right]. \tag{G-298}$$

Intrinsic Device Charge for CAPMOD = 2 (Which Was CAPMOD = 3 in BSIM3)

We saw in the CAPMOD = 1 equations that there are several basic components to the gate, bulk, source, and drain charges: Q_{acc}, $Q_{sub,0}$, δQ_{sub}, and Q_{inv}. Q_D and Q_S are basically some kind of partition of Q_{inv}; their sum adds up to Q_{inv}. Q_G and Q_B are some combination of the four basic components. Of these four components, Q_{acc} and $Q_{sub,0}$ are associated with the accumulation and depletion regions. They are most important when V_{GS} is below the threshold voltage value. Due to the smoothing properties, they remain finite (but fairly close to zero) above the threshold. In contrast, δQ_{sub} and Q_{inv} are charges associated with the inversion channel. These latter components are significant only at V_{GS} greater than the threshold voltage, although they are also finite below threshold.

In CAPMOD = 2, BSIM4 decides that there are really two effective oxide thicknesses to work with. One thickness matters most to the calculation of charges in the accumulation and the depletion region, while another, the strong inversion region. Consequently, the effective oxide thickness for the calculations of Q_{acc}, $Q_{sub,0}$, differs from the effective oxide thickness used to calculate Q_{inv} and δQ_{sub}.

For the calculation of Q_{acc} and $Q_{sub,0}$:

$$C'_{ox,\text{eff}} = \frac{\epsilon_{ox}}{\text{TOXP}} // \frac{\epsilon_s}{X_{DC,\text{eff}}}, \tag{G-299}$$

where

$$X_{DC,\text{eff}} = X_{DC,\max} - \frac{(X_{DC,\max} - X_{DC} - \delta_x) + \sqrt{(x_{DC,\max} - X_{DC} - \delta_x)^2 + 4\delta_x X_{DC,\max}}}{2}.$$
$$\tag{G-300}$$

APPENDIX G

The various terms inside the $X_{DC,\text{eff}}$ are given by

$$X_{DC} = \frac{L_{Debye}}{3} \exp\left[\text{ACDE}\left(\frac{\text{NDEP}}{2 \times 10^{16}}\right)^{-1/4} \cdot \frac{V_{GB,\text{eff}} - V_{fb,zb}}{10^8 \times \text{TOXP}}\right]. \quad \text{(G-301)}$$

$$L_{Debye} = \sqrt{\frac{\epsilon_s \cdot k[\text{TNOM} + 273.15]/q}{q\text{NDEP} \cdot 10^6}}. \quad \text{(G-302)}$$

$$X_{DC,\max} = \frac{L_{Debye}}{3}. \quad \text{(G-303)}$$

$$\delta_x = 10^{-3}\text{TOXP}. \quad \text{(G-304)}$$

The $X_{DC,\text{eff}}$ calculated above is used only for the calculation of accumation charge in the accumulation region.

$$C_{ox,\text{eff}} = C'_{ox,\text{eff}} \times W_{\text{eff,CV}} \times L_{\text{eff,CV}} \times \text{NF}. \quad \text{(G-305)}$$

$$Q_{acc} = C_{ox,\text{eff}}(V_{FB,\text{effCV}} - V_{FB,\text{CV}}). \quad \text{(G-306)}$$

$$Q_{sub,0} = C_{ox,\text{eff}}\left(\text{K1}\frac{\text{TOXE}}{\text{TOXM}}\right)\sqrt{\phi_{s,\text{dep}}}. \quad \text{(G-307)}$$

For the calculation of Q_{inv} and δQ_{sub}: The moment the transistor enters sub-threshold/depletion or inversion region, another X_{DC} is sought.

$$C'_{ox,\text{inv}} = \frac{\epsilon_{ox}}{\text{TOXP}} // \frac{\epsilon_s}{X_{DC,\text{inv}}}, \quad \text{(G-308)}$$

where

$$X_{DC,\text{inv}} = \begin{cases} \dfrac{1.9 \times 10^{-9}}{\left[1 + \dfrac{V_{GST,\text{effCV}} + 4(\text{VTH0} - V_{FB} - 2\phi_f)}{2 \times 10^8 \cdot \text{TOXP}}\right]^{0.7}} & \textit{if } (\text{VTH0} - V_{FB} - 2\phi_f) \geq 0; \\ \dfrac{1.9 \times 10^{-9}}{\left[1 + \dfrac{V_{GST,\text{effCV}}}{2 \times 10^8 \cdot \text{TOXP}}\right]^{0.7}} & \textit{if } (\text{VTH0} - V_{FB} - 2\phi_f) < 0. \end{cases} \quad \text{(G-309)}$$

It appears that $X_{DC,\text{inv}}$ is identical to $X_{DC,\text{IV}}$ (used for the I-V calculation). A closer examination shows that $X_{DC,\text{inv}}$ utilizes $V_{GST,\text{effCV}}$, while $X_{DC,\text{IV}}$ uses $V_{GST\text{eff}}$.

$$C_{ox,\text{inv}} = C'_{ox,\text{inv}} \times W_{\text{eff,CV}} L_{\text{eff,CV}} \times \text{NF}. \quad \text{(G-310)}$$

In addition to the new X_{DC}, BSIM3 also accounts for the fact that the surface potential does not get fixed at $2\phi_f$ in the strong inversion region.

$$\phi_\delta = \begin{cases} \dfrac{kT}{q} \ln\left[\dfrac{1 + V_{GST,\text{effCV}}(V_{GST,\text{effCV}} + 2 \cdot \text{K1} \cdot (\text{TOX/TOXM}) \cdot \sqrt{2\phi_f})}{\text{MOIN} \cdot \text{K1}^2(\text{TOX/TOXM})^2 \cdot (kT/q)}\right] & \text{if K1} > 0; \\[2ex] \dfrac{kT}{q} \ln\left[1 + \dfrac{V_{GST,\text{effCV}}(V_{GST,\text{effCV}} + \sqrt{2\phi_f})}{0.25 \times \text{MOIN}(kT/q)}\right] & \text{if K1} \le 0. \end{cases}$$

(G-311)

$$V_{DS,\text{satCV}} = \frac{V_{GST,\text{effCV}} - \phi_\delta}{A_{bulk,\text{CV}}}. \tag{G-312}$$

$$V_{DS,\text{effCV}} = V_{DS,\text{satCV}} - \frac{(V_{DS,\text{satCV}} - V_{DS} - \delta_4)}{2} - \frac{\sqrt{(V_{DS,\text{satCV}} - V_{DS} - \delta_4)^2 + 4\delta_4 V_{DS,\text{satCV}}}}{2}$$

δ_4 is fixed at 0.02. (G-313)

A substitution of $V_{DS} = 0$ into the equations involving the evaluation of $V_{DS,\text{effCV}}$ will prove that $V_{DS,\text{effCV}} = 0$ when $V_{DS} = 0$. However, just due to rounding errors, a computer may evaluate $V_{DS,\text{effCV}}$ to be a finite (though very small) value at $V_{DS} = 0$. In order to make sure $V_{DS,\text{effCV}}$ is zero when V_{DS} is zero, BSIM4 specifically adds an if–then statement that sets $V_{DS,\text{effCV}}$ to zero when V_{DS} is precisely zero.

$$Q_{inv} = -C_{ox,inv}\left[\left(V_{GST,\text{effCV}} - \phi_\delta - \frac{A_{bulk,\text{CV}} V_{DS,\text{effCV}}}{2}\right) + \frac{A_{bulk,\text{CV}}^2 V_{DS,\text{effCV}}^2}{12(V_{GST,\text{effCV}} - \phi_\delta - A_{bulk,\text{CV}} V_{DS,\text{effCV}}/2 + 10^{-20})}\right]. \tag{G-314}$$

The factor 10^{-20} is added in the denominator, mainly to prevents the term from going to a negative value when the term without the factor is negative and close to 0.

$$\delta Q_{sub} = C_{ox,inv}\left[\frac{1 - A_{bulk,\text{CV}}}{2} V_{Ds,\text{effCV}} - \frac{(1 - A_{bulk,\text{CV}})A_{bulk,\text{CV}} V_{DS,\text{effCV}}^2}{12(V_{GST,\text{effCV}} - \phi_\delta - A_{bulk,\text{CV}} V_{DS,\text{effCV}}/2 + 10^{-20})}\right]. \tag{G-315}$$

After all four charge components are found (with different effective oxide thicknesses), the terminal charges are given by

$$Q_G = -Q_{inv} - \delta Q_{sub} + Q_{acc} + Q_{sub,0}; \tag{G-316}$$

$$Q_B = \delta Q_{sub} - Q_{acc} - Q_{sub,0}. \tag{G-317}$$

If XPART < 0.5 (40/60 charge partition; the 10^{-20} factor is dropped to simplify expression), then

$$Q_S = -\frac{C_{ox,\text{inv}}}{2\left(V_{GST,\text{effCV}} - \phi_\delta - \dfrac{A_{bulk,\text{CV}}V_{DS,\text{effCV}}}{2}\right)^2}$$
$$\times [(V_{GST,\text{effCV}} - \phi_\delta)^3$$
$$- \tfrac{4}{3}(V_{GST,\text{effCV}} - \phi_\delta)^2 A_{bulk,\text{CV}}V_{Ds,\text{effCV}}$$
$$+ \tfrac{2}{3}(V_{GST,\text{effCV}} - \phi_\delta)A_{bulk,\text{CV}}^2 V_{DS,\text{effCV}}^2$$
$$- \tfrac{2}{15} A_{bulk,\text{CV}}^3 V_{DS,\text{effCV}}^3]. \tag{G-318}$$

$$Q_D = -\frac{C_{ox,\text{inv}}}{2\left(V_{GST,\text{effCV}} - \phi_\delta - \dfrac{A_{bulk,\text{CV}}V_{DS,\text{effCV}}}{2}\right)^2}$$
$$\times [(V_{GST,\text{effCV}} - \phi_\delta)^3$$
$$- \tfrac{5}{3}(V_{GST,\text{effCV}} - \phi_\delta)^2 A_{bulk,\text{CV}}V_{DS,\text{effCV}}$$
$$+ (V_{GST,\text{effCV}} - \phi_\delta)A_{bulkCV}^2 V_{DS,\text{effCV}}^2$$
$$- \tfrac{1}{5} A_{bulk,\text{CV}}^3 V_{DS,\text{effCV}}^3]. \tag{G-319}$$

If XPART = 0.5 (50/50 charge partition), then

$$Q_S = Q_D = \frac{Q_{inv}}{2}. \tag{G-320}$$

If XPART > 0.5 (0/100 charge partition), then

$$Q_S = -C_{ox,\text{inv}} \left[\frac{V_{GST,\text{effCV}} - \phi_\delta}{2} + \frac{A_{bulk,\text{CV}}V_{DS,\text{effCV}}}{4} \right.$$
$$\left. - \frac{A_{bulk,\text{CV}}^2 V_{DS,\text{effCV}}^2}{24(V_{GST,\text{effCV}} - \phi_\delta - A_{bulk,\text{CV}}V_{DS,\text{effCV}}/2 + 10^{-20})} \right]; \tag{G-321}$$

$$Q_D = -C_{ox,\text{inv}} \left[\frac{V_{GST,\text{effCV}} - \phi_\delta}{2} - \frac{3A_{bulk,\text{CV}}V_{DS,\text{effCV}}}{4} \right.$$
$$\left. + \frac{A_{bulk,\text{CV}}^2 V_{DS,\text{effCV}}^2}{8(V_{GST,\text{effCV}} - \phi_\delta - A_{bulk,\text{CV}}V_{Ds,\text{effCV}}/2 + 10^{-20})} \right]. \tag{G-322}$$

APPENDIX G   571

Intrinsic Device Capacitances

The device capacitances are obtained from chain rule. Although BSIM4 works out the exact expressions, they are quite complicated. We will just express the definition.

$$C_{xy} = \delta_{xy} \frac{\partial Q_x}{\partial V_y}. \qquad (G\text{-}323)$$

$\delta_{xy} = 1$ if $x = y$, and -1 if $x \neq y$. For example, $C_{gg} = +\partial Q_G / \partial V_G$, and $C_{gd} = -\partial Q_G / \partial V_D$.

Source–Bulk Junction Charge and Capacitance

If $V_{BS} < 0$, then

$$C_{j,SB} = \frac{\text{CJS}}{\left(1 - \frac{V_{BS}}{\text{PBS}}\right)^{\text{MJS}}} AS_{\text{eff}} + \frac{\text{CJSWS}}{\left(1 - \frac{V_{BS}}{\text{PBSWS}}\right)^{\text{MJSWS}}} PS_{\text{eff}}$$

$$+ \frac{\text{CJSWGS}}{\left(1 - \frac{V_{BS}}{\text{PBSWGS}}\right)^{\text{MJSWGS}}} W_{\text{eff,CJ}} \times \text{NF}. \qquad (G\text{-}324)$$

$$Q_{j,SB} = \int_0^{V_{BS}} C_{j,SB}(V'_{BS}) dV'_{BS}$$

$$= \frac{\text{CJS} \cdot AS_{\text{eff}} \text{PBS}}{1 - \text{MJS}} \left[1 - \left(1 - \frac{V_{BS}}{\text{PBS}}\right)^{1-\text{MJS}} \right]$$

$$+ \frac{\text{CJSWS} \cdot PS_{\text{eff}} \text{PBSWS}}{1 - \text{MJSWS}} \left[1 - \left(1 - \frac{V_{BS}}{\text{PBSWS}}\right)^{1-\text{MJSWS}} \right]$$

$$+ \frac{\text{CJSWGS} \cdot W_{\text{eff,CJ}} \times \text{NF} \cdot \text{PBSWGS}}{1 - \text{MJSWGS}} \left[1 - \left(1 - \frac{V_{BS}}{\text{PBSWGS}}\right)^{1-\text{MJSWGS}} \right].$$

$$(G\text{-}325)$$

If $V_{BS} \geq 0$, then

$$C_{j,SB} = \text{CJS}\left(1 + \text{MJS}\frac{V_{BS}}{\text{PBS}}\right) AS_{\text{eff}} + \text{CJSWS}\left(1 + \text{MJSWS}\frac{V_{BS}}{\text{PBSWS}}\right) PS_{\text{eff}}$$

$$+ \text{CJSWGS}\left(1 + \text{MJSWGS}\frac{V_{BS}}{\text{PBSWGS}}\right) W_{\text{eff,CJ}} \times \text{NF}. \qquad (G\text{-}326)$$

$$Q_{J,SB} = \text{CJS} \cdot AS_{\text{eff}}\left(V_{BS} + \frac{\text{MJS}}{\text{PBS}}\frac{V_{BS}^2}{2}\right) + \text{CJSWS} \cdot PS_{\text{eff}}\left(V_{BS} + \frac{\text{MJSWS}}{\text{PBSWS}}\frac{V_{BS}^2}{2}\right)$$

$$+ \text{CJSWGS} \cdot W_{\text{eff,CJ}} \times \text{NF}\left(V_{BS} + \frac{\text{MJSWGS}}{\text{PBSWGS}}\frac{V_{BS}^2}{2}\right). \qquad (G\text{-}327)$$

Drain–Bulk Junction Charge

If $V_{BD} < 0$, then

$$C_{j,\text{DB}} = \frac{\text{CJD}}{\left(1 - \dfrac{V_{BD}}{\text{PBD}}\right)^{\text{MJD}}} AD_{\textit{eff}} + \frac{\text{CJSWD}}{\left(1 - \dfrac{V_{BD}}{\text{PBSWD}}\right)^{\text{MJSWD}}} PD_{\textit{eff}}$$

$$+ \frac{\text{CJSWGD}}{\left(1 - \dfrac{V_{BD}}{\text{PBSWGD}}\right)^{\text{MJSWGD}}} W_{\textit{eff},\text{CJ}} \times \text{NF}. \quad (\text{G-328})$$

$$Q_{j,\text{DB}} = \int_0^{V_{BD}} C_{j,\text{DB}}(V'_{BD}) dV'_{BD}$$

$$= \frac{\text{CJD} \cdot AD_{\textit{eff}} \text{PBD}}{1 - \text{MJD}} \left[1 - \left(1 - \frac{V_{BD}}{\text{PBD}}\right)^{1-\text{MJD}} \right]$$

$$+ \frac{\text{CJSWD} \cdot PD_{\textit{eff}} \text{PBSWD}}{1 - \text{MJSWD}} \left[1 - \left(1 - \frac{V_{BD}}{\text{PBSWD}}\right)^{1-\text{MJSWD}} \right]$$

$$+ \frac{\text{CJSWGD} \cdot W_{\textit{eff},\text{CJ}} \times \text{NF} \cdot \text{PBSWGD}}{1 - \text{MJSWGD}}$$

$$\times \left[1 - \left(1 - \frac{V_{BD}}{\text{PBSWGD}}\right)^{1-\text{MJSWGD}} \right]. \quad (\text{G-329})$$

If $V_{BD} \geq 0$, then

$$C_{j,\text{DB}} = \text{CJD}\left(1 + \text{MJD}\frac{V_{BD}}{\text{PBD}}\right) AD_{\textit{eff}} + \text{CJSWD}\left(1 + \text{MJSWD}\frac{V_{BD}}{\text{PBSWD}}\right) PD_{\textit{eff}}$$

$$+ \text{CJSWGD}\left(1 + \text{MJSWGD}\frac{V_{BD}}{\text{PBSWGD}}\right) W_{\textit{eff},\text{CJ}} \times \text{NF}. \quad (\text{G-330})$$

$$Q_{j,\text{DB}} = \text{CJD} \cdot AD_{\textit{eff}} \left(V_{BD} + \frac{\text{MJD}}{\text{PBD}} \frac{V_{BD}^2}{2} \right) + \text{CJSWD} \cdot PD_{\textit{eff}} \left(V_{BD} + \frac{\text{MJSWD}}{\text{PBSWD}} \frac{V_{BD}^2}{2} \right)$$

$$+ \text{CJSWGD} \cdot W_{\textit{eff},\text{CJ}} \times \text{NF} \left(V_{BD} + \frac{\text{MJSWGD}}{\text{PBSWGD}} \frac{V_{BD}^2}{2} \right). \quad (\text{G-331})$$

Gate–Source Overlap Capacitance and Charge

$$V_{GS,\text{overlap}} = \frac{1}{2}\left[V_{GS} + \delta_1 - \sqrt{(V_{GS} + \delta_1)^2 + 4\delta_1}\right]; \quad \delta_1 = 0.02. \tag{G-332}$$

$$Q_{ov,GS} = \text{NF} \times W_{\text{eff,CV}} \times \left\{ \text{CGS0} \cdot V_{GS} + \text{CGS1}\left[V_{GS} - V_{GS,\text{overlap}}\right.\right.$$
$$\left.\left. - \frac{\text{CKAPPAS}}{2}\left(-1 + \sqrt{1 - \frac{4 \cdot V_{GS,\text{overlap}}}{\text{CKAPPAS}}}\right)\right]\right\}. \tag{G-333}$$

$$C_{ov,GS} = \frac{\partial Q_{ov,GS}}{\partial V_{GS}} = \text{NF} \times W_{\text{eff,CV}} \times \left[\text{CGS0} + \text{CGS1}\right.$$
$$\left. -\text{CGS1}\left(1 - \frac{1}{\sqrt{1 - \frac{4 \cdot V_{Gs,\text{overlap}}}{\text{CKAPPAS}}}}\right)\left(\frac{\frac{1}{2} - (V_{GS} + \delta_1)}{2\sqrt{(V_{GS} + \delta_1)^2 + 4\delta_1}}\right)\right]. \tag{G-334}$$

Gate–Drain Overlap Capacitance and Charge

$$V_{GD,\text{overlap}} = \frac{1}{2}\left[V_{GD} + \delta_1 - \sqrt{(V_{GD} + \delta_1)^2 + 4\delta_1}\right]; \quad \delta_1 = 0.02. \tag{G-335}$$

$$Q_{ov,GD} = \text{NF} \times W_{\text{eff,CV}} \times \left\{ \text{CGD0} \cdot V_{GD} + \text{CGD1}\left[V_{GD} - V_{GD,\text{overlap}}\right.\right.$$
$$\left.\left. - \frac{\text{CKAPPAD}}{2}\left(-1 + \sqrt{1 - \frac{4 \cdot V_{GD,\text{overlap}}}{\text{CKAPPAD}}}\right)\right]\right\}. \tag{G-336}$$

$$C_{ov,GD} = \frac{\partial Q_{ov,GD}}{\partial V_{GD}} = \text{NF} \times W_{\text{eff,CV}} \times \left[(\text{CGD0} + \text{CGD1})\right.$$
$$\left. -\text{CGD1}\left(1 - \frac{1}{\sqrt{1 - \frac{4 \cdot V_{GD,\text{overlap}}}{\text{CKAPPAD}}}}\right)\left(\frac{1}{2} - \frac{(V_{GD} + \delta_1)}{2\sqrt{(V_{GD} + \delta_1)^2 + 4\delta_1}}\right)\right]. \tag{G-337}$$

Fringing Capacitance and Charge

$$C_f = \text{NF} \times W_{\!e\!f\!f,\text{CV}} \cdot \text{CF}. \tag{G-338}$$

$$Q_{f,\text{GS}} = \int_0^{V_{GS}} C_f dV'_{GS} = \text{NF} \times W_{\!e\!f\!f,\text{CV}} \cdot \text{CF} \cdot V_{GS}. \tag{G-339}$$

$$Q_{f,\text{GD}} = \int_0^{V_{GD}} C_f dV'_{GD} = \text{NF} \times W_{\!e\!f\!f,\text{CV}} \cdot \text{CF} \cdot V_{GD}. \tag{G-340}$$

BSIM4 assumes that gate–source and gate–drain fringing capacitances are identical.

Parasitic Gate–Bulk Capacitance and Charge

$$C_{gb,0} = \text{NF} \times L_{\!e\!f\!f,\text{CV}} \cdot \text{CGB0}. \tag{G-341}$$

$$Q_{GB,\text{p}} = \int_0^{V_{GB}} C_{GB,\text{p}} dV'_{GB} = \text{NF} \times L_{\!e\!f\!f,\text{CV}} \cdot \text{CGB0} \cdot V_{GB}. \tag{G-342}$$

Total Terminal Charges (Intrinsic Plus Parasitic Charges)

$$Q_{G,\text{t}} = Q_G + Q_{ov,\text{GS}} + Q_{ov,\text{GD}} + Q_{f,\text{GS}} + Q_{f,\text{GD}} + Q_{GB,\text{p}}. \tag{G-343}$$

$$Q_{D,\text{t}} = Q_D - Q_{ov,\text{GD}} - Q_{f,\text{GD}} - Q_{j,\text{DB}}. \tag{G-344}$$

$$Q_{S,\text{t}} = Q_S - Q_{ov,\text{GS}} - Q_{f,\text{GS}} - Q_{j,\text{SB}}. \tag{G-345}$$

$$Q_{B,\text{t}} = Q_B + Q_{j,\text{SB}} + Q_{j,\text{DB}} - Q_{GB,\text{p}}. \tag{G-346}$$

Total Device Capacitances (Intrinsic Plus Parasitic Capacitances)

In the following equations, $C_{f,\text{GS}}$ and $C_{f,\text{GD}}$ are identical to the C_f given in Eq. (G-338). BSIM4 does not distinguish between the fringing capacitances at the source side and the drain side. We denote them separately to make the equations easier to understand. The overlap capacitances for the source side and the drain side ($C_{ov,\text{GS}}$

and $C_{ov,\text{GD}}$), in contrast, are different in BSIM4. They were given in Eqs. (G-334) and (G-337), respectively.

$$C_{gg,t} = C_{gg} + C_{ov,\text{GS}} + C_{f,\text{GS}} + C_{ov,\text{GD}} + C_{f,\text{GD}} + C_{gb,0}. \qquad (G\text{-}347)$$
$$C_{gd,t} = C_{gd} + C_{ov,\text{GD}} + C_{f,\text{GD}}. \qquad (G\text{-}348)$$
$$C_{gs,t} = C_{gs} + C_{ov,\text{GS}} + C_{f,\text{GS}}. \qquad (G\text{-}349)$$
$$C_{gb,t} = C_{gb} + C_{gb,0}. \qquad (G\text{-}350)$$
$$C_{dg,t} = C_{dg} + C_{ov,\text{GD}} + C_{f,\text{GD}}. \qquad (G\text{-}351)$$
$$C_{dd,t} = C_{dd} + C_{ov,\text{GD}} + C_{f,\text{GD}} + C_{j,\text{DB}}. \qquad (G\text{-}352)$$
$$C_{ds,t} = C_{ds}. \qquad (G\text{-}353)$$
$$C_{db,t} = C_{db} + C_{j,\text{DB}}. \qquad (G\text{-}354)$$
$$C_{sg,t} = C_{sg} + C_{ov,\text{GS}} + C_{f,\text{GS}}. \qquad (G\text{-}355)$$
$$C_{sd,t} = C_{sd}. \qquad (G\text{-}356)$$
$$C_{ss,t} = C_{ss} + C_{ov,\text{GS}} + C_{f,\text{GS}} + C_{j,\text{SB}}. \qquad (G\text{-}357)$$
$$C_{sb,t} = C_{sb} + C_{j,\text{SB}}. \qquad (G\text{-}358)$$
$$C_{bg,t} = C_{bg} + C_{gb,0}. \qquad (G\text{-}359)$$
$$C_{bd,t} = C_{bd} + C_{j,\text{DB}}. \qquad (G\text{-}360)$$
$$C_{bs,t} = C_{bs} + C_{j,\text{SB}}. \qquad (G\text{-}361)$$
$$C_{bb,t} = C_{bb} + C_{j,\text{SB}} + C_{j,\text{DB}} + C_{gb,0}. \qquad (G\text{-}362)$$

The following transcapacitances are not really used in the code, but it is convenient for us to list them out anyway.

$$C_m = C_{dg,t} - C_{gd,t}. \qquad (G\text{-}363)$$
$$C_{mb} = C_{db,t} - C_{bd,t}. \qquad (G\text{-}364)$$
$$C_{mx} = C_{bg,t} - C_{gb,t}. \qquad (G\text{-}365)$$

Transient Current Equations

$$i_G(t) = C_{gg,t} \cdot \frac{dv_{GS}}{dt} - C_{gd,t} \cdot \frac{dv_{DS}}{dt} - C_{gb,t} \cdot \frac{dv_{BS}}{dt}. \qquad (G\text{-}366)$$
$$i_D(t) = I_{DS} + I_{sub} - I_{j,\text{DB}} - C_{dg,t} \cdot \frac{dv_{GS}}{dt} + C_{dd,t} \cdot \frac{dv_{DS}}{dt} - C_{db,t} \cdot \frac{dv_{BS}}{dt}. \qquad (G\text{-}367)$$
$$i_B(t) = -I_{sub} + I_{j,\text{DB}} + I_{j,\text{SB}} - C_{bg,t} \cdot \frac{dv_{GS}}{dt} - C_{bd,t} \cdot \frac{dv_{DS}}{dt} + C_{bb,t} \cdot \frac{dv_{BS}}{dt}. \qquad (G\text{-}368)$$
$$i_S(t) = -i_G(t) - i_D(t) - i_B(t). \qquad (G\text{-}369)$$

Small-Signal Current Equations

$$i_g = j\omega C_{gs} v_{gs} + j\omega C_{gd} v_{gd} + j\omega C_{gb} v_{gb}. \tag{G-370}$$

$$i_d = g_m v_{gs} + g_d v_{ds} + g_{mb} v_{bs} + G_{bg} v_{gs} + G_{bd} v_{ds} + G_{bb} v_{bs} + g_{j,\text{DB}} v_{db}$$
$$+ j\omega C_{sd,t} v_{ds} + j\omega C_{gd,t} v_{dg} + j\omega C_{bd,t} v_{db} - j\omega C_m v_{gs} - j\omega C_{mb} v_{bs}. \tag{G-371}$$

$$i_b = -G_{bg} v_{gs} - G_{bd} v_{ds} - G_{bb} v_{bs} + g_{j,\text{DB}} v_{bd} + g_{j,\text{SB}} b_{bs}$$
$$+ j\omega C_{bs,t} v_{bs} + j\omega C_{bd,t} v_{bd} + j\omega C_{gb,t} v_{gb} - j\omega C_{mx} v_{bg}. \tag{G-372}$$

$$i_s(t) = -i_g(t) - i_d(t) - i_b(t). \tag{G-373}$$

Temperature Modeling

If the device temperature is not equal to TNOM, then various quantities need to be substituted with the results given below. See the entry TNOM in Section 3.2 for more details.

$$V_T(T_{device}) = V_T(\text{TNOM}) + \left[\text{KT1} + \frac{\text{KT1L}}{L_{eff}} + \text{KT2} V_{BS,\text{eff}} \right]$$
$$\times \left[\frac{T_{device} + 273.15}{\text{TNOM} + 273.15} - 1 \right]. \tag{G-374}$$

$$\text{VSAT}(T_{device}) = \text{VSAT}(\text{TNOM}) - \text{AT} \times \left[\frac{T_{device} + 273.15}{\text{TNOM} + 273.15} - 1 \right]. \tag{G-375}$$

Suppose that the device temperature is high enough such that $\text{VSAT}(T_{device})$ given above decreases to a negative value. BSIM4 will catch this unphysical event and issues a fatal error. There is another BSIM4 checking mechanism, which requires the parameter PARAMCHK to be turned on. When PARAMCHK is set to 1, BSIM4 will start to issue a warning when $\text{VSAT}(T_{device})$ decreases below 1×10^3 m/s.

$$\text{U0}(T_{device}) = \text{U0}(\text{TNOM}) \times \left(\frac{T_{device} + 273.15}{\text{TNOM} + 273.15} \right)^{\text{UTE}}. \tag{G-376}$$

BSIM4 checks whether $\text{U0}(T_{device})$ is less than 0. If so, BSIM4 issues a fatal warning since the mobility must be a positive number. It is clear from the above equation that the only way to make $U0(T_{device}) < 0$ is by having nonphysical $U0(TNOM)$.

$$UA(T_{device}) = UA(TNOM) + UA1 \times \left[\frac{T_{device} + 273.15}{TNOM + 273.15} - 1\right]. \quad (G\text{-}377)$$

$$UB(T_{device}) = UB(TNOM) + UB1 \times \left[\frac{T_{device} + 273.15}{TNOM + 273.15} - 1\right]. \quad (G\text{-}378)$$

$$UC(T_{device}) = UC(TNOM) + UC1 \times \left[\frac{T_{device} + 273.15}{TNOM + 273.15} - 1\right]. \quad (G\text{-}379)$$

UA, UB, and UC are coefficients that modify the carrier mobility. They can be negative or positive, as far as calculating a numerical value for the carrier mobility is concerned. BSIM4 does not check whether they are positive or negative.

$$RDSW(T_{device}) = RDSW(TNOM) + PRT \times \left[\frac{T_{device} + 273.15}{TNOM + 273.15} - 1\right]. \quad (G\text{-}380)$$

If the calculated $RDSW(T_{device})$ is negative, then BSIM4 issues a warning and sets $RDSW(T_{device})$ to 0.

$$RDSWMIN(T_{device}) = RDSWMIN(TNOM) + PRT \times \left[\frac{T_{device} + 273.15}{TNOM + 273.15} - 1\right]. \quad (G\text{-}381)$$

If the calculated $RDSWMIN(T_{device})$ is negative, then BSIM4 issues a warning and sets $RDSWMIN(T_{device})$ to 0.

$$RSW(T_{device}) = RSW(TNOM) + PRT \times \left[\frac{T_{device} + 273.15}{TNOM + 273.15} - 1\right]. \quad (G\text{-}382)$$

If the calculated $RSW(T_{device})$ is negative, then BSIM4 issues a warning and sets $RSW(T_{device})$ to 0.

$$RSWMIN(T_{device}) = RSWMIN(TNOM) + PRT \times \left[\frac{T_{device} + 273.15}{TNOM + 273.15} - 1\right]. \quad (G\text{-}383)$$

If the calculated $RSWMIN(T_{device})$ is negative, then BSIM4 issues a warning and sets $RSWMIN(T_{device})$ to 0.

$$RDW(T_{device}) = RDW(TNOM) + PRT \times \left[\frac{T_{device} + 273.15}{TNOM + 273.15} - 1\right]. \quad (G\text{-}384)$$

If the calculated RDW(T_{device}) is negative, then BSIM4 issues a warning and sets RDW(T_{device}) to 0.

$$\text{RDWMIN}(T_{device}) = \text{RDWMIN(TNOM)} + \text{PRT} \times \left[\frac{T_{device} + 273.15}{\text{TNOM} + 273.15} - 1\right]. \quad \text{(G-385)}$$

If the calculated RDWMIN(T_{device}) is negative, then BSIM4 issues a warning and sets RDWMIN(T_{device}) to 0.

$$\text{CJ}(T_{device}) = \text{CJ(TNOM)} \times [1 + \text{TCJ} \cdot (T_{device} - \text{TNOM})]. \quad \text{(G-386)}$$

If $\text{TCJ} \cdot (T_{device} - \text{TNOM})$ is less than -1, then BSIM4 issues a warning and sets CJ(T_{device}) to 0.

$$\text{PB}(T_{device}) = \text{PB(TNOM)} - \text{TPB} \cdot (T_{device} - \text{TNOM}). \quad \text{(G-387)}$$

If the calculated PB(T_{device}) is less than 0.01, then BSIM4 issues a warning and sets PB(T_{device}) to 0.01.

$$\text{CJSW}(T_{device}) = \text{CJSW(TNOM)} \times [1 + \text{TCJSW} \cdot (T_{device} - \text{TNOM})]. \quad \text{(G-388)}$$

If $\text{TCJSW} \cdot (T_{device} - \text{TNOM}$ is less than -1, then BSIM4 issues a warning and sets CJSW(T_{device}) to 0.

$$\text{PBSW}(T_{device}) = \text{PBSW(TNOM)} - \text{TPBSW} \cdot (T_{device} - \text{TNOM}). \quad \text{(G-389)}$$

If the calculated PBSW(T_{device}) is less than 0.01, then BSIM4 issues a warning and sets PBSW(T_{device}) to 0.01.

$$\text{CJSWG}(T_{device}) = \text{CJSWG(TNOM)} \times [1 + \text{TCJSWG} \cdot (T_{device} - \text{TNOM})]. \quad \text{(G-390)}$$

If $\text{TCJSWG} \cdot (T_{device} - \text{TNOM})$ is less than -1, then BSIM4 issues a warning and sets CJSWG(T_{device}) to 0.

$$\text{PBSWG}(T_{device}) = \text{PBSWG(TNOM)} - \text{TPBSWG} \cdot (T_{device} - \text{TNOM}). \quad \text{(G-391)}$$

If the calculated PBSWG(T_{device}) is less than 0.01, then BSIM4 issues a warning and sets PBSWG(T_{device}) to 0.01.

$$\text{JSS}(T_{device}) = \text{JSS}(\text{TNOM}) \left[\exp\left(\frac{E_g(T_{device})}{\text{NJS} \cdot kT_{device}} - \frac{E_g(\text{TNOM})}{\text{NJS} \cdot k\,\text{TNOM}} \right) \right]$$
$$\times \left(\frac{T_{device}}{\text{TNOM}} \right)^{\text{XTIS/NJS}} ; \qquad (G\text{-}392)$$

$$\text{JSD}(T_{device}) = \text{JSD}(\text{TNOM}) \left[\exp\left(\frac{E_g(T_{device})}{\text{NJD} \cdot kT_{device}} - \frac{E_g(\text{TNOM})}{\text{NJD} \cdot k\,\text{TNOM}} \right) \right]$$
$$\times \left(\frac{T_{device}}{\text{TNOM}} \right)^{\text{XTID/NJD}} \qquad (G\text{-}393)$$

$$\text{JSWS}(T_{device}) = \text{JSWS}(\text{TNOM}) \left[\exp\left(\frac{E_g(T_{device})}{\text{NJS} \cdot kT_{device}} - \frac{E_g(\text{TNOM})}{\text{NJS} \cdot k\,\text{TNOM}} \right) \right]$$
$$\times \left(\frac{T_{device}}{\text{TNOM}} \right)^{\text{XTIS/NJS}} ; \qquad (G\text{-}394)$$

$$\text{JSWD}(T_{device}) = \text{JSWD}(\text{TNOM}) \left[\exp\left(\frac{E_g(T_{device})}{\text{NJD} \cdot kT_{device}} - \frac{E_g(\text{TNOM})}{\text{NJD} \cdot k\,\text{TNOM}} \right) \right]$$
$$\times \left(\frac{T_{device}}{\text{TNOM}} \right)^{\text{XTID/NJD}} ; \qquad (G\text{-}395)$$

$$\text{JSWGS}(T_{device}) = \text{JSWGS}(\text{TNOM}) \left[\exp\left(\frac{E_g(T_{device})}{\text{NJS} \cdot kT_{device}} - \frac{E_g(\text{TNOM})}{\text{NJD} \cdot k\,\text{TNOM}} \right) \right]$$
$$\times \left(\frac{T_{device}}{\text{TNOM}} \right)^{\text{XTIS/NJS}} ; \qquad (G\text{-}396)$$

$$\text{JSWGD}(T_{device}) = \text{JSWGD}(\text{TNOM}) \left[\exp\left(\frac{E_g(T_{device})}{\text{NJD} \cdot kT_{device}} - \frac{E_g(\text{TNOM})}{\text{NJD} \cdot k\,\text{TNOM}} \right) \right]$$
$$\times \left(\frac{T_{device}}{\text{TNOM}} \right)^{\text{XTID/NJD}} ; \qquad (G\text{-}397)$$

where

$$E_g \text{ (in eV)} = 1.16 - \frac{7.02 \times 10^{-4} T_{device}^2}{T_{device} + 1108} \qquad (T_{device} \text{ is in Kelvin}). \qquad (G\text{-}398)$$

Thermal Noise Modeling

In BSIM4, there are basically two thermal noise models. TNOIMOD = 0 resembles the model used in BSIM3. TNOIMOD = 1 is touted the holistic thermal noise model, since all the short-channel effects and velocity saturation effects that are incorporated in the I-V model are automatically included in this thermal noise model.

When TNOIMOD = 0 (whether RDSMOD = 1 or 0), then

$$\frac{\overline{i_d^2}}{\Delta f} = \frac{4kT \cdot \text{NTNOI}}{R_{DS} + L_{\text{eff}}^2/(\mu_{\text{eff}}|Q_{inv}|)}. \qquad (G\text{-}399)$$

When TNOIMOD = 1 (whether RDSMOD = 1 or 0; see the noise equivalent circuits in Fig. 5-17), then

$$\frac{\overline{i_d^2}}{\Delta f} = 4kT \frac{V_{DS,\text{eff}}}{I_{DS}} (\beta g_m + \beta g_{mb} + g_d)^2 - 4kTR_x(g_m + g_{mb} + g_d)^2; \qquad (G\text{-}400)$$

$$\frac{\overline{i_{RS}^2}}{\Delta f} = 4kT \frac{R_x + R_{S,\text{eltd}} + R_{S,\text{bias-dep}}}{(R_{S,\text{eltd}} + R_{S,\text{bias-dep}})^2}; \qquad (G\text{-}401)$$

where

$$R_x = \theta^2 \frac{V_{DS,\text{eff}}}{I_{DS}}; \qquad (G\text{-}402)$$

$$\beta = 0.577 \times \left[1 + \text{TNOIA} \cdot L_{\text{eff}} \left(\frac{V_{GST,\text{eff}}}{\varepsilon_{sat} L_{\text{eff}}} \right)^2 \right], \qquad (G\text{-}403)$$

$$\theta = 0.37 \times \left[1 + \text{TNOIB} \cdot L_{\text{eff}} \left(\frac{V_{GST,\text{eff}}}{\varepsilon_{sat} L_{\text{eff}}} \right)^2 \right]. \qquad (G\text{-}404)$$

ε_{sat}, the saturation electric field, is defined to be $2 \cdot \text{VSAT}/\mu_{\text{eff}}$, as given in Eq. (G-73). The following equations apply to only to TNOIMOD = 0; they were not needed for TNOIMOD = 1 because Eq. (G-401) incorporates their noise sources already.

$$\frac{\overline{i_{RS,\text{eltd}}^2}}{\Delta f} = \frac{4kT}{R_{S,\text{eltd}}}. \qquad (G\text{-}405)$$

$$\frac{\overline{i_{RS,\text{bias-dep}}^2}}{\Delta f} = \frac{4kT}{R_{S,\text{bias-dep}}}. \qquad (G\text{-}406)$$

The following equations apply to both TNOIMOD = 0 and 1:

$$\frac{\overline{i^2_{RD,\text{eltd}}}}{\Delta f} = \frac{4kT}{R_{D,\text{eltd}}}. \tag{G-407}$$

$$\frac{\overline{i^2_{RD,\text{bias-dep}}}}{\Delta f} = \frac{4kT}{R_{D,\text{bias-dep}}}. \tag{G-408}$$

$$\frac{\overline{i^2_{RG,\text{eltd}}}}{\Delta f} = \frac{4kT}{R_{G,\text{eltd}}}. \tag{G-409}$$

$$\frac{\overline{i^2_{RB,\text{pb}}}}{\Delta f} = \frac{4kT}{R_{B,\text{pb}}}. \tag{G-410}$$

$$\frac{\overline{i^2_{RB,\text{pd}}}}{\Delta f} = \frac{4kT}{R_{B,\text{pd}}}. \tag{G-411}$$

$$\frac{\overline{i^2_{RB,\text{ps}}}}{\Delta f} = \frac{4kT}{R_{B,\text{ps}}}. \tag{G-412}$$

$$\frac{\overline{i^2_{RB,\text{db}}}}{\Delta f} = \frac{4kT}{R_{B,\text{db}}}. \tag{G-413}$$

$$\frac{\overline{i^2_{RB,\text{sb}}}}{\Delta f} = \frac{4kT}{R_{B,\text{sb}}}. \tag{G-414}$$

Flicker Noise Modeling

When FNOIMOD = 0, a simple Flicker noise model is used.

$$\frac{\overline{i^2_d}}{\Delta f} = \text{KF} \times \frac{|I_{DS}|^{\text{AF}}}{C'_{ox,e} L^2_{\text{eff}} f^{\text{EF}}}. \tag{G-415}$$

When FNOIMOD = 1, a rather complicated Flicker noise model is used.

$$\frac{\overline{i^2_d}}{\Delta f} = \frac{S_{\text{inversion}} \times S_{\text{subthreshold}}}{S_{\text{inversion}} + S_{\text{subthreshold}}}; \tag{G-416}$$

$$S_{\text{inversion}} = \frac{q^2 kT \mu_{\text{eff}} I_{DS}}{C'_{ox,e} L^2_{\text{eff}} A_{bulk} f^{\text{EF}} 10^{10}} \left[\text{NOIA} \ln\left(\frac{N_0 + N^*}{N_1 + N^*}\right) + \text{NOIB}(N_0 - N_1) \right.$$

$$\left. + \frac{\text{NOIC}}{2}(N_0^2 - N_1^2) \right]$$

$$+ \frac{kT/q I^2_{DS} \Delta L_{clm}}{W_{\text{eff,special}} L^2_{\text{eff}} f^{\text{EF}} 10^{10}} \frac{\text{NOIA} + \text{NOIB} \cdot N_1 + \text{NOIC} \cdot N_1^2}{(N_1 + N^*)^2}; \tag{G-417}$$

$$S_{\text{subthreshold}} = \frac{\text{NOIA} \, kT/q I^2_{DS}}{W_{\text{eff,special}} L_{\text{eff}} (f^{\text{EF}}) \cdot (N^*)^2 \cdot 10^{10}}; \tag{G-418}$$

where

$$N^* = \frac{kT}{q^2}(C'_{ox,e} + C'_{dep} + \text{CIT}); \qquad (G\text{-}419)$$

where C'_{dep} was defined in Eq. (G-53).

$$N_0 = C'_{ox,e} V_{GST,\text{eff}}/q; \qquad (G\text{-}420)$$

$$N_1 = \frac{C'_{ox,e} V_{GST,\text{eff}}}{q}\left(1 - \frac{A_{bulk} V_{DS,\text{eff}}}{V_{GST,\text{eff}} + 2kT/q}\right); \qquad (G\text{-}421)$$

$$\Delta L_{clm} = L_{itl} \ln\left[\frac{(V_{DS} - V_{DS,\text{eff}})/L_{itl} + \text{EM}}{\varepsilon_{sat}}\right]; \qquad (G\text{-}422)$$

where L_{itl} was defined in Eq. (G-102), and ε_{sat} in Eq. (G-73).

In the noise models, it is $L_{\textit{eff}}$, not $L_{\textit{eff},\text{CV}}$, in all of the above equations.

INDEX

∞-differentiability, 29

A_{bulk}, see Bulk-charge coefficient
Analog model, Section 1.3
Asymmetrical MOSFET, 340, 343, 345, 393
Asymmetry, see Symmetry property
Avalanche breakdown, 103, 170, 242

Bandgap energy, 205, 423
Berkeley SPICE, 1, 2, 159
Binning, 90, Section 1.12, 161–162, 172, 396
Body effect, 192, 266, 403–404
Body-effect coefficient (γ), 27, 101, 206. See also Bulk-charge factor
Body-effect factors (GAMMA1, GAMM2, K1, K2), 200–201, 205–207
BSIM, BSIM2 models, 5, 6
BSIM3, 6, 26, Appendix A
  backward compatibility, 259, 399
  BSIM3v3.1, BSIM3v3.2, BSIM3v3.2.1, BSIM3v3.2, see BSIM3, versions
  BSIM3v3.3, 6, 99, 360, 399, 437
  calculation flow diagram, 275–282
  capacitance simulation artifacts, 25, Sections 4.8 to 4.10
  C–V model, 102, 120, 173–175, 184–186, 347, 353. See also CAPMOD
  derivatives, model vs. numerical, 37–38
  diode model, see Diode
  impact–ionization model, see Impact–ionization
  industry usage, 284, 399
  I–V model, 104–109
  large-signal (transient) model, Section 2.4
  noise model, 103, Section 2.6, 187. See also Noise modeling
  parameters, Section 3.1, Section 3.2
  parameter check, 234–235
  predictive modeling, 247, 251, 390–391, 413
  simulation artifacts, Chapter 4
  small-signal model, Section 2.5
  summary, Section 2.1
  unofficial, 159, 162
  versions, 6, 162, 174, 235, 258–260, 333, 343, 399
  version test, 260
  y-parameters, Section 2.3
BSIM4, 7, 399, Chapter 5, Appendix G
  capacitance issues, 406–407
  predictive modeling, 405, 413, 461
Built-in voltage, 191, 235, 393
Bulk-charge coefficient (A_{bulk}), 163, 167, 171, 185, 211, 272, 401. See also Bulk-charge factor
Bulk-charge factor (δ), 27, 49
  relation to A_{bulk}, 49, 163. See also Body-effect coefficient
Bulk-referencing, 72, 129
Bump in output voltage, 132, 273, 287, 299

Capacitance, see also BSIM3, Capacitor, C–V characteristics
  0/100 vs. 40/60 charge partition, 273, 507
  C_{dg}'s effect on quasi-static analysis, 132, 273, 287, 302, 314
  $C_{gd} \neq C_{gs}$ at $V_{DS} = 0$, see Symmetry property

583

584   INDEX

Capacitance (*continued*)
  C_{gs}, $C_{gs,p}$, and $C_{gs,t}$, distinction, *see* Notation convention
  $C_{xy} = C_{yx}$ (reciprocity), 23, 61, 116–117. See also Meyer capacitance model
  $C_{xy} \neq C_{yx}$ (non-reciprocity), 61
  definition, 64
  diffusion, 123-124
  fringing, *see* Fringing capacitance
  intrinsic device, 120, Appendix B
  junction, *see* Diode, C–V model
  linear relationships in device capacitances, 65–66
  measurement, 81–82, 84–85
  negative, *see* Negative capacitance
  nonlinear, 51
  non-quasi-static, 453
  overlap, *see* Overlap capacitance
  theoretical expressions, Appendix B
  velocity saturation effect, 185–186, 353
Capacitor
  MOSFET, 341
  parallel plate, 23, 60
CAPMOD (capacitance model), 34, 165, 173–175, 341
Chain rule, 36, Section 1.5, 125
Channel charge,
  governing equation, 39, 68
  quasi-static solution, 40–41
  small-signal solution, 69, Appendix C
  static solution, 40
Channel doping, 218–220
Channel length modulation, 236, 242
Channel reflected gate resistance, *see* Gate resistance
Channel resistance (r_{ch}), 5 Section 4.3, 327
Charge,
  0/100 vs. 40/60, 492–493, 496, 567–568, 571
  MOSFET (Q_G, Q_B, Q_S, Q_D, Q_{inv}), 45–49, 79, 187, 357, Appendix B
  numerical evaluation, 80
  SPICE calculation, 52
Charge-based model, 2, 53, 55, 123
Charge conservation, 5 Section 1.8, 102
Charge front, 43
Charge non-conservation, 51, 57
  in linear capacitor, 55
  test circuits, 57, 59
Charge partition, Section 1.7, 85, 272–274, 286–287, 375
  0/100 vs. 40/60, 286, 300, 314
  misconception, 48, 273
  non-quasi-static model, 47, 519

Charge sharing, 102, 190–192, 266, 390, 403–404
Charge-thickness capacitance model, 165–167, 175, 401
Circuits
  charge-pump, 57
  current mirror, 110
  induced gate noise, 365
  input-impedance matched amplifier, 374, 522, 525
  Killer NOR gate, Section 4.2
  NMOS discharging capacitor, 131, 288, 296
  noise correlation verification, 443, 444
  noise figure calculation, 368, 370
  parallel branches of R–C, 309
  passgate, 293
  switch-capacitor, 59
  two-stage amplifier, 376
Common-bulk, 127
  vs. bulk-referencing, 129
Common-source, 126, 136
Compact–Model Council, 7, 399
Conductance, 63, *see also* g_m/I_D
  $1/(5 g_m)$, 307, 309
  definition 13, 77
  $g_m^\dagger$, 137–139, 304, Section 4.4, 371, 375, 453
  GMIN, *see* GMIN
  model vs. numerical, 37
  modification by source/drain resistance, 113
  negative, *see* Negative conductance
Conformal mapping, 347, Appendix D
Convergence, 4, 9–11, 18, 54, 207, 273. See also Singular matrix, Tolerance
Cubic Spline, 78
Cutoff frequency, 290. See also Transit time, intrinsic-channel
C–V characteristics, 20–26
  BSIM3 simulated, 338–341
  BSIM3v3.1 and BS1M3v3.2, comparison, 335–337
  C_{gg} turn-on characteristics, 228, 342, 343–345
  MEDICI simulation, 349–355
  negative, *see* Negative capacitance
  polydepletion effects, 225
  VOFFCV, NOFF, effects on, 343

Debye length, 166
De-embedding, 83
δ, *see* Bulk-charge factor

INDEX **585**

Δ, 30
DELTA, 30, 187–188
Depletion mode device, 101, 250
Diffusion capacitance, *see* Capacitance, diffusion
Digital model, Section 1.3
Diode, 102–103, 105–106, 160
   breakdown, 124, 170, 392, Section 5.9
   built-in voltage, *see* Built-in voltage
   C–V model, 118–121, 155–156, 181–183, 393, Section 5.8, 466
   I–V model, 152–155, 201–205, Section 5.9
   junction capacitance, effects on circuits, 302, 314, 366, 453
   large-signal model, 123
   source-bulk/drain-bulk symmetry, 181, 392–393, 418
Discontinuity in simulated characteristics, Section 1.3, 29, 31
Drain-bulk diode, *see* Diode
Drain current, 104, 125, 126, 127, 201, 241
   BSIM4, 409–412, 426
   finite current artifact at $t < \tau_{tr}$, 42-44, 78, 297
Drain-induced barrier-lowering (DIBL), 76, 102, 189, 198, 236, 266, 382
Drain-induced gate noise, 103
Drain-induced threshold shift, 411

Early voltage, 189, 236–239, 272, 402, 410–411
Effective channel length, 101, 188, 214–217
Effective channel width, 101, 195–197, 268–270
   special note, 269, 360
Effective junction width, 414, 431
EKV model, 6, 72, 75
Electric field
   lateral, 248
   normal, 248–249, 413
Elmore constant, 197, 520
Equivalent circuit
   BSIM3, Chapter 2
   BSIM4, 425, 428, 433, 434, 439–442, 449, 463
Error voltage, 293

First-generation models, 7, 24
First-moment technique, 46
Flatband voltage, *see also* Smoothing function, $V_{FB,\text{effCV}}$

C–V calculation, 34, 173, 175, 259–260, 261, 333, 342
I–V calculation, 261
Fortran programming, 1, 298–299
Fringing capacitance, 102, 116–121, 176-177, 302, Appendix D
   inner, 119, 345-346

Gate-bulk parasitic capacitance, 116, 177–178
Gate-induced drain leakage (GIDL), 103, 151, Section 4.16, Section 5.10
Gate leakage current, 103, 104, 318, 395. *See also* Tunneling current.
Gate material, polysilicon or metal, 224, 261, 267, 316
Gate resistance,
   1/3 vs. 1/12 factor, 317–318, 431
   BSIM4 equivalent circuit, 433, 440, 442
   channel-reflected resistance, 432, 452
   distributed resistance, 105–106, 149, Section 4.5, Section 5.12
   sheet resistance/resistivity, 316, 431
Gauss law, 248
GBMIN, 435
g_m/I_D, 15–18, 341, 388, 409
$g_m^\dagger$, *see* Conductance.
GMIN, 18, 103, 106, 121, 149, 202, 384, 418
   effect on off current, 150

Holistic thermal noise model, 439, Section 5.15
   verification with van der Ziel representation, 442–449
Hot carrier, 108
HSPICE, 2, 159. *See also* Level 28

Ideality factor, 34, 221–223
   V_{BS} dependence, 34
Impact-ionization, 103, 106–111, 150, 160, 241, 384, 396
   BSIM3, 168–170
   conductances, 121
   large-signal model, 124
   NMOS vs. PMOS, 108, 110
   noise, 147
Impedance matching, 319, 373, 522
Inverse modeling, 73, Section 1.10, 149

Junction, isolated, merged, shared, 467

Killer–NOR gate, 48, 273, Section 4.2
Kink, *see* Digital model
Kirchoff current law, 4, 8, 62, 132, 288

Layout, transistor, 394, 432, Section 5.19
LCR meter, 84–85
LEVEL, 7, 159, 213
Level 1, 2, 3 models, 5, 18, 24, 60
Level 28, 8
Linear region, 29, 40
Liu, xii, 99

Maximum operating voltage, 104
MEDICI, 3, 85, Section 4.1, 340, 348–355
MESFET, 55, 81, 101
Meyer capacitance model, 24, 61
Mobility, 247–256, Section 5.7
   model predictability, 251, 413
   temperature dependency, 251, 256
   universal mobility, 249, 413
MOS Model 9, 6

Narrow width effect, 208–210, 266
Negative capacitance
   C_{ds}, C_{sd} (physical), 340
   C_{dd}, 340
   C_{gb}, 337
   C_{gd}, 339, 406–407
   C_{gs}, 259–260, 335, 337, 340, 344–345
Negative conductance, 5, 9, Section 4.15
   g_m, 380–382
   g_{mb}, 193, 207, 212, 376
Newton–Raphson algorithm, 3, 4, 8–10, 526
NF, number of finger, 410, 415, 429, 465, 467. *See also* Noise figure
Number of instances, difference, 410, 465
Noise analysis, 144
   calculation with correlation coefficient, 364, 443, 447
Noise figure, Section 4.13, Appendix F
   0 dB simulation artifact, 372–375, 446
   calculation using SPICE, 368
   definition, 369
   frequency dependency, 370
   measured results, 367
   minimum, 372, Appendix F
Noise modeling, 19, Section 2.6. *See also* Holistic thermal noise model.
   correlation coefficient, 364

distributed gate resistance, 319
excess noise factor (γ), 358, 361, 437–438
Flicker noise, 146, 229–230, 361, 370, Section 5.16
impact ionization, 147
induced-gate noise, 146, 149, Section 4.12
junction diode, 149
kink in NOIMOD=4, 357
noiseless resistor, 359, 365, 368
resistor, 142–143
shot noise, 146, 387, 439, 464
thermal noise, 141, 146, 230–231, Section 4.11, Section 5.15
tunneling currents, 439, 441, 464
van der Ziel representation, 363, 440, 442–439
white noise, 143
Non-quasi-static model, Section 1.6, Section 4.1, 432
   in BSIM3, 38, 102, 197, 232–234, 296–299, Appendix E
   in BSIM4, Section 5.17, Section 5.18
   small-signal equivalent circuit, 308
Notation convention,
   capacitance, intrinsic, (such as C_{gs}) 70
   capacitance, parasitic, (such as $C_{gs,p}$) 116
   capacitance, sign, 64
   capacitance, total, (such as $C_{gs,t}$) 120
   current direction, 125
   noise current sources, 364
   prime, 39
   static vs. instantaneous, 38
Number of instances, *see* NF.

Off (drain-leakage) current, 103, 150, 223, 409
Overlap capacitance, 102, 104, 116–121
   effects on circuit, 302, 307, 324
   voltage dependency, 178–180, 345, Section 5.14
Oxide thickness, 165–166
   electrical, 246, Section 5.2
   measured, 247, 396, 400
   physical, Section 5.2, 402

Paired parameters, 191, 195, 199, 238, 244
Parasitic bipolar transistor, 103, 108, 110, 169
Parasitics, *see also* Diode, Fringing capacitance, Gate-bulk capacitance,

INDEX **587**

Overlap capacitance, Resistance, Source/drain capacitance
Partition, *see* Charge partition
Path-independent, 80
PISCES, *see* MEDICI
Pocket implant, 118, 413
  BSIM4 modeling, 404–407, 411
Polydepletion, 224
Predictive modeling, *see* BSIM3, BSIM4
PSPICE, 2, 159

Quantum mechanical effects, 103, 165, 175
Quasi-static
  assumption validity, 44, 138, 285
  comparison with non-quasi-static model, Section 4.1
  model, Section 1.6, Chapter 2, Appendix B
  sectioning transistor, *see* Sectioning transistor

Regional Model, 15, 29, 174, 187
Relaxation time, 452, 520
Reliability, 396
Resistance,
  access resistance, 395
  channel, *see* Channel resistance
  gate, *see* Gate resistance
  measurement, 316–317
  source/drain, *see* Source/drain resistance
  substrate, *see* Substrate distributed resistance
Reverse short-channel effects, 102, 226, 266, 390, 403–404
Root model, 81

Saturation index, 40
Saturation region, 29
Saturation velocity, 171, 184, 264, 313. *See also* Velocity saturation
  temperature dependence, 171
Saturation voltage ($V_{DS,\text{sat}}$), 27–29, 32, 185, 401
Scalable model, Section 1.12, 101, 435
Second-generation models, 7
Sectioning transistor (transistor breakup technique)
  distributed resistance simulation, 319
  quasi-static simulation, 291–296

Self-heating, 103, 242, 396
Single equation, 29, 31. *See also* Smoothing function
Singular matrix, 435
Small-size effect, 193, 266
Smoothing function, 30, Section 1.4
  $V_{BS,\text{eff}}$, 257–258
  $V_{DS,\text{eff}}$, 27, 30, 187, 242, 258
  $V_{FB,\text{effCV}}$, 34, 258
  $V_{GST,\text{eff}}$, 32, 228, 258, 408
  $V_{GST,\text{effCV}}$, 227
Snap-back, *see* Parasitic bipolar transistor
Source-bulk diode, *see* Diode
Source/drain capacitance, 119, 347
Source/drain resistance, 104, 110, 244
  absorbed/lumped approaches, 112–116, 160, 165, 243–245, 360
  BSIM4, Section 5.11, 437, 440, 442
  conductance modification, 113
  problems of the absorbed approach, 113–114, 147, Section 4.19
  sheet resistance, 245
  temperature dependence, 239
  y-parameters, effects on, 122, 140
Source-drain swapping, *see* Inverse modeling
Source-referencing, 72, Section 1.10, 122, 129
s-parameters, 79
  measured, 312, 324–327
  measurement consideration, 82–84, 327
  parameter extraction, 323
SPECTRE, 2
SPICE, 158
  acronym, 1
  history, Section 1.1
  models, Section 1.1
  netlist, 130, 158
  parameters, 92, 159, Section 3.2
  simulator, Section 1.1, 435, 452
SPICE special-function circuits
  $C \cdot dv/dt$, 130
  excess noise correction, 358–359
  GIDL correction, 385
  $g_m^{\dagger}$ correction, 315
  noise analysis, 143–145
  noise figure calculation, 370, Appendix F
  small-signal parameters, 139–140
Static assumption, 40
Strong inversion region, 431
Substrate distributed resistance 83, 149, Section 4.6, Section 5.13
  analytical formula, 328
  BSIM4 equivalent circuit, 440, 442

Subthreshold, 150
  capacitance, 21, 23
  discontinuity in model, 18
  slope, 104
  subthreshold-slope ratio, 18, Section 4.17
Subthreshold region, 31
Surface potential ($2\phi_f$), 27, 175, 198, 402–403, 459
  ϕ_δ, 166–167
Symmetry property (or lack of), 103, Section 4.7
  Capacitance, 70, 76, 102, 129, 329
  Current, 71–73, 331

Table-lookup model, 78, Section 1.11, 297
τ_1, 137, 304, 313
τ_2, 306
Temperature effects, 102, 245–246
  junction capacitance, 183
  junction leakage, 204, 423
  mobility, 251
  saturation velocity, 171
  source-drain resistance, 239
  threshold voltage, 213
Third-generation models, 7
Threshold voltage, 224, 265–267, Section 5.4
  difference between I–V and C–V, 24, 76, 173
  measured result, 227
  rollup, Section 4.18, 405, 407
  temperature dependency, 212
T.I. SPICE, 2
Tolerance, 10, 53, 54, 57–58

Transcapacitance, 55–56, 126
Transistor breakup technique, *see* Sectioning transistor
Transit time, intrinsic-channel, 42, 44, 139, 285
Tunneling current, 103, 395, Section 5.18
  noise, 439, 441, 464

Velocity overshoot, 265
Velocity saturation, effect on capacitance, 185–186, 353

ω_0, 44, 137, 285

XPART, 272–274. *See also* Charge partition

y-parameters
  BSIM3, Section 2.3
  common-source, common-bulk configurations, 68, 80
  conversion from s-parameters, 79
  definition, 67
  including parasitics, 120, 122, 140
  intrinsic transistor, 120–121
  linearly dependent relationships, 67, 121
  measured results, 81
  non-quasi-static, 69, 306, Appendix C
  quasi-static, 69–70, 79
  y_{gg}, purely imaginary, 319, 322, 360, 428
  y_{gs}, 306